Introduction to
WILDLIFE AND FISHERIES

Introduction to WILDLIFE AND FISHERIES
An Integrated Approach

SECOND EDITION

David W. Willis ■ Charles G. Scalet ■ Lester D. Flake

Department of Wildlife and Fisheries Sciences
South Dakota State University

W. H. Freeman and Company / *New York*

Publisher: Sara Tenney
Acquisitions Editor: Jerry Correa
Associate Director of Marketing: Debbie Clare
Associate Project Manager: Hannah Thonet
Photo Editor: Ted Szczepanski
Cover and Text Designer: Blake Logan
Project Editor: Jane O'Neill
Illustrations: Network Graphics
Senior Illustration Coordinator: Bill Page
Production Coordinator: Lawrence Guerra
Composition: Preparé
Printing and Binding: Hamilton Printing Company

Chapter opening photo credits appear on pages 434–435.

Library of Congress Control Number: 2007939065
ISBN-13: 978-1-4641-0913-3

Printed in the United States of America

First printing

W. H. Freeman and Company
41 Madison Avenue
New York, NY 10010
Houndmills, Basingstoke
RG21 6XS, England
www.whfreeman.com

To Ginger, Marcia, and Susan for their constant support, forbearance, and love

Contents

Preface to the Second Edition

Introduction to Wildlife and Fisheries, first published in 1996, is the first and only book to integrate the topics of wildlife and fisheries into a single introductory textbook. In writing the second edition, we incorporated many technological, scientific, and legislative advances of the past 12 years while maintaining our ultimate goal of offering students an introductory but comprehensive view of all aspects of wildlife and fisheries.

As with the first edition, we continue to believe that this book fills a void in available textbooks by integrating introductory information from both the wildlife and fisheries disciplines. Although there are numerous texts in each area, they are often intended for upper-level courses, and none incorporate both fisheries and wildlife in an even manner. We expect that the primary audience for this book will be students taking their first course in wildlife or fisheries or both. The text is designed as an introduction to the principles involved in wildlife and fisheries and how these principles relate to the organisms in question, their habitats, and their users. As in the first edition, our intent is to cover breadth rather than depth, leaving detailed information to more advanced texts and courses, although more depth was added to acomodate recent advances.

We had two primary objectives for this revision. First, we wanted to provide updated information and new, relevant examples. Updates include technological advances in geographic information systems (GIS) and biotelemetry; preferred structures for fish aging; information concerning diseases such as chronic wasting disease (CWD), avian flu, West Nile virus, viral hemorrhagic septicemia (VHS), and whirling disease; Pacific salmon stock sustainability; charismatic megafauna; behavioral ecology; and the effects of invasive species coupled with the loss of native species. The list above includes just a few of the many additions to the new edition.

Second, we made concerted efforts to broaden the scale and focus of this textbook while streamlining the existing content to maintain book length. Some of these efforts involved the addition of new sections; other efforts involved substantial section expansion; and still others involved deletion of material. Both the wildlife and fisheries professions are evolving toward larger-scale concepts and management. Thus, we specifically looked for opportunities to enhance or add information concerning entire communities (assemblages), adaptive management, biomanipulation, landscape ecology, watershed management, and ecosystem management. We did, however, maintain a focus on population assessment and management, because conservation efforts are often based on understanding populations before understanding communities, landscapes, and ecosystems. Examples of added sections include community structure, biodiversity, the scientific method, adaptations to heat and cold, and future aspects of the professions.

We hope to educate students who take only an introductory course in this area and those who use an introductory course as a foundation for later, more in-depth study. In using this text, students may expect to gain an understanding of the terminology specific to these fields, the ecological concepts involved with fisheries and wildlife, the application of these concepts to practice, contemporary issues facing wildlife and fisheries, and numerous other building-block elements essential to basic understanding.

To ensure readability, scientific names were seldom used in the text. The appendix contains the current scientific names of all fish, bird, mammal, and other vertebrate species used as examples in the text.

In addition, each chapter contains a variety of words and terms that appear in boldface. These terms or words are generally defined or described within their context of use the first time they appear in the text, as well as in the alphabetized glossary. This complete glossary includes many basic words or terms that are commonly not defined or are insufficiently defined in other publications.

We are pleased to recognize the improved quality of visual images in the second edition. Given improved print technology and digital imaging, the images in the second edition are of substantially higher resolution than those in the first edition. Many people contributed images; they are acknowledged elsewhere.

The first acknowledgment we would like to make is to the hundreds of students with whom we have been associated. Their need for an integrated text led a wildlife biologist and two fisheries biologists to work very closely together—a learning experience that caused the authors to recognize some of their own biases and preconceived notions.

We would also like to recognize the following people for their assistance with the first or second edition: Dr. Ira Adelman, Dr. Richard Anderson, Dr. Charles Berry, Dr. Michael Brown, Dr. Steven Chipps, Dr. John Connelly, Dr. Shelli Dubay, Dr. Walter Duffy, Dr. Jerran Flinders, Kurt Forman, Dr. Donald Garling, Dale Gates, Dr. Michael Hansen, Dr. Kenneth Higgins, Dr. Daniel Hubbard, Dr. Jonathan Jenks, Dr. Kent Jensen, Jack Kinnear, Dr. Robert Klaver, Dr. Bruce Leopold, Dr. David Manuwal, Dr. George Mitchell, Dr. Brian Murphy, Dr. Ray Oglesby, Dr. Pete Pekins, David Phillips, Dr. Harold Picton, Dan Plut, Robert Prieksat, Dr. William Shelton, Eileen Dowd Stukel, Dr. Trent Sutton, George Vandel, and Dr. Samantha Wisely. We especially extend thanks to Dr. David Philipp (first edition) and Dr. Eric Hallerman (second edition) for their assistance with Chapter 4 and to Dr. Larry Gigliotti for his assistance with Chapter 16. We also thank Dr. Scott Craven for his information on chronic wasting disease in Wisconsin. We thank Beth Staehle and Dr. Harry Hogdon for coordinating permissions from the American Fisheries Society and The Wildlife Society. We also want to thank the multitude of colleagues and peers who over the course of the last 12 years have made substantive comments that helped us improve the book.

We would also like to express our appreciation to the numerous people associated with W. H. Freeman and Company who assisted with various phases and editions of this textbook: Jeff Ciprioni, Debbie Clare, Jerry Correa, Maia Gil'Adi, Lawrence Guerra, Blake Logan, Jane O'Neill, Bill Page, Ted Szczepanski, Sara Tenney, Hannah Thonet, and Sandro Vitaglione.

The authors welcome you to the study of wildlife and fisheries. We hope that you will find these important fields as fascinating as we do.

CHAPTER 1

Defining Wildlife and Fisheries

What is it about wildlife and fisheries resources that interests and attracts people? Why or how are people motivated to pursue these areas as either a vocation or an avocation? Answers to these questions are probably as varied as the people who work in fisheries and wildlife or are interested in these fields. Experience tells us that interest in fisheries and wildlife is not necessarily associated with where one is raised. For example, one author of this text grew up on the Oregon coast, another came from the windswept plains of central North Dakota, and another is from the streets of Chicago. What all three have in common, and what most people who are oriented toward these resources have in common, is a desire to protect, conserve, and learn more about wild organisms.

This chapter is intended to introduce basic terminology used in wildlife and fisheries, to discuss the components of a fishery or a wildlife system, to describe some of the major differences between wildlife and fisheries, to provide a brief history of the fields, to address the scientific method, and to provide additional introductory information, including changes in the professions.

1.1 Wildlife, Fish, Fishes, Fishery, and Fisheries: What Are They?

It would seem that defining which animals come under the purview of wildlife and which are in the fisheries realm would be a simple matter. There are, however, differences of opinion. One of the authors of this text is a wildlife biologist, two are fisheries biologists. It became readily apparent during the writing of this book that each perceives things somewhat differently; each sees through the lens of his own background, education, and experience. Whales are mammals, but they live their entire lives in water—are they a wildlife species or do they belong in fisheries? How can a lobster, a crustacean, be the primary or target animal in a fishery? Issues such as these and others require further discussion.

What are **wild animals**? A dictionary definition of wild is "something living or growing in its original, natural state." Therefore, something wild cannot be domesticated or cultivated. Unfortunately, such a dictionary definition is not necessarily applicable from a wildlife or fisheries perspective. Domesticated livestock such as cattle, sheep, or pigs are not wild. But what about domesticated pigs that escape human control? Wild pigs in areas such as Arizona and Arkansas are descendants of domesticated pigs; anyone who has been near a wild pig quickly realizes that it is anything but domesticated. A white-tailed deer in the Florida Everglades is wild. But what about a white-tailed deer that a game rancher owns? The captive deer surely is not in a natural state, nor is it publicly owned like its Everglades counterpart. In addition, in most states a wild white-tailed deer is under the control of the wildlife agency in that state, while a privately owned white-tailed deer is often under the control of a board that regulates domesticated livestock. Similarly, in fisheries, a channel catfish in a river in Mississippi is wild, but a channel catfish a few meters away in the pond of a fish farmer is not wild. Wild brook trout in California and wild gray partridges in Montana are species that have been introduced into those areas. They are not in their original natural state, but we consider them to be wild.

Because of these difficulties in determining whether or not an animal is wild, legal means are usually necessary to determine the state of wildness of particular animals in particular situations. In this text the term *wild* refers to those organisms that are usually legally defined as being wild. Whatever the legal description, these organisms are **renewable resources**—they replenish naturally through biological processes.

Early in the development of these fields the term **wildlife** was considered to be inclusive of all wild animals, whether they were fishes, birds, mammals, insects, or other animals. Most biologists now take a more restrictive view—the view we will follow in this book—and consider wildlife primarily to be wild terrestrial and partly terrestrial vertebrate animals. Therefore, such organisms as mountain bluebirds, greater sage-grouse, Dall's sheep, thirteen-lined ground squirrels, smooth green snakes, and other totally terrestrial wild vertebrates are considered to be wildlife. There are also numerous aquatic mammals, such as walruses, common muskrats, American beavers, and hooded seals, that are categorized as wildlife even though a considerable portion of their lives may be spent in water. The same is true of numerous birds like **waterfowl** (ducks, geese, and swans), **shorebirds** (small wading birds, most of which feed along shores and mudflats), and **seabirds** (birds that are associated with oceans and, at times, their shorelines). Many amphibians and reptiles tend to be associated with both terrestrial and aquatic systems and can pose a problem with regard to whether they are considered to be in wildlife or fisheries. In addition, invertebrates such as butterflies and spiders are often categorized as wildlife. For the purposes of this book we shall limit our focus of wildlife examples mostly to birds and mammals, but this should not imply that other animals, such as those mentioned above, are not considered wildlife.

In addition, wildlife is not limited only to game or recreationally important species. **Game animals** are species that are harvested for recreational purposes; **nongame animals** are the remaining majority of species that are not pursued for consumptive recreational purposes. Species taken for commercial purposes are also at times placed in the game category, but there are many exceptions. In reality, whether an animal is considered to be a game or nongame species depends on legal definitions. For example, in some states where mourning dove hunting is legal, doves are game birds. In states where mourning dove hunting is illegal, they are nongame animals. The terms *game* and *nongame* will not be used often in this text; most of the topics covered have relevance to all wild animal populations.

The term **fish** refers to both jawless forms, such as silver lampreys and black hagfish, and jawed forms, such as thresher sharks, bluntnose minnows, and bull trout. Some believe that the singular form, fish, should be used when referring to a single species of fish or an individual fish, and the plural form, **fishes**, should be used when more than one species is involved. With this format, one alligator gar or 200 alligator gars are fish and one alligator gar and one spotted gar are fishes. Alternatively, some believe that either term can be used as the plural form. The term **fishery** is also complicated. A fishery can either involve a single target or featured species, such as the chain pickerel fishery in New Jersey, or cover all the fishes in a particular area, such as the fishery of Beaver Lake, Arkansas. **Fisheries** is the plural form and refers to more than one fishery. Not only can a fish be the target or featured organism in a fishery, but so can a number of other aquatic organisms. The term *lobster fishery* may sound like an oxymoron, but lobsters, which are crustaceans and not fishes, can be the biotic focal point of a fishery. Fisheries can include a wide range of living things, from lake trout to sturgeon chubs, from snapping turtles to American bullfrogs, and from clams to crayfish. In addition, some aquatic mammals that essentially spend their entire lives in water, such as short-finned pilot whales and bottlenose dolphins, are often considered to be in the fisheries realm. As with wildlife, some people think of only game or commercially important animals as the focal point of a fishery. This perception is outdated; a broad array of aquatic animals can be the focal point of a fishery.

1.2 What Constitutes a Fisheries or a Wildlife System?

A fishery is composed of three interactive components: **biota**, **habitat**, and **human users** (Figure 1.1). No comparable term is used in wildlife to describe the sum of these three components, so in this text we will instead use the term *wildlife system* as the counterpart term to *fishery*. Each of the three components of a fishery or a wildlife system influences the way it functions, and knowledge of each component is essential for understanding the entire system.

Biota includes the fishes, birds, mammals, reptiles, amphibians, other animals, and some of the plants in a fishery or a wildlife system. Often there will be a specific target or featured organism in a

FIGURE 1.1 The three equally important, interactive components of a fisheries or a wildlife system are the biota, the habitat, and the human users. *[Biota image courtesy of G. Kramer, U.S. Fish and Wildlife Service; habitat image courtesy of K. Jensen; human users image courtesy of M. Brown.]*

fishery or a wildlife system, as in the example of the chain pickerel fishery in New Jersey used earlier, but the biota component also includes all the animals and some of the plants that interact with the chain pickerel. These animals and plants may function as its food, they may be parasites, or they may interact with chain pickerel in a variety of other ways. Some consider all living organisms in a system to be part of the biota component (Nielsen and Lackey 1980). However, we believe that some living organisms, especially many plants, are more appropriately placed in the habitat component of a fishery or a wildlife system.

Habitat is where an organism lives. **Abiotic** (nonliving) components, such as rocks, water, wind, soil, humidity, salinity, oxygen level, stream gradient, water depth, and water **turbidity** (water opaqueness due to suspended particulate matter), constitute a portion of the habitat of an organism. We also include some **biotic** (living) elements as habitat components. For most animals, organisms such as trees, grasses, and large aquatic plants represent important habitat components. It should also be remembered that habitats are not just natural settings—humans create habitats, as in urban settings, that affect organisms.

Any fisheries or wildlife system also has a human user component. These human users represent what is often referred to as the **human dimensions** of a system. Human dimensions involve how people value wildlife and fisheries, how they want these resources managed, and how they affect or are affected by wildlife and fisheries management decisions. Human users can be categorized as **direct users** or **indirect users**. Direct users are anglers, wildlife and fisheries law enforcement officers, bait dealers, or any other humans who contribute to, use, or directly benefit from a fishery. They are also bird-watchers, binocular manufacturers, people who sell birdseed, or any others who contribute to, use, or directly benefit from a wildlife system. Humans can also be indirect users, such as the people who use or manage some aspect of the habitat for another purpose and in doing so affect a fishery or a wildlife system. Examples would be those who harvest trees, those who use water from streams, or those who graze domesticated livestock. These people may not actually use the fisheries or wildlife resource in question, but their activities affect the system. Some people are **consumptive users**, those who harvest a deer, a tree, or a flower, while others are **nonconsumptive users**, those who canoe a river, climb rocks, photograph wildlife, or view autumn foliage. Even people who are nonconsumptive users are consumptive in the sense that they are using the biota or habitat in some way. Nonconsumptive users can affect consumptive users; the opposite is also true. Additionally, direct users can affect indirect users, and the converse also occurs.

Human dimensions are now viewed as a more important component of a fishery or a wildlife system than has been the case in the past. As human populations increase and change locations, their impact on natural resources become more pronounced. As a result, biologist attention to aspects of human dimensions is rapidly expanding. All human users, whether consumptive or nonconsumptive and direct or indirect, have substantial effects on a fishery or a wildlife system and must therefore be viewed as portions of this component. Table 1.1 provides more in-depth examples of some potential direct, indirect, consumptive, and nonconsumptive users of a specific natural resource.

1.3 Fisheries and Wildlife: Similar, but Different

In many ways the study of fisheries and the study of wildlife are similar; that is why this text includes both. For example, the principles involving the use of fish scales or mammal teeth to determine age, the principles of the techniques used to capture aquatic and terrestrial animals, and the analysis of food habits are all similar for animals associated with wildlife systems and fisheries. A comprehensive list of similarities would be long; the similarities of the two fields far outnumber their differences. However, a fishery and a wildlife system can differ appreciably, and understanding some of the basic differences between them is important in understanding both systems.

Fishes generally exhibit **indeterminate growth**, while birds and mammals generally exhibit **determinate growth**—that is, most fishes continue to grow throughout their lives, while most bird and mammal species reach a certain adult size and essentially stop growing. A white perch increases in size as it increases in age. This growth will be determined by food supply, among other things, and as long as conditions for growth are satisfactory, the fish will continue to increase in length and weight throughout its

TABLE 1.1 Examples of some potential direct, indirect, consumptive, and nonconsumptive users of a small pond in Ohio.

Direct Users	Indirect Users	Consumptive Users	Nonconsumptive Users
Someone who fishes in the pond.	Someone who raises corn near the pond and applies fertilizer that runs into the pond.	Someone who removes fish from the pond.	Someone who canoes on the pond.
Someone who sells bait to the person who is fishing in the pond.	Someone who cuts and removes trees from near the pond, which results in increased wind action on the pond.	Someone who removes an American beaver from the pond.	Someone who scuba dives in the pond.
Someone who traps common muskrats from the pond.	An aerial sprayer who sprays crops near the pond causing some of the spray to drift into the pond.	Someone who collects cattails from the pond for floral arrangements.	Someone who picnics on the pond bank.
Someone who rents a motel room to someone using the pond.	Someone who drains a wetland near the pond, which affects pond water depth.	Someone who removes ducks from the pond.	Someone who fishes on the pond but catches nothing.
Someone who sells gasoline used on the pond or to get to or from the pond.	Someone who grazes domesticated livestock around the pond.	Someone who uses water from the pond to irrigate a field.	Someone who photographs the pond.
Someone who views shorebirds on the pond.	Someone who serves on a local board that decides zoning regulations near the pond.	Someone who cuts and removes trees from near the pond.	Someone who uses the pond for swimming.
Someone who sells binoculars to the person viewing birds on the pond.		Someone who uses water from the pond for domesticated livestock watering.	
Someone who sells a license for someone to fish on the pond.			
Someone who manages the fishery resources of the pond.			
Someone who enforces game and fish laws on the pond.			

life. If there are insufficient resources for growth, the fish will stop growing but will not necessarily die. If resources become sufficient at some time in the future, the fish can resume growing. Bird and mammal species tend to have a more uniform or specific adult size; animals with determinate growth do not get taller and longer with age. This does not mean that a three-year-old American badger or trumpeter swan may not be larger and heavier than a two-year-old badger or trumpeter swan, just that growth is not directly incremental with age and environmental conditions.

Mammals and birds are **endothermic** (**homeothermic**), while fishes are **ectothermic** (**poikilothermic**). That is, birds and mammals maintain their body temperatures at or near a specific level that is required for them to remain alive, regardless of the temperature of the environment in which they live. Their primary source of heat comes from within their bodies. Fishes and other ectotherms generally exist at or near the temperature of the surrounding environment, which is referred to as the **ambient temperature**. Their primary source of heat comes from outside their bodies. This does not mean that ectotherms have no control over their body temperatures. They can increase their body temperatures by moving to an area with a higher ambient temperature, and some species can even increase body temperature above ambient temperature by activity, such as rapid swimming. However, ectotherms do not maintain a specific body temperature under all environmental conditions. Endotherms must expend energy to maintain their body temperatures; ectotherms do not face this energy expenditure. This difference has direct ramifications for processes such as growth and metabolic rates of the two types of organisms. In the past the terms *warm-blooded* and *cold-blooded* were often used to differentiate between endotherms and ectotherms. *Warm-blooded* and *cold-blooded* are terms that are much too inaccurate to be useful. For example, a Red River pupfish living in 39°C water is anything but cold-blooded.

Fishes live in an aquatic environment, while most birds and mammals are primarily terrestrial. The two environments differ in many ways. The atmosphere of the earth contains approximately 19% oxygen; this amount is essentially unchanging. Well-oxygenated water contains only 10–15 milligrams of oxygen per liter (0.0010%–0.0015%), and this amount can vary by time of day, season, or location within a water body. This difference affects the amounts of oxygen available to terrestrial and most aquatic organisms, how they extract it, their respective metabolic rates, and other facets of their lives.

Gravity has a greater effect on terrestrial organisms than on aquatic ones. Most fishes have the ability to adjust their body density to that of the surrounding water by using the **gas bladder**. Fishes can increase or decrease the internal gas content of this organ and thus their body density. Even in fish species that lack a gas bladder, body density is similar to that of water. Thus, fishes can remain suspended in their environment with little expenditure of energy. Terrestrial organisms, in contrast, are much denser than air. Differences in characteristics such as the amount of bone structure needed for support, the specific density of the body, and mechanisms and energy expenditures for maintaining positions in the environment are a result of the differing effects of gravity on terrestrial and aquatic organisms. Fishes and other aquatic animals, however, have a problem that terrestrial organisms do not have: more energy is required to move water than air because of the greater density of water. This difference affects facets of life such as obtaining oxygen and movement.

Aquatic systems are generally more stable, in the short term, than terrestrial systems. Terrestrial organisms often must adapt to ambient temperature changes of 20°C or more during a 24-hour period. Aquatic ambient temperature changes during the same period are normally a couple of degrees at most. Terrestrial organisms, especially endotherms, can therefore usually adjust more easily to rapid environmental changes than aquatic organisms.

The concept of day and night is quite different in the terrestrial and aquatic environments. Water quality characteristics, such as depth and turbidity, affect the ability of light to pass through water. The aquatic environment is very different from the terrestrial in terms of light intensity, duration, and color.

Terrestrial and aquatic organisms also differ in their mobility. How they are able to move from one place to another is affected by the medium in which they live. Freshwater fishes live in "islands" of water surrounded by land. Their movement is generally restricted to water areas. This restriction of movement can be compared with the situation of bird and mammal species found on islands in the middle of oceans. Most wildlife species are confronted by less formidable barriers to movement than the absence of the medium in which they live. For example, the overland movement of birds or mammals may be limited by barriers such as mountain ranges, forests, or deserts but not by the lack of atmosphere. In general it is more difficult for freshwater fishes to move to new areas than it is for most terrestrial species. Marine fish mobility is more similar to that of terrestrial organisms than to that of freshwater fishes. While water may connect all oceans, barriers such as landmasses, wide expanses of deep water, water temperature and salinity differences, and other factors restrict movements of many marine fishes.

Another substantial difference between wildlife and fisheries is that fishes tend to exist in a more three-dimensional world than do most wildlife species. Fishes live in a world of length, width, and depth, whereas most wildlife species live primarily in a length and width, or two-dimensional, environment. That is not to say that wildlife species are relegated only to the surface of the earth. Examples of wildlife species that are **fossorial** (burrow into the earth), **aerial** (have flight capabilities), or **arboreal** (live in trees) are numerous, but most wildlife species are fairly narrowly associated with the surface-to-atmosphere interface of the earth for most, if not all, of their lives. Even organisms using different horizontal layers of a forest, such as birds, are essentially associated with the surface of the earth.

Wildlife and fishes also differ greatly with regard to what organisms are generally managed for consumptive purposes as the target or featured species of a fishery or a wildlife system. Most of the fishes that are consumptively managed eat other animals, while most of the consumptively managed wildlife species eat plants. For example, fishes such as the muskellunge, smallmouth bass, and walleye generally eat other animals, while wildlife species such as the mule deer, Ross' goose, and scaled quail eat plants. There are obvious exceptions to this generalization, such as the plant-eating grass carp and the animal-eating mink, but overall, most past management and conservation activities associated with consumptively used animals have been directed toward animal-eating fish and plant-eating wildlife species.

An additional difference between fishes and most wildlife species involves the measurement or enumeration of their respective populations. Birds and mammals are usually counted in numbers per unit of area—for example, the number of pronghorns or American robins found per square kilometer or hectare. These numbers are usually referred to as the **population density** of a wildlife species. Fishes are sometimes counted in numbers per unit of area but are most often measured in terms of **biomass**—for example, the number of kilograms of banded pygmy sunfish per surface hectare. One of the primary reasons for this difference in enumeration is the previously mentioned difference between determinate and indeterminate growth. For example, if a count involved killdeers in an area, a count would most appropriately be made of all of the members of that species because they would all be the same approximate size. There might be a particular biomass of killdeers present, but killdeer numbers in one population would be about the same as the numbers in another killdeer population of similar biomass. If, however, a small pond had a bluegill population with a total biomass of 150 kilograms of this species, there could be a wide variety of bluegill numbers present. There could be 1,500 0.10-kilogram bluegills, 300 0.5-kilogram bluegills, 6 25-kilogram bluegills (which could be exciting for anglers and swimmers), or any combination of numbers that totaled 150 kilograms, depending on the sizes of fish present. Because of their three-dimensional aquatic environment, aquatic organisms can also be measured in numbers per unit of volume. Population number and biomass per unit of area or volume can be used as quantitative measures of a resource. The term **abundance** is often equated with these quantitative measures.

Most fish and wildlife species also differ markedly in their reproductive potential. This difference affects their mortality and survival rates, behavior patterns, reproduction strategies, how rapidly they can repopulate an area, and how they are managed. Fishes generally produce many more offspring than bird or mammal species. Some fishes, such as the common carp, produce large numbers of eggs; a large female carp can produce in excess of 2 million eggs during one reproductive season. Even a fish such as the cutthroat trout, which is considered to have a moderately low rate of egg production, usually produces up to 4,500 eggs per mature female per year. When these fish egg production numbers are compared with the relatively small numbers of offspring that birds or mammals produce per year (usually fewer than a dozen), one readily sees that there are major differences between fishes and wildlife species with regard to reproductive potential. These differences are reflected in the use of different strategies for successful reproduction. Fishes tend to produce large numbers of offspring with low survival rates; mammals and birds tend to produce smaller numbers of offspring with higher survival rates.

Another difference between wildlife and fisheries is that human-created wildlife habitat is usually more fleeting than human-created fisheries habitat. Particular types of vegetative communities developed or used as wildlife habitat tend to be more easily changed. The removal of trees and shrubs from around a building can be accomplished in a short time; these changes quickly affect wildlife species. Such rapid changes can also occur over larger areas. An example of such a change involves the Conservation Reserve Program (CRP), instituted in 1985, which shifts land use on erodible land from agricultural row crops to cover vegetation (see section 18.9). This CRP habitat, which covers many millions of hectares, could be radically changed very quickly if these lands reverted to traditional agricultural crops. With such a change, large numbers and kinds of terrestrial wildlife species would be rapidly affected. On a short-term basis, most aquatic habitats tend to be more stable than terrestrial systems.

In addition, human-made water bodies are generally produced at the expense of wildlife habitat. The flooding of a river bottom for a reservoir or the building of a farm pond usually reduces some types of wildlife habitat while increasing aquatic habitat. This kind of change should not be thought of as a total shift, because construction of water bodies also produces habitat for such wildlife species as shorebirds, waterfowl, and common muskrats, but it does shift the area away from a totally terrestrial system. However, one should not think of aquatic habitat building as a one-way street. Some of the ponds and reservoirs we build today are the wetlands of tomorrow and then potentially will be terrestrial areas again. While such a succession of changes may take hundreds of years, this is a relatively short period of time in a biological sense. Humans have a tendency to think of time in relation to the length of a human life span; this clouds our perception of time.

Finally, humans perceive most wildlife species differently than they perceive fishes and other species less similar to themselves. **Anthropomorphism** is the attributing of human shape or characteristics to, among other things, other animals. Anthropomorphism is more often directed at birds and mammals than at fishes. Birds and mammals are endotherms and are therefore warm, often have eyelids and therefore blink, usually have hair or feathers and are therefore dry and smooth, have appendages such as legs that humans more closely identify with, and make sounds that humans can hear. They are more similar to humans in these ways than are ectothermic, nonblinking, scaled, wet (slimy), legless, and seemingly soundless fishes. In addition, humans can usually observe the activities of birds and mammals more readily than those of fishes, because most wildlife lives in the same terrestrial environment as humans. We also perceive ourselves to be more like these other terrestrial organisms because we relate more readily to their sensory receptors and the way they perceive their environment. We generally think we know more of what a bird sees or what a mammal smells than what a fish sees or smells or how it uses a sensory structure such as its **lateral line system,** a mechanical sound reception system that humans do not possess. Humans also tend to be **anthropocentric,** that is, believe that they are the center of the universe. If something seems more like us, it acquires added value in our eyes. Our perception of other mammals and birds as being more like us than most fishes and other less advanced organisms influences the way we react and relate to them and how we perceive their relative importance. Yet there is a growing awareness of and subscription to the fundamental **biocentric** view that all creatures in an ecosystem are equally important. The human tendency to be anthropocentric is reflected in our division of wildlife systems and fisheries into three interactive components: biota, habitat, and human users. While we reserve for ourselves a special place in this triad, we should recognize that we are really a portion of the biota.

These are some of the most important differences between wildlife and fisheries. Realizing that differences exist allows one to understand the reasons why the study of fisheries and the study of wildlife systems may differ. There are other differences, as well as many similarities, between the two fields of study, many of which appear throughout this text.

1.4 A Brief History of Wildlife and Fisheries

Humans have always interacted closely with wildlife and fisheries resources. Early humans living along the seashore or on lakes or rivers developed life rhythms that coincided with seasonal rhythms of fish movements and behavior. Nomadic hunter-gatherer societies were so closely intertwined with the wild organisms they required for life that they often viewed some of those organisms as deities. For example, North American Plains Indians were dependent on American bison for food, shelter, and many other products. It is no wonder that the bison is a focal point of their lives and important in their religious beliefs.

Humans needed knowledge about the organisms on which they depended to survive. Over time, some of the knowledge they gained allowed them to change or modify their relationships with these wild organisms. In some cases it led to domestication, while in others it led to human activities that changed habitats in favor of preferred wild organisms. The burning of grasslands to promote the return of wildlife to an area, the construction of shallow water areas in which to trap fishes, and the clearing of forest patches to attract animals are all examples of early attempts at modifying the relationship between humans and wild organisms.

While human knowledge about wildlife and fisheries spans many thousands of years, our knowledge concerning what early humans knew about wildlife and fisheries is primarily limited to the last few thousand years, because only this time period is represented by a written record. Early written records are primarily Chinese, Greek, and Roman; most of these are oriented toward the natural history of organisms.

Early work (from the late eighteenth century to the late nineteenth century) on wildlife and fisheries resources in the United States was primarily the result of efforts by two groups. One group is represented by people educated in the sciences whose work primarily involved classifying the organisms present, describing what they looked like, and determining where they were found. The second group is represented by writers, artists, politicians, philosophers, and others who championed the cause of saving wild organisms and unique natural areas.

At the end of the nineteenth century and the start of the twentieth century, the goals of protecting and preserving natural resources in the United States reached fruition and the pace of natural resource conservation activities quickened. Numerous U.S. government natural resource agencies were initially developed during this period, as was the scientific basis for many of our present-day activities directed toward natural resource conservation and management.

If the beginning of a comprehensive written account of the principles and practices involved in managing wild animals of North America could be identified, one would have to point to the 1933 publication *Game Management* by Aldo Leopold. Leopold outlined current practices of wildlife management and also consolidated the theories and concepts used to support those practices. Using the Leopold wildlife book as a model, Carl Hubbs and Ralph Eschmeyer published a book in 1938 on fisheries management. This publication, *The Improvement of Lakes for Fishing,* set the stage for fisheries management as a field of study. These two publications laid the foundations for the variety of wildlife and fisheries work that has continued to the present time. In addition to those who worked specifically in wildlife and fisheries, there has been a long succession of plant and animal ecologists, limnologists, geneticists, physiologists, ethologists, and people in other fields who have contributed to or forged foundations for the fields of wildlife and fisheries.

Not all of the seminal work and publications contributing to wildlife and fisheries were of a purely scientific nature. Few books have had the impact of Leopold's *A Sand County Almanac,* in which he philosophized on the land ethic, or *Silent Spring,* by Rachel Carson, which drew attention to the effects of humans on the environment. Because these books are basic to understanding the relationship of humans to their environment, each should be required reading for everyone, not just those who work in natural resource fields.

Any attempt to recognize all the individuals who made important contributions to wildlife and fisheries would be difficult. One could start with Fan Lee around 2500 BP (before present), who wrote a book on Chinese fish culture. Other names, like Aristotle, Carolus Linnaeus, Thomas Malthus, and Charles Darwin, would need to be included, as would many others. For the sake of brevity, only North American contributors are listed in Table 1.2; even on this list most contemporary contributors are omitted.

1.5 Successes and Failures

The history of wildlife and fisheries is replete with examples of successes and failures with regard to increases or decreases in population levels. There have been successes in which native species that had been

TABLE 1.2 A general chronological listing of some North Americans who have made important contributions to wildlife, fisheries, and related fields.

Alexander Wilson	Late 1700s to early 1800s	The father of U.S. ornithology.
John James Audubon	Early 1800s	Noted wildlife artist and naturalist.
Constantine Rafinesque	Early 1800s	Early ichthyologist and natural historian who worked in the Ohio River area while it was still a frontier area.
George Catlin	Early to mid-1800s	Naturalist, artist, and writer who chronicled the early American West.
George Perkins Marsh	Mid-1800s	Naturalist and writer who demonstrated how human activities affected the environment.
Louis Agassiz	Mid-1800s	Biologist and many-faceted naturalist who gained fame in Europe before coming to the United States. He conducted studies in a variety of areas, including ichthyology.
Theodatus Garlick	Mid-1800s	Father of American fish culture and also one of the founders of the American Fish Culturists' Association, which later became the American Fisheries Society.
Spencer Baird	Mid-1800s	Commissioner of the first U.S. government fisheries agency.
John Muir	Mid- to late 1800s	Noted naturalist and writer who supported the preservationist view of conservation. He also founded the Sierra Club.
Theodore Roosevelt	Late 1800s to early 1900s	U.S. president noted for his conservation efforts. He initiated the development of the National Wildlife Refuge system, greatly expanded the National Forest system, and established a presidential mechanism for saving unique resource areas as National Monuments, one of which was the Grand Canyon.
Gifford Pinchot	Late 1800s to early 1900s	Forester who helped establish the American Society of Foresters, he was director of the U.S. Department of Agriculture, Division of Forestry, and promoted wise use of resources on a sustained yield basis. He also first used the term *conservation* as it applies to natural resources.
William Hornaday	Late 1800s to early 1900s	Biologist and naturalist whose efforts helped to save the American bison from extinction. He also promoted reforms of state wildlife conservation agencies.
C. Hart Merriam	Late 1800s to early 1900s	Head of the first U.S. government wildlife agency who also developed the life-zone concept of ecological classification.
Stephen A. Forbes	Late 1800s to early 1900s	Ecologist who described the lake community as an ecosystem.
George Bird Grinnell	Late 1800s to early 1900s	Naturalist and founder of the National Association of Audubon Societies.
Edward Birge	Late 1800s to early 1900s	Limnologist who collaborated with Juday and worked primarily on Wisconsin lakes. His work centered on zooplankton.
Ernest Seton	Late 1800s to early 1900s	Artist and mammalogist who is particularly noted for his natural history work on game animals.
David Starr Jordan	Late 1800s to early 1900s	Author of hundreds of ichthyological publications from 1874 to 1931, including the monumental work *Fishes of North and Middle America*.

Stephen Mather	Early 1900s	First director of the National Park Service; he was especially adept at public relations.
Chancey Juday	Early 1900s	Limnologist who collaborated with Birge on the study of Wisconsin lakes.
Arthur Bent	Early 1900s	Author of a large series of books on the life histories of birds.
Victor Shelford	Early to mid-1900s	Plant ecologist who helped develop the study of communities.
Frederich Clements	Early to mid-1900s	Plant ecologist who collaborated with Shelford on the concept of biomes.
Robert Marshall	Early to mid-1900s	Noted forester and founder of the Wilderness Society.
Aldo Leopold	Mid-1900s	Viewed as the founder of American wildlife management. Wrote the 1933 book *Game Management.* He was also a philosopher and developer of the land ethic concept, which he described in his 1949 book *A Sand County Almanac.*
J. N. "Ding" Darling	Mid-1900s	Writer, naturalist, and promoter of numerous wildlife causes. He helped persuade Congress to pass the Duck Stamp Act, was instrumental in the development of federal wildlife cooperative units at universities, helped establish the Wildlife Management Institute and the National Wildlife Federation, and was director of the Bureau of Biological Survey, the forerunner of the present-day U.S. Fish and Wildlife Service.
Carl Hubbs	Mid-1900s	Scientist who worked in many areas of ichthyology and fisheries management. Coauthored *The Improvement of Lakes for Fishing* in 1938, which inventoried contemporary theories on fisheries management.
Ralph Eschmeyer	Mid-1900s	Early fisheries management worker who coauthored the 1938 publication on fisheries management with Hubbs.
Homer Swingle	Mid-1900s	Early fisheries manager who demonstrated fish population dynamics in small ponds.
H. Albert Hochbaum	Mid-1900s	Longtime director of the Delta Waterfowl Research Station in Canada; he was also noted for his work on waterfowl ecology.
Paul Errington	Mid-1900s	Both a biologist and writer, he was particularly noted for his work on the population dynamics of furbearers.
William Ricker	Mid- to late 1900s	Developer of many concepts of fish population dynamics, such as stock recruitment relationships; he also worked on computational techniques for fisheries data.
Durward Allen	Mid- to late 1900s	Biologist, conservationist, and author noted for his wolf-moose predator-prey relationship work in Isle Royal National Park.
Stephen Jay Gould	Mid- to late 1900s	Paleontologist, biologist, and author who was especially adept at conveying science to the general public.
Robert MacArther	Mid- to late 1900s	Best known as a geographical ecologist who primarily worked with birds.
Edward O. Wilson	Mid- to late 1900s to present	Biologist and author of numerous important works on biodiversity and sociobiology.

greatly reduced in numbers are now more numerous. Bald eagles were once thought to be on the road to extinction, but they are now flourishing (Figure 1.2A). Atlantic salmon, once severely reduced, are now found in areas from which they disappeared decades ago (Figure 1.2B). White-tailed deer, after years of depletion and scarcity, currently have reached populations that probably exceed their numbers at the time of North American settlement by Europeans (Figure 1.2C). There are many other success stories, such as northern fur seals, wood ducks, wild turkeys, American alligators, and pronghorns. Depletion of these animals was usually a result of some human activity, such as uncontrolled harvest, pollution, or habitat degradation. However, their recoveries were also a result of human activities. These wildlife and fisheries success stories are therefore the result of positive corrective actions by humans.

Not all success stories involve increased populations of native species. There have also been introductions of nonnative animals that have had positive results and are generally viewed as successes. The ring-necked pheasant, originally native to Asia, has populated new habitats in North America created by agricultural activity (Figure 1.3A). Native species that were historically present, such as the sharp-tailed grouse and greater prairie-chicken, could not withstand the agriculturally induced habitat changes in many areas, but ring-necked pheasants were able to survive in these new landscapes. The nonnative brown trout, because it could withstand higher water temperatures, expanded into areas that could not support native trouts (Figure 1.3B). The brown trout introduction has had positive aspects, especially in rela-

A

B

C

FIGURE 1.2 Three North American animals that were once greatly reduced in numbers but are now increasing. *(A)* The bald eagle, once predicted to be on the road to extinction, is now increasing in numbers. *(B)* The Atlantic salmon, a native of northwestern North America, the North Atlantic, and Western Europe, while still small in numbers, has responded positively to recovery efforts. *(C)* The white-tailed deer, once reduced to approximately 350,000 animals, now exceeds 12 million. *[Part A image courtesy of F. Deines, U.S. Fish and Wildlife Service; part B image courtesy of D. Pugh; part C image courtesy of J. Stehn, U.S. Fish and Wildlife Service.]*

A

B

FIGURE 1.3 Some introductions of nonnative animals into North America have generally been viewed as having positive results. *(A)* The ring-necked pheasant has populated new habitats in North America created by agricultural activity. *(B)* The brown trout, because it can withstand warmer water temperatures, has expanded into areas that could not support native trout. *[Part A image courtesy of South Dakota Tourism; part B image courtesy of M. Anderson.]*

tion to recreational fisheries, but has also had some drawbacks because of the negative effects of brown trout on the native species they can displace. The introduction of some species of **Pacific salmon** (pink, chum, coho, sockeye, and Chinook salmon) into the Great Lakes of North America is also generally viewed as positive.

Unfortunately, for every nonnative introduction that is viewed as having primarily positive effects, there are multiple examples of nonnative introductions that were unfortunately too successful. Many of these organisms are now found in large numbers, and their effects on native species, crops, and even buildings have been such that these introductions must be categorized as failures. Some of these were purposeful introductions; others were accidental. The common carp was a purposeful introduction into North America (Figure 1.4A). However, it has often displaced native fishes and has also had negative effects on habitats and on other organisms, such as waterfowl. House sparrows and European starlings (Figure 1.4B) were purposeful introductions that have displaced and reduced native bird species. Coypu (previously known as nutria) were both accidental and purposeful introductions (Figure 1.4C). Originally brought into North America for fur purposes, many escaped or were released when they could not be sold. They were also purposefully released to control problem aquatic vegetation. They have displaced native animals and also cause damage to agricultural crops, trees, roadbeds, and other human-made structures. The round goby was an accidental introduction. It was probably introduced with **ballast water** (water in the holds of ships) from ships originating in Europe. This species is now widespread in the Great Lakes basin and represents a threat to native aquatic communities. Any positive features of these nonnative introductions are far outweighed by their negative effects.

The list of problem introduced species continues to grow as human activities promote, either purposefully or accidentally, the movement of organisms. Animals such as zebra muscles and various species of Asian carp as well as plants such as purple loosestrife are just a few of the species that have the potential to cause or are causing large-scale problems. The list of these problem organisms is long and rapidly increasing.

Potential success stories that are still in the formative stages are the attempts by humans to maintain species that have been reduced to small numbers. The ongoing efforts to save whooping cranes, Colorado

A

B

C

FIGURE 1.4 Some nonnative introductions have been too successful and have had many negative effects. *(A)* The common carp now ranges over almost all of North America in a wide variety of habitats. *(B)* European starlings were purposefully introduced in small numbers but are now found in most areas of North America. *(C)* The coypu (formerly called nutria), while having some positive attributes, is responsible for much damage to native organisms and to the environments in which it lives. *[Part A image courtesy of J. Jolley; part B image courtesy of L. Karney, U.S. Fish and Wildlife Service; part C image courtesy of J. and K. Hollingsworth, U.S. Fish and Wildlife Service.]*

pikeminnows, black-footed ferrets, and a variety of other endangered and threatened species are potential successes (Figure 1.5); they are also potential failures (see Chapter 11).

Efforts to save some species came either too late or not at all. These are failures in the truest sense, because these species are gone and nothing can be done to ameliorate their loss. The heath hen, the Carolina parakeet, the blue pike (Figure 1.6A), the passenger pigeon (Figure 1.6B), and others are gone forever. Their loss is irreversible and unfortunate. How the loss of numerous plant and animal species, primarily as a result of human activities, will affect the integrity and functioning of the ecosystems of the world is currently unknown. Aldo Leopold compared the human experiment of modifying the earth to a tinkerer taking apart a machine (Leopold 1966). A good tinkerer does not throw away a part because its function or position in the machine is unknown. Humans continue to throw away parts of ecosystems at an ever-increasing rate. In addition, to further compound problems, humans increasingly, both intentionally and unintentionally, are adding parts (new species) into areas in which they are not native.

There have been some magnificent successes with regard to wildlife and fisheries populations, and there have also been many abysmal failures. Even with the successes taken into account, the future of many species is not bright. For every population that is flourishing, there are many others that are just holding their own or declining. All too often humans are concerned only with large species, such as American bison or California condors, or economically important species, such as canvasbacks or Pacific salmon,

A

B

C

FIGURE 1.5 Species that humans are attempting to save from extinction represent potential successes or failures. *(A)* Whooping cranes once had been reduced to 14 birds but now number over 400. *(B)* Colorado pikeminnows have been diminished in numbers primarily as a result of dam construction and nonnative introductions, but their numbers are now increasing. *(C)* Because of their small numbers, all known black-footed ferrets were captured in an attempt to increase their numbers through captive breeding programs. *[Part A image courtesy of S. Hillebrand, U.S. Fish and Wildlife Service; part B image courtesy of Colorado Division of Wildlife; part C image courtesy of L. Parker, U.S. Fish and Wildlife Service.]*

while a myriad of small and seemingly unimportant species, some known, some unknown, are relegated to inattention and benign neglect. All organisms, whether large or small, in great numbers or small numbers, economically important or unknown, represent a portion of the web of life—all are interconnected in some way. It would seem that the overall success or failure of the human species in the end will depend on the success or failure of many other species; their future and our future are intertwined.

Successes and failures do not just pertain to individual species. Protecting small areas of habitat, larger landscapes or ecosystems, and assemblages of organisms are also issues of importance.

Saving a small area of unique habitat is more likely to be successful than saving a large area. For example, a small **fen** (a type of wetland) can be preserved if the area upon which the fen is dependent is protected. However, protecting larger landscapes and ecosystems is a much different situation. Rain forests, large grassland tracts, and coastal wetlands are just a few of the large-scale areas on the planet that are being lost at alarming rates—their protection is much more complicated and a less successful undertaking.

Preserving unique assemblages of organisms is also usually an issue of scale. It is easier (note that the word *easy* was not used) to protect a group of organisms found in a small area rather than those in a large area. For example, saving a group of organisms found in a small desert spring is more likely to be successful than our current abysmal efforts directed at protecting valuable marine fishes or Great Lakes fish assemblages.

FIGURE 1.6 For some species, any attempts to save them came too late; they are now extinct. *(A)* The blue pike, variously regarded as a color phase of the walleye, a walleye subspecies, or a species, was commercially important in Lake Erie up to the 1950s. It is believed to have become virtually extinct by the early 1970s. *(B)* The population of passenger pigeons at one time numbered approximately 5 billion birds and ranged from Canada to Mexico. The last individual died in 1914. *[Part A image courtesy of W. Carrick.]*

1.6 Wildlife and Fisheries and the Scientific Method

Garton et al. (2005) summarized the scientific method as "a circular process in which previous information is synthesized into a theory, predictions are deduced from the theory, the predictions are stated explicitly in the form of hypotheses, hypotheses are tested through an investigation involving experimentation, observation, or quantitative models, the theory is supported, modified, or expanded on the basis of results of these tests, and the process starts again." Some essential characteristics of science are that it is guided by natural laws, it is explanatory by reference to natural laws, it is testable against the empirical world, and its conclusions are tentative. Hypothesis testing is central to science and the scientific method.

A **hypothesis** is a proposition or supposition regarding a theory that is tentatively accepted to explain known phenomena or to provide a basis for further investigation. A working hypothesis generally represents what is initially considered to be the most likely explanation of a phenomenon. Usually several alternative research hypotheses are formulated; these are based on a variety of possible explanations or reasons for facts observed.

In the scientific method, hypotheses are rejected or fail to be rejected depending on study results. When data support a hypothesis, it cannot be concluded that the hypothesis is true, only that it has not been rejected. In addition, a hypothesis is generally only rejected or not rejected at some statistical level. The scientific method can never prove that a hypothesis or theory is correct, only that alternative hypotheses or theories are rejected. The potential correctness of a hypothesis or theory increases as alternative hypotheses are rejected.

It is common to hear statements about theories such as "there is no scientific proof," "it has been proven scientifically," or "the hypothesis was accepted." Such statements indicate a lack of understanding about the scientific method. A hypothesis can only be rejected or not rejected; that is all.

The primary drawback to the scientific method is that it can never prove that theories involving issues such as evolution, gravity, or causes of global climate change are correct. However, we can continue to test hypotheses associated with these theories and, from the results, strengthen or weaken the theories. The primary strength of the scientific method is that when studies are correctly designed, implemented, and interpreted, human biases and beliefs are eliminated from the equation. There is a great difference between human belief and human knowledge; the scientific method provides knowledge-based information. Theories or hypotheses that cannot be tested with the scientific method are not really theories or hypotheses—they are beliefs.

Many facets of the study of wildlife and fisheries are not and probably never will be like the pure sciences. In pure sciences, such as chemistry, there is a greater ability to control and modify variables associated with an experiment. This ability to control and modify variables results in experiment repeatability under the same conditions to confirm previous results. In wildlife and fisheries, or any biological science, especially those involving field work, it is difficult to control and modify variables. For example, the population dynamics of cactus wrens in a certain locale may be determined. These determinations could be correctly made and then not be repeatable (either at a different location or even at the same location) because of differences in variables such as death rates, growth characteristics, or hatching success resulting from variations in weather conditions, habitat modifications, or any number of other changeable factors. Some variables associated with the study of wildlife and fisheries are controllable in laboratory situations, but such information is not necessarily applicable to nonlaboratory settings.

Much of our knowledge of wildlife and fisheries is based on observations and conclusions derived from associations rather than from experimental tests of hypotheses (Romesburg 1981). Such knowledge is correlative rather than indicative of cause and effect; thus, interpretation may be inaccurate and misleading. Romesburg (1981, 1991) and others have been critical of this aspect of wildlife and fisheries as science. An example of knowledge obtained through association (often called a descriptive-type study) would be sampling a fish species from a number of locations, examining their stomach contents, and reporting what was eaten. The study is descriptive because it does not ask the "why" question, i.e., why are fish eating what was found, and a manipulative study was not designed to test a prediction. While descriptive-type studies are still conducted in wildlife and fisheries, there has been a strong trend in the last couple of decades to test specific hypotheses and produce research results that are not solely based on associations. However, information based on an association still has value. Many hypotheses have been developed based on information obtained from descriptive studies, and many untested hypotheses will evolve in the future from such information. However, hypothesis-testing research usually provides more reliable information and typically is the next step after completing a descriptive study.

The primary factors that have allowed wildlife and fisheries biologists to progress from association-type studies to hypothesis-testing studies are the advent of improved statistical tests and the almost universal use of computers. Increased complexity and power of statistical tests and constantly improving technology associated with computer hardware and software have combined to allow for greater research rigor in wildlife and fisheries.

Our current knowledge of wildlife and fisheries is incomplete. Because of this, biologists are often faced with the dilemma of making decisions based on insufficient information. This dilemma leads to questions such as "Is the controlled burning of this piece of grassland at this time really the best manner of managing this resource?" or "Are you sure that this is the most effective way to protect this fish assemblage?" The answers to such questions are not "yes" or "no"; the answer should be "based on our current knowledge, this is the best practice that can be implemented." As our knowledge base increases, we often learn that the best practice at some previous time was really not the most appropriate solution. Although this modification of solutions has been called waffling by some, and the term is used to diminish credibility, this approach is unavoidable in science, where the objective is to find truth. The general public is often confused when information previously thought to be accurate turns out to be inaccurate. This confusion is primarily a result of not understanding the scientific method, especially that the scientific method cannot be used to "prove" something. Under the circumstances, all a biologist can do is use the best information currently available and then strive to increase knowledge using the scientific method.

1.7 Biology, Science, Management, Conservation, and Ecology: What Are They?

The terms *fisheries* and *wildlife* are often followed by words such as *biology*, *science*, *management*, *conservation*, and *ecology*. For example, the phrases "wildlife conservation and management," "fisheries science," and "fisheries biology" are just a few commonly used combinations. Unfortunately, books and papers using these terms often define the same terms differently or do not define them at all. This leads to unnecessary confusion. In an attempt to reduce this confusion, we will give our definitions for various terms. Some people will disagree with our definitions, but these definitions are as correct as any.

Wildlife management and **fisheries management** are the art and science of manipulating biota, habitat, or human users to produce some desired end result. For example, habitat modification efforts to increase Cape Fear shiner numbers and rules implemented to reduce harvest of American black ducks fall into the category of management. **Fisheries science** or

wildlife science involve the process of obtaining knowledge about and studying fisheries or wildlife. For example, development of stock recruitment curves in fisheries and the rationale and mechanics involved in the use of anesthetics in wildlife represent facets of science. In other words, science concerns the process of obtaining knowledge about something, while management represents application of that knowledge.

Conservation is the effective management and use of natural resources. It thus follows that **fisheries conservation** is the effective management and use of fisheries resources and **wildlife conservation** is the effective management and use of wildlife resources. **Ecology** is the study of the interrelationships of organisms with other organisms and the environment. Therefore, **fisheries ecology** is the study of the interrelationships of organisms that are considered to be a part of the fisheries realm with other organisms and the environment; **wildlife ecology** is the study of the interrelationships of those organisms generally considered to be under the purview of wildlife with other organisms and the environment. **Biology** is the study of living things; thus, **fisheries biology** is the study of fisheries and **wildlife biology** is the study of a wildlife system. It should be apparent that the definitions of *management* and *science* are fairly specific and quite different from each other, while the definitions of the other terms are broad and overlap with one another as well as with the terms *science* and *management*.

These commonly used terms can also be combined in other ways. For instance, **conservation biology** is "the applied science of maintaining the earth's biological diversity" (Hunter 2002).

Definitions change with time. Leopold (1933) defined game management as "the art of making land produce sustained annual crops of wild game for recreational use." While it is not specified, most people assume that the "game" of which Leopold writes are animals that are harvested for recreational purposes. Current perceptions of wildlife management would not be as restrictive; more than game animals would be involved, and few people would restrict wildlife management to only those animals used for consumptive recreational purposes. Interestingly, Leopold said nothing of science but described management as an "art."

Rounsefell and Everhart (1953) began the preface of their book by defining fisheries management as "the application of scientific knowledge concerning fish populations to the problems of obtaining the maximum production of fisheries products, whether stated in tons of factory material or in hours of angling pressure." The definition is of fisheries management, but their book title is *Fishery Science: Its Methods and Applications*; they provide no definition of *fisheries science*. Their definition of *management* is directed only at consumptively used fisheries resources, which is quite different than would be the case today. In addition, the term *maximum production* is used in their definition. In current definitions, *optimum* would replace *maximum*. There is a great deal of difference between maximum and optimum, as the reader will learn elsewhere in this book.

Whereas *science* or *management* was generally thought to be sufficient to describe activities in wildlife and fisheries in the past, changing terminology has often led to the more common inclusion of terms such as *ecology* and *conservation* in the lexicon of the fields. Excellent wildlife or fisheries books by Bolen and Robinson (2003), Sinclair et al. (2006), and Walters and Martell (2004) all include *management* in their titles along with *ecology* and, in one case, *conservation*. It is obvious that *ecology* is added to book titles to indicate a broader scope of content and because ecological knowledge is indispensable in management. Management is often described as applied ecology, so the inclusion of *ecology* is appropriate from this perspective also. *Conservation* might be added under similar circumstances because it implies a concern for wildlife and fish resources. *Biology* could be added because of the positive connotations of that term; biology is typically perceived as being scientific. Reasons for using such terms are probably as numerous as the people who use them, with each having his or her own particular rationale. What these additions of terms most likely represent is change and growth in language as it pertains to wildlife and fisheries. Word and term usage changes through time, and that has occurred and is occurring in wildlife and fisheries.

It should be noted that the title of this book contains the words *wildlife* and *fisheries*, but not *conservation*, *ecology*, *biology*, *management*, or *science*. The book contains elements of all of the latter terms, and a title that included all would be too long—they were purposefully omitted.

1.8 Use of Scientific Names

Common names of organisms are generally used in this text, but this does not mean that scientific names are unimportant. Scientific names allow us to be certain what organism is being discussed regardless of the language being used. Whether a paper or a book is written in English, German, Russian, Japanese, or any other language, the currently accepted scientific name for an organism is the same. Common names often vary depending on location; this is not the case with scientific names. When standardized scientific names are used, there can be no confusion as to the organism under consideration. The appendix contains the scientific names of vertebrates used as examples in this text.

The scientific name of an organism generally consists of generic and specific names. For example, the current scientific name of the smallmouth bass is *Micropterus dolomieu*; *Micropterus* is the genus and *dolomieu* is the species. People can sometimes be confused by the way the word *species* is commonly used. For example, consider the question "What species is this?" If the organism is a smallmouth bass, the answer is *Micropterus dolomieu*. The smallmouth bass is considered to be a species, but both the generic and specific portions of the scientific name are required to identify the fish. The plural form of *species* is *species*, while the plural form of *genus* is *genera*. In a biological sense there really is no singular form of *species*, because a species name does not represent a single organism but instead represents all of the individual organisms that constitute that species.

Scientific names can include more than just the genus and species. In situations in which subspecies have been described, the subspecies name follows the species. In addition, the person who named the organism and when it was described may be included. For example, the lake sturgeon has the name Rafinesque and the year 1817 following the species name. Its scientific name would appear as *Acipenser fulvescens* Rafinesque, 1817. This means that this species was described by Rafinesque in 1817. If the species was originally described with another scientific name and that name has been changed, the name of the person who originally described the organism and the date of that description are placed in parentheses. For example, the scientific name of the shovelnose sturgeon has been changed; therefore, it appears as *Scaphirhynchus platorynchus* (Rafinesque, 1820).

The genus, species, and, if present, subspecies names should be printed in italics; if that is not possible, they should be underlined. The first letter of the generic name should be capitalized; no letters in the specific or subspecific portion of the name should be capitalized.

The term **taxon** is a general expression for a taxonomic group, whatever its rank. Therefore, a genus, species, family, or any of the other groupings used in the hierarchy of taxonomy is referred to as a taxon.

1.9 The Professions—The Future

The authors of this text have among them almost 100 years of professional experience since completing their terminal university degrees. During that time much has changed in the wildlife and fisheries professions. Thirty years ago almost all activity in these fields was directed at organisms and habitat, but changes in these fields have necessitated increased exposure to many other areas. Greatly expanded efforts directed at skills involving human dimensions (as mentioned earlier in this chapter), teamwork, communications, statistics, computers, critical thinking, globalization, and ethics have been needed. These new areas have essentially been additive to what professionals in these fields need to know, and information available about organisms and habitat has expanded exponentially.

For wildlife and fisheries professionals to be most effective in the future, it will be essential that they possess these additional skills in ever-increasing depth. One cannot successfully work with people without having effective oral, written, and listening communication skills. Working well in groups, including interdisciplinary groups, requires teamwork and communication skills. Technical skills involving computer and statistical applications are mandatory. Solving complicated problems requires clear and critical thought processes. Being able to put what one does into a global and ethical context is also essential.

The pace of change has been rapid, and it continues to accelerate. Envision the future problems for natural resources caused by ever-expanding human populations. Ponder the massive biological ramifications of global climate change on wild organisms, habitat, and people. Visualize a world with decreasing ocean fisheries resources. Reflect upon future changes in energy use and production and how these changes will affect wildlife and fisheries resources. Consider the impacts of new and spreading disease outbreaks on organisms, including humans.

Contemplate the increased information that will be needed as we continue to shift our thinking emphasis from individual populations to ecosystem, landscape, and global levels. Think about the effects of future world geopolitical upheaval on natural resources. Speculate about the effects on ecosystems caused by our continued movement of organisms to new areas in which they did not evolve and the continued loss of organisms that were there. Weigh the impacts of continued urbanization on many countries. Wonder about the changes that we cannot even speculate about because we cannot foresee them. These are only a few of the daunting tasks that must be faced.

One thing should be obvious: change will continue. Biologists of the future will need to be in a constant continuing education mode and must understand the systems and people with which they work better than their current counterparts.

SUMMARY

Wild animals are organisms that are legally defined as being wild. Terrestrial and partly terrestrial vertebrate animals are generally categorized as wildlife. Fishes are considered to be in the fisheries realm, but a variety of other organisms are also usually included. A fishery or a wildlife system is composed of three interacting components: biota, habitat, and human users. Biota consists of the target or featured organism of a fishery or a wildlife system and most other living organisms in the system. Habitat consists of abiotic components and some biotic components, primarily large plants. Human users can be direct users, indirect users, consumptive users, or nonconsumptive users.

There are some basic differences between the target animals of a fishery and those of a wildlife system that make the study of the two fields different. Aspects of these animals such as growth patterns, body temperature regulation, the environment in which they live, their mobility, the stability of their habitats, how their abundance is measured, their reproductive potential, and the way humans view them are all quite different.

The history of wildlife and fisheries involves a variety of naturalists, philosophers, artists, politicians, and biologists who have made contributions. The past successes and failures of wildlife and fisheries revolve around (1) successes involving native species that were reduced in numbers but are now flourishing, (2) successes of some nonnative introduced organisms that have had positive effects, (3) failures resulting from nonnative introduced organisms that have had negative effects, (4) potential successes or failures involving species currently small in numbers, and (5) failures resulting from species that have become extinct.

The scientific method is important in wildlife and fisheries research. It is imperative to remember that the scientific method cannot prove that a theory or hypothesis is correct. It can only reject or not reject hypotheses.

Wildlife science and fisheries science involve the process of obtaining knowledge about and studying wildlife or fisheries. Wildlife management and fisheries management involve manipulation of biota, habitat, and human users. The terms *ecology*, *conservation*, and *biology* are also commonly used in reference to these fields. Scientific names allow us to communicate with people around the world with no confusion as to the organism in question.

Change in these fields will continue to occur at an ever-increasing pace. Biologists in the future will be in a constant continuing education mode.

PRACTICE QUESTIONS

1. Why is it difficult to define which animals are wild and which are not?

2. What are the three components of a fisheries or a wildlife system? Select a fisheries or wildlife resource in your region and provide examples of each component.

3. Choose a specific fisheries or wildlife system in your region and then develop a list of six of each of its direct, indirect, consumptive, and nonconsumptive users.

4. What is the primary difference between population density and biomass?

5. Why do humans relate more closely to birds and mammals than to fishes?

6. With regard to their ability to move from one area to another, how are marine fishes and land mammals and birds similar?

7. Describe two major written contributions of Aldo Leopold.

8. What is the primary aspect of the scientific method that most people do not understand?

9. List a number of possible benefits of introducing nonnative organisms and then list a number of negative consequences associated with nonnative introductions.

10. What is science and what is management?

SELECTED READINGS

Anderson, S. H. 1991. *Managing our wildlife resources.* 2nd ed. Prentice Hall, Englewood Cliffs, N.J.

Carson, R. 1962. *Silent spring.* Houghton Mifflin, Boston.

Gould, S. J. 1996. *Full house.* Harmony Books, New York.

Kallman, H., chief ed. 1987. *Restoring America's wildlife 1937–1987.* U.S. Department of the Interior, Fish and Wildlife Service, Washington, D.C.

Krausman, P. R. 2002. *Introduction to wildlife management: The basics.* Prentice Hall, Upper Saddle River, N.J.

Leopold, A. 1949. *A Sand County almanac and sketches here and there.* Oxford University Press, London.

Nielsen, L. A. 1999. History of inland fisheries management in North America. Pages 3–30 *in* C. C. Kohler and W. A. Hubert, eds. *Inland fisheries management in North America.* 2nd ed. American Fisheries Society, Bethesda, Md.

Podulka, S., M. Eckhardt, and D. Otis. 2004. Birds and humans: A historical perspective. Pages H.1–H.42 *in* S. Podulka, R. W. Rohrbaugh, and R. Bonney, eds. *Handbook of bird biology.* 2nd ed. Cornell Lab of Ornithology, Ithaca, N.Y.

Taber, R. D., and N. F. Payne. 2003. *Wildlife, conservation, and human welfare.* Krieger Publishing Company, Malabar, Fla.

Trefethen, J. B. 1975. *An American crusade for wildlife.* Winchester Press, New York.

Wilson, E. O. 1998. *Consilience.* Alfred A. Knopf, New York.

Wilson, E. O. 1992. *The diversity of life.* Harvard University Press, Cambridge, Mass.

LITERATURE CITED

Bolen, E. G., and W. L. Robinson. 2003. *Wildlife ecology and management.* 5th ed. Prentice Hall, Upper Saddle River, N.J.

Garton, E. O., J. T. Ratti, and J. H. Giudice. 2005. Research and experimental design. Pages 43–71 *in* C. E. Braun, ed. *Techniques for wildlife investigations and management.* 6th ed. The Wildlife Society, Bethesda, Md.

Hunter, M. L. 2002. *Fundamentals of conservation biology.* 2nd ed. Blackwell Science, Inc., Williston, Vt.

Leopold, A. 1933. *Game management.* Charles Scribner's Sons, New York.

Leopold, A. 1966. *A Sand County almanac with other essays on conservation from Round River.* Oxford University Press, New York.

Nielsen, L. A., and R. T. Lackey. 1980. Introduction. Pages 3–14 *in* R. T. Lackey and L. A. Nielsen, eds. *Fisheries management.* John Wiley & Sons, New York.

Romesburg, H. C. 1981. Wildlife science: Gaining reliable knowledge. *Journal of Wildlife Management* 45:293–313.

Romesburg, H. C. 1991. On improving the natural resources and environmental sciences. *Journal of Wildlife Management* 55:744–756.

Rounsefell, G. A., and W. H. Everhart. 1953. *Fishery science: Its methods and applications.* John Wiley & Sons, New York.

Sinclair, A. R. E., J. M. Fryxell, and G. Caughley. 2006. *Wildlife ecology, conservation, and management.* 2nd ed. Blackwell Publishing, Malden, Mass.

Walters, C. J., and S. J. D. Martell. 2004. *Fisheries ecology and management.* Princeton University Press, Princeton, N.J.

CHAPTER 2

Ecological Concepts

Working with populations and communities of living organisms is important in wildlife and fisheries. To understand these organisms it is necessary to understand the ecological principles that govern various facets of their lives, such as where they live, what roles they play, and how they interact with the abiotic and biotic components of their environment. Much of the work conducted by wildlife and fisheries biologists involves the application of this knowledge of ecological principles.

The decisions of wildlife and fisheries biologists often have far-reaching implications for an ecosystem, several ecosystems, or even the entire biosphere. For example, a wildlife biologist may be involved in evaluating and prescribing timber harvest practices on private or public lands in North America. These practices may directly influence nesting sites, available food sources, and other needs of several migrant bird species, some of which winter in tropical portions of South America. Think of the myriad effects that a fisheries biologist could have while working with a large and important group such as migratory **salmonids** (members of the fish family Salmonidae; trout and salmon). These fishes have ecological importance on a global scale. An overall ecosystem and landscape approach to conservation is being increasingly emphasized, particularly on public lands.

The purposes of this chapter are to present a variety of ecological concepts, to illustrate the importance of these concepts with examples, and to introduce some basic ecological terminology. Many of these concepts will be referred to in later chapters.

2.1 Ecosystems, Biotic Communities, and Populations

An **ecosystem** is an interacting system of biotic and abiotic components in a particular area or place. Many different types of ecosystems can be identified. Ecosystem classification is generally based on some major characteristic, such as the dominant plant in terrestrial ecosystems or the water body type (pond, stream, or lake) in aquatic ecosystems. Ecosystem classification can be applied at broad levels, such as grasslands or mountain lakes, or applied more specifically to areas with the same species of dominant trees (oak-hickory forest), animals (coral reef), or other recognizable characteristics. Even a pile of cow manure in a pasture or a pool of rainwater in a natural rock bowl could be considered small temporary ecosystems.

A water body such as a pond or lake is an aquatic ecosystem with more or less clearly defined boundaries and with aquatic organisms in large part tied to those boundaries. However, other more mobile species that are part of this aquatic ecosystem, such as little blue herons, southern watersnakes, and raccoons, regularly traverse the boundaries of the ecosystem. These animal species remove energy and nutrients from the system in excess of what they may replace. Thus, the influence of such an ecosystem, particularly if it is highly productive, reaches far beyond its shoreline.

Ecosystems in general have influences reaching beyond their geographic boundaries. For instance, the northern regions of Canada and Alaska provide nesting and rearing areas for many species of shorebirds that spend the remainder of the year in southern temperate or tropical climates. The American golden-plover, for example, breeds in the Arctic and winters in southern South America. As these birds move between and through ecosystems, nutrients are transferred great distances and the animal life of two or more widely separated ecosystems becomes interdependent. Likewise, the nutrients obtained by many Pacific salmon species in the marine ecosystem are important to the **productivity** (the rate at which energy is captured and stored in an ecosystem) of the freshwater and terrestrial ecosystems where they migrate to spawn and die during their life cycle. Oxygen produced in marine ecosystems and tropical forests is important to the atmosphere of the earth and to the survival of living organisms. Thus, even though we identify specific ecosystems, the influences of each ecosystem extend to other ecosystems and to the entire **biosphere** (the layer around the earth in which all living things are found).

Sometimes extensive regions of similar vegetation are identified as **biomes** (see section 12.1). Biomes are major terrestrial communities or major ecosystems that often recur on the various continents. A system for classification and mapping of ecologically similar systems into **ecoregions** (Bailey 1998) has recently become widely used, especially by groups such as the World Wildlife Foundation in their Global Conservation Program. In this system, climate types, vegetation associations, soil types, and other factors are used in a hierarchical classification to group ecologically similar areas. An attractive feature of the ecoregion approach is that it can be used to classify and map both terrestrial and aquatic regions, although the applicability of the World Wildlife Foundation's ecoregion approach in global conservation efforts has received criticism because the classification system may not be as repeatable or as explicit as classification systems that are already in place and have proven reliable (Jepson and Whittaker 2002).

The living portion of an ecosystem is called a **biotic community**; it is usually just referred to as a **community**. This interacting assemblage of living organisms is sometimes subdivided into plant and animal communities. At times the terms *community* and *ecosystem* are incorrectly used as though they were synonymous. Ecosystem is a broader term because it includes the interaction of the biotic and abiotic components of the system, while a community includes only the biotic component.

Ecosystems are often studied in terms of their structure and function. **Ecosystem structure** refers to the composition of the system with regard to species of plants, animals, and microbes, the nutrients available, the availability of water, and so forth, or generally a "grocery list" of what is there. **Ecosystem function** refers to how the ecosystem works in its many interacting ways, particularly how energy and nutrients move through and within the system.

In studies of ecosystems or their plant and animal communities, the number of species in a particular community is often called **species richness.** Species richness is sometimes used as a rough measure of **species diversity.** Species diversity, however, includes both the number of species present and some measure of **species equity** or **evenness.** Species equity or evenness refers to the proportional abundance of the various species in a community. For example, two forests (A and B) with 20 tree species occurring in each would have the same tree species richness. However, if the most common tree species in forest A accounted for only 10% of all the trees present while the most common tree in forest B accounted for 90% of all the trees present, forest A would have a higher species diversity than forest B. The greater the chance that a random tree selected from one of the forests would be of the same species as a previously selected tree, the lower the species diversity.

More recently, terms similar to *species richness* and *species diversity* have become widely used by both biologists and the general public. These terms, **biodiversity** and **biological diversity**, are synonymous, and they are often used in situations in which other terms, such as *species diversity* or *species richness*, would be more appropriate. The U.S. Congress Office of Technology Assessment (1987) defined biological diversity as the variety and variability among living organisms and the ecological complexes in which they occur. Primack (2002) divided biological diversity into three components: (1) genetic diversity within populations, (2) species diversity within ecosystems, and (3) community and ecosystem diversity across the landscapes of entire regions. Levin (2001) presented an extensive collection of interesting articles addressing various aspects of biodiversity in aquatic and terrestrial ecosystems.

Species diversity is being eroded by increasing human populations and their effects on natural resources and ecosystems (Figure 2.1). Losses of species and species diversity can have many negative influences on humans. For example, insect populations have many checks and balances that in part depend on their interrelationships with other organisms within an ecosystem. Reductions or losses of other species may cause some insects, once kept under control by natural processes, to increase in abundance, causing economic damage to crops or becoming a nuisance to humans. By losing species diversity, ecosystems become inherently less stable. There are many other reasons for maintaining species diversity; some of these are discussed in Chapter 11. Biologists, realizing the importance of species diversity, attempt to maintain populations of species and to restore those that are endangered, threatened, or declining.

A **population** is defined as all the individuals of a species within a specified area at a given time. The area specified defines the scope of the population in question. A population can include all the individuals of a species in the world, for example, all the skipjack herrings. A population can also include only certain segments of a species, for example, the rock sandpipers in western Washington or the lake chubs in Lake Michigan.

At times the term *population* is used loosely to refer to more than one species, such as all the fish species in a pond or all the small mammals in a given area. This use of the term *population* is incorrect; multispecies assemblages are communities.

A population existing as a set of geographically separate subpopulations but with some genetic interchange among them is sometimes referred to as a **metapopulation**. Greater prairie-chickens, for example, that have minimal genetic interchange currently exist in several subpopulations throughout the Interior Plains of North America; these subpopulations together could be considered a metapopulation. Likewise, the subpopulations of mountain goats associated with different mountains or mountain ranges in western Canada could be considered a metapopulation. Coho salmon have many subpopulations associated with different rivers; these subpopulations compose a metapopulation.

A

B

FIGURE 2.1 Human encroachment can reduce species diversity and cause substantial changes in ecosystems. *(A)* Urban development has caused extensive loss of traditional winter range for mule deer and other indigenous ungulates in the intermountain west; such developments also increase human–wildlife conflicts. *(B)* This native prairie site is being tilled and converted to cropland and will soon be void of native plants that provide habitat for numerous prairie birds and other animal species. *[Part A image courtesy of T. Benzon, South Dakota Department of Game, Fish, and Parks; part B image courtesy of B. Schulz, U.S. Fish and Wildlife Service.]*

2.2 Biogeochemical Cycles and Energy Flow

In any ecosystem, energy and the essential elements of living cells must be available to support the system. Living organisms obtain essential elements from their environment, including the soil, water, air, and other organisms. In this way these elements are cycled through the ecosystem, thus the term **biogeochemical cycles**. Cycles of elements essential to the growth of living organisms may also be called **nutrient cycles**.

The elements essential as nutrients to most plants include carbon (C), hydrogen (H), oxygen (O), phosphorus (P), potassium (K), iodine (I), nitrogen (N), sulfur (S), calcium (Ca), iron (Fe), magnesium (Mg), chlorine (Cl), copper (Cu), boron (B), manganese (Mn), zinc (Zn), and molybdenum (Mo). Essential elements for animals include those used by plants, with the exception of boron. Additional elements required by animals include selenium (Se), cobalt (Co), fluoride (F), chromium (Cr), and silicon (Si). Essential elements for animals are further discussed in section 5.1. Understanding how these elements cycle through ecosystems is important to understanding ecosystem function. Let us consider nitrogen, carbon dioxide, and sulfur as examples of element cycling.

The Nitrogen Cycle

Nitrogen (N_2) occurs as an inert gas in the atmosphere, but it is unusable by most plants in this form. The conversion of inert, gaseous nitrogen to ammonia (NH_3), nitrites (NO_2^-), or nitrates (NO_3^-) is termed **nitrogen fixation**. The nitrogen cycle (Figure 2.2) is especially

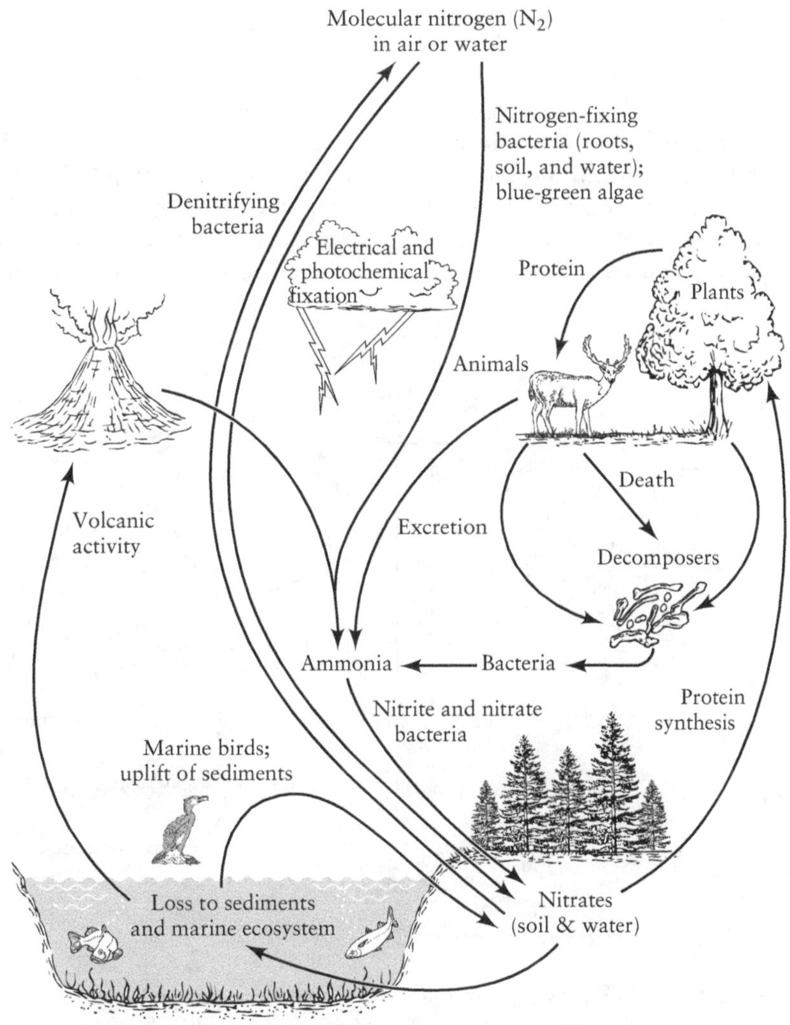

FIGURE 2.2 The nitrogen cycle. Nitrogen-fixing bacteria are the primary mechanism by which inert nitrogen is made available to plants and, through plants, to other living organisms.

interesting because certain plants, particularly legumes, in association with **symbiotic** bacteria in their root nodules (a symbiotic relationship is one in which interacting species are closely and permanently dependent on each other), can fix nitrogen in the form of ammonia (NH_3), making it available to plants and animals. Some free-living **aerobic** (living in the presence of oxygen) bacteria in the soil are also able to make nitrogen available to the biota, as can blue-green algae in aquatic systems.

Biological nitrogen fixation produces ammonia by splitting the nitrogen molecule and combining it with hydrogen; this process requires energy. Most fixation of nitrogen is biological, although nonbiological fixation, the formation of nitrates from sources such as lightning and volcanoes, also occurs. Other parts of the cycle involve additional types of bacteria that convert ammonia to nitrites (nitrite bacteria) and nitrites to nitrates (nitrate bacteria), as well as denitrifying bacteria living under **anaerobic** conditions (in the absence of oxygen) that release inert nitrogen back into the atmosphere. The amount of nitrogen fixed daily in the biosphere and the amount returned to the atmosphere are roughly equal. Collapse of the nitrogen cycle would quickly place most of the organisms in the world, including ourselves, in jeopardy of a very short existence.

The Carbon Cycle

The carbon (C) cycle (Figure 2.3) is currently receiving much attention because of its possible role in global warming (Stern et al. 1992; Ricklefs 2001) (see section 14.1). Carbon dioxide (CO_2) is required by plants for photosynthesis (see below). It is produced by plants and animals through normal processes of cellular respiration. Relatively small concentrations of CO_2 occur in the atmosphere and water, yet these concentrations are critical to ecosystems, and changes in these concentrations are dangerous in terms of possible perturbations to the climate of the planet. Carbon is in a dynamic equilibrium among the forms present in the oceans and other waters (bicarbonates and carbonates), in sediments and fossil fuels, and in the atmosphere, plants, and soils.

The Sulfur Cycle

Sulfur (S) is essential to living organisms primarily for the formation of certain amino acids and enzymes. Sulfur occurs in **organic** (carbon-containing compounds, including living or dead organisms) and **inorganic** (nonorganic minerals) deposits and is generally present in adequate supply for plants and animals. Sulfur at the surface of the earth is released

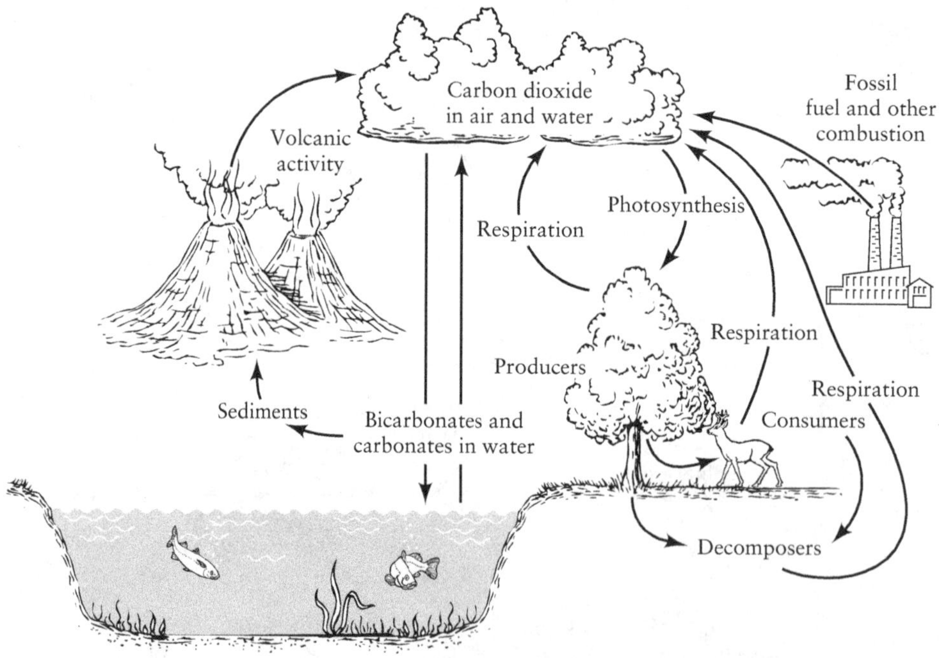

FIGURE 2.3 The carbon cycle. Carbon dioxide is required by plants for photosynthesis and thus is closely tied to energy flow in ecosystems. The ocean and terrestrial biomass serve as nutrient sinks by storing over half of the total CO_2 released into the atmosphere by various sources.

into the air, water, and soil by decomposition of organic matter, including that in stored deposits such as peat, and by surface weathering, erosion, and volcanic eruptions (Figure 2.4). Sulfur is also released in gaseous form and in salt spray from the surface of the ocean. Sulfur occurs in various forms, including hydrogen sulfide (H_2S) and sulfur dioxide (SO_2), in the atmosphere and water, and as sulfate (SO_4) in water and soil. Sulfur in the atmosphere first appears as toxic hydrogen sulfide but is quickly converted to

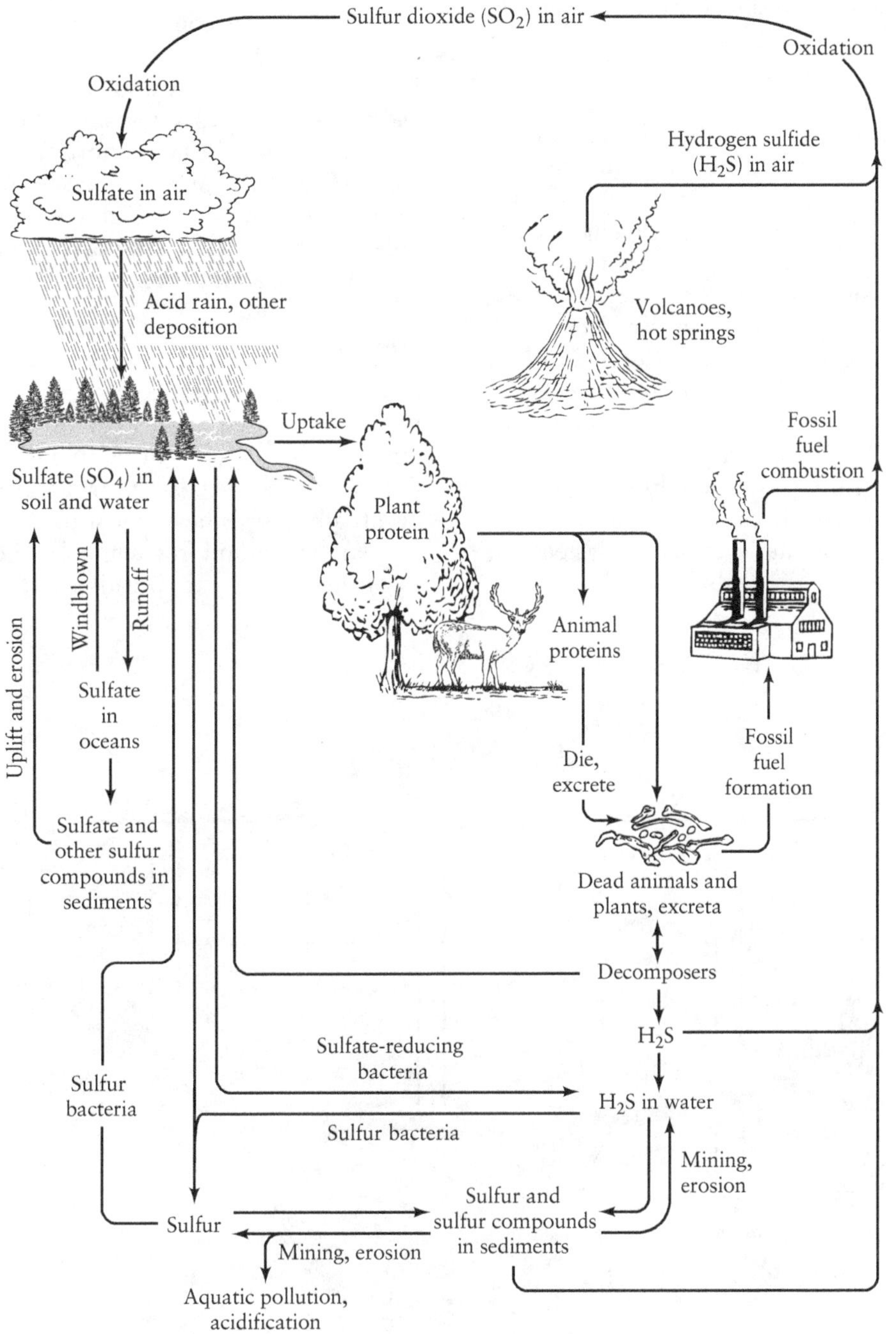

FIGURE 2.4 The sulfur cycle. Sulfur is essential for living organisms, but excess amounts released by the burning of fossil fuels and from acidic drainage of coal mines have caused environmental damage.

sulfur dioxide. Sulfur dioxide is soluble in water and forms weak sulfuric acid (H_2SO_4) in precipitation, which, upon reaching the surface of the earth, again becomes available to plants. Bacteria and fungi that act as **decomposers** (organisms that obtain energy from dead organisms and waste products) release sulfur from organic material. Certain bacteria are also able to use inorganic sulfur in the soil and make it available to the ecosystem. Plants obtain sulfur primarily as sulfate from soil and water, although some plants are able to use sulfur dioxide directly. Animals obtain adequate sulfur by eating plants or other animals.

Exposure of sulfur in mining deposits, particularly coal deposits, has in some cases led to acidification of water by sulfuric acid. The release of sulfur into the atmosphere through industrial processes, particularly the burning of high-sulfur coal or associated high-sulfur sediments, results in hydrogen sulfide formation in the atmosphere. The volatile gases containing sulfur combine with water to increase the level of acidity in rain, snow, dew, and fog, causing acidic deposition, which can alter the acidity of streams, lakes, and soils and damage ecosystems (see section 14.1).

The Water Cycle

Water also cycles through ecosystems (Figure 2.5). The oceans provide moisture to the continents in this **hydrologic cycle** (water cycle). Most precipitation that falls on land evaporates, is stored as surface water or groundwater, or is returned to the oceans via river flow. Water composes only a small proportion of the atmosphere, and the turnover rate is high. Human-induced or natural changes in the hydrologic cycle can affect ecosystems in a broad way. Human activities that influence runoff or evaporation rates, use surface water or groundwater, or reduce natural recharge of **aquifers** (underground water in permeable formations of sand, gravel, or rock) can influence this important cycle.

Energy Flow

A major aspect of any ecosystem is **energy flow**. The word *flow* is used because energy comes into the system, usually as sunlight, is fixed by plants through photosynthesis, is passed through the system, and is eventually lost as heat. Most of the energy captured by plants is bound in chemical bonds through photosynthesis, and part of this energy is harvested by

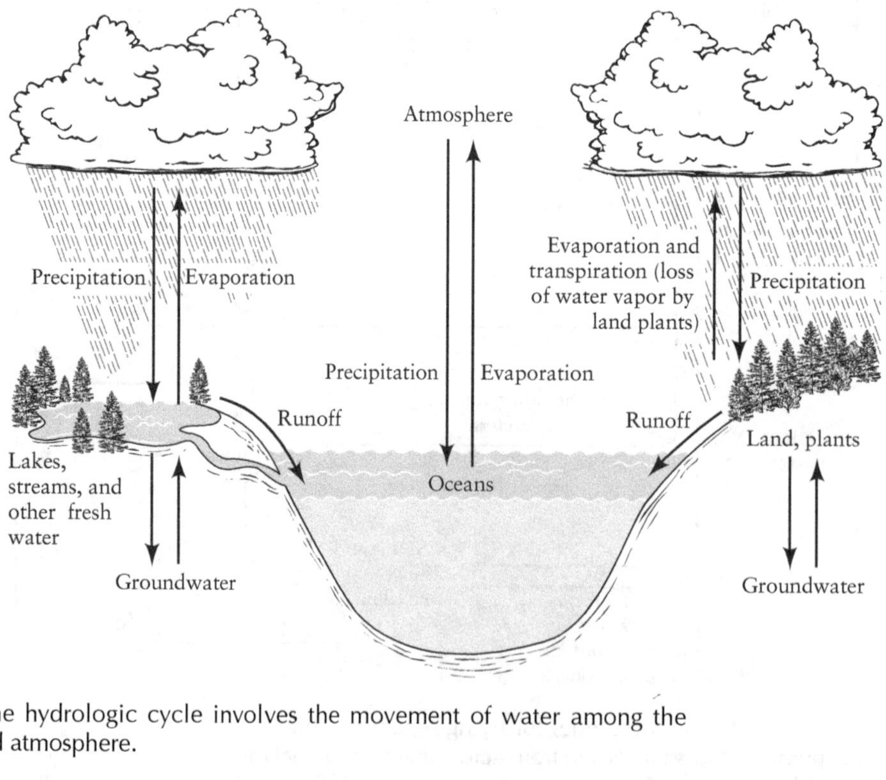

FIGURE 2.5 The hydrologic cycle involves the movement of water among the oceans, land, and atmosphere.

other organisms and passed through the system. Because energy captured by the ecosystem through photosynthesis eventually leaves the system, the process is called energy flow rather than cycling.

Photosynthesis involves the use of carbon dioxide and water to form complex organic molecules. The reaction requires light energy, which is captured by chlorophyll molecules in the leaves or sometimes in the stems or other parts of plants. The primary products of photosynthesis are sugars and oxygen, but other organic products, such as fatty acids, fats, vitamins, and proteins, are also produced by plants. The products of photosynthesis may vary by plant species and even with the age or part of the plant; for example, young leaves on a shrub may produce much more protein than older leaves. The basic chemical reaction of photosynthesis is:

$$\underset{\text{Carbon + Water}}{6CO_2 : 6H_2O} \xrightarrow{\text{Sunlight}} \underset{\text{Glucose + Oxygen dioxide}}{C_6H_{12}O_6 \cdot 6O_2}$$

This formula shows the basic products and raw materials of photosynthesis. The photosynthetic process involves a light phase, in which energy is captured and water is split into oxygen and hydrogen ions; oxygen is produced in this reaction. It also involves a dark phase, in which carbon dioxide and hydrogen are combined to form a variety of organic compounds, including glucose. Photosynthetic pathways and mechanisms for the capture of energy and formation of sugars are presented in greater detail in Smith and Smith (2001).

While photosynthesis uses energy from the sun to build complex organic molecules and produce oxygen, decomposition breaks down these organic compounds, releases carbon dioxide and water, and returns nutrients to an inorganic state. Decomposition of plant and animal matter is accomplished primarily by **microflora** such as fungi and bacteria, along with invertebrates and other organisms that feed directly on **detritus** (dead organic matter). Invertebrates that feed on litter and break it into fragments are usually called **detritivores**, while bacteria and fungi involved in decomposition are called decomposers. However, microfloral decomposers and invertebrates that serve as detritivores are sometimes grouped together as decomposers because they are all important in the decomposition process. Fungal and bacterial decomposers and detritivores obtain nutrients and energy in the process of breaking down dead organic matter. At the same time, these decomposers and detritivores release the energy bound in these materials and return nutrients to the inorganic state. Other larger organisms also break down detritus by their physical activities or by ingestion. Even the effects of precipitation in leaching nutrients from detritus and of weather in breaking down detritus, primarily plant matter, into smaller fragments are important in decomposition.

The primary decomposers of animal matter are bacteria, while fungi are more important in the decomposition of plant litter. Microflora secrete enzymes into plant detritus to assist in the decomposition process. Many species of bacteria can function in both aerobic and anaerobic environments. Bacteria that function in anaerobic environments are particularly important in the decomposition of organic matter that collects at the bottoms of lakes or slow-moving streams.

Mites, protozoans, nematodes, caddisfly larvae, stonefly nymphs, earthworms, and millipedes are all examples of detritivores. Detritivores play an important role in the decomposition process by fragmenting organic material and inoculating it with bacteria and fungi. The composition of the microflora and detritivores changes with the stage in decomposition. At one stage organisms that live on simple sugars may be successful, while at a later stage the remaining decomposers and detritivores may be those that specialize in the use of more complex carbohydrates, such as cellulose and lignins in the woody cell walls of plants. Another group of organisms, the **microbivores**, feed on the nutrients and energy in bacteria and fungi. Examples of microbivores include larval beetles, flies, mites, and amoebas.

While decomposers may use nutrients and keep them out of circulation in their own body tissues (**nutrient immobilization**), their role in decomposing detritus and releasing inorganic nutrients (**mineralization**) is invaluable to the ecosystem. Microflora and detritivores are essential to the release of nutrients and energy bound in the ecosystem and thus to maintaining ecosystem productivity.

2.3 Trophic Levels, Food Chains and Webs, and Pyramids of Energy

Energy is transferred from plants to animals when the animals consume the plants. Subsequently, this energy is passed through additional feeding or **trophic levels** as animals are eaten by other animals. **Autotrophs** are organisms, such as plants, that can produce their own food; **heterotrophs** must utilize autotrophs or other heterotrophs to obtain food. Energy flows through the ecosystem as organisms eat and are eaten. This path of energy flow can be diagrammed as a **food chain** or, more accurately, as a **food web** (Figure 2.6). Food chains are simple examples often used to introduce this topic, but nature really functions in food webs. A chain implies linearity, but that is not the way in which these trophic interactions generally occur.

The trophic levels, in the order of energy flow, include **primary producers** (autotrophs or plants), **primary consumers** (**herbivores** that eat primarily autotrophs), and **carnivores** (organisms that feed primarily on other heterotrophs). The term **secondary consumer** is often used for carnivores such as the cougar that feed primarily on primary consumers. The term **tertiary consumer** can be used for carnivores such as the white shark that feed primarily on secondary consumers. The term **predator** is also used for organisms that attack, kill, and eat other animals, which are known as their **prey**.

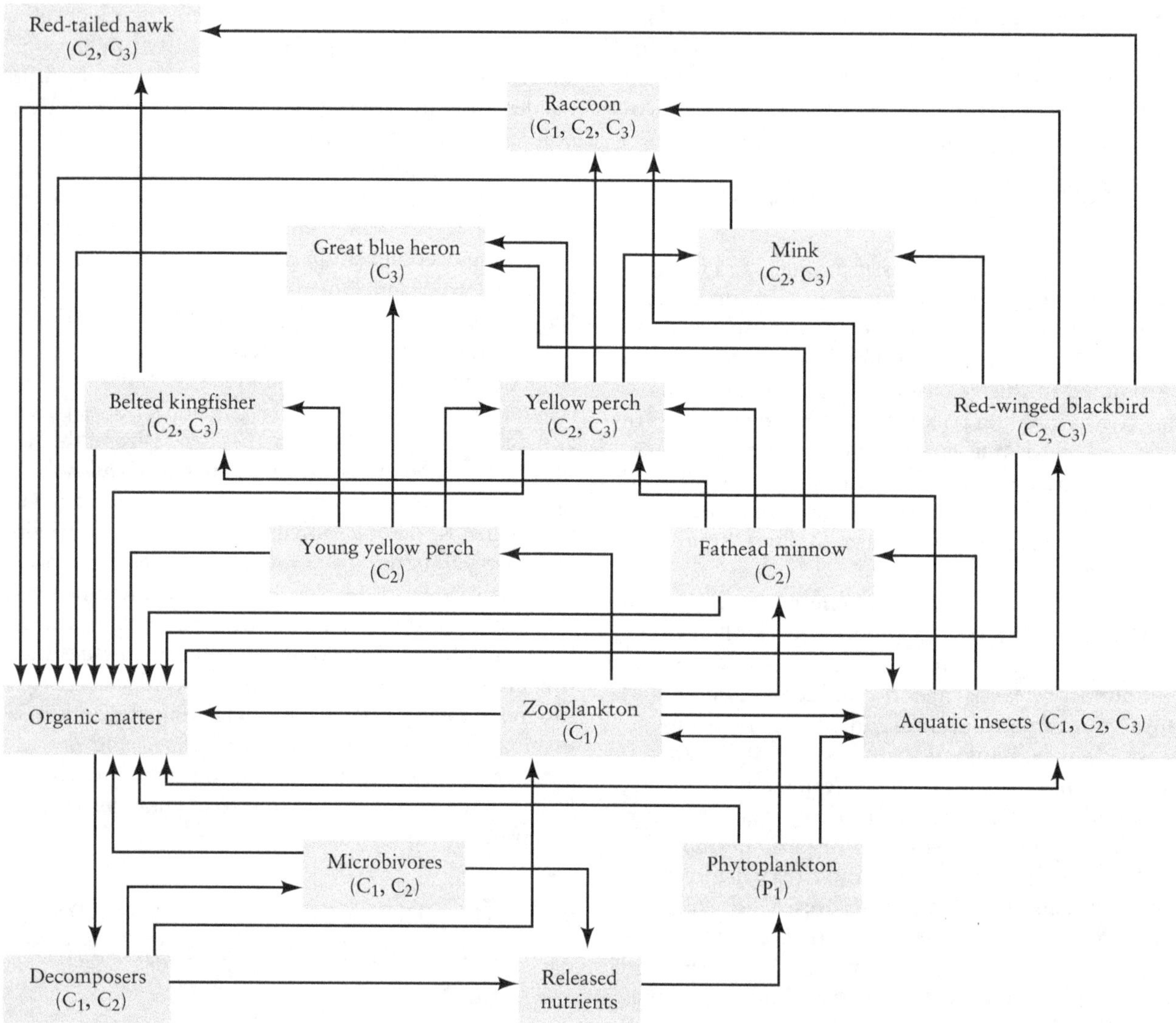

FIGURE 2.6 A simplified food web for a small farm pond in the Midwest. The arrows show the paths of energy flow and nutrient cycling among the various trophic levels. P_1, primary producer; C_1, primary consumer; C_2, secondary consumer; C_3, tertiary consumer. Many other organisms could be added to such a food web.

Other feeding levels are also used to group organisms. Animals such as American black bears, channel catfish, and raccoons, like humans, readily eat both autotrophs and heterotrophs and are called **omnivores.** Omnivorous species fill a mixed role as both primary consumers and carnivores (secondary and/or tertiary consumers). Also contributing to the food web are the decomposers, detritivores, and microbivores discussed in section 2.2. Depending on what they feed on (plant or animal matter), these organisms can also be grouped as primary consumers or carnivores. Larger organisms that obtain much of their food from dead animal matter, such as black vultures, are often termed **scavengers;** they are also carnivores. **Saprophytes** are nonanimals, primarily fungi, which feed mainly on dead plant material; they are also appropriately called decomposers. **Parasites,** such as fleas, sea lampreys, and intestinal nematodes, obtain nutrients and energy by living inside of or on other organisms.

Plants, whether they are trees, grasses, shrubs, or **phytoplankton** (tiny aquatic plants that float or drift in the water), are primary producers. They are the living base of food webs. Organisms such as greater white-fronted geese, grass carp, mule deer, common muskrats, and **zooplankton** (tiny aquatic animals that primarily float or drift in the water) are, for the most part, primary consumers. A qualifying statement is needed here, because an animal such as the common muskrat will eat animal material, including other muskrats, if the need and opportunity arise, but such material constitutes only a small portion of its diet. Numerous other species alternate as primary consumers and secondary consumers depending on their nutrient needs. For instance, female ducks, such as northern pintails and mallards, are largely herbivorous much of the year but feed heavily on invertebrates, such as aquatic insects, fairy shrimp, snails, and earthworms, during late winter, spring, and early summer when preparing to reproduce and during egg laying. Ducklings also depend on the high-protein content of invertebrates, especially in the first several weeks after hatching. The young of **gallinaceous** birds (members of the order Galliformes, such as the sharp-tailed grouse, California quail, and gray partridge) exhibit a similar age dependence on invertebrate foods. These examples illustrate why food webs are a more appropriate metaphor than food chains for the flow of energy through an ecosystem.

Numerous organisms also shift between being secondary consumers and tertiary consumers. Predatory great barracuda feed primarily on invertebrates as secondary consumers in the early portion of their lives but shift to a fish diet and a tertiary consumer role as they grow. Thus, even though trophic levels are often used to categorize species, these levels may be dynamic within a species in relation to age, reproductive status, and opportunity to exploit new food sources. For species that switch trophic levels seasonally or with age, it works best to think of them at seasonal or age-specific trophic levels instead of assigning them one overall trophic level designation.

Energy is transferred through the various trophic levels. Although energy cannot be created or destroyed (the **first law of thermodynamics**), a considerable loss of energy occurs with each conversion in the food web. Producers fix a given amount of energy from the sun in various chemical bonds. Primary consumers use this energy through herbivory and digestion, but much of it is dispersed in unusable forms such as heat. A reasonable rule of thumb is a 90% loss of energy at each conversion. In other words, if 1,000 kilocalories of plant material (primary producers) were consumed by primary consumers, only about 100 kilocalories would be converted to primary consumer biomass. This loss of energy from one trophic level to the next follows the **second law of thermodynamics.** According to this law, when energy is transferred or transformed, much is lost or dispersed in unusable forms, primarily heat. Thus, species at higher trophic levels, such as muskellunge or bobcats, are limited in population biomass by the amount of energy they can receive from lower trophic levels. By the laws of energy flow within ecosystems, a population of a **piscivorous** (fish-eating) fish species such as the northern pike is incapable of reaching the biomass that shorthead redhorses, bluegills, or other fishes at lower trophic levels can reach. Pyramids of energy illustrate the manner in which energy is transferred and lost as it flows through trophic levels (Figure 2.7). Other similar **ecological pyramids** include those for numbers and biomass.

In fisheries, the primary game and food fishes are secondary and tertiary consumers, while almost all game wildlife species are at lower trophic levels, filling the role of primary consumers. Some predators managed as **furbearers** (mammals commonly

FIGURE 2.7 A generalized pyramid of energy for a terrestrial ecosystem showing the transfer of energy through different trophic levels. Primary producers include not only aboveground but also belowground plant matter (roots) as well as dead plant materials. Organisms involved in decomposing dead plant or animal material are grouped with the appropriate consumers.

harvested for their hides, such as American mink and red foxes) fill roles at higher trophic levels and, as in the fisheries example above, cannot by the laws of energy flow reach the population biomass of a primary consumer species such as the common muskrat or white-footed deermouse.

In some cases, consumer organisms can have a large influence on other trophic levels or organisms within their own trophic level, causing what has been termed a **trophic cascade**. For example, overabundant elk (all North American elk are subspecies of red deer) in Yellowstone National Park caused damage to quaking aspen and willows to such an extent that the appearance, makeup, and productivity of the plant and associated animal community were greatly changed. After wolf introduction, elk numbers in high-wolf-use areas declined due to elk avoidance of wolves and wolf predation on the elk. Quaking aspen and willows in high-wolf-use areas responded to reduced browsing by elk with improved growth and reestablishment (Ripple et al. 2001). A similar trophic cascade occurred with lesser snow geese along coastal areas in much of their eastern Arctic nesting area. Large-scale increases in lesser snow goose populations have caused long-term damage to fragile plant communities in the nesting area through goose foraging activities; increases in goose populations were likely related to warming of the Arctic and the abundance of agricultural food in goose migration and wintering areas. Lesser snow goose destruction and damage to Arctic plant communities has in turn been detrimental to several other species, particularly other birds that nest and rear young in the same coastal ecosystems. In a trophic cascade in the Black Sea, decreases in predatory fishes and increases in fishes that prey on zooplankton have been linked to extensive increases in phytoplankton and deteriorating environmental conditions (Daskalov 2002).

Ecologists sometimes use the terms **top-down control** and **bottom-up control** when referring to influences of trophic levels on numbers or biomass of organisms in other trophic levels. Top-down control occurs when higher trophic levels, predators, for example, control the size of one or more lower trophic levels. If wolves and grizzly bears are depressing the numbers and biomass of reindeer (North American subspecies are commonly called caribou) and moose in an area, the situation would represent top-down control. However, if the numbers of reindeer and moose were more strongly influenced by long-term changes in primary production, then the situation would be considered bottom-up control.

2.4 Ecological Succession

Ecological succession involves the processes that occur when an ecosystem changes from an early, immature, transient stage, such as that found at a newly disturbed site, to a more mature, relatively stable stage. This more mature and stable stage has often been termed a *climax community*, but this term has been replaced by many ecologists in recent years with the terms *shifting-mosaic stage* or **shifting-mosaic steady state**; we will use the latter. Multiple factors influence the direction of succession, including soils, moisture, wind, seasonal temperatures, seed sources, and even disturbances such as fire. Examples of more mature and somewhat stable communities include old-growth rain forests in Brazil, old-growth oak-hickory forests in the southeastern United States, and desert areas dominated by giant saguaro and palo verde in southern Arizona. Researchers have pointed out that even the most mature and stable plant communities are constantly replacing small portions of the overall community through many cycles of succession at various stages. Thus, a mature rain forest or oak-hickory forest includes within the community a mosaic of small patches at different stages, from early to late succession. The ability of a shifting-mosaic steady-state plant community to self-perpetuate depends on the balancing effects of the many successional cycles occurring within the overall plant community. A shifting-mosaic steady-state plant community may be a relatively permanent stage, but in reality it often will not reproduce itself without the continuous array of successional cycles occurring in disturbed areas (openings, wind-damaged areas, replacement of dead trees, etc.) within the overall community—thus the term *shifting-mosaic steady state*.

Ecological succession involves changes in the structure and function of the ecosystem and is reasonably predictable in terms of the dominant plants and other associated plants and animals that will occur in the various successional stages. In succession, the biotic community modifies the environment and exerts much control over the direction of succession. The physical environment of the general area also exerts a strong influence. The transient but recognizable stages in succession are called **seral stages**, while the entire sequence of seral stages from early to late succession is termed a **sere**. Succession occurring on sites unchanged by living organisms is called **primary succession.** An example of primary succession would be the development of a biotic community on a recently cooled lava flow. Succession occurring on sites that were previously occupied by other organisms, as in an abandoned agricultural field or burned forest, is **secondary succession.**

A ponderosa pine forest burned by a wildfire in which all the trees and many shrubs were killed provides an example of succession. The earliest phase of succession in this sequence might be a grass-forb seral stage (a **forb** is a nonwoody plant such as poison ivy that is not a grass or grasslike), followed by grass-forb-shrub, shrub–pine sapling, and other intermediate seral stages, until the mature ponderosa pine forest with its associated **understory** (plants growing on forest floors under trees) and animal life is eventually reached in 200 or 300 years (Figure 2.8). The ponderosa pines form an **overstory** or **canopy** above the understory. During this sequence the species composition and energy flow within the system change, and the ecosystem itself, along with the physical environment, dictates the nature of the system. As noted previously, the mature ponderosa pine forest includes within its boundaries a mosaic of seral stages forming numerous small patches within the overall community.

In general, the structure and function of an ecosystem become increasingly more complex as succession progresses; this complexity imparts greater stability to the ecosystem. A mature rain forest is a relatively stable ecosystem due to its great complexity in terms of species diversity, energy flow, and nutrient cycling. Its many checks and balances resist perturbations to the system. The rain forest system would probably be very resistant, for example, to attacks of harmful insects due to checks and balances in the form of a broad array of **insectivores** (insect eaters). The diversity of plants in such an ecosystem also adds to its stability. The loss or reduction of a single plant species due to insects or a disease, for example, would tend to have less of an effect on an ecosystem with a great diversity of plant species than on a simpler ecosystem with few plant species.

A B

FIGURE 2.8 Secondary succession from a burned-over forest (grass-shrub-sapling seral stage) *(A)* to an old-growth forest dominated by large ponderosa pines *(B)* will take from 200 to 300 years. The areas shown in these images are in the northern Black Hills in Wyoming.

Succession is not confined to terrestrial ecosystems; it also occurs in aquatic systems. For example, over time a river progresses through various successional stages that can be illustrated by looking at river sections of different ages. The headwaters of rivers typify an early successional stage. Headwaters are characterized by V-shaped valleys, colder and clearer water, lower fertility (concentrations of soluble minerals that are important as nutrients in the ecosystem), rock and gravel bottoms, and lower species diversity. Downstream successional stages in older sections of the river are characterized by having U-shaped valleys, water that is warmer and more turbid, increased fertility, more silted bottoms, and increased species diversity. Standing water bodies such as lakes also move through successional stages. Early successional-stage lakes are deeper and have more oxygen, more inorganic bottoms, better water transparency, lower fertility, and generally less vegetation than do older lakes. As these characteristics change in water bodies, so do the fish and other animal and plant populations.

One of the reasons biologists are interested in succession is the direct effect it has on the occurrence and abundance of organisms. Post-fire succession in forested regions from grass or grass-shrub communities to shifting-mosaic steady-state communities is accompanied by predictable changes in animal communities. For example, early forest clearing by means of plow, axe, and fire in central Wisconsin transformed once-forested regions into excellent sharp-tailed grouse and greater prairie-chicken habitat. These grouse species thrived temporarily until the grass, early shrub, and agricultural communities were replaced by young quaking aspens and eventually by coniferous forest. Many other wildlife species also fluctuated in abundance and even occurrence as succession progressed from grassland-shrub-agriculture to quaking aspen and eventually back to the shifting-mosaic steady-state forest.

Biologists are aware of the effects of succession and use these principles to manage species associated with various seral or shifting-mosaic steady-state stages. A biologist attempting to maintain summer breeding habitat for the endangered Kirtland's warbler in northern Michigan may use burning to maintain an early successional stand of jack pines. Ruffed grouse are managed by maintaining different-aged stands of quaking aspen, an early successional-stage plant species. Other animals, such as northern spotted owls in Oregon or Sitka black-tailed deer in southeastern Alaska, directly benefit from maintenance of old-growth coniferous forest. These are examples of management for individual species; however, management for an individual species through manipulation of seral stages influences the entire animal community, not just the target species. Thus, the emphasis on the northern spotted owl in relation to old-growth forests in the Pacific Northwest has been broadened to include the variety of other animal species and subspecies that depend on old-growth forests, such as the American marten, pileated woodpecker, marbled murrelet, northern goshawk, and others. Plants dependent on the environment created by these old-growth forests are also being considered.

It may be best to provide a variety of seral stages to increase the diversity of wildlife in an area. Additionally, many species require several widely divergent seral stages. White-tailed deer and elk, for example, readily use clearings for feeding but seek nearby forest cover for escape and protection from the elements. Even within woodland areas, thick stands of small trees may provide hiding cover, while more mature forests with extensive overstory can be important as thermal protection in both winter and summer.

Newly constructed reservoirs provide an excellent example of aquatic succession and its effects on fish populations. When Lake Oahe, a Missouri River reservoir in North and South Dakota, was being filled, northern pike were abundant. As the reservoir filled with water and for a few years after filling, the flooded soils and terrestrial vegetation released abundant nutrients that led to high levels of productivity; the vegetation also provided reproductive habitat for the northern pike. After the reservoir had been filled for some time, productivity declined, and flooded terrestrial vegetation was no longer available for northern pike reproduction. The northern pike population declined, and this species was replaced by other fishes, such as walleyes, that were better adapted to the reservoir at this later successional stage. Continued succession in Lake Oahe can be expected in the future.

2.5 Niche, Habitat, and Environment

The role or function of an organism in the biotic community, its "occupation," has been defined as its **niche** (Elton 1927). This definition emphasizes the way an organism functions in an ecosystem in terms of obtaining energy. For example, the Mexican long-nosed bat is a primary consumer that specializes in feeding on nectar and pollen. The term **multidimensional niche**, proposed by Hutchinson (1957), can be defined as the variety of niche components that constitute the total niche of an organism. This term is often used because it better explains the breadth and complexity of the niche concept. The multidimensional niche concept was first developed in relation to studies of competition between organisms that occupy similar niches. Hutchinson (1978) viewed the niche of an organism as comprising a variety of components, including trophic level (functional role), where it seeks food (for example, forest understory or the rocky bottom of a stream), when it feeds, what size food it eats, and where it reproduces. The multidimensional niche concept emphasizes the role the organism plays in an ecosystem and is broader than, but inclusive of, the definition provided by Elton (1927). For example, the niche of a cliff swallow, an aerial feeder, might be viewed as two-dimensional, comprised of size of insect prey, and, during the nesting season, distance to nest support structures (Figure 2.9); density of insect prey or distance to water could add a third dimension to this multidimensional niche.

The niche of any species involves many dimensions and can be an abstract and difficult concept. However, most studies of niches target from one to three dimensions. For example, a study comparing the niches of mule deer and elk on winter ranges could target overlap in food habits and spatial distribution in relation to escape cover such as dense forest or shrub cover. A study examining possible niche overlap in nesting red-tailed and Swainson's hawks might target prey size as one dimension and nest substrate as a second dimension. Researchers examining possible niche overlap in largemouth, smallmouth, and spotted bass might focus on habitat used, diversity in prey items, and the sizes of prey items.

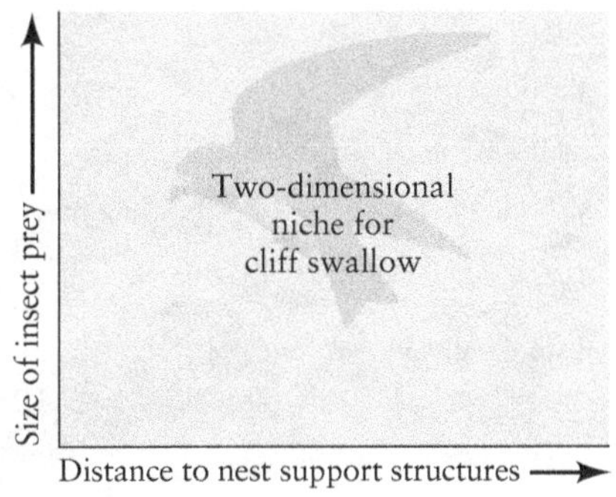

FIGURE 2.9 Size of insect prey and distance to nest support structures (bridges, rock cliffs, etc.) are two potential dimensions of the cliff swallow's niche during the breeding season.

In section 1.2 we defined habitat as the place where an organism lives (Odum 1971). That term needs to be further discussed. Brewer (1994) defined habitat as the specific set of environmental conditions under which an individual, species, or community exists. Perhaps it should be emphasized that, for many animal species, the place where an animal lives or exists need not be used for a large portion of its life. When animals migrate between areas of regular seasonal use, they may spend only a few hours, days, or weeks at any particular site. For example, reindeer are a highly mobile species. All the areas used seasonally by reindeer herds, such as calving grounds, windy hilltops and cooler slopes used to escape insects in the summer, wintering areas, and areas they migrate through, are needed to support reindeer and constitute habitat for that species. Likewise, on its route to Arctic nesting grounds the American golden-plover migrates from southern South America through the middle of North America. The mudflats, pastures, and fields where American golden-plovers stop to feed and rest during migration qualify as migration habitat, even though the birds may remain at such sites for only a few hours or days. Habitats used by migrating animals to rest and replenish food supplies are sometimes referred to as staging areas, particularly for birds. An organism moving rapidly through an area without using the resources present would not necessarily be in habitat suitable for that organism. For example, a coyote that is just crossing an extensive stretch of frozen lake is not in coyote habitat. The airspace occupied by migrating Arctic terns over the Atlantic Ocean is not tern habitat. However, the stream through which a Chinook salmon migrates provides resting and escape habitat even in areas where reproduction may not occur and is considered a type of migrational habitat.

Even on a localized basis, year-round resident species may shift habitats seasonally. During winter in the northern Great Plains, white-tailed deer and ring-necked pheasants move into areas with dense winter cover, such as cattail or willow stands, to seek protection from winds and extreme cold. Fish species such as lake trout shift from deep water habitat to shallower areas as water temperatures cool during the fall.

The habitats of organisms are usually described by the general characteristics of the area being used. For example, the breeding habitat of pied-billed grebes

FIGURE 2.10 Wild turkeys usually place their nests in microhabitats that provide overhead and lateral concealment.

is a shallow body of water that has an interspersion of emergent vegetation and open water. **Emergent vegetation** refers to rooted aquatic plants that can grow through the water column and above its surface; cattails are an example. The water and emergent vegetation are emphasized as the dominant characteristics of the habitat, but other aspects, including prey availability, size of the water body, escape cover, and humidity, may also be important parts of the habitat.

The term **environment** includes the habitat of the organism, but it is much broader in scope. We define *environment* as all the surroundings of an organism, including various conditions or influences that may occur irregularly. Aspects such as acidic deposition, unusual flooding, hurricanes, volcanoes, and extreme temperatures are not generally used to describe habitat but are part of the environment of an organism.

Microhabitat is defined in a variety of ways but is generally viewed as a small area of intensive use within the overall habitat of an organism. For example, a wild turkey usually selects a nest site microhabitat that provides some type of visual obstruction, such as shrub or sapling limbs, around or over the nest (Figure 2.10).

The term **guild** is sometimes used to refer to a group of organisms that use resources, such as food sources, in similar patterns and thus often have considerable niche overlap. For example, fishes that feed on organisms on the bottom of a stream, such as spotted suckers and suckermouth minnows, could be considered part of the guild of "bottom feeders." Birds that catch insects primarily in the air, such as common nighthawks and cliff swallows, could be considered a guild. Small mammals, such as the little pocket mouse and Ord's kangaroo rat, would belong to a seed-eating guild.

2.6 Exotics, Transplants, Invasives, and Restorations

Prior to defining the terms introduced in this section, we must first discuss the zoogeographic regions of the world. The distributions of animals are often described in terms of the zoogeographic regions first defined by Alfred Russel Wallace (1876). These zoogeographic regions, or realms, represent continents or portions of continents that are relatively isolated from one another in terms of the dispersal of animals (Figure 2.11). One zoogeographic region, the **Australian** (Australia, New Zealand, and New Guinea), has long been separated from other regions by extensive areas of ocean and has a unique assemblage of organisms. The **Neotropical region** (South America, Central America, and the West Indies) was also separated from other regions until the development of a land bridge in the last few million years; this isolation plus the variety of habitats and the stability of the climate have contributed to a diverse and unique assemblage of organisms in this region. Both the **Oriental** (India, Burma, Indochina, Malaysia, Sumatra, Java, Borneo, and the Philippines) and **Ethiopian** (Africa south of the Sahara, southern Arabia, and Madagascar) regions, because of their variety of habitats, suitable climates, and long period of sta-

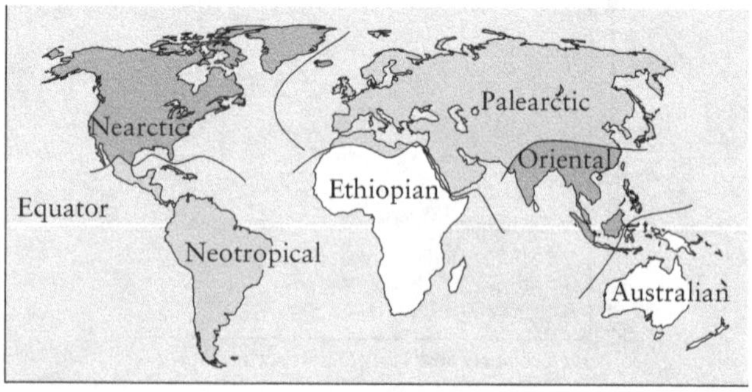

FIGURE 2.11 The major zoogeographic regions of the world.

bility compared with glaciated regions, also have considerable species diversity. The northern zoogeographic regions, the **Nearctic** (North America and Greenland) and **Palearctic** (Europe and Asia north of the Himalayas), have less species diversity than the southern regions; this lack of species diversity may be due to recent glaciation and colder climates, particularly in the northern portions of these two zoogeographic regions.

Because of the activities of humans, many organisms, both plants and animals, have been moved to areas where they did not naturally occur. A variety of terms are used to describe these organisms. These terms should be standardized for legislative, enforcement, and scientific purposes, but unfortunately they are not. We define an **exotic** as an organism introduced from another zoogeographic region. Exotics are sometimes defined as organisms introduced from a foreign land; however, this latter definition seems inadequate in light of changing national boundaries and the lack of relationship between zoogeography and national boundaries. The European starling is an example of an exotic species in North America (Nearctic region), because it was introduced from another zoogeographic region, the Palearctic.

A **transplant** is an organism moved outside its native range but within the same zoogeographic region. The **native range** of an organism is the geographic area in which the species is **indigenous** (originally found in the area). The rainbow trout, a native of the Pacific coast areas of the Nearctic region, is a transplant when introduced into areas of North America outside its native range. For example, rainbow trout in New York or Manitoba are transplants. Rainbow trout introduced into New Zealand or Europe are exotics. Both exotics and transplants can be referred to as **introduced organisms** (**introductions**). The term **invasive species** refers to introduced organisms that rapidly expand their ranges; we will present numerous examples throughout this book, as invasive species are the cause of some of our most pressing biological and social problems.

Once animal species have been introduced into an area they may or may not establish a naturally reproducing population there. Many introduced species, such as rainbow trout and ring-necked pheasants, reproduce in their new areas after initial introduction. Naturally occurring offspring produced by exotics or transplants are called **naturalized organisms**.

There is another type of organism classification that should be added: **feral animals.** These are animals that have reverted to the wild state after having been domesticated. Examples of such organisms are feral horses and feral pigs in the Nearctic region.

Species eliminated from portions of their native range and then successfully released and reestablished on these vacated portions of their native range are **restored species**; the act of reestablishing a native population is called a **restoration.** The reestablishment of breeding populations of giant Canada geese is an example of a successful restoration. Restoration biology is a topic of growing importance because of its relationship to species diversity and richness.

The various terminologies used for organisms moved by humans, either accidentally or purposefully, are a subject of debate. Some of these terms are used interchangeably, and this leads to unnecessary confusion. Throughout this text the definitions provided above will be used in reference to "moved" species.

2.7 Interspecific and Intraspecific Competition

The niche concept is closely associated with interspecific competition. **Interspecific** means between or among different species; **intraspecific** means within a species. **Competition** occurs when two or more organisms living in the same area have overlapping niche requirements for a resource that is in limited supply. For example, there could be competition for nesting sites, such as **secondary tree cavities** (cavities that have already been constructed), or for a specific food resource. In North America the eastern bluebird has declined in part because of interspecific competition for secondary nesting cavities with two bird species introduced from Europe, the European starling and the house sparrow. However, two or more species may overlap almost completely on several dimensions of a niche and still not be in competition. For example, great horned owls and red-tailed hawks are found in the same geographic areas and habitats and have some of the same niche requirements. However, the owl generally hunts at night, while the hawk is active during the day. This difference in temporal (time) activity patterns, a single niche dimension, reduces the competition between these two species. The tendency for species that live in the same area and require similar resources to have niche requirements that differ in one or more dimensions is called **niche segregation**. The **competitive exclusion principle** (**Gause's principle**) addresses the matter by stating that no two species can simultaneously and completely occupy the same niche for an indefinite period of time. This principle has also been succinctly stated as "complete competitors cannot coexist."

A good example of the flexibility of organisms with regard to interspecific competition occurs in the **sunfishes** (members of the fish family Centrarchidae). The pumpkinseed eats a broad variety of invertebrates when it is the only sunfish species in a pond. However, if bluegills and green sunfish live in the same pond, pumpkinseeds will feed extensively on snails. Likewise, where yellow-headed and red-winged blackbirds occur in the same area, the two species compete for overwater nesting sites in emergent vegetation. When nesting, the yellow-headed blackbirds will displace the red-winged blackbirds from the emergent vegetation over the deeper parts of the water body; the red-winged blackbirds will be forced to set up territories in the shallower peripheral areas of the water body or on adjacent upland habitats. Red-winged blackbirds will establish territories and nests over the deeper parts of water bodies where yellow-headed blackbirds are absent. Mule and white-tailed deer in southeastern Arizona were found to have considerable overlap in their food habits but to have relatively little overlap in local distribution due to differences in habitat selection (Anthony and Smith 1977). In all of these examples some niche overlap occurs, but not to the extent of elimination of the other species. Some niche overlap between species that live in the same area is common. The **fundamental niche** of an organism is the niche occupied by that organism when there is no competition from other species. The **realized niche** of an organism is the niche occupied by that organism when some competition with other species is occurring. Most organisms occupy a realized niche instead of a fundamental niche because competition with other organisms is usually occurring to some degree. Predation, like competition, can dictate the realized niche of species. Predation by wolves has been shown to limit the habitats and change the movements of elk in several studies.

Interspecific competition can also be human induced. Introductions of trout into areas and systems lacking native trout species, such as the Black Hills in South Dakota and Wyoming and the streams of New Zealand, have provided considerable potential for competition with native fishes. Brown, rainbow, and brook trout were all introduced into the troutless streams of the Black Hills. Native fish species in those streams, such as the mountain sucker and creek chub, although not endangered, must now compete for prey with the introduced trout species. The brown trout introduced into New Zealand are well known for their rapid growth rates, but they compete with native piscivorous fishes. If competition from introduced species is sufficient, native species may be reduced in numbers, eliminated in some areas, or possibly brought to extinction.

Invasive species also create interspecific competition because they are capable of reproducing successfully and spreading rapidly into new habitats with subsequent damage to native communities and ecosystems. Several invasive species in North America, including common carp, house sparrows, coypu, and European starlings, were described in section 1.5. Few questions were asked about possible interspecific competition with native species when these exotic species were intentionally brought to North America. The above species and many others have spread rapidly to fill available habitat and compete with native species. Other recent invasives, such as various species of Asian carps and snakeheads, pose threats to native species in water bodies where they become established.

In addition to animal species, plant species such as downy brome (cheat grass) have become established and dominant across much of the sagebrush desert in the western United States. Cheat grass has little value to grazing animals and dries quickly during the summer, providing fuel for unwanted fires. The negative repercussions of cheat grass on sagebrush ecosystems and on sagebrush-dependent species such as greater sage-grouse and pygmy rabbits is of much concern. In a similar manner, invasive plants such as kudzu, purple loosestrife, Russian olive, and tamarisk are disrupting native vegetation and ecosystems in areas of North America where they have become well established or even dominant. The invasive species problem is worldwide and growing in extent and severity; invasive species will continue to be of major importance to biologists, the public, and even national governments in the future.

In some cases a native species may increase greatly in numbers and range and can have the same harmful or nuisance effects as an invasive species. Such increases are usually due to landscape changes caused by humans or a natural change, such as changing climate. For example, pied currawong were native to forest habitats in southeastern Australia but have adapted and spread to urban habitats in the past 40 years; their behaviors such as large flock formation, calling loudly and frequently, and predation on nestling songbirds have made them unpopular with human residents.

Biologists should be especially concerned about possible competition between organisms that are native and those that are introduced into an area. Animals found on islands are particularly susceptible to competition and possible elimination after the introduction of new species. For example, the mallard, introduced from Europe and North America, appears to be competing strongly with and increasing in relation to the native Pacific black duck in New Zealand (Caithness et al. 1991); the two species also interbreed.

Intraspecific competition is competition for limited resources among individuals of the same species. The strongest competitor of any individual organism is another organism of the same species, because these organisms have essentially the same niche requirements. In some cases niche requirements vary by gender or age categories, and this variation can reduce direct intraspecific competition. For example, males of most hawk species are generally markedly smaller than females and may exploit different-sized prey, reducing intraspecific competition and expanding the ability of a pair to feed their young. Wild turkeys segregate by gender, and in males, by age, during much of the year and in this manner may reduce intraspecific competition for limited resources. **Sexual segregation** (differential use of space by gender within a species during a portion of the year) is common in many animal species and likely influences intraspecific competition.

Intraspecific competition is sometimes divided into **exploitative competition** and **contest competition** (**interference competition**). In exploitative competition the competing individuals divide the resource somewhat equally, and all suffer similar effects, such as death or reduced growth or reproductive rates, from resource shortages. For example, in a stream with a high density of adult yellow bullheads, the fish are competing for a limited food supply. The result of competition is that all the yellow bullheads may have reduced growth rates, but all are basically influenced in the same manner by the lack of food. Contest or interference competition occurs when some individuals compete more successfully for the resource at the expense of other competing individuals. Nesting barn owls often hatch several young, and because they initiate incubation as soon as they lay their first egg, these nestlings are of different ages and sizes. If the food supply during the rearing stage is limited, the adults may not be able to obtain adequate food for all of the nestlings. In this case the older, larger nestlings will dominate the "contest" for food brought to the nest and thus be more likely to survive.

The consequences of intraspecific competition can also be observed in reduced reproductive rates in some species. For example, when a northern bobwhite population is high, the number of young produced per adult is generally lower than in years when the population is comparatively low. Elk on overcrowded ranges will similarly reflect competition with lowered survival rates of young and reduced birth rates. In some species, younger individuals delay entry into the breeding population by a year or more where competition for resources, including space, is severe.

2.8 Home Range and Territory

Some animals, particularly vertebrates, tend to establish a normal daily area of use as individuals or groups. The area within which an individual animal normally travels in its daily activities is termed its **home range**. Home range should not be confused with the **geographic range** of a species, which defines the geographic boundaries within which the species occurs (see section 2.6). Geographic range is often just referred to as the **range** of a species.

Within an annual period, the home range of an animal normally changes in response to factors such as reproductive condition, weather and habitat conditions, and food availability. For year-round resident species, shorter-term home ranges, such as those used in the nesting season, can be identified, as can longer-term or annual home ranges. Measuring home ranges during different periods of the year can give more realistic information on animal needs in terms of habitat than a single annual measurement. For example, the annual home range for a black-backed woodpecker would include the general area used during courtship and nesting during spring and early summer as well as other periods of the year, such as winter or fall. A biologist would want to know specific habitat needs during shorter periods of time, such as the nesting season, to provide the most useful information. Likewise, large northern pike may have home ranges in the warmer near-shore zone of a lake during spring and move to deeper, cooler portions of the lake in midsummer as water temperature increases. Putting these seasonal home ranges together would mask much important information.

All organisms have space requirements, even if just for a place to attach for life, as with a barnacle. However, many organisms, especially vertebrates, reflect their spatial needs through territorial behavior. **Territory** can be defined as a portion of the home range defended against others of the same or sometimes closely related species. Thus, a large rainbow trout will defend part of a stream against other rainbow trout that would compete for that space as a resting site, feeding site, or for other uses. Likewise, one would expect a snapping turtle or American black bear to defend certain areas against others of the same species. During the breeding season birds often proclaim their nesting territories through songs, calls, and displays. Other types of territories, such as feeding territories and wintering territories, have been identified for various animals. Even a long-domesticated animal such as the dog retains the territorial traits exhibited by its wild kin. The smallest dog can have a big bark and a hard bite in its own territory; its aggressiveness wanes rapidly when it is removed from that territory.

Wolves defend pack territories and act as a group. Reproduction within the pack is restricted to the dominant male and female or at the least to a few dominant individuals, and the entire pack helps in rearing the young. The pack territories are marked by scent and extend over a large area that often exceeds 100 square kilometers. The reason for such large territories is clear when the extensive prey needs of wolves are noted. A large territory assures a food supply throughout the year. Wolves can suffer severe population losses due to starvation during winters, when their primary food sources are scarce. Ignoring pack boundaries can lead to injury or even death for trespassing individuals should the pack not accept them.

Male largemouth bass are highly defensive toward other largemouth bass, as well other fish species, around their nest sites. Thus, largemouth bass nests are widely spaced. In contrast, bluegills commonly nest in colonies, with each male bluegill defending only a small area around his individual nest site.

Some species group together at certain times of the year, or even for most of the year, and appear to show little evidence of territoriality. However, many of these species have an **individual distance** requirement even when found in groups. Observe Bohemian waxwings crowded into a small tree, red-winged blackbirds perched together in the early spring or fall, or migrating birds such as bank swallows on a power line (Figure 2.12). Each individual maintains a fixed, although minimal, distance between itself and the next bird. Birds crossing that distance do what is apparently socially unacceptable to the next bird. Many fish and mammal species behave in a similar fashion.

FIGURE 2.12 Birds in flocks, such as these Bohemian waxwings, still maintain a space, or individual distance, between themselves and the nearest bird or birds. *[Image courtesy of J. Iron.]*

2.9 Limiting Factors and the Law of Tolerance

Organisms, both plant and animal, have a variety of needs that must be fulfilled by the environment. These needs can include food, water, escape cover, habitat for reproduction or nesting, habitat for rearing young, and many others. Populations may be limited by the lack of sufficient amounts of these items. A single component that limits population growth is considered to be a **limiting factor** (**limiting resource**). Populations may also be limited by a combination or interaction of factors.

Biologists attempt to identify the most important limiting factors that affect a population. A smallmouth bass or spottail shiner population normally remains at a certain biomass level in a specific water body. That level is often determined by a limiting factor that could be, for example, prey availability in one area or lack of reproductive habitat in another. Bald eagle breeding populations may be limited by the availability of large nesting trees near water, human disturbance, or an adequate prey base. If human populations and activity in an area are too great, nesting bald eagles may be unable to use potentially valuable habitat. If food sources are lacking, bald eagles will not be able to use large trees isolated from human activity that otherwise would provide suitable nest sites. Determination of limiting factors allows biologists to develop management plans to increase populations of target organisms. If food is the limiting factor, for example, and it is increased, the population can expand, but at some point a new limiting factor will restrict population growth. This discussion could be about mayflies, white-footed deermice, western hog-nosed snakes, or humans—they are all influenced by limiting factors (see section 3.5).

Available space can be a limiting factor for a territorial species. Usually biologists must work with the given space requirements of the species, realizing that the upper limit is set by the behavior of the organism and, ultimately, by limited resources. However, for some species, territory size is at least partly related to the availability of other resources or even to the degree to which territorial individuals are visible to each other; in such cases, management to increase populations may be possible. For example, some fish species are highly territorial around their nests during the breeding season. If that territoriality were dependent on visual contact with other breeding fish at other nest sites, the use of visual obstructions could enhance breeding densities of that species of fish.

For some species, the biologist's goal is to manipulate limiting factors to reduce populations. Rock pigeons and house sparrows are introduced species that are immensely successful in North America. Most city officials would gladly accept no or fewer rock pigeons and house sparrows instead of the abundant populations that often exist, but some buildings provide perfectly engineered habitat for these troublesome species. Engineers could manipulate limiting factors for these species by designing buildings with few sites for rock pigeon roosts or house sparrow nests. Conversely, barn and cliff swallows, which are desirable in most situations, are limited by the availability of clifflike nesting sites protected from excessive rain. Humans have accidentally influenced a limiting factor for these species with the construction of bridges and other structures that provide protected sites for attachment of their mud nests (Figure 2.13).

Within the geographic range of a species one finds regions where the species thrives and other areas where it may occur only sporadically. Often, conditions near

FIGURE 2.13 The construction of bridges has removed a limiting factor by providing nesting habitat for cliff swallows in areas of North America once unsuitable for their nesting populations.

the edge of the geographic range of a species approach the limits of the species' tolerance with regard to one or more environmental conditions; this is the **law of tolerance**, a concept attributed to the ecologist V. E. Shelford (1913). Odum (1971) stated the law of tolerance in the following manner: "Absence or failure of an organism can be controlled by the qualitative or quantitative deficiency or excess with respect to any one of several factors that may approach the limits of tolerance for that organism." Thus, species tolerate ecological minimums and maximums with regard to various conditions. For instance, greenback cutthroat trout living in streams on the eastern slope of the Rocky Mountains in Colorado can exist only under environmental conditions amenable to that species. If one moves downstream along a hypothetical Rocky Mountain river with a greenback cutthroat trout population, summer water temperatures will at some point exceed the limits of tolerance for that species and it will no longer occur or will occur only sporadically. For black bullheads the range of tolerance for embryonic fish has been clearly defined: black bullhead embryos can survive only in water between 18° and 28°C, with an optimal temperature between 20° and 22°C (Stuber 1982) (Figure 2.14).

FIGURE 2.14 The suitability of water for embryonic development of black bullheads is dependent on water temperature. The suitability index is directly proportional to population abundance. Minimum and maximum temperatures provide an example of the range of tolerance during embryonic development for this species. Optimal temperatures appear to fall between 20 and 22°C. *[Adapted from Stuber 1982.]*

Northern bobwhites are found across most of the eastern, midwestern, and southern United States. To the north they reach into southern Minnesota, Wisconsin, and South Dakota, but their numbers fluctuate greatly there with severity of winters and snow cover. It appears that northern bobwhites can withstand a limited duration and depth of snow cover but cannot withstand continuous snow along with extreme cold for more than a few days. Thus, they reach the limits of tolerance for snow cover and temperature in the northern parts of their geographic range. The variation in tolerance for minimum temperatures among small birds is well illustrated by the ability of the horned lark to winter in the harsh cold of the northern Great Plains while many similar-sized species retire to tropical or subtropical areas to winter.

2.10 Dispersion Patterns and Dispersal

Each species has a geographic range over which it is distributed. For example, the geographic range of the Mountain cottontail covers most of the sagebrush deserts found in the intermountain region of western North America. Geographic ranges vary from very small areas to almost worldwide. A species' geographic range reflects factors such as the mobility of the organism, the distribution of habitat conditions appropriate for the species, its ability to tolerate ranges of conditions, the distribution of other species that might exclude the organism, and other physical or behavioral barriers to movement.

Individuals within populations of plants and animals are dispersed over the surface of the earth in various patterns. **Dispersion** refers to the spatial pattern or distribution pattern of individuals in a population. These patterns are sometimes categorized as **random, uniform (spaced)**, or **clumped dispersion** patterns (Figure 2.15). Random dispersion of individuals in animal populations may occur in species that lack social tendencies toward clumping and where there is little variation in the environment. In a random dispersion the position of an organism is independent of the positions of other like organisms. Uniform dispersion of animals is often related to territoriality or other strong intraspecific competition. The dispersion of most animals can be described as clumped and reflects the distribution of suitable habitats as well as social tendencies in various species. A

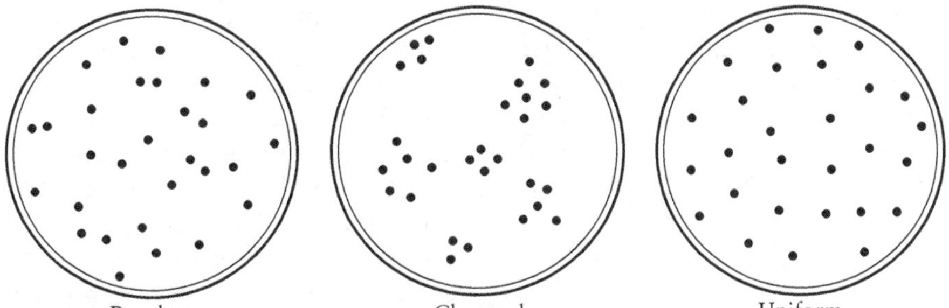

FIGURE 2.15 Individuals within populations may be dispersed across their geographic range in various patterns.

school of bleeding shiners or a herd of muskoxen represent clumped dispersion patterns. The dispersion patterns of animal species often reflect seasonal tendencies toward formation of large groups, pairs, family groups, or other social conglomerations. Dispersion of pairs or other groupings of animals, such as coveys of quail or pods of whales, can also be categorized as random, uniform, or clumped. For example, packs of wolves or territorial pairs of passerine birds would probably be uniformly dispersed if the habitat were somewhat uniform. **Passerine birds** are members of the largest order of birds, the Passeriformes, many of which have the ability to produce complex bird songs, such as those of song sparrows and western tanagers.

Dispersal is defined as outward movements of individuals away from their established areas of activity. For example, when a moose is occasionally found in an Iowa cornfield, it is likely to be a dispersing individual, usually a young male who has left his **natal area** (place of birth or hatching). The terms *dispersal* and *dispersion* are often confused.

Migration is defined as a two-way movement to and from an area that occurs with characteristic regularity or with changes in life history stage; it is often related to reproduction, feeding, or movement between summer and winter ranges. One-way movement out of an area or population is **emigration**; one-way movement into an area or population is **immigration**. Sometimes, as with Pacific salmon, migration involves a return to the natal area to reproduce, with death of the adult occurring after reproduction.

Both dispersal and migration reflect the mobility of an organism. Birds tend to be highly mobile, but their degree of mobility varies greatly even if flightless birds are excluded from comparison. A northern pintail will sometimes cross the Pacific Ocean to reach the Hawaiian Islands, while ruffed grouse will seldom cross even a few hundred meters of water to reach an adjacent shoreline. Birds are the best dispersers in the animal kingdom, although some species of fishes, mammals, and insects can and do make extensive dispersal movements.

Various barriers inhibit movements of animal species, depending on the species' mobility. The movement and dispersal of birds and mammals are greatly influenced by physical barriers such as mountain ranges, deserts, and large bodies of water. A waterfall can limit the movements and distribution of Atlantic salmon during migration. Fishes may be limited in their dispersal by water temperature or salinity beyond the range of tolerance of a species. Animals may be limited in their distribution by biological barriers such as lack of prey species, changes in vegetation, and competition with other closely related species. A forested area may be a major barrier to a species adapted to grasslands; the converse is also true.

Some species are highly adaptable and by nature are distributed across broad geographic regions. The white-breasted nuthatch breeds and winters from Mexico and the southern United States to the southern forests of Canada. North American deermice are found over most of North America. Likewise, the range of the channel catfish extends from southern Canada to Mexico. Other species are much less adaptable and are more limited in their geographic ranges. Seaside sparrows are limited to specific coastal wetlands of the eastern and southern United States. The Colorado pikeminnow, an endangered species, is limited to the turbid river habitat of free-flowing portions of the Colorado River system and is not adaptable to other water types. Most endangered species lack the ability to adapt to changes in their habitats, including human-induced changes. Species with a narrow range of tolerance for various environmental conditions, if those conditions are limited as to the geographic areas in which they occur, are often the species that reach threatened or endangered status.

2.11 Landscape Ecology, Patch Size, Corridors, and Ecotones

Landscape ecology is an area of study that spans several disciplines and is tied directly to ecosystems (Turner et al. 2001). Landscape ecology is defined as the study of structure, function, and change in landscapes. A **landscape** is a heterogeneous land area composed of interacting ecosystems that are repeated in similar patterns throughout an area or region. For our purposes, land area should be interpreted to include aquatic ecosystems such as streams and lakes as well as terrestrial ecosystems. For instance, an agricultural landscape in eastern Kansas might include various crops (winter wheat, milo, soybeans) interspersed with hedgerows, shelterbelts, stream bottoms, and constructed ponds, each of which might be described as a type of ecosystem. Such a system is highly disturbed from its original state but is now a new landscape with characteristic habitat patterns, plants, and animals. This agricultural landscape pattern would repeat itself in similar form throughout a region with similar agricultural potential.

When the needs of individual organisms and populations of organisms are viewed in terms of the minimum-sized living spaces required to meet those needs, one begins to see the importance of landscapes. For instance, forest **patch size**, a landscape characteristic, as well as other characteristics of the forest, such as age, tree diameter, and understory, are related to the success of various plant and animal species. A **patch** is a nonlinear (not long and narrow) surface area within a landscape that differs from surrounding areas. For example, a continuous area of mature ponderosa pine covering 10 hectares (patch size) surrounded by grassland is a patch, as would be a similar-sized area of grassland surrounded by tilled agricultural land or with urban habitat on all sides. If a patch is isolated from other patches of the same habitat or vegetation type, it could be described as having poor connectivity. **Connectivity** is the continuity or connectedness of a habitat or vegetation type across the landscape (Turner et al. 2001).

The decline in the number of northern spotted owls in the Pacific Northwest requires consideration at a landscape level and is related to concepts such as patch size and connectivity. Populations of the northern spotted owl are thought to be partially dependent on the patch size of old-growth or shifting-mosaic steady-state forest. Remaining patches of old-growth forest are also prime logging habitat and are thus susceptible to **habitat fragmentation** (the breaking of larger patches of habitat into smaller, disconnected patches) or elimination. Logging and associated lumber industries are the major sources of employment in many areas of the Pacific Northwest. Logging restrictions to protect remaining old-growth forest are extremely unpopular with segments of the public dependent on logging for their income. Many people have already lost their logging-related jobs, and local economies have in some cases been seriously impaired. Often those dependent on logging place the blame for their economic losses on the northern spotted owl; this is by no means the only, or even the most important, reason for economic losses in that region. Understandably, northern spotted owls have become a politically sensitive issue because they were used as a primary reason to reduce logging on old-growth timber. Other segments of the public want the remaining old-growth forest on public lands preserved for aesthetic reasons, for tourism, and as habitat for the northern spotted owl and other species. The debate concerning the northern spotted owl, its abundance and population status, and its dependence on old-growth forest is still continuing.

Patch size is an important concept in the management of many other species as well. The occurrence and abundance of western meadowlarks and a host of other grassland birds are very much tied to grassland patch size. The success of the reintroduction of the endangered black-footed ferret, whose primary prey is the black-tailed prairie dog, is closely tied to the availability of patches of prairie dog towns of some suitable size. Reduction of patch size will result at some point in the area no longer being able to support a viable black-footed ferret population, regardless of how the ferrets are otherwise protected.

The concept of patch size has seen little use in fisheries, but the principles are applicable. Areas of submergent vegetation provide escape cover and may determine the survival rates of some prey fishes. **Submergent vegetation** is aquatic vegetation that generally does not extend above the water surface; examples would be kelp and sago pondweed. Perhaps the total area of submergent vegetation as well as the size of submergent vegetation blocks (patch size) would relate well to survival rates of prey species such as fathead minnows, golden shiners, or brook sticklebacks. Coral reefs nicely tie into the idea of patch size with regard to the types of organisms that find a particular-sized reef to be suitable habitat.

The concept of **corridors** has application to both wildlife and fisheries. A terrestrial corridor is a relatively narrow strip of land that differs, usually in terms of dominant vegetation form (such as forest or grassland), from surrounding areas. The same concept can be applied to rivers, chains of lakes, or other linear arrangements of aquatic systems. A narrow strip of trees extending across a large meadow and connecting extensive patches of forest on either side of that meadow is a corridor. Similarly, a narrow stretch of grassland running through a forested region or a series of periodically flooded shallow water habitats in close proximity to one another are also types of corridors. Woodlands on the banks and numerous islands of the Platte River in Nebraska serve as a corridor for daily and seasonal movements of Cooper's hawks, small woodland birds, and a variety of other species (Figure 2.16). Extensive timber removal by explorers, settlers, and railroad workers in the mid- to late 1800s destroyed much of this woodland corridor on the Platte River, but reforestation and spread of woodlands has since occurred due to reduction of timber cutting, reduced stream flows, and lack of prairie fires (Johnson and Boettcher 2000). Loss of woodland, shrubland, or other corridor types can influence the movements, vulnerability to predators, and many other aspects related to the welfare of species dependent on such corridors.

Various fish species use streams and rivers as corridors of movement. Humans have in many instances altered fish corridors with dams, dredging, timber operations, or other activities that cause changes such as siltation and temperature increases. A good example of corridor disruption with resultant negative effects on fish populations is the series of dams on the Columbia River and its tributaries; these dams have negatively affected Pacific salmon migrations. In addition, sections of rivers may become too polluted for some aquatic species, thus interrupting a corridor.

The extensive reduction of shallow water habitats and their associated vegetation (such as bulrushes, spike rushes, and cattails) in North America and many other areas of the world can be viewed as a corridor and patch problem; it is also a form of habitat fragmentation. As shallow water habitats are drained, the distance between them increases and the number of such habitats per unit of area is reduced. How are organisms such as tiger salamanders, southern leopard frogs, and other seldom-studied species affected by the densities and linkages of these shallow water habitats? The decline of frogs in many areas may be related to loss of wet areas and thus the landscapes in which the frogs once thrived. Other factors, such as pesticides, that could affect populations of such organisms should not be ignored. Nevertheless, losses of shallow water habitats to urbanization, agricultural drainage, and

FIGURE 2.16 Explorers from the early to mid-1800s described countless timber-covered islands along the Platte River in Nebraska, providing evidence that this prairie river has served as a corridor of closely linked woodland patches for hundreds if not thousands of years. *[Photograph courtesy of C. Johnson.]*

other causes are closely related to patch and corridor concepts and the welfare of a great variety of organisms. In some states over 90% of the original shallow water habitats have been drained.

Some corridors, if too narrow, can be viewed as predator traps for species using them. For example, ground-nesting birds are more vulnerable to predation by red foxes, raccoons, and other mammalian predators if nesting areas are reduced to narrow corridors that can be easily traversed and searched by these predators. A fenceline habitat with cropland on either side can be a dangerous nesting area for a blue-winged teal. However, a small increase in corridor width could reduce predation rates. Unfortunately, in areas where most of the land is tilled for crops, the only nesting areas remaining for ground-nesting birds are often narrow corridors or strips of grass along field or fenceline edges. In most agricultural landscapes the tendency has been to reduce the width of these corridors. Species adapted to extensive and continuous stands of forest may be at greater risk of predation if the forest is reduced to patches connected by thin strips of forest. The relationship of corridor width to predation has received little study and deserves more consideration.

The transition area between two different ecosystems is called an **ecotone** and is also known as an **edge**. For instance, the area where a mixed-grass prairie and a burr oak woodland come together is an ecotone. Such transition areas are often associated with greater species diversity, because the two habitats in juxtaposition provide an increased array of trophic, spatial, and habitat possibilities.

In fisheries the word *ecotone* is not commonly used, but there are transition zones in aquatic environments to which it could be applied. For example, the transition zone of a lake bed between a shallow water area and a deep water area is an ecotone. The structural change at the edge of a coral reef is an ecotone. Even the interface between beds of emergent or submergent vegetation and open water fits this description. A variety of species, both predators and prey, find transition areas attractive.

Some ecosystems are rich in species in part because they tend to be narrow and transitional in nature. A **riparian zone** is the habitat associated with the edges of rivers and streams. The riparian zone is really a transition from the stream or river to the adjacent upland shrubland, grassland, or other community. Because of the varieties of habitat at these edges, riparian zones have high species diversity and are important to the reproduction, movements, and other activities of a disproportionately large number of species.

Finally, **scale** represents a critical concept in landscape ecology and ecological research in general (Turner et al. 2001). Scale should be considered in both space and time. Spatial scale and sampling approaches, for example, would differ greatly between a study evaluating habitats and landscapes needed by a northern pocket gopher population compared to that for a herd of migratory reindeer. From a time perspective, results from a study of burning effects on a forested area likely would be interpreted quite differently 2 years after treatment compared to 40 years after treatment.

2.12 Predator–Prey Interactions

As previously described, anything that hunts, kills, and eats other animals to obtain most of its energy and nutrients is a predator; the organism eaten is the prey. Animals such as the red-tailed hawk, cougar, and muskellunge normally come to mind when the word *predator* is used. One does not normally think of eastern kingbirds or blacktail shiners as predators, but they fit this description because they hunt and kill other animals (e.g., insects).

Predator–prey interactions are not as simple as they may seem; the lines between predators and prey are often blurred. The interaction between largemouth bass and bluegills is a good example. The largemouth bass is usually viewed as the predator, with the bluegill assuming the prey role. However, bluegills can eat largemouth bass eggs or young. In addition, a juvenile largemouth bass may be a predator of small bluegills but can be a prey item for a larger predator such as a walleye, bowfin, or northern pike.

Predation is also affected by humans because of our influence in fragmenting or eliminating habitats. Grassland nesting birds, including species such as northern pintails and long-billed curlews, face reduced habitat along with increased predation when land-use practices fragment nesting cover. In situations where livestock are allowed to overgraze and damage grasslands, predation on nests and nesting females of various bird species may also increase. Humans also change the predator composition through various land-use practices such as planting trees in grassland regions or removing once-dominant predators that controlled smaller predator species. Predators such as raccoons and red fox appear to thrive in human-disturbed habitats and in this manner add to predation on native bird species.

Wolves are predators that specialize in taking large **ungulates** (hoofed animals), though they may also seek smaller prey, such as American beavers. The predator–prey interactions of wolves have been studied intensively. When wolves feed on deer, moose, or reindeer, they are sometimes seen as competitors with humans. The effects of wolves on large ungulate populations have received and continue to receive much attention from researchers, managers, and the public. When packs of wolves roamed the prairies in areas such as eastern Montana and North Dakota, both wolves and ungulates seemed in ample supply. The ungulates thrived on the abundant grasses and forbs, and wolves as well as grizzly bears, each limited by intraspecific competition (territories), likely attained relatively stable population levels. A quote by Meriwether Lewis describing the grasslands of north central Montana paints an early picture of predator and prey in the northern plains: "One vast plain, in which innumerable herds of buffalo were seen, attended by their constant shepherds, the wolves" (Dillon 1965).

Work in northern Minnesota indicated that wolves depend on white-tailed deer as a primary food source (Fuller 1989). In areas with abundant white-tailed deer the effect of wolves on the deer population may be negligible, but if an area within the home range of a wolf pack has a low deer density, the possibility of an effect becomes greater. Some researchers have indicated that the likelihood of wolf effects on a prey population is closely tied to the predator-to-prey ratio in terms of biomass. Where white-tailed deer or other ungulate populations are reduced due to severe winter weather or other reasons, the wolf population may further depress prey populations. However, reproduction and thus population levels in northern Minnesota white-tailed deer are also greatly influenced by vegetation changes related to forest management and succession. Thus, wolves are just one factor in a complex relationship between white-tailed deer and their environment.

It has been suggested that wolves maintain reindeer populations at low levels (top-down control) in areas where there is a low number of prey per predator. Under such conditions wolf populations likely could be reduced by human intervention to allow reindeer recovery, and that wolf control would eventually allow both prey and predator to thrive once the ratio of prey to predator increased. This argument was used in Alaska, where some reindeer and moose populations were at low levels and experiencing poor calf survival due, in part, to predation from wolves and grizzly bears (Gasaway et al. 1983). A large-scale study involving wolf control in the southwestern Yukon indicated that wolf populations strongly limited successful reproduction (recruitment) in reindeer and survival of adult moose (Hayes et al. 2003). Other factors, such as long-term changes in habitat, weather patterns, and normal cycles or fluctuations in ungulate populations, also need to be considered in attempting to understand changes in abundance of reindeer and moose (Boutin 1992). Mech and Peterson (2003) noted that the relationship between wolf populations and their ungulate prey is complicated by the availability of alternative prey (reindeer plus moose, for example). The influence of other large predators such as grizzly bears, productivity of the habitat and of prey populations, human effects on predator and prey, snow conditions, and other interacting factors also affect resulting predator–prey relations. The idea that predation can limit ungulates to a low-density equilibrium remains a matter of debate and of ongoing research.

Fisheries biologists look at predator–prey relations in different ways depending on the type of fish community being managed. For example, largemouth bass can be managed to create high-quality panfisheries in small impoundments (Gabelhouse 1984). **Panfish** are small game fishes such as bluegills, black crappies, white perch, and yellow perch. A high density of small (20–30 centimeter) largemouth bass can reduce overabundant young of some panfish species so that the surviving panfishes experience decreased competition. These panfishes grow quickly and reach sizes of interest to anglers. However, northern pike and yellow perch can have a very different predator–prey relationship. When northern pike density in Horseshoe Lake, Minnesota, became too high, the number and sizes of yellow perch in the lake declined (Anderson and Schupp 1986). Larger northern pike preferred to feed on large prey, represented by the larger yellow perch. Thus, northern pike actually competed with humans for large yellow perch.

Certain outbreaks of insects in forest areas demonstrate an additional economic aspect of predator–prey interactions. For instance, some insectivorous birds are highly mobile and seek areas with abundant insects. In areas experiencing an outbreak of an insect causing damage to timber, this relationship of predator and prey is of potential economic interest to humans. Biologists may have an additional interest in what happens to the insectivorous birds in terms of the effect of abundant prey on reproduction and survival of young. For instance, the increased availability of prey in an area with an insect outbreak may remove food as a limiting factor for one or more bird species.

Thus, in predator–prey interactions it is important to understand what happens to the predator as well as to the prey. Both prey and predator are part of the interaction, and prey effects on the predator population can be as substantial as predator effects on the prey. When prey is scarce, animals such as snowy owls, Canadian lynx, and northern hawk owls (Figure 2.17) reduce their reproductive efforts and may suffer increased losses of young. Prey shortages commonly influence reproductive success in predators.

FIGURE 2.17 Abundance of prey influences numbers of eggs and young produced by species such as these northern hawk owls. These young have left the nest but are not yet capable of flight. *[Image courtesy of R. MacDonald, U.S. Fish and Wildlife Service.]*

The effect of prey shortages on a predator also depends on the degree to which the predator is a **generalist** or a **specialist**. Specialists, which require a more specific prey base, are more at risk with fluctuations of their primary prey; generalists can more readily switch to other prey sources. For example, a northern grasshopper mouse will eat a great variety of insects, other invertebrates, seeds, and even other mice—it is a generalist. In comparison, the black-footed ferret is a species strongly dependent on black-tailed prairie dogs for its prey—it is a specialist.

SUMMARY

Ecosystems represent interacting systems of organisms and their abiotic environment and are characterized by dominant vegetation or some other dominant feature. Ecosystems have a structure in terms of the types of organisms and abiotic materials present. Ecosystem function relates to nutrient (mineral) cycling and energy flow. Nutrients and energy move through ecosystems in a complex manner that can be represented as a food web. Energy is captured by photosynthesis by primary producers, but much is lost in unusable forms at each trophic level as animals consume primary producers or other animals. Energy pyramids illustrate this loss of energy.

Ecological succession relates closely to what organisms are present and to the energy available to those organisms. Wildlife and, to a lesser extent, fisheries management and conservation practices directed at improving habitat are often designed to manipulate succession to create seral stages favorable to particular target organisms or communities.

The role of an organism in an ecosystem is termed its niche; this complex concept can be illustrated using the multidimensional niche concept. Competition can occur when resources are in short supply and some aspects of the niches of two or more organisms living in the same geographic area overlap. Many introduced species, planned or unplanned, have reproduced successfully and greatly expanded their populations into suitable habitats; these species, especially if they cause damage to native organisms and ecosystems, are considered invasive species.

Animals establish daily patterns within an area defined as their home range. Some animals establish territories to defend an area of habitat, usually against others of the same species. The strongest competitor of an organism is usually another organism of the same species. Habitat must often be described on a seasonal or functional basis for animals that migrate or otherwise make seasonal changes in habitat use.

Animal populations can be limited by a single factor (limiting factor) or by more than one factor acting in concert. The conditions near the edge of the geographic range of a species usually approach the limits of tolerance of that species for one or more environmental factors.

Animals may be dispersed in random, uniform (spaced), or clumped patterns. Animals moving away from established home ranges are dispersing, while migration is a two-way movement to and from an area occurring on an annual or periodic basis. Animals introduced into a new area from a different zoogeographic realm are exotics, while those introduced into new areas within the same zoogeographic realm are transplants.

Landscape ecology is a field of study that examines patterns that exist within a heterogeneous land area. Animal populations are often influenced by landscape characteristics such as patch size and habitat fragmentation. Animal movements may be dependent on corridors of certain types of habitat. The transition zone between two ecosystems is an ecotone and may be particularly rich in plant and animal species.

Predator–prey interactions influence both prey and predator populations. The effects of predators on prey are complicated by other factors, such as habitat quality, that influence reproduction and survival of the prey.

PRACTICE QUESTIONS

1. Select two ecosystems with which you are familiar and describe their dominant features.

2. Select two landscapes with which you are familiar and describe those landscapes in terms of interacting ecosystems.

3. Give two specific examples in which widely separated ecosystems (terrestrial, marine, freshwater, or any combination) are interrelated.

4. How is species diversity different from species richness? Give an example in conceptual terms using a taxon and ecosystem familiar to you.

5. Give two examples, other than those provided in the chapter, of populations and metapopulations.

6. Diagram the primary aspects of the carbon, sulfur, and nitrogen cycles.

7. Develop a food web example of your own. Show routes of energy flow and categorize species by trophic level.

8. Develop an energy pyramid for the food web in question 7.

9. Give an example of top-down and bottom-up control of trophic levels.

10. Give one example of a trophic cascade and explain the possible repercussions.

11. Give an example of a shifting-mosaic steady-state community other than the examples provided in this chapter. Why does your example fit the definition? Name two animal species primarily associated with that community stage. Describe an earlier seral stage of this community and name two animal species characteristic of that stage.

12. Give two dimensions of a niche for an organism with which you are familiar. Relate each dimension to the niche of the animal.

13. Describe the habitat of two animal species. Indicate whether the habitat is seasonal.

14. Name three invasive species found in the area where you live. How might one of these species compete with a native species?

15. Give your own examples of contest competition and exploitative competition.

16. What are some possible advantages and disadvantages of territorial behavior? What are some of the ways that animals mark or otherwise advertise territory boundaries?

17. Give four of your own examples of possible limiting factors for different aquatic or terrestrial animal populations.

18. How might fisheries biologists view predation differently in terms of management and conservation than might wildlife biologists?

19. How have humans influenced predation on ground nesting birds?

SELECTED READINGS

Bissonette, J. A., and I. Storch, eds. 2003. *Landscape ecology and resource management: Linking theory with practice*. Island Press, Washington, D.C.

Diana, J. S. 2004. *Biology and ecology of fishes*. 2nd ed. Biological Sciences Press, Carmel, Ind.

Errington, P. L. 1967. *Of predation and life*. Iowa State University Press, Ames, Iowa.

Levin, S. A., ed. 2001. *Encyclopedia of biodiversity*. vols. 1–5. Academic Press, San Diego.

McCullough, D. R., ed. 1996. *Metapopulations and wildlife conservation*. Island Press, Washington, D.C.

Mech, L. D., and L. Boitani, eds. 2003. *Wolves: Behavior, ecology, and conservation*. University of Chicago Press, Chicago.

Ricklefs, R. E. 2001. *The economy of nature*. 5th ed. W. H. Freeman and Co., New York.

Ricklefs, R. E., and G. Miller. 2000. *Ecology*. 4th ed. W H. Freeman and Co., New York.

Smith, R. L., and T. M. Smith. 2001. *Ecology and field biology*. 6th ed. Benjamin Cummings / Addison Wesley Longman, Inc., San Francisco.

Stroud, R. H., and H. Clepper, eds. 1979. *Predator-prey systems in fisheries management*. Sport Fishing Institute, Washington, D.C.

Wiens, J. A., and M. R. Moss, eds. 2005. *Issues and perspectives in landscape ecology*. Cambridge University Press, Cambridge, U.K.

LITERATURE CITED

Anderson, D. W., and D. H. Schupp. 1986. *Fish community responses to northern pike stocking in Horseshoe Lake, Minnesota*. Investigational Report No. 387. Minnesota Department of Natural Resources, St. Paul, Minn.

Anthony, R. G., and N. S. Smith. 1977. Ecological relationships between mule deer and white-tailed deer in southeastern Arizona. *Ecological Monographs* 47:255–277.

Bailey, R. G. 1998. *Ecoregions: The ecosystem geography of the oceans and continents*. Springer-Verlag, New York.

Boutin, S. 1992. Predation and moose population dynamics: a critique. *Journal of Wildlife Management* 56:116–127.

Brewer, R. 1994. *The science of ecology*. 2nd ed. Saunders College Publishing, Fort Worth, Tex.

Caithness, T., M. Williams, and J. D. Nichols. 1991. Survival and band recovery rates of sympatric grey ducks and mallards in New Zealand. *Journal of Wildlife Management* 55:111–118.

Daskalov, G. M. 2002. Overfishing drives a trophic cascade in the Black Sea. *Marine Ecology-Progress Series* 225:53-63.

Dillon, R. H. 1965. *Meriwether Lewis: A biography*. Coward-McCann, Inc., New York.

Elton, C. S. 1927. *Animal ecology*. Macmillan, New York.

Fuller, T. K. 1989. Population dynamics of wolves in north-central Minnesota. *Wildlife Monographs* 105.

Gabelhouse, D. W., Jr. 1984. An assessment of crappie stocks in small midwestern private impoundments. *North American Journal of Fisheries Management* 4:371–384.

Gasaway, W. C., R. O. Stephenson, J. L. Davis, P. E. K. Shepherd, and O. E. Burris. 1983. Interrelationships of wolves, prey, and man in interior Alaska. *Wildlife Monographs* 84.

Hayes, R. D., R. Farnell, R. M. P. Ward, J. Carey, M. Dehn, W. Gerald, A. M. Baer, C. L. Gardner, and M. O'Donoghue. 2003. Experimental reduction of wolves in the Yukon: Ungulate responses and management implications. *Wildlife Monographs* 152:1-35.

Hutchinson, G. E. 1957. *A treatise on limnology*. Vol. 2 of *Introduction to lake biology and the limnoplankton*. John Wiley & Sons, New York.

Hutchinson, G. E. 1978. *An introduction to population ecology*. Yale University Press, New Haven, Conn.

Jepson, P., and R. J. Whittaker. 2002. Ecoregions in context: A critique with special reference to Indonesia. *Conservation Biology* 16:42-57.

Johnson, W. C., and S. E. Boettcher. 2000. The presettlement Platte: Wooded or prairie river. *Great Plains Research* 10:39-68.

Levin, S. A., ed. 2001. *Encyclopedia of biodiversity*. vols. 1–5. Academic Press, San Diego.

Mech, L. D., and R. O. Peterson. 2003. Wolf-prey relations. Pages 131-160 *in* L. D. Mech and L. Boitani, eds. *Wolves: Behavior, ecology, and conservation*. University of Chicago Press, Chicago.

Odum, E. P. 1971. *Fundamentals of ecology.* 3rd ed. W. B. Saunders, Philadelphia.

Primack, R. B. 2002. *Essentials of conservation biology.* 3rd ed. Sinauer Associates, Sunderland, Mass.

Ricklefs, R. E. 2001. *The economy of nature.* 5th ed. W. H. Freeman, New York.

Ripple, W. J., E. J. Larsen, R. A. Renkin, and D. W. Smith. 2001. Trophic cascades among wolves, elk and aspen on Yellowstone National Park's northern range. *Biological Conservation* 102:227–234.

Shelford, V. E. 1913. *Animal communities in temperate America.* University of Chicago Press, Chicago.

Smith, R. L., and T. M. Smith. 2001. *Ecology and field biology.* 6th ed. Benjamin Cummings/Addison Wesley Longman, Inc., San Francisco.

Stern, P. C., O. R. Young, and D. Druckman, eds. 1992. *Global environmental change: Understanding the human dimensions.* National Academy Press, Washington, D.C.

Stuber, R. J. 1982. *Habitat suitability index models: Black bullhead.* FWS/OBS-82/10.14. U.S. Fish and Wildlife Service, Washington, D.C.

Turner, M. G., R. H. Gardner, and R. V. O'Neill. 2001. *Landscape ecology in theory and practice: Pattern and process.* Springer-Verlag, New York.

U.S. Congress, Office of Technology Assessment. 1987. *Technologies to maintain biological diversity.* OTA-F-330. Washington, D.C.

Wallace, A. R. 1876. *The geographical distribution of animals.* 2 vols. Harper and Brothers, New York.

CHAPTER 3

Population Dynamics

Before one can understand how communities and ecosystems function, it is necessary to understand how populations function. When a biologist considers the dynamics and structure of a population, specific concepts come to mind. **Population dynamics** is the study of changes in numbers or weight (biomass) of organisms in populations and of the factors influencing those changes.

Populations are affected by three dynamic rate functions: natality or recruitment, growth, and mortality. Natality and recruitment add new individuals to a population, growth adds biomass to a population, and mortality removes individuals from a population. These three dynamic rate functions interact to influence a population and determine **population structure.** Population structure generally is assessed as size structure, age structure, or sex ratio. Population structure can also be called **population form,** and population dynamics is sometimes referred to as **population function.** This chapter introduces the basic concepts of population dynamics and describes how the interaction of the three dynamic rate functions influences population structure, population regulation, production, and yield. Finally, the chapter will address **community structure,** which is a result of interactions among populations. Community structure is most often defined as the composition (species composition and abundance) of an assemblage of organisms.

3.1 Natality/Recruitment

Natality refers to birth or hatching and is commonly expressed as the number of young individuals born or hatched within a specified period of time. An example would be the number of big brown bats added to a population through birth in one year. Most often, wildlife biologists express natality rate as the number of offspring per 1,000 females. Natality can also be used in reference to fishes, although most fishes are hatched, not born. In contrast, fisheries biologists consider **recruitment** more often than natality. Recruitment can be defined in numerous ways; however, fisheries biologists often define recruitment as the number of fish hatched or born in any year that survive to reproductive size. Thus, fisheries biologists are not so concerned with the number of fish that hatch; they are more concerned with the number that survive to reach adulthood and enter the reproductive pool.

Why are fisheries biologists more concerned with recruitment than with natality? The answer has to do with differences in reproductive strategies (see sections 1.3 and 6.10). Many wildlife species have low numbers of offspring. For example, white-tailed deer usually have only one or two fawns per year. Species that produce few young generally provide them with substantial parental care, and a large percentage of the young survive. Conversely, most fishes tend to produce many eggs and invest much less energy in parental care. For example, a large female common carp can produce 1 or 2 million eggs per year. These adhesive eggs are released onto flooded vegetation, and the parents swim away and let the eggs and newly hatched offspring fend for themselves. For most fishes, as egg numbers increase, parental care decreases; the converse is also true.

Only a fraction of the fish eggs that are produced will actually hatch, and only a fraction of the newly hatched fish will be recruited into the adult population. The following is a hypothetical, but realistic, example using sockeye salmon. These fish attain most of their growth as adult fish in the ocean but return to fresh water to **spawn** (release their eggs and sperm). Sockeye salmon use their body reserves to get to the spawning stream and die after spawning. A female sockeye salmon might contain 3,000 eggs. Of those 3,000 eggs:

In a balanced system, only two offspring from a pair of adults need to return to the stream and reproduce to continue this age-old cycle.

Although it is not as strongly stressed as in fisheries work, wildlife biologists are also concerned with recruitment. Monitoring of fawn-to-doe ratios in white-tailed deer and counts of the young of songbirds, waterfowl, and upland game birds are common examples. **Upland game** is an artificial category describing the many birds and small mammals, such as ruffed grouse and eastern fox squirrels, which are found in terrestrial systems and used for recreational hunting.

Natality and recruitment are affected by both **density-dependent** and **density-independent factors.** A density-dependent factor is an environmental factor that affects a population based on its density. For example, when the size of an elk herd reaches the point at which the animals are competing for a limited food supply, a lower percentage of cow elk will produce calves. A density-independent factor is an environmental factor that acts independently of population density. An example in a fishery could be the influence of a major flood on a stream. Flood conditions could result in no or reduced survival of young creek chubs, regardless of the population size. In other words, the survival of one young fish would not be dependent on how many others were present; the environmental factor (the flood) would have killed many of the young creek chubs, regardless of population density.

Density-dependent influences on natality and recruitment are commonly discussed by biologists. Wildlife biologists often refer to the concept of **inversity.** When the density of adult white-tailed deer is low, the

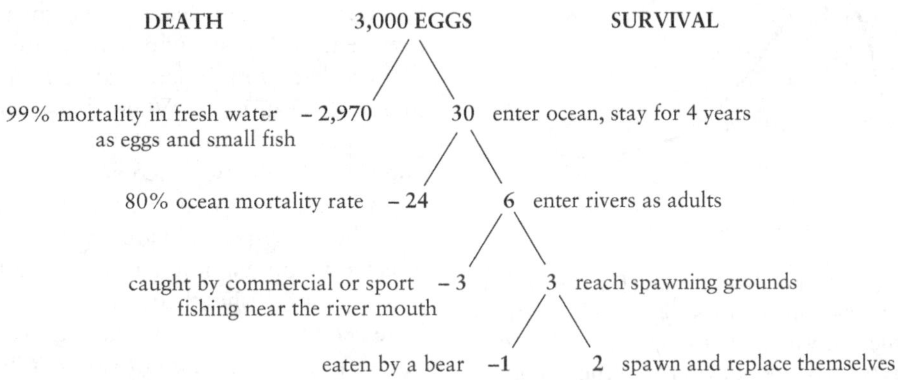

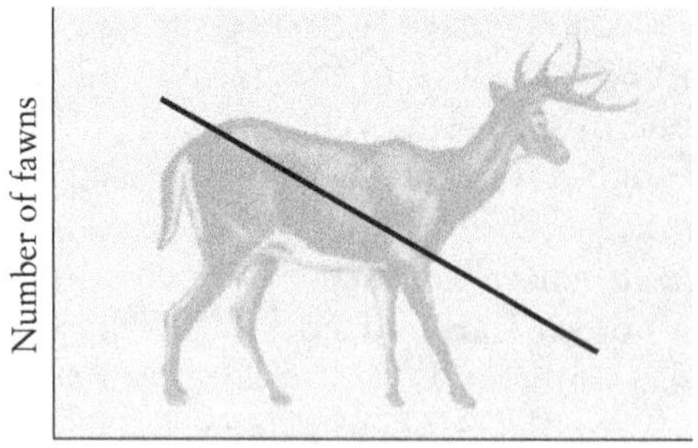

FIGURE 3.1 The concept of inversity is illustrated by the relationship of natality in a white-tailed deer population to adult population density. As the number of adult white-tailed deer in a population increases, the number of fawns produced decreases.

proportion of yearling does that bear fawns increases and adult does often have twins or even triplets. At such times there tends to be less competition for resources such as food, and well-nourished white-tailed deer are more likely to breed at a younger age and produce larger numbers of offspring. When the density of adult white-tailed deer is high, food may be a limiting factor and the nutritional status of the deer may be poor. As a result, few yearling white-tailed does bear fawns, and adult does usually have only a single fawn (Figure 3.1).

Fisheries biologists typically are more concerned with changes in recruitment than with changes in reproductive capability. **Stock-recruitment relationships** demonstrate the influence of parental abundance on subsequent recruitment (Figure 3.2). A **stock** is a group of organisms with a common ancestry or parentage that is adapted to a particular environment; *stock* is primarily a fisheries term (see section 4.2). Wildlife biologists generally use the term *population*, but at times they use *stock*. Chinook salmon that are spawned in different Alaskan and Canadian rivers will intermingle in the ocean. However, when they return to the rivers to spawn, each stock will return to the place where it was spawned.

When the density of an adult stock is high, there is room for little recruitment because adults are abundant (point *z*, Figure 3.2). As adult density decreases, recruitment increases until a point is reached at which recruitment is at its maximum (point *y*, Figure 3.2). If adult density decreases below this point, recruitment decreases because there are too few adults left to replace themselves (point *x*, Figure 3.2); the stock then "collapses." Stock-recruitment relationships have most often been developed for commercially harvested fishes in the ocean or large lakes, but such relationships have also been noted for wildlife species such as white-tailed deer (McCullough 1979) and inland recreational fishes such as walleye (Hansen et al. 1998; Beard et al. 2003).

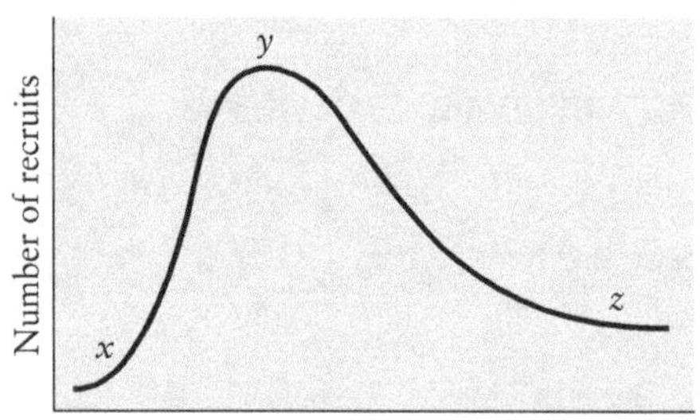

FIGURE 3.2 The greatest number of young fish is recruited to a fish population when the biomass of adults is intermediate (*y*). At extremely low adult biomass, there are too few adults to quickly rebuild the population (*x*). When biomass of adults is high, fewer young fish will be able to recruit to the adult population because of density-dependent factors such as competition for food resources (*z*).

3.2 Growth

Most birds and mammals exhibit determinate growth, while fishes exhibit indeterminate growth (see section 1.3). Organisms with determinate growth tend to grow quickly and reach their approximate adult size in a year or two. For organisms with indeterminate growth, growth patterns are more variable.

Fish growth is very much density dependent. Indeterminate growth allows the development of high-density, slow-growing fish populations that are often referred to as **stunted populations.** Yellow perch growth can be used as an example. In a high-density, slow-growing population, yellow perch may reach 125 millimeters after four years (a weight of 20 grams). In low-density, fast-growing populations, yellow perch can reach 260 millimeters after four years (approximately 290 grams). When density is high, there is much competition for a limiting factor such as food and growth is slow. When density is low, food is more abundant and growth rates are substantially faster.

The growth of individual wildlife organisms can be affected by population density, but not to the same extent as that of fishes. Because most birds and mammals have determinate growth, stunting to the extent possible in fish populations cannot occur. While there may be some density-dependent slowing of growth in birds and mammals, such a problem usually changes to density-dependent mortality, as we will discuss in the next section.

3.3 Mortality

Mortality is the death of an individual, but it is often expressed as the percentage of a population that dies in one year. Mortality can be categorized as density dependent or density independent. An example of mortality that is primarily density independent occurs in ring-necked pheasant populations in the northern Great Plains. An extensive winter blizzard can result in high mortality in these birds; that is, a certain percentage of the ring-necked pheasants will be killed due to the blizzard, regardless of the density of the birds. This does not mean that all storm-related mortality is necessarily density independent; a density-dependent factor such as competition for winter cover might also influence mortality from the blizzard.

Density-dependent mortality caused by factors such as predation or limited food supplies can occur in animal populations. When common muskrat populations are high, they may outstrip the supply of available food. An increase in mortality may occur that reduces the population to a level that the available habitat can support. Another example of density-dependent mortality involves walleye predation on yellow perch. In a year when young yellow perch are abundant in a lake, walleye predation may account for a substantial portion of total mortality for the perch. Conversely, in years when young yellow perch density is low, the walleyes are likely to switch to other, more abundant fishes as primary prey items, and walleye predation will account for a smaller portion of the mortality of young perch.

Wildlife biologists monitor both natural and hunting mortality for game species, while fisheries biologists monitor both natural and fishing mortality for game and commercial fishes. **Natural mortality** is mortality caused by predation, starvation, diseases, accidents, or other natural causes. **Harvest mortality** results from human activities directed at taking organisms for food or recreational purposes and can include harvest related to activities such as hunting, fishing, or trapping. Harvest mortality is an anthropocentric term; this mortality is actually one more type of predation, and it could be included in the definition of natural mortality. **Total mortality** is the sum of natural and harvest mortality. Natural mortality is closely tied to the growth rates and longevity of animals. A longer-lived species such as the walleye may have natural mortality rates near 30% per year, while a shorter-lived species such as the white crappie may have natural mortality rates closer to 50%. Similarly, shorter-lived wildlife species such as eastern cottontail rabbits have higher natural mortality rates than longer-lived species such as American black bears.

Because fishes are ectotherms, growing season length affects fish growth rates and ultimately also influences natural mortality rates. In general, slower-growing individuals of a fish species tend to live longer than faster-growing individuals of that species. There certainly are exceptions, but this general rule often holds. For example, a 45-centimeter walleye in a Kansas reservoir is likely to be 3 years old. Such populations may have natural mortality rates near 40% per year. A 45-centimeter walleye in a northern Canada lake may be as old as 15 years; natural mortality rates for such populations are likely to be less than 10% per year.

The term *natural* in *natural mortality* can be misleading. Natural mortality can be influenced by humans, especially in situations in which human activities affect habitat. If human disturbances affect the quantity or quality of the habitat an organism needs to reproduce or survive, then natural mortality rates can be increased.

A population will experience a certain level of natural mortality whether or not harvest mortality occurs. For example, approximately 70% of a northern bobwhite population will die in a normal year, regardless of whether hunting mortality occurs. Due in part to density-dependent mortality factors, removal or addition of mortality factors will likely have little influence on total annual mortality rates. Therefore, a **harvestable surplus** of animals can be removed, within limits, without adversely affecting a population. The harvestable surplus is that portion of the population that can be taken by humans without affecting subsequent generations of that organism. For example, if hunting accounted for 35% of the annual mortality of northern bobwhites in a given area, then 35% mortality would occur from natural causes and total mortality would equal approximately 70%. If hunting mortality accounted for 15% of the mortality, then natural mortality would account for the remaining 55% and total mortality would remain at 70%. For the animal population, what caused the total mortality would be essentially irrelevant. In such a situation, hunting mortality is said to be **compensatory mortality**. That is, the 35% or the 15% hunting mortality replaces part of the natural mortality that would have occurred even without hunting mortality. Compensatory mortality is not just a phenomenon associated with harvest mortality. If a particular species had an annual total mortality of 50%, with 25% of that mortality being contributed by predation and 25% by disease, compensatory mortality could occur. For example, if disease mortality increased to 40% and predation mortality decreased to 10%, compensatory mortality would have occurred.

The question of when harvest mortality is compensatory mortality and when it is **additive mortality** has stirred controversy. Harvest mortality is additive when it results in a total mortality that exceeds what would have occurred due to natural mortality alone. If a portion of harvest mortality is additive, then population density may be reduced. In the northern bobwhite population discussed above, harvest mortality would be compensatory only up to a certain level of harvest. If 60% of the northern bobwhites were harvested by hunters, total mortality would very likely exceed 70%. The amount by which total mortality exceeded 70% (the expected total mortality in a nonhunted population) would be additive mortality.

Even when additive mortality due to harvest is possible in a population, that level of mortality may not occur. There are a number of reasons why additive mortality may not occur. Human interest in harvest may simply decline if animal population density decreases. With fewer organisms present, hunting success may be reduced to the point at which harvest ceases simply because the organisms are too difficult to find. Also, human interest can decline if the animals become elusive. For example, sharp-tailed grouse become increasingly wary after being hunted, and they soon begin to flush (rise from the ground) out of shooting range. In either case, the law of diminishing returns enters the picture and hunters stay home because success becomes increasingly unlikely.

There also may be no additive mortality because of harvest regulation. For example, pronghorns are quite vulnerable to hunting because of their tendency to use open habitat, their highly visible coloration, and the accuracy of rifles with telescopic sights. However, additive mortality sufficient to cause population declines in this species can be avoided by proper harvest regulation. That is, the state, provincial, or territorial conservation agency issues only the number of hunting permits that ensures that harvest mortality is not additive.

Conversely, it is possible for harvest mortality to become additive if harvest regulations are inappropriate (most often this means too liberal) or if additive mortality is a management goal. Additive mortality can be a purposeful management strategy. For example, white-tailed deer populations have been increasing in many areas during the last two decades. These increased populations have caused increased damage to crops, urban area problems, and car–deer accidents. Thus, many management agencies have increased the number of white-tailed deer permits issued in an attempt to reduce the deer population density. This reduction would not harm the ability of the species to maintain a population; therefore, some additive mortality would not necessarily be negative.

Other factors complicate an assessment of whether harvest mortality is additive or compensatory in a particular situation. Some species respond to increased mortality by increasing natality or recruitment (remember the discussion of inversity in section 3.1). For example, the reproductive rate of coyotes may increase as mortality rates increase. Also, movements of animals such as ruffed grouse and northern bobwhites from lightly hunted surrounding areas into a heavily hunted area can complicate an assessment of mortality. Estimates of total mortality for northern bobwhites can exceed 100% in some hunting areas. Obviously, mortality in excess of 100% is not possible. Movement of birds from surrounding lands onto the hunted lands can result in a harvest greater than the original number of birds present.

North American duck populations fluctuate widely in abundance over time, with declines occurring most recently in the early 2000s. The causes of these declines are complex, and they have been attributed to various factors, including habitat loss and potential additive hunting mortality. Most studies have indicated that waterfowl harvest at current levels is compensatory; however, the compensatory versus additive mortality question in waterfowl management will continue to remain a point of controversy.

The issue of when and whether harvest mortality is additive or compensatory is further complicated because mortality rates are typically assessed based on "expected" or "normal" annual mortality rates. Mortality rates are not necessarily stable or normal. As an example, consider the previously described population of northern bobwhites. Through regulations these birds may be managed in an area to maintain a certain population level. Great care may be taken to ensure that harvest mortality is not additive. All of these efforts may be to no avail if a density-independent factor, such as a severe blizzard or a substantial loss of reproductive habitat, reduces the northern bobwhite population to some exceedingly low level. All of the management efforts over a number of years to ensure the maintenance of the population at some level would be negated by that single storm or habitat loss event. The topic of compensatory and additive mortality for harvested species will remain a point of contention. It is an important aspect of the human dimensions in wildlife and fisheries, a subject addressed later in this text.

In natural systems, surplus organisms in a population die from natural causes, such as predation. For example, cougars were a common predator on mule deer in the southwestern United States before humans reduced cougar numbers. In perturbed systems, there may be no predators other than humans. Once cougar populations were reduced by human activities, regulated hunting became the primary source of predation for the mule deer. In highly perturbed systems, such as urban areas, this becomes a special problem. With no or reduced predation and the unavailability of traditional human harvest methods (hunting), increasing numbers of some wildlife species have become a major concern. In many areas, species such as Canada geese and white-tailed deer often reach problem population levels.

To conclude this section on mortality, let's review an analysis of the identifiable causes of mortality for female Merriam's wild turkeys in Arizona, Montana, and South Dakota (Rumble et al. 2003). In all three locations, the primary cause of annual mortality was predation (60%–90% of the annual mortality), primarily from coyotes and avian predators. Poaching mortality ranged from 5% to 25% among locales, while legal hunting mortality ranged from 5% to 9%. Other measurable causes of mortality included disease and collisions with vehicles.

3.4 Interactions of the Dynamic Rate Functions

Natality or recruitment, growth, and mortality interact to determine the nature of a population (Figure 3.3). Consider a population of white catfish in a South Carolina river: recruitment adds new individuals, growth adds biomass, and mortality reduces the number of fish in the stock. In a stable system, the inputs (recruitment and growth) are balanced by outputs (harvest and natural mortality).

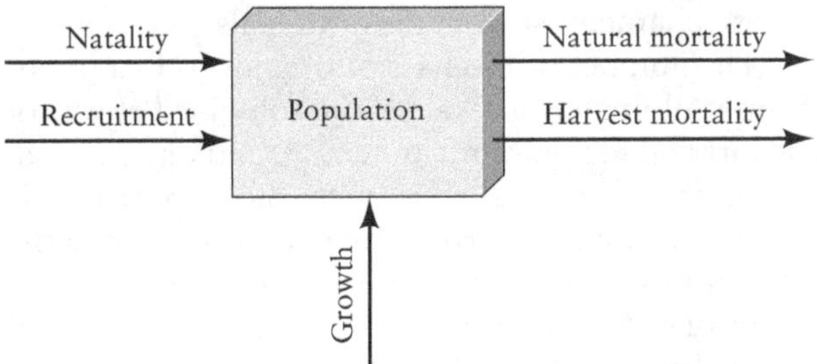

FIGURE 3.3 The three dynamic rate functions (natality/recruitment, growth, and mortality) interact to determine the nature of an animal population. Natality and recruitment add new individuals to the population, growth adds biomass to the population, and mortality results in a loss of both individuals and biomass.

Few systems are stable, however, especially those that have been perturbed by humans. If mortality exceeds natality or recruitment, then the population density will decrease. The converse is also true. If natality or recruitment exceeds mortality, then the population density will increase. However, only a certain biomass can be supported by any particular environment, and that biomass cannot be exceeded for any extended time period.

3.5 Population Growth and Regulation

A species in a new and favorable environment often increases at an exponential rate. This type of population growth curve is termed a **J-curve** (Figure 3.4A). However, this exponential increase in numbers can continue only as long as no limiting factors are reached. For example, as numbers increase, there could be increasing competition for food. At some point there would be insufficient food for new individuals to be added to the population. Thus, the J-curve flattens at the upper end and becomes an **S-curve**, also known as a **logistic growth curve** (Figure 3.4B). Think back to the white-tailed deer inversity example in section 3.1. When white-tailed deer density is low, the reproductive rate is high. As adult density increases, the reproductive rate (offspring per adult) decreases.

The level at which the logistic growth curve flattens is generally close to **carrying capacity** (K) (Figure 3.4B). Carrying capacity typically is defined as the maximum biomass of a population that can be sustained within a defined area throughout a specified period of time. Very often, the period of time used by biologists is an entire year. Thus, a small Iowa pond might have a carrying capacity of 50 kilograms per hectare for largemouth bass. The same pond might have a carrying capacity of 150 kilograms per hectare for bluegills, which are at a lower trophic level (see section 2.3). In an Alabama pond, which is less fertile than the Iowa pond because the soils there are highly leached by abundant rainfall and the waters tend to be less productive, carrying capacity may only be 15 kilograms per hectare for largemouth bass and 60 kilograms per hectare for bluegills. Thus, the trophic level of a particular fish species and the fertility of a water body both influence carrying capacity. In wildlife, carrying capacity is generally given in numbers of animals rather than biomass. In addition, wildlife biologists generally consider carrying capacity to be the

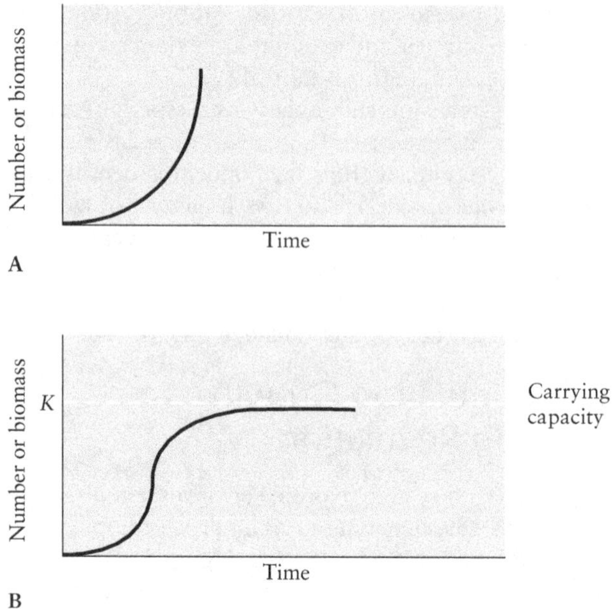

FIGURE 3.4 *(A)* When a species is first introduced into a new and favorable environment, the resulting increase in population resembles a J-curve. *(B)* Population growth cannot continue unchecked, however, because the population will reach carrying capacity (*K*), often defined as the maximum biomass for that population that can be supported over an entire year. Thus, population growth slows and the J-curve becomes an S-curve (also known as the logistic growth curve).

maximum population number that can be supported without causing habitat damage. Krausman (2002) summarized carrying capacity by providing seven different definitions for this term. Consider a county in southern Ohio. The white-tailed deer in that county have an ecologically based carrying capacity, using the term as we have defined it. However, that population also has a minimum-impact carrying capacity (or wildlife acceptance capacity) that is based on a desire by humans in that county to reduce car–deer accidents by maintaining a lower population abundance for white-tailed deer.

Management practices can be directed at either increasing or decreasing carrying capacity for a species, depending on management objectives for that particular area. For example, the carrying capacity for Baird's sparrows on a section of grassland can be quickly changed by the manner in which the grassland habitat is managed: mow or heavily graze the area and the carrying capacity for the species declines; enhance the grass species on which the Baird's sparrows depend and the carrying capacity increases. Management practices to increase or decrease the carrying capacity for a target population affect other nontarget species in that area as well, some favorably and others unfavorably. That is, nontarget species may increase or decrease in numbers depending on the extent to which the management practice affects them and their habitat.

Unfortunately, carrying capacity is not measurable; it represents a potential. In fact, biologists do not even agree on a standard definition for carrying capacity (MacNab 1985). Although carrying capacity is neither measurable nor well defined, it is a necessary concept for understanding population dynamics.

Standing stock (**standing crop**), unlike carrying capacity, is measurable. Standing stock is the abundance of organisms present at a given time in a given area and does not include the year-round portion of the carrying capacity definition. It represents the amount, usually by weight for fishes and number for birds and mammals, of a given species or complex of species present in a specific habitat or area at a specific moment. In reality, the term is commonly used in fisheries but seldom by wildlife biologists. Fisheries biologists commonly refer to standing stock for a single fish species or to the total standing stock for the entire fish community. Wildlife biologists usually refer to population density rather than standing stock, but the concept is certainly the same.

Assume that the standing stock for redear sunfish in a body of water is 50 kilograms per hectare. The population biomass can be the same whether the population contains 500 or 1,000 individuals. In the 1,000-fish population, mean weight per individual is one-half of that in the 500-fish population. Stable biomass with changing numbers of individuals is possible in fish populations because of the indeterminate growth characteristics of fishes. For most wildlife populations, a decline in population number results in a decline in biomass because of the determinate growth characteristics of these animals.

The number or biomass of organisms in a population is ultimately a result of the combined effects of one internal factor, **biotic potential**, and one external factor, **environmental resistance**. Biotic potential is synonymous with reproductive potential and is the maximum rate of population increase under ideal conditions. As shown in Figure 3.5, biotic potential can be thought of as a "force" that pushes up on the population growth curve.

In the natural world, a series of environmental factors (both biotic and abiotic) interact to limit the abundance of a population; these factors constitute environmental resistance. Environmental resistance acts to hold the rate of population increase below the level theoretically possible as a result of biotic potential. As shown in Figure 3.5, this environmental resistance is the "force" that pushes down on the population growth curve. Environmental resistance might include predation, limited sites for nesting or reproduction, climatic conditions, or competition for a limited supply of food. Carrying capacity is the theoretical balance between biotic potential and environmental resistance. Standing stock is the actual balance, in numbers or biomass, between the two "forces" at a particular moment in time.

Figure 3.5 is also useful as an illustration of population dynamics. Few populations in the natural world remain stable for very long. Changes in environmental resistance allow changes in population density. For example, if predation is reduced, population density may increase. If food supplies decline, population density may decrease.

Populations may increase and decrease on either a regular or an irregular basis. A **cyclic population** is one in which abundance increases or decreases on a regular basis. Cycles may be short, like those of Siberian brown lemmings, which have a 3- to 5-year cycle of abundance, or longer, as for snowshoe hares, which seem to have a 9- to 11-year cycle. Black-tailed jackrabbits in the shrub deserts of western North America have populations that are **irruptive**, here defined as populations that dramatically increase and decrease at irregular intervals. Some biologists consider black-tailed jack rabbits to have cyclic populations and argue that they increase and decrease at somewhat regular intervals in an approximately 7-year cycle. Whether their population dynamics are cyclic or irruptive, studies of black-tailed jackrabbit populations indicate that predation, primarily by coyotes, is not what causes the initial population downturn that occurs when jackrabbit populations reach a peak. At the peak of black-tailed jackrabbit abundance, jackrabbits per predator are well beyond the level that the predators can control. At the time when the black-tailed jackrabbit population starts to decline, the coyotes are abundant because of the abundant food that has been available to them over the past few years. As the black-tailed jackrabbit population continues to decline, the prey-per-predator ratio becomes so low that predation depresses the population of jackrabbits further. Possible delays in recovery of the black-tailed jackrabbit population may be related to predators continuing to depress the prey population. At some time in the future, the vegetation recovers and the prey species again begins to increase. By this time, coyote populations have declined to low levels along with those of the black-tailed jackrabbits. The coyotes, because of their dependence on black-tailed jackrabbits, have adjusted their reproduction to prey availability. Thus, whether predator or prey is regulating the cycle is unclear. In reality, the regulating agent changes depending on the time in the cycle. In fact, weather, vegetation, predators, and prey can all have an influence at some point.

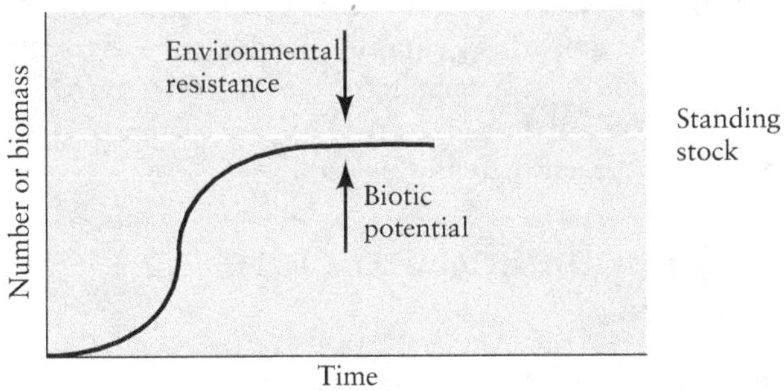

FIGURE 3.5 Biotic potential (reproductive potential) can be thought of as a "force" that pushes upward on the population growth curve. Environmental resistance can be thought of as a "force" composed of factors such as predation and competition that limits abundance. The balance between these two forces results in standing stock, which is the population abundance at any given moment.

Black and white crappies are often referred to as cyclic species because their recruitment (year-class strength) can be variable, resulting in substantial fluctuations in population abundance. However, this terminology is incorrect because changes in their abundance are not regular. Crappie populations in many waters commonly experience high variation in recruitment among years, often in response to environmental variables such as temperature and precipitation. Thus, their abundance is irregular rather than cyclic; irruptive would probably be a better descriptor.

3.6 Production and Yield

Production is defined as the biomass accumulated by a population during a year or some other specific period, and it includes both living organisms and those that died during that period. **Surplus production** is that portion of production that can be removed from a population by natural causes or human harvest without adversely affecting future population levels. Surplus production results in the harvestable surplus discussed in section 3.3. In commercial fisheries, surplus production is synonymous with sustainable yield (Ricker 1975). That portion of the population taken by humans is referred to as **yield** and is just one of the several fates of production.

Wildlife biologists generally consider surplus production in terms of numbers of animals (see section 1.3). An example might be a harvest of a certain number of giant Canada geese. Fisheries biologists, however, generally consider surplus production in terms of weight, not number. A Georgia farm pond will support a certain biomass of largemouth bass, but that biomass may be many small fish, fewer large fish, or moderate numbers of small, medium, and large fish. Fisheries biologists often attempt to influence the size structure of fish populations through harvest regulations (see Chapter 17).

In fisheries, **maximum sustained yield (MSY)** is a traditional management philosophy with a goal of achieving maximum harvest from a population over a series of years; that is, the maximum biomass that can be harvested without affecting future harvest. As a simple example, commercial ocean fisheries might allow harvest of 60% of the adult fish and leave 40% of the adults for reproduction. These percentages can vary tremendously depending on species and location. The problem with MSY is that it does not take into account the quality (size) of the organism being harvested. The maximum harvest of walleyes from a Minnesota lake, in terms of biomass, might result from a 70% exploitation rate, with the mean length of walleyes harvested being only 33 centimeters. Few anglers would be satisfied with that type of fishery. Thus, biologists now tend to manage for **optimum sustained yield (OSY)**, which can be defined as a management philosophy with objectives that consider ecological and socioeconomic factors. Biologists consider such factors as user desires and expectations and cost–benefit relationships—the human dimensions of a fishery. Such a philosophy pertains to both commercial and recreational activities. In contrast to the above MSY example using walleyes, one management objective under an OSY strategy might be to produce a mean length of 45 centimeters for walleyes in the angler harvest.

If a wildlife population, such as elk, was managed under MSY, biologists would generally attempt to keep the population at the inflection point on the S-curve (Figure 3.6). However, MSY is a dangerous management strategy, because biologists seldom know exactly where this point lies on the S-curve and because an event such as an extremely harsh winter or an outbreak of disease could push the population below this point, causing a substantial population decline. Thus, wildlife biologists also usually manage at somewhere between the inflection point on the S-curve and K in Figure 3.6 to avoid population catastrophes. If a population was managed for trophy hunting, it would be managed under OSY rather than MSY. In Michigan and Tennessee studies, biologists determined that MSY for white-tailed deer occurred when the herd was reduced to 56%–61% of the prehunt population (McCullough 1987; Jenks et al. 2002).

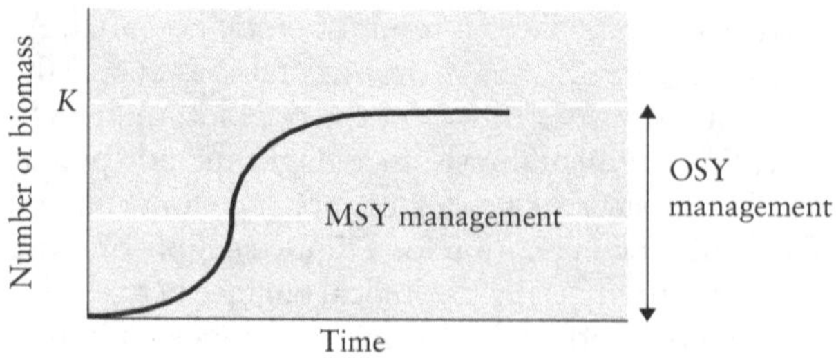

FIGURE 3.6 Maximum sustained yield (MSY) is a management target using a strategy that typically seeks to produce the maximum harvest without endangering future reproduction and recruitment; K represents carrying capacity. Optimum sustained yield (OSY) represents a target that can occur anywhere along the population curve, depending on management objectives. In fact, OSY may also occur at a point below MSY in some circumstances (see section 3.6).

Can you think of a situation in which a biologist might want to set an OSY strategy that results in the decline of a particular organism? Waterfowl biologists often struggle with common carp populations in waters intended primarily for duck production (see section 10.2). In such a situation, the OSY strategy for common carp likely would be to reduce the population rather than to maintain or increase the population abundance.

While the MSY and OSY philosophies may seem simple, they are not (e.g., Cooper 1983; Mangel 2000). It is difficult to determine sustainable levels for populations. To be assured of sustainability, much more information is generally needed than is available. Therefore, a determination of the OSY or MSY for a particular population is at most a best guess based on currently available data.

3.7 Population Structure

Population structure is, as previously mentioned, the result of the three interacting rate functions. Biologists are concerned with both **age structure** and **size structure**, the age and size distributions of a population.

Consider the age structure of harvested buck white-tailed deer in two different states (Figure 3.7). Which of these harvests likely occurred where the state conservation agency operated under a management

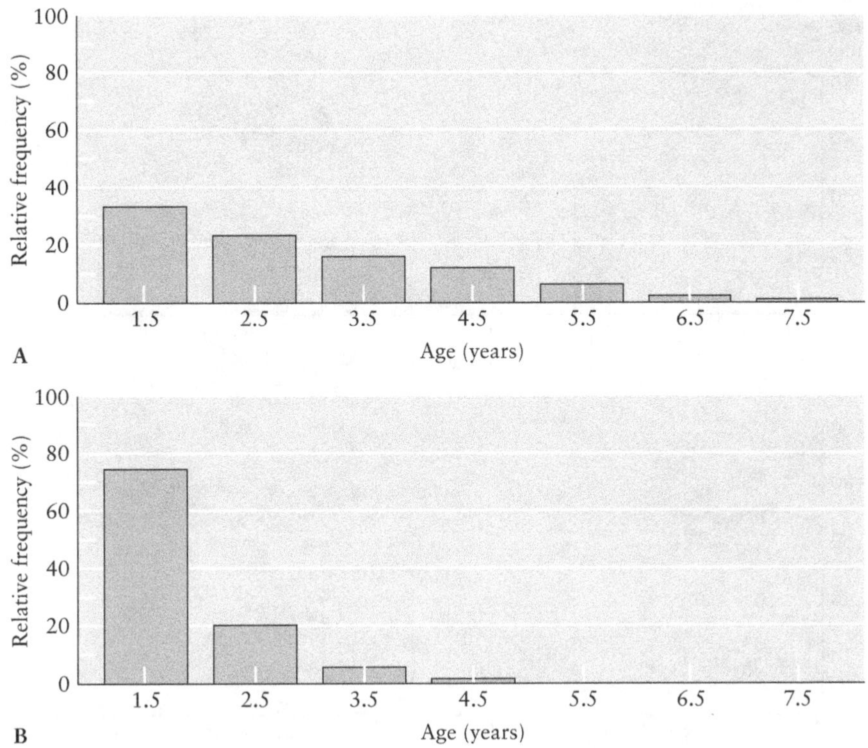

FIGURE 3.7 Age structures for buck white-tailed deer harvested from two populations that are under different management strategies. *(A)* This population is managed with limited buck licenses so that at least 20% of the harvest will be composed of bucks that are 4.5 years old and older. *(B)* This population is managed with unlimited buck license sales to allow maximum hunting opportunity.

philosophy seeking to maximize hunting opportunity and thus did not limit the number of licenses for antlered white-tailed deer? Which of these two age structures likely represents the harvest in a state that limits the sale of buck licenses to maintain a certain level of quality in the buck white-tailed deer harvested—perhaps 20% of harvested bucks being at least 4.5 years old? High mortality of white-tailed deer bucks caused by unlimited buck harvest results in an age structure that is dominated by young deer, with few bucks older than 2.5 years being harvested (Figure 3.7B). Conversely, lower mortality of bucks under limited harvest results in a greater number of older bucks being harvested (Figure 3.7A). Either management strategy can be defended, but the age structures of the two populations are certainly different.

Fisheries biologists also consider age structure. Which of the two black crappie populations represented in Figure 3.8 has consistent recruitment, and which has inconsistent recruitment? Black crappies in Lake Mitchell, South Dakota, successfully recruited every year for a 6-year period (Figure 3.8A). In contrast, the black crappies in Roy Lake, South Dakota, successfully recruited only three times in a similar 6-year period (Figure 3.8B). Anglers are more likely to consistently catch black crappies in a population such as that in Lake Mitchell than in one like Roy Lake's. Black crappie recruitment in South Dakota was found to be more consistent in artificial impoundments than in natural lakes, apparently because of physical differences in the amount of shoreline irregularity and the degree of water level fluctuation (Guy and Willis 1995). Lake Mitchell is an impoundment, while Roy Lake is a natural lake.

Fisheries biologists rely heavily on assessments of fish population size structure, while wildlife managers seldom are concerned with size structure. Bluegills are popular recreational fish in much of the United States, but they have the tendency to overpopulate and stunt. Two different bluegill populations are depicted in Figure 3.9. Population B is at or

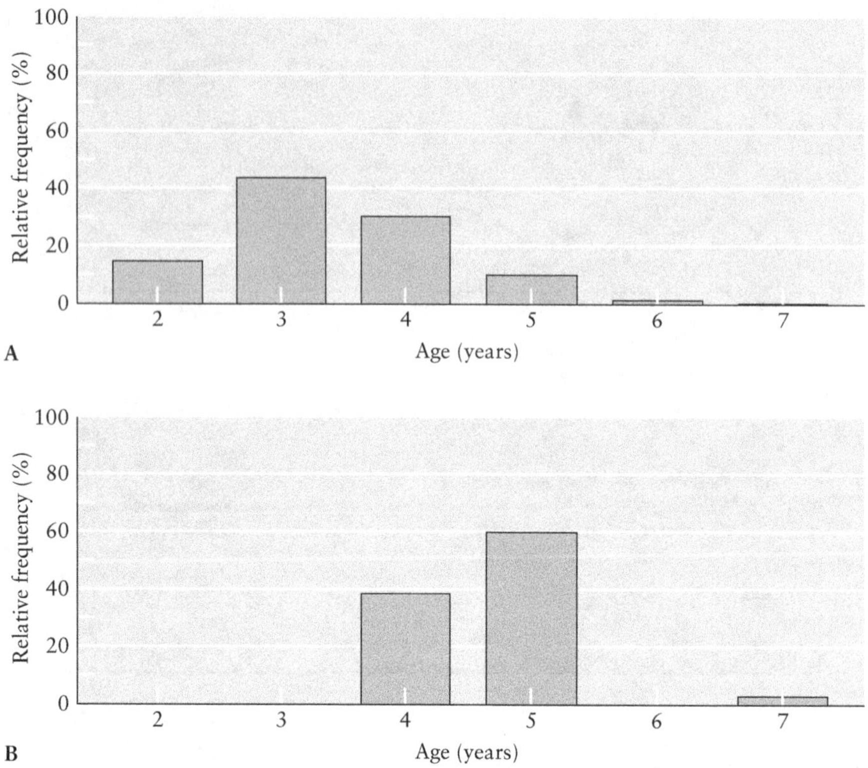

FIGURE 3.8 Age structure of black crappies sampled from *(A)* Lake Mitchell, South Dakota, and *(B)* Roy Lake, South Dakota. Recruitment is inconsistent in the Roy Lake population; no individuals from age groups 2, 3, and 6 were collected in this sample.

near carrying capacity, and growth is slow. This high-density, slow-growing population is a classic example of a stunted bluegill population—size structure is said to be small, low, or poor. Population A is below the theoretical carrying capacity for that water body that might be expected given the available nutrients. As a result, growth is faster and there are larger individuals in the population. The size structure depicts a much higher quality for this particular population. Given a choice, most anglers would prefer to fish for bluegills such as those represented in Figure 3.9A rather than those in Figure 3.9B.

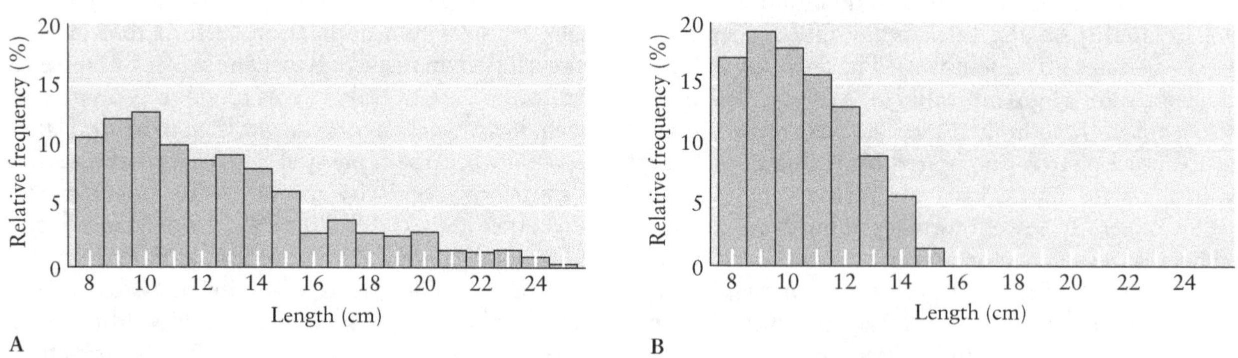

FIGURE 3.9 Size structures of bluegill populations in two different water bodies. *(A)* Bluegills in this population are faster growing because their standing stock is below the theoretical carrying capacity that could be supported by nutrients available in that water body, and they reach 17 centimeters (cm) by age 4. *(B)* Bluegills in this population are slower growing because of high population density and only reach 12 centimeters by age 4.

3.8 Community Structure

Community structure is the composition (species composition and abundance) of an assemblage of organisms. Generally, ecologists consider two components: the number of species present in a community (richness; see Chapter 2) and the abundance of the various species in a community (evenness or equity). All of the previous discussion in this chapter involved populations rather than communities; however, numerous populations of organisms constitute communities. How individual populations are affected by interactions with other populations is highly variable and of importance to ecologists (see section 2.7).

Fish communities are often structured through predator–prey relations. A simple and common example occurs for largemouth bass (predator) and bluegill (prey) in small warmwater impoundments. When largemouth bass density is high (e.g., when they are near carrying capacity), then predation on small bluegills will be high. As a result, bluegill density will be low and growth will be fast. The bluegill size structure will be dominated by larger fish (Figure 3.10). If a biologist changes the size structure of a largemouth bass population through the use of a particular length limit (section 17.5), then the size structure of a prey population (e.g., bluegill) will also change. For example, if a biologist changes a largemouth bass population in a small impoundment from a higher-density population dominated by small individuals to a lower-density population with larger individuals present, then there will be an accompanying change in the bluegill population. Because fewer largemouth bass are now present in such a population, there will be less predation on the small bluegills. As a result, bluegill abundance will increase, growth will slow, and fewer large individuals will be present in the population (e.g., Eder 1984).

FIGURE 3.10 A high-density, slow-growing largemouth bass population will prey extensively on small bluegills in small, warmwater impoundments. As a result, bluegill growth tends to be fast and many bluegills will reach larger sizes that will interest recreational anglers. *[Image courtesy of J. Jolley.]*

The diversity of fish species within a community can also be related to predator–prey relations. Consider the pallid sturgeon, which is native to the Missouri River ecosystem. This fish was placed on the federal endangered species list in 1990, and most biologists believe that **anthropogenic** (i.e., human-induced) changes in the river system caused by dam construction (e.g., blockage of spawning movements) resulted in nearly complete loss of recruitment for pallid sturgeon. However, Gerrity et al. (2006) suggested an alternative explanation for the lack of recruitment and population decline. They found that hatchery-reared, subadult pallid sturgeon that were released into the river above the uppermost dam in Montana fed extensively on fishes. In fact, most of their diet was composed of sicklefin chubs and sturgeon chubs, both of which also are species of concern and whose populations have declined over most of their ranges. The authors suggested that the anthropogenic changes in the Missouri River ecosystem caused by dam construction may have altered the prey fish base available to pallid sturgeon, and the modified fish community structure potentially could be a valid alternative explanation for sturgeon population declines.

Predator–prey interactions and resulting community structure become more complex as one moves from overly simplistic food chains to more realistic food webs. For example, what happens when coyotes and cougars have access to both mule deer and white-tailed deer as prey at a location within the Rocky Mountains? Typically, the causes of predation will differ in populations of the two deer species, with cougars preying more selectively on mule deer and coyotes preying more on white-tailed deer. As more and more components of a food web are evaluated, the links become more and more complex.

SUMMARY

Three dynamic rate functions (natality or recruitment, growth, and mortality) form the basis for the population dynamics of any species. Natality and recruitment add new individuals to a population, growth adds biomass to a population, and mortality results in a loss of both individuals and biomass. Density-dependent and density-independent factors affect all three rate functions. While stable populations would exhibit a balance among these three rate functions, most wild populations are not stable but instead fluctuate in response to factors such as variations in environmental conditions. If recruitment or natality exceeds mortality, the population will increase; if mortality exceeds recruitment or natality, the population will decline.

When a species is introduced into a new and favorable environment, the population will increase in number and biomass. At first, this increase in abundance can be exponential. However, such rapid increases do not continue, because the carrying capacity of the system is reached. Populations cannot be sustained beyond the capacity of a system to provide for their needs. Biotic potential, or reproductive potential, can be thought of as a "force" that is driving the population to increase. Environmental resistance is the sum of the biotic and abiotic environmental factors that interact to limit populations.

Production is the biomass that is accumulated by a population during a specific time period, most often one year. Surplus production results in a harvestable surplus of many recreational and commercial fish and wildlife species, and the portion of the population taken by humans is termed *yield*. The maximum sustained yield philosophy has as its goal obtaining maximum yield without affecting future yield but does not consider the quality of the harvested organisms. Today, most biologists operate under a philosophy of optimum sustained yield, whereby ecological and socioeconomic factors, rather than simply number or biomass to be harvested, are considered.

The interactions of the dynamic rate functions are expressed in population structure. Biologists assess both age and size structure for populations. Assessment of age structure can provide information concerning irregular recruitment or excessive mortality. Fisheries biologists in particular assess size structure, which provides information on the quality of the fish population to anglers. Interactions of populations create community structure.

PRACTICE QUESTIONS

1. What is the primary reason why wildlife biologists use natality and fisheries biologists use recruitment to adult size when considering how many individuals are added to a population?

2. What is inversity?

3. Consider two brook trout populations in beaver ponds in Wyoming; neither has been fished by anglers. In one pond, brook trout density is high and few individuals exceed 20 centimeters. In the other, density is low and the majority of adult brook trout are longer than 20 centimeters. In which pond would you expect faster growth rates? How would intraspecific competition relate to expected growth rates in the two ponds?

4. Consider two mammal species that reside in your geographic area. Choose one species that is short lived and in which most individuals do not live more than one or two years. Examples of the shorter-lived species might include Arizona pocket mice and pygmy rabbits. Choose a second species at the other end of the longevity spectrum, in which individuals have the capacity to live for five, ten, or more years. Longer-lived species might include Dall's sheep, brown bears, moose, or one of the deer species. Contrast the rates of reproduction/recruitment and mortality for the two species you have chosen.

5. Choose an animal species with which you are familiar. Make a list of the various biotic and abiotic factors that are part of the environmental resistance for a population of this species. Now decide which of these factors are density dependent, which are density independent, and which might be both.

6. Why is carrying capacity not a measurable number?

7. Wildlife and fisheries biologists typically consider surplus production as being a harvestable surplus for game or commercial species. Think back to the lessons from ecology in Chapter 2. In general, will harvestable surplus be greater for a species at a low trophic level or a high trophic level? In other words, is biomass available for harvest generally greater for a herbivore or a carnivore? Why?

8. How do determinate growth and indeterminate growth affect surplus production?

9. Describe how total mortality, natural mortality, harvest mortality, compensatory mortality, and additive mortality may interact in an animal population.

10. Consider competition for nesting sites for two bird species in the same community. How might this competition affect resulting community structure?

SELECTED READINGS

Begon, M., and M. Mortimer. 1986. *Population ecology: A unified study of animals and plants*. 2nd ed. Blackwell Scientific Publications, Oxford, U.K.

Braun, C. E., ed. 2005. *Techniques for wildlife investigations and management*. 6th ed. The Wildlife Society, Bethesda, Md.

Busacker, G. P., I. R. Adelman, and E. M. Goolish. 1990. Growth. Pages 363–387 *in* C. B. Schreck and P. B. Moyle, eds. *Methods for fish biology*. American Fisheries Society, Bethesda, Md.

Caughley, G. 1977. *Analysis of vertebrate populations*. John Wiley & Sons, New York.

Getz, W. M., and R. G. Haight. 1989. *Population harvesting: Demographic models of fish, forest, and animal resources*. Monographs in Population Biology 27. Princeton University Press, Princeton, N.J.

Guy, C. S., and M. L. Brown. 2007. *Analysis and interpretation of freshwater fisheries data*. American Fisheries Society, Bethesda, Md.

Hilborn, R., and C. J. Walters. 1992. *Quantitative fisheries stock assessment: Choice, dynamics, and uncertainty*. Chapman and Hall, New York.

Ricklefs, R. E., and G. L. Miller. 2000. *Ecology*. 4th ed. W. H. Freeman and Company, New York.

Van Den Avyle, M. J., and R. S. Hayward. 1999. Dynamics of exploited fish populations. Pages 127–166 *in* C. C. Kohler and W. A. Hubert, eds. *Inland fisheries management in North America*. 2nd ed. American Fisheries Society, Bethesda, Md.

LITERATURE CITED

Beard, T. D., Jr., M. J. Hansen, and S. R. Carpenter. 2003. Development of a regional stock-recruitment model for understanding factors affecting walleye recruitment in northern Wisconsin lakes. *Transactions of the American Fisheries Society* 132:382–391.

Cooper, C. F. 1983. Carbon storage in managed forests. *Canadian Journal of Forest Research* 13:155–166.

Eder, S. 1984. Effectiveness of an imposed slot length limit of 12.0-14.9 inches on largemouth bass. *North American Journal of Fisheries Management* 4:469–478.

Gerrity, P. C., C. S. Guy, and W. M. Gardner. 2006. Juvenile pallid sturgeon are piscivorous: A call for conserving native cyprinids. *Transactions of the American Fisheries Society* 135:604–609.

Guy, C. S., and D. W. Willis. 1995. Population characteristics of black crappies in South Dakota waters: A case for ecosystem-specific management. *North American Journal of Fisheries Management* 15:754–765.

Hansen, M. J., M. A. Bozek, J. R. Newby, S. P. Newman, and M. D. Staggs. 1998. Factors affecting recruitment of walleyes in Escanaba Lake, Wisconsin, 1958-1996. *North American Journal of Fisheries Management* 18:764–774.

Jenks, J. A., W. P. Smith, and C. S. DePerno. 2002. Maximum sustained yield harvest versus trophy management. *Journal of Wildlife Management* 66:528–535.

Krausman, P. R. 2002. *Introduction to wildlife management: The basics*. Prentice-Hall, Upper Saddle River, N.J.

MacNab, J. 1985. Carrying capacity and related slippery shibboleths. *Wildlife Society Bulletin* 13:403–410.

Mangel, M. 2000. Trade-offs between fish habitat and fishing mortality and the role of reserves. *Bulletin of Marine Science* 66:663–674.

McCullough, D. R. 1979. *The George Reserve deer herd: Population ecology of a* K*-selected species*. University of Michigan Press, Ann Arbor.

McCullough, D. R. 1987. The theory and management of *Odocoileus* populations. Pages 535–549 in C. M. Wemmer, ed. *Biology and management of the Cervidae*. Smithsonian Institution Press, Washington, D.C.

Ricker, W. E. 1975. *Computation and interpretation of biological statistics of fish populations*. Bulletin 191. Fisheries Research Board of Canada, Ottawa, Canada.

Rumble, M. A., B. F. Wakeling, and L. D. Flake. 2003. Factors affecting survival and recruitment in female Merriam's turkeys. *Intermountain Journal of Science* 9:26–37.

CHAPTER 4

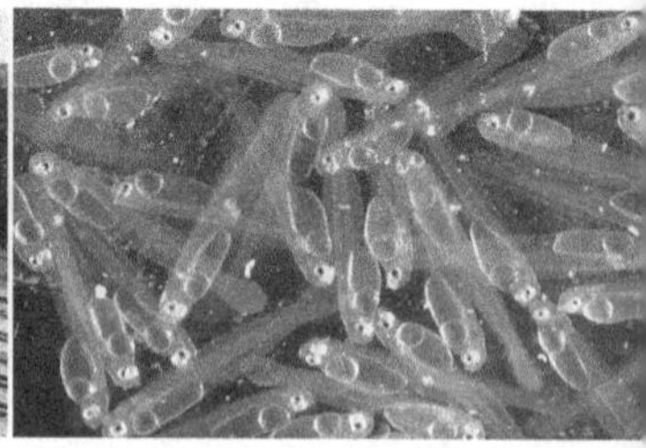

Genetics in Wildlife and Fisheries

Any discussion of fisheries and wildlife requires coverage of genetics, because understanding inheritance of traits is basic to understanding organisms, populations, communities, and ecosystems. **Genetic diversity** includes all genetic variation within a species and occurs at different levels; there is genetic diversity among individuals within populations, and there is genetic diversity among different populations of a species. Greater genetic diversity in populations results in greater genetic diversity in the communities to which they belong. Understanding how genetic diversity is transformed into differences within and among populations is fundamental to understanding evolution.

While it has always played an important role in fisheries and wildlife, genetics is coming more to the forefront in these fields as greater attention is given to the implications of genetics. Genetic modifications purposefully made by humans, genetic effects of adding organisms to or subtracting them from populations and communities, using genetic information for law enforcement, marking organisms and populations via their genetic material, and problems with captive breeding programs are just a few examples of topics of heightened interest. This chapter will address some basic genetic concepts and briefly describe how they relate to speciation, natural selection and adaptation, artificial selection, inbreeding and outbreeding, and other genetics-related topics; it will also cover some newer applications of genetics in wildlife and fisheries and address issues associated with population genetics.

4.1 Basic Genetics

As in all organisms, traits in species associated with wildlife and fisheries resources are transferred from one generation to the next by **genes**. Genes are segments of hereditary material that are positioned on **chromosomes** within cells. Chromosomes are structures that carry genes and are found in the nuclei of cells. Genes are heritable and determine the characteristics of an organism. Some traits are transmitted by single genes. For example, in rainbow trout, expression of normal skin color versus colorless skin (albino) is controlled by a single gene. Other traits are controlled by multiple genes and are called **polygenic traits**. For example, in rainbow trout, growth is controlled by multiple genes present at a variety of locations on chromosomes. Genes that encode amino acid sequences or the structure of proteins are called **structural genes**, while those that control expression of structural genes during development or among different tissues are called **regulatory genes**.

Most vertebrates are **diploid** organisms; that is, they have two complete sets of chromosomes. Each individual organism has two copies of each gene, usually one received from its mother and one from its father. These copies are called **alleles** and may be alternative forms of the same gene. Alleles for the same gene occur at the same **locus**. A locus is the location of a particular gene on a chromosome. Two chromosomes that each carry the same set of genes are called **homologous chromosomes**. The specific genetic information possessed by an individual is called its **genotype**; the actual physical expression of traits is referred to as its **phenotype**.

The phenotype of an individual is a product of its genotype interacting with its environment. Two organisms can have the same genotype for a particular trait but have different phenotypes. For example, two individuals of a species may have the same genotype for growth but be of different sizes because one receives more or less food than the other. The converse can also be true; two individuals of a species can have different genotypes for a particular trait but, because of environmental conditions, have the same phenotype. For example, one individual may have a genotype for faster growth than another individual, but the two may grow at the same rate if the individual with the genotype for faster growth receives less food than its counterpart with a genotype for slower growth.

Individuals with two copies of the same allele at a particular locus are **homozygous**; individuals with two different alleles at a particular locus are **heterozygous**. An individual may have two **dominant alleles**, two **recessive alleles**, or one of each at a particular locus. Dominant alleles are usually expressed over recessive alleles in a heterozygous individual, although in some cases, both alleles are expressed phenotypically but to different degrees; this is called **incomplete dominance**. Both alleles at a locus may also be expressed to the same degree; this situation is called **codominance**.

Genes and their alleles determine not only what characteristics are expressed in an individual but also how well an individual or population is adapted to its environment. The **gene pool** of a population is composed of all alleles present within all individuals constituting that population. If little genetic variability is present in the gene pool, a population will be less able to adjust to changes in environmental conditions. The opposite is also true: the greater the genetic variability, the greater the ability to adapt or change.

The terms **heterozygosity** and **homozygosity** are commonly used in referring to both individuals and populations. Heterozygosity is the condition of having dissimilar alleles at one or more loci; homozygosity is the condition of having identical alleles at one or more loci. Use of these terms differs somewhat depending on whether individuals or populations are being discussed. With regard to an individual organism, heterozygosity and homozygosity can refer to the frequencies of such genotypes across all genes in the individual, but most often it refers to the probability of heterozygosity or homozygosity for any one gene. When used in reference to populations, heterozygosity or homozygosity may refer to the proportion of the population that is homozygous or heterozygous at a particular locus or across all loci.

As a very simplified example of inheritance, we can consider pigmentation in rainbow trout. Albinism, or lack of normal pigmentation, is controlled by a single locus with two alleles; the double recessive genotype gives rise to albinism (Bridges and von Limbach 1972). If a fish population had only alleles for normal pigmentation, then only pigmented fish would occur in that population. If only albinistic alleles were present in another population, then only albino individuals would occur in that population. Neither of these populations would have any heterozygosity for pigmentation. Therefore, no within-population variation would be present for this trait, but a difference between the two populations would exist. If, however, a third population had both pigmented and albinistic alleles in its gene pool, then within-population variation would exist, and this population would have some level of heterozygosity. Various levels of heterozygosity can be present among different populations depending on the frequency of the alleles in question. When pigmentation is combined with the myriad of other genetically determined traits an organism can possess, one can readily see the endless variety of combinations that can be available to produce genetic similarities and differences.

4.2 Species, Subspecies, Populations, and Strains

Evolution is the theory that pertains to the process of continuous and generally gradual transformation of lines of descent from a common ancestor. It involves changes in the gene pool of a population that over time lead to adaptation to the environment in which the population lives. The products of evolution are genetically new organisms. A **species** is a naturally occurring group of organisms that can successfully interbreed and produce fertile offspring; they usually are unable to successfully interbreed with organisms outside that group, or if they do so they produce fewer viable offspring. However, students should be aware that the term *species* can be quite perplexing—it is defined differently by different people. **Subspecies** (**races**) are formally recognized geographic subdivisions of a species whose members resemble one another to some extent and can interbreed to produce fertile offspring, but which differ from other subdivisions in some recognizable way. **Populations** (**stocks**) are groups within a species that actively interbreed and that are adapted to similar environments. **Strains** are cultured populations that reliably express a suite of characteristics.

Subspecies and strains differ from other members of a species in genetic composition and in the phenotypic expression of that composition. For example, there are a number of subspecies of cutthroat trout in the Rocky Mountain region. These subspecies are separated by barriers that have not allowed them to intermingle for some time. Although they are all the same species, differing conditions in which they live have resulted in allelic frequencies in each population's gene pool that are somewhat different from that of the next. Another example involves subspecies of cackling goose. Pairs mate for life; they winter at traditional sites with their young from the previous summer and return to the same nesting area with the young birds in the spring. These behavioral characteristics have led to the evolution of different subspecies of the cackling goose that interbreed only to a limited extent.

Formally named fish strains are often developed by humans for aquaculture purposes. **Aquaculture** is the culture and husbandry of aquatic organisms. Rainbow trout have been cultured for a long period, and sufficient records of parentage exist to allow development of named strains with specific characteristics. For example, a rainbow trout strain that was initially developed at the fish hatchery in Manchester, Iowa, is called the Manchester strain. Although different strains of rainbow trout are all the same species, they can differ markedly in spawning time, growth rate, vulnerability to angling, susceptibility to diseases, and a wide range of other traits. With the advent of **game ranching** and **game farming**, specific strains of various wildlife species have been developed. Game ranching is the husbandry of native animals, usually large ungulates, for the production of meat and other products. It is a variation of range management that in North America involves species such as elk and mule deer, rather than domesticated sheep or cattle. At times game ranching is also used to refer to pay-for-hunting enterprises in which animals, primarily exotic ungulates, are harvested for recreational purposes. Game farming usually involves the commercial production of smaller organisms such as ring-necked pheasants and northern bobwhites. These animals can be used for food or recreational purposes.

Similarly, exposure of natural populations to varying environmental conditions has given rise to different stocks, each with its own specific traits. For example, a variety of stocks of Pacific salmon have developed naturally on the west coast of North America and the east coast of Asia.

4.3 Isolating Mechanisms

Populations can have their gene pools become segregated by a variety of **isolating mechanisms**. Isolating mechanisms are mechanisms that tend to keep organisms from interbreeding. **Speciation** (the process of species formation) cannot occur without some form of isolation among groups. If populations are **allopatric** (having nonoverlapping geographic ranges), they normally do not interbreed unless changes occur that mix them in a common environment (Figure 4.1). When organisms are separated in this way, each group exists in a somewhat different habitat. Over a long period

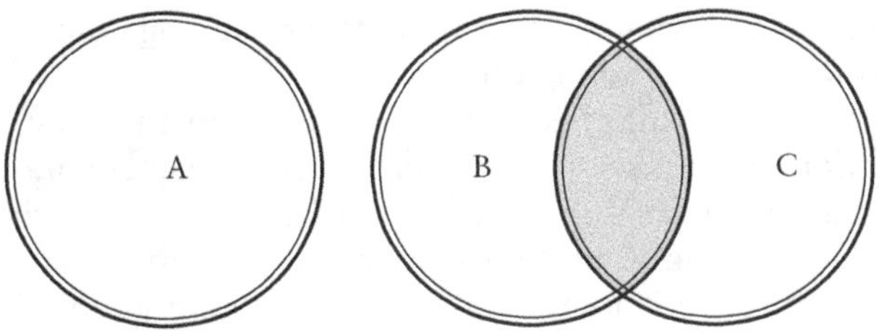

FIGURE 4.1 A diagram of the ranges of allopatric and sympatric populations. Population A is allopatric with populations B and C; it is effectively isolated from the others, at least by space. Populations B and C have overlapping ranges (shaded area) and are, therefore, sympatric.

of time, these differences in habitat can result in both populations evolving and inheriting genetic traits suitable to their respective environments. If they become different enough, they may no longer have the capacity to interbreed. Thus, the spatial separation of populations allows for genetic differentiation and **allopatric speciation**.

Even if populations are currently **sympatric** (their geographic ranges coincide or overlap), isolating mechanisms may evolve that prevent them from interbreeding (see Figure 4.1). Often several of these mechanisms will work in concert. Premating isolating mechanisms, such as sexual or other behavioral differences, physiological differences, anatomical differences, and mating time differences, can work to reduce interbreeding. Even if interbreeding occurs, several postmating isolating mechanisms, such as spontaneous abortion and hybrid sterility, can prevent the successful establishment of interbred populations. **Sympatric speciation**, which is much rarer than allopatric speciation, has been observed in whitefishes in North America and Eurasia and especially in **cichlid** (members of the family Cichlidae) fishes in African rift lakes.

The ultimate test of whether speciation has occurred happens when two previously allopatric populations come back into contact. For example, the red-shafted and yellow-shafted races of northern flickers have been brought back into contact through forest plantings by humans in the Great Plains. Where their ranges now overlap, the two groups interbreed freely, producing fertile offspring, indicating that they have not been isolated long enough to form separate species.

4.4 Natural Selection and Adaptation

Natural selection and **adaptation** are closely interrelated concepts. The theory of natural selection is an attempt to account for the adaptation of organisms to their environment. The major tenet of natural selection is that organisms of different genotypes in a population contribute differently to the gene pool of succeeding generations. Those organisms most successful in an environment will survive better and produce more offspring with the characteristics that led to their success. Differences between individuals with regard to survival and subsequent reproduction result in genotypic differences between generations. In this way, natural selection promotes adaptation in populations of organisms. A simple definition of *adaptation* is the change in structure, physiology, behavior, or life history that allows species to adjust to their environment.

Consider the responses of populations to selective pressure. Ricker (1981) considered 24 or more years of data on size, age, and state of maturity for five species of Pacific salmon on the west coast of North America. Many stocks of these species were subject to size-selective harvest, generally with removal of the largest and hence later-maturing individuals. Changes in average size and age of fish varied among species and fisheries. Chinook salmon subject to fishing pressure decreased greatly in size and age as individuals that matured at later ages were harvested more intensively and age at maturity declined over time. Coho and pink salmon exhibited decreases in average size over time because gill-net fisheries removed fish of larger than average size. Thus, through selection imposed by the fisheries, the populations adapted to the changing environmental conditions. The above is an example of adaptation (the process in which allele frequencies of a population are subjected to selection by environmental factors, thereby increasing the ability of the population to survive and reproduce in that environment). This adaptation within the population would not necessarily mean that alleles for rapid growth and later maturation would be completely lost from the population; the phenotypic or genotypic incidence of rapid growth and later maturation would just be reduced. Continued heterozygosity for rapid growth and later maturation would be beneficial because it would allow for future adaptation if environmental conditions were to change again. Envision what would happen to this population if some heterozygosity for growth rate and age of maturation was retained and at some time in the future harvest regulations (e.g., mesh size of nets) were changed.

An example such as this does not have to portray an either/or situation. Incomplete dominance or codominance could result in a range of growth rates and ages of maturation. In addition, multiple genes controlling a trait could also result in a range of phenotypic expression of these traits.

The above example is human induced, but a natural example of local adaptation is exhibited by the Mexican tetra. This fish species has cave-dwelling populations that are eyeless, normal populations living outside of caves that have well-developed eyes, and populations that live near the mouths of caves that have intermediate-sized eyes (Mitchell et al. 1977). These fish were able to exhibit local adaptation to their environments because the genetic compositions of their populations responded to natural selection.

On a long-term basis, evolutionary lineages can display striking adaptive change or evolution. The different types of forelimbs exhibited by mammals are an example. Adaptation is reflected in the shape and purpose of these limbs: in common chimpanzees the forelimbs are adapted for climbing, in pallid bats for flying, in eastern moles for digging, in killer whales for swimming, in coyotes for walking, and in humans for handling objects.

The striking effects of natural selection and adaptation are also seen in examples of **convergent evolution**. Such evolution occurs when species with different origins and histories develop similar traits and characteristics that allow them to occupy similar niches. Often these species are widely separated by geographic location and by their positions in evolutionary history. For example, several bird species have evolved into flightless, running organisms. This characteristic is exhibited by birds such as dwarf cassowaries, greater rheas, ostriches, and emus. There are some advantages to flightlessness. Flight is an energetically expensive activity. If a bird can avoid predation by **nocturnal** (pertaining to the night) or secretive activity patterns, keen vision, rapid running, or aggressiveness rather than by flight, it can be evolutionarily successful. The point is that flightless, running birds have developed in widely separated areas of the world; this convergent evolution is a result of natural selection and adaptation to similar ecological niches.

Another excellent example of convergent evolution involves the sunfishes of the Nearctic region and the cichlids of the Ethiopian region (Figure 4.2). In their respective regions each family has evolved to fill a wide variety of freshwater ecological niches. Both families have species that are prey organisms, while

A

B

FIGURE 4.2 The Nile tilapia is an African cichlid *(A)*, and the bluegill is a North American sunfish *(B)*. They are examples of fishes from families that exhibit convergent evolution. Note the similarity of features such as general body shape, fin size and location, and mouth size and location. These characteristics and others are indicative of the fact that the two families to which these species belong exhibit convergent evolution. *[Part A image courtesy of M. Brown; part B image courtesy of C. Paukert.]*

others are predators; some eat primarily snails, while others eat small soft-bodied bottom invertebrates, hatching insects, or a number of other food organisms; some live in rocky areas, while others live in vegetated areas (Barlow 2000). In many cases a cichlid and a sunfish that have adapted to similar ecological niches closely resemble each other in characteristics such as body shape, fin location, and mouth size. They have evolved to fill similar niches, but on two different continents.

4.5 Artificial Selection

Just as natural factors can impose selection, so can artificial or human-induced factors. In **artificial selection** humans choose the parents of each generation based upon phenotypic traits that they want to produce. Such selective breeding has produced many kinds of domesticated organisms that differ markedly in physical, physiological, and behavioral characteristics from their wild ancestors. The variety of different shapes, sizes, and colors of dogs, cattle, and chickens that exists today is the result of artificial selection.

Artificial selection, like natural selection, is a **systematic process**, that is, a genetic process that changes allelic frequencies in a population in some predictable, nonrandom fashion. While artificial selection has benefits in domestication, such selection is usually not beneficial to wild populations. A bulldog may do quite well living in your house, and a Holstein cow may survive well on a dairy farm, but such organisms likely would not survive or maintain their current forms in a natural ecosystem.

The rainbow trout is cultured throughout the world for both commercial and recreational purposes. Through selective breeding, a population of rainbow trout can be developed to have some trait desired by humans, such as rapid growth, resistance to a key disease, or reddish fillet coloration. Such selection may be beneficial only for those individuals raised in an artificial situation while being prepared for market. The same fish would be unlikely to compete and grow well in a natural ecosystem. In fact, it would probably be at a distinct disadvantage because of its lack of a variety of genetic traits of use in the wild. Its introduction might also have negative effects on local wild populations if it survived and reproduced, because genetic material unsuited for that environment might be added to the gene pool.

As mentioned above, artificial selection is not practiced only in fisheries. As game ranching and game farming become more prevalent, so does artificial selection in farmed and ranched "wild" animals. For example, white-tailed deer are being selectively bred to produce qualities such as large antlers. Records of which males and females have genotypes for large antler production in male offspring are kept, and those animals are used for breeding purposes. However, offspring of such animals may lose other beneficial genetic traits needed for survival in a wild situation.

Even when the intent is to raise a "wild" organism for stocking, unintentional **domestication selection** can often occur. By rearing or holding organisms in a hatchery or some other artificial situation, humans inadvertently select for eggs, young, or adults that survive well in that particular artificial situation. Individuals that have the genetic composition to survive such an experience will live to pass on their genetic information; those that do not do well in such an environment may not. This differential survival and reproductive success changes the genetic composition of future generations.

4.6 Inbreeding and Outbreeding

When populations are small, either in natural or captive situations, the probability that individuals in the population will mate with closely related individuals increases. Such mating is called **inbreeding**. Both the individual and the population can be affected by the loss of genetic variability that results. Inbreeding is most often a problem in fish culture and other artificial situations and in small wild populations. Inbreeding is a **dispersive process**, meaning that it causes random change in allelic frequencies and a loss of genetic diversity in populations over time. **Inbreeding depression**, a reduction in **fitness** or vigor due to increased homozygosity, is the result. Fitness is a measure of survival and reproductive success of individuals or populations. While inbreeding depression is difficult to demonstrate in natural populations, it is a common phenomenon in domesticated animals and laboratory populations that are extremely inbred.

In natural populations, the **founder effect** occurs when a few individuals leave a large population to establish a new population and changes in allele frequencies result from random genetic drift (see section 4.7). Rare alleles found in the large population will be lost if no individuals in the new population carry that genetic material. Also, the smaller the initial number of individuals that found a population, the greater the risk of inbreeding depression. It is important to remember that once inbreeding has occurred, its effects remain, even if the population increases to a much greater number; individuals in the population are still the offspring of relatively few parents and thus have the genetic diversity of only those few.

It might be assumed that if inbreeding leads to decreased genetic variability (homozygosity) and is not desirable, then the best situation would be for organisms to mate with their most distant relatives. This is not necessarily the case. Such distant mating is termed **outbreeding**. Populations often have local adaptations; while outbreeding increases heterozygosity, it can also result in the breakup of coadapted gene complexes, that is, of combinations of alleles conferring genetic fitness for specific conditions. Outbreeding is usually not a problem in natural situations because members of distantly related populations do not normally meet for mating. However, it can be a major problem when humans move organisms from place to place.

4.7 Mutation, Random Genetic Drift, and Migration

Selection, inbreeding, and outbreeding are all ways in which genotypic change can occur. **Mutation** of genes, a change in gene **DNA** (**deoxyribonucleic acid**), is another type of genotypic change. DNA is the chemical used to store genetic information in most organisms. Gene mutation occurs when the DNA of a gene permanently and spontaneously changes, resulting in a new allele. An allele produced by mutation may be entirely new to a population. If the mutation in some way makes the organism more successful, the trait may be passed to the next generation and have an increasing frequency in the population. Therefore, the mutation can lead to different genotypic and phenotypic frequencies in future generations; eventually it can also lead to a totally new species, especially if isolating mechanisms separate populations. However, it is actually much more likely that the mutation will have a negative or neutral effect and ultimately be lost from a population.

Mutation is a process that increases genetic diversity in populations and counteracts the loss of genetic diversity caused by dispersive processes. It is a normal phenomenon in nature. In general, mutation occurs at random or in an undirected manner. Genetic materials supplied by mutation provide the building blocks that allow for natural selection and, therefore, evolution.

Another mutational mechanism of genetic alteration is chromosomal change resulting from **chromosome aberrations**. Such changes can occur through **deletion**, **duplication**, **inversion**, or **translocation**. Deletion occurs when a chromosome breaks, then reunites, but loses a segment in the process. Duplication occurs when a segment of a chromosome is replicated; this results in that segment appearing twice in the genetic composition of the organism. Inversion occurs when the sequence of genes on a chromosome becomes inverted. Translocation occurs when a segment from one chromosome becomes attached to a **nonhomologous chromosome**. Nonhomologous chromosomes are chromosomes that carry genes for different characters. Each of these modes of genetic alteration can cause genotypic change in an organism and can, therefore, affect populations. In rare instances, whole chromosome sets can be duplicated, such that an individual organism has four or more sets (**polyploidy**). Such polyploidization events have occurred in the evolutionary history of certain shrimps, salmons, sturgeons, carps, suckers, and salamanders. Polyploidization is relatively common in plants and often drives speciation events.

Genetic drift is another mode of genotypic change. Genetic drift occurs due to random changes in allelic frequencies caused by sampling errors that occur in each generation. In this situation, sampling refers to the small sample of parental eggs and sperm that actually develops into offspring. For example, if an event causes the death of organisms in a population and the cause has nothing to do with the fitness of those organisms, genetic drift could occur. Assume that there is a Baltimore oriole population with only eight males, all with a different genotype for a trait such as bill color. Also assume that the five males with the darkest bills are killed for reasons unrelated to their genetic composition. The three remaining males would be the only ones that could mate, and therefore, gene frequency in the population would change in the direction of their genotypes—in this case, a lighter bill color. Genetic drift removes genes from the population but cannot add them. The chance of genetic drift increases as population size decreases, and it is usually associated with small populations. Genetic drift, like inbreeding, is a dispersive process, so it tends to increase homozygosity.

Genetically effective migration also can cause shifts in genetic composition. The movement of an organism to a new population that is genetically different from its original population and its subsequent reproduction can cause genetic change in the receiving population. Migration can have a much more immediate effect if the receiving population is small.

4.8 Some Genetic Applications in Wildlife and Fisheries

The many ways in which genetics is applied in wildlife and fisheries are too numerous to discuss in this broad introduction. However, a brief discussion of some newer applications follows. For deeper discussions, see Hallerman (2003) for fisheries applications and Frankham et al. (2002) for wildlife applications.

Genetic forensics is the use of genetics to make inferences regarding past events. In the context of this book it primarily relates to questions relevant to wildlife and fisheries law enforcement. The use of genetic information can allow for species, population, and even individual organism identification. For example, one could use the genetic makeup of a fish fillet to identify the species or perhaps even the genetic stock from which it came. Conservation law enforcement officers can use this information as evidence of illegal activity.

Genetic marking involves purposefully changing allele frequencies in a group of organisms to identify them or their descendants. The ability to identify descendants is a particularly powerful tool and has numerous applications. For example, by genetically marking a cultured fish stock one can detect the contribution of alleles from that cultured stock to a receiving wild population. Increasing the frequency of rare, naturally occurring alleles can also be used for identification purposes. Genetic marking and the use of genetic marks will continue to increase as new genetic tools are developed.

Management of migratory species may be complicated because distinct populations may be harvested together. For example, subadult salmon originating from different rivers may be harvested in mixed-stock marine fisheries. Geneticists observe and analyze allele frequency data from source populations and harvest to estimate the contribution of each source population to the fishery (Utter and Ryman 1993). Management of mixed-stock harvests involves restricting harvest in situations where the total allowable catch has been taken from demographically weak stocks.

Genetics can also be used to analyze **population viability** (the ability of a population to sustain itself demographically). How many organisms are needed in a population for that population to exist? By using genetic information one can work toward solving the issue of minimum viable population size. That is, determinations can be made concerning what population size is needed to provide a given probability of the persistence of a population for a given time period. Such a tool is especially applicable to species or populations that have decreased to very low numbers, such as endangered species or populations.

When population or species numbers become too low, genetic diversity is threatened and organisms have reduced capabilities for adapting to change. **Conservation genetics**, which involves both natural and captive organisms, seeks to devise methods to maintain as high a degree of genetic variability as possible in populations, especially those near extinction. An example is the case of the Florida panther. This population resides in south Florida and had been reduced to approximately 30 to 50 adults. Because of its low numbers, inbreeding appeared to be an issue, as did genetic bottlenecking. A **genetic bottleneck** or the **bottleneck effect** is a loss of genetic variability resulting from a restricted number of individuals reproducing in a population. In addition, this population was essentially isolated from other populations, thus reducing gene flow from the outside. These problems appeared to be leading to genetically related physical abnormalities in Florida panthers and a reduction in their abundance. In an effort to increase genetic diversity and numbers in the Florida panther population, cougars from Texas were introduced into the Florida panther's range in 1995. Do not become confused by the names; this big cat is variously called cougar, panther, puma, catamount, or mountain lion, but all are the same species (*Puma concolor*), and currently no North American subspecies are recognized. Was this stocking done without controversy? No! Has the stocking been successful? Recent evaluations are mixed, and the debate will continue (Maehr et al. 2006; Pimm et al. 2006a;

Pimm et al. 2006b). Will this conservation action result in the maintenance of a south Florida panther population with a unique gene pool or just a population of panthers in that region? The two results are quite different from a genetic standpoint. Many other questions remain. However, as our genetic tools improve there should be more effective means of addressing such problems in the future.

Another topic that is receiving much greater attention is the genetics of captive animals. Specific protocols for facilities, such as fish hatcheries, have been developed to try to ameliorate the genetic effects of captive-rearing programs. These protocols are just another indication of the growing importance of genetics in effectively managing and conserving organisms.

4.9 Population Genetics—Peril and Promise

There are many current perils and promises associated with population genetics. This section will address only a few of the more important ones. Remember that there are always potential negative and positive outcomes when humans intervene in biological systems.

Genetic engineering is a source of genetic change. Potential hazards and benefits of the ability of humans to create "new" organisms is currently a topic of both scientific and public debate. This debate is not just a biological one. Genetic engineering relates to economics, ethics, and a variety of other social issues.

This discipline involves **gene transfer**, induced polyploidy, and other artificial methods of producing "new" organisms. Gene transfer involves insertion by humans of genetic information from one species into the DNA of another. The result is the construction of an individual bearing a novel genotype and is called a **transgenic organism**. The new gene or genes can be obtained from any other species and may confer some specific beneficial trait to improve the recipient organism, such as disease resistance, faster growth, or greater tolerance for some environmental condition.

Polyploidy is a condition of an organism having more than two sets of chromosomes. As discussed earlier, some organisms are naturally occurring polyploids, but this condition can also be artificially induced in organisms in numerous ways. Polyploidy, especially **triploidy** (having three sets of chromosomes), is induced in some species, such as grass carp, to produce fish that cannot reproduce. This practice reduces the potential of this exotic species becoming established where it is stocked. In other cases, for example, in rainbow trout, triploidy is induced because triploid fish are generally sterile and can direct energy intake into body growth, rather than gonadal development and reproductive behavior.

Other genetic engineering methods are currently being used in many areas, including aquaculture. Such efforts include production of **monosex populations**, **gynogenesis**, and **androgenesis**. Monosex populations are populations that consist of only one sex. These may be useful because one sex may be more valuable than the other; for example, females of the species may grow more rapidly than males. Another potential benefit of monosex populations is that reproduction can be controlled. Monosex populations are usually produced by crossing two closely related species, by using hormones to artificially reverse the sex of organisms, or by induced gynogenesis. Gynogenesis is the production of viable young that have genetic material only from their mother; androgenesis is the production of viable young that have genetic material only from their father. Both gynogenesis and androgenesis, which involve reproduction without fertilization, can be used to produce diploid monosex or polyploid fishes.

With all forms of genetic engineering a question arises: Do the potential positives outweigh the potential negatives? Only the results of well-designed applied research will provide the answers to this question.

Artificial selection is another area of population genetics that holds peril and promise. The problem of artificial selection, whether intentional or unintentional, faces the profession whenever an organism is produced in a controlled or artificial environment. Such organisms may lack fitness. Fitness is a trait that is the product of many genes and the environment interacting throughout the lifetime of an individual organism. When we raise organisms in artificial conditions, the offspring will not necessarily have the same genetic makeup as the offspring of their wild counterparts (Figure 4.3). What is the effect, for example, on the genetic composition and fitness of mallards, elk, or razorback suckers when they are raised in captivity? Should these captive-reared organisms be released into natural environments? If so, under what conditions and in what situations? Domesticated turkeys have lost much of their poten-

FIGURE 4.3 A trout hatchery *(A)* and a natural stream *(B)* are dissimilar habitats that will produce dissimilar organisms, each adapted to its particular environment.

tial to survive in the wild due to genotypic change; if they breed with wild turkeys, the inherent wildness and survival ability of wild birds could be reduced. Some pen-reared ring-necked pheasants are bred for large size and are poor flyers, yet many of them are released into the wild. If they interbreed with wild birds, how do they affect the ring-necked pheasant population?

Whenever mixed-stock populations result from human activities, this question of **genetic integrity** arises. Genetic integrity refers to the genetic characteristics of a group of organisms that distinguish it from other groups or populations of the species. Little is known concerning the effects of loss of genetic integrity. Unfortunately, there are many populations, both native and introduced, in which genetic integrity is really no longer a relevant issue because human activities have already altered them so much.

In some situations the effects that humans have on a population due to captive holding or rearing are unavoidable. Black-footed ferrets and California condors were so low in numbers that all known individuals were captured and held for captive breeding. Obviously the intent of doing this was to increase their numbers and then later move them back to the wild, and this has been done. However, we cannot know what genetic consequences resulted from their time in captivity.

The issue of population genetics in relation to human conservation and management activities raises many questions. Is a walleye stocking program as effective as it might have been if fish with greater genetic diversity had been introduced? Would using parental walleyes from a nearby, similar lake within the watershed produce a better fishery? Could the introduction of the exotic gray partridge into North America have produced a better wildlife system if greater attention had been paid to genetic diversity and if larger numbers of partridge had been originally stocked? It is difficult to incorporate genetic diversity into a large population once the population has been established; for example, adding new gray partridge genetic material to an existing large population would be difficult at best. Such new genetic information is usually overwhelmed by that of the population already present. It might be possible in this situation to have an effect, but it might take many generations. What will be the genetic diversity of wolf populations resulting from current restoration efforts in the western United States? Wolf restorations are usually made with only a few animals, often captured from the same location. Will these restored populations be less successful than hoped for—or unsuccessful—because of the specific animals used in the restoration effort? Inbreeding and small population size must be a consideration in such efforts.

The maintenance of gene pools and genetic diversity is important if we are to have long-term success with populations that are managed or maintained by adding organisms to the population or by reestablishing populations. If gene pools and genetic variability are not maintained, these populations, no matter how large they become, will always be less adaptable to changing conditions than most naturally occurring populations. The loss of within-population variability through inbreeding and the loss of among-population variability through outbreeding both must be considered, especially when small populations are involved.

Hybridization is another area of peril and promise to population genetics. Piper et al. (1982) defined **hybrids** as the offspring resulting from mating of parents that are genetically unlike. In most cases hybrids represent crosses between different species, subspecies, or strains. In some situations hybridization is a natural occurrence, while in others it is human induced. An increase in some phenotypic value (growth rate, for example) of a hybrid relative to its parents is called **hybrid vigor** (**heterosis**). Hybrid vigor is usually attributed to increased heterozygosity. In some situations hybrids are unable to reproduce, while in other cases they can produce offspring.

Some human practices have led to negative population effects due to hybridization. The cutthroat trout has been eliminated or reduced in parts of its native range because of hybridization with introduced rainbow trout. After several generations of interbreeding, the rainbow trout phenotype often becomes dominant. Such a result is called **introgression**. This shift to rainbow trout is obviously not beneficial to native cutthroat trout, because they are reduced in numbers or eliminated. Various subspecies of wild turkeys have been introduced into new regions in North America. In some cases, introductions of wild turkeys of mixed-subspecies ancestry or introductions of two subspecies into the same area have occurred. The possible loss of identifiable subspecies and their gene pools resulting from such practices should be considered prior to introduction or reintroduction programs.

Hybrids constructed and used in a controlled fashion can have positive effects. The decision to use hybrids depends on the management objectives and whether those objectives can be accomplished with little or acceptable risk to other populations. A commonly used hybrid in fisheries work is the striped bass–white bass hybrid. This hybrid apparently does not require a coolwater refuge during the summer, something that large striped bass need. However, the hybrid retains some preferred attributes of the striped bass, especially with regard to size, in that they grow larger than white bass. The risk of using this hybrid is its potential adverse genetic and ecological effects on pure-species populations of white bass and striped bass.

One of the greatest threats to population genetics is the total loss of species, subspecies, strains, or populations. Total loss can have important ramifications on communities and species; there are really no positives involved. An example of the downside of complete loss is the case of the Audubon bighorn sheep. This bighorn sheep subspecies was historically found in the badlands and Missouri River breaks of North and South Dakota, but it became extinct by the early twentieth century. Since then, introduction of other bighorn sheep subspecies into that area has met with limited success. It is likely that the Audubon subspecies was adapted to local environmental conditions, and if that subspecies were still in existence today, restoration efforts would be more successful.

It should be obvious that there are multiple perils and promises associated with population genetics. Anytime we move (either intentionally or unintentionally), domesticate, hold captive, genetically modify, or in some other way affect organisms or populations, we introduce the potential for negative and positive consequences. Those effects on populations then influence communities. Having the collective wisdom to make correct and informed decisions is the key to reducing the perilous aspects of these endeavors and elevating the promising aspects.

4.10 Results of Evolution

Evolution results from genetic change in populations. Because genetic modifications provide the building blocks for evolution, understanding the way in which inherited variations occur is basic to our understanding of the mechanisms of evolution. Much of what has been discussed in this chapter relates directly to the evolution of organisms. As the genetic composition of a species changes, the phenotypes of the individuals and populations change and evolve. A combination of the individual's genetic composition and the selective pressure of environmental change acts to determine how successful an individual or population is at surviving.

The time scales over which genetic change occurs are highly variable. Some organisms are able to speciate or change in only a few generations, while others remain apparently unchanged or little differentiated over hundreds of thousands or millions of years. The genetic composition of organisms such as the horseshoe crab, American alligator, platypus, brown kiwi, and paddlefish has allowed these organisms to survive in relatively unchanged form for long periods of time and through many environmental changes. The genetic composition of groups such as minnows, warblers, and mice has allowed them to develop multiple species in many new habitats as these habitats became available and changes occurred. Many contemporaries of these species or groups did not have the genetic composition necessary to adapt to or confront change; they have become extinct and are just fossil memories.

SUMMARY

Genes transfer traits from one generation to the next. Some traits are transmitted by single genes, while others are transmitted by multiple genes. Most vertebrates are diploid organisms that have two complete sets of chromosomes, one from their mother and one from their father. The specific genetic information possessed by an organism is called its genotype; the actual physical expression of traits is called its phenotype. The phenotype is a product of the genotype interacting with the environment. Alleles are alternate forms of a gene. An organism with two of the same alleles at a given locus is homozygous at that locus. If the alleles are different, the organism is heterozygous at that locus. The amount of heterozygosity or homozygosity present in a population determines its ability to change or adapt. Genetic diversity occurs within and among populations.

Species are naturally occurring groups of organisms that can successfully interbreed and produce fertile offspring. Subspecies are formally recognized geographical subdivisions of species whose members can interbreed but which differ from other subdivisions in some recognizable way. Populations (stocks) are groups within a species that actively interbreed and that are adapted to similar environments. Strains are cultured populations that reliably express a suite of characteristics.

Isolating mechanisms tend to keep organisms from interbreeding. Without some form of isolation, speciation is almost impossible. Natural selection and adaptation are closely intertwined. The theory of natural selection is an attempt to account for the adaptation of organisms to their environments.

Artificial selection is a process in which humans select for some desired phenotypic trait. Artificial selection, whether intentional or unintentional, occurs in both wildlife and fish populations.

Inbreeding is a process that is usually associated with small populations and often results in increased homozygosity. Outbreeding, while promoting heterozygosity, may be detrimental if it reduces the capacity of an organism to survive in a particular environment.

Mutation is a systematic process that increases genetic diversity. Genetic drift, usually a process affecting only small populations, is a result of chance fluctuations in allelic frequencies. It is a dispersive process that tends to increase homozygosity.

There are numerous newer genetic applications in wildlife and fisheries. Included are genetic forensics, genetic marking, conservation genetics, and genetic management of captive breeding programs.

There are many perils and promises associated with population genetics. A few include moving organisms from place to place, losing species or populations, maintaining captive populations, domestication, and genetically modifying organisms.

PRACTICE QUESTIONS

1. How can two organisms with the same genotype for a trait have different phenotypes for that trait?

2. How can two organisms with different genotypes for a trait have the same phenotype for that trait?

3. Define a species.

4. Select two closely related sympatric organisms and describe some of the isolating mechanisms that might keep them from interbreeding.

5. Excluding examples used in this text, develop a hypothetical organism and set of environmental conditions that would lead to some specific scenario of natural selection in that organism.

6. Think of conditions that a particular organism would experience in a human-controlled artificial environment, then list a dozen ways in which that artificial environment could select for particular genetic traits.

7. Why is inbreeding a problem even after a population founded by a small number of individuals grows large?

8. What are some potential negative ramifications concerning population genetics when "domesticated" white-tailed deer escape into a wild population?

9. Describe a hypothetical mutation that would be beneficial to an organism, then describe how that characteristic could lead to the formation of a new species.

10. Why and for what reasons could a variety of different people be supportive of or against genetic engineering?

11. In section 4.8 there are three citations concerning the issue of Florida panthers. Read the papers and draw some of your own conclusions about this issue. In addition, search for more recent papers on this issue.

SELECTED READINGS

Allendorf, F. W., and M. M. Ferguson. 1990. Genetics. Pages 35–63 *in* C. B. Schreck and P. B. Moyle, eds. *Methods for fish biology.* American Fisheries Society, Bethesda, Md.

Avise, J. C., and J. L. Hamrick, eds. 1996. *Conservation genetics: Case histories from nature.* Chapman and Hall, New York.

Frankham, R., J. D. Ballou, and D. A. Briscoe. 2002. *Introduction to conservation genetics.* Cambridge University Press, Cambridge, U.K.

Hallerman, E. M., ed. 2003. *Population genetics: Principles and applications for fisheries scientists.* American Fisheries Society, Bethesda, Md.

Hunter, M. L. 2002. *Fundamentals of conservation biology.* 2nd ed. Blackwell Publishing, Williston, Vt.

Pennell, W., E. D. Lane, and F. Dalziel. 2001. Open systems: The culture of fish for release into natural systems. Pages 187–239 *in* G. A. Wedemeyer, ed. *Fish hatchery management.* 2nd ed. American Fisheries Society, Bethesda, Md.

Oyler-McCance, S. J., and P. L. Leberg. 2005. Conservation genetics in wildlife biology. Pages 632–657 *in* C. E. Braun, ed. *Techniques for wildlife investigations and management.* 6th ed. The Wildlife Society, Bethesda, Md.

LITERATURE CITED

Barlow, G. 2000. *The cichlid fishes: Nature's grand experiment in evolution.* Perseus Publishing, Cambridge, Mass.

Bridges, W. R., and B. von Limbach. 1972. Inheritance of albinism in rainbow trout. *Journal of Heredity* 62:152–153.

Frankham, R., J. D. Ballou, and D. A. Briscoe. 2002. *Introduction to conservation genetics.* Cambridge University Press, Cambridge, U.K.

Hallerman, E. M., ed. 2003. *Population genetics: Principles and applications for fisheries scientists.* American Fisheries Society, Bethesda, Md.

Maehr, D. S., P. Crowley, J. J. Cox, M. J. Lacki, J. L. Larkin, T. S. Hoctor, L. D. Harris, and P. M. Hall. 2006. Of cats and Haruspices*: Genetic intervention in the Florida panther. Response to Pimm et al. (2006). *Animal Conservation* 9:127–132.

Mitchell, R. W., W. H. Russell, and W. R. Elliott. 1977. *Mexican eyeless characin fishes, genus Astyanax: Environment, distribution, and evolution.* Special Publication 12. Texas Tech Museum, Lubbock.

Pimm, S. L., O. L. Bass, and L. Dollar. 2006a. Ockham and Garp. Reply to Maehr et al.'s (2006) response to Pimm et al. (2006). *Animal Conservation* 9:133–134.

Pimm, S. L., L. Dollar, and O. L. Bass. 2006b. The genetic rescue of the Florida panther. *Animal Conservation* 9:115–122.

Piper, R. G., I. B. McElwain, L. E. Orme, J. P. McCraren, L. G. Fowler, and J. R. Leonard. 1982. *Fish hatchery management.* U.S. Department of the Interior, Fish and Wildlife Service, Washington, D.C.

Ricker, W. E. 1981. Changes in the average size and average age of Pacific salmon. *Canadian Journal of Fisheries and Aquatic Sciences* 38:1636–1656.

Utter, F., and N. Ryman. 1993. Genetic markers and mixed stock fisheries. *Fisheries* 18:11–21.

CHAPTER 5

Nutrition and Environmental Physiology

The ways in which organisms obtain nutrients and the ways in which they are adapted to their particular environments are important aspects of animal and plant populations and communities. The nutritional needs of wild animals are met primarily through feeding patterns that have evolved to promote the survival and reproduction of each species. Animals must obtain an array of mineral elements, vitamins, proteins, fats, and carbohydrates in their diets. While broad similarities in nutritional needs exist among animals in the same guilds, there are interesting differences among species; differences also often exist within species by season, gender, and age. In exploiting available food sources, vertebrate species have evolved various structural and functional differences in their digestive tracts that allow the processing of widely differing food materials. Many animals, such as ungulates, even enlist the help of an array of bacteria, protozoans, and other microbes in extracting needed nutrients from their diets.

The physiology of a species determines the environment in which it can survive and reproduce. Endotherms and ectotherms live in environments that are both static and changeable. Organisms must be able to adapt to a variety of conditions if they are to be successful. For example, aquatic ectotherms are adapted to an environment with low oxygen concentrations. Aquatic ectotherms and some endotherms must be able to adapt to changing salinities. All organisms must be able to adapt to changing temperatures, but such adaptations are most developed in endotherms, because they must maintain their body temperatures within a narrow range.

A single chapter on topics as extensive as nutrition and environmental physiology cannot be all inclusive or in depth. Nevertheless, because of their importance to understanding how organisms function, nutritional and physiological comparisons can and should be made between and among different animals. The objectives of this chapter are to provide an overview of the nutritional needs of fish and wildlife species and to discuss some selected physiological adaptations of these species to their environments.

5.1 Minerals and Other Elements

Both fish and wildlife species need a number of inorganic elements in large amounts (**macroelements**), as well as several that are needed in only trace amounts (**trace elements** or **microelements**). Several additional elements are required by captive animals in sterile environments and on purified diets; these **ultratrace elements** are not known to be lacking under natural conditions (Robbins 1993). Macroelements, microelements, and ultratrace elements are also referred to as macronutrients, micronutrients, and ultratrace nutrients, respectively.

Macroelements, microelements, and ultratrace elements occur in soil and water as minerals (Table 5.1). Other elements that are needed in large quantities, but which are usually classified as nonmineral elements and thus are not listed with the macroelements, include oxygen, hydrogen, carbon, and nitrogen. The seven macroelements—calcium, phosphorus, sodium, potassium, magnesium, chloride, and sulfur—constitute approximately 4% of the body weight of vertebrates, while the microelements contribute less than 1%. Oxygen, carbon, hydrogen, and nitrogen constitute about 96% of the body weight of vertebrates.

Calcium is involved with activities such as muscle contraction, nerve impulse transmission, bone formation, eggshell formation in birds, blood clotting, and acid–base balance. As much as 99% of the calcium in the body may be found in the **hydroxyapatite matrix** (mineral matrix of calcium phosphate), of which vertebrate bones and teeth are primarily composed (Kincaid 1988; Schmidt-Nielsen 1997). The growing young of fish and wildlife species need high levels of calcium. During egg laying in birds, the intestine increases its absorption of calcium; this is reflected by increased levels of calcium in the blood and increased calcium deposition in the bones. Female birds use additional calcium from diets and some stored bone calcium for eggshell formation. Eggshells are about 94% calcium carbonate. Calcium is also mobilized from bones during **lactation** (milk secretion) in mammals. Antlers contain about 22% calcium and create a substantial calcium demand in many ungulates during the late spring and early summer, when new antler growth occurs (Robbins 1993).

Calcium is obtained through the diet, but wild animals sometimes encounter deficiencies. Young carnivores must consume adequate amounts of bone in the diet, because animal flesh generally does not provide adequate amounts of calcium. Various bird species, from ring-necked pheasants to shorebirds, ingest bits of calcium-containing items such as snails and small pieces of bone. **Raptors** (predatory birds such as hawks and owls) and their young may at times ingest rather large bones, and rodents may chew on bones or shed antlers. Fishes often obtain calcium and several other elements through the gills (see section 5.8) as well as obtaining them in food.

Phosphorus is involved in a wide array of metabolic activities, including muscle contraction, nerve metabolism, and metabolism of fats, carbohydrates, and amino

TABLE 5.1 Macroelements, trace elements (microelements), and ultratrace elements that occur as minerals and are needed as nutrients by animals (Robbins 1993).

Macroelements	Trace Elements	Ultratrace Elements[a]
Calcium (Ca)	Iron (Fe)	Silicon (Si)
Phosphorus (P)	Zinc (Zn)	Tin (Sn)
Sodium (Na)	Manganese (Mn)	Boron (B)
Potassium (K)	Copper (Cu)	Bromine (Br)
Magnesium (Mg)	Molybdenum (Mo)	Cadmium (Cd)
Chlorine (Cl)	Iodine (I)	Lead (Pb)
Sulfur (S)	Selenium (Se)	Lithium (Li)
	Cobalt (Co)	Vanadium (V)
	Fluorine (F)	Nickel (Ni)
	Chromium (Cr)	Arsenic (As)

[a] Captive animals in sterile environments and on diets completely lacking these elements may show deficiencies. These elements are rarely deficient in the diets of wild animals.

acids. Phosphorus is extremely important as a mineral element in combination with calcium in the hydroxyapatite matrix of bone (Hays and Swenson 1993). Bone includes approximately 360 grams of calcium, 170 grams of phosphorus, and 10 grams of magnesium per kilogram (McDonald et al. 1981). Phosphorus is also used in small amounts in eggshell formation and composes about 11% of the weight of antlers (Robbins 1993). Like calcium, additional phosphorus can be obtained in the diet through ingestion of bone.

Sodium is the primary **cation** (positively charged ion) needed for maintaining the cells of vertebrates in a saline fluid environment. It is also important in nerve impulse transmission and for general growth and reproduction. Carnivorous aquatic and terrestrial vertebrates generally obtain adequate sodium in the tissues of their prey. Herbivorous vertebrates may have difficulty obtaining sodium because it is not essential to plants. Plants that do contain high sodium levels, such as submergent vegetation, may be sought as sodium sources by herbivores such as the moose. Ungulates such as moose and elk may seek areas of soil or rock with concentrations of sodium and other minerals to supplement mineral requirements that are not being supplied by foraging alone (Figure 5.1) (Ayotte et al. 2006). Ingestion of soil or other earthy material by animals to obtain minerals is termed **geophagia**. Such mineral sites may be used over long periods of time by various ungulate species. African elephants in Wankie National Park in Zimbabwe excavate scrapes and pits in areas with unusually high concentrations of soluble sodium (Weir 1969). Termite mounds contain elevated concentrations of soluble sodium and may also be used as licks by African elephants in areas where sodium concentrations cannot be located near the ground surface. Salt, usually in block form, can be placed on the ground for use by domesticated livestock and may also be used by wild ungulates or other animals.

Both potassium and magnesium are readily available in plant and animal food sources. Potassium is critical for processes such as nerve impulse transmission, muscle contraction, maintenance of cell osmotic pressures, body metabolism, and enzyme activation. Potassium is normally available in adequate amounts in vegetation or in ingested animal tissue, so deficiencies are rare in wild animals. However, high levels of potassium in ingested vegetation can reduce the retention of other elements, such as sodium or magnesium, and thus contribute to the deficiency of other minerals (Robbins 1993).

FIGURE 5.1 Ungulates use traditional licks to ingest soils high in supplemental sources of minerals that cannot be acquired by foraging alone. This moose is visiting a wet-lick site used by moose and elk in northern British Columbia; dry-lick sites can also be important to many ungulates. *[Image courtesy of M. Gillingham.]*

Magnesium is critical to functions such as enzyme activation and energy metabolism and is also needed for bone and tooth formation; about 70% of the magnesium found in animals occurs in bone (Goff 2004). Magnesium is present in ample amounts in plant and animal tissue, and deficiencies are rare in wild animals.

Sulfur is needed for the production of sulfur-containing amino acids, while chlorine functions as the primary **anion** (negatively charged ion) in the blood and body fluids. Chlorine is also important in maintaining body pH and in hydrochloric acid formation in the stomach. Chlorine is obtained by fishes, in large part, through the gills.

All of these macroelements must be obtained in relatively large amounts; in some cases feeding preferences have been related to animals seeking needed macroelements. Table 5.2 describes some of the macroelement requirements in prepared diets of captive animals.

Trace elements are used in lesser amounts than macroelements but are still necessary to normal body functions in animals. Geophagia observed in wildlife may be related to trace element requirements. Though these elements are needed only in small amounts by wildlife and fish species, they are still important. For instance, iron and manganese are both trace elements, but they are essential to animals, the former in red blood cells and the latter for bone formation and energy metabolism

TABLE 5.2 Nutrient requirements for selected macroelements in captive animals, expressed as a percentage of dry matter in the diet.

Macroelement	Species	Life Stage	% of Dry Diet
Calcium	Ring-necked pheasants; northern bobwhites	Growth	0.5–1.0
	Ring-necked pheasants; northern bobwhites	Breeding female	2.5
	Mallards	Growth	0.6
	Mallards	Breeding	2.8
	American mink; red foxes		0.3–0.6
	Common carp		0.3
	Atlantic salmon		None required in seawater
Phosphorus	Ring-necked pheasants; northern bobwhites; mallards		0.3–0.7
	American mink; red foxes	All stages, but needs highest at lactation	0.3–0.6
	Rainbow trout		0.5–0.8
Sodium	Ring-necked pheasants; northern bobwhites		0.15
Salt (NaCl)	American mink; red foxes		0.5
Potassium	Japanese quail		0.4
	Chinook salmon	Juvenile in fresh water	0.6–0.8
Magnesium	Ring-necked pheasants; northern bobwhites; mallards		0.03–0.05
	Domesticated guinea pigs	Growth	0.06
	Rainbow trout; common carp; channel catfish		0.04–0.06

Sources: National Research Council 1982, 1993, 1994; Robbins 1993.
Note: Mineral requirements in fishes are difficult to study, because in addition to those obtained in the diet, several elements can be taken in through the gills, through the skin, or by drinking water.

as well as other functions. Iodine, as in humans, must be present for normal thyroid gland function and growth. A lack of selenium, though it is toxic in high amounts, will cause muscle degeneration and labored breathing. Even the lack of fluoride and chromium can cause reduced growth and other problems. Most of the trace elements needed by fish and wildlife species are found in sufficient quantities in natural diets, and deficiencies are uncommon except in captive situations.

5.2 Vitamins

Vitamins are organic substances required by animals for normal body functions. Vitamins are of two types, water soluble and fat soluble. Fat-soluble vitamins are often stored in large quantities in the bodies of animals; these stores can be used when dietary amounts are inadequate. Water-soluble vitamins, with some exceptions, are generally not stored in large amounts in the body and must be consistently available in the diet. The fat-soluble vitamins include vitamins A, D, E, and K, while the water-soluble vitamins include thiamin (B_1), riboflavin (B_2), nicotinic acid or niacin, pyridoxine (B_6), pantothenic acid, folic acid, cyanocobalamin (B_{12}), choline, biotin, and ascorbic acid (C). The vitamin needs of fishes have been largely determined through feeding experiments with cultured species such as salmon, trout, channel catfish, and common carp. Likewise, the vitamin requirements of wildlife species have largely been determined from studies and observations on a variety of captive animals.

TABLE 5.3 Vitamins and their major functions.

Vitamin	Known Functions
FAT SOLUBLE	
A	A constituent of visual pigment (rhodopsin); maintenance and growth of epithelial tissue; synthesis of glycoprotein; bone and tooth development
D	Calcium absorption from intestine and resorption from bone; calcium metabolism; promotion of bone growth
E	Maintains functional integrity of cell membranes; prevents breakdown of vitamins A and C in gut
K	Electron transport role in body energetics; needed in formation of blood clots
WATER SOLUBLE	
Thiamin (B_1)	Formation of connective tissue; needed in use of iron; coenzyme in carbohydrate metabolism
Riboflavin (B_2)	Functions with coenzymes flavin adenine dinucleotide (FAD) and flavin mononucleotide (FMN) in breakdown of glucose and other carbohydrates
Niacin (nicotinic acid)	Functions with coenzyme nicotinamide adenine dinucleotide (NAD^+) and nicotinamide adenine dinucleotide phosphate ($NADP^+$) in breakdown of glucose and other carbohydrates
Pyridoxine (B_6)	Coenzyme role in protein metabolism
Pantothenic acid	Coenzyme role in carbohydrate, fat, and amino acid metabolism
Folic acid	Coenzyme role in amino acid and nucleic acid metabolism
Cyanocobalamin (B_{12})	Coenzyme role in carbohydrate metabolism; also important in nucleic acid metabolism; linked with folic acid in synthesis of methionine.
Biotin	Coenzyme role in fat and glycogen formation and in amino acid metabolism
Choline	Needed in nervous system (component of acetylcholine); component of phospholipids in cells and tissues
C (ascorbic acid)	Needed in bone, teeth, and collagen formation and carbohydrate metabolism; may reduce infections

Sources: Adapted from Robbins 1993; Pond et al. 2005.

In general, the small quantities of vitamins needed are found in adequate proportions in the diets of wild animals. Problems occur when fishes or wildlife are reared in captivity and needed vitamins are inadequate in prepared diets. Vitamins can also be lost from prepared diets during feed production or if the feed is stored too long. Some fat-soluble vitamins, such as A and D, can be highly toxic when excessive amounts are included in the diet. For example, the livers of some carnivores, such as polar bears and various seal species, accumulate large quantities of vitamin A and can be toxic if fed to other animals. At times some animals eat feces (**coprophagia**). In the case of some rabbit species, this is done to obtain and conserve vitamins produced by bacteria in the rabbits' digestive tracts.

Vitamins perform a variety of functions in animals (Table 5.3). Lack of various vitamins can cause serious consequences, such as disturbances in the normal calcification of growing bones, a lack of appetite and growth, nervous disorders, hair loss, difficulty in blood clotting, gill disease, anemia, hemorrhagic kidneys, or even sudden death (Robbins 1993). Many of the effects of vitamin deficiencies in hatchery-reared fishes and captive-reared wildlife were thought to be caused by parasites, diseases, or other factors until the role of vitamins in diets was demonstrated.

5.3 Amino Acids

Amino acids are the building blocks of proteins. They can be obtained by animals in two ways. Amino acids that an animal can manufacture within its own body are called **nonessential amino acids.** Others, called **essential amino acids,** must be obtained in the diet, either because the animal cannot manufacture them or the quantities it can manufacture are insufficient to provide for its needs. Do not be confused by these terms; all of these amino acids can be essential to an animal. The essential versus nonessential designation refers to how they are generally obtained.

Most fishes and terrestrial vertebrates with simple stomachs have similar lists of essential amino acids, which include arginine, histidine, isoleucine, leucine, lysine, methionine, phenylalanine, threonine, tryptophan, and valine. Nonessential amino acids for most fishes and other vertebrates include alanine, aspartic acid, citrulline, cystine, glutamic acid, glycine, norleucine, proline, serine, and tyrosine. Essential amino acid requirements are reduced or eliminated in **ruminants,** such as mountain goats and white-tailed deer, which have four-chambered stomachs where microbes break down plant matter (see section 5.5). Essential amino acid requirements are also reduced or eliminated in dusky grouse (the species formerly known as blue grouse has been divided into a dusky and a sooty grouse), spruce grouse, African elephants, and other birds and mammals with well-developed **caeca** (see Figure 5.11 on page 91). These caeca are outpocketings of the intestine in which microbial populations break down complex plant matter and improve digestive efficiency. The bacteria involved in these processes are themselves digested, providing an important source of amino acids.

5.4 Proteins, Fats, and Carbohydrates

Body metabolism can be divided into **anabolic processes,** which build up tissue, maintain body tissue, and store energy, and **catabolic processes,** which break down substances and release energy. Fish and wildlife species get their energy from proteins, fats, and carbohydrates in the diet. These food sources undergo chemical breakdown (are **catabolized**) through the various processes in the tricarboxylic acid cycle (Krebs or citric acid cycle) to produce the energy needed for growth, body heat, and muscular contraction (Frandson et al. 2003). Carbohydrates obtained in the diet include simple sugars, such as glucose (a **monosaccharide**), as well as more complex molecules, such as **oligosaccharides** (a short chain of two or more simple sugars) or **polysaccharides** (a straight or branched chain of hundreds or thousands of sugar molecules). Glucose is the primary carbohydrate found in the blood and other tissue fluids, and it can be readily obtained from plants or from the breakdown of other carbohydrates.

When blood glucose and other carbohydrates are in short supply, as during fasting, glucose can be synthesized from noncarbohydrate sources, such as fats and proteins, in a process occurring primarily in the liver that is called **gluconeogenesis.** Use of proteins in gluconeogenesis requires catabolism of amino acids and **deamination** (removal of an amino group [NH_2] from the amino acid); deaminized amino acids can also be used directly for energy without being converted to glucose. Fats represent a stored form of potential energy and can be broken down by hydrolysis (a catabolic process known as **ketogenesis**) to enter the tricarboxylic acid cycle.

Protein is more important for building body tissues, particularly muscle, than as an energy source. The protein component of bone, including antlers, is substantial at all stages of life, but particularly during growth. Vertebrates catabolize ingested proteins into amino acids and **peptide groups** (combinations of two or more amino acids) and later **anabolize** (construct) proteins as needed by the animal. In many cases, the most efficient manner of obtaining needed proteins would be to eat other animals of the same species, something many fish species do on a regular basis. Even a rogue common muskrat, normally an herbivore, obtains added amino acid benefits from eating its own kind.

Protein can be obtained from vertebrate muscle tissue, invertebrates, and new vegetation growth. Fats and carbohydrates are normally ingested in sufficient amounts by carnivores, which tend to eat the entire prey organism, including fat deposits. It should be noted, however, that animals contain a low proportion of carbohydrates compared to plants. Herbivores and omnivores may seek plant material particularly high in fat or carbohydrates, such as seeds and fruits.

Protein requirements vary among vertebrates by species and within species by age, gender, and season of the year. Vertebrates feeding on the flesh of other animals consume high levels of protein during all adult life stages. In terrestrial carnivores, protein requirements for maintenance in adults often exceed 25% of the diet. Carnivores, unlike herbivores and most omnivores, are adapted to obtain most of their energy requirements from proteins. Carnivores have enzymes in their livers that are used to deaminate proteins and that have higher activity levels than similar enzymes found in herbivores and omnivores. In comparison, ruminants have protein requirements of 5%–9% of the diet; their protein requirements during growth are commonly 13%–20% but are still less than those of adult or juvenile carnivores (Robbins 1993). As noted earlier, the ruminants, unlike carnivores, supplement their amino acid intake with proteins from the bodies of microorganisms in their complex stomachs.

Fishes exhibit great species differences in the amounts of proteins, fats, and carbohydrates that they need or can tolerate in the diet. Carbohydrate needs in most fishes are generally quite low. If trout or salmon in captivity are fed a high-fat diet, they develop fat-infiltrated livers, while common carp thrive on the same diet (Lagler et al. 1977). However, carnivorous fish such as salmon must obtain and store considerable fat for energy needs and accomplish this through predation on smaller fishes, squids, or other organisms with high lipid stores (Moyle and Cech 2004). In general, protein levels below 6% of the diet are too low to sustain even herbivorous fishes. Many piscivorous fishes ingest protein levels exceeding 40% of the diet, depending on water temperature and the physiological needs of the species.

Young birds and mammals can be expected to require higher percentages of protein in the diet than at most times later in the life cycle (Figure 5.2). American wigeon ducklings require large amounts of invertebrates in the diet, particularly in the first few weeks of life. The young of greater white-fronted geese graze on new-growth sedges and other Arctic plants that are better sources of protein than mature plants. Young mammals such as nursing Ord's kangaroo rats or desert cottontails obtain their protein needs from the lactating female. In fishes that are herbivorous as adults, such as grass carp, the young initially feed on zooplankton, then switch to herbivory as they grow. The protein needs of young white-tailed deer are higher than those of adults, even after weaning, because they are growing rapidly.

Reproduction in fishes and terrestrial vertebrates is often accompanied by increased protein demands in the females. In mammals, **gestation** (growth of the embryo in the uterus) and lactation require extra dietary protein. Likewise, the production of bird eggs requires considerable parental investment in terms of proteins and fats in the yolk and proteins in the albumen. Several species of waterfowl and gallinaceous birds will attempt to renest if the original **clutch** (group of eggs) is destroyed or otherwise terminated. Each clutch may weigh an appreciable amount, in some cases more than the female, so the costs to the female in proteins, fats, and calcium (for the eggshell) are high (Figure 5.3). Additionally, the hen must stay on the nest for prolonged periods of time during incubation, which limits her opportunities to feed.

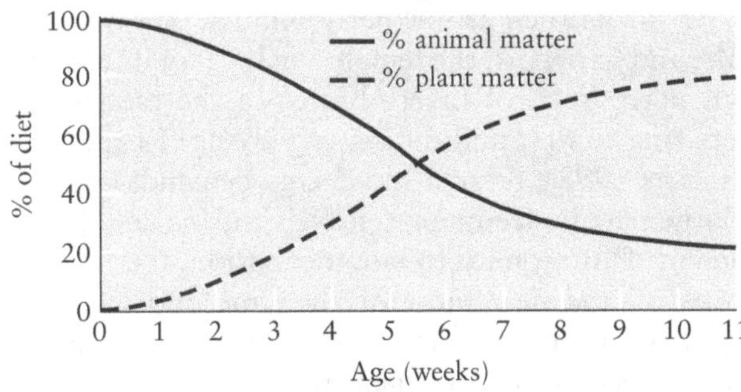

FIGURE 5.2 Percentage of animal matter (primarily invertebrates) and plant matter in diets of young gallinaceous birds and surface-feeding ducks.

FIGURE 5.3 Egg production can be energetically demanding, particularly in species that produce large clutches like this one, laid by a mallard.

In ducks such as northern pintails, blue-winged teals, and gadwalls, the females include an unusually high percentage of invertebrates in the diet while preparing to nest and during egg laying (Krapu and Reinecke 1992). Protein and energy demands are met in large part by feeding intensively during egg development. During mid- to late incubation, the hen remains on the nest most of the time and has few opportunities to feed; therefore, some body protein may be needed to supplement the diet. Female ducks also use stored body fat and carbohydrates during incubation. However, the resulting weight loss in these female ducks is not as severe as in some other species, such as most Arctic-nesting geese.

Arctic-nesting geese, such as lesser snow geese, forage extensively on corn and other waste grains as well as winter wheat during their slow spring migration. Prior to the extensive development of agriculture, migrating lesser snow geese probably sought roots and other nutritious parts of native plants in wetlands and associated uplands. During this northward migration, lesser snow geese store large supplies of fats, carbohydrates, and proteins for use during the nesting season. On the Arctic nesting grounds, nutrient-rich food in the form of new-growth grasses and sedges is especially limited early in the year, and the reproductive season is short. Nesting must be initiated soon after arrival by the already mated females to allow time for incubation and rearing of the young to flight stage before fall migration.

If the year is colder than normal, snow may cover potential nesting sites, and many lesser snow goose pairs may not attempt to nest that year. The short season does not allow adequate time for renesting if initial nesting fails. Because avian and mammalian predators are quick to take advantage of unprotected clutches of lesser snow goose eggs, females have evolved highly persistent incubation habits. Successful females seldom leave the nest, so most of their protein, as well as energy, must come from body stores (Figure 5.4). The female uses protein from her breast and leg muscles as well as from internal organs such as the gizzard and the once-enlarged oviduct. After an incubation period of about 23 days, the female lesser snow goose has lost over 25% of her body weight and may approach a point (about 40% weight loss) at which further weight loss could mean death (Ankney and MacInnes 1978). A small number actually die on the nest, so precipitous is this loss of energy and associated decline in body condition.

FIGURE 5.4 Female lesser snow geese remain on the nest throughout incubation, with only brief periods of feeding. During the incubation period, they generally lose 25% or more of their body weight. *[Image courtesy of D. Ankney.]*

Even the male lesser snow goose, while guarding the female from other males, is required to use more protein and energy than he can replace by feeding during the nesting period (Ankney 1977). However, weight loss by paired males is not nearly as substantial as that of nesting females, and the males are able to accept increased parenting duties after the young hatch. Females are in an emaciated condition by that time and enter into a period of **hyperphagy** (rapid and lengthy feeding) to replenish the lost supply of proteins, fats, and carbohydrates.

Migration of salmon species native to the Pacific Ocean (Pacific salmon) could be viewed as being similar to the ordeal of some of the Arctic-nesting geese, except that these parents die soon after reproduction. In the months before they enter spawning streams, Pacific salmon undergo a period of hyperphagy. While still in the ocean, the female needs protein and energy to form the large egg masses. These fishes undertake a long, arduous upstream journey that requires a large amount of energy. In sockeye salmon making spawning migrations of 596–1,023 kilometers, females depleted over 90% of their fat reserves and 55%–61% of their body protein by the time of death; male loss of fat and protein was also severe but not as great as that of females (Burgner 1991). The change from a marine to a freshwater environment is an added stress to migrating salmon. Because Pacific salmon species, while in the ocean, feed on

FIGURE 5.5 Pacific salmon migrate up freshwater rivers and streams, then spawn and die; their nutrients enrich the stream and thus indirectly nourish young fish. Brown bears and many other predators and scavengers feed heavily on adult salmon during the spawning run. *[Image courtesy of S. Hillebrand, U.S. Fish and Wildlife Service.]*

fishes or invertebrates, they have a high-protein diet in comparison to herbivores.

Death of the parents immediately after reproduction is a successful reproductive strategy in Pacific salmon because the eggs do not require post-spawning care; it would not work in lesser snow geese. In a sense, however, the dead Pacific salmon continue to feed their young. By dying in the home stream, they provide a nutrient source that enriches the stream and eventually feeds invertebrates and other stream organisms, which then in turn may be eaten by the young salmon. Even those Pacific salmon eaten by brown bears have carried nutrients upstream that will eventually enter the stream and nourish young salmon (Figure 5.5).

Nutritional needs and body condition can vary greatly through the seasons of the year in white-tailed deer. Forage with adequate protein and energy content is available to the deer from mid-spring or early summer through early fall in most areas. White-tailed deer feed heavily during summer and fall to increase their body fat and nutrient reserves. Energy demands due to factors such as lactation, growth, and **rut** (the period associated with male breeding activity in some ungulates) delay or reduce the accumulation of fat and nutrient reserves. These reserves are essential because of the poor quality (low protein content) of the food usually available during the winter months. For white-tailed deer does, the winter season is also the time when protein and energy are needed for gestation. Individuals that cannot accumulate adequate reserves are faced with a greater risk of death during the winter (Figure 5.6) (Mautz 1978).

As a general principle, nutritionally fit animals are much more likely to be reproductively successful. In terrestrial vertebrates, especially mammals, the relationship between body condition and reproduction is also quite clear. Ungulates that are in good condition are more likely to become pregnant at a younger age and to have multiple births. White-tailed deer with diets high in protein and energy, such as those in the agricultural regions of the upper Midwest, reproduce at a younger age and have more twins and triplets than do populations in nutrient-poor environments; for

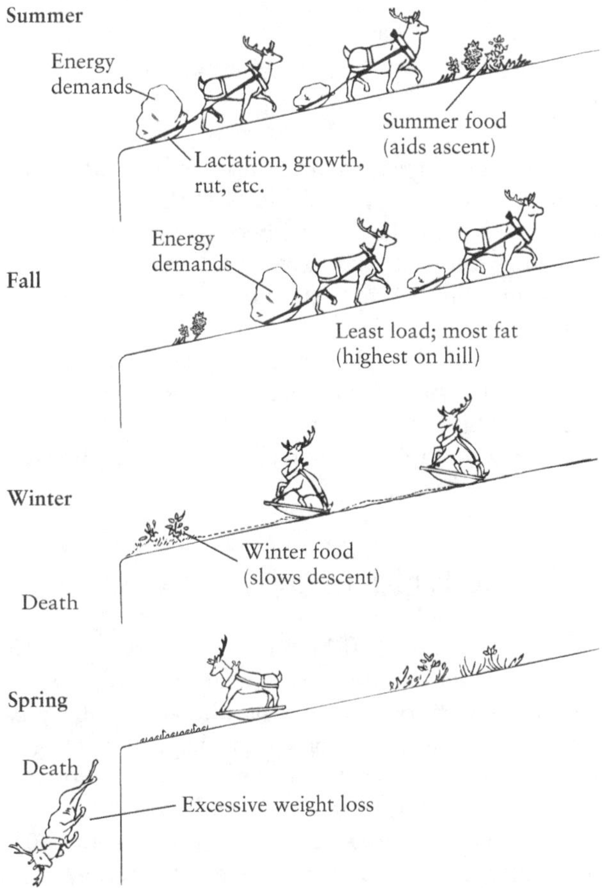

FIGURE 5.6 An analogy of the yearly body fat cycle in deer. The uphill climb represents fat accumulation; the downhill slide represents fat catabolism or weight loss. *[Adapted from Mautz 1978.]*

FIGURE 5.7 White-tailed deer does in good nutritional condition often have twins or triplets. *[Image courtesy of D. Linde.]*

example, in nutrient-rich portions of the northern Great Plains approximately 70% of the white-tailed deer does born in late spring become pregnant in the following fall and have fawns the next spring, and about one-third of those does may carry two fetuses (Petersen 1984) (Figure 5.7). Carnivores, such as Canadian lynx, have larger litters in years when prey is abundant than during years of food scarcity. A number of bird species respond to improved diets with more eggs, multiple clutches, and more young **fledged** (reaching flight stage).

5.5 The Digestive Tract

The digestive tract of fishes provides the basic design found throughout the vertebrates. The system includes a mouth, oral cavity (mouth cavity), **pharynx** (between the oral cavity and the esophagus), **esophagus** (connecting the pharynx to the stomach), stomach, small intestine, large intestine, and anus.

In fishes, the mouth cavity, pharynx, esophagus, and stomach are lined with a mucous membrane containing glands that secrete mucus that lubricates food items being swallowed. Fishes lack salivary glands, with a few exceptions, such as the parasitic lampreys (Lagler et al. 1977). The digestive tract in general is highly elastic and allows the passage of large food items. If a prey item can fit into the mouth of a fish, it will rarely be too large for the digestive tract. Occasionally, however, a fish with spines, such as the black bullhead, can become lodged in the anterior part of the digestive tract of a predatory fish.

Birds, like fishes, with the exception of many aquatic bird species, have glands that secrete mucus in the mouth cavity, pharyngeal region, and esophagus. As in fishes, the mucus aids the passage of food items to the stomach. Birds generally have salivary glands in the pharynx; these glands are particularly abundant in seed-eating birds and assist in initiating the breakdown of starches into sugars. Some birds, such as chimney swifts, also use salivary secretions to cement nesting material together.

Like fishes, many birds have a highly elastic pharynx and esophagus that allow the passage of large food items or the storage of food items in the esophagus. In birds, the swallowing of large food items is made possible by a lower jaw that hinges with movable bones on the skull (quadrates), giving the bird wide gaping abilities. Reptiles, particularly snakes, are especially adept at gaping and have a mobile articulation of the jaw to the skull similar to that of birds. An eastern diamondbacked rattlesnake, for example, can capture and swallow an eastern cottontail that appears much too large for the snake to swallow. After the rabbit is swallowed, a lump that is larger than the normal diameter of the snake can be observed as the prey is slowly digested.

In many birds that eat seeds, such as mourning doves, the esophagus has an expanded area, known as the **crop,** where food can be stored until space is available in the stomach. In the case of the mourning dove and other species of doves and pigeons, the lining of the crop in both genders contains specialized cells that produce a rich milk-like material (**pigeon milk**) that is fed to the young.

Most mammals have abundant salivary and mucous glands in the mouth cavity and pharyngeal region but lack mucous glands in the esophagus. Unlike that of fishes, reptiles, and birds, the jaw articulation of mammals does not allow wide gaping, so mammals generally do not swallow large prey whole. The mammalian esophagus also is not as distensible as that of fishes, reptiles, and birds.

The stomach in fishes is single chambered but can vary in shape from elongate to small or even be absent; in the latter case the esophagus connects directly to the intestine. In some fishes, such as the gizzard shad, the stomach is modified into a muscular **gizzard.** The lining of the gizzard is greatly thickened and muscularized for grinding food material, but it lacks digestive glands.

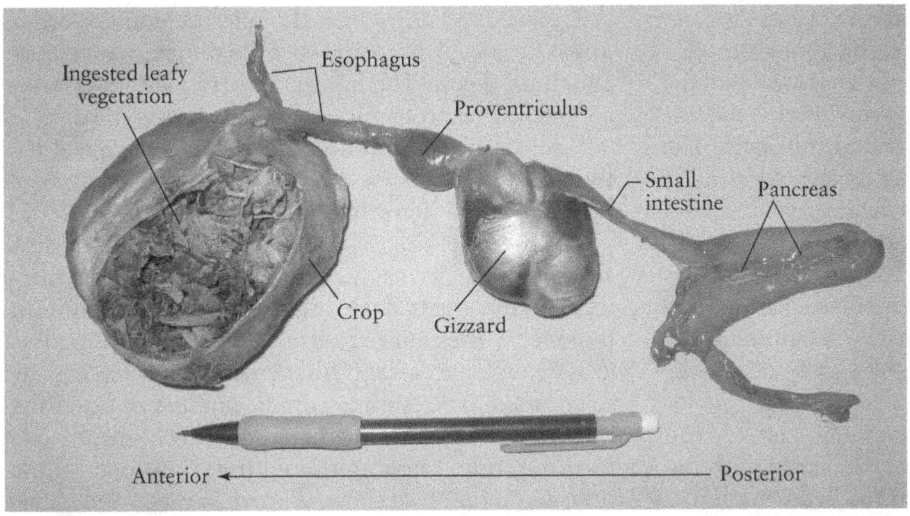

FIGURE 5.8 Illustration of a portion of the digestive tract of a ring-necked pheasant showing the crop, proventriculus (glandular stomach), muscular gizzard, and pancreas. The crop of this wintering pheasant is full of leafy vegetation from wetlands; the crop is normally filled with corn or other high-energy seeds in winter if they are available.

Stomach glands are common in other fishes, particularly those that are carnivorous. As in higher vertebrates such as birds and mammals, these glands secrete hydrochloric acid and pepsinogen, which help to catabolize protein molecules for digestion. Hydrochloric acid also is important in dissolving bones.

Birds have an anterior **glandular stomach** (**proventriculus**) and a posterior muscular stomach or gizzard that generally lacks glands (Figure 5.8). The birds with the best-developed glandular stomachs include predatory species such as American white pelicans, sharp-shinned hawks, and snowy owls. In these species the gizzard is weakly developed. Birds that eat foods that require crushing, such as seeds, snails, or small clams, have well-developed muscular gizzards. The lining of the gizzard in these birds has hardened ridges that assist in the grinding of food. Birds also eat grit (sand and small stones), which is deposited in the gizzard to assist in the grinding of seeds and other hard food objects. Some reptiles also have gizzards.

Although the enzymes associated with digestion are similar among vertebrates, mammalian stomachs vary with food type and can be more complex than those of other vertebrates. Many mammals, such as humans and northern grasshopper mice, have simple stomachs consisting of one chamber. Other mammals, particularly the ruminants, have complex stomachs consisting of multiple compartments. Ruminants are ungulates with an even number of toes (such as the two toes on each foot of a giraffe or mule deer) that regurgitate and chew their food repeatedly (chew their cud). Ruminants have stomachs that generally consist of four chambers, the **rumen, reticulum, omasum,** and **abomasum** (Figure 5.9).

Plant material eaten by the animal enters the rumen, where it is moistened and mixed into a mass. It then passes into the reticulum, where the coarser material is formed into small cuds or masses that are regurgitated and rechewed while the animal rests. After chewing, the animal swallows the regurgitated material again, and it enters the rumen for reprocessing. The finer plant materials and the digested materials from the rumen pass into the omasum. In the omasum the plant

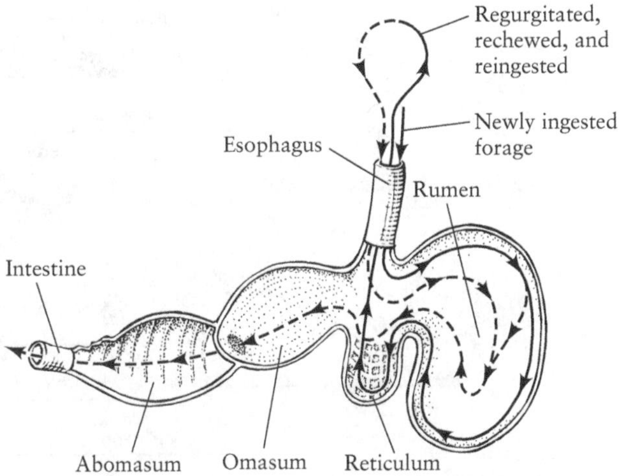

FIGURE 5.9 The stomach of a ruminant, illustrating the arrangement of the three nonglandular stomachs (rumen, reticulum, and omasum) and the glandular abomasum. The solid line indicates the flow of newly ingested material; the dashed line, the flow of most regurgitated, rechewed, and reingested material. Some forage may be regurgitated more than once.

material is subjected to further churning before it is released into the glandular stomach, the abomasum, for digestion. Small fingerlike projections on the walls of the rumen and leaflike folds in the omasum increase the surface area for absorption of water and nutrients. The reticulum, as its name implies, has a reticulated (netlike) pattern of absorptive tissue on its inner surface.

The rumen is a small ecosystem that contains abundant bacteria, protozoans, fungi, and other microorganisms capable of living in an anaerobic environment. These microorganisms break down plant material through a fermentation process. The rumen is inoculated with microorganisms from sources such as ingested plant material, animal-to-animal contact, and feces. In the young, inoculation of the rumen with microorganisms from the parent or through other animal-to-animal contact may be essential; this inoculation can occur through mechanisms such as exchange of cud or salivary transfer. The microorganisms living in the rumen are involved in a symbiotic relationship with the ruminant. The ruminant provides the physiological conditions necessary for the microorganisms to survive and reproduce. In return, the microorganisms break down fibrous plant matter that the ruminant otherwise could not digest. The bodies of these microorganisms also serve as a protein source for the host ruminant.

Although the ruminant stomach is designed for the digestion of coarse food materials, the capacity of ruminants for digesting coarse or fibrous material varies considerably by species. Species that are less selective feeders, such as bighorn sheep and American bison, must have highly advanced ruminant stomachs; these species are sometimes grouped as **grass/roughage eaters.** Other species, the **concentrate selectors,** have a simple rumen capable of handling only vegetation with low fiber concentrations; species in this group, such as the white-tailed deer, mule deer, and moose, are more selective in their food habits (Hofmann 1988). Many other ruminants, such as pronghorns, reindeer, and elk, are grouped as intermediate-type feeders and have a ruminant stomach intermediate between the simple and highly advanced forms (Figure 5.10). Mammals with ruminant stomachs are sometimes referred to as **foregut fermenters.**

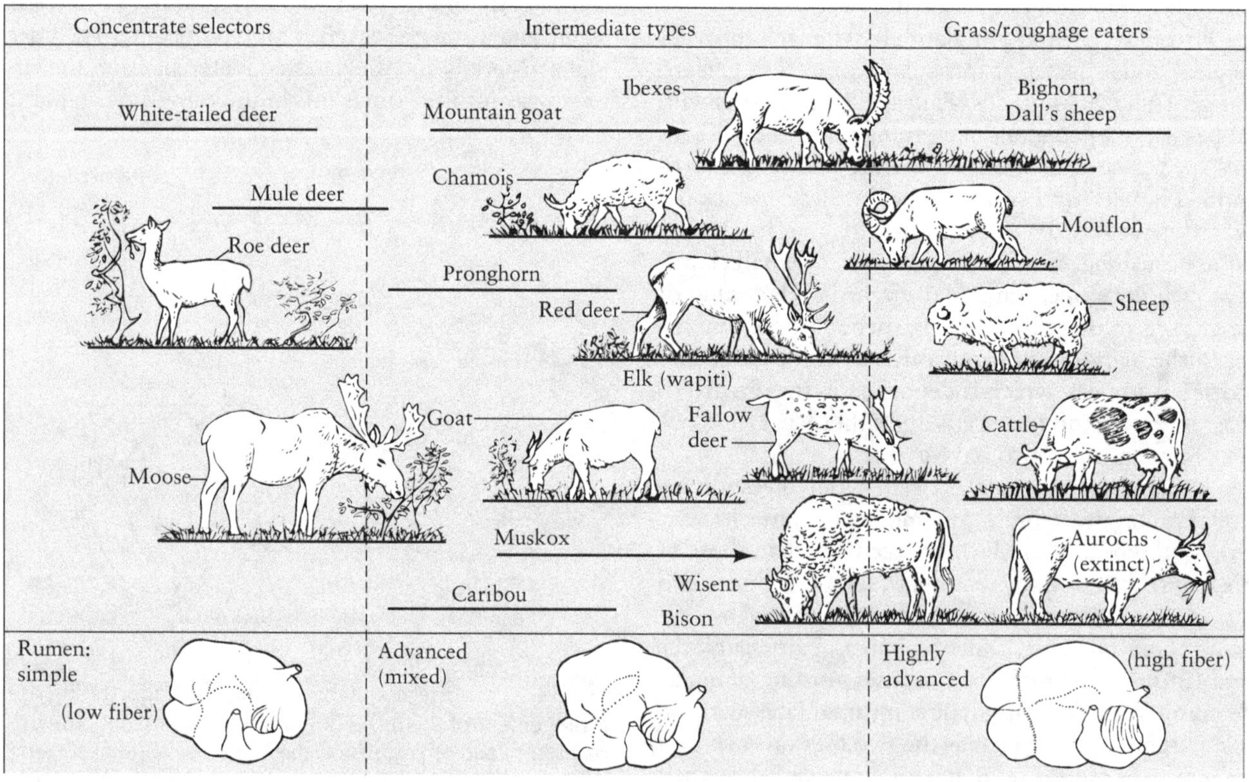

FIGURE 5.10 Positions of European and North American ruminant species within the system of feeding types. The farther the baseline for a species extends to the right, the greater its ability to digest fiber in the rumen. An arrow on the baseline indicates that it extends farther to the right. *[Adapted from Hofmann 1988.]*

Many vertebrate species have caeca, or outpocketings of the digestive system, that are also involved in digestion. Most fishes have **pyloric caeca** that occur at the junction of the stomach and intestine; these may number from one to more than one hundred. They probably function in both digestion and absorption. Many reptiles have a caecum arising at the junction of the small and large intestine. Birds with caeca have two, which are located at the junction of the large and small intestine. The length and development of the caeca are highly variable, with the most highly developed caeca rivaling the length of the intestine. Caeca are well developed in birds that eat coarse plant material, such as lesser snow geese, spruce grouse, and dusky grouse (Figure 5.11). Most mammals have a single caecum at the junction of the large and small intestine. Even ruminants have a caecum. The caecum is highly developed in some of the large herbivores that lack ruminant stomachs. For example, the white rhinoceros and Asian elephant lack ruminant stomachs; microbial fermentation of fibrous plant material in these species occurs in the greatly enlarged caecum and in the large intestine (Owen-Smith 1988). Species in which fermentation occurs primarily in the caecum and intestine are sometimes grouped as **hindgut fermenters.**

Among endotherms and ectotherms, primary consumers tend to ingest a higher percentage of their body weight daily than do organisms at higher trophic levels. This is particularly true of organisms that require a diet of vegetable matter other than high-energy seeds. Therefore, a grass carp or a greater sage-grouse, both herbivores, could be expected to ingest a greater percentage of their body weight daily than a white shark or a bobcat. Vertebrate species that feed on vegetation, particularly green or leafy matter, tend to have much longer small intestines than do predators; this allows for more efficient processing of ingested plant material. Therefore, one would expect a predatory Pacific barracuda to have a shorter intestine in relation to its body length than a grass carp. In general, vegetable matter is much more difficult to digest than animal matter.

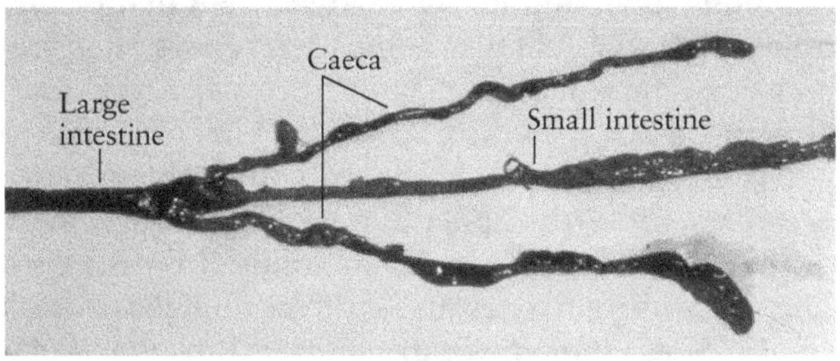

FIGURE 5.11 The caeca branch anteriorly from the junction of the small and large intestines and are important in digestion of coarse plant material in birds such as pheasants, grouse, and waterfowl (shown).

Some generalizations can be made concerning the digestive secretions and hormones released into the small intestine in vertebrates. Digestive secretions from the pancreas and liver are released into the anterior portion of the small intestine. The pancreas releases enzymes important in catabolizing carbohydrates, fats, and proteins in the intestine. The pancreas also releases a bicarbonate that helps to buffer the hydrochloric acid reaching the intestine from the stomach. Specialized cells in the pancreas (**alpha cells**) also produce **glucagon**, a hormone that causes stored polysaccharides and amino acids in the liver to be converted to glucose, thus increasing blood glucose levels. When blood glucose levels are high, **beta cells** in the pancreas produce **insulin,** a hormone that stimulates the liver, muscles, and fat tissues to take up glucose and thus reduces blood glucose levels. The liver releases bile into the small intestine through the bile duct that helps to **emulsify** fat (to finely divide globules of fat), to break down the fat with enzymes, and to encourage absorption of fat into the blood.

5.6 Body Temperature, Metabolic Rate, and Body Size

Most ectotherms, including fishes, have body temperatures that are close to the ambient temperature, and their temperatures change with the ambient temperature. Endotherms, in contrast, maintain nearly constant body temperature levels. Small changes in internal temperatures that would cause severe stress or death to most bird and mammal species can be easily tolerated by ectotherms. Most ectotherms, however, are not as well adapted to withstand wide ranges of or rapid changes in ambient temperature as endotherms.

The relatively slower and generally narrower temperature changes in aquatic environments as compared with terrestrial ones ideally fit the needs of aquatic ectotherms (see section 1.3). The geographic ranges of ectotherms are often limited by high or low temperatures that are lethal to particular species. For example, threadfin shad cannot exist in waters that become cold enough to allow ice formation; conversely, Arctic char cannot survive in the warm waters found in an Oklahoma reservoir during midsummer. Freshwater fish species are commonly categorized as **warmwater, coolwater,** or **coldwater species** based on the general optimal water temperatures at which they survive, grow, and reproduce. Common warmwater species include largemouth bass, bluegills, striped bass, channel catfish, red shiners, and brook silversides. Common coolwater fishes include walleyes, yellow perch, northern pike, and brook sticklebacks. Common coldwater fishes include lake trout, brook trout, coho salmon, Arctic graylings, mountain whitefish, and spoonhead sculpins.

Ectotherms sometimes use behavioral methods to warm their bodies above the ambient temperature. A western fence lizard, for instance, can warm itself by using a warm rock and the heat-absorptive capacity of its body. Some fishes, such as the albacore, can maintain body temperatures considerably above the ambient temperature by physiological mechanisms (Carey 1982). Vigorous swimming activity generates metabolic heat, and these fishes' systems of blood vessels keep much of that heat from reaching the body surface, where it would be lost. These fishes also have a special hemoglobin that, unlike the hemoglobin of most other fishes, increases its oxygen-holding capacity with warming (Cech et al. 1984). Fishes that maintain their body temperatures above the ambient temperature in this manner are sometimes grouped as **heterotherms**.

Endotherms can survive great variations in ambient temperature but must maintain their body temperatures within a narrow range. **Homeostasis** is the tendency to maintain relatively stable conditions within the body. To maintain body temperature homeostasis, endotherms must expend energy. This energy expenditure is often accompanied by increased oxygen consumption. Within a certain range of ambient temperatures, called the **thermoneutral zone,** no additional oxygen beyond that needed for normal body metabolism is required to maintain body temperature (Figure 5.12). At ambient temperatures above or below the thermoneutral zone, below the **lower critical temperature** or above the **upper critical temperature,** the animal must increase its oxygen consumption to maintain temperature homeostasis. The temperature zones within which the animal must expend additional energy to maintain homeostasis are termed the **upper** and **lower zones of metabolic compensation.** The lower zone of metabolic compensation represents a wider range of ambient temperatures

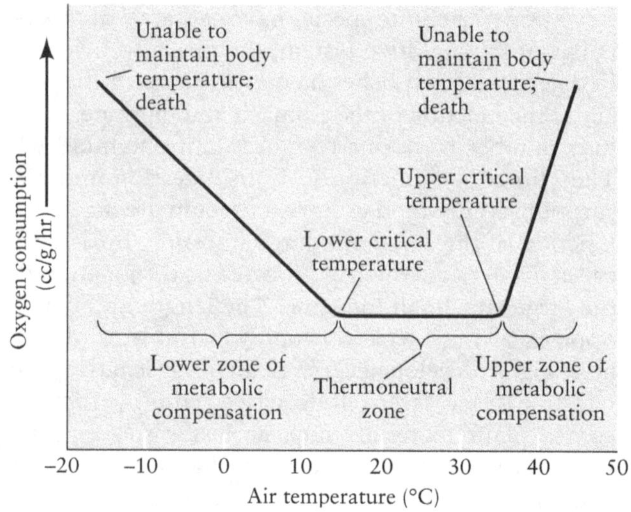

FIGURE 5.12 The relationship between energy used (as measured by oxygen consumption) by a hypothetical bird in maintaining body temperature homeostasis and the ambient air temperature. The thermoneutral zone is the temperature range within which no extra work is required to maintain body temperature; upper and lower critical temperatures are the points at which oxygen consumption begins to increase.

than does the narrower upper zone of metabolic compensation for almost all species. At some point on the temperature scale, at both low and high temperatures, the animal is no longer able to maintain its body temperature, and it dies.

The breadth of the thermoneutral zone is highly variable among species and reflects their adaptation to a particular environment. A snowy owl in winter has a broad thermoneutral zone with the lower critical temperature set at a considerably colder temperature than that for the migratory ring-billed gull; the upper critical temperature is probably also lower in the snowy owl, but not much lower than that of the ring-billed gull. A similar comparison could be made between cold-adapted muskoxen in the Arctic and pronghorns on the North American Great Plains.

When endotherms are faced with ambient temperatures above the upper critical temperature, they can use several strategies to maintain homeostasis. They may move to sites out of direct sunlight or otherwise choose more favorable locations to avoid high ambient temperatures. Many desert species remain hidden during daylight hours and become active only at night. Panting in birds and mammals, fluttering of the **gular sac** (a distended membrane in the throat area) in cormorants and pelicans, sweating in horses and some other species that have sweat glands, and exposure of the skin to the surrounding air to allow heat loss by convection (transfer of heat by movement of heated liquid or gas) in birds are all adaptations to maintain homeostasis when ambient temperatures are above the upper critical level. Additional adaptations for cooling the body include loss of water directly through the skin in animals lacking sweat glands, dilation of blood vessels in the lower legs and feet of wading birds to increase circulation to the extremities and thus dispersal of heat into the environment, and seasonal losses in the coat of insulative fur or feathers.

When endotherms are faced with temperatures below the lower critical level, the strategies they use include shivering; fluffing of the fur or feathers to increase insulating layers of stable air around the body; increases in insulating fur, feathers, or layers of fat during the colder months; shunting of blood away from the extremities; and exchange of heat between warmer blood traveling to the extremities and cooler blood returning from the extremities. Many behaviors, such as sleeping in aggregations by small birds, use of roosts in snow tunnels by ruffed and other grouse, curling up to reduce exposed surface area by most mammals, use of tree cavities for roosting, seeking habitats protected from the wind, seasonal migration, and seeking warm background areas and direct sunlight, are used as means of reducing heat loss in endotherms.

As a generalization, birds and mammals must consume considerable quantities of high-energy food during winter to support endothermy and maintain their body weight. In cold weather, it is not unusual for small birds or mammals to ingest 25% of their body weight or more in food daily; at the same temperatures, larger birds or mammals eating energetically similar foods ingest a markedly lower percentage. Actual intake of food depends greatly on the energy and protein content of the food; it also varies within species with factors such as age and activity, and it varies greatly among species. The **metabolic rate** (the rate at which chemical reactions occur in an organism) of ectotherms, such as fishes, amphibians, and reptiles, is reduced during cold periods, and their energy demands are also low. For example, during the summer, bluegills may consume about 5% of their body weight daily, while at colder winter temperatures they consume 1% or less of their weight in food.

Small endotherms have special problems in cold weather as they attempt to maintain body temperature homeostasis. Small endotherms have a high surface area-to-body mass ratio, and they lose body heat proportionally faster than do larger endotherms. Because of their higher surface-to-mass ratio, smaller animals also tend to be cooled more rapidly by wind through convection (Figure 5.13). Thus, wind combined with cold environmental temperatures would be much more stressful to a 7-month-old mule deer than to a mature mule deer with twice the body mass. Small endotherms must also consume proportionally more energy in relation to their body weight than larger endotherms because of the higher metabolic rates and greater energy demands of small bodies (Randall et al. 2002).

These relationships between body size and heat loss are encapsulated in **Bergmann's rule.** Bergmann's rule states that species living in colder climates often have larger body sizes than subspecies or closely related species living in warmer regions. The range in size of white-tailed deer from the large northern subspecies to the smaller southern subspecies well illustrates Bergmann's rule. **Allen's rule** states that ears, bills, tails, and other extremities tend to be shorter in the colder portions of the range of a species; these adaptations would also serve to reduce the surface area-to-body mass ratio.

FIGURE 5.13 Subspecies of Canada geese are able to winter in colder regions than subspecies of cackling geese because of their larger body size, lower surface area-to-body mass ratio, and lower energy use per unit of body weight. The cackling goose in the foreground is in close proximity to the Canada goose. *[Image courtesy of J. Iron.]*

Even though larger endotherms require less energy per kilogram of body mass than smaller ones, they still require more total food per individual. Thus, an African elephant must take in more total energy in food than a blue wildebeest, even though the wildebeest requires more energy per unit of body mass. However, because of the lower metabolic rates of larger animals, less energy would be required to support 100,000 kilograms of African elephants than 100,000 kilograms of blue wildebeests. Likewise, 10,000 kilograms of herbivorous mice would use more energy than an equal mass of moose.

Large herbivores, particularly **megaherbivores** (those that weigh over 1,000 kilograms), must take in large amounts of forage to satisfy their energy needs. Among the ruminants, the proportion of crude protein in the diet generally declines with body size, indicating that larger animals are better adapted to processing poorer-quality, more abundant forage (Owen-Smith 1988). However, some megaherbivores, such as giraffes, are able to obtain a diet higher in protein than would be expected based on their body size (Owen-Smith 1988). Giraffes, which are ruminants, are able to move rapidly between clumped food sources and have relatively high crude protein levels in their diet (Figure 5.14). If an herbivore is highly selective, it is dependent on foods that are in more limited supply in the environment than if it were a generalist.

Because of their lower metabolic rates, large endotherms can survive on an equivalent proportion of body energy stores for longer than small endotherms. Thus, a wild turkey could survive longer on stored energy than a horned lark, and an American bison longer than a mule deer. Mobility also tends to be less costly in terms of energy used (oxygen used per unit of body weight) in larger mammals (Schmidt-Nielsen 1997). Body size and mobility can also be related to the habitat requirements of an animal. For example, mule deer may not adapt as easily to logging debris or deep snow as elk because locomotion per unit of body weight is more costly in the smaller animal. In reality, differences in the costs of locomotion between mule deer and elk would interact with additional factors such as average bite size, food consumption rate, and forage access in influencing energetics and habitat needs in these species.

FIGURE 5.14 Giraffes are megaherbivores with ruminant stomachs and are able to obtain relatively nutritious vegetation by moving between clumped food sources. *[PhotoDisc]*

The heart in endotherms, particularly birds, is a larger, more powerful organ than that in ectotherms. Unlike the two-chambered heart of fishes and the three-chambered hearts of amphibians and most reptiles, the endotherm heart has four chambers, with completely separated left and right ventricles. The separation of the heart into four distinct chambers segregates newly oxygenated blood returning from the lungs from oxygen-depleted blood returning from the rest of the body. Thus, the endotherm is able to pump a blood supply higher in oxygen content through the body. Endotherms require higher oxygen levels in the blood than do ectotherms to support their higher level of metabolic activity.

The benefits of endothermy, such as the capacity for increased mobility, especially flight, and the capacity to remain active in cold environments, have proved successful for a large number of bird and mammal species. Flight in birds and bats, for example, is an energy-intensive activity that requires the high metabolic rates found in endotherms. Many species of birds migrate to Arctic regions for breeding each summer, a feat that would not be possible in their ectothermic cousins, the contemporary reptiles. Ancient flying reptiles, the Pterosaurs, were probably endothermic. Low species richness of reptiles and amphibians in Arctic regions compared to bird and mammal species demonstrates the advantage of endothermy.

Among endotherms, some have seemingly reached back in time and borrowed certain ectothermic strategies. Hummingbirds are too small to maintain their high body temperatures throughout cool nights; to conserve energy, these birds go into a type of daily **torpor** (lethargy or sluggishness) or **hypothermy**. Hypothermy is defined here as the lowering of body temperature in endotherms to save energy during periods of stress. **Hibernation** has been defined as a type of hypothermy in which the body temperature drops to approximately that of the surroundings of the animal (Schmidt-Nielsen 1997). Willmer et al. (2000) referred to "deep" and "shallow" torpor in terms of body temperature drop but also noted that there is a continuum of temperatures between these categories. Most daily hypothermy fits the shallow category rather than seasonal, which is often a deep hypothermy. As one would expect, deep hypothermy is preceded by considerable weight gain in the animal. Periods of torpor or hypothermy during the summer are often referred to as **estivation**. Because of a lack of precise definitions, hibernation, estivation, and torpor are all referred to as hypothermy in this text.

Common poorwills in the southwestern United States go into a winter hypothermy, and they may remain in this state for a month or more until insect prey populations warrant their return to an active lifestyle. Mammals such as the thirteen-lined ground squirrel go into a summer hypothermy in mid- to late summer, then, usually after a brief emergence from this inactive period, go into winter hypothermy. Numerous species, such as small rodents and bats, go into daily or seasonal hypothermy to conserve energy (Schmidt-Nielsen 1997; Feldhamer et al. 2004). In small mammals exhibiting seasonal hypothermy, periodic arousal or awakening commonly occurs; this explains why thirteen-lined ground squirrels may at times be seen on warm late fall or winter days. Thus, the benefits of ectothermy in saving energy, especially in cold periods or periods without food, can be partially matched by these endothermic species.

Some larger animals, such as bears, American beavers, and skunks, experience a period of winter sleep that involves only a mild degree of hypothermy. In this shallow hypothermy, a bear's body temperature drops from a normal 37°C to about 31°C; an animal as large as an American black bear cannot afford the energy to awake from body temperatures observed in mammals during deep hypothermy. Marmots, such as the woodchuck, are the largest mammals that go into a deep hypothermy; their body temperature drops to between 5°C and 15°C. (Randall et al. 2002; Feldhamer et al. 2004).

5.7 Obtaining Oxygen in Terrestrial and Aquatic Environments

Fishes differ greatly from endotherms and terrestrial ectotherms in their capabilities for obtaining and circulating oxygen. In section 1.3, it was noted that the terrestrial atmosphere consists of approximately 19% oxygen, while well-oxygenated water may have only 10–15 milligrams per liter (0.0010%–0.0015%) available to organisms. Oxygen levels in aquatic environments also fluctuate much more than those in terrestrial systems. For example, oxygen in aquatic habitats can fluctuate greatly on a seasonal basis and may also fluctuate on a 24-hour basis. Terrestrial organisms obviously have a more readily accessible and constant supply of oxygen available than do totally aquatic organisms. When one thinks of limiting factors, oxygen can be very important to aquatic species but is seldom mentioned as a limiting factor for terrestrial species.

Aquatic ectotherms, such as fishes, have gills or other specialized structures that they use to extract oxygen from water. Terrestrial vertebrates typically use lungs for respiration. All of the aquatic endotherms, such as whales and seals, use lungs and breathe atmospheric oxygen. Given the small amounts of oxygen available in water, an endothermic animal would not be able to obtain sufficient oxygen from the water to satisfy the needs of endothermy. To exacerbate problems for aquatic ectotherms, the capacity of water to hold dissolved oxygen decreases as water temperature increases.

Aquatic ectotherms, therefore, often face decreased oxygen availability as well as the need for increased oxygen consumption as water temperature increases.

Another difference between endotherms or other terrestrial vertebrates and aquatic ectotherms is the energy expenditure needed to pass the oxygen-bearing medium (air or water) over their respiratory surfaces. Water is 800 times denser than air; therefore, it requires much more energy to pass water over gills than air over lungs. Add this to the different levels of oxygen available in the two media, and one can readily see that aquatic ectotherms must work harder to extract oxygen than do endotherms and other terrestrial vertebrates.

Birds and mammals have lungs that differ in their basic structure and function. In mammalian lungs, the air enters a dead-end sac and gas exchange with the blood occurs in tiny pockets called **alveoli.** Because each lung is basically a single sac, the used air is never completely exhaled and newly inhaled air mixes with the air from previous inhalations. This mixing reduces oxygen uptake efficiency. In bird lungs, the air flows in a one-way direction, avoiding the problem of stale air in the lungs. The respiratory systems of birds feature several air sacs, usually one single and four paired, which lack blood vessels and are not involved in gas exchange; air flow occurs between the lungs and air sacs. A volume of air first enters the more posterior air sacs on inhalation, on the following exhalation progresses into the lungs for gas exchange in the **parabronchi,** flows into the more anterior air sacs on the next inhalation, and exits the body on the following exhalation (Figure 5.15). Bird lungs are highly efficient in oxygen uptake.

As we have seen, fishes have to work harder than endotherms to obtain oxygen, but they do have some strategies for obtaining oxygen more efficiently and reducing oxygen demand. Fishes make use of a countercurrent system of gas exchange between the water and the blood in their gills to increase their efficiency in obtaining oxygen and ridding the body of excess carbon dioxide (Randall et al. 2002). Most fishes are also able to save energy by maintaining a neutral buoyancy by using the gas bladder. Gas can be moved into the gas bladder in two ways: through swallowing, as is sometimes the case in fishes in which the gas bladder opens to the digestive tract (**physostomous fishes**), or through a complex system of blood vessels and gas exchange, as is usually the case in fishes in which the gas bladder lacks a connection to the digestive tract in adult fishes (**physoclistous fishes**). Neutral buoyancy creates an effect similar to weightlessness and clearly results in energy savings. The gas bladder is one of the reasons why swimming fishes use less energy than flying birds or running mammals per unit of distance traveled (per kilogram of body weight) despite the high density of water (Schmidt-Nielsen 1997).

Differences between endotherms and aquatic ectotherms also exist in the delivery of oxygenated

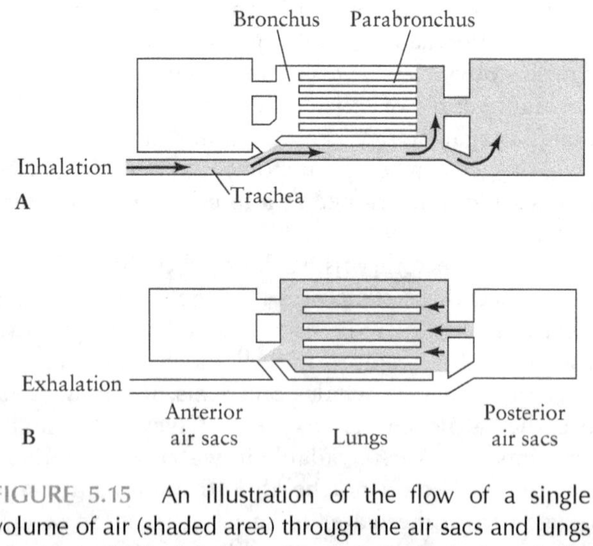

FIGURE 5.15 An illustration of the flow of a single volume of air (shaded area) through the air sacs and lungs of a bird. *(A)* An inhaled volume of air travels through the trachea and bronchus, expanding the posterior air sacs with fresh air. *(B)* On the following exhalation, the volume of air enters the lungs and passes through the parabronchi for gas exchange. *(C)* On the next inhalation, the volume of air passes into the anterior air sacs. *(D)* On the following exhalation, the volume of air is expired from the body. *[Adapted from Schmidt-Nielsen 1971.]*

blood to the body tissues. As mentioned in section 5.6, birds, mammals, and a few reptiles have four-chambered hearts, while fishes have two-chambered hearts. In organisms with four-chambered hearts, deoxygenated blood is pumped to the lungs, where it undergoes oxygenation. To be oxygenated, the blood must be passed through a fine capillary network where oxygen uptake (and carbon dioxide elimination) can be accomplished. This passage of blood through the lungs reduces blood pressure and is the reason why blood must go back to the heart for further pumping. The added pumping stage ensures that the oxygenated blood is rapidly distributed to the body. In animals with three-chambered hearts, such as amphibians and most reptiles, part or most of the blood is also pumped through the lungs and returned to the heart before being pumped to the body. In fishes, blood is pumped from the heart to the gills. The blood is oxygenated in the gills and flows directly to the body. No secondary pump is used; therefore, blood pressure is low and oxygen distribution is less efficient (Figure 5.16).

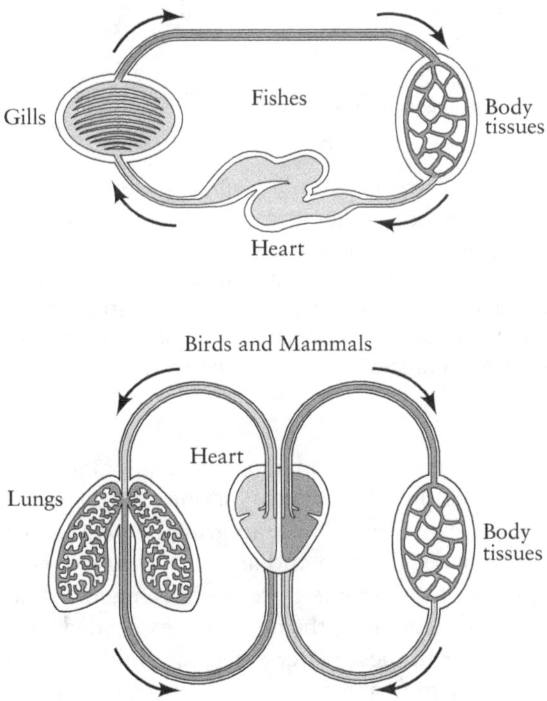

FIGURE 5.16 In fishes, blood is pumped to the gills and then flows directly to the body tissues. Delivery of blood through the blood vessels in fishes is therefore slower, at lower pressure, and less efficient than that in other vertebrates in which oxygenated blood returns to the heart before being pumped to the body tissues.

Some birds and mammals are able to dive underwater and remain completely submerged for considerable periods of time. The sperm whale can dive to over 900 meters and remain underwater for over an hour; other whales and seals are also highly adapted for diving and staying submerged for long periods. Among birds, common murres may reach depths in excess of 150 meters, razorbills 120 meters (Piatt and Nettleship 1985), and emperor penguins 265 meters (Kooyman et al. 1971). In terms of respiration, these endotherms are adapted to diving in several ways. Deep-diving birds and mammals can withstand much higher levels of carbon dioxide in the blood than can their terrestrial relatives. Diving birds can store considerably more oxygen in the blood and muscle than can nondiving birds, and they can recirculate air in the air sacs through the lungs. The heart rate slows greatly in diving birds and mammals when they are submerged, and blood is shunted to critical areas such as the nervous system and heart. In addition, their muscles can operate anaerobically and incur an oxygen debt that is repaid after the animal resurfaces.

5.8 Adapting to Freshwater and Marine Environments

Fishes live in aquatic habitats and, as such, face very challenging physiological problems with regard to water and mineral balance (**osmotic balance**) in their body fluids and cells. Some fishes live in marine (saltwater) habitats, others live in freshwater systems, and some move between both; in each case fishes face different osmotic problems because of the differing levels of salinity they encounter. In addition, salinity does not remain constant. As water evaporates from water bodies, the ions that cause salinity remain. Thus, salinity can increase with evaporation or decrease with precipitation. To further add to the problem, some inland waters are very saline, and waters in coastal areas where freshwater and marine environments meet can vary daily in their salt content.

In freshwater bony fishes, the blood and other body fluids contain higher concentrations of minerals and organic materials than does the surrounding water. Bony fishes are fishes with skeletons primarily composed of bone, in contrast to cartilaginous fishes, such as sharks and rays, which have skeletons made of cartilage. The higher **osmotic pressure** of the blood and body fluids in freshwater bony fishes leads to constant movement of water from the environment into the body. These fishes are said to be **hyperosmotic.** The rate at which water moves into the blood and body tissues is reduced by a partially impervious layer of skin and sometimes scales, but water easily crosses selectively permeable membranes such as the gills, mouth lining, and intestinal surface. Sharks and rays are also hyperosmotic.

Freshwater bony fishes use several strategies in their constant battle against water intrusion from the surrounding medium. The kidneys of freshwater bony fishes are specially adapted to remove water from their bodies. Freshwater bony fishes excrete copious amounts of dilute urine in order to maintain osmotic balance. In fact, although the main function of the kidneys in most vertebrates is the removal of nitrogenous wastes, the kidneys of most fishes are primarily used for maintaining osmotic balance; most nitrogenous wastes are eliminated through the gills. In general, freshwater bony fishes have many more and larger kidney **glomeruli** (conglomerations of blood vessels associated with the functional units of a kidney) than marine bony fishes; this reflects their increased ability and need to excrete water. These freshwater species excrete little salt because they need to maintain body salts against the osmotic gradient. They compensate for salt losses to the environment by **active transport** of salts from the surrounding medium across the gills (Figure 5.17A). Active transport involves the use of body energy to transport salts across selectively permeable membranes and hold them against the osmotic gradient.

Marine bony fishes face the problem of an environment with a higher osmotic pressure than their bodies, which results in a constant loss of body fluids and gain of salts. These marine bony fishes are said to be **hyposmotic.** Salts are taken in continuously with food and water. Most of these salts, along with 90% of nitrogenous waste, are excreted through the gills. The excretion of salts through the gills against an osmotic gradient requires active transport, which is in part associated with cells called **chloride cells** (Figure 5.17B). Because osmotic pressure favors loss of fluids in marine bony fishes, their kidneys are adapted to reabsorb water.

Primitive fishes, such as the Pacific hagfish, have body fluids similar in osmotic pressure to the surrounding water (they are **isosmotic**); they do not have substantial problems with osmotic balance. Marine sharks and rays regulate their osmotic balance differently than marine bony fishes by retaining large quantities of organic salts. The osmotic balance system of marine sharks and rays is therefore more like that of most freshwater bony fishes than that of marine bony fishes.

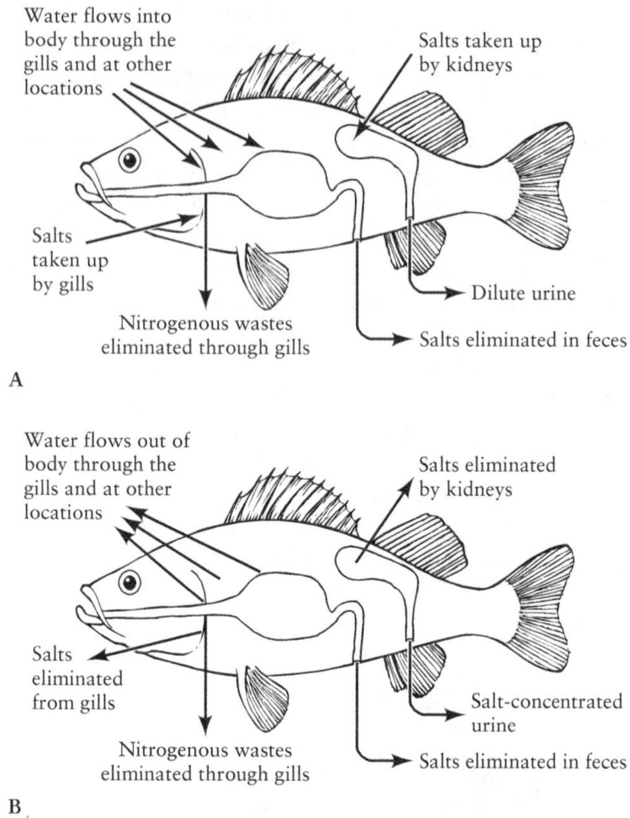

FIGURE 5.17 Net inflow and outflow of salts and water in *(A)* freshwater bony fishes that are hyperosmotic to the surrounding medium and *(B)* marine bony fishes that are hyposmotic to the surrounding medium. In freshwater bony fishes, chloride cells in the gills actively transport salts into the body. In marine bony fishes, the chloride cells in the gills actively transport salts out of the body.

Some fishes are able to migrate between marine and freshwater systems and are able to adjust kidney and gill function to these extremes of changing osmotic pressure. Considering the adaptations needed to survive in either of these water types, the ability to switch from marine to freshwater environments, or vice versa, is truly remarkable. These changes in the structure and function of the glomeruli of the kidneys and the gills involve the active transport system. Apparently, the chloride cells in the gills of some fishes, such as the chum salmon, can change their function from transferring salts into the body in fresh water to removing salts in marine environments. The adaptability of some of these fishes has been surprising to biologists. Chinook and pink salmon were once thought to need both freshwater and saltwater life phases, but they have been able to adapt to totally freshwater environments, such as in the Great Lakes. In a similar case, the Santee-Cooper Dam in South Carolina blocked the return routes of a striped bass population to the ocean, but biologists soon found that these fish could complete their life cycle in a large freshwater lake.

Birds associated with marine or highly saline environments do not have a serious problem of water intrusion into the body fluids because of their relatively impermeable skin. However, many bird species commonly drink marine water. Marine water has a concentration of approximately 3% salt, compared with 1% in the blood; the urine of birds contains only about 0.3% salt. Their solution to this problem involves the active transport of salts out of the bloodstream through a gland above the eye known as the **salt gland.** In birds with salt glands, the size of the gland varies with the degree to which the birds are drinking marine or other saline waters or feeding on highly saline foods. The salty exudate (approximately 5% salt) secreted from the salt gland drips from the bill or nasal area of marine birds and the marine iguana (Figure 5.18). A similar salt removal system is found in marine turtles, which secrete salt from an opening near the back corner of the eye.

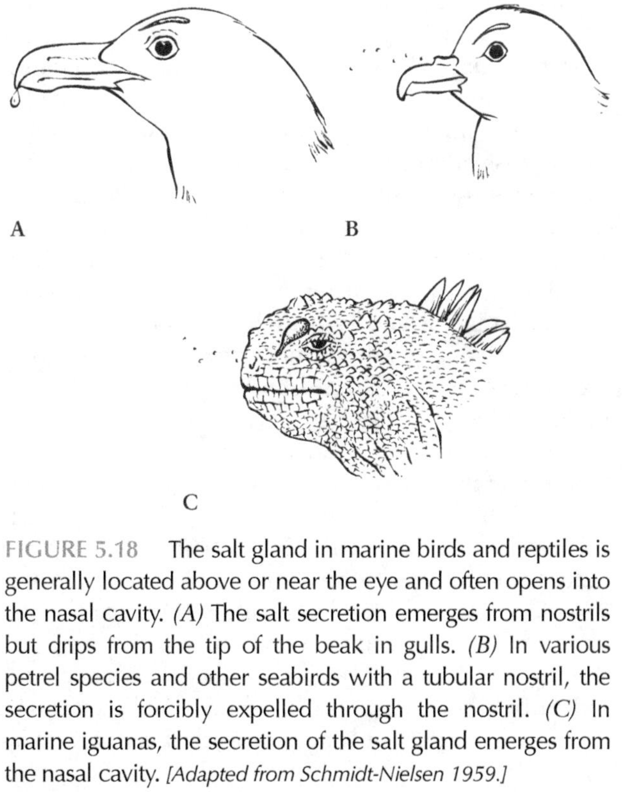

FIGURE 5.18 The salt gland in marine birds and reptiles is generally located above or near the eye and often opens into the nasal cavity. *(A)* The salt secretion emerges from nostrils but drips from the tip of the beak in gulls. *(B)* In various petrel species and other seabirds with a tubular nostril, the secretion is forcibly expelled through the nostril. *(C)* In marine iguanas, the secretion of the salt gland emerges from the nasal cavity. *[Adapted from Schmidt-Nielsen 1959.]*

Marine mammals such as seals and whales live in an environment where they cannot obtain fresh water for drinking. These mammals have highly efficient kidneys that produce urine more concentrated than marine water (Schmidt-Nielsen 1997). Additionally, a labyrinth-like proliferation and arrangement of membranes in the nasal passages are specially adapted to reduce water loss (Randall et al. 2002). Water needs are largely supplied by **metabolic water** and fluid from their prey. Metabolic water is a by-product, along with carbon dioxide, of the oxidation process in cells (respiration) wherein carbohydrates, lipids, or proteins are metabolized.

5.9 Adaptations to Cold and Heat

Many fishes are adapted to extremely cold water; however, having a body temperature at or near the surrounding ambient temperature is normal for ectotherms. Some mammals, such as certain seals, dolphins, and whales, and some birds, such as king penguins, also can live in extremely cold water, but they must maintain a body temperature very different from that of their surroundings. Heavy layers of fat underlying the fur or feathers are especially critical to protecting these marine mammals and birds from heat loss. Their appendages project into an extremely cold environment; these appendages can allow a desirable loss of heat during periods of high metabolic activity or an undesirable loss of heat at other times.

In marine mammals, wading birds, and many other endotherms, particularly those that live in cold climates, the arterial blood going to the extremities (legs, wings, flippers, and ears) comes in close contact with returning venous blood. In many cases the veins are specifically arranged to facilitate the flow of heat from the adjacent arteries into these veins. Thus, there is a **countercurrent exchange** that results in the warming of blood returning to the core of the body and the cooling of blood going to the extremities. This countercurrent exchange reduces heat loss in the extremities as well as the energetic costs of heating venous blood returning from the extremities. Similar systems are widely used by humans to reduce energy costs associated with heating and cooling in various industrial processes. In the flippers of seals and other marine mammals and in the legs (in the tibiotarsus or drumstick area) of some wading birds, there is a special netlike assemblage of blood vessels called the **rete mirabile** (meaning "wonderful net") that brings large numbers of small arteries and veins into close contact to increase countercurrent exchange.

Animals with countercurrent exchange systems are also able to shunt blood away from the extremities when required to conserve heat or to dilate the blood vessels to increase blood flow to the extremities to dissipate heat. A Canada goose, mallard, or glaucous gull, for instance, can stand on ice for hours in extremely cold temperatures. The temperature of the feet may be at 0–5°C while the body remains at approximately 38°C; the flow of blood to the feet is reduced during this period (Figure 5.19). The noses, ears, and legs of terrestrial mammals may also show a temperature gradient, with the extremities often close to ambient temperature; freezing of body extremities is unusual but occasionally occurs in nature. These adaptations for heat conservation provide substantial energy savings, without which these organisms could not survive in a cold environment.

FIGURE 5.19 Glaucous, ivory, Iceland, and other northern gulls can stand for long periods on ice or in extremely cold water because of their ability to maintain cold temperatures and reduced blood flow in their lower legs and feet. The temperature of this gull's feet is below 5°C, while the internal body temperature is approximately 38°C. Waterfowl such as Canada geese and mallards are similarly adapted. *[From Irving 1966.]*

At other times, increased blood flow to the extremities is necessary to dissipate excess metabolic heat. In running black-tailed jackrabbits, their long ears can provide considerable heat dispersal from the body to keep core temperatures from adversely increasing (Willmer et al. 2000). Large ears on elephants are important instruments for release of excess heat. Even marine mammals in cold waters must dissipate excess metabolic heat through their flippers during active periods. Because they have a thick insulating layer of fat under the skin, seals, sea lions, whales, and some other marine mammals generate excess metabolic heat during active periods or when water is warm. Part of that heat can be released by dilation of the blood vessels going to the appendages. In addition, these marine mammals also have blood vessels in the fat layer that can dilate and carry heat to the body surface for release into the environment or, if conserving heat, shunt blood away from the surface. These marine mammals differ from terrestrial mammals because their primary insulator (fat layer) is located below the surface from which heat is dissipated. In most terrestrial endotherms the primary insulator (hair or feathers) is external to the skin surface (Schmidt-Nielsen 1997).

In regions with cold winters, most terrestrial mammals are adapted to seasonal temperature changes through changes in the thickness and insulative qualities of their fur. Resident birds in these same regions are likewise adapted to the changing temperatures by growing more feathers for the winter. Birds and terrestrial mammals use erection (or depression) of the feathers (**ptiloerection**) or fur (**piloerection**) as a means of reducing heat loss or retaining heat. Birds can expose the skin surface by feather erection to increase heat loss through evaporation, convection, radiation, and conduction. Birds can trap increased amounts of air in the feathers through partial ptiloerection and thus also use this as a means to reduce heat loss. Terrestrial mammals also use piloerection to reduce heat loss by fluffing the fur and increasing the amount of trapped air (Randall et al. 2002). Mammals are generally less effective at controlling hair erection than are birds in controlling feather erection. Desert animals sometimes use entrapment of air in the hair or plumage as a means to reduce conduction, convection, and radiation of heat into the body (Willmer et al. 2000). Loss or gain of heat in terrestrial vertebrates is also influenced by use of shade, sunny warm sites, burrowing, wetting of plumage or fur (enhanced heat conductance), nocturnal or diurnal activity, migration, panting, and other behaviors. Even seasonal changes in fur or plumage color can be important in heat exchange through radiation.

Some desert animals are adapted to withstand higher than normal body temperatures during the hot periods of the day (Willmer et al. 2000). This adaptation allows animals such as various species of camels, oryx, and elands to have daytime body temperatures several degrees higher than their nighttime body temperatures. Even some small desert mammals, such as Harris's antelope squirrel, can increase their body temperature several degrees above normal during diurnal feeding and then return to a cool burrow to "unload" this excess heat. This adaptation to variable temperatures allows these mammals to survive in hot, dry climates without the need for much water for evaporative cooling (Randall et al. 2002).

5.10 Adaptations of Terrestrial Wildlife to Water Scarcity

Terrestrial vertebrates living in desert ecosystems have adapted to a scarcity of water through a variety of means. Any of the adaptations to heat (see section 5.9), such as burrowing or nocturnal activity, that allow animals to avoid heat and reduce the need for evaporative cooling also represent adaptation to water scarcity. Various species of kangaroo rats are adapted to desert living and are notable for their ability to meet virtually all of their water needs from metabolic water; they do not need drinking water. Many desert rodents are also highly efficient at conserving water loss because of specialized kidneys that are adapted to release unusually concentrated urine. Loss of water due to evapotranspiration is reduced in many desert mammals because of the countercurrent flow of warm, moist air from the lungs against nasal membranes kept cool at night or while in a burrow; water condenses on these nasal membranse and is conserved. Seeds harvested by various kangaroo rats are cached in belowground burrows where the seeds take up water due to the higher relative humidity there than aboveground, thus adding water to diets of these desert species (Willmer et al. 2000).

Many vertebrate species in dry regions or during dry periods become selective for plants with high moisture content. In addition, vertebrates in deserts or during dry periods in normally moist ecosystems commonly obtain water from early-morning dew. Excretion of nitrogen as uric acid in a pastelike form is an adaptation for water conservation widely found in birds and reptiles. The amount of water in the excrement of desert-adapted birds is commonly considerably less than in birds in ecosystems with greater water availability. Lipids in the skin of many desert birds reduce cutaneous water loss (Williams and Tieleman 2005). Some birds even have special adaptations for bringing water to the young. In African desert regions, birds such as the chestnut-bellied sandgrouse and black-faced sandgrouse have specially adapted belly feathers that can be used to carry water to nestlings over many kilometers. The adaptation and evolution of many vertebrates that allows them to thrive in the driest of desert conditions is an impressive accomplishment.

SUMMARY

Animal species require an array of inorganic elements in their diets. The elements needed in the greatest amounts are referred to as macroelements, while those needed in trace amounts are the trace elements or microelements. Terrestrial vertebrates obtain essential inorganic elements primarily from food, although some animals locate and ingest soil with concentrations of elements such as sodium. Fishes obtain most of the essential elements through food or directly from the water.

Vitamins are grouped as fat soluble and water soluble. They serve a variety of functions in animals. Most vitamins are obtained in adequate quantities in the diet. Lack of vitamins may be manifested in a variety of symptoms, depending on the vitamin.

Amino acids are grouped into those that animals can manufacture within their own bodies (nonessential amino acids) and those that must be obtained, at least in part, from the diet (essential amino acids). The need for essential amino acids may be reduced or eliminated in animals with multiple-chambered stomachs or caeca in which microbial populations contribute to the digestive process.

Animals obtain their energy from ingested proteins, fats, and carbohydrates. Energy is normally obtained by breaking down carbohydrates into simple sugars such as glucose. When blood glucose and other carbohydrates are in short supply, glucose can be synthesized from fats and proteins. Protein, however, is more important for building muscle tissue than as an emergency source of energy. Protein requirements vary by species, with carnivores having the highest requirements; protein requirements within species may also vary by season, gender, and age. Nutritionally fit animals are much more likely to reproduce successfully than those with deficiencies.

Broad similarities exist in the digestive enzymes and morphology of the digestive tracts of vertebrates. The stomachs of most birds are divided into a glandular stomach and a separate muscular stomach (gizzard) for grinding. Ruminant mammals have complex stomachs in which microorganisms break down plant materials through a fermentation process. Many vertebrate species have caeca in which digestion of coarse materials occurs through fermentation.

High metabolic rates are characteristic of endotherms. Within the thermoneutral zone, little energy expenditure beyond that required for normal body metabolism is needed to maintain body temperature. Above or below the thermoneutral zone, endotherms use a variety of behavioral, physiological, and morphological adaptations to help them maintain temperature homeostasis. Smaller endotherms have a larger surface area-to-body mass ratio and higher metabolic rates than do endotherms with greater mass. Because of their lower metabolic rates, larger endotherms can generally survive on less energy per unit of body mass than can smaller endotherms. Larger endotherms can also normally walk or fly more efficiently than can smaller endotherms.

Oxygen is in much more abundant and constant supply in terrestrial ecosystems than in aquatic ecosystems. Fishes must extract oxygen from a medium with extremely low and fluctuating oxygen concentrations. Endotherms could not obtain sufficient oxygen from water to survive. Because of their neutral buoyancy, fishes can move more efficiently through water than can birds through air or mammals overland. The lungs of birds are more efficient than those of mammals because of their one-way air flow pattern and the resulting reduction of stale air remaining in the lungs after exhalation.

Fishes face challenges in maintaining the osmotic balance of their body fluids. Freshwater bony fishes are hyperosmotic to their environment and must contend with the loss of salts from the body and the gain of water. Freshwater bony fishes have kidneys especially adapted for the secretion of water from the body; the gills use active transport to pump salts into the body against the osmotic gradient. Marine bony fishes are hyposmotic to their environment and use active transport to move salts out of the body (through the gills); their kidneys are adapted to reabsorb water. Birds that drink salt water and eat saline foods often have salt glands that secrete salt through the nostrils and bill.

Endotherms living in cold aquatic environments must have mechanisms for heat conservation. They must also be able to dissipate metabolic heat during periods of activity. Loss of heat through the extremities is controlled by a countercurrent system that cools blood going to the extremities and warms blood returning to the body. Dilation or constriction of blood vessels leading to the extremities or body surface can also control heat loss. Special networks of blood vessels in the flippers or legs of some species enhance heat exchange and assist in reducing heat loss and blood flow to the extremities. Terrestrial endotherms also have a variety of mechanisms important in conserving or dissipating heat. Some desert animals are able to tolerate a higher than normal body temperature during the day and then dissipate that heat load and cool the body during the night. Other aspects such as erection of fur or feathers, seasonal changes in feather or fur thickness, burrowing, finding sun or shade, panting, wading in cool water, and migration are some of the adaptations and behaviors important in adapting to changing temperatures.

Vertebrates that dwell in deserts often have special adaptations to conserve water. Some species never need drinking water, with much of their water budget coming from metabolic water. Reduction in water loss from excretion (nitrogenous waste) and adaptations that reduce evaporative water loss allow vertebrates to live in deserts.

PRACTICE QUESTIONS

1. List four macroelements and a function of each in animals.

2. What is geophagia?

3. Provide examples of three fat-soluble vitamins and three water-soluble vitamins and list important functions of each.

4. What is the definition of a nonessential amino acid? What is the definition of an essential amino acid?

5. Why are protein requirements in some species related to season, age, and gender?

6. Compare the reproductive energetics of northern pintails and lesser snow geese.

7. How do Pacific salmon, in a sense, nourish their young even after the adults are dead?

8. How is nutritional condition related to reproductive success in fish and wildlife species?

9. Describe the form and function of a foregut fermenter and of a hindgut fermenter.

10. Compare the digestive tract of a California quail with that of a northern grasshopper mouse.

11. What is temperature homeostasis in endotherms? How do endotherms maintain temperature homeostasis when ambient temperatures are below the lower critical temperature or above the upper critical temperature?

12. Compare the changes in energy demands of house sparrows and bluegills from summer to winter.

13. Relate metabolic rates and energy requirements to body mass in endotherms.

14. What is hypothermy and what advantages does it offer an animal species? Give an example.

15. Discuss oxygen availability and the difficulties aquatic ectotherms face in obtaining oxygen from water.

16. How does the heart and basic blood circulation plan differ between fishes and mammals?

17. Compare the ways in which fishes in marine and freshwater ecosystems are adapted to these different osmotic environments.

18. How are some birds adapted for drinking salt water and eating highly saline foods?

19. What are some methods that seals in the Arctic Ocean might use to reduce or promote loss of metabolic heat?

20. What are four ways that desert kangaroo rats are adapted to living in desert ecosystems?

SELECTED READINGS

Barton, M. 2007. *Bond's biology of fishes.* 3rd ed. Thomson Brooks / Cole, Belmont, Calif.

Halver, J. E. 2002. *Fish nutrition.* 3rd ed. Academic Press, New York.

Moyle, P. B., and J. J. Cech Jr. 2004. *Fishes: An introduction to ichthyology.* 5th ed. Prentice-Hall, Upper Saddle River, N.J.

Owen-Smith, R. N. 1988. *Megaherbivores: The influence of very large body size on ecology.* Cambridge University Press, Cambridge, U.K.

Randall, D., W. Burggren, and K. French. 2002. *Ekert animal physiology: Mechanisms and adaptations.* W. H. Freeman, New York.

Robbins, C. T. 1993. *Wildlife feeding and nutrition.* 2nd ed. Academic Press, San Diego.

Schmidt-Nielsen, K. 1997. *Animal physiology: Adaptation and environment.* 5th ed. Cambridge University Press, Cambridge, U.K.

Willmer, P., G. Stone, and I. Johnston. 2000. *Environmental physiology of animals.* Blackwell Science, Ltd., London.

LITERATURE CITED

Ankney, C. D. 1977. The use of nutrient reserves by breeding male lesser snow geese *Chen caerulescens caerulescens. Canadian Journal of Zoology* 55:1984–1987.

Ankney, C. D., and C. D. MacInnes. 1978. Nutrient reserves and reproductive performance of female lesser snow geese. *Auk* 95:459–571.

Ayotte, J. B., K. L. Parker, J. M. Arocena, and M. P. Gillingham. 2006. Chemical composition of lick soils: Functions of soil ingestion by four ungulate species. *Journal of Mammalogy* 87:878-888.

Burgner, R. L. 1991. Life history of sockeye salmon *(Oncorhynchus nerka)*. Pages 3–117 *in* C. Groot and L. Margolis, eds. *Pacific salmon life histories.* University of British Columbia Press, Vancouver, Canada.

Carey, F. G. 1982. Warm fish. Pages 213–233 *in* C. R. Taylor, K. Johansen, and L. Bolis, eds. *A companion to animal physiology.* Cambridge University Press, Cambridge, U.K.

Cech, J. J., Jr., R. M. Laurs, and J. B. Graham. 1984. Temperature-induced changes in blood gas equilibria in the albacore, *Thunnus alalunga*, a warm-bodied tuna. *Journal of Experimental Biology* 109:21–34.

Feldhamer, G. A., L. C. Drickamer, S. H. Vessey, and J. F. Merritt. 2004. *Mammalogy: Adaptation, diversity, ecology.* 2nd ed. McGraw Hill, New York.

Frandson, R. D., W. L. Wilke, and A. D. Fails. 2003. *Anatomy and physiology of farm animals.* 6th ed. Lippincott, Williams, and Wilkins, Philadelphia.

Goff, J. P. 2004. Minerals. Pages 575–599 *in* W. O. Reece, ed. *Dukes' physiology of domestic animals.* 12th ed. Cornell University Press, Ithaca, N.Y.

Hays, V. W., and M. J. Swenson. 1993. Minerals and bones. Pages 517–535 *in* M. J. Swenson and W. O. Reece, eds. *Dukes' physiology of domestic animals.* 11th ed. Cornell University Press, Ithaca, N.Y.

Hofmann, R. R. 1988. Anatomy of the gastro-intestinal tract. Pages 14–43 *in* D. C. Church, ed. *The ruminant animal: Digestive physiology and nutrition.* Prentice-Hall, Englewood Cliffs, N.J.

Irving, L. 1966. Adaptations to cold. *Scientific American* 214:94–101.

Kincaid, R. 1988. Macro elements for ruminants. Pages 326–341 *in* D. C. Church, ed. *The ruminant animal: Digestive physiology and nutrition.* Prentice-Hall, Englewood Cliffs, N.J.

Kooyman, G. L., C. M. Drabek, R. Elsner, and W. B. Campbell. 1971. Diving behavior of the emperor penguin, *Aptenodytes forsteri. Auk* 88:775–795.

Krapu, G. L., and K. J. Reinecke. 1992. Foraging ecology and nutrition. Pages 1–29 *in* B. D. Batt, A. D. Afton, M. G. Anderson, C. D. Ankney, D. H. Johnson, J. A. Kadlek, and G. L. Krapu, eds. *Ecology and management of breeding waterfowl.* University of Minnesota Press, Minneapolis.

Lagler, K. F., J. E. Bardach, R. R. Miller, and D. R. M. Passino. 1977. *Ichthyology.* 2nd ed. John Wiley & Sons, New York.

Mautz, W. W. 1978. Sledding on a bushy hillside: The fat cycle in deer. *Wildlife Society Bulletin* 6:88–90.

McDonald, P., R. A. Edwards, and J. F. D. Greenhalgh. 1981. *Animal nutrition.* 3rd ed. Longman, London.

Moyle, P. B., and J. J. Cech Jr. 2004. *Fishes: An introduction to ichthyology.* 5th ed. Prentice Hall, Upper Saddle River, N.J.

National Research Council. 1982. *Nutrient requirements of mink and foxes.* 2nd rev. ed. National Academy Press, Washington, D.C.

National Research Council. 1993. *Nutrient requirements of fish.* National Academy Press, Washington, D.C.

National Research Council. 1994. *Nutrient requirements of poultry.* 9th rev. ed. National Academy Press, Washington, D.C.

Owen-Smith, R. N. 1988. *Megaherbivores: The influence of very large body size on ecology.* Cambridge University Press, Cambridge, U.K.

Petersen, L. E. 1984. Northern plains. Pages 441–448 *in* L. K. Halls, ed. *White-tailed deer ecology and management.* Stackpole Books, Harrisburg, Pa.

Piatt, J. F., and D. N. Nettleship. 1985. Diving depths of four alcids. *Auk* 102:293–297.

Pond, W. G., D. C. Church, K. R. Pond, and P. A. Schoknecht. 2005. *Basic animal nutrition and feeding.* 5th ed. John Wiley & Sons, Inc., Hoboken, N.J.

Randall, D., W. Burggren, and K. French. 2002. *Ekert animal physiology: Mechanisms and adaptations.* 5th ed. W. H. Freeman, New York.

Robbins, C. T. 1993. *Wildlife feeding and nutrition.* 2nd ed. Academic Press, San Diego.

Schmidt-Nielsen, K. 1959. Salt glands. *Scientific American* 200:109–116.

Schmidt-Nielsen, K. 1971. How birds breathe. *Scientific American* 225:72–79.

Schmidt-Nielsen, K. 1997. *Animal physiology: Adaptation and environment.* 5th ed. Cambridge University Press, Cambridge, U.K.

Weir, J. S. 1969. Chemical properties and occurrence on Kalahari sand of salt licks created by elephants. *Journal of Zoology* 158:293–310.

Williams, J. B., and B. I. Tieleman. 2005. Physiological adaptation in desert birds. *Bioscience* 55:416–425.

Willmer, P., G. Stone, and I. Johnston. 2000. *Environmental physiology of animals.* Blackwell Science, Ltd., London.

CHAPTER 6

Behavior

When an organism responds to its environment, it often does so by performing specific behaviors. The behavior of individuals or groups of animals is closely tied to sensory perception; therefore, an understanding of behavior requires some general knowledge of sensory structures. In addition, because animals often respond to their environment through behavior, any modification of that environment can affect their behavior. Animal behavior is influenced by the genetic constraints of the organism, by natural processes that affect the environment, and by human modifications of the environment, both intentional and unintentional.

Succession (see section 2.4) is an excellent example of the way naturally occurring environmental change can affect an animal's behavioral patterns. Terrestrial animals whose behaviors are adapted to open areas are affected when trees or shrubs appear at a later seral stage. Fishes that are dependent on clear water are behaviorally affected by succession that results in reduced water clarity.

Human-induced modification of the environment can also affect animal behavior patterns. Plowing or urbanization of grassland may not be done with the intent to modify the habitats of an organism, but its effects may be substantial. For example, human effects on the grassland habitat of the lesser prairie-chicken can eliminate its reproductive habitat. Habitat, and therefore behavior, can also be intentionally affected. For example, if there are no suitable locations for hunting perches or nesting sites for ferruginous hawks in a particular area, such structures can be provided. Thus, an understanding of behavior may be a necessary prerequisite for programs that are intended to increase or decrease the abundance of a particular species or groups of species. Knowledge of animal behavior is also important with regard to activities such as animal sampling. A biologist cannot sample effectively without knowing when and where a species is vulnerable to a sampling design or method.

This chapter first provides an overview of the senses used by organisms in their behavioral activities, and then a variety of individual and group behaviors are addressed.

6.1 Senses

When humans think of senses, we typically think of sight, hearing, touch, smell, balance, and taste. We interact with our surroundings through the use of these senses. Terrestrial and aquatic animals use these same senses, but they often differ from humans in the level of their capabilities. For example, the range of sounds heard by a mammal such as the red fox is broader than the range of sounds heard by humans. Night vision is better developed in a bird such as the barn owl than it is in humans. Many rare sea turtle species will not nest at night on a beach if it is lit. Color vision is well developed in humans but poorly developed in many other mammals. In addition, some animals have sensory structures that humans do not possess, such as the lateral line system in fishes. This sensory structure is used for detecting sound, but it is quite different from the "typical" ears of birds, mammals, or even fishes themselves.

The eyes of different fish species vary in effectiveness depending on the habitat in which they live and how they have incorporated sight into their sensory repertoire. Some species, such as the predatory chain pickerel, have acute vision. Others, such as the southern cavefish, have no eyes; they rely on cues from their other senses to compensate for their inability to see. A good deal of what a fish sees is influenced by the medium in which it lives. The intensity of light is reduced as the light passes through water. The rate of reduction depends on water transparency as well as depth. Light reaching the bottom of a clear mountain stream is of a totally different intensity than is the light reaching the same depth in a muddy lake. In addition, light passing through water loses different portions of the color spectrum depending on water transparency. For example, red light penetrates deeper than blue light when water transparency is low, while the opposite is true when water transparency is high (Horne and Goldman 1994).

Birds generally gain much information from their surroundings by sight. Depending on the species, they can detect direction, distance, size, shape, brightness, color, intensity, and motion. Most mammals have poorer sight but a good sense of smell (**olfactory** capabilities), probably because mammals initially evolved as nocturnal organisms (active during periods of darkness). In the past, biologists believed that mammals possessed "smell brains" because their brains have large olfactory lobes, while birds possessed "eye brains" because of their large optic lobes. More recently, we have come to realize that some birds possess a sense of smell and can detect odors as well as mammals (Young 2005). In mammals, color vision is primarily found in the primates, a group that has relatively well-developed vision. For many mammal species, the ability to detect motion appears to be more important than the ability to distinguish colors. Another indication that color vision is less important to most mammals is their own rather drab coloration compared with that of birds, fishes, and many other animals.

Most birds are sight feeders, although some, such as turkey vultures and brown kiwis, are quite sensitive to smell. Air can be a difficult medium through which to follow a scent, depending on the amount of air movement, turbulence, and humidity. In contrast, the terrestrial surface environment where most mammals live usually retains odors for a longer time than air. Although mammals, like other organisms, can smell odors that have **volatilized** (vaporized at a relatively low temperature), the source of the odor often remains in place and produces odor for an extended time period.

Most terrestrial animals use their sense of taste for short-range perception of chemical stimuli and their sense of smell for long-range perception. Aquatic animals, on the other hand, can use both senses for long-range perception because of the different medium in which they live. Chemical cues for both taste and smell must pass through a moist membrane to be perceived. For a terrestrial animal to taste something, the material usually must be present in its mouth; this is not the case for an organism such as a fish. The external body surface of terrestrial organisms is not moist; the external body surface of fishes is.

In the aquatic environment, odors can remain in solution for a long time. Fishes perceive primarily organic odors, which can be either smelled or tasted. They use both an olfactory apparatus contained in a nasal sac and their taste buds to detect these odors. Some fishes, such as Pacific salmon and many sharks, have developed an acute sense of smell. In fishes, the sense of smell is usually more acute than the taste sense.

Most fishes have taste buds in the mouth, lip, and throat regions, but some, like the channel catfish, have taste buds over all of their bodies. The common carp, like mammals, can distinguish salty, sweet, bitter, and acid tastes. Birds have few taste buds compared with other vertebrates. For example, a mallard has only 375 taste buds, while a human has 9,000 and a rabbit has 17,000 (Welty and Baptista 1988). A black bullhead can have 100,000 taste buds located over all of its body (Lagler et al. 1977).

Hearing is the sense that allows animals to detect sound waves. The hearing apparatus is relatively simple in fishes and becomes progressively more refined in reptiles, birds, and mammals.

The inner ear portion of the sound receptor in fishes has no opening to the outside of the fish. Because the density of fish flesh and the density of water are similar, sound passes through the body of the fish, strikes structures with differing densities such as bones or the gas bladder, and is transmitted to the inner ear.

The lateral line system is an additional system for sensing sound waves that is unique to fishes and some amphibians. The most obvious portion of the lateral line system found in most fish species is located in a canal that runs along the side of the fish and has numerous openings to the outside of the fish (Figure 6.1). In scaled fishes, pores through the scales on the lateral line allow water to enter the canal; in scaleless fishes, the pores do not need to pass through scales. Within the canal are structures called **neuromasts** that can detect displacement of water and are thus sensitive to mechanical sound waves and motion around the fish. In some fishes that have particularly sensitive lateral line reception, no canal is present and the neuromasts are on the exterior of the fish; the Ozark cavefish is an example of a species that has many exposed neuromasts on the skin surface. Finally, the lateral line system is not limited to the side of a fish. Most fishes also have a portion of the lateral line system on their heads.

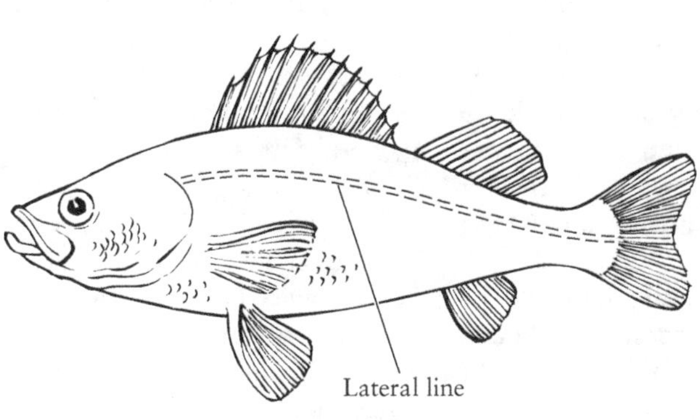

FIGURE 6.1 A portion of the lateral line system of many fishes is located along the side of the body. The line is actually a series of pores that allow sensory cells inside the lateral line canal to respond to water displacement.

The lateral line system has numerous uses. For example, a predatory fish can use the lateral line system to detect prey, schooling fish can use it to detect one another, and fishes can also detect nonmoving objects when water currents "bounce" off the objects. This "distant touch" organ therefore allows location of both moving and nonmoving objects.

Mammals and birds have ears that are divided into three parts: the external ear, middle ear, and inner ear. In most birds, the external ear is simply an opening, often shaped to enhance sound gathering, which allows sound waves to enter. Some birds, such as owls, have feathers associated with ear openings that may direct sound into the ear. Most mammals have an external ear (**pinna**) that helps to intercept and direct sound waves into the ear. Different species or groups of species can sense different frequencies of sound. Some birds are capable of hearing notes in their songs that humans cannot hear because they occur too rapidly. Bats have developed a sound production and reception system for very high frequencies that helps them locate prey. They produce a high-pitched sound wave, then use their large pinnae to monitor return signals (echoes) from prey organisms such as flying insects. This same system allows them to fly in areas such as caves without striking the walls.

The middle ear of birds and mammals transmits sound waves from outside the animal to the inner ear. A single elongated bone comprises the middle ear of birds, while mammals have three connected bones. The inner ear is the location where sound is converted from mechanical energy to electrical energy (the form in which it travels through the nervous system). In fishes, **otoliths** (earstones) located in fluid-filled sacs in the inner ear stimulate sensory cells lining the sacs that convert mechanical to electrical energy. In birds and mammals, vibrations passing through the middle ear are transmitted to the viscous fluid in the membrane-surrounded chambers of the middle ear. Vibrations in the fluid of the inner ear are then received by tiny sensory cells and transmitted as electrical signals to the brain.

In vertebrates, the sense of hearing is closely tied to the sense of equilibrium. The inner ear arose in the earliest vertebrates as a mechanism for maintaining equilibrium. In mammals, birds, and fishes, the **semicircular canals** of the inner ear and otoliths can both contribute to the sense of equilibrium.

Birds and mammals both have touch receptors in the skin, but these receptors are much better developed in mammals. There is little evidence of such receptors in most fish species. The skin is also the location where many organisms can sense temperature. The skin of birds and mammals contains nerve endings that can detect heat and cold. Higher bony fishes also have nerve endings in the skin and can detect temperature changes as small as 0.03°C (Lagler et al. 1977).

Various fishes have electroreceptors that can sense electrical energy (Bleckmann 1993). In addition, some fishes have specialized organs that can produce electrical fields. Some catfish species have electroreceptors, termed **ampullary organs,** that are also sensitive to pressure, touch, temperature, and chemical stimuli. While many amphibians also have this electrical sense, the platypus and short-beaked echidna are the only mammals known to possess electroreceptors. There is much that humans do not know or understand about the ability of animals to produce or detect electrical stimuli.

There are numerous other sensory capabilities that animals possess. Humans again do not always have a good understanding of these capabilities, because such senses often have no externally obvious sensory reception area, such as an eye or ear. We are also less aware of some of these senses because the inputs they provide are not as distinct as those provided by senses such as sight or hearing. For example, many birds apparently can detect magnetic fields. In fact, many of the birds that migrate long distances probably use the magnetic field of the earth for at least part of their navigation. Determination of direction using the sun and stars also appears to be important in animal migration, as do other senses involving lunar and tidal perception. Additionally, little is known concerning the function of the **pineal gland** (**pineal organ**). This structure, which is located in the brain, is sensitive to light and may function in many animals as an ultrasensitive light receptor.

6.2 Innate and Learned Behavior

Behavior displayed by an organism can be **innate** (**inborn**) **behavior** or **learned behavior.** These are not always easy to differentiate; biologists commonly debate whether various behaviors are innate or learned. For example, does a scarlet tanager instinctively know how to build a nest, or is part or all of that behavior learned by observation? In truth, many behavior patterns are likely to be a blend of innate and learned behaviors.

At the simplest level, innate behavior can consist of **autonomic responses** that act independently of volition. Common examples of autonomic responses include breathing and the response of the pupil of the eye to changes in environmental brightness. Another simple type of innate behavior is a **reflex**, which is an autonomic response to a stimulus. The knee-jerk reflex of a human during a medical examination is a familiar example. If a house mouse touches a hot surface, quick reflex movement away from the surface is its immediate response. A **taxis**, another innate behavior, is the oriented movement of an organism toward (positive taxis) or away from (negative taxis) a stimulus. A taxis may occur in response to such environmental stimuli as light, air or water currents, or soil or water salinity. For example, many moth species are positively phototactic, meaning that they move toward light, while a brown bullhead is negatively phototactic, meaning that it moves away from light. An **instinct** is an innate, stereotyped behavior that is characteristic of a given species. A newly hatched northern pike has a holdfast structure on its upper snout. It instinctively "knows" that it can use the holdfast to attach to a piece of vegetation while it absorbs its yolk sac. Young birds respond to touch or shaking of the nest, or even shadows, by gaping for food. Adult male mallards and green-winged teal perform stereotyped and species-specific courtship activities toward hens in the late winter and spring. All of these behaviors are examples of instincts.

Learned behaviors are those that have been shaped as a result of experience. American goldfinches learn to feed at thistle feeders in the backyards of birdwatchers. Some largemouth bass that have been caught and released by anglers apparently learn from this experience; they may be less vulnerable to angling after an initial capture. Similarly, coyotes that have escaped from a leghold trap are much more wary with regard to traps. Conversely, some animals may actually become "trap-happy." Waterfowl biologists banding ducks during the summer on the Canadian prairie and using grain as bait often capture the same duck day after day. Has this bird learned that the grain is available and that it will be released unharmed, or has it failed to learn to avoid the trap? Questions such as these are difficult to answer. Learning capacity is higher for some bird species, such as American crows and blue jays, than for others. Fish and mammal species also vary considerably in ability to learn.

The simplest form of learning is **habituation learning,** which means learning not to respond to meaningless stimuli. A bird nesting near a construction site, for example, must learn not to respond to frequent, familiar loud noises. **Trial-and-error learning** involves sorting through several nonrewarding actions to find an action that will provide a reward or avoid a punishment. A western kingbird may feed on many types of insects. Perhaps some insects taste better than others, or perhaps some might even make the bird sick. Trial-and-error learning allows the bird to learn which insects should be avoided and which should be sought. **Insight learning** is a much higher form of learning, and it requires that an animal have the ability to perceive relationships. It needs to have the capacity to relate a similar experience to the one at hand. Some biologists consider the use of sticks to remove ants from burrows by common chimpanzees to be an example of insight learning, although this is arguable. Humans obviously use insight learning.

As mentioned earlier, most behaviors are likely to be a blend of innate and learned behavior. For example, the basic template of a bird song is innate, but each individual needs learning and practice to perfect its song. A young male western meadowlark hears adult male meadowlarks and perfects its own song. A northern mockingbird innately knows how to sing, but it increases its song complexity as it hears and incorporates elements of the songs of other birds.

6.3 Mobility in Early Life

At hatching or birth, young organisms can be classified as either **precocial** or **altricial**, depending on their particular behavioral pattern and state of development. A precocial bird, such as the American golden-plover or sharp-tailed grouse, hatches with down, is alert and well developed, and is mobile soon after hatching; in most cases, precocial young are soon able to fend for themselves with respect to activities such as feeding or avoiding predators. Mammals that bear precocial young include black-tailed jackrabbits and pronghorns. Although this term is seldom used in fisheries work, it is still applicable. A newly hatched walleye is free-swimming and can be called precocial.

Altricial birds, such as mourning doves, are typically naked, blind, and weak at the time of hatching. They require more intensive parental care to survive than do precocial species. The same is true for the altricial young of eastern cottontails and North American deer mice, which are naked and blind at birth. Although they are not born naked, the young of coyotes and tigers are also altricial, relying on intensive parental care for their survival. Newly hatched northern pike can also be termed altricial. After hatching, they struggle to attach themselves to stems of vegetation. They remain attached for several days while they absorb their yolk sac. The energy provided by the yolk sac allows them to develop further so that they will be free-swimming and able to feed when they detach from the vegetation.

Whether its offspring are precocial or altricial, each species has evolved a particular reproductive strategy. Parental investment is needed to enhance the likelihood of offspring survival, and these two strategies simply vary the time of parental investment. At first, it might seem that parents of altricial offspring have to expend more energy because they have to feed and care for their young. However, they typically do not invest as much energy in the actual production of an egg or young animal as do parents of precocial offspring. Even though parents of precocial offspring do not need to invest as much time in feeding and caring for their young, they generally invest more energy in the development of young before hatching or birth.

6.4 Rhythmic Behavior

The activity patterns of individual organisms often involve **rhythmic behavior** associated with daily or seasonal changes. Commonly recognized examples include **diel (circadian) behaviors**, which change throughout the day (diel refers to the 24-hour period), and **circannual behaviors**, which change throughout the year.

Mourning doves have a typical diel pattern of activity. For example, Howe and Flake (1989) found that mourning doves in Idaho sought water by visiting a pond most frequently at midmorning and late evening during early September (Figure 6.2).

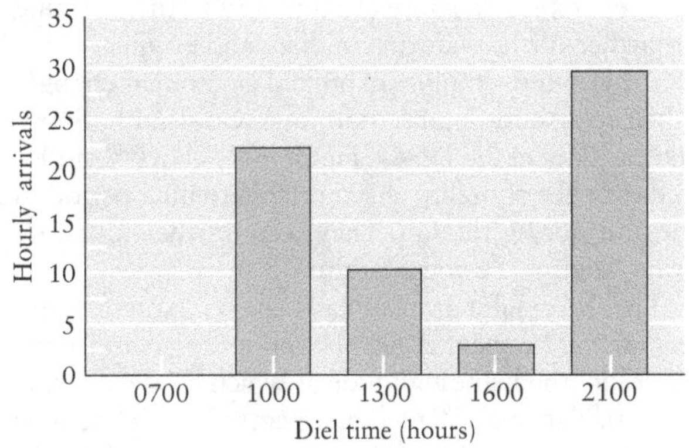

FIGURE 6.2 Arrivals of mourning doves at various times of the day during early September at a pond located at the Idaho National Engineering Laboratory. *[Adapted from Howe and Flake 1989.]*

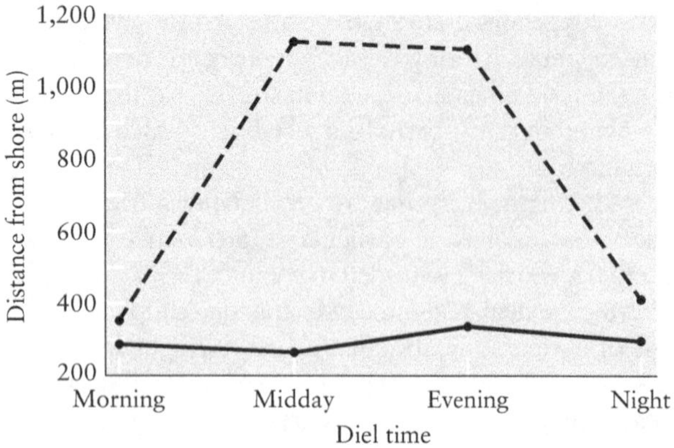

FIGURE 6.3 Diel activity patterns for northern pike in Lake Thompson, South Dakota, during August. Small northern pike (solid line; mean total length = 53 centimeters) remained approximately the same distance from shore throughout the day, while large pike (dotted line; mean total length = 84 centimeters) moved farther offshore during the warmer midday and evening periods. *[Adapted from Neumann 1994.]*

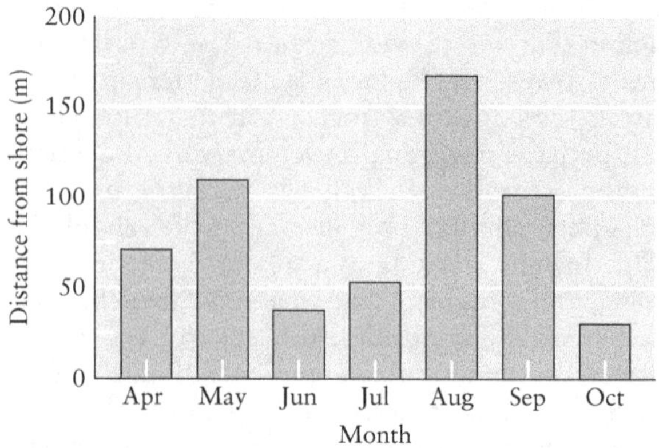

FIGURE 6.4 Average distance from shore for adult white crappies in Lake Goldsmith, South Dakota, at different times of the year. Fish were closest to shore during spawning in June and again in the fall and farthest from shore during the warm-weather period in August. *[Adapted from Guy et al. 1994.]*

Diel behavior can vary among species. For example, although many bats are nocturnal, there is substantial variation in activity patterns at various times of the night among bat species (Erkert 1982). Many insectivorous bats are most active near sunset, while many bats that feed on fruit or nectar are most active just prior to sunrise.

Diel behavior can also vary within a species. For example, Neumann (1994) found that the activity patterns of small and large northern pike differed in Lake Thompson, South Dakota, during August (Figure 6.3). Small northern pike remained about the same distance from shore throughout the day, while large northern pike were closer to shore at night and in the morning but farther offshore during midday and evening.

Adult white crappies exhibited circannual changes in their distribution pattern in Lake Goldsmith, South Dakota (Guy et al. 1994) (Figure 6.4). They were located near the shoreline during the spawning period in June and during the fall. They were farthest offshore during August, the warmest time of the year.

Elk also exhibit a circannual behavior pattern. Summers tend to be spent at high elevations, even above the **timberline** (the highest altitude at which trees can grow in a particular area). Winters are spent at lower elevations that are warmer and provide more access to food. Other examples of circannual behavior are the various migrations exhibited by some organisms (see section 6.8).

Most biologists believe that many circannual behaviors are related to **photoperiod**, the proportion of light and dark in a 24-hour period. Increasing day length during the spring may be an important cue for species, such as northern hog suckers and white-breasted nuthatches, that breed at that time of year. Similarly, decreasing day length in the fall probably cues reproductive behavior in species such as brown trout and elk.

6.5 Early Social Behavior

As soon as most animals are born or hatched, they must participate in a variety of social behaviors. These can be intraspecific or interspecific interactions and can take numerous forms. Immediately upon hatching, a blue-winged teal duckling must interact with its siblings and parent. It will have to communicate in order to survive. A young red fox will need to interact with others of its species, but it will also need to learn to interact with prey species, such as rodents and waterfowl, and with competitors and potential predators, such as the coyote.

Social behavior may begin just prior to hatching in the young of many precocial birds. Some researchers believe that the calls of the parent prior to and during **pipping** (breaking of the eggshell during hatching) in precocial species are important in early bonding between the parent and its young. During the first 10–20 hours after hatching, the young become strongly attached to the parent in a process known as **filial imprinting.** Imprinting is critical to the survival of precocial young that must follow their parent soon after hatching. Interestingly, if they are exposed to a moving object (e.g., a wooden decoy, an orange toy truck, or even a human) during this imprinting period, precocial birds will often become imprinted to the substitute parent. Following a parent through grass and over obstacles actually increases the filial bond. Clucking sounds similar to those made by the natural parent will enhance the imprinting process. Once the brief imprinting period passes, the young will no longer accept a substitute parent.

Filial imprinting does not occur to the same extent for altricial young as it does for precocial young. Apparently, the most likely imprinting of altricial young is habitat imprinting or sexual imprinting, which occur somewhat later in life.

6.6 Social Organization

Sociality is the tendency to gather in organized groups. Social organization is highly developed in many animals. Even invertebrates such as some bees and ants have developed high levels of social organization. Many social animals establish a **hierarchal order** (order of dominance) to help them deal with social gatherings. This hierarchy is often referred to as a **pecking order**. A flock of domesticated chickens will establish an order of dominance, in which dominant individuals will dominate subordinate individuals, often by pecking them. Such hierarchies allow animals to avoid conflict that requires energy expenditure. Rather than having continual, energy-consuming fights over day-to-day matters, the social hierarchy determines which individual has dominance, and there is little need for fighting. The position of an animal in a hierarchy is generally established through fighting or threat and is influenced by such things as breeding status, age, size, and gender. The hierarchy of dominance may be linear but is often more complicated, with dominance relationships resembling a polygon (Figure 6.5).

Examples of social hierarchies are plentiful. A wildlife biologist may want to capture wild turkeys during winter for relocation. Often, this is accomplished by placing a bait such as corn or other grain under a large net. The net is dropped when a sufficient number of birds are under the net. However, the dominant wild turkeys, generally older males, will sometimes keep other turkeys from feeding on the bait, complicating the task of the watching biologist, who would usually prefer to capture more than just the older males. Wolves and eastern gorillas also operate under social hierarchy systems. South

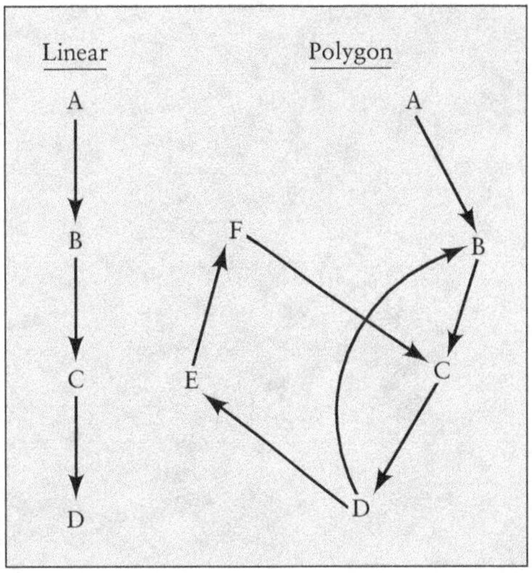

FIGURE 6.5 Dominance hierarchies may have a simple linear pattern (left), but they are typically more complicated and often resemble complex polygons (right). In the complex polygon shown here, individual A is dominant over B, B over C, C over D, D over B, D over E, E over F, and F over C.

American cichlid fishes, such as the firemouth cichlid and Jack Dempsey, will quickly establish social hierarchies when moved into an aquarium. Wild rainbow trout in a stream maintain specific feeding locations, with the dominant fish maintaining the best locations.

6.7 Communication

Most communication in the animal world occurs within or between species by sound, sight, smell, and taste. Electrical communication is rare, but it does occur in some organisms (see section 6.1). Touch can also play a role in animal communication.

Sound has several advantages over visual displays or scents for communicating: a wide range of vocal frequencies and intensities is possible; little energy expenditure is needed; sound can penetrate visual barriers; and communication is possible in the dark. Sound can be used for spacing individuals of the same species in a given environment, keeping a group together when visibility is limited, conveying information about predators, or indicating the presence of food. Sound is also important for reproductive functions, such as the proclamation of gender, development of pair bonds, and mate identification.

Birds use sound as a communication tool in their calls and songs. The “oak-a-lee” song of a male

FIGURE 6.6 The song of a male red-winged blackbird announces his establishment of a territory during the breeding season. *[Image courtesy of D. Dewhurst, U.S. Fish and Wildlife Service.]*

red-winged blackbird in a cattail marsh during the spring is a clear message to other males and potential mates that the bird has chosen a territory (Figure 6.6). Similarly, a northern bobwhite covey that has been separated can regroup through calls. Not all sound communication in birds is by vocalization. The male ruffed grouse, for example, uses rapid wingbeats to produce its characteristic "drumming" sound during the spring mating season (Figure 6.7).

Numerous mammal species use sound for similar functions. Some mammals use little sound communication, while others, such as coyotes, are great vocalizers. The sounds produced by mammals vary from the bugling of an elk (accomplished with specialized teeth) to the distress screams of an injured white-tailed jackrabbit. However, most mammals are poor vocalizers compared with birds. An exception would be the bats, which use sound for such tasks as individual identification, echo-location of food, and finding roost sites.

Sound also plays an important role in aquatic ecosystems. Humans hear poorly in the underwater environment, so we label it the "silent world." Nothing could be further from the truth; we are just not physically adapted to hearing underwater sound. Sound waves actually travel farther and faster in water than in air. Many fish species use sound as a form of communication. An example is the freshwater drum, which produces a drumming sound using its gas bladder. The songs of the humpback whale are well known and also serve as a means of communication. With devices designed to detect underwater sounds, humans find that the underwater world is far from a silent place. In reality, it is a rather noisy place.

Visual forms of communication, some relating to color and others to posture or structure, are common in animals. Visual forms of communication have the advantage of being visible over long distances in open habitat. Color, which is a function of vision, is commonly used by aquatic and terrestrial animals for activities such as courtship and species recognition. Removal of the moustache (black or red in color, depending on the race) from a male northern flicker will confuse sex recognition by females. The display of colorful body parts is a major aspect of courtship for many species. The males of most ducks, such as wood ducks, mallards, and ruddy ducks, display bright breeding plumage, while the females have drab coloration. The colorful buttocks of Hamadryas baboons provide information on female reproductive status and also on dominance status. Even in humans, who have the most highly developed verbal communication skills, visual cues are very important during interpersonal communication.

Many North American sunfish species are reproductively isolated by visual cues associated with color. In the wild, a female green sunfish generally will not breed with a male bluegill. However, bluegill × green sunfish hybrids can be produced in a fish culture pond.

FIGURE 6.7 During the breeding season, male ruffed grouse produce a drumming sound by means of rapid wing movements, not by vocalization. *[Image courtesy of the Ruffed Grouse Society.]*

FIGURE 6.8 Male bluegills have a large, dark tab on the gill flap. Female bluegills may use the tab as one visual cue to help ensure that they breed with a male bluegill and not a male of another sunfish species. *[Image courtesy of the Nebraska Game and Parks Commission.]*

FIGURE 6.9 Pronghorns have a highly visible white rump patch that can provide a danger signal to other pronghorns across substantial distances in the open habitat that this species typically inhabits. *[Image courtesy of Yellowstone National Park.]*

If the fish culturist removes the dark tab on the gill flap of a male bluegill (Figure 6.8), a female green sunfish will be more likely to accept him as a mate. The tab is probably a visual cue that serves as one mechanism that aids in reproductive isolation; another such mechanism is sound (Gerald 1971). Hybrid sunfish in the wild are often found where submerged aquatic vegetation is very dense and visual cues are obscured.

Another type of visual signal common to many animals is a signal displayed as a sign of danger. Examples include the white rump of a pronghorn (Figure 6.9) and the white underside of the tail in white-tailed deer. Many prairie birds, such as western meadowlarks and vesper sparrows, which are vulnerable to predators, have white outer tail feathers. These white feathers flash during escape maneuvers and may serve as a distraction to predators, a danger signal to **conspecifics** (others of the same species), or both.

Visual communication can also involve the position or posture an animal assumes or the use of specially designed structures. For example, when a rock bass is disturbed, it erects the front portions of its dorsal and anal fins (the spined parts); this gives the fish the appearance of being larger. The **dewlap** (fold of skin on the neck) of some lizards can be enlarged for use in visual communication during reproduction. The tails of male greater sage-grouse and wild turkeys are opened or fanned during courtship behavior. Many such structures are conspicuously colored when displayed. In addition, particular types of movements may be used. During reproduction, brook trout males display in a sideways fashion directly in front of females with whom they are trying to mate. A lizard that extends its dewlap may also perform a series of "push-ups" or "bobs." Male and female western grebes run along the water surface during their highly stylized courtship displays.

Visual communication does not necessarily mean drawing attention to the organism; in fact, it can work in the opposite manner. Cryptic or camouflage coloration patterns can make an organism difficult to see. Female ducks, such as wood ducks, mallards, and ruddy ducks, are drab in color, probably because it makes them less conspicuous while nesting. This form of communication conveys the message, "I am not here." Coloration patterns may also be used to send false messages. For example, some fishes have large black spots on their tails. This makes the tail look like the head of a large fish (the large spots resemble the eyes). In this case, two false messages are conveyed, "This is my head" and "I am big." In some cases, coloration is combined with body shape to hide an organism. This form of communication sends the message, "I am a tree leaf," "I am a twig," or other similar messages.

In many animals, agonistic displays involve both sight and sound. **Agonistic behavior** is behavior involving aggression or threat displays. A wolf pack has a social hierarchy, and visual displays as well as sounds (growls) are a part of keeping this order with little actual fighting. The curled-lip aggressive stance that displays the teeth of a wolf is well recognized within the pack. Lowering of the tail and exposure of the vulnerable head or neck regions are submissive responses to aggressive displays. Such submissive responses usually serve to repress any escalation of aggressive behavior, such as attack, by the dominant animal.

Animals can also use chemical forms of communication by means of their olfaction and taste senses. Chemical messages are sometimes lumped together into a group of compounds called **pheromones** (Grier 1984). Although some biologists use a more specific definition of pheromones, we will use the term in this more general sense. One common form of chemical communication involving pheromones is the territorial marking or signposting of an area by male mammals. Anyone who has ever watched a male dog proceed around its territory (which often corresponds to the property boundaries of its owner), while periodically stopping to sniff and urinate, is viewing chemical communication between dogs. Territorial marking is also commonly used by animals such as deer, rabbits, and wolves. In most mammal species, males ascertain the reproductive status and receptiveness of females by scent. However, the extent of communication involved can vary substantially from organism to organism. For example, North American elk (McCullough 1969) and mule deer (Bowyer 1986) use scent marking for both male–male and male–female communication, while other cervids such as moose (Bowyer et al. 1994) and reindeer (Adams et al. 2001) primarily use scent marking for male–female communication. Chemical communication is also important for some fish species. For example, female blue and channel catfish use odors (sexual pheromones) to be certain of the species of a potential mate (Todd 1971). In addition, chemical substances produced by some fishes, especially minnows, can warn nearby conspecifics of danger.

6.8 Movements and Migrations

Migration is normally defined as a two-way movement to and from an area with characteristic regularity or with changes in life history stage (see section 2.10). **Movements** can be defined as changes in position or posture, and they are not necessarily regular in nature. Most migrations are associated with food supply, reproduction, or changes in ambient temperature. They are often cued by photoperiod and can be examples of rhythmic group behavior (see section 6.4).

Some fishes are **diadromous**; that is, they migrate between salt and fresh water. **Anadromous** fishes spawn in fresh water but live most of their lives in the ocean. The various species of Pacific salmon spawn in freshwater lakes and rivers in northwestern North America, and their offspring migrate to the Pacific Ocean, where they grow to adult size. Juvenile salmon are able to use the sun's position and the earth's magnetic field for orientation, while adult salmon rely on olfactory cues to return to the stream in which they hatched (Dittman and Quinn 1996). The striped bass, native to the Atlantic coast of North America, is another example of an anadromous fish. **Catadromous** fishes spawn in the ocean and return to fresh water to accomplish much of their growth. For example, American eels that live and grow in the Mississippi River system migrate to the Sargasso Sea area of the Atlantic Ocean (east of southern Florida) to reproduce.

Fish migrations can also occur strictly within fresh or salt water. For example, a strain of rainbow trout that once lived in Lake McConaughy in western Nebraska migrated hundreds of kilometers up the North Platte River and actually spawned in Wyoming. Other fish movements may be shorter in distance. For example, male black crappies spend much of the year in the open water of a lake but migrate to the shoreline at spawning time to build their solitary nests.

Both elk and mule deer in the Rocky Mountains migrate from summer feeding areas at high elevations to wintering areas at lower elevations. Reindeer migrate over tremendous distances in Alaska and northern Canada. They feed far to the north in the summer, then migrate south for wintering. Mountain goats exhibit an opposite pattern, because they seasonally migrate up in elevation to feed on windblown slopes and exposed areas during the winter.

Birds are some of the most famous of all animal migrants. For example, American golden-plovers breed along the northern coast of Alaska and across northern Canada, then winter in the southeastern part of South America, migrating a distance of 12,000 kilometers or more. Most North American waterfowl species migrate north during the spring and summer to areas with abundant and productive shallow water. Such areas include the extensive prairie pothole region (Figure 6.10) as well as northern river deltas and floodplains. In this way, waterfowl can take advantage of the high productivity of these water bodies during the summer while avoiding the harsh winter. Many snow geese nest north of the Arctic Circle and then migrate south during the winter, and many spend the winter on the Gulf of Mexico coast. Birds may rely on a variety of cues for

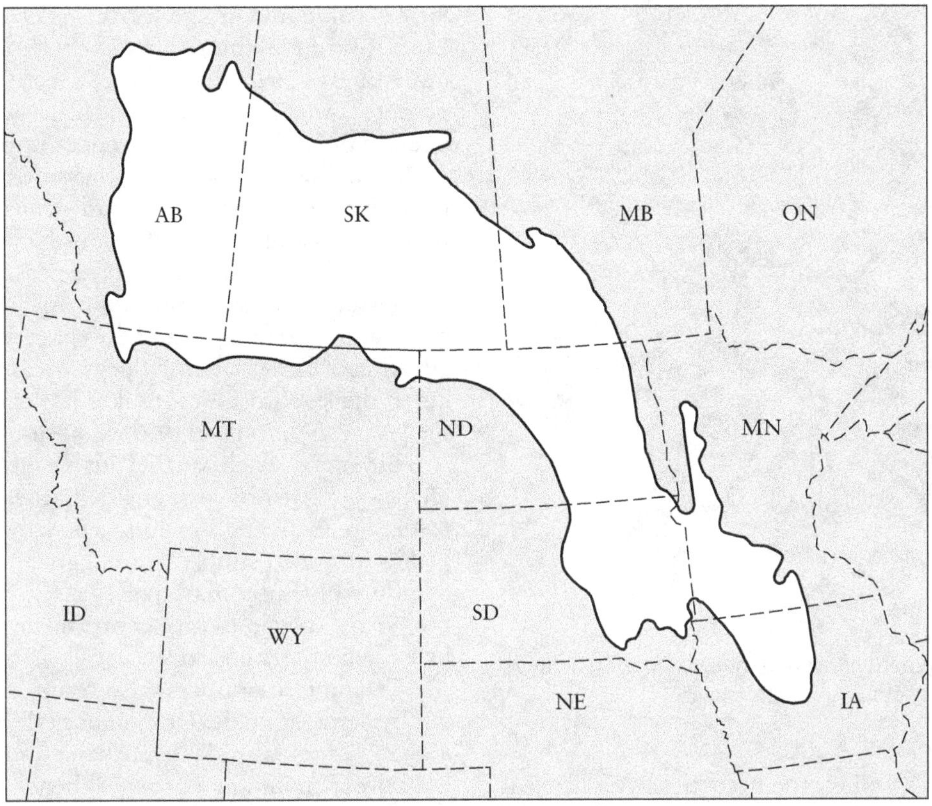

FIGURE 6.10 The prairie pothole region of North America extends from northwestern Iowa to southeastern Alberta.

navigation during these long migrations, such as the positions of celestial bodies, the magnetic field of the earth, and landscape features.

Homing is the ability of an animal to return to a familiar site that is outside the range of the direct senses, such as vision and hearing. Homing is useful for finding past successful breeding sites, wintering areas, feeding sites, and so forth. Pacific salmon home to the site where they hatched, probably with the aid of their sense of smell (**chemical imprinting**). Little brown myotis in New England home as far as 300 kilometers to a wintering cave (Davis and Hitchcock 1965). Giant Canada geese frequently return to breed at the location where they fledged, as do many other bird species.

6.9 Flocking, Herding, and Schooling

Terrestrial and aquatic animals may gather for a variety of reasons. Social groups may demonstrate flocking, herding, schooling, or other grouping behavior. A fish school is defined here as a group of fish that is together because of social attraction for one another and displays coordinated movement. Bird flocks also are social assemblages, as are mammal herds.

Many social assemblages are the result of predator–prey interactions. For fishes that are prey species, schooling can provide protection from predators. A closely packed school may appear to be a larger organism and thus deter predation. In addition, a predator is more likely to be observed by at least one of the many individuals in a school than by a lone fish. Some predator species are known to take more prey items when the items are presented individually rather than in a group. It is also possible that there is less likelihood of a compact mass being found than widely scattered individuals. A predator may simply be confused by the high level of motion in a school and not be able to strike as effectively. Some biologists have suggested that schooling provides a greater chance of group survival through the sacrifice of some individuals in the group. The more individuals there are in the group, the greater the chance that

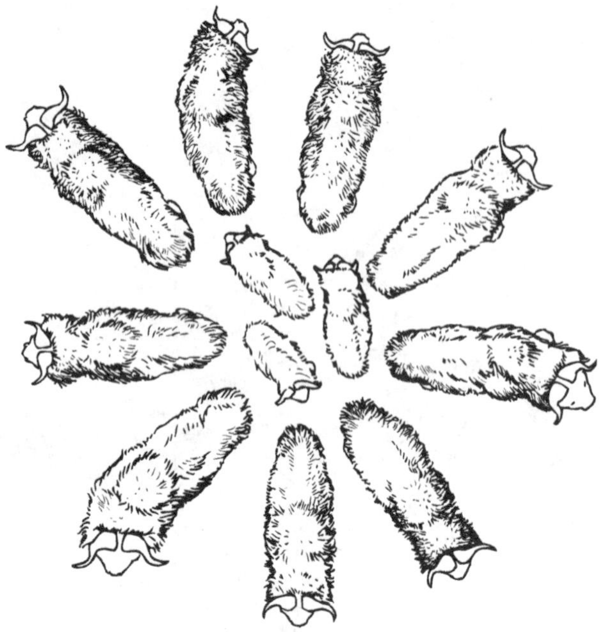

FIGURE 6.11 When threatened, muskoxen form a circle, with young herd members in the center.

some individuals will escape predation. A school of newly hatched smallmouth bass is typically composed of siblings; therefore, the sacrifice of some individuals in the school may increase the chances of survival for related individuals. Most fish species school as juveniles, but only about 20% school as adults (Bond 1979).

Birds may or may not flock with parents and siblings. For example, newly hatched great horned owls only stay in family groups until they are old enough to be independent. In contrast, Canada geese tend to flock and migrate in parental/sibling groups. Snow geese and other Arctic-nesting geese also remain in family groups and form large flocks during migration. The chances of survival of the young are probably increased when they flock with adults. Increased survival may result from direct protection by the adults, or the experience of the older birds may lead to higher survival rates of the young during migration.

Large ungulates of the open plains, such as pronghorns, commonly form herds as a defense against predators. Muskoxen form a circle when threatened, with young animals in the center (Figure 6.11).

The predators themselves may also operate in social assemblages. Wolves tend to group in packs of genetically related individuals. Packs apparently offer a feeding advantage for this species; the wolves can capture prey more effectively in groups than as individuals. Striped bass are effective open-water predators that feed in schools. Killer whales also cooperate in a group for more effective predation.

Schooling fishes and flocking birds may also gain hydrodynamic or aerodynamic advantages. Fish schools pass through water more efficiently than do individuals, as the lead fish allows for easier passage by the following individuals. Similarly, many birds flock together during migration and thus gain an aerodynamic advantage. In snow goose flocks, birds take turns flying in the lead position. It takes less energy to follow the leader than to be in the lead position.

Birds may flock, or they may simply be in short-term aggregations. Songbirds gathering at a bird feeder are in an aggregation, primarily seeking food. Fishes may also simply be in aggregations. A group of adult bluegills may aggregate for feeding purposes, or they may concentrate near submerged brush when seeking cover.

Breeding seasons can also result in social grouping by some animals. Dominant male elk will guard a group of females. Wild turkey gobblers will similarly try to maintain a group of hens.

Some birds form colonies during breeding seasons. Both double-crested cormorants and great blue herons form breeding colonies, often in large, dead trees or in live trees near water (Figure 6.12). The trees provide nest sites that are relatively safe from predators, and the adults use the trees as a nighttime

FIGURE 6.12 Great blue herons nest in colonies in either live or dead trees over or near water. *[Image courtesy of Yellowstone National Park.]*

roosting site. Nesting in a colony also helps to synchronize breeding, and information about food sources can apparently also be shared. Colonial nesting birds generally require some minimal number in the colony to successfully reproduce, and larger colonies often have a higher percentage of active breeders.

Many birds are intolerant of conspecifics during the breeding season but may flock together for the remainder of the year. The red-winged blackbird is a good example of such a bird. During the breeding season, a red-winged blackbird is highly territorial. During the remainder of the year, it may join other red-winged blackbirds, yellow-headed blackbirds, European starlings, common grackles, and other birds in mixed-species flocks. Such mixed flocks typically join together for feeding and roosting; the flock may also be a good defense against predation.

6.10 Reproductive Behavior

Mating systems in higher animals are generally categorized as **promiscuous, monogamous**, or **polygamous.** Promiscuous species generally form no pair bonds between males and females. Rainbow smelt in the Great Lakes are promiscuous breeders. They spawn in large groups, with no prior development of pair bonds. Two or more males maintain positions beside a female and release **milt** (sperm-bearing fluid) as the female releases her adhesive eggs. Ruby-throated hummingbirds also have a promiscuous mating system.

Monogamous mating systems involve a pair bond between one male and one female that can last for as short as one mating season or for the life span of the individual. Both the gray partridge and the field sparrow are monogamous species. Monogamy is rare in mammals, but some mammals, such as the red fox, have a monogamous mating system. Monogamy is even rarer in fishes. Mating systems that are mostly monogamous occur in some species, such as some of the African cichlids, in which both parents provide care for the young. However, even among these fishes the monogamous system is not strictly followed, as a male often accepts a second female.

In a polygamous system, a single individual of one gender mates with numerous individuals of the opposite gender. A **polygynous** species is one in which a single male mates with multiple females. Common examples include ring-necked pheasants, elk, and mule deer. Less common are **polyandrous** species, in which a single female mates with multiple males. Female Wilson's phalaropes mate with multiple males. In this species, the female is larger than the male and has brighter plumage. The male builds the nest, incubates the eggs, and rears the young alone.

Most animals have developed reproductive strategies that can be categorized along a continuum from production of large numbers of offspring that receive little parental care to production of few offspring that receive a great deal of parental care. This range of reproductive strategies is referred to as the ***r–K* continuum.** ***r*-selected species** use a strategy of low energy expenditure for production of gametes or care of offspring but produce large numbers of offspring. ***K*-selected species** use a strategy of high energy expenditure for embryo or egg development and care of offspring and produce small numbers of offspring. Organisms can be placed at a variety of points along the *r–K* continuum. That is, some *K*-selected species or *r*-selected species may be farther toward the ends or the middle of the continuum than others. In birds and mammals, *K*-selected species tend to have longer life spans than *r*-selected species; this is generally not true for fishes.

The *r*-selected female common carp simply broadcasts her adhesive eggs over flooded vegetation. The eggs, which are produced in large numbers (as many as 2 million in 1 year), are left to hatch, and the young must survive with no parental care. The common carp invests her energy in developing a large number of eggs; thus, only a fraction of a percent of the eggs needs to survive to perpetuate the species. Toward the other end of the continuum in fishes, smallmouth bass are nest builders, and their nests commonly contain several thousand eggs. The male guards the eggs after spawning and also guards the school of young after hatching.

The California condor, a *K*-selected bird, does not become sexually mature until 6–8 years of age, and then it incubates only one egg every other year. Development is unusually slow, with the young learning to fly at 5 months of age. Considerable energy is invested in incubation and care of the young by both parents over a period of about 10 months. The mourning dove is a more *r*-selected bird species. Although only two eggs are laid per clutch, mourning doves can hatch up to five clutches over the course of the summer.

The *r*-selected house mouse reaches sexual maturity 35 days after being born and has a **gestation period** (developmental period in the uterus) of 19 days. Litters can contain up to twelve young, and a single female can have as many as six litters in a year. In contrast, *K*-selected African elephants have a maturation period of 8–16 years, have a gestation period of over 20 months, and have a single calf in one cycle.

Fishes have evolved a variety of reproductive strategies, ranging from egg laying (longnose suckers, golden trout, and wedgespot shiners) to bearing live young (guppies and western mosquitofish). There is tremendous diversity within the egg-laying fishes. Some build and guard nests (black crappies, black bullheads, and brook sticklebacks), some build nests but do not protect the eggs (brook trout and pink salmon), and others broadcast eggs onto vegetation (common carp and muskellunge) or gravel (white bass and walleyes) and provide no care thereafter. Striped bass have semi-buoyant eggs that will hatch while being carried by a river current. Freshwater drums have eggs that are nearly neutral in buoyancy and hatch while suspended in water. There are also numerous fish species that carry developing eggs in their mouths or attached to or in some other portion of their bodies.

Male bluegill sunfish actually have alternative reproductive behaviors within a population. Typically considered a colonial nesting species, most males exhibit a "parental male" life history. Such males typically delay maturation until they reach larger sizes, often 15 or 20 cm. Thus, most energy is used for body growth until maturation. These males then prepare nests, attract a female, guard the nest and eggs, and even guard newly hatched fry for a short time. In contrast, there are also "cuckold" males that mature at a small size. These males often sneak into a spawning bout between a parental male and a female and fertilize some eggs before being driven off by the parental male. Other cuckolds are female mimics. They will assume the coloration of a female rather than a male; as a parental male mates with a female in the nest, the female mimic will join them and fertilize eggs. Because of the female coloration, the parental male does not drive the female mimic from the nest.

Birds may nest in isolation (green-winged teal and upland sandpipers) or in colonies (double-crested cormorants, king penguins, and bank swallows). They may have single (gray partridge and red-tailed hawks) or multiple (mourning doves) broods within a given year. Many species that have single broods, such as most ducks and gallinaceous birds, have the capability to renest if the first nest is destroyed. Some birds also will parasitize the nests of other bird species. The brown-headed cowbird does not build its own nests; instead, the females lay their eggs in the nests of various other birds and allow these other species to raise the brown-headed cowbird young. The young brown-headed cowbirds are aggressive and often are the most successful young in the host bird's nest. Some birds, such as the redhead, are flexible in their parasitism; they can either lay and incubate their own eggs or deposit their eggs in other nests.

Mammals generally bear live young. The nutritious milk provided by most female mammals aids in the development of the young, but at a substantial energetic cost to the female. Once again, there is a great deal of variation in the length of time that mammals may nurse. For example, young Dall's sheep begin to feed on grasses and are at least partially weaned within one month of birth. Conversely, brown bear cubs nurse from the time of their birth (January to March) and are not fully weaned until October to December.

Fishes generally reach reproductive maturity based on their size rather than their age. For example, male walleyes typically are mature by the time they reach 38 centimeters, while females mature by 45 centimeters. In Kansas, a 38-centimeter male is likely to be 2 years old, while a 45-centimeter female is likely to be 3 years old. In northern Canada, where walleye growth rates are much slower, these fish may not mature until they are 6–8 years old or even older. Conversely, mammals and birds typically reach reproductive maturity based on age. Age at reproduction can still be influenced by factors such as size and nutritional status, but not nearly to the same extent as in fishes. Most common terns do not begin nesting until they reach 3 years of age; American robins, many species of warblers, and many species of ducks nest at 1 year of age; California condors first nest at 6–8 years of age. Female montane voles can breed at an age of only 21 days, while most female mountain goats do not breed until they are 2.5 years of age or older.

6.11 Feeding Behavior

There are several aspects to consider when describing the feeding behavior of organisms. What an organism eats is referred to as its **food habits**. Food habits vary from species to species and represent information that is important if an organism or its habitat is to be correctly managed. Food habits are covered in more detail in section 9.6.

Feeding behavior involves when, where, and how an organism obtains food, and it is also referred to as **feeding habits**. The "when" can refer either to time of year, because feeding habits can vary by season, or to time of day. Some animals are **diurnal** feeders; they consume food during daylight hours. The term *diurnal* refers to any activity that occurs during daylight hours. A central stoneroller slowly moving along the bottom of a stream during midday and scraping food organisms from rocks, a scissor-tailed flycatcher capturing flying insects on a warm summer afternoon, and a Franklin's ground squirrel foraging for seeds and insects during the day are all practicing diurnal feeding. Some organisms feed at night and are termed nocturnal feeders. Channel catfish, barn owls, and raccoons are all primarily nocturnal feeders. Other animals are termed **crepuscular** feeders; examples include walleyes and common nighthawks. The crepuscular period is the twilight hours of dawn and dusk.

The "where" of feeding strategies is also an aspect of feeding behavior. Some fishes feed in shallow water, while others feed in deep water. Some fishes are surface feeders, some feed in midwater areas, and yet others are **benthic** (bottom) feeders. Much information about the feeding behavior of a fish can be ascertained by looking at the size and location of its mouth (Figure 6.13).

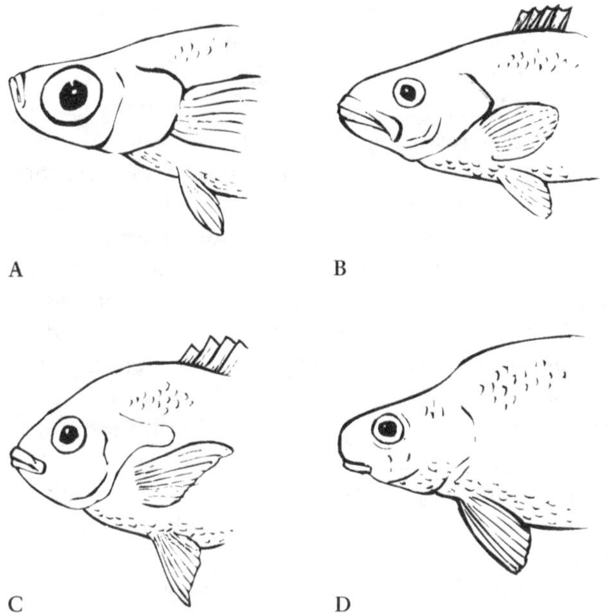

FIGURE 6.13 Mouth location in fish species is related to feeding behavior. The starhead topminnow *(A)* has a supraterminal (or superior) mouth for capturing organisms located above the fish. The smallmouth bass *(B)* has a terminal mouth that is large enough to capture large prey, such as fishes and crayfishes. The redbreast sunfish *(C)* also has a terminal mouth, but it is small and well suited to feeding on insects. Both of these species with terminal mouths primarily feed on items found in front of them. The river carpsucker *(D)* has a subterminal (or inferior) mouth that is suited to its benthic-feeding behavior.

Birds also vary greatly with regard to where they feed. Some feed on the ground, others dive underwater to obtain food, some capture their food in midair, and some creep up or down the trunks of trees while feeding. The variety of feeding locations is almost endless. The same can be said for mammals. Moose feed on aquatic plants in marshes, lakes, and rivers; American mink feed on aquatic or terrestrial organisms along the periphery of water bodies; and some bat species feed in midair on flying insects.

The "how" aspect of feeding behavior is also variable. Some animals actively pursue their prey, while others lie in wait for prey to come to them. Striped bass pursue their prey, often feeding in schools. Flathead catfish are also predators but typically lay motionless, waiting for prey to come close enough to be captured. Some predators swallow their food whole, while others bite or tear off chunks of food. Most predatory fishes and many owls ingest prey in one piece (usually headfirst); many hawks and most mammalian predators do not, depending on the size of the prey organisms.

Most organisms need to be as efficient as possible in their food intake. The **optimal foraging theory** is applicable to the feeding behavior of both aquatic and terrestrial animals. This theory is based on the premise that natural selection shapes the behavior of an animal in such a way that it does not expend more energy than necessary to obtain food. The most efficient foraging strategy may be dictated by prey type, location, size, or nutrient content. For example, as a northern pike grows larger, it will tend to select increasingly larger prey items. Prey availability and vulnerability may also be important. A large predator may feed on abundant small prey if they are easier to catch than a single large prey item.

The feeding strategies of birds and mammals can also be discussed in terms of the optimal foraging theory. American robins may select earthworms over smaller invertebrates, and downy woodpeckers may concentrate on insect larvae of a particular size to maximize energy intake and minimize energy expenditure. Moose need large amounts of sodium in their diets, and aquatic plants provide good sources of sodium. However, aquatic plants are available only during the summer, and they supply less energy than do terrestrial plants. Thus, moose have to "determine," probably through natural selection processes, how much of their summer diet will be composed of sodium-rich aquatic plants and how much will be composed of energy-rich terrestrial plants (Stephens and Krebs 1986).

Subtle differences in feeding behavior between closely related species can relate back to potential intraspecific competition (section 2.7). For example, Swainson's and red-tailed hawks can overlap in geographic nesting locations. In a study in Idaho (Hansen and Flake 1995), both hawk species preyed upon black-tailed jackrabbits. However, Swainson's hawks preyed more commonly on birds than did the red-tailed hawks. The red-tailed hawks, in turn, preyed more upon ground squirrels and reptiles than did the Swainson's hawks.

6.12 Sociobiology

Sociobiology is the study of the biological basis of social behavior and the organization of societies. Sociobiology came to the forefront of behavioral science with the publication of a book entitled *Sociobiology: The New Synthesis* (Wilson 1975). Sociobiologists specifically add the genetic component to the assessment of behavior; most behaviors have a genetic basis and are therefore subject to natural selection. The genes underlying a behavior that enhances the likelihood of survival will be passed on to future generations, but the genes underlying an unsuccessful behavior will be lost from a population because an unsuccessful individual is less likely to reproduce.

Grier (1984) suggested that the major taxonomic groups of organisms can be assessed for the incidence and extent of sociality using seven criteria. The use of these criteria allows a biologist to more thoroughly understand the group behavior of a particular species. The first criterion is related to the number of organisms of the same species in a group: the larger the number, the more social the organism. The second criterion is the length of time or part of the life cycle during which the group remains together. Sibling smallmouth bass remain in a school for no more than the first few weeks of their lives. Conversely, snow geese often remain in family groups from hatching in the spring through the fall migration, wintering, and spring migration. The third criterion relates to how much time or energy is spent on social behavior. Some organisms may actually interact relatively little in a group, while others have substantial and complex interactions. Consider the highly advanced social behavior of the common chimpanzee, in which much time is spent on social interactions such as grooming. The fourth criterion concerns whether reciprocal communication is necessary to keep the group together. Honeybees convey messages, but typically without reciprocation. The communication is one way; honeybee scouts use a dance to tell other bees the direction and distance to a food source, but there is no return message from the honeybees at the hive. More highly evolved social species, such as bighorn sheep, have much more reciprocal communication. When a ram displays an agonistic behavior to another ram during the breeding season, the second ram will either communicate its willingness to fight or display a submissive behavior to avoid a fight. The fifth criterion relates to the existence of social structures in which different individuals have different roles. For example, various species of ants have clearly developed such roles. Some individuals may be workers, others may be defenders, and yet others may be responsible for reproduction. The sixth criterion relates to the existence of an overlap of generations in which families or parts of families remain together. Wolves have such a social structure; even "relatives" such as "uncles" may remain with the pack.

Finally, the criterion least commonly met is the highest level of social behavior: **altruism**. Altruistic behavior is aid-giving behavior that has a cost to the aid-giving individual. This final aspect of social behavior has been a point of contention among sociobiologists. The most extreme example of altruism occurs when an individual sacrifices its life for another. From a genetic or evolutionary perspective, such an extreme level of altruism should result in the loss of the genetic material that allowed such a sacrifice, because that individual would not live to reproduce and replace itself. In some instances, however, **kin selection** has been proposed as a justification for altruistic behavior. The more closely related two individuals are, the more genetic material they share. If individual A lives in a family group and sacrifices its life for the remaining family members, this altruistic act will increase the chances that some of the genetic material individual A shares with its relatives will survive in those relatives and be passed to future generations. The closer the kinship, the greater the tendency for animals to cooperate. A common example of kin selection might involve an "uncle" in a wolf pack who is not involved in breeding but who helps to feed and protect the pups in his family group. Because he is related to at least one of the parents of the pups, his altruistic behavior increases the chances that some of his genes will be passed to future generations. In contrast, there is evidence (Komdeur 1992) that birds typically will act in their own best interest. When habitat is limited it may be in their best interest to stay at home and help the family group raise closely related individuals, but when open habitat is available they will set up territories, attract mates, and nest.

The concept of sociobiology has led to differences of opinion among behavioral scientists. At the simplest level, some biologists believe that it is just a new term applied to an established science. Some of the controversy concerning sociobiology has arisen as sociobiologists have applied their science to human behavior. Often, such application conflicts with religious beliefs; similar conflicts arise with respect to other scientific concepts, such as evolution. In addition, some individuals have argued that the ideas proposed by Wilson concerning genetic determinism in humans could be used to justify racism, sexism, imperialism, and war (Grier 1984), although they were not intended to imply such justifications.

SUMMARY

The behavior of an individual organism is largely dependent upon its senses. The senses of sight, smell, taste, hearing, balance, and touch are commonly understood by humans. Less well understood are the senses that relate to magnetic fields, electrical fields, and solar and lunar influences.

An organism may display innate or learned behaviors. Simple behaviors in simple organisms are almost certainly innate. Many behaviors are actually a blend of innate and learned behaviors.

Precocial offspring are capable of moving and feeding soon after birth or hatching. Altricial offspring are born or hatched in a much more helpless state and require more parental care at this stage of life than do precocial individuals.

Individual organisms commonly display rhythmic behaviors. Behaviors that vary throughout the day are termed diel or circadian. Circannual behaviors are those that change seasonally or throughout the year. Photoperiod, the proportion of light and dark in a 24-hour period, cues many of the rhythmic behaviors.

Social interactions can be intraspecific or interspecific. Intraspecific interactions are often regulated by dominance hierarchies. If individuals in the social group know their place in a hierarchy, conflicts can be minimized.

Most animals communicate by sight, smell, sound, or taste. Vocalizations can be effective means of communication even in habitats with poor visibility. Visual displays can be effective over long distances in open habitat. Chemical communication has an advantage in that its message can be "sent" for an extended time period.

Migrations are two-way movements with characteristic regularity. Schools, flocks, and herds are social groups, and these groups show coordinated movements. Social assemblages such as these should not be confused with aggregations, which are generally gatherings of a less formal and shorter-term nature.

Mating behaviors are obviously social behaviors. Common mating systems include promiscuous systems, in which no pair bonds are formed; monogamous systems, in which one male and one female form a pair bond; and polygamous systems, in which one individual of one sex mates with multiple members of the opposite sex.

Reproductive strategies are highly variable and fall along an *r–K* continuum. *K*-selected species tend to produce few offspring and provide substantial parental care. At the other end of the continuum, *r*-selected species tend to produce many offspring but provide little or no parental care.

Feeding habits involve when, where, and how organisms obtain food. They should not be confused with food habits, which are defined as what an organism eats. Diurnal feeders feed during the day, nocturnal feeders feed at night, and crepuscular feeders are those that feed during the twilight periods. The optimal foraging theory indicates that an organism will obtain food in the most efficient manner possible.

Sociobiology is the study of the biological basis of social behavior and the organization of societies. Sociobiologists specifically add the genetic component to the assessment of behavior. Biologists can use several criteria to assess the degree of sociality in a species.

PRACTICE QUESTIONS

1. Make three lists of organisms with which you are familiar that commonly use (1) vocal communication, (2) visual communication, and (3) chemical communication. Be sure to include a variety of fishes, birds, and mammals on each list.

2. Is competition for limited breeding sites among male yellow-headed blackbirds an example of intraspecific or interspecific competition? Is competition between red foxes and coyotes preying upon meadow voles an example of intraspecific or interspecific competition?

3. Are learned behaviors more likely to occur in an insect such as a mayfly or in a common chimpanzee? Why?

4. What is the relationship between parental care, as suggested by the *r–K* continuum, and the state of development in early life (altricial or precocial) for a species?

5. Choose an animal species with which you are familiar. How do both the diel and circannual behaviors of an individual of that species vary?

6. What is the difference between a school of fish and an aggregation of fish? Give an example of each.

7. How can a dominance hierarchy result in less agonistic behavior? What are the values of reducing agonistic behavior?

8. Why do northern pintails have an annual migration that places them in northern latitudes on the North American continent during summer and in southern latitudes during winter?

9. Why is it to the advantage of the various species of Pacific salmon to migrate from the ocean to fresh water for reproduction?

10. Parental male bluegills have a reproductive strategy that results from maturing later in life after attaining a large body size. Their nesting behavior as well as care of eggs and newly hatched fry come with a substantial energetic cost. Cuckold males attempt to "steal" fertilizations and provide no parental care. What is the likelihood that parental males might be able to recognize their own offspring and perhaps adjust their parental care based on relatedness? After making your best guess, see the Neff and Sherman (2005) citation listed below.

11. How is sociobiology the study of individual behavior? How is it the study of group behavior?

SELECTED READINGS

Alcock, J. 2001. *The triumph of sociobiology*. Oxford University Press, New York.

Alcock, J. 2005. *Animal behavior: An evolutionary approach*. 8th ed. Sinauer Associates Incorporated, Sunderland, Mass.

Beard, T. D., Jr., M. T. Drake, J. E. Breck, and N. A. Nate. 1997. Effects of simulated angling regulations on stunting in bluegill populations. *North American Journal of Fisheries Management* 17:525–532.

Drake, M. T., J. E. Claussen, D. P. Philipp, and D. L. Pereira. 1997. A comparison of bluegill reproductive strategies and growth among lakes with different fishing intensities. *North American Journal of Fisheries Management* 17:496–507.

Festa-Bianchet, M., and M. Apollonio, eds. 2003. *Animal behavior and wildlife conservation*. Island Press, Washington, D.C.

Groot, C., and L. Margolis, eds. 1991. *Pacific salmon life histories*. University of British Columbia Press, Vancouver.

Gross, M. R., and E. L. Charnov. 1980. Alternative male life histories in bluegill sunfish. *Proceedings of the National Academy of Sciences* 77:6937–6940.

MacArthur, R. H., and E. O. Wilson. 1967. *The theory of island biogeography*. Princeton University Press, Princeton, N.J.

Martin, K. 1998. The role of animal behavior studies in wildlife sciences. *Wildlife Society Bulletin* 26:911–920.

Morse, D. H. 1970. Ecological aspects of some mixed-species foraging flocks of birds. *Ecological Monographs* 40:119–168.

Nevitt, G. 1999. Foraging by seabirds on an olfactory landscape. *American Scientist* 87:46–53.

Pianka, E. R. 1970. On *r*- and *K*-selection. *American Naturalist* 104:592–597.

Pitcher, T. J., ed. 1993. *Behaviour of teleost fishes.* 2nd ed. Chapman and Hall, London.

Quinn, T. P. 2005. *The behavior and ecology of Pacific salmon and trout.* University of Washington Press, Seattle.

Smith, J. M. 1964. Group selection and kin selection. *Nature* 201:1145–1147.

Wilson, E. O. 1975. *Sociobiology: The new synthesis.* Harvard University Press, Cambridge, Mass.

Wilson, E. O. 1992. *The diversity of life.* Harvard University Press, Cambridge, Mass.

Witherington, B. E. 1997. The problem of photo-pollution for sea turtles and other nocturnal animals. Pages 303–328 *in* J. R. Clemmons and R. Buchholz, eds. *Behavioral approaches to conservation in the wild.* Cambridge University Press, Cambridge, U.K.

LITERATURE CITED

Adams, C. A., R. T. Bowyer, J. E. Rowell, W. E. Hauer, and J. A. Jenks. 2001. Scent marking by male caribou: An experimental test of rubbing behavior. *Rangifer* 21:21–27.

Bleckmann, H. 1993. Role of the lateral line in fish behaviour. Pages 201–246 *in* T. J. Pitcher, ed. *Behaviour of teleost fishes.* 2nd ed. Chapman and Hall, London.

Bond, C. E. 1979. *Biology of fishes.* W. B. Saunders, Philadelphia.

Bowyer, R. T. 1986. Antler characteristics as related to social status of male southern mule deer. *Southwestern Naturalist* 31:289–298.

Bowyer, R. T., V. Van Ballenberghe, and K. R. Rock. 1994. Scent marking by Alaska moose: Characteristics and spatial distribution of rubbed trees. *Canadian Journal of Zoology* 72:2186–2192.

Davis, W. H., and H. B. Hitchcock. 1965. Biology and migration of the bat, *Myotis lucifugus,* in New England. *Journal of Mammalogy* 46:296–313.

Dittman, A. H., and T. P. Quinn. 1996. Homing in Pacific salmon: Mechanisms and ecological basis. *The Journal of Experimental Biology* 199:83–91.

Erkert, H. G. 1982. Ecological aspects of bat activity rhythms. Pages 201–242 *in* T. H. Kunz, ed. *Ecology of bats.* Plenum Press, New York.

Gerald, J. W. 1971. Sound production during courtship in six species of sunfish (Centrarchidae). *Evolution* 25:75–87.

Grier, J. W. 1984. *Biology of animal behavior.* Times Mirror/Mosby College Publishing, St. Louis.

Guy, C. S., D. W. Willis, and J. J. Jackson. 1994. Biotelemetry of white crappies in a South Dakota glacial lake. *Transactions of the American Fisheries Society* 123:63–70.

Hansen, R. W., and L. D. Flake. 1995. Ecological relationships between nesting Swainson's and red-tailed hawks in southeastern Idaho. *The Journal of Raptor Research* 29:166–171.

Horne, A. J., and C. R. Goldman. 1994. *Limnology.* 2nd ed. McGraw-Hill, New York.

Howe, F. P., and L. D. Flake. 1989. Mourning dove use of man-made ponds in a cold-desert ecosystem in Idaho. *Great Basin Naturalist* 49:627–631.

Komdeur, J. 1992. Importance of habitat saturation and territory quality for evolution of cooperative breeding in the Seychelles warbler. *Nature* 358:493–495.

Lagler, K. F., J. E. Bardach, R. R. Miller, and D. R. M. Passino. 1977. *Ichthyology.* 2nd edition. John Wiley & Sons, New York.

McCullough, D. R. 1969. *The tule elk: Its history, behavior, and ecology.* University of California Publications in Zoology 88, University of California Press. Berkeley and Los Angeles.

Neff, B. D., and P. W. Sherman. 2005. In vitro fertilization reveals offspring recognition via self-referencing in a fish with paternal care and cuckoldry. *Ethology* 111:425–438.

Neumann, R. M. 1994. "Growth, distribution, and movement of northern pike in a South Dakota natural lake, with sampling considerations." PhD dissertation, South Dakota State University, Brookings.

Stephens, D. W., and J. R. Krebs. 1986. *Foraging theory.* Princeton University Press, Princeton, N.J.

Todd, J. H. 1971. The chemical languages of fishes. *Scientific American* 224:98–108.

Welty, J. C., and L. Baptista. 1988. *The life of birds.* 4th ed. Saunders College Publishing, Fort Worth, Tex.

Wilson, E. O. 1975. *Sociobiology: The new synthesis.* Harvard University Press, Cambridge, Mass.

Young, J. R. 2005. Animal behavior: Its role in wildlife biology. Pages 616–631 *in* C. E. Braun, ed. *Techniques for wildlife investigations and management.* 6th ed. The Wildlife Society, Bethesda, Md.

CHAPTER 7

Sampling the Biota

To study wildlife and fisheries resources, biologists often need to sample organisms. A variety of devices and methods are employed for this purpose. Sampling devices vary in overall shape and size, in the mechanics used to attract or retain the animal, in mesh size if netting or screening is involved, in construction material, and in many other characteristics. The permutations are endless and depend only on the needs of the biologist and the imagination and ingenuity that he or she uses. Fisheries biologists refer to capture devices as **gear**; wildlife biologists have no such all-encompassing term for the capture devices they use.

This chapter contains more information concerning capture of aquatic organisms than terrestrial organisms because capture devices used in fisheries and aquatic work are more numerous and are generally used more often than those employed in the wildlife field. There are a number of possible reasons for this disparity. Terrestrial species can be seen or heard by humans more readily than aquatic animals and therefore can frequently be monitored without being captured. Because aquatic animals are seen or heard less often, more of the information on those species is based on capture. Also, because of indeterminate growth, a greater variety of sizes of a fish species must usually be captured in order to sample the species. In addition, aquatic animals live in a more three-dimensional world than terrestrial animals, and this difference affects capture methodology.

This chapter describes the most common methods employed to sample fishes, birds, and mammals, as well as some methods used to sample smaller aquatic and terrestrial organisms, such as plankton and insects. There are many less commonly employed methods and devices that are not included.

7.1 Sampling: Purposes and Problems

Seldom can an entire population be counted or captured for some particular purpose; samples usually must be taken. A **sample** represents a subset or portion of the total number of organisms in a population. In a broader sense, a sample is a subset or portion of the whole. Samples can be obtained by capturing organisms. For example, a sample of a fish population can be obtained by using a specific kind of net or trap. Samples can also be obtained by utilizing noncapture methods. For example, a sample of birds present in an area can be obtained by listening to and identifying bird songs at sampling sites.

There are a multitude of reasons why we may need to sample organisms. We may need to sample organisms to estimate population number or to index population abundance. We may need to sample organisms to determine information such as food eaten, food availability, survival rates, parasites present, health or condition, habitat use, movements and home range, or for a variety of other purposes. We may need to capture an animal in order to transport it and release it in a new area. We may need to capture it so that a radio or sonar device can be attached or implanted, or so that the animal can be tagged or marked in some other fashion.

Some sampling methods are lethal. Other sampling methods are nonlethal, making it possible to sample animals with little or no damage to them. The fate of animals sampled by nonlethal means is variable. Some may be captured and immediately returned to the wild after data such as length or weight are obtained or after being marked or tagged. Others may be live-captured and then transported to a laboratory or other holding facility for further study. Some of these animals may be returned to the wild at the conclusion of the study. Some animals may be captured, transported to a laboratory or other holding facility, and killed for the purpose of the study. Study objectives determine how the animal is sampled as well as its subsequent fate.

Even those sampling methods that are nonlethal often result in **stress** to an organism. Stress results from biotic and abiotic forces that place pressure on the internal stabilizing mechanisms of an organism beyond its ability to maintain homeostasis. In other words, stress is an internal response by an organism to confront external stressing forces. Because stress can affect such important processes as reproduction, growth, and resistance to disease, it is important that it be minimized. Physical injury can also result from nonlethal capture methods. For example, a capture device that damages the fins of a fish or causes the loss of most tail feathers of a bird affects that organism. Stress and physical injury should be kept to a minimum in sampling.

There are ways to reduce physical damage and stress resulting from sampling. For example, head coverings or blindfolds can be placed on large mammals or birds. Many species can be anesthetized upon capture. Damage can also be reduced by avoiding the use of a sampling device in an area containing large numbers of nontarget species that are susceptible to that particular device. Stress and physical damage problems can be encountered in the use of many of the sampling methods and devices discussed in this chapter.

Many of the devices and methods biologists use to sample animals are illegal, except with special sampling permits. State, provincial, territorial, and federal governments all may be involved in regulating who can sample; how, when, and where sampling can be done; and what can be sampled. Biologists must be aware of these regulations.

The device or method used greatly influences what organisms are sampled. All sampling devices and methods are in some way selective; that is, no matter what the device or method, some organisms are going to be sampled more or less effectively than other organisms. In some situations this selectivity is positive; we can choose methods or devices that are either effective or ineffective for particular organisms. For example, a particular type of net with a specific mesh size may be chosen because it will be effective in sampling a particular species and will also be effective in capturing only a particular size of that species. Conversely, a particular kind of device may be chosen because it will be ineffective at sampling an organism that we wish to avoid capturing. Selectivity can also be negative in that **bias** may result. Bias is defined as the distance of an estimated value or parameter from the actual value or true parameter. The farther the estimate is from the actual value, the more biased the information. Bias, especially bias of which we are unaware, can greatly affect the utility of information. For example, while a certain type of net may be quite effective for sampling dollar sunfish, it may be much less effective for sampling largemouth bass. Using a sample obtained with that net to estimate relative species composition of both fishes would result in biased information. The dollar sunfish would appear more abundant, which would not necessarily be true.

Susceptibility to capture, and therefore bias, may depend on more than just species; it can also be affected by size, gender, behavior, anatomical characteristics, season, or a variety of other variables associated with organisms. For example, adult orangebelly darters may be readily sampled with one type of gear, while young orangebelly darters may require a totally different type of gear for effective sampling. There are a wide array of devices used specifically to sample fish eggs and young; these gears are usually ineffective for capturing adult fishes. A device designed to capture waterfowl during flightless molting periods would be totally different with regard to effectiveness if it were employed when the birds can fly.

There are multiple reasons why sampling devices or methods can be selective. A net with openings, or meshes, of a size that allows some members of the target population to pass through will result in a selective sample. Placing a device where it will capture one gender more often than the other can also give a selective sample. Placing a device in an area only during the day when most of the target organisms are there only at night is another example. Using a capture device that remains stationary when the target organism is not actively moving will be inefficient. In any wildlife or fisheries work that involves sampling, the selectivity or effectiveness of the specific method or device used, in relation to the specific animal to be sampled, must be considered.

7.2 Passive Entanglement Devices

Passive capture devices are put into place and essentially remain stationary; movement or behavior of the target organism results in its capture. Passive capture devices can involve either passive entanglement or passive entrapment, which we will discuss in the next section. **Passive entanglement devices** are stationary devices that snare and tangle the animals encountering the device. Examples of such devices are **gill nets** and **trammel nets** (Figure 7.1A) used to capture fishes, **mist nets** (Figure 7.1B) used to capture birds, and **drive nets** used to capture some mammal and bird species. These nets are usually nonlethal to the target organisms. Passive entanglement nets for fishes are suspended in water. To make these nets most effective, a **float line** (a buoyant line attached to the top of the net) and a **lead line** (a weighted line attached to the bottom of the net) are used to keep the net from collapsing while in water. Maintaining these nets in an open position ensures that the greatest possible amount of net area is available for fish capture. Mist nets and drive nets are set on land and are held open by tying them to trees, bushes, or moveable poles.

Moving organisms encounter these nets and become entangled in them. Gill, mist, and drive nets have a single panel of netting. When an animal encounters the net, a part of its body may penetrate the mesh, but the entire animal cannot pass through the net. Thus, gill coverings, fins, spines, feathers, wings,

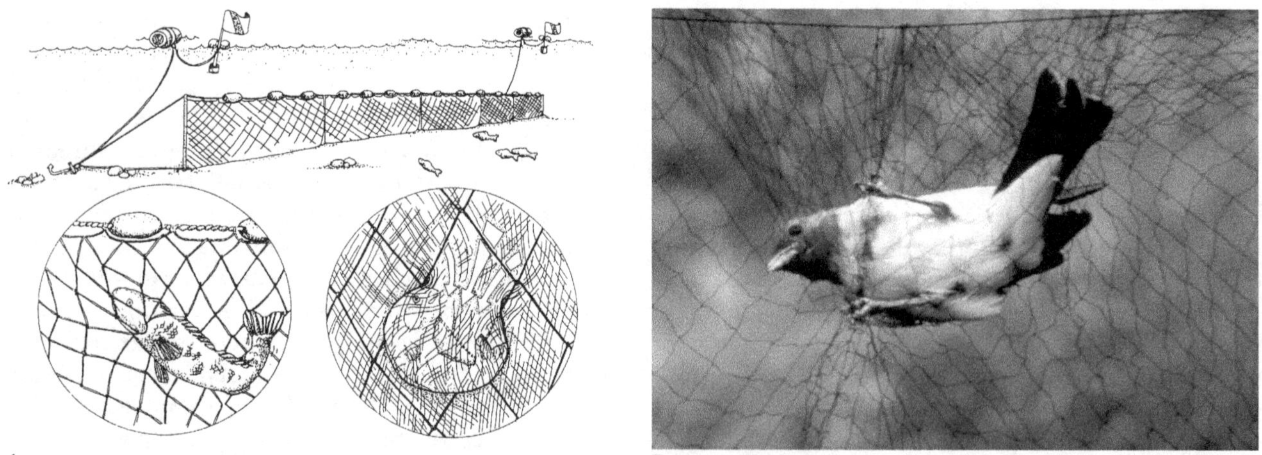

A B

FIGURE 7.1 Passive entanglement devices remain stationary, and moving organisms become entangled in them: *(A)* gill net (left), trammel net (right); *(B)* mist net with entangled western tanager. *[Part A drawings courtesy of the American Fisheries Society; part B image courtesy of K. Jensen.]*

scales, legs, feet, or other body extensions of the animal may become entangled in the net.

In fisheries work, gill nets are sometimes made with numerous sections of different-sized meshes. These nets are called **experimental gill nets** and are used to capture a wider length range of target organisms than is possible with nets of only one mesh size. In addition to net size and mesh size, elasticity, strength, flexibility of twine, net visibility, and time and manner fished can affect what kinds, what sizes, and how many fishes are captured in gill nets (Hubert 1996).

Trammel nets function on a slightly different principle. They usually consist of three panels of netting, with the inner panel having a smaller mesh size than the outer two; the middle panel also has greater depth and hangs loosely between the two outer panels. Fishes encountering the trammel net pass through the first outer panel, are entangled by the inner panel, and usually pull some of this inner panel of mesh through the second outer panel. The organism is thus captured in a "pouch" or "bag" of mesh.

Caution must be exercised in the use of passive entanglement devices if mortality is to be minimized. Organisms may be damaged or killed if they remain in the net too long or if they are mishandled after capture. Damage and mortality can be reduced or eliminated by frequent emptying of the nets and careful handling and disentanglement procedures. At times, however, studies require the mortality of the animal.

Passive entanglement devices can be constructed of any net and mesh size, depending on the target organism and purpose of the study. Some nets may be hundreds of meters long, while others may be only 10 meters or less. Similar variation is found in net height. Nets can be constructed of cotton or linen (seldom used anymore), multifilament nylon, monofilament nylon, or any other appropriate material. Netting made of synthetic material is much more resistant to rotting and tearing than the older cotton or linen nets. Mesh size is the primary factor that determines the size of the animals such nets capture. Smaller meshes are generally more effective at capturing smaller organisms, and larger meshes generally capture larger organisms. In most cases, especially for fishes, the closer the mesh size is to the diameter of the target organism, the more effective the net will be. Additionally, if the correct mesh size is selected, fewer nontarget animals will be captured.

Knowledge of target organism behavior also influences the effectiveness of passive entanglement devices. If low-flying birds are the target organisms, mist nets must be set near the ground. If deep-water fishes are the targets, gill or trammel nets need to be set in deep rather than shallow water. Target organism behavior will also dictate whether these nets are set during daylight or at night, in what season they are most effective, and how long they need to remain in place.

At times, the effectiveness of these passive entanglement devices can be enhanced by scaring or driving target organisms in the direction of the nets. Large mammals are sometimes directed into drive nets by humans on foot or horseback or in trucks, snowmobiles, aircraft, or other motorized vehicles. Some birds can be captured by walking through shrubs or other vegetation to drive them into a mist net. It is also possible to capture fishes in this manner. An excellent method of capturing adult paddlefish with a gill net is to place the net near the sand bars where these fish congregate, then scare the paddlefish toward the net using a motorized boat.

Another type of passive entanglement device is used primarily by wildlife biologists to capture individual animals; fisheries workers have no comparable passive capture devices that are employed to take individual fish. A variety of **leghold traps, snap traps** (like the familiar mousetrap), and **snares** are in this passive capture category (Figure 7.2). These three capture devices are categorized as entanglement devices because they fit better in this category than in the category of passive entrapment devices, which capture animals by enclosing them in a specific space (see section 7.3). Leghold traps and snares can be used to capture animals without injury if properly used; snap traps kill the animal captured. Leghold and snap traps come in a variety of sizes, shapes, and construction materials. Padded-jaw leghold traps have been developed to minimize damage to captured animals. The target animal triggers its own capture by stepping upon or in some other way releasing the mechanism that traps it. Snares are usually constructed of steel wires or ropes and are set to capture animals by the leg; they can also be set to capture animals around the neck. When the leg or head of the animal enters the looped end of the snare, the movement of the animal decreases the size of the loop. When the loop is pulled tight, the animal is caught. A checking device can be placed on a snare to keep the loop from becoming too tight.

Oftentimes with passive capture devices it is necessary to determine how many devices need to be used, in what arrangement they must be placed, and how often the devices need to be checked. Such information is necessary if sampling is to be effective.

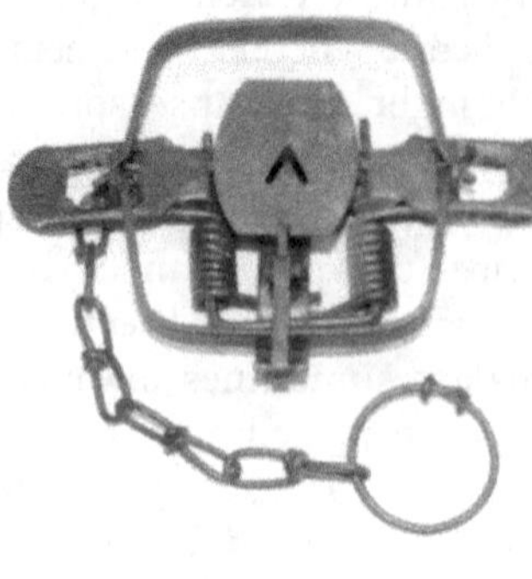

A

B

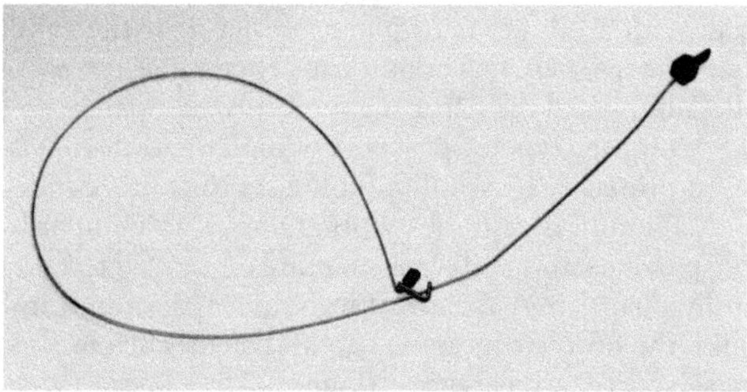

C

FIGURE 7.2 Some passive entanglement devices are used for capturing individual animals: *(A)* leghold trap; *(B)* snap trap; *(C)* snare.

It is also imperative that behavior and habits of target organisms be understood if these devices are to be successful. Setting them in particular places, and in particular ways, will often result in capture of the desired species to the exclusion of others. Individuals who are experienced with traps and snares not only know correct placement but also know how to use items such as sticks, stones, or logs to direct an animal to the capture device. Some people are sufficiently skilled to capture the target animal by a particular leg.

Often, the effectiveness of traps and snares can be enhanced by providing baits, lures, decoys, or scents to attract the target species. Some of these items, such as fermented eggs or striped skunk musk, can be very odoriferous to humans, while others, such as peanut butter, are relatively innocuous.

A quite different type of snare is composed of a series of monofilament loops set atop a post to snare birds by the feet. This type of entanglement device, when set above a cage containing a prey organism such as a mouse, is called a **bal-chatri trap**. It can be particularly effective for capturing raptors.

7.3 Passive Entrapment Devices

Passive entrapment devices capture organisms that move into an enclosed area by retaining them in that area. Like passive entanglement devices, they remain in a stationary position, thus the term *passive*. Examples of these devices are **hoop, fyke, modified-fyke,** and **trap nets**, which are used by fisheries biologists (Figure 7.3A, B, C, D); **funnel traps**, which are used to capture organisms such as molting waterfowl, wild turkeys, mourning doves, or gray partridge (Figure 7.3E); **box traps** (sometimes called **live traps**), which are used to capture a variety of birds and mammals (Figure 7.3F); **Clover traps**, which are generally used to capture deer (Figure 7.3G); **pit traps**, which, depending upon their size, can be used to capture everything from snakes to tigers (Figure 7.3H); and **corral traps**, which are used on a variety of organisms (Figure 7.3I). All of these passive entrapment devices are intended to be nonlethal to captured animals.

The effectiveness of such passive entrapment devices often depends on the behavior of the organism.

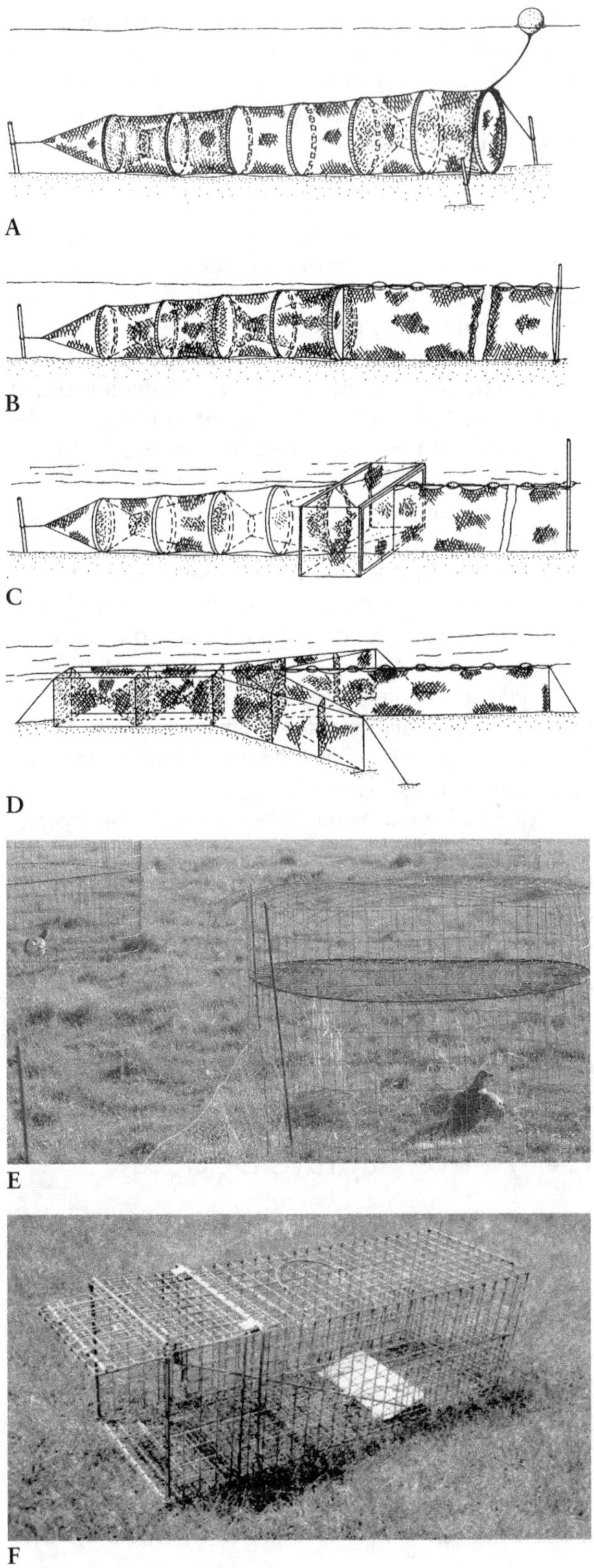

FIGURE 7.3 Passive entrapment devices are used to capture a wide variety of aquatic and terrestrial animals: *(A)* hoop net; *(B)* fyke net; *(C)* modified-fyke net; *(D)* trap net; *(E)* funnel trap (with sharp-tailed grouse); *(F)* box trap; *(G)* Clover trap; *(H)* pit trap; *(I)* corral trap (with molting lesser snow geese). *[Part A, B, C, and D drawings courtesy of the American Fisheries Society; part E image courtesy of K. Jensen; part G image courtesy of R. Sweitzer; part H image courtesy of P. Johnson.]*

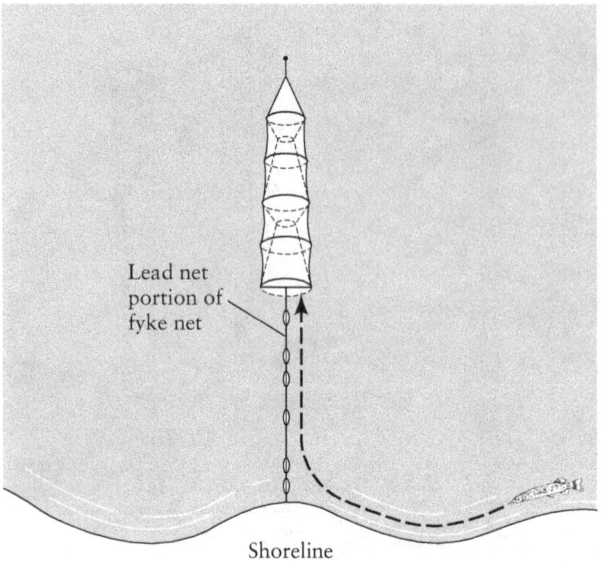

FIGURE 7.4 This arrangement of a fyke net and its lead net placed near the shoreline is especially effective for capturing northern pike during their spring reproductive movements.

For example, fyke netting northern pike in the spring is more successful if the net is placed near the shoreline of a water body, where the **lead net** portion of the fyke net will direct the fish into the capture part of the device (Figure 7.4). In spring, northern pike move along the shoreline seeking inflowing streams or flooded terrestrial vegetation for spawning. With correct fyke net placement, fish are more readily captured. Similarly, a corral trap set for elk works best if placed in an area where natural structures such as rocks or water bodies narrow the escape route; lead nets can also be placed to direct these animals into the trap.

In some situations baits, lures, scents, or decoys can be used to attract target organisms into a passive entrapment device. Clover traps baited with apples can be an effective means of capturing white-tailed deer. Hanging a bag of stink bait or cheese by-products in a hoop net set for channel catfish can greatly increase net effectiveness. A live female duck that serves as a decoy can entice male ducks into a funnel trap during the spring breeding season. A mirror in a box trap can enhance the capture of drumming ruffed grouse. Other attractants that can be used are calls, including chick and prey distress calls, stuffed decoys, and live prey.

It may be necessary for humans to drive or direct target animals to the trap in some situations. This strategy is more effective for some animals than for others. **Big game** (large animals hunted for recreation, such as elk and bighorn sheep), molting geese, and feral horses are often directed to traps in this manner.

Once in the trap, animals must be retained by the device. There are numerous methods of accomplishing their retention. A net or other blocking device can be put in place, either manually or remotely, to block the exit. In some traps, the opening to the device is designed so that animals can enter easily but find it difficult to exit. This type of arrangement is demonstrated by the funnel or baffle arrangement of netting in the hoop, fyke, and trap nets used by fisheries biologists and in the various funnel traps used by wildlife biologists. An organism easily follows the large funnel to the small opening into the trap but rarely finds the opening again from the inside of the trap. Another technique to keep the animal in the trap once it has entered is to have the captured animal trip a release that closes a door or panel on the trap, as in the various box trap designs. Pit traps retain captured animals because they are unable to climb out of a pit into which they have fallen. Pit traps are often constructed with smooth metal sides to increase capture retention.

All passive entrapment devices function by getting the organisms into an enclosed area from which they cannot escape. A fish trap designed to capture a marine fish species may have many lead nets and a containment area that covers many hectares, but it works on the same principle as a small minnow trap. Additionally, a pit trap dug for an animal the size of a tiger works on the same principle as a small pit trap designed for capturing long-nosed leopard lizards.

7.4 Active Capture Devices

Active capture devices require human or mechanical action to move the device; this human-aided movement results in the capture of the target organism. The active capture devices described in this section are generally nonlethal to target animals.

A common type of active capture gear used in fisheries work is the **seine** (Figure 7.5A). Seines have a float line and a lead line, and the ends, especially on smaller seines, are attached to **brails** (poles or rods), which allow workers to hold onto and pull the net. Larger seines are often pulled by tractors, other

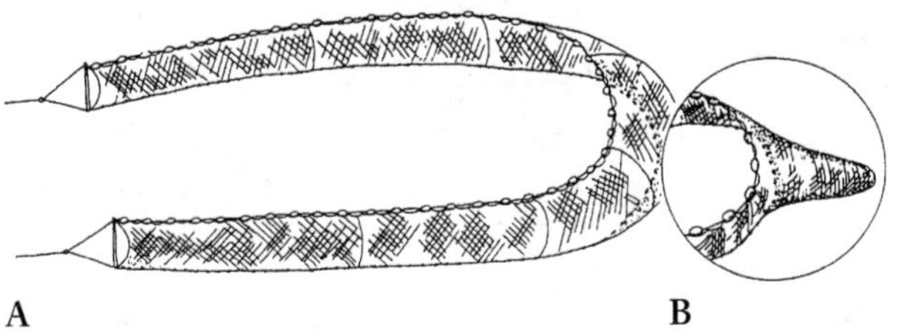

FIGURE 7.5 *(A)* The seine is an effective device for capturing fishes in shallow water areas. *(B)* Adding a bag to a seine can increase its sampling effectiveness. *[Drawings courtesy of the American Fisheries Society.]*

types of motorized vehicles, or winches. Seines come in a variety of shapes and sizes. Sometimes a bag, constructed of the same netting as the seine, is added to enhance its capture effectiveness; such seines are called **bag seines** (Figure 7.5B). Seines are primarily shallow water sampling devices and thus are often called **beach seines.** For these devices to be most successful, the lead line must remain on the bottom of the water body and the float line must not go below the water surface while the seine is pulled to the shore.

Purse seines are used when the net cannot be moved to shore for capture purposes (Figure 7.6). They are primarily used in deep open water, such as large lakes or the ocean, and are put in place with the use of boats. The purse seine is placed in the water in a circular arrangement. Along its weighted bottom is a purse line, which when drawn together reduces the size of the bottom opening; the top of the purse seine is attached to a float line. When the purse line is completely drawn together, any fishes within the seine are prevented from escaping out the bottom of the net. The enclosed area is then reduced in size by pulling in the seine, and the fishes are ready for removal.

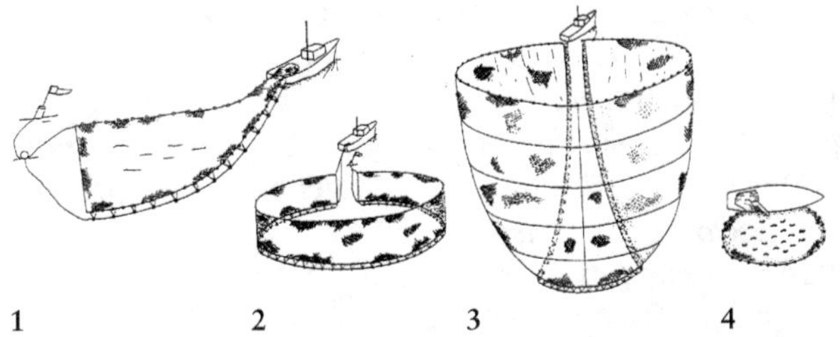

FIGURE 7.6 Purse seines are used in deeper water areas for capturing aquatic animals. Stages 1 through 4 indicate the order in which the purse seine is set and retrieved. *[Drawings courtesy of the American Fisheries Society.]*

The variety of trawls used in fisheries work constitute another type of active capture gear. Trawls are bag- or funnel-shaped nets that are pulled through the water, usually by means of one or more motorized boats. After the trawl has been pulled for some distance, it is brought to the boat and the captured animals are removed through the cod end of the net (the small end opposite the mouth). The mouth or front end of the trawl is held open by various means. **Beam trawls** have a rigid beam across the mouth end of the net (Figure 7.7A). **Otter trawls** have weighted boards (otter boards) that are angled so as to keep the trawl mouth open as it is pulled through the water (Figure 7.7B). Trawls vary in shape and size. Some are used to capture surface and midwater fishes, others are used to capture benthic fishes, and still others are used to capture aquatic animals other than fishes, such as shrimp. Knowledge of target animal behavior, especially location, movement patterns, and gear avoidance behavior, is very important for successful use of these gears.

Dredges are active capture devices used in fisheries work to capture a variety of benthic organisms (Figure 7.8). Some dredges are designed specifically to capture mollusks, others are used for crustacean capture, and others are used for a variety of benthic animals. These devices are designed to be dragged over the bottom and usually have blades, rakelike teeth, or other structures that enhance the capture of target animals.

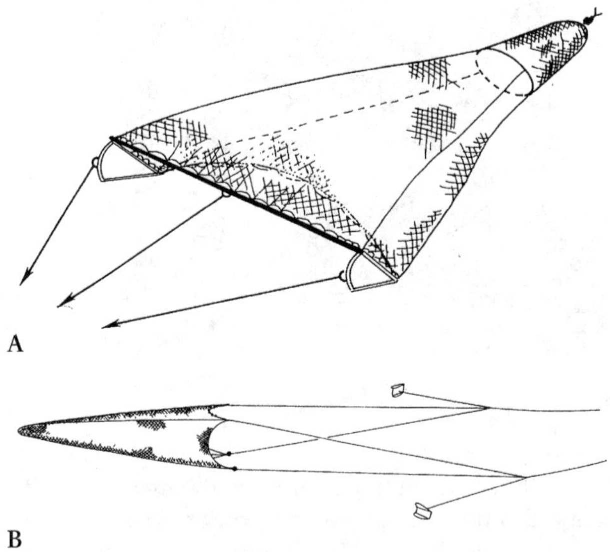

FIGURE 7.7 Trawls can be used to capture a variety of aquatic animals at various water depths: *(A)* beam trawl; *(B)* otter trawl. *[Drawings courtesy of the American Fisheries Society.]*

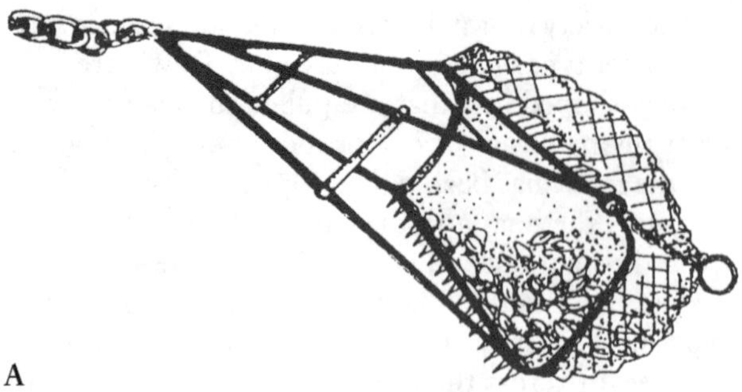

A

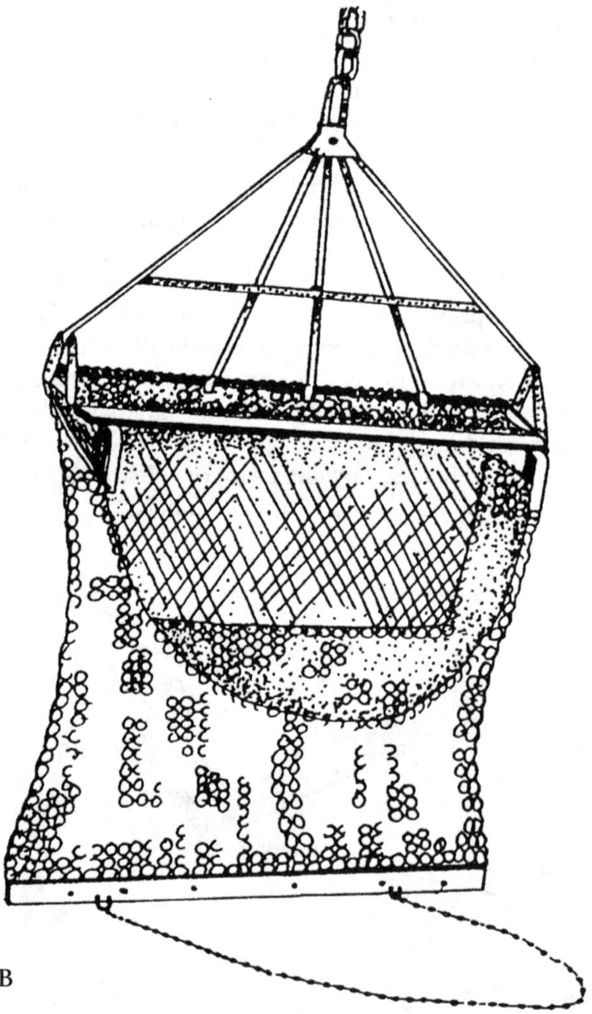

B

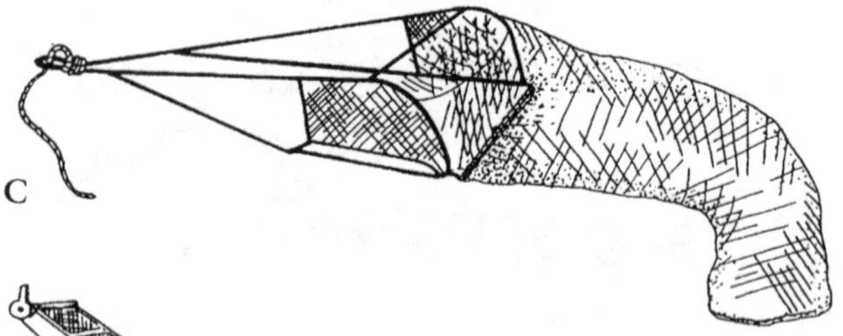

C

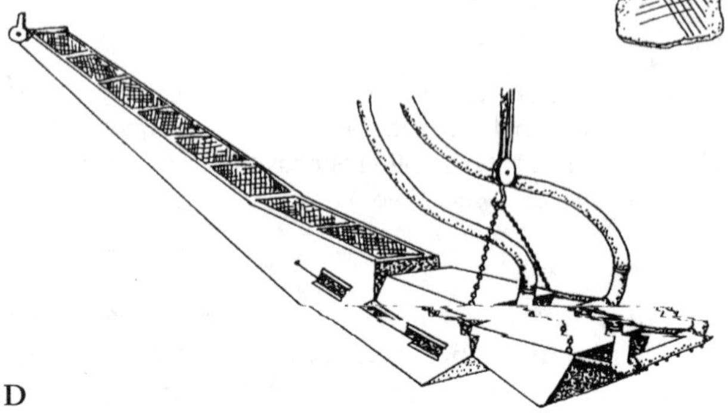

D

FIGURE 7.8 Dredges can be used to capture a variety of bottom-dwelling animals: *(A)* oyster dredge; *(B)* scallop dredge; *(C)* crab scrape; *(D)* hydraulic or jet dredge. *[Drawings courtesy of the American Fisheries Society.]*

Some active capture devices are also used in wildlife work. Examples are **cannon nets, rocket nets, drop nets,** and **net guns** (Figure 7.9). Cannon, rocket, and drop nets are used to capture a variety of bird and mammal species; their success is often enhanced by baits or decoys that attract target organisms to the proximity of the net. Cannon nets are lightweight nets that are carried over a group of animals by mortar-type projectiles. Rocket nets are propelled by small rocket-type projectiles. A drop net is suspended above the ground, and when target animals are under the net, it is dropped. Net guns fire projectiles that are attached to a net. Such guns are usually fired from the shoulder or are mounted on helicopters. Net guns have been used effectively on a variety of bird and mammal species.

In addition to these active capture devices, wildlife and fisheries personnel use a variety of handheld nets. These nets, which have variable handle lengths, net diameters, and mesh sizes, can be used from vehicles, boats, or on foot. Such nets are employed to capture a variety of animals, from dip nets used on fishes to lightweight, large-diameter handheld nets used to capture ground-roosting birds. Vehicle- or backpack-powered lights can be used if the captures take place at night. Loud music has even been used to distract some upland game birds during nighttime lighting and netting operations so that they may be more effectively captured.

Terms such as *passive*, *active*, *entanglement*, and *entrapment* have always been used in reference to fisheries gears. The terms have not generally been applied to capture devices used in the wildlife field. However, these terms do apply to the ways that wildlife sampling devices and methods work. Therefore, it seems appropriate to use these terms in relation to wildlife sampling as well.

A

B

C

FIGURE 7.9 Some active capture devices can be used to sample wildlife: *(A)* rocket net; *(B)* drop net; *(C)* net gun. *[Part A image courtesy of the National Wild Turkey Federation; part B image courtesy of K. Jensen; part C image courtesy of J. Ratti.]*

7.5 Anesthetics for Sampling Wildlife

Anesthetics are most commonly used by wildlife biologists for live capture of large organisms. Anesthetics are used in fisheries work also, but usually after capture. Wildlife anesthetics can be delivered in several ways. The most common method of delivery for larger mammals is a **dart gun** (Figure 7.10A). The darts are syringelike projectiles designed so that a predetermined amount of anesthetic is injected when the animal is struck (Figure 7.10B). The projectile gun is usually an air gun or a powder-charged gun, but a bow and arrow, blowgun, or even the end of a pole or stick can be used to deliver the dart or injection.

When using anesthetics to capture animals, care must be taken to ensure that target animals are not overdosed or in some other way harmed. Television and other media describing wildlife capture activities often depict the darting anesthetization of large animals as a rather simple and foolproof activity. It is seldom mentioned that injury or mortality is not uncommon with such a capture method. Anesthetization is not as simple as it sounds, because animal body size must usually be estimated to determine the proper drug dosage and because there is considerable variation in the response to drugs by individual animals, even within the same species. Even if proper drugs and dosage levels are used, injury and mortality can still occur.

A variety of drugs are used to immobilize wildlife species, with each having a particular effectiveness depending on the target organism in question. Most of the drugs used are legally controlled substances and can be utilized only by licensed users. Drugs can be general anesthetics, neuromuscular blockers, tranquilizers and sedatives, or antagonists (reverses the effects of other drugs). When using drugs it is important to be aware of how they can be administered, concentration levels needed, their side effects, drugs (if available) that can

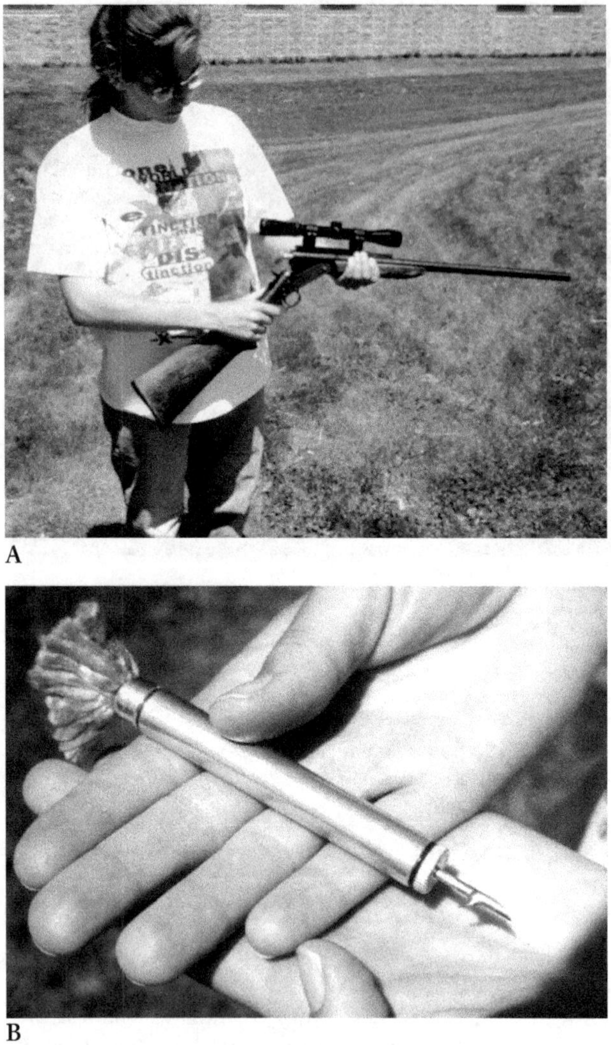

FIGURE 7.10 A dart gun *(A)* and dart *(B)* used to anesthetize animals. Care must be taken with this device to prevent injury to the target animal.

be used to reverse their action, and other aspects of their usage. It is also a good policy to have a qualified veterinarian as a member of any capture team using anesthetics—in some cases it is required.

Oral drugs can also be used to capture animals by placing the drug in food. Once the target animal eats the treated food and is anesthetized, the drugged animal is available for study. The drug, alpha-chloralose, has been used to capture a variety of birds (Schemnitz 2005). Additionally, some radio transmitters attached to animals (see section 9.14) include anesthetic devices that can be remotely triggered.

7.6 Toxicants for Sampling Fishes

Toxicants, or poisons, called **piscicides** (**ichthyocides**) because they affect fishes, are used in fisheries work; wildlife biologists have no comparable means of capture. Piscicides are toxicants that, when added to water, result in immobilization of fishes. This is primarily a lethal capture method.

Toxicant use is one of only two methods available for sampling all fishes present in a body of water. The other method available for the complete sampling of fishes is draining a water body.

A variety of piscicides have been used over the years. Most, because of their harmful or potentially harmful effects on the environment and on humans, are no longer legally usable. Currently, the most common legally approved piscicides are **rotenone** and **antimycin**. These are general-purpose toxicants used on a variety of fish species. Rotenone is a naturally occurring material found in some plants and is sold under a variety of brand names. It can be purchased in liquid or powder form and comes in various concentration levels. Rotenone blocks oxygen uptake in the gills of fishes. The effectiveness of rotenone is dependent upon fish size and species and upon water temperature, pH, oxygen concentration, amount of organic matter, and turbidity. It is used at parts-per-million concentration levels. Rotenone affects all gill-breathing animals in aquatic environments.

Antimycin, an antibiotic, also inhibits oxygen uptake of fishes, but at the tissue level. It is sold under the brand name Fintrol. The effects of antimycin are modified by fish size and species, water temperature and pH, sunlight, and fish metabolic activity. It is used at parts-per-billion concentration levels because it is more toxic to fishes than rotenone.

Both rotenone and antimycin can be used somewhat selectively for some fish species in the presence of other fishes, because not all species have the same susceptibility to the same concentration levels. Both piscicides can be inactivated by potassium permanganate; however, potassium permanganate is also toxic if concentration levels are too high.

While rotenone and antimycin are general-purpose piscicides, there are other piscicides that are used for more specific purposes. Two such toxicants, **TFM** (**3-trifluoromethyl-4-nitrophenol**) and **Bayluscide**, are used for sea lamprey control. **Lampricides** (kills lampreys) are used primarily on larval sea lampreys in Great Lakes tributary streams. Their primary purpose is to reduce sea lamprey populations in or-

FIGURE 7.11 Piscicides can be applied in numerous ways. Shown here is application with a handheld spray.

FIGURE 7.12 After a piscicide has been applied, the fishes must be collected. This is usually done with handheld dip nets.

der to reduce the negative effects of this species on other fishes, such as lake trout.

Searches continue for piscicides that are selective for target species and not harmful to nontarget species. Development of toxicants that would selectively remove undesirable fishes, such as the various species of snakeheads and Asian carps (all of which are exotic to North America), would be beneficial.

Toxicants can be applied in a variety of ways, ranging from boat-mounted devices to handheld sprayers (Figure 7.11). Boat-mounted devices are used in large water bodies; handheld sprayers can be used in very small, shallow water bodies and in difficult-to-reach portions of large water bodies. At times, even helicopters have been used to apply the toxicant. After the fishes are immobilized by the toxicant, they are quickly collected with handheld nets (Figure 7.12). Some of those that sink before collection will float to the surface in a day or two.

The use of toxicants has declined in recent years. Problems with toxicant effects on nontarget species, potential effects on human users, aversion to large quantities of dead fishes for aesthetic and ethical reasons, and cost have all played roles in reducing piscicide usage. However, under proper circumstances and when correctly used, toxicants are an effective method for sampling fishes, eliminating fish populations or communities, or manipulating biological systems.

Chemical use in fisheries (and wildlife) work is a constantly changing arena. Depending upon the chemical and its use, its control in the United States may come under the jurisdiction of the Environmental Protection Agency or the Food and Drug Administration. There are state and federal programs to register and re-register these chemicals for use. Regulations regarding which chemicals can be legally employed, who can apply them, and the animals on which they can be used change from year to year. Biologists must be aware of all current regulations with regard to these chemicals. The chemicals listed as being applicable in fisheries work in this section of the book were legal at the time of publication; they may not be legal at some point in the future. The same is true of chemicals utilized in wildlife work (see section 7.5).

7.7 Electrofishing

Use of electricity to capture animals is a technique primarily employed in fisheries work. Electrofishing devices take many forms. Electrofishing units employed in large water bodies are usually mounted on boats (Figure 7.13A). Backpack electrofishing units are usually employed in small water bodies, such as streams (Figure 7.13B). Electric seines are also used in some situations for fish capture. There have been substantial improvements in electrofishing technology in recent years. Most improvement efforts are directed at making these devices more effective while reducing potential physical damage to fishes and human operators.

Electrofishing devices function by disrupting the nervous systems of fishes. When correctly used, they can temporarily immobilize a fish without causing lasting damage. After the fish is immobilized by the electric current, it can be captured with a handheld

FIGURE 7.13 Electrofishing devices are employed as a nonlethal method of capture: *(A)* boat-mounted unit; *(B)* backpack unit. *[Part B image courtesy of T. Modde.]*

net. Electrofishing is a nonlethal capture method; however, if too much current is applied or if a fish is too close to an electrode (the terminal of the electrical source), ruptured blood vessels, broken vertebrae, or other damage can occur. Such damage may or may not result in fish mortality. The currents most commonly employed are alternating current (AC), direct current (DC), and pulsed DC. Fish respond to AC with forced swimming without orientation—that is, they may move toward or away from the electrode. In contrast, DC attracts fishes to the positive electrode; pulsed DC extends the duration of attraction to the positive electrode. With most newer units, the operator has control over voltage, amperage, and other electrical field characteristics. This control allows for the selection of electrical characteristics that are particularly effective on various fish species and sizes. Electricity is usually produced by either a gasoline-powered generator or batteries, but hand-operated generators have also been used.

Among the variables that influence the effectiveness of electrofishing devices are differences in water conductivity (its capacity to conduct electricity), fish size, fish behavior, water temperature, and the method of electrical delivery. In addition, fish species, either because of their physiological characteristics or because of the habitat in which they dwell, differ in their susceptibility to electrical currents. Electrofishing devices are primarily used for shallow water sampling.

Use of electrofishing gear continues to raise issues in regard to damage to target and nontarget organisms. Substantial research on injury and mortality rates has been conducted to develop the most appropriate electrofishing methods to minimize damage. Salmonids are especially sensitive to muscle damage and spinal injury. Most of the warmwater and coolwater fishes are much less susceptible to injury when appropriate electrical fields are used. The issue of injury and death to sampled organisms remains of importance to biologists, who have a responsibility to minimize those rates. However, electrofishing remains a very useful sampling tool for many species and in specific habitats.

7.8 Sampling Small Terrestrial and Aquatic Organisms

In addition to large terrestrial and aquatic biota, small organisms such as insects or plankton may also need to be sampled. Many different methods must be employed because of the various habitats in which these organisms are found and their differing behavior patterns. Some burrow into terrestrial soil or aquatic system bottoms. Others fly or swim. Some crawl or walk on the land surface or in aquatic systems. Some inhabit terrestrial or aquatic vegetation. Others live in water for most of their lives before entering a terrestrial life stage.

FIGURE 7.14 Burlese-Tullgren funnels are used to separate small animals from material such as soil, grass, or leaves.

Terrestrial invertebrates are important to terrestrial wildlife species but can also influence aquatic animals. For example, terrestrial invertebrates constitute a major portion of the diets of some stream fishes. Similarly, potential effects of aquatic invertebrates on fisheries resources are numerous, but these aquatic animals also influence water-associated wildlife species such as waterfowl and raccoons. In addition, many aquatic animals have terrestrial life phases; these animals can be important to a variety of animals, including totally terrestrial ones.

Many terrestrial invertebrates can be sampled by taking a specific measured amount of material, such as soil, grass, or tree leaves, and counting the animals present in that material. The animals can be hand-separated from the material, but this is a tedious procedure. More commonly, separation is accomplished with a **Burlese-Tullgren funnel** (Figure 7.14). The soil or other medium containing the invertebrates is placed on a screen located in the funnel portion of the device. A heat source (such as a light bulb) at the top of the funnel forces the invertebrates out of the sample and into a collecting jar. Numerous modifications can be made to this device; for example, other methods of forcing or attracting animals into the sampling jar can be used.

Pit traps (see section 7.3) can be used for capturing a variety of small terrestrial animals that move along the soil surface. Pit traps can be provided with attractants or even killing agents and preservatives to increase trapping effectiveness.

The kinds and numbers of terrestrial invertebrates found on vegetation can be determined by using tubes, boxes, or nets that totally enclose an item such as a tree branch or a patch of grass. After total enclosure, the animals must be separated from the vegetative material. They can be removed by hand picking or with the funnel device already mentioned. They can also be taken directly from the environment with a device such as a **D-vac sampler** (Figure 7.15). This sampler is essentially a large field vacuum cleaner that collects the animals in a cloth bag located in the collection head.

Terrestrial invertebrates can also be collected with **sweep nets**, which are similar in structure and function to the handheld nets used to capture fishes, birds, and mammals. Many people fondly remember the hours of childhood enjoyment associated with using these "butterfly" nets.

At times, an abundance index for invertebrates found in trees is needed. One method of accomplishing this sampling is by striking tree branches with a pole or stick and collecting the dislodged animals on canvas sheets or in funnel devices located under the trees. Pesticides can also be used to dislodge such organisms.

Aerial insects can also be sampled with **suction traps**. These traps draw in a specific volume of air so that estimates of the total number of collected organisms can be quantified. There are numerous other devices that can be used to capture aerial invertebrates, including traps that contain sticky substances and traps that use lights, baits, or other attractants.

FIGURE 7.15 D-vac samplers, which are essentially field vacuum cleaners, can be used to collect terrestrial invertebrates.

FIGURE 7.16 Emergence traps are designed to capture invertebrates as they emerge from an aquatic life stage to begin a terrestrial existence or as they emerge from the ground. The organisms are collected in the jar.

Many invertebrates spend all of their lives as terrestrial forms; others spend the early stages of their lives in aquatic systems and become terrestrial in a later life stage. These organisms can be collected with some of the devices already mentioned in this section. They can also be captured as they emerge from the ground or the water to begin their next life stage. There are a number of different types of such **emergence traps** that are used to capture both terrestrial invertebrates and those that are emerging from aquatic life stages (Figure 7.16). Some work best in flowing water, while others are designed to capture animals emerging from standing water. Some of the devices float on the water surface; others are submerged. In addition, some use vacuum devices or sticky substances to increase their effectiveness.

There are many other invertebrate capture devices that could be mentioned. For example, two additional traps designed specifically for capturing flying invertebrates are the **flight-intercept trap** (Figure 7.17A), which captures and holds organisms below the trap in special collection devices, and the **ultraviolet survey trap** (Figure 7.17B), which attracts night-flying insects with its light. Both are used for terrestrial animals and aquatic emergents.

A

FIGURE 7.17 Specifically designed traps are used to capture both terrestrial and emergent aquatic flying organisms: *(A)* flight-intercept trap; *(B)* ultraviolet survey trap. *[Part A image courtesy of P. Johnson.]*

B

A

B

FIGURE 7.18 Fine mesh nets are used to capture smaller organisms: *(A)* Wisconsin plankton net; *(B)* meter net.

Some aquatic organisms are quite small and require specially designed sampling devices. The **Wisconsin plankton net** is a device commonly employed for capturing plankton (Figure 7.18A), but others, such as the **Juday net** and the **Clarke-Bumpus sampler**, are also used. These devices, because of the sizes of the organisms captured, are smaller and have finer mesh sizes than the sampling nets previously mentioned in this chapter. While mesh sizes for fish and wildlife capture devices are often measured in centimeters, one centimeter of plankton net could have hundreds of openings or meshes. Even at such fine mesh sizes, mesh size still influences what is captured. Most phytoplankton require a smaller mesh size than do generally larger zooplankton. Some plankters are too small for even the smallest net meshes and must be extracted from water samples using filter paper or by centrifuging. Plankton nets can be pulled by hand or by boat. Some have metering devices to measure how much water has passed through the net. Some devices are used primarily to sample fish larvae and eggs. For example, **meter nets** (Figure 7.18B) are commonly used for fish eggs and larvae.

If the aquatic biota being sampled are not suspended in the water column, other devices need to be employed. A variety of devices can be used to sample benthic organisms or organisms that are attached to submerged objects such as logs, rocks, or vegetation. For sampling benthic organisms in soft bottoms such as mud, sand, or small rocks, a variety of grab samplers can be used; two common ones are **Ekman** and **Petersen grabs** (Figure 7.19). These devices, when lowered to the bottom, grab a portion of the bottom material,

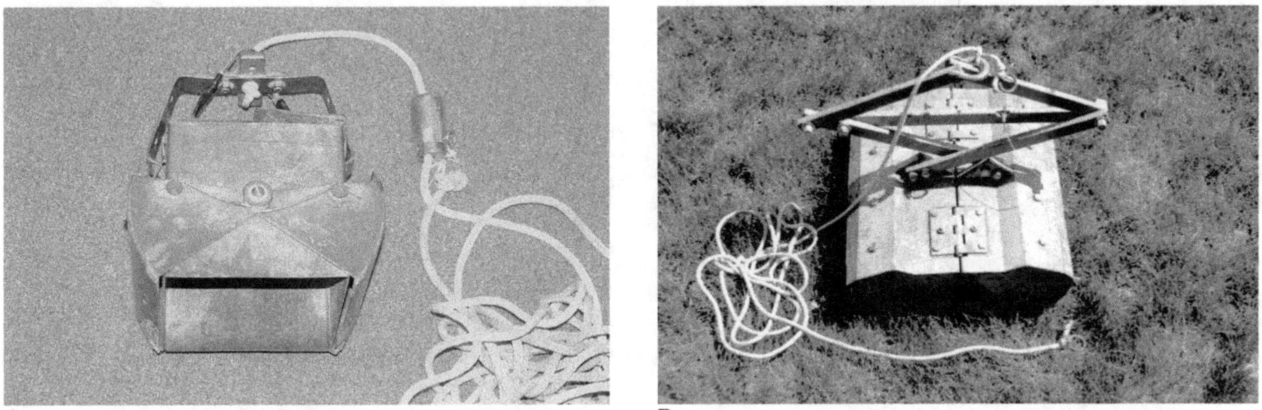

A B

FIGURE 7.19 Several types of grabs can be used to sample bottom material and the organisms associated with that material: *(A)* Ekman grab; *(B)* Petersen grab.

FIGURE 7.20 Surber stream samplers are used to collect larger benthic invertebrates in flowing-water environments.

which can then be sifted or strained to separate the benthic organisms from unwanted organic and inorganic material. Other devices, such as the **Surber stream sampler**, are used to collect larger organisms attached to bottom objects in flowing-water environments (Figure 7.20). The organisms are manually dislodged from rocks, vegetation, or other objects and are then carried by the water current into the sampler, which has been placed downstream. Organisms that are attached to larger objects and cannot be captured by grabs or other samplers can be manually removed from the submerged objects. In addition, biologists can place artificial substrates in the water body, wait for organisms to colonize them, and then remove the artificial substrates for assessment.

7.9 Combining Sampling Devices

There are times when sampling devices are best used in combination. In sampling with electrofishing gear, there are situations in which portions of the water body being sampled are blocked with nets to reduce fish escapement. Nets are also commonly used with piscicides to define the sampling area and prevent fish entry or escape. Fishes captured in purse seines are often removed from the seine using handheld or mechanically operated dip nets. Brown bears may initially be captured using a snare and then anesthetized. House finches may be initially captured in a funnel trap and then removed from the trap with a handheld net.

7.10 Noncapture Sampling

Not all sampling requires the actual capture of organisms. **Drag lines** are ropes or cables that can be pulled by two people or two vehicles across fields to flush ground-nesting birds, which can then be counted or their nests examined (Figure 7.21). These lines are specifically designed to eliminate damage to any nests they may encounter. In water, photoelectric cells can be used to count fishes as they pass through a confined area. **Hydroacoustics** are also used in fisheries work and involve the use of sonar. Some hydroacoustic devices are very technologically advanced, and use of hydroacoustics for activities such as fish enumeration and behavioral studies has increased as these devices have become better developed. Video or film photography, either normal or infrared, can also be used to sample organisms. Cameras can be manual, time-lapse, or activated by the target animal. This activity is even popular with the lay public who may be interested in what animals are using their backyards and hunters who want to know what kinds of deer are using particular places. Birds, mammals, and even fishes can be viewed with optical equipment, such as binoculars and spotting scopes, or the unaided eye. Some optical equipment can be quite sophisticated, such as **starlight** or **night-vision scopes** (Figure 7.22). Fishes can be observed with a variety of underwater viewing devices or observed by scuba or snorkel-equipped divers.

Counts of some animals can be made from vehicles, aircraft, or with satellite imagery or radar. A common practice, especially in the wildlife profession, is to sample with the use of lights. Using lights at night to find and count animals such as white-tailed deer, American alligators, swift foxes, and coyotes is a common sampling method. Sampling may involve seeing the whole animal or just the light reflecting from its eyes.

FIGURE 7.21 Drag lines are used to flush ground-nesting birds so that nests can be located or birds counted. *[Image courtesy of the U.S. Fish and Wildlife Service.]*

FIGURE 7.22 Night-vision or starlight scopes can be used to see animals under low-light conditions.

Birds and mammals can be counted in a variety of ways that do not even require the actual viewing of the animal. For example, counts or indices of abundance can be obtained from bird songs or calls, ruffed grouse drumming, coyotes calling, elk bugling, or white-tailed deer droppings (**scats**, **fecal pellets**, or **pellets**). In addition, information on some birds and mammals can be obtained from their tracks. Counting of tracks is often enhanced by smoothing dirt and adding scent attractants to the area, as in work with coyotes or red foxes. Track counting under snow cover conditions can also be an effective sampling method (Figure 7.23). For smaller organisms such as mice, carbon-blackened paper attached to trees or located in other areas the mice inhabit can be used to record tracks. When a mouse moves across the paper, it leaves footprints in the carbon black.

FIGURE 7.23 Tracks can be used as an indicator of abundance for various animals. These are ring-necked pheasant tracks in snow. *[Image courtesy of South Dakota Tourism.]*

7.11 Additional Sampling Devices

Not all sampling devices are designed and made specifically for scientific or commercial use. Under certain circumstances, regular fishing gear, used by the biologist with either natural baits or artificial lures, can be effective for fish capture. Even spearguns used in conjunction with scuba or snorkel diving can be effective. In addition, recreationally and commercially captured fishes can yield a variety of valuable data. Useful data can also be obtained from wildlife species taken by hunters, trappers, or biologists with shotguns, rifles, bows, traps, or other legal devices. Use of recreationally harvested fisheries and wildlife biota for data collection has the added benefit of reducing the cost of sampling. Even road-killed organisms can provide useful information. For example, in many states it is common to check road-killed deer to determine the number of young that pregnant females are carrying. Such information is helpful in determining the health of the deer herd and potential population trends.

Numerous other devices and methods not mentioned in this chapter have been employed to capture animals involved in wildlife and fisheries work. These vary from use of explosives in fisheries work to use of handheld snare poles to capture wildlife. In addition, sampling of terrestrial and aquatic invertebrates as well as fish larvae and eggs is carried out by means of devices and methods too varied and too numerous to be discussed here. We reiterate, however, that all capture methods are selective and can introduce bias into the information obtained. We must also reiterate the importance of our responsibilities to these animals for humane care and handling. Various professional societies have developed protocols and procedures for correct animal handling. The topic of animal welfare is discussed in section 16.6. In addition, there are laws and regulations, at various government levels, concerning the welfare of captured and held animals.

7.12 Human Safety

Throughout this chapter, examples were provided concerning various ways that sampling devices could be harmful to target and nontarget organisms. Most of these devices also have the potential to harm humans who are employing them. Each has its own set of personal risks, some more so than others. For example, the use of electricity in proximity to water requires many safety precautions to protect the operator. Special equipment, such as rubber gloves and boots, life jackets, and dip nets with insulated handles, must be used. Circuit-breaker switches on control units and foot-operated pedals are usually designed as a dual-safety arrangement into boat electrofishing systems. Electrofishing units are dangerous pieces of equipment if improperly handled.

In most cases special training on the correct use of sampling devices is essential to reduce risk. There are specific protocols that must be followed to improve safety for the operator. No equipment can be totally safe, but injuries can be minimized if proper care and techniques are used.

SUMMARY

A sample is a subset of the organisms in a population. Who can sample; how, where, and when sampling can be done; and what animals can be sampled is regulated. Many of the specialized devices used in wildlife and fisheries for sampling or capturing animals require special permits.

All sampling methods are selective and can introduce bias into the information obtained. Sampling devices can be selective for such variables as species, size, gender, and age. Regardless of whether a sampling device is lethal or nonlethal, animal welfare must be considered during any sampling activity.

Sampling devices tend to be either passive or active. Passive devices are put into place and target animals move or are driven into the devices. Active devices are themselves moved, and this movement results in animal capture. Passive entanglement devices capture animals by entangling some body part such as a leg, wing, or fin. Passive entrapment devices capture animals by enclosing the entire animal in a confined area from which it cannot escape.

Anesthetics are primarily used by wildlife biologists for the live capture of animals. Anesthetizing drugs are usually delivered with syringelike projectiles delivered by a dart gun. Use of toxicants for sampling is primarily a fisheries activity. General-purpose toxicants such as rotenone and antimycin are used, as are toxicants specific to sea lampreys. Use of chemicals in wildlife and fisheries work is controlled by a variety of regulations, which change frequently. Biologists must be aware of current regulations.

Electrofishing is primarily used in fisheries work. This method is generally used in shallow water and results in the live capture of fishes.

Specific devices and methods are used to sample small aquatic and terrestrial organisms, such as insects and plankton. These range from nets that strain small organisms from water to vacuum devices used to sample insects from trees.

Not all sampling requires that animals actually be captured. At times all that is needed for sampling is to see or hear an animal or observe some evidence that it is or was present.

User safety is always an issue when using different types of equipment. Proper training can minimize hazards to users.

PRACTICE QUESTIONS

1. Identify a dozen reasons why a fisheries biologist might need to capture a plains minnow.

2. Why might the capture of a young wild turkey require a different sampling method or strategy than the capture of an adult wild turkey?

3. List four types of passive entanglement devices and describe how they capture an animal.

4. List four types of passive entrapment devices and describe how each may be selective for a particular organism.

5. List four types of active capture devices and describe how they capture an animal.

6. Why is the use of anesthetics delivered by dart guns unpredictable in regard to the effects on the target animal?

7. How do AC and DC differ in their effects on fishes?

8. Why has the use of piscicides as a sampling technique decreased in recent years?

9. Describe a dozen different ways that animals can be sampled without capturing them.

10. Describe four devices used to capture or sample aquatic invertebrates.

11. Describe four devices that can be used to capture or sample terrestrial invertebrates.

SELECTED READINGS

Berry, C. R. 1996. Safety in fisheries work. Pages 63–81 *in* B. R. Murphy and D. W. Willis, eds. *Fisheries techniques*. 2nd ed. American Fisheries Society, Bethesda, Md.

Bettoli, P. W., and M. J. Maceina. 1996. Sampling with toxicants. Pages 303–333 *in* B. R. Murphy and D. W. Willis, eds. *Fisheries techniques*. 2nd ed. American Fisheries Society, Bethesda, Md.

Brandt, S. B. 1996. Acoustic assessment of fish abundance and distribution. Pages 385–432 *in* B. R. Murphy and D. W. Willis, eds. *Fisheries techniques*. 2nd ed. American Fisheries Society, Bethesda, Md.

Dolloff, A., J. Kershner, and R. Thurow. 1996. Underwater observation. Pages 533–554 *in* B. R. Murphy and D. W. Willis, eds. *Fisheries techniques*. 2nd ed. American Fisheries Society, Bethesda, Md.

Hayes, D. B., C. P. Ferreri, and W. M. Taylor. 1996. Active fish capture methods. Pages 193–220 *in* B. R. Murphy and D. W. Willis, eds. *Fisheries techniques*. 2nd ed. American Fisheries Society, Bethesda, Md.

Hubert, W. A. 1996. Passive capture techniques. Pages 157–181 *in* B. R. Murphy and D. W. Willis, eds. *Fisheries techniques*. 2nd ed. American Fisheries Society, Bethesda, Md.

Kelso, W. E., and D. A. Rutherford. 1996. Collection, preservation, and identification of fish eggs and larvae. Pages 255–302 *in* B. R. Murphy and D. W. Willis, eds. *Fisheries techniques*. 2nd ed. American Fisheries Society, Bethesda, Md.

Larkin, R. P. 2005. Radar techniques for wildlife biology. Pages 448–464 *in* C. E. Braun, ed. *Techniques for wildlife investigations and management*. 6th ed. The Wildlife Society, Bethesda, Md.

Rabeni, C. F. 1996. Invertebrates. Pages 335–352 *in* B. R. Murphy and D. W. Willis, eds. *Fisheries techniques*. 2nd ed. American Fisheries Society, Bethesda, Md.

Reynolds, J. B. 1996. Electrofishing. Pages 221–253 *in* B. R. Murphy and D. W. Willis, eds. *Fisheries techniques*. 2nd ed. American Fisheries Society, Bethesda, Md.

Roffe, T. J., S. J. Sweeney, and K. E. Aune. 2005. Chemical immobilization of North American mammals. Pages 286–302 *in* C. E. Braun, ed. *Techniques for wildlife investigations and management*. 6th ed. The Wildlife Society, Bethesda, Md.

Schemnitz, S. D. 2005. Capturing and handling wild animals. Pages 239–285 *in* C. E. Braun, ed. *Techniques for wildlife investigations and management*. 6th ed. The Wildlife Society, Bethesda, Md.

Winter, J. 1996. Advances in underwater biotelemetry. Pages 555–590 *in* B. R. Murphy and D. W. Willis, eds. *Fisheries techniques*. 2nd ed. American Fisheries Society, Bethesda, Md.

LITERATURE CITED

Hubert, W. A. 1996. Passive capture techniques. Pages 157–181 *in* B. R. Murphy and D. W. Willis, eds. *Fisheries techniques*. 2nd ed. American Fisheries Society, Bethesda, Md.

Schemnitz, S. D. 2005. Capturing and handling wild animals. Pages 239–285 *in* C. E. Braun, ed. *Techniques for wildlife investigations and management*. 6th ed. The Wildlife Society, Bethesda, Md.

CHAPTER 8

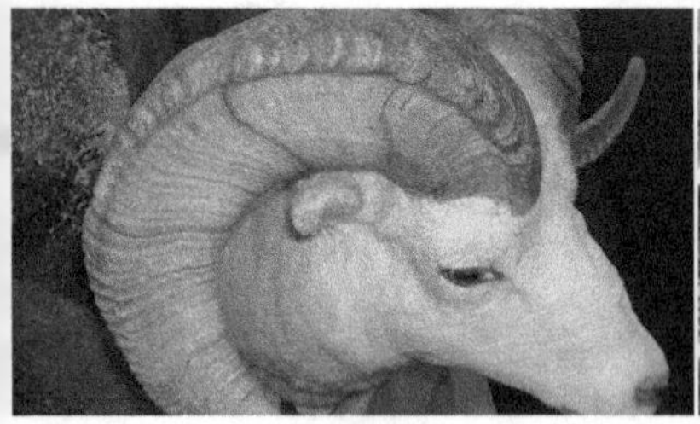

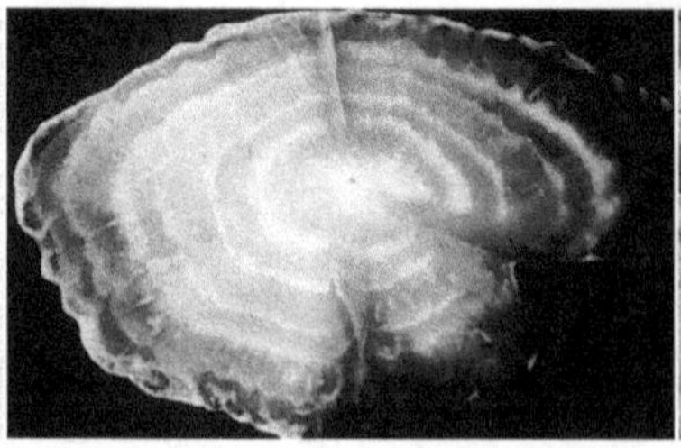

Determination and Use of Age, Growth, and Gender Information

Information on age, growth, and gender provides insight into the status and health of populations of wild animals. For example, age and growth information can be used to evaluate various fish harvest strategies. If a fish population is being overharvested, it will be characterized by young, rapidly growing fish and few large individuals. If age and growth data collected for a fishery reveal this pattern, biologists can then implement appropriate measures, such as harvest regulations (Chapter 17) or habitat enhancement (Chapter 15), to improve the quality of that fishery. Perhaps, while fishing, you helped to provide such data for evaluating a fish population when biologists checked your fish for length and weight and collected a few scales.

Wildlife biologists can determine the relative productivity of various populations, such as American wigeon, eastern fox squirrels, or elk, by determining the proportion of the population that occurs in various age categories. You may have assisted in collecting such data by mailing waterfowl wings to the U.S. Fish and Wildlife Service or deer or elk incisor teeth to a state, provincial, or territorial natural resource agency. The coloration and shape of various feathers on duck wings in the fall can be used to determine whether a duck was hatched the preceding summer or more than a year earlier. Likewise, growth rings on the roots of deer or elk incisors can be used to estimate the age of the elk or deer. The pre- and posthunting season gender composition of harvested animal populations, such as ring-necked pheasants in which males are selectively harvested, can be used to estimate the percentage of the population killed by hunters. In projects involving banding, marking, tagging, or biotelemetry (Chapter 9), it is often necessary to determine the gender or age of an animal at the time of capture.

Age, growth, and gender information can be used in a variety of ways by biologists to understand and conserve wildlife and fisheries resources. This chapter is a review of some of the more common methods used in obtaining age, growth, and gender data and includes discussions of the application of such information.

8.1 Age Category Terminology

The terminology used in grouping animals by age can be confusing, and there are differences among the terms used by wildlife and by fisheries biologists. When fish population age structure is evaluated, biologists generally place fishes in either age groups or year classes. **Age groups** represent the number of years a fish has lived. A fish in its first year of life is a member of age group 0, a fish in its second year of life is a member of age group 1, and so on. Thus, the age group of an individual fish changes each year. Commonly the *group* portion of the term is deleted; thus, fishes in age group 1 are usually called age-1 fishes. All fishes are assumed to have a 1 January "birthday" (more properly "hatch-day" for most fishes). In other words, regardless of whether a fish was hatched in March or August of a year, it will become an age-1 fish on 1 January of the next year in the northern hemisphere. Alternatively, fisheries biologists can place fishes in **year classes** instead of age groups. If a fish was hatched in 2001, then that fish would be a member of the 2001 year class. In 2 years, 4 years, or 9 years, it would still be a member of the 2001 year class.

The term **young-of-the-year**, which refers to animals in the early portions of their lives, is commonly used in both fisheries and wildlife work. In fisheries, there is an effort to replace the term *young-of-the-year* with the term *age 0*. The term **subyearling** is also used in fisheries to denote a fish of this age. A variety of other terms, such as **fry** and **fingerling**, are commonly used to refer to young fishes. However, these terms are variously defined, often have different meanings to different people, and should not be used unless they are specifically defined. Piper et al. (1982) defined fry as fishes between the time of hatching and the time at which they reach 25 millimeters in length and fingerlings as fishes between 25 millimeters and the length at age 1. Newly hatched fishes are also often referred to as **larval fishes**. Various terminologies are used for phases of larval fish development (Kelso and Rutherford 1996), but no one system has been universally accepted.

In addition, specific terminology is used for young members of the salmon family. **Parr** refers to the salmonid life stage from the time of first feeding until fishes become sufficiently pigmented to obliterate their parr marks. **Parr marks** are vertical pigmentation bands found on the sides of young salmonids (Figure 8.1). Some fish species other than salmonids also have parr marks. Salmonids that migrate from fresh water to the ocean become **smolts** at the time of physiological adaptation to life in the marine environment. The parr marks are typically lost at this time.

FIGURE 8.1 Parr marks (arrows) are characteristic of young salmonids such as this Rio Grande cutthroat trout. Parr marks on this fish will fade as it matures. *[Image courtesy of L. Hazzard, U.S. Fish and Wildlife Service.]*

The age group system used on fishes, with the assumed 1 January birthday, is not normally applied to wildlife, although that system would work for species whose age can be determined in years. Most often in wildlife, the term **age class** is used to refer to chronological age in years and to various age or life stage groupings; the term has a very broad usage. For example, individuals less than 1 year of age may be grouped as age class 0–1, age class 0, age class 0.5, or age class 1. To further confuse the issue, the term *year class* in wildlife is sometimes used interchangeably with the term *age class*. However, the year-class system is also sometimes used for wildlife in a manner identical to that in fisheries. For example, one may find reference to the 1989 year class of elk, meaning the elk that were born in that year. In this text, *year class* will be used as defined in fisheries work and will not be used interchangeably with other terms.

The term **cohort** is sometimes used synonymously with *age class*, *year class*, and *age group*. However, while a cohort can be a group of individuals in a population born or hatched during a 1-year period, a cohort can also be defined on the basis of shorter or longer time periods. For example, larval fish cohorts may be defined as narrowly as those hatched on 1 day.

The terms **immature** or **juvenile**, **subadult**, and **adult** are also used in both wildlife and fisheries, further illustrating the broad usage of a variety of terms

FIGURE 8.2 During fall migration, subadult and adult (adult pair in image) greater white-fronted geese have black patches on the lower breast and belly and white on the forehead, while juveniles hatched that summer (HY birds) lack these markings. *[Image courtesy of D. Menke, U.S. Fish and Wildlife Service.]*

having to do with age classification. In wildlife, the terms *immature* and *juvenile* are used interchangeably for animals that are too young to breed and can be distinguished from adults based on external characteristics. For example, a 3-year-old bald eagle lacks a white head and tail and is easily distinguishable from an adult; thus, it is classed as a juvenile or immature bird. In the fall, a greater white-fronted goose that hatched in late June of that year lacks the dark markings on the breast and belly and the white on the front of the head found on subadults (1 and 2 years old) and adults; thus, it is classed as a juvenile or immature (Figure 8.2). In wildlife, the term *subadult* is used specifically for individuals that are too young to breed but are externally indistinguishable from adults. For example, 2-year-old Canada geese are classed as subadults because they look like adults, but they usually do not breed until 3 or 4 years of age, at which time they are adults. Two-year-old American black bears are subadults because they look like adults but are not reproductively active. In fisheries, the terms *immature*, *juvenile*, and *subadult* tend to be used interchangeably. In fisheries and wildlife the adult age classification includes animals that have reached normal breeding age or size.

In ungulates and wild turkeys, animals in their second year of life are sometimes called **yearlings**, a term that is also an age classification. For example, a female elk in its second fall or winter may be referred to as a yearling. The term *yearling* is also used in fisheries to designate an age-1 fish. Even **prebreeding** and **postbreeding** categories are sometimes used as age classifications.

Natal down is the short, fluffy feathering on newly hatched birds. It is especially abundant on newly hatched waterfowl, gallinaceous birds, and other precocial birds. Some altricial birds, particularly hawks and owls, also have abundant natal down. The term **juvenal plumage**, with the *-al* suffix, is used to describe the plumage following natal down. A rose-breasted grosbeak hatched in early June would have a drab brownish plumage in July that is properly called juvenal plumage; this bird would be classed as a juvenile or immature bird because it has not reached sexual maturity and can be distinguished from the adult.

In birds, the terms **hatching year (HY)** and **after-hatching year (AHY)** are commonly used as age classifications. The category HY includes all birds younger than 1 year old (hatched that year), while birds 1 year old and older are grouped as AHY. For species such as northern bobwhites, which reproduce or attempt to reproduce at 1 year of age, the terms *HY* and *juvenile* or *immature* can be used interchangeably to describe those that have not reached breeding age; the terms *AHY* and *adult* can be used after they reach sexual maturity, at approximately 1 year. The term *HY* is more specific than *juvenile* because some birds are still juveniles after 1 year.

8.2 Determination of Age

The most common methods of age determination are those that use hard body parts such as scales, feathers, and bones or soft body parts such as reproductive organs. Age determination for fishes and many mammals is usually possible to the nearest year. In most birds, the methods of age determination are primarily effective in separating individuals that have not reached reproductive maturity from those that are adults (reproductively mature). An overview of age determination techniques is provided in this chapter, but biologists should consult more advanced books and recent scientific journals to determine the most appropriate methods of age determination for particular species. It is quite common for particular aging methods to vary in effectiveness from one species to the next.

In both wildlife and fisheries, it is often difficult to correctly determine the age of an organism. Regardless of which structure is chosen for aging, biologists must validate the **accuracy** of the technique on a case-by-case basis (Beamish and McFarlane 1983). Accuracy is the closeness of a measure to the true value (see section 9.1). One way in which biologists validate accuracy is by marking known-aged animals so that they may be identified later and compared to animals whose ages have been determined through the technique in question. Animals reared in captivity are also used to validate aging techniques.

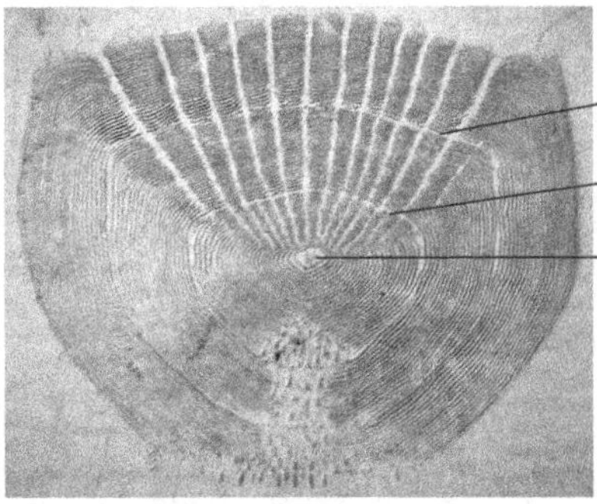

FIGURE 8.3 Impression of a white crappie scale on an acetate slide. This fish was age 2 when collected in the fall, after completing most of its third-year growth. Thus, the scale has two visible annuli.

8.3 Aging Methods

Scales and bones in fishes, teeth and bones in mammals, and feathers and bones in birds are all hard body parts that can be used to estimate age. Fishes are sometimes aged using the annual growth rings on scales (**scale annuli**) (Figure 8.3). Scales have concentric markings (**circuli**) around the center (**focus**). Most fish species living in temperate regions grow more rapidly during the warmer periods of the year, when the fish feed more and their body temperatures and metabolic rates are higher. These periods of greater growth are characterized by widely spaced scale circuli. When growth is slow, usually during the colder periods of the year, the scale circuli become crowded and form an annual growth ring (**annulus**). Therefore, a fish captured in the summer of 2005 that had three annuli on its scales would be placed in age group 3 (an age-3 fish) and in the 2002 year class. Some fishes can be aged accurately by scale annuli, but most often the method is used for younger fish within a species. Older fish cannot be aged with certainty from scales. **False annuli** or **spawning checks** (scale growth variations caused by periods of environmental stress or at spawning time) can make aging difficult. Such false annuli can also be found on some other hard parts used to age animals. Annuli on older fish tend to crowd toward the scale edge, making it difficult to accurately estimate age. One of the primary advantages of using scales is that they can be removed with little or no damage to the fish; scales are regenerated in the area from which they are removed.

A variety of other hard body parts can also be used for fish aging. The bones of the inner ear (otoliths; Figure 8.4) are used for some species, while other hard structures such as spines, **cleithra** (part of the pectoral fin support), and **branchiostegal rays** (part of the hyoid, or throat, support system) are also used. Ages estimated from these hard structures often are more accurate than those determined from fish scales. For example, consider the ages estimated

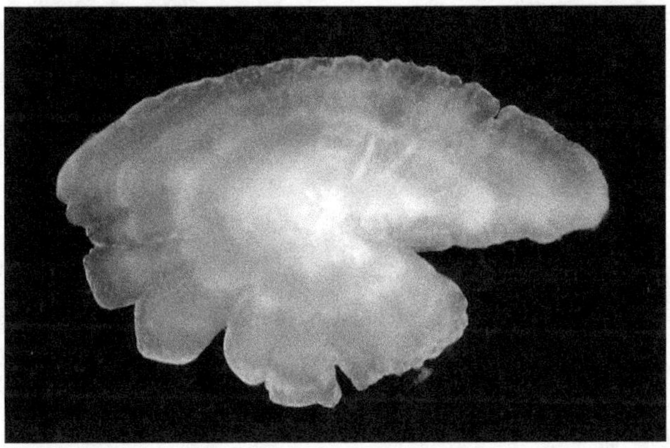

A

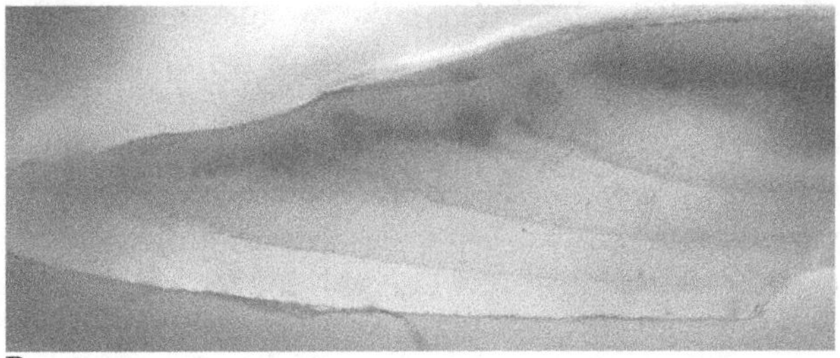

B

FIGURE 8.4 *(A)* Whole-view otolith for an age-4 bluegill captured in June. *(B)* Cracked (half) and sanded otolith for an age-4 white bass captured in May. *[Image B courtesy of C. Soupir.]*

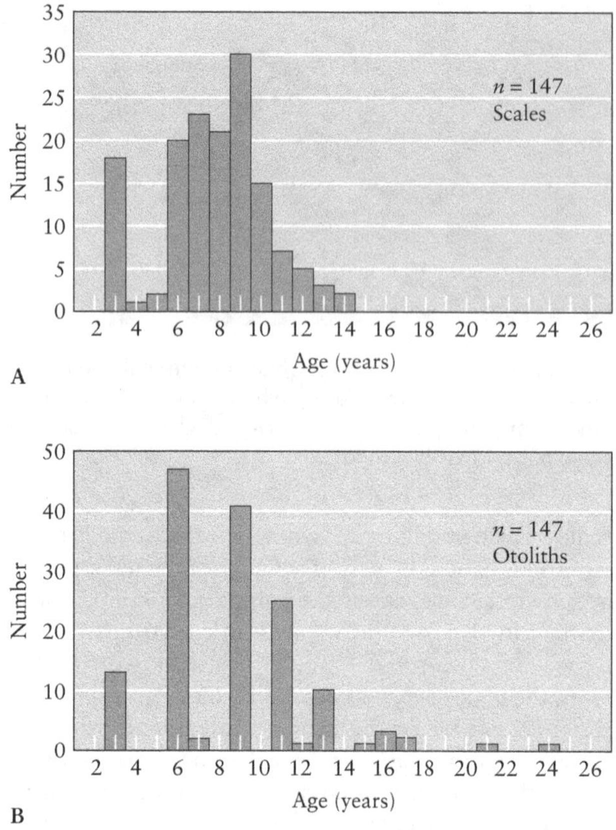

FIGURE 8.5 Age structure for 147 common carp based on scales *(A)* and otoliths *(B)*. Scales tend to underestimate ages of older fish and thus tend to provide less accurate age assignments than otoliths.

for the 147 common carp in Figure 8.5. Brown et al. (2004) validated the use of otoliths for accurate aging of common carp. Ages assigned to scales tend to be underestimated for larger individuals of most fish species, which was true for common carp (Phelps et al. 2007). The oldest fish in the sample according to otolith aging was 24, while the scale from that same fish was assigned an age of only 14. In addition, because the ages assigned to scales are less accurate, there is less apparent variability in abundance of the various age groups when scales are used for age estimation. A biologist would believe that recruitment was more erratic, given the otolith sample age structure in Figure 8.5 compared with the scale sample.

In paddlefish and sturgeons, the ossified and cartilaginous parts of the jaw and skull have growth rings that can be used for age estimation. Bone or cartilage density differences between periods of fast growth and slow growth are indicative of different growth periods. By viewing sections of these hard body parts with reflected or passed-through light, bone or cartilage density differences can be seen and ages determined. In most cases the use of bones requires killing the fish, although in catfish species the annual rings on a cross section of the pectoral spine can be used to determine age without killing the fish. Another disadvantage of the use of bones for aging is that it often takes longer to use them than scales, as they must be sectioned and in other ways prepared for viewing. While most of these methods yield information on fish age in years, some hard parts of young fish, such as otoliths, can also be used to determine days of age. Again, validation is necessary.

In a number of mammal species, annual deposits (**cementum annuli**) on the teeth provide an aging method somewhat similar to otolith aging in fishes (Figure 8.6). In deer, for instance, the roots of the incisors (front teeth) can be sectioned, stained, placed on slides, and the age estimated; a lower canine or premolar can be used for species such as brown bears and coyotes (Schroeder and Robb 2005). Differences in cementum deposition between the periods of slowest growth (winter) and the remainder of the year cause the development of growth rings, or annuli, that can be counted. Cementum annuli closest to the dentine of

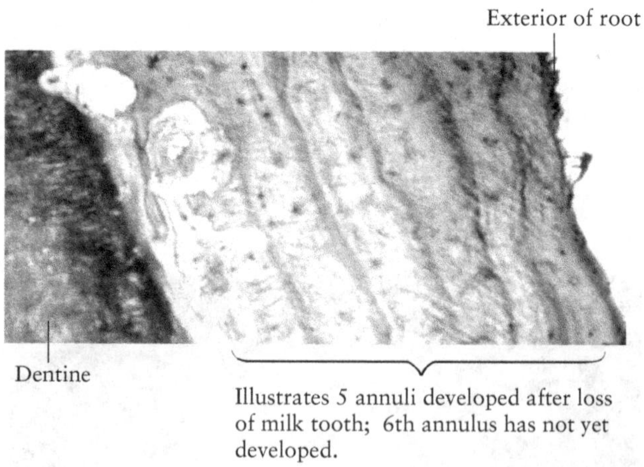

FIGURE 8.6 Longitudinal section of the root of an incisor tooth of a white-tailed deer, illustrating the cementum annuli of a 6.5-year-old deer. The oldest annulus (band of cementum) is nearest the dentine, and the most recently formed annulus is nearest the exterior of the root. This age calculation takes into account 1 year prior to milk tooth loss. *[Image courtesy of L. Rice.]*

the tooth root are the oldest, and those nearest the exterior of the root are the most recently formed. This method seems simple but in reality is quite complex due to various aberrations, such as false annuli, that require experience to decipher. Similar problems occur in aging other hard parts of animals.

Another common mammalian aging technique is the use of annual rings on the horns of mountain goats and bighorn sheep. These growth rings, which are visible on the surface of the horn, also reflect slow versus rapid growth periods during the year (Figure 8.7). The rings closest to the skull are the most recently formed. This method works well for several years, but in older animals, 9 years of age, for example, the rings become difficult to decipher near the skull. Also, the end portions of the horns may be broken or worn, making age determination difficult.

In some species, such as American mink, raccoons, and long-tailed weasels, there is a bone in the penis (**baculum**) that grows with age and can be used to estimate age in years. In these species, collections of bacula from pen-reared or marked known-aged animals can provide validation for estimating the ages of individuals from wild populations.

Studies of captive deer have clearly demonstrated variation in antler development due to nutrition as well as genetics; therefore, antler size is generally not a very dependable source of age information. For example, in some areas where the

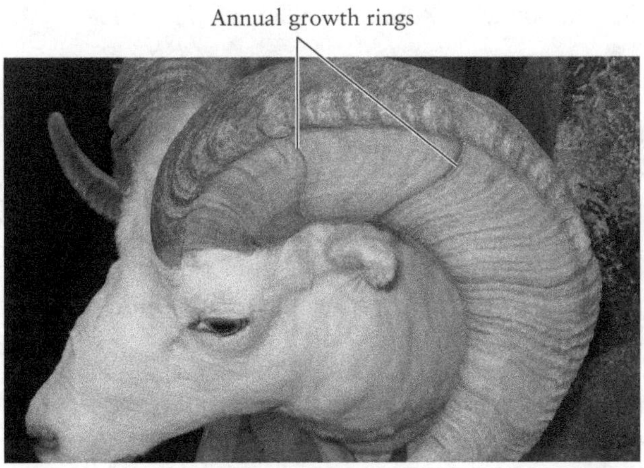

FIGURE 8.7 Annual rings can be used in age determination for various sheep species, such as this Dall's sheep from Alaska. Rings near the skull are newest but become crowded and more difficult to distinguish in older males.

quality of the forage is high, 1.5-year-old white-tailed deer males may have four points (**tines**) on each antler, and 2.5-year-old males can have impressive **beams** (the main trunk of the antlers) with lengthy tines that hunters would define as trophy quality. In other areas with poorer nutrient levels in the soil and forage, 1.5-year-old males seldom have more than two tines on each antler and have very little antler mass; 2.5-year-old males may have three or four tines per side and still have little antler mass or size. In addition, injury to the growing antlers may give rise to extra tines. A biologist who is familiar with age and growth characteristics of a particular deer population may be able to infer something about age from antler size. However, in general, antler growth is not an acceptable method for estimating deer age.

Considerable variation can also occur in antler growth rates in species such as elk and moose. Age of elk males, however, is generally evident through the first 3 years of life to a person familiar with antler growth in a particular population. For example, in most elk herds, 1.5-year-old males have single spikes or poorly developed forks for antlers. A **raghorn**, or a 2.5-year-old male elk, often has four tines per antler but can usually be recognized because its antler structure is much lighter and smaller than that of an older elk. Male elk at 3.5 years or older cannot be aged accurately from the antlers. Moose can be aged accurately from the antlers only through the yearling stage.

Milk teeth (the first set of teeth in mammals) are replaced and permanent teeth erupt at predictable and characteristic times in mammals. Tooth eruption and replacement charts can be used to determine the ages of many ungulates to within an accuracy of several months. For example, the presence of three-cusped fourth premolars instead of adult, two-cusped fourth premolars reveals that a deer is less than 18 months of age (Figure 8.8). Likewise, the two middle incisors (anteriormost teeth of the lower jaw) are replaced by larger adult incisors by 6 months of age (Severinghaus 1949; Schroeder and Robb 2005). Wear on the crests of the molars and premolars and the development of **dentine lakes** (areas where surface enamel is worn away and dentine is exposed) can provide useful information for age determination in deer, elk, and other ungulates (Figure 8.9). In using the tooth wear method, the biologist must recognize that local

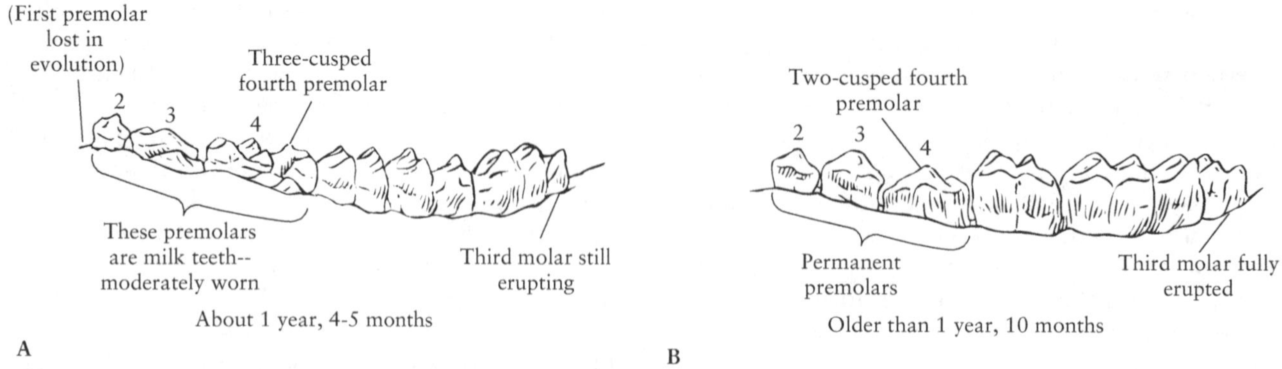

FIGURE 8.8 A three-cusped fourth premolar and an erupting third molar on the lower jaw distinguish a white-tailed deer approximately 1 year, 4–5 months of age *(A)* from a deer older than 1 year, 8–10 months *(B)*, which has a two-cusped permanent fourth premolar and a fully erupted third molar. *[Adapted from Larson and Taber 1980.]*

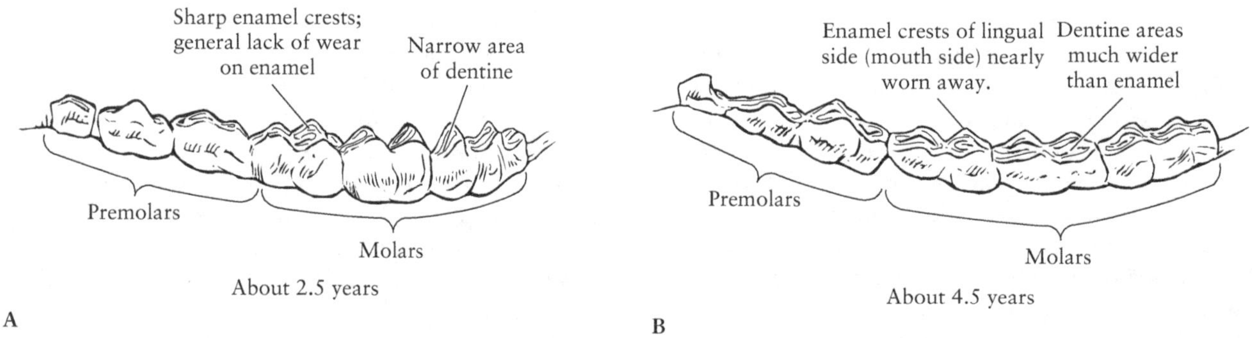

FIGURE 8.9 Enamel is less worn and dentine lakes (dark areas on dorsal surface) are smaller on premolars and molars (lateral view) of a 2.5-year-old white-tailed deer *(A)* compared with those of a 4.5-year-old deer *(B)*. Enamel crests and tooth height also wear rapidly with increased age. This method of aging is subject to many inaccuracies because of regional differences in tooth wear caused by diet and soil type. *[Adapted from Larson and Taber 1980.]*

and even gender-specific differences may occur due to differences in diet abrasiveness that can affect tooth wear. Tooth wear and eruption patterns are particularly useful if jaws from known-aged animals are available for comparison. Some of the advantages of using tooth eruption and wear data for aging include the availability at road checks of harvested animals and in other field situations, their usefulness on live-trapped animals, and the value of immediate age estimates as a public relations tool.

In young mammals, the long bones, such as the femur and humerus, are characterized by a cartilaginous zone near the ends of the bone where growth occurs. The caps or ends of these long bones are ossified, as is most of the remainder of the bone. When a long bone is X-rayed, this **epiphyseal cartilage** separating the **epiphysis** (the end of a long bone) from the **diaphysis** (the shaft of a long bone) is visible as a distinct line in the young of mammals such as raccoons, eastern gray squirrels, and eastern cottontails (Figure 8.10). Adults of these species exhibit no visible cartilage line in the long bones.

Wing feathers commonly used for age determination are the **primary feathers (primaries)**, **secondary feathers (secondaries)**, and **coverts**; coverts include the **primary coverts**; the **greater**, **middle**, and **lesser secondary coverts**; and the **marginal coverts.** Figure 8.11 illustrates the locations of these feathers on the wing of a duck. Aging methods related to feather color or shape are commonly used to separate

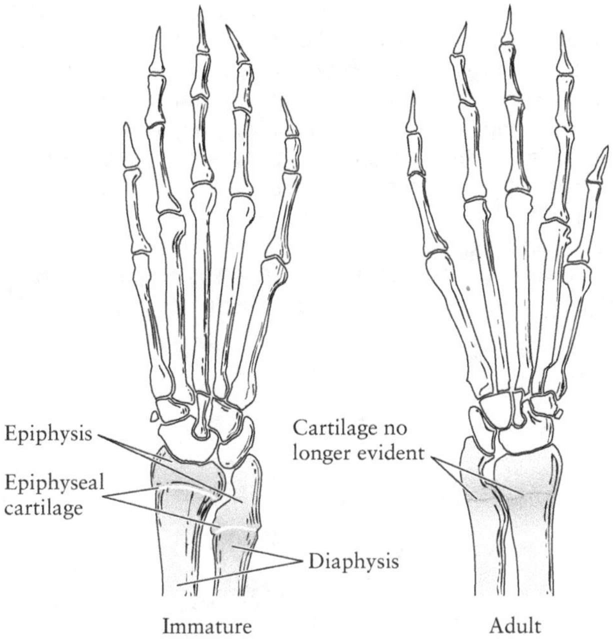

FIGURE 8.10 The epiphyseal cartilage can be seen in the forelimb of an immature raccoon but is absent in an adult raccoon. Drawings are from X rays. *[Adapted from Sanderson 1950.]*

HY and AHY birds of many species (U.S. Department of Interior and Canadian Wildlife Service 1977; Schroeder and Robb 2005). The most recent information on age and gender determination in birds is available through Birds of North America Online (http://bna.birds.cornell.edu/BNA/; Cornell Lab of Ornithology, Ithaca, N.Y.).

In HY (juvenile) gallinaceous birds, with the exception of pheasant species, the first eight primary feathers (P) of the wing are replaced during **molt** (loss of feathers in the process of renewal) from the one closest to the body (P1) out to P8; the last two juvenal primaries (P9 and P10) are retained through the yearling stage until late the following summer or early fall (16–17 months of age). When birds such as greater sage grouse molt P9 and P10 during the yearling stage, these pointed, faded, and worn juvenal flight feathers are then replaced by adult primaries with a more rounded tip, and the birds at that stage are classed as adults (Figure 8.12). Female yearling grouse (second summer) can be distinguished from female HY grouse until mid- to late September, prior to replacement of P9 and P10, by the fading and increased wear on these two flight feathers. Yearling male grouse generally molt P9 and P10 about 3–4 weeks earlier than yearling females.

In addition to differences in shape, the ratio of the diameter of the **calamus** (the bare shaft of the feather below the feathering) of primary feathers P9 and P8 (the diameter of P9 divided by the diameter of P8) is lower in gallinaceous birds such as sharp-tailed grouse before they molt the outer two primaries at 16–17 months than in adults (Caldwell 1980). At bag check stations for sharp-tailed grouse and other grouse species, biologists sometimes look for blood in the base

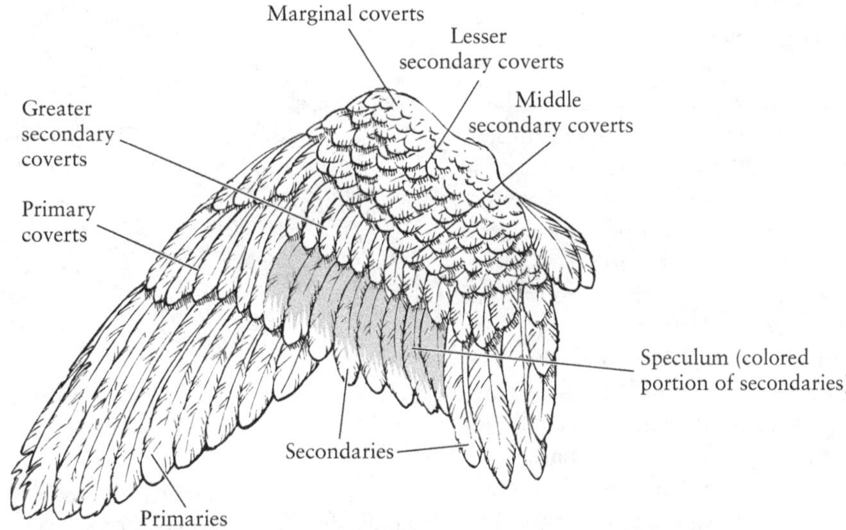

FIGURE 8.11 Feather groups identified on this duck wing are commonly used in age and gender determination. *[Adapted from Carney 1992.]*

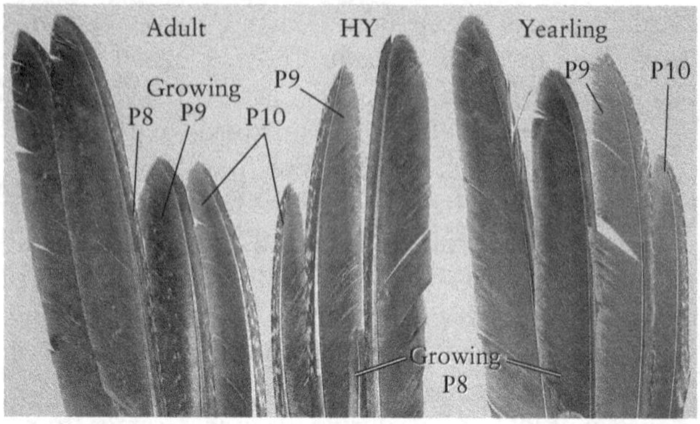

FIGURE 8.12 Shape, wear, and fading on the two most distal primaries (P9 and P10) can be used to determine age in grouse. The soon-to-be-replaced P10 (P9 is being replaced) on the adult (left) is comparatively rounded at the tip, indicating it is over 2 years of age. The tips of the outer primaries on the HY (center) and yearling (right) grouse are both relatively pointed (until replaced at about 16–17 months); yearlings at this stage will soon replace the faded and worn juvenal P9 and P10 with adult primaries and are thereafter classed as adults. Wings in this image were from females collected from hunters in September.

of the calamus of P9 and P10, which indicates a newly grown primary. There will be no blood in the calamus of P9 and P10 for HY birds or yearling grouse that have not molted P9 or P10. Grouse with blooded P9 and/or P10 are in the process of replacing or have just replaced juvenal primaries with adult primaries.

In addition to the other methods used for gallinaceous birds, male wild turkeys are often aged from the **rectrices** (large tail feathers). The rectrices of AHY wild turkeys are all of similar length, while the central three to four pairs of rectrices are noticeably longer than the other rectrices in HY turkeys after the prealternate molt in midwinter (Figure 8.13); these differences can be observed from a distance. Adult male wild turkeys also have longer spurs, generally a longer **beard** (a specialized group of hairlike feathers on the breast), and a broader and more colorful patch of greater secondary coverts than do HY males.

Feather characteristics can be used to age other bird species; a few more examples follow. In HY northern bobwhites, scaled quail, and Gambel's quail, the primary coverts are buffy or white tipped, more ragged at the tip, and less rounded than those of adults. Immature gray-cheeked thrushes have terminal buff spots on the greater secondary coverts, while adults lack these spots. The secondary coverts in an immature black-and-white warbler are brown, whereas in adults they are black.

Waterfowl biologists can use a combination of wing feather characteristics (shape and color) to categorize ducks as HY or AHY individuals. Ducks do not replace most of their HY flight feathers (primaries and secondaries) until the second summer of life. However, the most **proximal** (near the body) secondaries of HY birds, which are usually longer and differently shaped and colored than the other secondaries, are usually

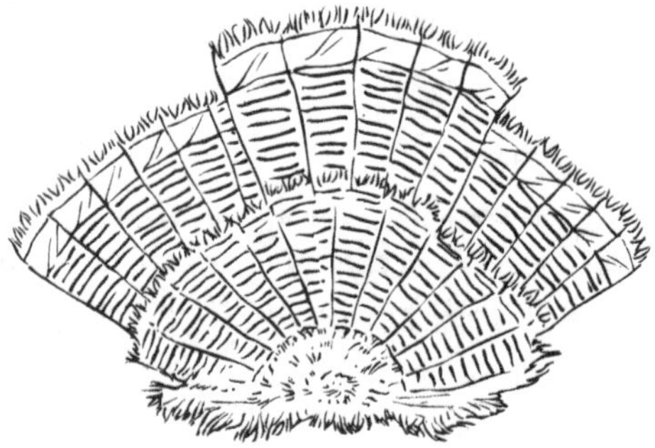

A

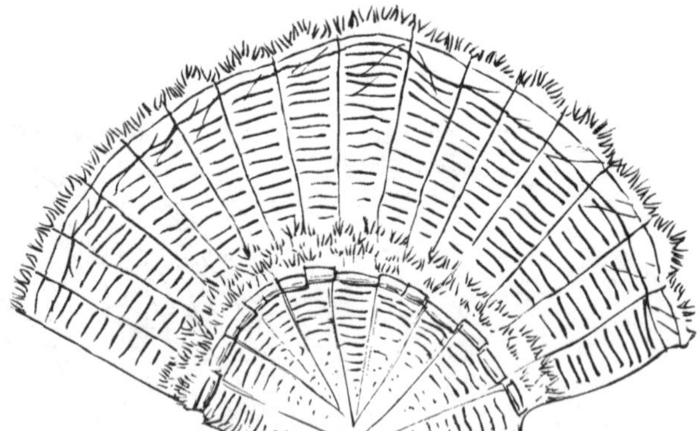

B

FIGURE 8.13 Central rectrices of HY wild turkeys *(A)* are noticeably longer than other rectrices after replacement in early winter, while those of AHY birds *(B)* are all of equal length.

molted with the body feathers during the first fall (Carney 1992). These proximal secondaries, because of their distinctive shape, length, and often color, can be used to separate HY from AHY ducks until they are molted in midfall. The juvenal wing coverts (referring to the primary and various secondary coverts) are replaced in the first fall, winter, or spring and, because of age-related differences in color and shape, may also be useful in determining age.

In addition, the rectrices of HY ducks and geese are notched until they are replaced sometime in the fall or winter (Figure 8.14). The notching in the rectrices of HY waterfowl is caused by abrasion or wearing of the downy tips of these feathers. Wear on the rectrices of adult waterfowl can sometimes cause the feathers to look notched like those of HY birds.

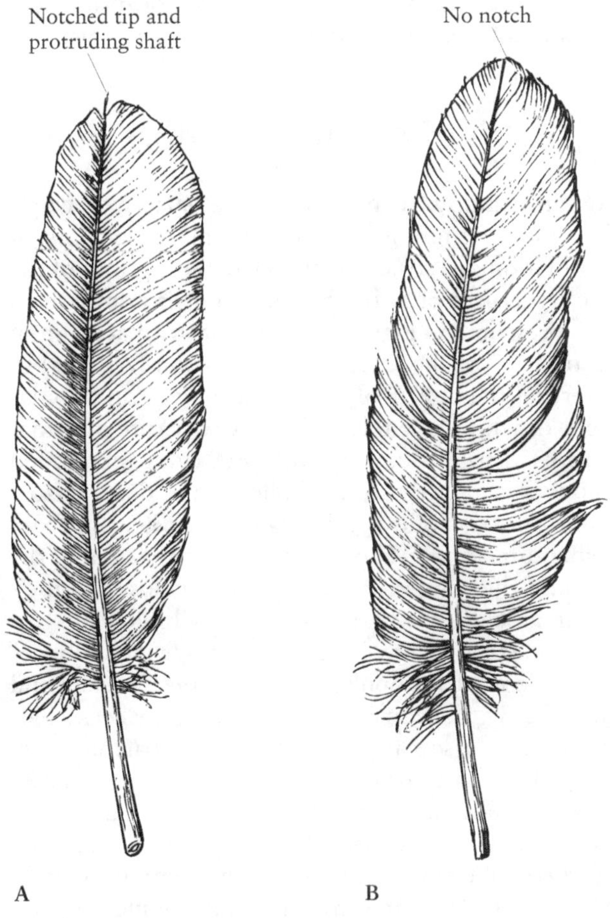

FIGURE 8.14 *(A)* Rectrices in the tail of a juvenile (HY) Canada goose illustrate the typical notched tip and protruding shaft. *(B)* Rectrices of an AHY Canada goose lack the notched tip.

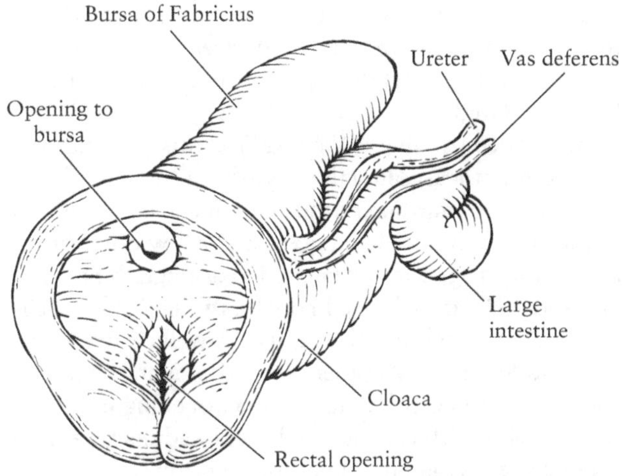

FIGURE 8.15 The bursa of Fabricius is a dorsal pocket in the cloaca of birds that is present in HY birds into late fall or winter, but it is absent or much shallower in adults. *[Adapted from Godin 1960.]*

Various other hard features, including additional tooth eruption patterns, external wear patterns on the incisors of several carnivores such as the wolf, gum line recession in carnivores, molt patterns on the pelts of mammals such as common muskrats, growth marks on the skeleton of sea turtles, spur characteristics, and additional feather characteristics, may also provide age information on wildlife species (Schroeder and Robb 2005). The fissures between many skull bones in juvenile carnivores, for example, are much more visible than in adults. In addition, in HY ring-necked pheasants, the lower bill will bend or break more easily than in AHY birds.

Some age determination methods depend on soft body parts. The weight of the eye lens, for example, increases with age in mammals. If lenses are collected from recently killed specimens, their dry weight can be used as an indication of age, particularly in medium-sized mammals such as the raccoon. The protein tyrosine increases in eye lenses throughout life and has been used to age both large and small mammals.

In birds, a dorsal pocket in the **cloaca** (the common chamber where the intestinal, kidney, and reproductive canals empty) termed the **bursa of Fabricius** is present in the young but disappears or becomes shallow in adults (Figure 8.15). The bursa is thought to be important to the immune system of young birds. The bursa can be used to estimate HY or AHY status for several months after fledging. For example, the bursa in a juvenile ring-necked pheasant male through the first November of life is normally greater than 10 millimeters in depth, while the bursa in an adult male is less than 8 millimeters. Bursal depth is also an effective means of determining HY or AHY status in ducks and geese until at least late fall or winter.

The shape and size of the penis, located on the wall of the cloaca in ducks and geese and a few other birds, changes with age and can be used to distinguish between HY and AHY male birds. In ducks, the age difference is very marked, and the method works well on live or recently killed birds. Another method of aging that uses soft body parts involves the oviduct opening in female waterfowl, which is easier to detect in adults that have produced eggs than in immature waterfowl.

In some animals, the developmental stage of the gonads, either ovaries or testes, can indicate age. For example, an immature brook trout will have undeveloped gonads, while a mature brook trout, even if similar in body length to the immature fish, will have developed gonads. This difference in gonadal development is especially apparent in or near the breeding season. Similar differences in gonad development between immature and adult mammals and birds also occur.

Other aging methods are also used. Because fishes exhibit indeterminate growth, a procedure called **length-frequency analysis** can be used in fisheries. Fish of similar age in a particular water body tend to be of approximately the same size. If the lengths of a large number of fish from a particular population are plotted on a graph (Figure 8.16), there will be a tendency for like-aged fish to occur within similar length ranges, although the length range tends to increase for older age groups. This method is more effective for some species than for others, and it is usually most effective in distinguishing younger fish (perhaps up to age 1 or 2).

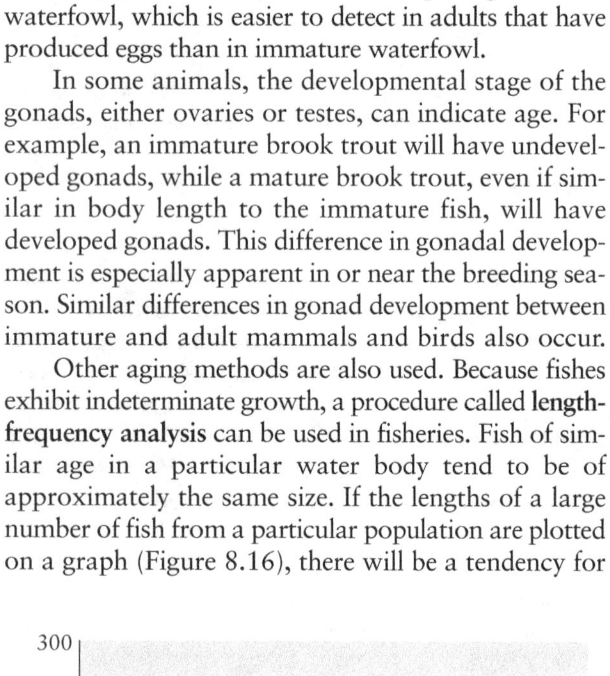

FIGURE 8.16 Length frequency for a black crappie sample obtained during spring using a modified-fyke net. Horizontal lines indicate the length range for each age group (1–5). As the fish age, there is more overlap in length between age groups (cm = centimeters).

Many additional characteristics can be mentioned that separate the young of a species from more mature individuals. The spots on a deer fawn or young American robin or the off-white coloration of a HY white-phase lesser snow goose are examples. Other examples of such characteristics are parr marks on young salmonids, eye (iris) and mouth lining color changes in some birds, development of the scrotum or teats in small mammals, the **breeding tubercles** (bumplike, epidermally derived skin structures usually used during reproductive behavior) found on fishes such as the male hornyhead chub, and a variety of other structural or color differences exhibited by either mature or young organisms.

8.4 Use of Age Information

Age information can be used in a variety of ways. The most common uses involve measurements of population age structure and growth. A graph of the age structures of two white-tailed deer populations was presented in Figure 3.7 along with an account in section 3.7 of the way that information was used to interpret the effects of harvest strategies on populations of white-tailed deer. The age structure information used to evaluate harvest strategies mentioned above was gained through collection of deer incisors from hunters followed by age determination from cementum annuli. The hunter records the area of the kill and the gender of the deer and encloses an incisor in an envelope that is mailed to a biologist.

Similarly, Figure 3.8 depicts the age structures of two different fish populations, one with consistent and the other with inconsistent recruitment. Fisheries biologists also assess mortality from age structure. Decreased proportions of older individuals can indicate excessive mortality. Figure 8.17A depicts the age structure of a sample collected from a lightly exploited northern pike population that might occur in southern Minnesota. The same hypothetical population, but with excessive angling mortality, is depicted in Figure 8.17B. Age determination alone is seldom used in fisheries work other than for assessment of age structure; in most cases growth information is also determined.

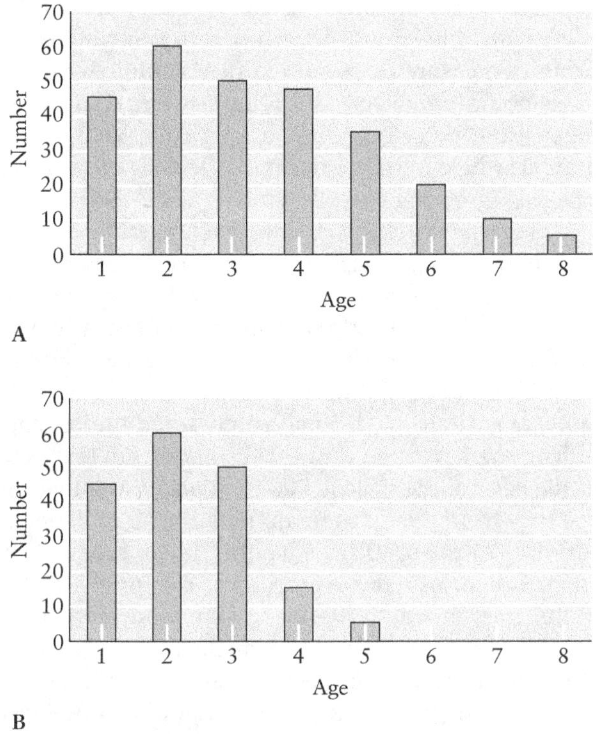

FIGURE 8.17 Age structures of *(A)* a lightly exploited population and *(B)* a heavily exploited population of northern pike.

Age structure data are also used in the management of populations of large animals such as cougars, American black bears, and American alligators. If such populations are being hunted excessively, their age structures are likely to exhibit a lack of older and larger animals in the harvest. Thus, hunting quality may be deteriorating. Of course, biologists managing such a population would also be closely observing recruitment rates.

Hunting regulations are sometimes based on visible age indicators. For example, in many bighorn sheep populations, only rams with a 3/4 curl or greater (horn curl equal to or greater than 3/4 of a full circle) are allowed to be harvested. Thus, biologists may base the number of permits on the number of 3/4-curl rams in the population to allow some harvest and still maintain an adequate number of older rams for aesthetic (watchable wildlife), breeding, and future harvest purposes.

Each year the U.S. Fish and Wildlife Service contacts a number of waterfowl hunters to request their assistance in saving and returning duck wings and goose rectrices for later identification, aging, and sexing (ducks only) by waterfowl biologists. These data provide information on the composition of the harvest by species as well as by age and gender. The ratio of young to adults, when corrected for age-related differences in vulnerability to hunting, provides an indication of annual recruitment. For most waterfowl species, recruitment information is also obtained through the annual aerial surveys undertaken by the U.S. Fish and Wildlife Service and the Canadian Wildlife Service.

For many species, such as eastern fox squirrels, ring-necked pheasants, eastern cottontails, and greater sage grouse, the long-term data obtained from harvested individuals have provided a general model of expected mortality or turnover rate in the population. The base of an **age pyramid**, representing animals less than 1 year of age, is generally broad, with a decline in subsequent layers or age classes (Figure 8.18). In age pyramids for birds, the age categories used are generally HY and AHY (just two tiers) because of the difficulty of estimating the ages of AHY birds to the nearest year. In most gallinaceous birds and in many small birds, such as red-winged blackbirds and American robins, 50%–80% of the total population in the fall is normally in the HY age category. Similar age pyramid characteristics are typical for a wide variety of short-lived species, including mourning doves, common muskrats, and eastern cottontails.

In terms of management and conservation value, information on age ratio (ratio of age class 0 individuals to all those in age class 1 or higher) may prove worthwhile in long-term studies relating environmental conditions to recruitment. There are, however, pitfalls in collecting age ratio data in harvested

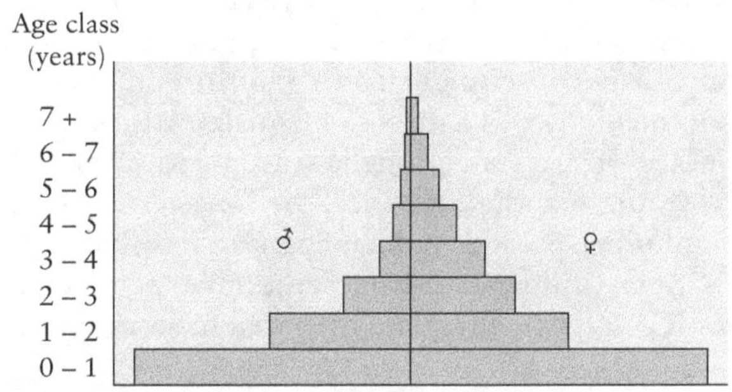

FIGURE 8.18 Age classes of white-tailed deer can be organized into a pyramid to illustrate the mortality rates in both the male and female segments of the population. Males have higher mortality than females after the first year because of hunter selection for antlered deer.

species because the young are generally more vulnerable to hunting or trapping than the adults. Similar biases may also exist in age ratio data obtained from live capture and release of game and nongame species. The data are still valuable, but the bias should be considered. In addition, if the adults have unusual and undetected spring or summer mortality so that few survive to reproduce in comparison to previous years, the ratio of young to adults might appear excellent and yet total recruitment could be low.

In a ring-necked pheasant population, the ratio of young to adults can be used to estimate the production per hen. Further, the ages of HY birds, based on primary molt progression, can be used to determine a **hatching curve** (plot of the number of clutches hatched in relation to the progression of the reproductive season). All of this information is biologically interesting. However, the benefits and costs of gathering annual age ratio data from some harvested species, such as the ring-necked pheasant, should be considered. Only male ring-necked pheasants are usually hunted. As the ring-necked pheasant hunting season progresses, the males become increasingly more difficult for hunters to flush within shooting range, and hunter effort and success usually decline. Adequate numbers of males will remain for breeding purposes in this polygynous species, regardless of season length or population numbers at the beginning of the season. The length of the season and bag limit are therefore of little consequence to year-to-year ring-necked pheasant populations in this naturally short-lived species. A biologist would be likely to recommend a long season and the same year-to-year bag limits (such as three per day) for a male-only season regardless of what population estimates in midsummer indicated.

Age ratios in sharp-tailed grouse and other upland game populations may have little practical value unless biologists have a specific purpose in mind. Such a purpose might be to inform hunters of recruitment rates as a means of explaining their success or lack of success during the season. However, if one is evaluating the influences of long-term weather conditions, possible habitat changes, or other factors that may influence recruitment, the expense of collecting age structure data may well be justified.

8.5 Measurement of Growth

Growth determination usually involves age data because growth is generally assessed per unit of time (per year, for example). Thus, an increase in weight or length per individual per unit of time is a measure of growth. Increase in length per unit of time is the primary measure of growth in fishes, while weight is more commonly used in wildlife species. Growth is most often determined for a particular year, but it may also be determined on a daily basis or, in some cases, by growing season.

Fisheries biologists can back-calculate body length at specific ages from hard body part lengths because the growth of fish body length and the growth of hard body part length are generally proportional (Figure 8.19). Assume that a redear sunfish is 150 millimeters in total length and has two annuli. When a scale from this fish is measured under magnification, it is found that the distance from the focus to the edge of the scale is 100 millimeters and that the distance from the focus to annulus 1 is 40 millimeters (or 40% of the total scale length). It can then be calculated that this redear sunfish was approximately 60 millimeters in length (0.40×150 millimeters) at the time the first annulus was formed. A variety of scale magnification techniques are used by biologists; scales are magnified so that the annuli can be more easily located and measurements more accurately made. The total scale length and length to annulus 1 are really much smaller than the 100- and 40-millimeter figures provided; these lengths occur because the scale is magnified. It does not make any difference what magnification is used because the proportional relationship between the two measurements is what matters. However, the same magnification must be used throughout the procedure.

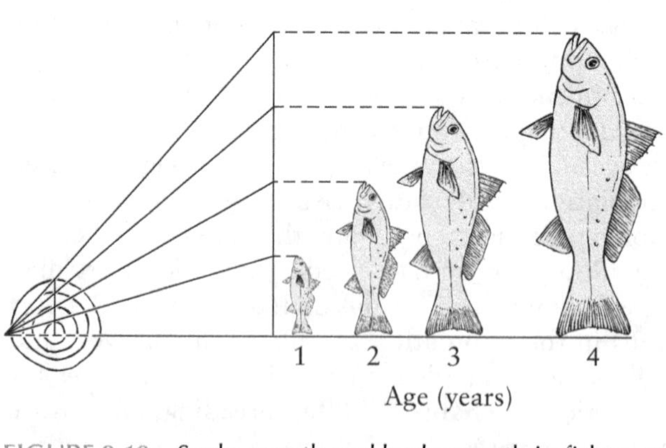

FIGURE 8.19 Scale growth and body growth in fishes are often directly proportional; therefore, body growth in length can be determined for past years using back-calculation from scale growth.

The methods used by fisheries biologists to back-calculate length at a specific age are actually more complex than this simple redear sunfish example would indicate. Body length and hard body part size are essentially directly proportional, but hard body parts do not necessarily start to grow or appear before the fish has developed some length. For example, a newly hatched fish has body length, but its scales may not yet be formed. This difference requires adjustments when back-calculating length at a particular age. A variety of length back-calculation methods take these differences into account, but they are beyond the scope of this text. Busacker et al. (1990) and DeVries and Frie (1996) described some of these methods. It is important to remember that the use of hard body parts in back-calculating fish lengths provides growth information with regard to prior fish length, not prior fish weight.

In wildlife, growth is usually determined by evaluating the weight of the animal in relation to age and gender. Captured small mammals, such as northern grasshopper mice, can be placed in a cloth bag and weighed; the weight can be determined by subtracting the bag weight. Likewise, captured ducks, songbirds, and many other small- to intermediate-sized animals can be weighed in this manner. In large ungulates, weight along with age and gender is sometimes obtained for field-dressed animals at bag check stations. Weights can be taken for any fish species, either by direct weighing or weighing in a water-filled container, with the weight of the container and water subtracted.

8.6 Use of Growth Information

Growth rate information is particularly important in fisheries and of lesser importance in wildlife because fishes, unlike wildlife, exhibit an indeterminate growth pattern (see section 1.3). In some fisheries applications, age is determined only in order to obtain growth data. The following are examples of the potential uses of growth information.

Fish growth is often a reflection of food availability in relation to population size. A decrease in food base (prey) or an increase in predator fish abundance can lead to predator populations with slower growth; the converse is also true. Biologists can collect annual growth rate information and, from these data, determine the adequacy of the prey base for the number of predator fish present. Sampling of prey fishes to determine their growth rates and age structure would be needed to more fully understand the status of the predator–prey relationship. If the predator fish population is exhibiting slow growth rates and prey species are found to be in low supply relative to the predator population, the biologist can modify harvest regulations or, if the predator is a stocked fish, modify future stocking rates. In addition, habitat modifications might be implemented to address such a problem.

Fisheries biologists commonly use comparative data to assess fish growth. Many conservation agencies have determined average growth rates for fish species in their region. In addition, regional summaries of growth data are available for many fish species (Carlander 1969, 1977, 1997). For example, smallmouth bass average 107, 202, 282, and 346 millimeters at annuli 1 through 4 in the southeastern United States. If a smallmouth bass population in a Tennessee lake were found to reach 390 millimeters at age 4, then by regional standards those fish would be considered fast growing.

Growth rates for wildlife species can also yield useful management information, particularly if comparisons can be made with populations from other areas or years. In wild ungulates, overpopulation can lead to poor nutrition and animals that lack the size and weight of similar-aged individuals under less crowded circumstances. For instance, weights of male elk in Yellowstone National Park might be compared with those of male elk of the same age in the Snowy Range near Laramie, Wyoming, to evaluate the growth rates of individuals within the two populations. Such comparisons can provide information on relative growth rates, which may reflect the availability of nutrients in particular habitats or during specific periods of time. Growth rate reductions in lesser snow geese, as reflected in adult size and weight, have been related to the depletion of forage plants in the Arctic nesting grounds of some populations of geese (Cooch et al. 1991). Thus, the biologist, by observing growth rates (or simply weights) and maintaining long-term records, can identify situations in which the habitat of a species is declining in quality.

Growth information on birds and mammals can be obtained from harvested animals, but this practice is not widespread. In general, body size is relatively constant within a species in a region. Thus, weight is not usually an important criterion for biologists or hunters and is usually not recorded. Weights in many birds and mammals, as in fishes, can vary markedly during the day depending on feeding patterns; weights can also vary during the year in relation to physiological changes and energy needs associated with events such as reproduction and migration.

8.7 Determination of Gender

Many animal species lack external **sexual dimorphism** (differences in form or shape due to gender) or **sexual dichromatism** (differences in color due to gender) during all or part of their lives. In such organisms, biologists may want to use genetic approaches based on DNA or other molecular differences that can be used to determine gender with noninvasive approaches. Genetic approaches to gender determination are particularly appropriate for work with endangered or threatened species and have been tested for a variety of birds and mammals. Use of DNA or other genetic material has also proven useful in determining the gender of sea otters, polar bears, and American beavers from harvested furs. In addition to gender determination, examination of genetic material in feces has been used to identify individual animals and monitor aspects of population dynamics in species such as coyotes and wolves (Lucchini et al. 2002; Prugh et al. 2005).

The gender of animal species lacking sexual dimorphism or sexual dichromatism can also be determined or confirmed through dissection to observe the presence of testes or ovaries, but this requires the availability of harvested, collected, or incidentally killed organisms. Even organisms that are not sexually mature can often be sexed based upon internal structures. The testes of fishes are usually elongate in shape, paired, and located in the upper portion of the body cavity. The ovaries of fishes are also dorsal, usually longitudinal (lengthwise direction), and paired. As a fish matures, the ovaries can expand to constitute as much as 70% (more typically 5%–15%) of the total weight in some species (Lagler et al. 1977). Mature testes in fishes seldom constitute more than 4%–5% of the total body weight. When the ovaries are mature and expanded, gross examination of the gonads will readily identify the gender of the fish (Figure 8.20). In sexually immature fishes or during nonbreeding periods when the gonads regress, microscopic examination may facilitate differentiation between ovaries and testes.

Unlike those of most mammals, bird testes are internal and may be cooled by their position in relation to the air sacs. The paired testes in birds are located dorsally near the anterior portion of the kidneys. They are whitish and generally shaped liked a kidney bean. Testes are readily identifiable during the breeding season, when they are enlarged. During the nonbreeding season the testes are small but can usually be located and identified internally when dissecting a bird that has recently died from disease or other causes.

FIGURE 8.20 The ovaries (arrows) in this coho salmon are filled with mature eggs and indicate it would have spawned in the next few weeks

The ovary in birds is composed of many **ovarian follicles**, each of which may develop into an ovum for release into the oviduct. Almost all female birds have only a single ovary and oviduct (on the left side), an adaptation that saves weight and may reduce cracking of egg shells in the lower oviduct. During the breeding season, the ovary in birds resembles a cluster of grapes (Figure 8.21); follicles from which ova have been released resemble a grape in which the skin has been split and the pulp removed. The oviduct is enlarged and easily identifiable during the reproductive season. During the nonbreeding season the ovary and oviduct are greatly reduced in size and are more difficult to identify; the quiescent ovary resembles a cluster of sand granules. By first locating the ovary in reproductively active birds, a biologist can obtain the experience needed to identify ovaries in reproductively inactive birds. In general, ovaries are more difficult to identify during the nonbreeding season than are testes.

In mammals, internal identification of gender is seldom necessary due to the dimorphism of external features. However, if needed for gender identification, the uterus in females is normally identifiable and is located just below the bladder in the abdominal region.

External sex organs can usually be employed to determine gender in mammals, although identification can

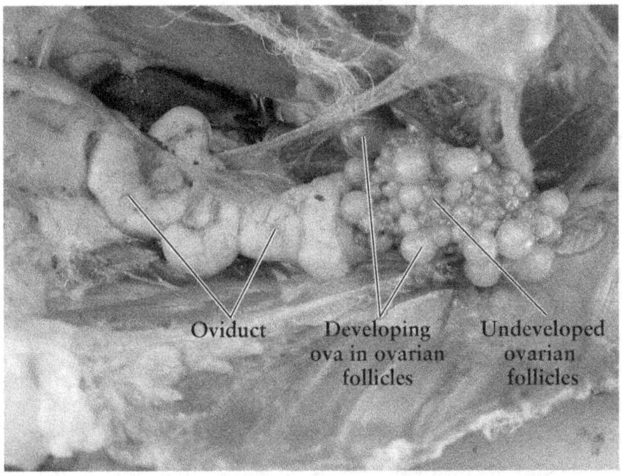

FIGURE 8.21 The ovary, composed primarily of ovarian follicles, and the oviduct of birds, such as this female mallard, become greatly enlarged during the breeding season. The same structures can be difficult to locate during nonbreeding periods.

be difficult during the nonbreeding season in some mammals, such as the North American deermouse. During the breeding season, the scrotum in most rodents swells greatly as the testes descend through the **inguinal canal** to enter the scrotum and expand to a large and easily identifiable size. In many mammals, such as raccoons and coyotes, there should be no difficulty in identifying the scrotum or the penis with its bony inclusion (baculum). In studies of a number of small mammal species, the condition of the scrotum has been used to identify the period of reproductive activity. Likewise, the enlarged teats of female mammals are evident during the reproductive season, especially during lactation.

There are a variety of other external characteristics that allow for gender determination in various mammal species. During much of the year, the presence or absence of antlers differentiates adult males from females and juveniles in most deer species. The size of the horns in pronghorns, much larger in males than in females, can be used to differentiate the sexes. The size and shape of horns also varies between male and female American bison.

Although not as common as in birds and fishes, sexual dichromatism is exhibited by some mammals. The dark neck patch and snout (mask) of male pronghorns, for example, is distinct from the lighter color and pattern of the female. The mandrill, a species of baboon found in equatorial West Africa, exhibits reds and blues in the area of the nose, face, and buttocks; these colors are more intense in adult males than in females.

Determination of gender by external appearance is possible in birds that display sexual dichromatism. However, winter plumage (basic plumage) and juvenal plumage can at times make it difficult or impossible to determine gender in many species. In some HY birds, differences in feather color or shape can provide clues as to gender. Duck wings, even if in drab juvenal plumage, can generally be identified by gender. Adult duck wings can be particularly useful in determining the bird's gender, based on differences in color, color pattern, or feather shape. For example, the color of the **speculum** (a colored, sometimes iridescent patch on the secondaries; see Figure 8.11) can be used to sex some duck species such as northern pintails. The color and shape of the elongated proximal secondaries can be used to determine gender in many duck species (Carney 1992).

Feather size and color can be used to determine gender in many species that are superficially similar in their plumage. In greater prairie-chickens (Figure 8.22) and sharp-tailed grouse, the barring patterns on the tail and crown (top of the head) can normally be used to determine gender. The **pinnae feathers** (elongated feathers on the neck) (see Figure 8.22) of greater and lesser prairie-chickens are longer in males than in females. The presence of two dots on the rump feathers of a ruffed grouse indicates a male; females have a single dot. Male ruffed grouse also have longer rectrices than similar-aged females. In northern bobwhites, the white on the head of

FIGURE 8.22 Pinnae feathers on the side of the neck are longer on greater prairie-chicken males than females; the tail (rectrices) and dark crown on the head of males normally lack the light and dark crossbarring characteristic of most females. In courtship, the male greater prairie-chicken often holds the elongated pinnae and the rectrices erect as shown here. *[Image courtesy of D. Backlund.]*

the male versus the buff on the female distinguishes the sexes. In mourning doves, the pinkish rosy hue on the facial and throat area and pinkish iridescence on the neck are usually more distinct during the breeding season in males than in females. There are numerous other methods for determination of gender in birds that have sexually similar plumages as adults or that must be sexed while in the juvenal plumage (U. S. Department of Interior and Canadian Wildlife Service 1977; Schroeder and Robb 2005) (also see Birds of North America Online; http://bna.birds.cornell.edu/BNA/; Cornell Lab of Ornithology, Ithaca, N.Y.).

Some birds also have other characteristics that can be used to determine sex, such as the beard, spur, and **wattle** (fleshy lobe on throat) of male wild turkeys, bill color in many ducks, or the external air sacs in greater prairie-chickens. The presence or absence of a penis in ducks and geese can be used to determine gender.

Many fishes likewise can be sexed by external appearance, especially during the breeding season. Male orangespotted sunfish have bright orange spots, while the spots of females are rusty orange in color. Mature fathead minnow males have dark coloration compared to females. In many salmonids, males develop prominent hooked jaws (**kype**) and a humped back during the spawning season. In some fishes, such as the channel catfish, bluegill, and muskellunge, differences in the shape, size, and direction of the urogenital opening can be used to determine gender. The swollen body shape of a female fish carrying eggs can also be used to determine gender in some situations. The presence of breeding tubercles on various parts of the body, especially during the breeding season, is characteristic of males in many fish species. In some fishes the size and shape of the **genital papillae** (fleshy protuberances through which eggs and sperm are discharged) also differ between males and females. Fin shape and size can also vary. The male fantail darter has knoblike structures on the spine tops of the dorsal fin, while the mature Arctic grayling male has a larger dorsal fin than the female. There are a variety of color, fin shape, body shape, and other sexual differences exhibited by numerous fish species.

8.8 Use of Gender Information

The ratio of antlered to antlerless individuals in the harvest of some large ungulates can provide important information under certain harvest regulation schemes. For instance, if biologists were concerned about overcrowding of white-tailed deer in an area, information on the antlerless kill under an any-deer permit system would be important. Do hunters, given the opportunity, take enough does to control or reduce recruitment the following year, or do they concentrate heavily on antlered deer, having little effect on population numbers? Biologists have found that any-deer permits do not result in the doe harvest level needed to control prolific deer populations. Antlered deer harvest in a deer season in which both antlered and antlerless deer are legal may exceed 70%, thus defeating the primary purpose of any-deer permits if population reduction is the goal.

Comparisons of pre- and postseason data on gender ratios have long been of value for making a change-in-ratio estimate of kill during seasons in which harvest is directed at a particular gender. As an example, if the preseason sex ratio in a ring-necked pheasant population were approximately 1 male:1 female and the post-season ratio were 1 male:5 females, the approximate percentage of kill on males could be estimated as follows: $([5 - 1] \div 5) \times 100 = 80\%$. The change in ratio is proportional to the kill of males, assuming that hunter kill of females is negligible. Appropriate corrections can be made to adjust for illegal hen kill.

Data on sex ratios in the harvest are of questionable value in many cases. Gender in many game species, such as the northern bobwhite, cannot be readily determined by hunters until the animals are harvested. In northern bobwhites, the sex ratios in the population and in hunter-killed birds should be near 1:1, and annual collection of these data would seem to be of little value. The same could be said for sex ratios in many other fish and wildlife species.

Sex ratio data have limited utility in fisheries work, except in culture situations. However, because fishes can vary by gender in regard to food habits, movements, size, growth, longevity, or other characteristics, there are situations in which information on gender can be useful. In many fishes, including northern pike, the females tend to be larger than similar-aged males.

Just being able to sex an organism can have much utility. Fisheries or wildlife biologists may specifically desire information on one gender. In biotelemetry studies, for example, it is usually important to identify the gender of an individual at the time of initial capture. Banding data from birds can be used to estimate gender-specific mortality if gender is identified at the time of banding. Hunting regulations that require the protection of one gender, such as male-only seasons, have been immensely successful; however, such regulations are possible only if the sexes are readily identifiable.

8.9 Implications of Age, Growth, and Gender Information

We have illustrated that information on growth rates, which depends on knowledge of age, is important in fisheries and less important in wildlife conservation and management. Determination of age and growth are both important in the evaluation and management of fish populations and communities (see section 9.4). In fishes, growth, movements, behavior, and susceptibility to anglers may be quite different by gender; therefore, biologists may also be interested in the gender composition of a fish population or the gender of individual fish being used in research.

Age and gender ratios as well as weights (growth information) can be important to wildlife biologists, but, as in fisheries, a specific objective should be identified before allotting time, equipment, and other resources toward the collection of such data. As previously indicated, the gender and age ratios in the harvest can be important in the management of large ungulates such as white-tailed or mule deer. However, sex ratio data for most smaller game species are usually not needed for their management. As in fisheries, knowing the gender and age of animals being studied in behavioral research, for example, in a radiotelemetry study, is important.

In many cases there are public relations benefits of sufficient importance to warrant collection of age, growth, and gender data on any animal population. A variety of age, growth, and gender information on wildlife and fish populations can be used in newspaper articles and other public forums to gain support for a management or conservation plan, such as adoption of a length limit for a fish species or the harvest of additional cow elk, which may not be initially accepted by the public. Face-to-face contact between biologists and users during the collection of such data can also serve as part of a public relations and law enforcement effort. Knowledge of how to sex and age fish and wildlife species can be used in other ways by professionals when working with the public. People are often interested in the age or gender of a fish, bird, mammal, or other wild organism, or how rapidly it is growing. The ability to explain and provide such information is a public relations asset. In addition, providing assistance in determining the gender and age of an animal to anglers, hunters, or members of the general public may gain added credibility for the professionalism of biologists. This credibility can be helpful to biologists when they address more important and controversial issues. Thus, it is important for wildlife and fisheries personnel to become competent in sexing and aging animals for more than just conservation and research purposes.

Biologists, particularly those in administrative positions, must define specific and measurable goals before committing resources to the collection of gender, age, and growth data on wildlife and fish populations (see section 20.6). This should not be construed to mean that there is little biological utility for such data; however, it does mean that it is not information that should be routinely gathered without thought to its purpose.

SUMMARY

Wildlife and fisheries biologists collect age, growth, and gender information from populations to provide information needed in management, research, and public relations. Fishes are generally grouped in age groups or year classes. Fishes in their first year of life, for example, would be in age group 0, while fishes in their second year of life would be in age group 1. The year class of a fish is simply the year in which it was hatched; year class is sometimes used in the same manner for wildlife. In addition, the term *age class* is often used in wildlife. This term can be used in a manner similar to the use of *age groups* in fisheries, but it is also used much more broadly in wildlife. In many wildlife species, particularly birds, the exact age in years after the first year cannot be accurately estimated. Therefore, age class designations such as juvenile and adult or, in birds, hatching year (HY) and after-hatching year (AHY) are used.

Age is generally determined from hard body parts such as scales, bones, spines, teeth, and feathers, or from soft body parts such as the bursa of Fabricius. Annuli on scales and annual rings on otoliths of fishes or the roots of mammal teeth are commonly used in age determination. In several mammal species the baculum changes in size and shape with age and can be used to determine age. Antlers are difficult to use in age determination because their development is strongly influenced by nutrition and genetics. Tooth replacement patterns and the presence or absence of the epiphyseal cartilage in long bones can be used in age assessment in mammals. In birds, the characteristics and replacement patterns of feathers are often used to determine age.

The bursa of Fabricius, a soft body part, is reduced in depth or disappears as birds mature. Some birds, such as waterfowl, have a penis on the wall of the cloaca that increases in size and changes in shape as the bird matures; it can also be used in gender determination. Colors in some mammals and many fishes and birds change with age.

Other characteristics, such as breeding tubercles and kypes in some fishes and enlarged external testes in small mammals, provide information on maturity and thus can be used for age assessment. In some water bodies, length-frequency analysis of a fish population can also be used to determine age for younger age groups.

Age data provide information on recruitment patterns in various populations. Such information may be valuable in evaluating the effects of various harvest or management plans on a population. Data on age in relatively stable populations can be used to develop age pyramids reflecting mortality rates in various age groups or classes. Age ratio data may be influenced by greater vulnerability of juveniles to various capture and harvest methods.

Growth is generally measured in terms of length or weight increase per unit of time. Growth in fishes is generally measured in terms of length, although weight is also sometimes used. Because fishes exhibit indeterminate growth, growth information is used much more often in fisheries work than in wildlife work. Growth data in terms of relative weights per age class can provide measures of the relative health of some wildlife populations, such as elk or lesser snow geese.

Determination of gender can be made from external differences in some species, especially during breeding seasons, or may need to be determined by internal examination or even by genetic approaches. Most small mammals have a readily visible scrotum, testes, or teats during the breeding season unless they are juveniles. Cloacal characteristics are commonly used in juvenile waterfowl to determine gender. Many birds are sexually dichromatic, and even juvenile birds, or birds in which both sexes are the same color as adults, may have certain feather color or body shape differences that can be used in sexing. Many fishes have color or shape differences, especially during the breeding season, that distinguish males from females.

Vertebrate species that have been harvested or otherwise killed can be dissected to determine or confirm gender. The testes, ovaries, and accessory structures (such as the oviduct in birds) are well developed during the reproductive season, but they are more difficult to locate during the nonbreeding season.

Biologists may need to determine the gender of target organisms because various species exhibit differences between sexes with regard to aspects such as movements, food habits, and habitat use. Biologists may also need to identify the sexes during marking or observation. Gender determination can also be important in the evaluation of certain harvest strategies designed to selectively harvest one gender.

For a biologist working with the public, it is important to be proficient in age, growth, and gender determination. Information on both game and nongame species regarding age, growth, and gender is often requested when a biologist interacts with human users.

PRACTICE QUESTIONS

1. Incorporate the terms *age group* and *year class* into one or two sentences about a fish to illustrate how these terms are used.

2. Define the terms *juvenile*, *subadult*, and *adult* and give examples of their use as applied to wildlife species. How is the term *juvenal* used?

3. What do the abbreviations HY and AHY mean?

4. How are the scales of many fish species useful in age and growth determination? Why are they used less for age interpretation of older fish compared to younger fish?

5. Describe how cementum annuli are used in age determination in mammals.

6. How can long bones in mammals be used in age determination?

7. Discuss how P9 and P10 feathers are useful in age determination in most gallinaceous birds.

8. What is the bursa of Fabricius and how is it used in age determination?

9. How can length-frequency analysis be used to obtain age information for a fish population?

10. List three ways in which information on the age structure of a population can be used by biologists.

11. Why can fish scales be used to back-calculate fish length at specific ages?

12. How can growth rate information be important in managing a fishery? How can it be important in managing a population of elk or other large ungulates?

13. Provide two examples each (other than those used in the text) of sexual dichromatism in fishes and birds.

14. Describe a situation where genetic determination of gender could be a useful tool.

15. How useful are antlers for age determination in white-tailed deer, mule deer, and elk? Why?

16. How can information on the gender of captured or harvested animals be useful in wildlife and fisheries?

17. How might age, growth, and gender information sometimes be of value in public relations?

18. Why is it important to validate age estimates?

SELECTED READINGS

Bagenal, T. B., ed. 1974. *The ageing of fish*. Gresham Press, Old Woking, U.K.

DeVries, D. R., and R. V. Frie. 1996. Determination of age and growth. Pages 483–512 *in* B. R. Murphy and D. W. Willis, eds. *Fisheries techniques*. 2nd ed. American Fisheries Society, Bethesda, Md.

Schroeder, M. A., and L. A. Robb. 2005. Criteria for gender and age. Pages 303–338 *in* C. E. Braun, ed. *Techniques for wildlife investigations and management*. 6th ed. The Wildlife Society, Bethesda, Md.

LITERATURE CITED

Beamish, R. J., and G. A. McFarlane. 1983. The forgotten requirement for age validation in fisheries biology. *Transactions of the American Fisheries Society* 112:735–743.

Brown, P., C. Green, K. P. Sivakumaran, D. Stoessel, and A. Giles. 2004. Validating otolith annuli for annual age determination of common carp. *Transactions of the American Fisheries Society* 133:190–196.

Busacker, G. P., I. R. Adelman, and E. M. Goolish. 1990. Growth. Pages 363–387 *in* C. B. Schreck and P. B. Moyle, eds. *Methods for fish biology*. American Fisheries Society, Bethesda, Md.

Caldwell, P. J. 1980. Primary shaft measurements in relation to age of sharp-tailed grouse. *The Journal of Wildlife Management* 44:202–204.

Carlander, K. D. 1969. *Handbook of freshwater fishery biology*. vol. 1. Iowa State University Press, Ames.

Carlander, K. D. 1977. *Handbook of freshwater fishery biology*. vol. 2. Iowa State University Press, Ames.

Carlander, K. D. 1997. *Handbook of freshwater fishery biology*. vol. 3. Iowa State University Press, Ames.

Carney, S. M. 1992. *Species, age, and sex identification of ducks using wing plumage*. U.S. Fish and Wildlife Service, Washington, D.C. Jamestown, ND: Northern Prairie Wildlife Research Center Online. http://www.npwrc.usgs.gov/resource/tools/duckplum/index.htm.

Cooch, E. G., D. B. Lank, R. F. Rockwell, and F. Cooke. 1991. Long-term decline in body size in a snow goose population: Evidence of environmental degradation. *Journal of Animal Ecology* 60:483– 496.

DeVries, D. R., and R. V. Frie. 1996. Determination of age and growth. Pages 483–512 *in* B. R. Murphy and D. W. Willis, eds. *Fisheries techniques*. 2nd ed. American Fisheries Society, Bethesda, Md.

Godin, A. J. 1960. A compilation of diagnostic characteristics used in aging and sexing game birds and mammals. MS thesis, University of Massachusetts, Amherst.

Kelso, W. E., and D. A. Rutherford. 1996. Collection, preservation, and identification of fish eggs and larvae. Pages 255–302 *in* B. R. Murphy and D. W. Willis, eds. *Fisheries techniques*. 2nd ed. American Fisheries Society, Bethesda, Md.

Lagler, K. F., J. E. Bardach, R. R. Miller, and D. R. M. Passino. 1977. *Ichthyology*. 2nd ed. John Wiley & Sons, New York.

Larson, J. S., and R. D. Taber. 1980. Criteria of sex and age. Pages 143–202 *in* S. D. Schemnitz, ed. *Wildlife management techniques manual*. 4th ed. The Wildlife Society, Bethesda, Md.

Lucchini, V., E. Fabbri, F. Marucco, S. Ricci, L. Boitani, and E. Randi. 2002. Noninvasive molecular tracking of colonizing wolf (*Canis lupus*) packs in the western Italian Alps. *Molecular Ecology* 11:857–868.

Phelps, Q. E., K. R. Edwards, and D. W. Willis. 2007. Precision of five structures for estimating age of common carp. *North American Journal of Fisheries Management* 27:103–105.

Piper, R. G., I. B. McElwain, L. E. Orme, J. P. McCraren, L. G. Fowler, and J. R. Leonard. 1982. *Fish hatchery management*. U.S. Department of Interior, Fish and Wildlife Service, Washington, D.C.

Prugh, L. R., C. E. Ritland, S. M. Arthur, and C. J. Krebs. 2005. Monitoring coyote population dynamics by genotyping feces. *Molecular Ecology* 14:1585–1596.

Sanderson, G. C. 1950. *Techniques for determining age of raccoons*. Biological Notes 45. Illinois Natural History Survey, Urbana.

Schroeder, M. A., and L. A. Robb. 2005. Criteria for gender and age. Pages 303–338 *in* C. E. Braun, ed. *Techniques for wildlife investigations and management*. 6th ed. The Wildlife Society, Bethesda, Md.

Severinghaus, C. W. 1949. Tooth development and wear as criteria of age in white-tailed deer. *The Journal of Wildlife Management* 13:195–216.

U.S. Department of Interior and Canadian Wildlife Service. 1977. *North American bird banding manual*. vol. II. U.S. Department of the Interior, Fish and Wildlife Service, Washington, D.C.

CHAPTER 9

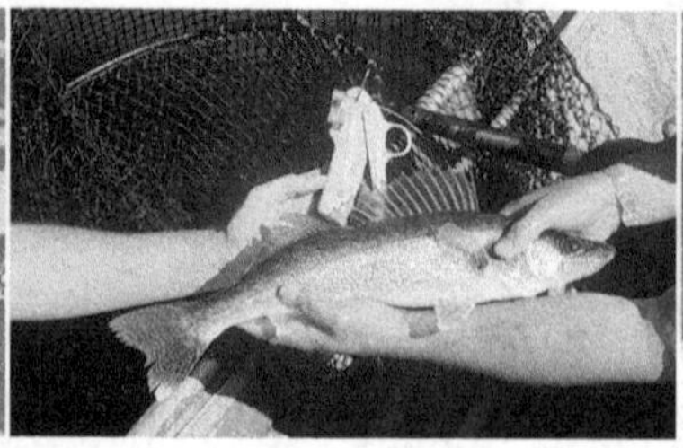

Population and Community Assessment

A biologist must understand certain aspects of a population or community before implementing any management or conservation strategy. Chapter 8 described the methods used to assess age, growth, and gender information for wildlife and fish populations. This chapter builds on that material by describing other population and community assessment techniques.

Biologists often are responsible for conserving a particular resource. This may involve a specific component of the biota, such as largemouth bass, black phoebes, or cougars, or it may involve entire animal communities or ecosystems. Regardless of the biota involved, some type of assessment of populations or communities is usually necessary. The type of assessment needed depends on the organisms involved and the information required. Even after initial assessment leads to the development and implementation of a management strategy, reassessment is needed to determine the success or failure of the implemented strategy (see section 20.6).

As an example, assume that you are working for a conservation agency as a biologist in charge of a private pond program. A pond owner is concerned because his or her pond contains only small largemouth bass. A corrective management strategy cannot be suggested until the population is understood. Does the pond contain a high density of largemouth bass that are slow growing (a stunted population)? If so, some small bass may need to be selectively removed so that the growth rates of the remaining fish will increase. Alternatively, the pond may contain a low density of small fish with fast growth rates because largemouth bass harvest by anglers has been excessive. In this case, the last thing the pond needs is further largemouth bass removal. Fish will need to be released after capture rather than being harvested, at least initially. Finally, after a period of time, the success of the chosen management strategy will have to be evaluated, and if the strategy was not successful, modifications will have to be made.

This chapter describes some of the more common techniques that are used to assess the biota and reviews some of the methods by which organisms are marked, tagged, or banded so that they can be identified during assessment. Most of the examples given involve population assessment. While biologists certainly are interested in communities and ecosystems, more assessment tools are available for assessing populations. Biologists must understand populations before they can understand communities, and understand communities before they can understand ecosystems.

9.1 Accuracy and Precision

Two characteristics of data describe their value: **accuracy** and **precision** (Figure 9.1). Accuracy refers to the closeness of a measure to the true value. The collection of accurate (unbiased) data requires knowledge and planning. For example, if knowledge concerning the size structure of a white crappie population in a Kansas reservoir is needed, fish can be sampled with a modified-fyke net. If the sampling is done during August, when most of the larger white crappies are offshore and are not likely to be caught with this near-shore gear, then the size structure and average size of white crappies in that reservoir will be underestimated. The sample will be biased rather than accurate.

Before proceeding to a discussion of precision, it will be useful to describe some basic statistical parameters that are often calculated for wildlife and fisheries data. Table 9.1 includes the weights measured for five adult female greater sage-grouse. The sigma (Σ) sign refers to the sum of the numbers.

The **range** for this data set is the difference between the largest and smallest observations in a sample.

$$\text{range} = 1.42 - 1.16 = 0.26 \text{ kilograms}$$

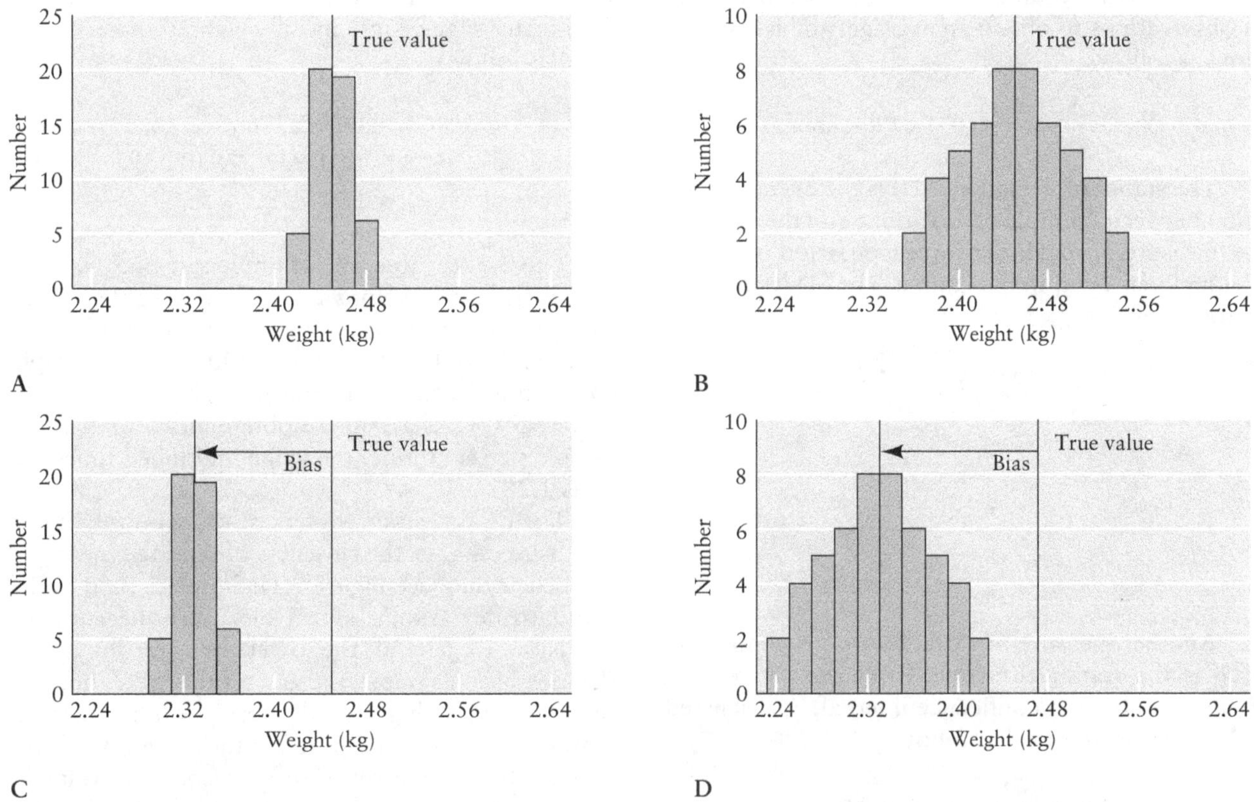

FIGURE 9.1 The difference between accuracy and precision. For each of the four graphs, assume that a different biologist made fifty repeated measures for a single variable, in this case the weight of an adult male greater sage-grouse that had an actual weight of 2.45 kilograms (kg). Each of the four biologists took the measurements at a different location, under different wind conditions, and used a different scale. Sample A is accurate and has high precision; B is accurate but has low precision; C is biased (inaccurate) but has high precision; D is biased (inaccurate) and also has low precision.

TABLE 9.1 Weights measured for five adult female greater sage-grouse and associated statistical parameters.

Sample (n)	Weight (kilograms) = Y	Y^2
1	1.16	1.35
2	1.27	1.61
3	1.31	1.72
4	1.39	1.93
5	1.42	2.02
$n = 5$	$\Sigma Y = 6.55$	$\Sigma(Y^2) = 8.63$

The range is a crude measure of variation, because it is solely dependent on extreme or outlying values.

The **arithmetic mean** ($\overline{Y}$) is obtained by summing all the observations and then dividing by the number of observations to obtain an average, and it is calculated as follows:

$$\overline{Y} = \frac{\Sigma Y}{n} = \frac{6.55}{5} = 1.31 \text{ kilograms}$$

The **standard deviation (SD)** is the average deviation between multiple observations and the mean. It is a measure of the dispersion or variation of the individual items within the sample. The SD is obtained as follows:

$$\text{SD} = \sqrt{\frac{\Sigma(Y^2) - \frac{(\Sigma Y)^2}{n}}{n - 1}}$$

$$= \sqrt{\frac{8.63 - \frac{(6.55)^2}{5}}{4}}$$

$$= 0.11$$

Another measure of variation that is commonly used is the **standard error** (**SE**) of the mean. It is needed to obtain a confidence interval, as discussed below, and is obtained as follows:

$$\text{SE} = \frac{\text{SD}}{\sqrt{n}} = \frac{0.11}{\sqrt{5}} = 0.05$$

A **confidence interval** (**CI**) can be calculated for many statistical parameters to assess their reliability. A CI is the interval within which a parameter has a certain probability of occurring. A 95% CI indicates that if a biologist takes 100 random samples from a population, 95 times the measured value will fall within the range indicated by the CI. The 95% CI for the greater sage-grouse mean weight above is calculated as follows:

$$95\%\text{CI} = \overline{Y} \pm 2.776\ (\text{SE})$$

$$= 1.31 \pm 2.776\ (0.05)$$

$$= 1.31 \pm 0.14$$

The 2.776 value for a sample size of 5 was obtained from a Student's t-distribution table, which can be found in any common statistics book (e.g., Sokal and Rohlf 1994; Moore 2003). When sample size is infinite, the value multiplied by the SE to obtain a 95% CI is 1.960. To obtain an 80% CI when sample size is infinite, the value multiplied times the SE is 1.282.

The term *precision* refers to the repeatability of a measure; that is, the closeness of repeated measures to one another (see Figure 9.1). Statistics such as the standard deviation about a mean are measures of precision. In general, the lower the variation about the mean for a sample, the greater the precision, and the more likely that a similar result will be obtained if another sample is collected in the same way. Compared with the data collected by scientists working in the controlled conditions of a laboratory, wildlife and fisheries field data tend to be quite variable. Thus, substantial sampling effort and a careful study design are often required if the level of precision is to be acceptable (see section 9.8).

One method by which the precision of a sample mean can be assessed is the use of confidence intervals. In the greater sage-grouse mean weight example (see Table 9.1), the 95% CI was 1.31 ± 0.14, or 1.17–1.45 kilograms. That is, if this same population were sampled 100 times, we would expect to obtain a mean weight between 1.17 and 1.45 kilograms 95 times out of 100. Only 5 times out of 100 would we expect to obtain a mean weight that was smaller or larger than the confidence interval. Biologists often use 95% CI values for research purposes, while 80% CIs may be used for management purposes. However, the actual confidence interval used may vary depending on study type and objectives.

9.2 Assessment of Population Number

Both wildlife and fisheries biologists commonly assess the number of individuals in a population. Wildlife biologists may estimate the number of mountain goats or lodgepole chipmunks in an area and summarize population density in terms of the number of individuals per square kilometer or some other unit of area. Fisheries biologists commonly discuss the number of fish per surface hectare of water or the number of fish in a particular portion of a stream. Assessments can also be accomplished for harvested organisms such as commercially captured fishes, shellfishes, or furbearing mammals. Some of the procedures for analyzing such data are quite complex. The type of procedure used depends on the methods of counting and the reasons for making the counts.

To assess a population or portion of a population, either a **census** or a **survey** can be conducted. The terms *census* and *survey* are often used interchangeably, which leads to needless confusion; these two terms have quite different meanings. A census is a complete count of the organisms present or harvested, while a survey is a partial count (sample). The results of a survey can be expanded into an estimate or may simply be used as an **index**. An index is a value, often a ratio, used in lieu of the actual number. For example, the number of longnose dace collected per hour of electrofishing in a stream might be used as an index to population density. An index can be obtained with less effort and expense than a population census. Both censuses and surveys can be conducted using either capture or noncapture techniques.

A census can be obtained for an entire population or for a portion of a population. Often, the harvested portion of a population is censused. If harvested organisms are involved, fisheries biologists undertake **creel censuses**, while wildlife biologists undertake **bag censuses**. Creel or bag censuses do not provide actual numbers of the population present, just numbers of organisms that are harvested. To do a creel or bag census, biologists obtain information from all anglers or hunters affecting the population in question over a defined time period. Common examples would include all fishes harvested from a reservoir in a 12-month period or the number of northern bobwhites harvested on a state-owned game production area during the fall hunting season. In addition to the total number of organisms harvested, a variety of information types, such as public opinion data, can be obtained through these censuses.

Censuses can also involve the counting of all organisms in a population, not just those harvested by hunters or anglers. A fisheries biologist may census a population of longnose gar by draining a body of water and counting all the individuals of that species. A wildlife biologist can census the population of whooping cranes in North America because the population size is small and the birds are highly visible and accessible on their wintering grounds; therefore, they can all be counted. At times, censuses of a portion of a population can be conducted. For example, biologists may census a particular breeding population of trumpeter swans.

Total censuses of animal populations are rarely undertaken because of the expense and logistical problems involved. However, censuses can provide valuable information on the actual number of organisms present in a population. Such data are often important for a thorough understanding of the biology of a species. For example, a census of brook trout populations in a series of Wyoming streams would allow a biologist to evaluate the environmental factors that are conducive to desirable population characteristics with much more confidence than would a partial count conducted by means of surveys.

Creel surveys or **bag surveys** are done more often than creel or bag censuses because they are less expensive and easier to conduct. Such surveys are typically expanded to an estimate of total harvest. Because surveys are only partial counts, biologists must use a variety of statistical methods to extrapolate an estimate of the total harvest. For example, assume that a technician is assigned to randomly check the number of ruffed grouse harvested by hunters on a public hunting area in northern Wisconsin. If the technician is at the checking station 1 out of every 4 days throughout the entire 60-day hunting season and finds that a total of 800 ruffed grouse were harvested on those 15 days (60 days ÷ 15 days = 4), then the total harvest for the entire season in that area was probably about 3,200 birds (4 × 800 = 3,200).

This ruffed grouse example is overly simple compared with most field situations. Hunting effort is generally not randomly distributed; for example, more hunting usually occurs on weekends than on weekdays. Thus, biologists often make separate estimates for weekend and weekday harvest. In this type of survey, **random-stratified samples** are collected from separate predetermined categories (stratifications). In this case, the stratification is the division of the survey into weekend and weekday periods; the random portion of the description refers to the randomly chosen sampling days within each of the two stratifications. Fisheries biologists also use random-stratified creel surveys. There are a variety of ways that samples can be stratified, including weekday–weekend, day–night, or shore anglers–boat anglers.

Regardless of the method used for such extrapolations, projected estimates should always include an associated range of values within which the probable true values occur. Most commonly, this takes the form of a confidence interval.

Relative abundance is based on information obtained from samples or observations and then used as an index to population abundance. Wildlife and fisheries biologists differ somewhat in their use of the term *index*. In the wildlife profession, an index often is a count or estimate of some aspect related to animal number. For example, to index common muskrat density, a biologist might count muskrat houses. The more muskrat houses, the higher the likely population density. Other examples of noncapture wildlife indices include mourning dove "coo" counts, deer track counts, coyote scent post surveys, and ring-necked pheasant brood counts.

Fisheries biologists commonly use the catch per unit of effort (CPUE) of fish captured with a particular sampling gear as an index to population density. Examples of such indices include the number of adult white catfish caught per net night in a hoop net, the number of age-0 walleyes caught per hour of night electrofishing during the fall, and the number of Utah chubs caught per hour a gill net is set. Thus, CPUE can provide an index to abundance. Relative abundance should not be confused with **relative species composition**, which is the proportional (percentage) numerical abundance of a species within a collection of species. Examples would include Topeka shiners composing 20% all fishes captured in a seining sample or common carp composing 90% by number of all fishes collected in a reservoir gill net sample.

Indices are not usually used to estimate total populations. Rather, they are used as comparative data over time or space. For example, a count of ring-necked pheasant broods along a particular road route at a particular time of day and year can be compared with brood counts done on that same route in previous years, or it can be compared with brood counts at other locations along a similar length of route. Appropriate use of relative abundance indices in fisheries requires that similar gears be used in similar locations at similar times of year if samples are to be compared over a series of annual samples. Even when indices are used, statistical methods should be employed to place a range of values about the average catch or observation so that sample precision can be assessed. Again, confidence intervals about a mean provide a common method of documenting the precision of such samples.

Some survey methods are designed to estimate the total population of organisms. **Population estimates** are made by obtaining a sample of the population and expanding those data into an estimate of total population number. Both capture and noncapture techniques are available. Wildlife and fisheries biologists each have somewhat different problems in observing the organisms for which they wish to make an estimate. Most wildlife species are mobile and can thus leave or enter a particular area, making it difficult to assess the population. Fish species are confined within the boundaries of water bodies but are usually not as visible to the human eye, making them difficult to count.

Population estimates can be obtained by capturing organisms. The simplest population estimates are single mark-and-recapture estimates. Consider a small plot of grassland that is surrounded by well-cultivated row crops on all four sides. The objective is to obtain a population estimate for meadow voles found on that piece of grassland. Four hundred box traps are placed in a square grid at 15-meter intervals over the entire field. The next day, seventy-seven meadow voles are removed from the traps, marked with a dye or paint, and released. After 3 days, meadow voles in the field are again trapped. Sixty-eight meadow voles are captured, and 21 of them bear marks that indicate they were among the 77 originally captured. The population estimate is obtained using the following formula:

$$N = \frac{(M)(C)}{R}$$

where

N = population estimate

M = number of meadow voles marked in the initial sample

C = total number of voles in the second sample

R = number of voles in the second sample that have marks (recaptures)

The estimated meadow vole population thus is $N = (77)(68) \div 21 = 249$. The 95% confidence interval for the population estimate is obtained using the following formula:

$$95\%\ \text{Cl} = N \pm 1.96\sqrt{\frac{(M^2)(C)(C - R)}{R^3}}$$

$$= 249 \pm 1.96\sqrt{\frac{(77^2)(68)(68 - 21)}{21^3}}$$

$$= 249 \pm (1.96)(45.2)$$

$$= 249 \pm 89$$

That is, if we sampled this population 100 times, 95 times we would expect the population estimate to be between 160 and 338 (249 ± 89). There are a number of important assumptions that must be met for such a population estimate to be valid (Van Den Avyle 1999).

The single mark-and-recapture population estimate described above is called the **Lincoln method** by wildlife biologists and the **Petersen method** by fisheries biologists. A more reliable population estimate can usually be obtained by using different methods to sample animals for the first and second captures. Consider the meadow vole example. If meadow voles are initially captured in box traps, those individuals may be less likely or more likely to enter the traps a second time, causing a bias in the population estimate. Thus, a wildlife biologist might use box traps initially to capture the meadow voles and pit traps for the second (recapture) sample.

A variety of more complex methods can be used to estimate population number (White et al. 1982; Van Den Avyle 1999; Lancia et al. 2005; Guy and Brown 2007). At times, these more complex methods may provide more precise and accurate estimates. The statistical techniques used for population estimation are chosen based on the type of method by which a particular population is sampled. For example, unique and complex estimation methods are typically used for commercially harvested marine fish species.

One example of a noncapture population estimation technique is **spot-mapping**, which is commonly used for songbirds that sing or call within established territories. This technique involves plotting the locations of individual birds on a gridded map during repeated visits by a biologist to a study area. Clusters of location sites then become apparent on the map. The total number of clusters in the study area equals the number of clusters completely inside the area plus the sum of the fractional cluster parts on the boundaries. For example, a cluster on the boundary that has 40% of observations inside the study area boundary and 60% outside would add 0.4 to the sum of fractional clusters. The estimated number of birds in the study population is then obtained by multiplying the number of clusters by the mean number of birds per cluster, which is commonly two if a breeding pair is present. As with population estimation techniques that involve captured organisms, a series of important assumptions must be met for an estimate obtained from the spot-mapping method to be valid. A variety of other noncapture population estimation techniques, such as drive counts, sighting probability models, and line transects, were described by Lancia et al. (2005).

Not all methods of population assessment are easily categorized as either a survey or a census. For example, a technique called cove rotenone sampling is sometimes used to sample reservoir fish populations in the southern United States. A small cove, often 1–2 hectares in surface area, is blocked with a net, and a piscicide such as rotenone is applied. Fishes are collected for 3 days after the addition of the toxicant. If all the fishes in the cove were collected, the sample would be a census of the fishes in that cove. However, recovery of fishes after piscicide treatment is generally incomplete. Some fishes do not float, others may be small and are therefore overlooked, and others may be eaten by animals such as American mink, turtles, or American crows before recovery can occur. If a correction factor for nonrecovered fishes is applied to the sample, it becomes an estimate and therefore would be categorized as a survey. If the fish numbers in a cove or coves were used to index or estimate populations for the entire reservoir, this count would also be a survey. Wildlife biologists also at times undertake complete censuses on multiple small areas and then extrapolate from these counts to obtain an estimate for a larger area.

Animal populations can also be estimated or counted by a variety of types of remote counting methods. In fisheries work, hydroacoustic techniques, which use sonar signals and the returning echoes from targets (fishes) struck by these sound waves, are used to count fishes. Wildlife population estimates or counts can be obtained by a variety of **remote sensing** methods, including visual surveys from aircraft, radar, remote photography, or thermal imagery (Lancia et al. 2005). In its broadest sense, remote sensing is the measurement or acquisition of information about some property of an object or phenomenon by a recording device that is not in physical or intimate contact with the object or phenomenon under study. The use of such methods for counting organisms is increasing. Other methods of sampling animals were discussed in Chapter 7; most of those methods can be used for population assessment purposes.

9.3 Assessment of Natality/Recruitment, Growth, and Mortality

Neither fisheries nor wildlife biologists are concerned only with the number of organisms in a population. They also measure the dynamic rate functions: natality/recruitment, growth, and mortality. Growth assessment was previously discussed in sections 8.5 and 8.6.

Wildlife biologists consider natality (birth or hatching rate) to be a measure of population health in many species. In general, more young are produced in healthy populations. Actual counts of young per female may be difficult to obtain because most wild animals bear (or hatch) and rear their young in relatively secluded locations. Thus, natality is often indexed as the number of young observed per female. Brood counts are undertaken by means of aerial surveys for many North American duck populations; roadside counts are also done for many upland game birds. Natality indices for large mammals such as white-tailed deer and elk can be obtained using doe:fawn or cow:calf ratios. Often such data are collected on standard survey routes during late summer, fall, or winter, when the animals are less secretive and more visible. Other methods of assessment include young:adult ratios in the harvest, examination of road-killed mammals for pregnancy rates, studies of nesting success, and biotelemetry studies to monitor reproduction by individual animals.

Fisheries biologists are usually concerned with the measurement of recruitment (individuals that survive to reproductive size) rather than natality. At times the number of eggs in ovaries is counted to obtain an estimate of **fecundity**. Fecundity is the potential number of reproductive products or units that can be produced by an organism. The term **year-class strength** is commonly used by fisheries biologists to refer to the abundance of a particular year class of fish. Year-class strength can serve as an index of recruitment. Because very young fish are likely to experience extremely high mortality rates, year-class strength is generally not determined until this period of extreme mortality has passed. Beyond this period, a fisheries biologist typically tries to determine the year-class strength of a fish species as early as possible in the life of that particular species, although year-class strength is set later in life for some fish species than for others. For example, walleye year-class strength is generally thought to be set by the late summer or fall of the first year of life. Thus, gill netting, bottom trawling, or night electrofishing in the late summer or fall are commonly used to determine an index to walleye recruitment (e.g., the number of age-0 walleyes caught per gill net night in a specific mesh size, the number of age-0 walleyes caught per distance or time a trawl is towed, or the number of age-0 walleyes caught per distance or time an electrofishing boat is operated). For other species, such as largemouth bass, year-class strength may not be established until spring, when fish have survived to reach age 1. Thus, the number of age-0 largemouth bass caught per seine haul in July may not serve as a reliable index to recruitment because these abundant young bass may still suffer very high or even complete mortality before the next spring. The number of age-1 largemouth bass caught per hour (or shoreline distance) of electrofishing in the spring is more likely to serve as a reliable index.

Mortality rates are determined by fisheries and wildlife biologists using similar techniques. However, wildlife biologists are more likely to calculate and report survival rates than mortality rates. In either case, remember that the mortality and survival rates al-

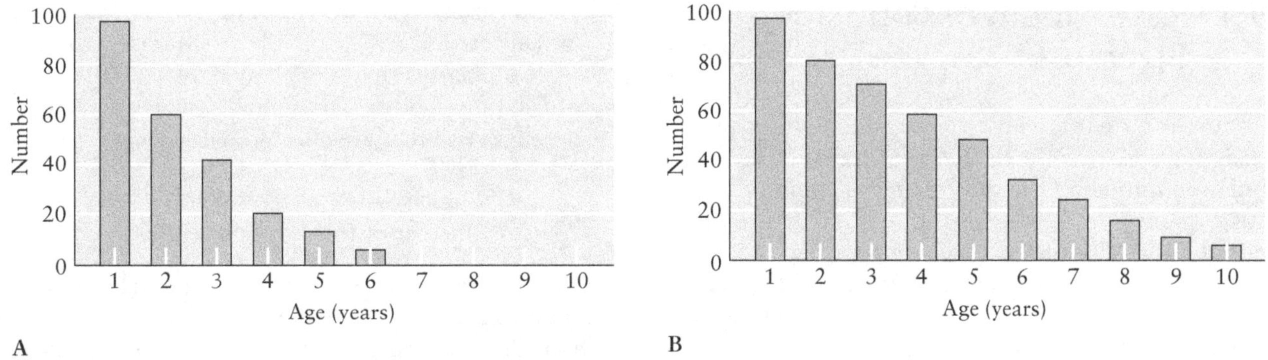

FIGURE 9.2 Age structures for samples from two populations of cutthroat trout. Population A has a higher mortality rate than population B, as evidenced by the more rapid decline in fish numbers for each age group.

ways add to 100%. That is, if total mortality is 40%, then 60% of the organisms must have survived.

The total mortality of an animal population can be determined by analyzing its age structure. Figure 9.2 depicts the age structures for samples from two populations of cutthroat trout. Population A has a higher mortality rate than population B, as evidenced by the more rapid decline in fish numbers for each age group. Mortality rates can be calculated from these data by the **catch-curve method** (Ricker 1975; Van Den Avyle 1999), based on the declining abundance of successive age groups. In Figure 9.2, the total annual mortality rate for population A is 44% (survival = 56%), meaning that on average, 44% of the population died in a 1-year period. The mortality rate for population B is only 27%.

Wildlife biologists use similar data to develop **life tables** (Caughley 1966; Dinsmore and Johnson 2005). A life table is simply a summary of the survivorship or mortality of a population from one age class to the next (Table 9.2).

Biologists also monitor natural and harvest mortality for harvested organisms. Natural mortality and harvest mortality together constitute total mortality for such organisms (see section 3.3). Information from marked animals can provide a basis for harvest mortality estimates. Wildlife biologists use banding analyses to determine harvest mortality for waterfowl and other migratory game birds. The relative return rates for bands also reflect survival to the years following banding and can be used to estimate survival rates for banded game and nongame birds (see section 9.10). Estimation of mortality or survival rates from banding data is challenging, and the statistical analysis involved is often complex. Similarly, if a fisheries biologist tags a certain number of the catchable-sized Arctic char in a population, the percentage of those tags returned after 1 year provides a minimum estimate of harvest mortality.

TABLE 9.2 Life table for a population of eastern fox squirrels.

Age (years)	Number[a]	Number of Deaths	Mortality Rate	Survival Rate
0–1	88	43	43/88 = 0.49	45/88 = 0.51
1–2	45	24	24/45 = 0.53	21/45 = 0.47
2–3	21	14	14/21 = 0.67	7/21 = 0.33
3–4	7	5	5/7 = 0.71	2/7 = 0.29
4–5	2	2	2/2 = 1.00	0/2 = 0.00

[a] Number of squirrels alive at the beginning of each age interval.

9.4 Assessment of Population Structure

Population quality can be interpreted either as the sizes and ages of animals in a population or total population number. Fisheries biologists commonly discuss quality in terms of the size structure of a population. Wildlife biologists most often use number as a measure of quality. Other aspects, such as antler size or body weight, can also be used as measures of quality for some animals.

Fish populations that consist primarily of small individuals are usually considered low-quality populations even if many fish are present. Fisheries biologists commonly have a management objective specifying that a certain percentage of the individuals in a recreationally or commercially harvested fish population be over a certain length. Thus, fisheries biologists managing under the concept of optimum sustained yield are generally concerned with the sizes of fish in a population as well as fish number.

Fisheries biologists often begin an assessment of population quality by determining population structure and plotting age structure (see Figure 9.2) as well as length frequency (Figure 9.3). In Figure 9.2, the cutthroat trout in population A have a higher mortality rate than those in population B. Therefore, population B has an older age structure than population A. The rock bass in population A of Figure 9.3 have a broader length range and reach larger sizes than those in population B, and most anglers would prefer to catch more of the larger fish found in population A.

Proportional size distribution (PSD) (Anderson 1976; Guy et al. 2007) is an index to the size structure of a fish population that is often used to assess the quality of fish populations. This index is the percentage of stock-length fish that are also of quality length.

$$\text{PSD} = \frac{\text{number of fish} \geq \text{quality length}}{\text{number of fish} \geq \text{stock length}} \times 100$$

The term *stock length* refers to the minimum length of a fish that can provide any recreational value to anglers. Fish smaller than this length are rarely caught by anglers. The term *quality length* refers to the minimum length of a fish species that anglers will harvest. Fish shorter than quality length are likely to be released after being caught. Stock and quality lengths are set by species and are based on standard percentages of world-record length for each fish species (Gabelhouse 1984). For example, bluegill stock length is 8 centimeters and quality length is 15 centimeters. For northern pike, stock length is 35 centimeters and quality length is 53 centimeters. Pallid sturgeon have a stock length of 33 centimeters and a quality length of 63 centimeters (Shuman et al. 2006).

Minimum stock length for rock bass is 10 centimeters, while minimum quality length is 18 centimeters. Thus, the PSD for population A in Figure 9.3 is 71; that is, there are 55 fish that are of stock length, and 39 of those are also of quality length [(39 ÷ 55) × 100 = 71]. The PSD for population B is 13 [(7 ÷ 52) × 100]. Populations with a high PSD are dominated by large individuals, while populations with a low PSD are dominated by small individuals.

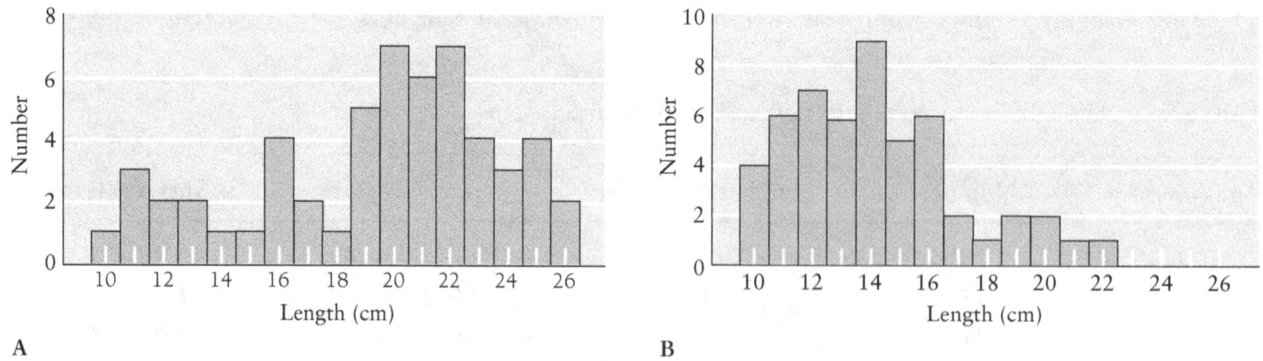

FIGURE 9.3 Length frequencies for samples from two rock bass populations. The proportional size distribution (PSD) for population A is 71, while the PSD for population B is 14. Minimum stock length for rock bass is 10 centimeters (cm), and minimum quality length is 18 centimeters.

Wildlife biologists less often discuss quality in terms of size structure, again reflecting a difference in philosophy between the fisheries and wildlife fields. Wildlife biologists more often equate quality with the number of organisms present. While there may be a desire by ring-necked pheasant hunters to bag a bird with long tail feathers, wildlife biologists do not manage populations with a strategy aimed at producing a certain percentage of birds with 55-centimeter tail feathers. Similarly, wildlife biologists do not manage Canada goose populations to ensure that 20% of the harvested geese exceed a particular weight.

Body size management may not be a common wildlife management objective, but it can be an indirect objective for some other management criterion, such as antler size. For example, in many geographic locations, yearling white-tailed deer bucks often have small antlers, such as single spikes or small forks, while older bucks often have three or more tines per antler. The antlers typically become heavier and larger during the first several years of a buck's life. While a conservation agency might manage white-tailed deer under a strategy that relates to age, the goal of that strategy would probably be to increase the number of deer with large antlers. The management objective for bucks might be a harvest composed of 60% yearlings (age class 1.5) and 40% older bucks (age class 2.5 and older). Of the 40% older bucks, the objective might be to harvest 25% that are age class 4.5 or older. Biologists would monitor the harvest by determining buck age from incisor teeth submitted by hunters or by using some other aging technique. If age class 2.5 and older bucks composed less than 40% of the harvest for a hunting season, the agency would be likely to decrease the number of buck licenses that are issued the next year.

9.5 Assessment of Physiological Status

Various physiological indices or indicators can be used to provide information for population assessment. For example, placental scars in some mammals, such as bobcats, appear as dark bands on the wall of the uterus. Because each fetus leaves a distinct scar, these marks can be used as a physiological index to reproduction. Likewise, a hormone-secreting body (**corpus luteum**) develops in the ovary and remains throughout pregnancy in mammals. Even after pregnancy, corpora lutea can be identified by slicing the preserved ovary into 1-millimeter-thick sections; they provide an index to pregnancy rates and numbers of fetuses per female in many ungulate populations, such as elk and mule deer. Measures of fat in the body cavity, in bone marrow, or around an organ such as the kidney can provide a physiological index to nutritional status in animals as diverse as white-tailed deer and rainbow trout. Analysis of urine samples collected from snow can be used to obtain information about the physiological condition of ungulates in the winter. Physiological indices can also be quite simple. For example, the mean weight of harvested yearling white-tailed deer bucks can provide a meaningful index to the nutritional status of the herd.

Blood characteristics can also reveal information about the nutritional status of an animal. A physiological index to stress can be obtained from blood chemistry or adrenal gland weight. Fish culturists can monitor blood chemistry to determine the health of fish being reared at a hatchery.

Fisheries biologists often use fish weight–length data to assess physiological status. Fish tend to become progressively heavier as length increases, which results in a curvilinear relationship between length and weight (Figure 9.4). Fish in a stunted population tend to weigh less at a given length than fish in a fast-growing population. Thus, fisheries biologists use fish weight–length data to calculate **condition** (relative plumpness). Several different types of condition

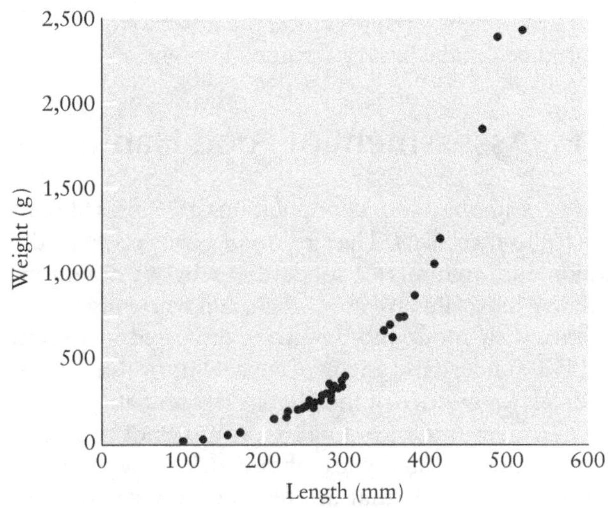

FIGURE 9.4 The relationship between weight in grams (g) and length in millimeters (mm) for a largemouth bass population sample.

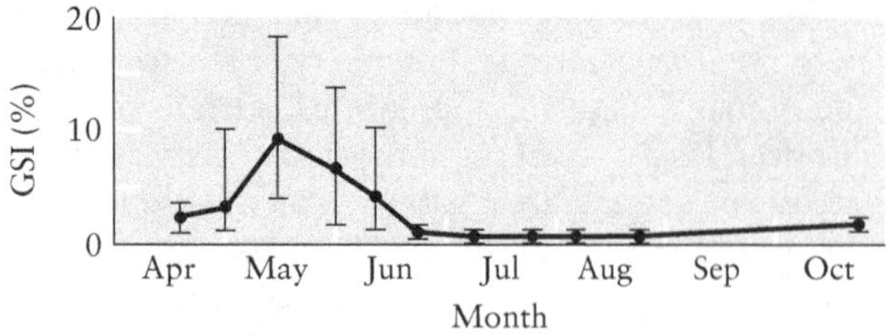

FIGURE 9.5 Gonadosomatic index (GSI) values for female gizzard shad collected from Melvern Reservoir, Kansas. The filled circle represents the sample mean; the vertical bar represents the range. *[Adapted from Willis 1987.]*

indices are used (Anderson and Neumann 1996). One commonly used condition index is **relative weight**, in which the actual weight of a fish is divided by a standard weight selected to represent an "optimal" weight for a fish of that species at that length and multiplied by 100. Fish with condition values of less than 100 are "thin," and those with values over 100 are "plump." Interpretation of condition indices is complicated by the fact that fish condition typically varies by season, by gender, and even by time since last feeding.

Seasonal changes in reproductive status in fishes can be monitored using the **gonadosomatic index (GSI)**. This index is obtained by dividing the weight of the ovaries or testes by the body weight of the fish and multiplying by 100. The GSI in most fishes increases rapidly just prior to spawning and declines after spawning (Figure 9.5). Seasonal changes in reproductive status for birds and mammals can also be monitored through changes in gonad size, weight, or appearance; the closer to the time of reproduction, the heavier the gonad weight.

9.6 Assessment of Food Habits

Very commonly, animal populations are assessed with regard to their diets. The term *food habits* refers to the kinds and amounts of foods eaten by an organism. Feeding habits involve how, when, and where organisms obtain their food, and they were discussed in section 6.11. To understand and manage wildlife or fisheries resources, knowledge of food habits is essential.

To determine food habits, samples of digestive tract contents usually are obtained; however, observation of animals that are feeding may provide information on the foods selected if an observer can view the animal at close quarters without disturbing it. In birds, food samples are normally obtained from the esophagus or crop, which are both located in the initial portion of the digestive tract (see Figure 5.8), to avoid biases caused by differential digestion rates for various food items. Another alternative is available in the case of raptors that regurgitate pellets containing remains of prey organisms. Raptor pellets commonly contain bones and other hard parts from the small animals the birds eat; these prey items can be identified by characteristics such as skulls, teeth, and feathers. Similarly, diet analyses for herbivores such as bighorn sheep can be conducted by analyzing their fecal pellets. However, fecal pellet analysis in ungulates is strongly biased toward the most indigestible items in the diet. The esophagus or crop of a bird can sometimes be flushed to obtain diet samples without killing the bird. The upper digestive tract of some shorebirds, such as pectoral sandpipers or long-billed dowitchers, can be pumped without injury to the bird. In some cases, as with small mammals such as Pacific jumping mice, black-eared deermice, or yellow-nosed cotton rats, the animal is usually killed to obtain its digestive tract contents.

Fish digestive tract contents can also be obtained by lethal or nonlethal techniques. Nonlethal techniques include inducing regurgitation, **gastric lavage** (flushing contents from the stomach with pumped water), or inserting glass or plastic tubes into the stomach and withdrawing most of the contents. These nonlethal methods do not always ensure that the total stomach contents are removed. Lethal techniques for obtaining fish digestive tract contents involve killing the organism so that all of the contents can be removed.

Foods eaten can vary by season, by habitat, by body size, and sometimes by time of day. Thus, one of the biggest challenges in completing a thorough food habits analysis for an animal population is to adequately sample individuals of all sizes, in all habitat types, at all times of the day and night, and in all seasons. A food habits analysis done in only one habitat type during one time of year is inadequate because it will provide only a partial documentation of food habits, or information on only a single window in time. For example, to study the food habits of flathead catfish in a reservoir, a biologist needs to sample all sizes of the fish, in deep and shallow water, in open water and along logs or boulders, throughout the entire year. Likewise, mule deer that occupy ponderosa pine forest areas in the summer have different diets from those summering in piñon–juniper forest areas; mule deer wintering in piñon–juniper forest areas are likely to have different diets from those living in the same area during the summer. Finally, food habits vary depending on food availability. For example, the most abundant prey of sand shiners in one stream may be quite different from that in another stream if the two streams have different prey availabilities.

Both wildlife and fisheries biologists tend to analyze food habits using similar methods. Food habits are generally described in terms of numerical analysis, frequency-of-occurrence analysis, gravimetric (weight) analysis, or volumetric analysis; each of these methods is described below.

In numerical analysis, the food items found in sampled animals are counted, and the percentage by number of each food type found in each animal or the entire collection of animals is then calculated. For example, 17% of all food items, by number, in a sample of digestive tract contents collected from brown trout might be mayflies. The weakness of this method is that the food most commonly ingested may not be the most important nutritionally. A rock bass may consume 99% zooplankton, but one ingested crayfish may provide more energy to the fish than all the zooplankton combined. Similarly, a wolf would obtain more energy from eating one snowshoe hare than one Arctic ground squirrel.

Frequency of occurrence is the percentage of the sampled organisms that consumed at least one of a particular food type. For example, a 25% frequency of occurrence for a particular kind of plant eaten by common muskrats means that 25% of the muskrats sampled contained that plant. This technique overemphasizes the use of persistent or highly abundant food items. For example, all golden trout in a California stream at a certain time of year may contain at least one caddisfly because those insects are extremely abundant at that time. However, the greatest proportion of the diet by weight or volume may actually be composed of large stoneflies that are hatching at the same time. The frequency-of-occurrence analysis might indicate that both food types are used equally, or even that the less-used food is found in more animals, but in reality they may have quite different levels of use.

In gravimetric analysis, food habits are expressed as the percentage of the total weight of food found in the animals that one food type composes. This method generally overemphasizes the use of large food items.

Finally, volumetric analysis expresses food habits as the percentage of total food volume that one food type composes. This method overemphasizes the use of large food items that are slowly digested.

Food habits studies generally need to be supplemented with information on the kinds and amounts of foods available, frequency of feeding, the nutritional value of the food eaten, and food preferences. In addition, food habits may change depending on the type and amount of competition that occurs or the vulnerability of prey to particular predators. More in-depth studies of food habits may examine daily food requirements, seasonal food requirements, efficiency of food conversion, stomach fullness, digestion rates, and indices that compare food eaten with food available.

9.7 Assessment of Community Structure

Community ecologists often face complex data sets and intricate mathematical analyses during measurement and characterization of community properties. The simplest level of community assessment involves species diversity (see section 2.1), which includes both the number of species present in a community (species richness) and the relative composition of species in that community (species evenness or equity). Thus, a logical first step in community assessment is to determine the number of species present and then assess the abundance of each species in proportion to the others (Figure 9.6). Quite often, such analyses are accomplished for a particular segment of the biotic community, such as the community of birds or trees. Community ecologists utilize various indices for determining diversity, such as the Shannon-Wiener index (Krebs 1999); the mathematical calculations involved in such indices are beyond the scope of this textbook.

Stream ecologists use an index of biotic integrity (IBI; Karr 1981) as a tool to assess the health of aquatic ecosystems. In effect, Karr used fishes as indicator organisms for biological monitoring. His original IBI involved the following parameters: number of fish species; presence of intolerant species; proportion of green sunfish; proportion of hybrid individuals; species richness and composition of darters (small minnowlike fishes in the perch family), suckers, and the sunfish other than green sunfish; number of individuals in a sample; proportion of omnivores; proportion of insectivorous minnows; proportion of top carnivores; and proportion of fishes with disease, tumors, fin damage, and other anomalies. These parameters accurately reflected the status of fish communities and the environment supporting them in Illinois streams. Since the original proposal of IBI, other authors have modified the index for use in different geographic locations (e.g., Fausch et al. 1984; Lyons et al. 1996).

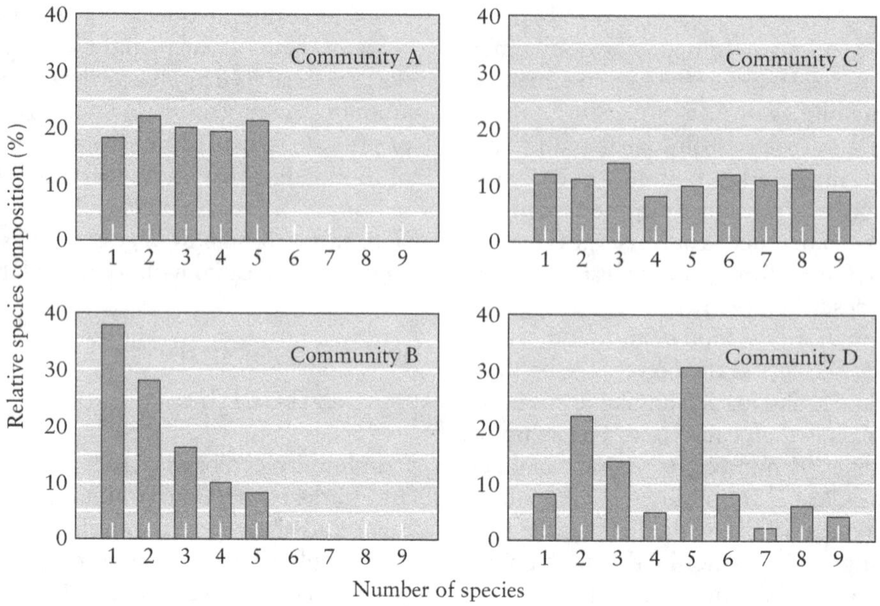

FIGURE 9.6 Community A and community B have lower species richness (n = 5 species) than community C and community D (n = 9). Community A has higher species evenness than community B, and community C has higher evenness than community D because the relative composition values for the species present are more similar in communities A and C compared with communities B and D. Note that relative species composition always sums to 100 for each community.

Other segments of the biotic community, such as invertebrates (e.g., Rabeni and Wang 2001), can likewise be used for bioassessment. Similarly, an index of plant community integrity was developed for assessing wetland plant communities (DeKeyser et al. 2003).

Many challenges remain for biologists wanting to assess community structure. Common examples include linking the geographic scale of an assessment to biological diversity (e.g., Beever et al. 2006) and measuring the success of attempted wildlife community restorations (e.g., Block et al. 2001; McCoy and Mushinsky 2002). These challenges are daunting from both a data collection and mathematical (statistical) perspective. As we move from populations to communities to ecosystems, and as we move from a small geographic scale to a large scale, assessment becomes more complex.

9.8 Sampling Design

Because field data tend to be variable, sample size and subsequent statistical validity are constant concerns for field biologists. Fisheries biologists using catch per unit of effort as an index to population density must use sufficient sampling effort to obtain reliable estimates. For example, compare the two gill net samples for American shad in Table 9.3. Both samples have a mean catch of 15 fish per net. However, the sample from population A has much more variation among individual net catches. The standard deviation for population A is 12.7, while it is only 3.7 for population B. The 95% confidence interval for the mean catch per net for population A is 15 ±13, while it is only 15 ± 4 for population B. As one more example, suppose that the same mean catch per net (15) and the same standard deviation (12.7) were obtained for population A, but that forty nets were used instead of six. The 95% confidence interval for such a sample would be 15 ± 4. As these examples demonstrate, the precision of samples is highest when the variation among individual net catches is low and when sampling effort is adequate.

Biologists are also concerned with **sample replication**. Replication can refer to repeated measures of samples or experimental units. However, Hurlbert (1984) warned that treating replicate samples as if they represented replicate experimental units is actually **pseudoreplication**. For example, assume that a study is designed to determine whether controlled burns affect

TABLE 9.3 Catches of American shad in experimental gill nets from two populations. The mean catch per net is 15 for both samples, but the sample for population A has much higher variation about the mean.

Net Number	Population A	Population B
1	29	10
2	1	20
3	24	15
4	2	18
5	8	12
6	26	15
Mean $(\overline{Y})$	15	15
Standard deviation (SD)	12.7	3.7
Standard error (SE)	5.2	1.5
95% confidence interval	±13	±4

the population density of chestnut-collared longspurs in grassland habitat. First, it will be necessary to obtain multiple samples from both burned and unburned areas so that the abundance of the birds, as well as the variation about that measure (such as the standard deviation or confidence interval), can be determined. However, the results of a study involving only a single burn would not be reliable. A biologist cannot simply obtain multiple samples on a single site prior to and after the burn. It is important that the burned areas and unburned areas be replicated at more than one site. If consistent trends in chestnut-collared longspur population density occur after burns are conducted on multiple study sites, and no increases in longspur density are observed at multiple unburned study sites, then biologists will be more confident that the effect was caused by the burns.

Because of the variation common in field sampling data, most biologists use **standardized sampling** procedures. Standardized sampling means that sampling devices are chosen that will effectively sample the target species and that the same sampling devices are used (as much as possible) at the same sites, at the same times of year, and from year to year. The proper method for selection of sampling sites is much debated. Some biologists subjectively choose sampling sites that they believe will allow effective sampling of the target species, while other biologists believe that sampling sites should be randomly selected. Even among those who prefer random site selection there is disagreement. Some biologists randomly choose the initial sampling sites, then return to those sites for samples in subsequent years. Still other biologists prefer to select sampling sites randomly each year.

The purpose of standardized sampling is to minimize variation due to sampling device efficiency and seasonal changes in sampling data. Long-term trends in population structure and dynamics can then be more reliably monitored. For example, wildlife biologists can monitor the relative abundance of coyotes using a scent post survey. Small, flat containers of scent are placed along dirt or gravel roads. All marks and tracks are smoothed from the soil surrounding the scent container so that coyotes that visit the site will leave recognizable tracks. In general, the more coyotes present in an area, the greater the number of scent sites that will be visited and the more visits per site. A wildlife biologist performing such a scent post survey would want to use the same scent, at the same locations, at the same time of year from year to year. In this way, trends in population abundance could be more reliably determined.

Fisheries and wildlife biologists realize that measurable objectives must be set for management strategies so that they can be evaluated (see section 20.6). Each management strategy should be developed with a desired result in mind. For example, if sampling data indicate that the PSD of a largemouth bass population in a lake is only 15, a fisheries biologist might recommend that a length limit be placed on bass in that lake. The intent of this management strategy is to improve the size structure of the largemouth bass population. Therefore, the biologist might set a measurable objective of improving PSD to 40–70. Measurable objectives can take many forms. Biologists may want to manage a white-tailed deer herd for a hunter success of 65%, a walleye population for an angling catch rate averaging 0.2 fish per hour, or an eastern bluebird population for densities reaching a certain number of birds per unit of habitat area. Unless objectives are measurable, the success or failure of a management strategy cannot be determined.

A long-standing practice in fisheries and wildlife has been the concept of a circular process for management. The management process involves setting of goals and objectives, problem identification, choice and implementation of actions, and evaluation of actions. Once actions have been evaluated, goals and objectives can be modified, new actions chosen if needed, and the process continues. This process results in feedback learning, which is a key component of the current movement toward **adaptive management.** Adaptive management has been defined as a structured process of "learning by doing" that involves more than ecological monitoring and changes in response to unexpected management impacts (Walters 1997). Adaptive management proponents begin with a concerted effort to integrate existing interdisciplinary experience and scientific information into models and then make predictions about the effects of alternative management strategies.

9.9 Modeling

A **model** is a conceptual or mathematical description used to portray a complex system. Modeling is an increasingly important tool in the assessment of animal populations and communities. The development of computer software and hardware capabilities has made modeling a common tool for biologists.

Wildlife biologists often use software such as RAMAS (Akçakaya et al. 1999) to model fluctuations in population number by age class. In this particular model, biologists must input initial abundance, emigration and immigration rates, fecundity, survival schedules, and density dependence information. The model then predicts abundance as a function of time, age class distributions for each generation, and so forth. Fisheries biologists can use the FAST (Fishery Analyses and Simulation Tools; Slipke and Maceina 2000) model to simulate the effects of various harvest mortality rates on the resulting age and size structure of fish populations. As another example, fish food habits can be related to fish growth using a bioenergetics model (Hanson et al. 1997).

Models are especially useful for predicting the likely effects of management strategies prior to their implementation. For example, Wall et al. (2004) used a **geographic information system (GIS)**, which is a system for the computerized mapping and analysis of geographic features using digitized data, to develop a model for endangered Topeka shiners. By using physical stream variables (e.g., size, slope, groundwater) and land cover (e.g., percent pasture), they predicted likely locations with appropriate habitat for this federally endangered (i.e., in the United States) minnow. Think of the amount of time that was saved in searches for this rare minnow because the fish-habitat modeling had previously been conducted to identify likely locations. Similarly, think of the value of this model to a biologist in charge of determining appropriate locations for habitat restoration to conserve this fish species. Thus, by using a computer model to identify appropriate management actions, a biologist can save both time and money and is more likely to be successful in management and conservation efforts. However, it is essential that computer models be validated by field testing (e.g., Chipps and Bennett 2002). This aspect of computer modeling is sometimes overlooked. Usually, models become more accurate over time as more complete data are entered into the model. The more accurate the data entered into the model, the better its predictive capabilities.

9.10 Purposes of Marking, Tagging, and Banding

Animals are marked, tagged, or banded by biologists for a variety of reasons. Any of these techniques can be used in mark-and-recapture studies or any other study that requires the identification of individual organisms or groups of organisms. For example, mortality/survival rate information is commonly calculated from marked, tagged, or banded animals. Mark, tag, or band returns can be used to obtain estimates of **exploitation** (harvest by humans) for organisms that are utilized for recreational or commercial purposes. If a certain number of fish are tagged in a lake, the percentage of those fish harvested by anglers in 1 year can provide an estimate of annual exploitation. This assumes that the tags have not influenced the natural mortality of the fish or the vulnerability of the fish to angling and that anglers return tags from all tagged fish. Many exploitation studies provide rewards for tag, band, or mark returns as an incentive for people to participate in the program.

The rates at which waterfowl and other migratory game bird bands are recovered over the years following banding can also be used to estimate both exploitation and overall survival rates. Band recovery rates (the proportion of bands recovered and reported to the banding agency) and band reporting rates (the proportion of recovered bands that hunters actually returned to the banding agency) can be used to estimate percentages of a population harvested by hunters. An estimation of harvest rate is obtained by dividing band recovery rates by band reporting rates in a particular year. Harvest rate is corrected using an estimate of nonretrieved birds killed by hunters to obtain the exploitation rate. Annual survival rates from all causes can also be estimated. The proportion of bands recovered from year to year from a particular banded population reflects the relative proportions of the population surviving and dying. For example, if 10% of all bands recovered from gadwalls banded as flightless ducklings in central Canada during a particular year were recovered during or after the fourth year post-banding, this would indicate that about 10% of the population was surviving to at least the fourth fall after hatching.

Banding, tagging, and marking programs can also provide information on animal movement. For example, an American avocet banded as a young bird by wildlife biologists in Minnesota and recaptured in Louisiana provides information on migration. A channel catfish tagged in the South Dakota portion of the Missouri River might be recaptured at the confluence of the Missouri and Mississippi rivers near St. Louis, Missouri. As another example, subadult male cougars generally disperse longer distances than subadult females, especially in higher-density populations. Because a radio transmitter (see section 9.14) was placed on a particular age-1 cougar, biologists were able to document an extensive dispersal when that individual was located on 3 September 2003 in the Black Hills of Wyoming and then killed by a collision with a train in Oklahoma on 27 May 2004 (Thompson and Jenks 2005). This 1,067-kilometer-minimum movement was the longest dispersal documented for cougars as of that time, and it would not have been known unless the tracking study was in place. The problem with movement information provided by most banding, marking, and tagging data is that only the minimum distance moved can be documented. Perhaps the channel catfish in the above example actually had been down the Mississippi River to New Orleans and back to St. Louis before it was recaptured.

Other information can also be obtained from tagging, banding, or marking. For example, preferred habitat can be determined, maximum life span can be estimated, and growth or age estimation methods can be validated.

9.11 Marking Methods

For the purposes of this text, the term *marking* will include both temporary and permanent marks placed on organisms. Examples of marking include toe clips on Mexican voles, fin clips on brindled madtoms, ear notches on mule deer, tattoos on bobcats, heat brands on white-tailed deer, freeze brands on walleyes, fluorescent pigments sprayed on fathead minnows, and subcutaneous latex injections on young sockeye salmon.

Marking often involves some form of mutilation, and biologists who use such techniques may be questioned by those who support animal rights (see section 16.6). In addition to ensuring that marking does not result in unnecessary damage or discomfort, a biologist needs to be certain that it does not influence the behavior of the marked organism. Such changes in behavior could bias the results of a study.

Marks can be used either to identify a group of organisms or to identify a relatively small number of individual organisms. Thus, marks could be used by a biologist to assess the performance of two strains of rainbow trout that are stocked into a stream. One strain could be marked with a fin clip or freeze brand on the left side of the body, and the other strain marked on the right side of the body. By clipping different fins or branding at different body locations, the biologist could even identify individual fish. However, the number of such mark combinations available for individual fish identification would be limited.

Fin clipping is the simplest method of externally marking a fish. When a fin is clipped near the body of a fish, the fin often does not regenerate. If the fin does regenerate, the mark can generally be recognized by the irregularities in the regenerated fin rays. Toes clipped from small mammals, such as Mexican voles, do not regenerate, and these marks are easily recognizable throughout the life of the animal.

Often, only short-term marks are needed. Fisheries biologists conducting a Petersen mark-and-recapture population estimate often use a handheld paper punch to mark fishes. This method creates a clean mark in a fish fin that is quick and easy to make, causes little damage to the fish, and is easy to recognize for a short time period.

Paints, dyes, and other substances are also used for marking organisms. Hair or feather dyes or patches of paint can be used for temporary marking of terrestrial animals, and dyes and stains can be used for temporary marking of fishes. Paints, dyes, and chemical substances are most often used for mass marking when it is not important to recognize individual organisms. Some of these marks are relatively short lived. Fluorescent pigments, embedded in the skin of fishes using compressed air and observed under ultraviolet light, can last up to a year. If fish are immersed in a chemical called oxytetracycline, the resulting mark can be detected for some time on scales; however, the oxytetracycline is eventually lost through exposure to sunlight. Some marks have longer durations. If fish are immersed for a sufficient time or fed oxytetracycline in a prepared diet, a permanent mark will be visible when bones such as vertebrae are sectioned and viewed under ultraviolet light. Finally, recall that fish can be recognized by genetic marks and marking, which were discussed in section 4.8. In fact, individual fish can be recognized by their unique mitochondrial DNA.

9.12 Tagging and Banding Methods

Tags and bands are differentiated from marks in that they involve the physical attachment of an artificial structure to the organism. Examples include neck collars on trumpeter swans (Figure 9.7), aluminum leg bands on American black ducks or house finches, ear tags on white-tailed deer (Figure 9.8), neck bands or collars on wolverines, and external dart or anchor tags on muskellunge. These examples are all external tags, but internal tags can also be used. Body cavity tags are simply small metal tags containing numbers, addresses, or other information that are inserted into the body cavity of an animal. Small coded wire tags are commonly inserted into the snouts of fishes; such tags can contain a substantial amount of information.

Tags can be used for either group or individual recognition. Neck collars placed on a group of Canada geese may be all of one color to allow recognition of a group of captured and released birds, and the collars may also be individually numbered so that individual birds can later be recognized. Coded wire tags can also be used to differentiate groups of fish. For example, if two groups of striped bass were reared to different sizes in a hatchery prior to stocking in a reservoir, two types of coded wire tags could be implanted in the snouts of the different groups. When striped bass were collected in subsequent years, a biologist could then determine which size of fish survived at a higher rate. However, a large number of individual codes can also be inscribed on this tiny tag; thus, individual fish can also be recognized if desired.

FIGURE 9.7 A trumpeter swan with a neck collar and leg bands. *[Image courtesy of the U.S. Fish and Wildlife Service.]*

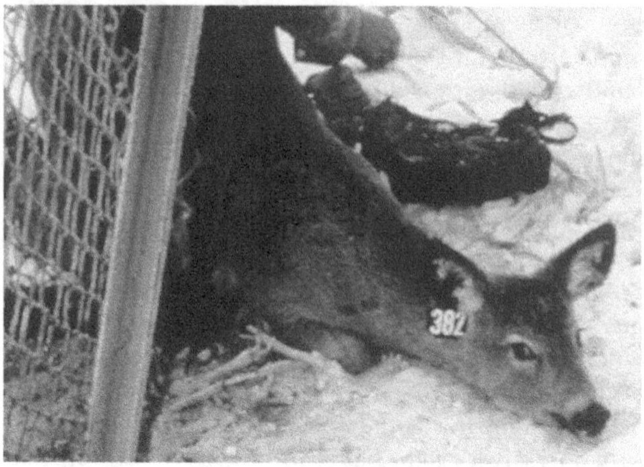

FIGURE 9.8 Numbered ear tags can be used to individually recognize white-tailed deer or other large ungulates. *[Image courtesy of D. Naugle.]*

Captured birds are often tagged by placing an aluminum band around a leg (Figure 9.9A). Bands generally are numbered and carry a return address. A band is properly fitted when it can move freely up and down the leg and turn smoothly but cannot be pulled down over the toes. Because many individuals and organizations are involved in bird banding, the U.S. Fish and Wildlife Service and the Canadian Wildlife Service issue banding permits and keep banding records for all migratory birds. Migratory birds include a variety of birds, ranging from waterfowl species to songbirds, that are **Neotropical migrants** (birds that migrate between the Nearctic and Neotropical regions). Birds can also be identified with back streamers, wing tags (Figure 9.9B), and nasal tags.

A variety of tags can be used for identifying fishes (Figure 9.10). These tags can be brightly colored to attract the attention of anglers and can carry a return address. If the tags are individually numbered, individual fish can also be monitored.

A similar array of external devices can be used on mammals. For example, neck collars are often used on large mammals, including types that are self-

A B

FIGURE 9.9 Birds can be identified with a band on a leg *(A)* or a patagial tag on the wing *(B)*. The patagial tag (number 74) on the wild turkey is visible from a substantial distance. *[Image courtesy of M. Rumble]*

adjusting for young ungulates. Ear tags are used on bats, and plastic ear tags are used on wild ungulates (see Figure 9.8) just as they are on domesticated livestock. Reflective tape can be placed on tags or collars to allow for better nocturnal observation.

9.13 Choosing Appropriate Marks, Tags, or Bands

Biologists must consider numerous factors when choosing marks, tags, or bands. A mark, tag, or band should not affect the behavior of an organism. While applying a temporary dye to a fish probably would not affect its swimming capability, clipping the entire tail fin could.

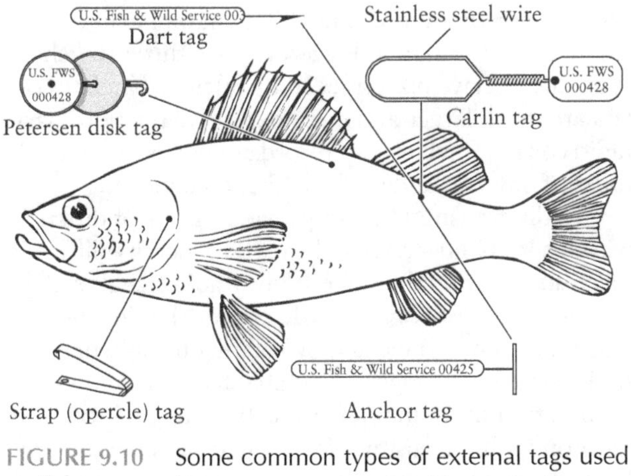

FIGURE 9.10 Some common types of external tags used on fishes.

As mentioned earlier, many marking, tagging, and banding programs are undertaken to determine mortality rates; thus, the chosen mark, tag, or band should not cause mortality or result in an increased mortality rate for the organism in question. For example, a tag or band that is too large for an organism could cause mortality. When fluorescent pigment is sprayed on rainbow trout, the air pressure must not be high enough to kill fish. A bright-orange ear tag placed on a marsh rabbit could be easily spotted by a biologist. However, it might also be easily spotted by an avian predator, and the tagged group of marsh rabbits might have higher predation rates than unmarked individuals. The chosen mark, tag, or band also should not affect the likelihood of recapture. For example, metal jaw tags on fish might make them more susceptible to capture in gill nets than untagged fish.

9.14 Biotelemetry

Biotelemetry involves electronic devices that are placed on or in an animal and relay information about that animal to biologists. Biotelemetry usually requires two devices, one placed in or on the animal to transmit information and the other retained by the biologist to receive the information. If the transmitters provide information on location, biologists can determine daily movement patterns, seasonal movement patterns, whether an animal establishes a home range, and a variety of other types of information. The biologist can follow all of the animal's movements, not just determine minimum distances moved as with tagged, marked, or banded organisms. Transmitters with unique capabilities that provide much information about animals are available. For example, some transmitters can send a motion signal that differs from the signal transmitted when the animal is at rest. By this means, biologists can determine not only animal locations but also activity patterns. Mortality-sensing collars, activated by a drop in body temperature, are also of value, particularly for predation studies. Some transmitters used for fishes can relay information on fish depth. More sophisticated transmitters can relay biological information such as body temperature, heart rate, or feeding activity. Archival transmitters can store data in the transmitter for later transmission to the receiver.

Biotelemetry allows for the study of organisms from sufficient distances to prevent disturbance of their normal activities. In fishes, biotelemetry also allows remote tracking of an organism that is not normally visible to humans. The same can be said for nocturnal terrestrial animals; however, even terrestrial animals active during daylight are often monitored using biotelemetry. Advances in biotelemetry equipment in recent years have been extraordinary. Increased miniaturization, more effective placement, longer transmitter life, and increased dependability have all added to its usefulness.

Wildlife biotelemetry studies most commonly use externally attached radio transmitters. These transmitters are usually battery powered, with life spans for most ranging from a few days to perhaps 2 years. Some transmitters are solar powered. Radio transmitters must have antennas; however, some antennas are quite small and internal and thus might not be easily seen. Transmitters used on large animals, such as black-tailed deer or collared peccaries, are generally attached with neck collars (Figure 9.11A). Radio transmitter systems that can remotely inject a drug into an animal are available, making it possible to anesthetize and recapture the animal at will. Transmitters for most birds are placed on the back or breast area and are partially covered by feathers (Figure 9.11B). They can be attached by a variety of methods, including harnesses around the body or wings, glue, suturing to the skin, and ponchos or necklaces fitted around the neck. Transmitters glued to tail feathers have proven valuable in bald and golden eagle studies.

An antenna and receiver are used to receive the radio signal from the transmitter; the antenna can be handheld (Figure 9.12A), mounted on a vehicle (Figure 9.12B), carried on or in an aircraft, or mounted on a stationary tower. Small transmitters used for songbirds typically have ranges from 200 meters up to 2 kilometers. Transmitters used on larger animals can be much more powerful. For example, transmitters used on white-tailed deer that are being tracked from the ground commonly have a range of 3–4 kilometers. Some animals, such as polar bears (Bethke et al. 1996), humpback whales (Mate et al. 1998), and loggerhead sea turtles (Nichols et al. 2000), have been tracked by satellite. Satellite collars are much larger and more expensive than ordinary radio collars. Satellites can record animal locations at all times of day and under all weather conditions.

A

B

FIGURE 9.11 Externally attached radio transmitters are used in determining animal locations and activity patterns in wildlife biotelemetry studies. *(A)* Collar on white-tailed deer; *(B)* antenna wire leading from the transmitter on an eastern meadowlark. *[Part A image courtesy of J. Jenks; part B image courtesy of D. Granfors.]*

Radio transmitters can provide signals of different wavelength bands that can be used to identify different individuals within the same study. Radio signals are directional, so a biologist can determine the direction in which an animal is located by rotating the antenna until the signal is strongest. This characteristic allows determination of animal locations by triangulation. If a raccoon with a transmitter is located within a particular area, biologists can obtain directional bearings from two or three different known locations, either simulta-

A B

FIGURE 9.12 Receiver antennas receive radio signals from transmitters and enable biologists to track and record the activities and locations of monitored animals. *(A)* Handheld unit; *(B)* vehicle-mounted unit. *[Part B image courtesy of D. Naugle.]*

neously or in a very short period of time before the animal moves. On a map of the area, lines are drawn from the known locations along the directional bearings obtained for the animal. The intersection of those lines provides the projected location of the animal.

Fish biotelemetry studies can be done with radio transmitters (Figure 9.13), but ultrasonic transmitters that send sound waves through water are also used. Ultrasonic transmitters are more effective than radio transmitters in waters that have high amounts of total dissolved solids, including salt water, and in deep water. Radio signals do not easily pass through such media. However, radio transmitters are commonly used in turbulent waters such as flowing rivers. The sounds created by the turbulence interfere with reception of ultrasonic transmitters.

FIGURE 9.13 Radio transmitter externally attached to a sauger; note the antenna trailing toward the posterior end of the fish. *[Image courtesy of B. Graeb.]*

Ultrasonic transmitters send sound waves, and underwater hydrophones (Figure 9.14) are used to receive their signals. The life expectancies and ranges

FIGURE 9.14 A hydrophone and receiver for an ultrasonic biotelemetry system.

of these transmitters vary with battery size and functions performed. Obviously, a larger fish can carry a larger transmitter with a larger power source than can a smaller fish. For general purposes, most ultrasonic transmitter signals cannot be received at distances of more than 1–2 kilometers.

Ultrasonic transmitters can be coded so that individual fish can be recognized. For example, the sound pulse pattern for one fish could be a series of four repeated signals, such as beep, beep, pause, beep. A second fish could be distinguished if its transmitter had a repeated pattern of beep, pause, beep, pause.

Ultrasonic signals are blocked or reduced by physical obstacles such as aquatic plants, lake features such as islands or points, areas of large temperature differences, turbulent waters, and outboard motors. Radio signals are better suited for waters with low levels of dissolved solids, shallow water, and turbulent water. Radio transmitters are well suited for use in large, shallow rivers and for fish species such as channel catfish or saugers that may make long movements. In such cases, fish can be located with an antenna that is handheld, mounted on a boat, or mounted on an aircraft that flies along the river. Radio signals can even be received through ice. Individual fish can be distinguished by different radio frequencies.

Both ultrasonic and radio transmitters are often implanted within the body cavity of a fish, but external attachment is also possible. Ultrasonic transmitters do not need an antenna, while radio transmitters must have some type of antenna. If the antenna can be extended outside the body of a fish, the range at which the signal can be received is greatly extended.

Fish locations can be obtained through triangulation. However, recent advances in technology have increased the techniques available for determining animal location. For example, if a boat is positioned over a fish with a transmitter, a **global positioning system (GPS)** can be used for location determination. Global positioning systems utilize earth-orbiting satellites for position location.

SUMMARY

To understand population and community assessment, some knowledge of basic statistics is essential. Accuracy refers to the closeness of a measure to its true value, while precision refers to the closeness of repeated measures to one another. Measures that are inaccurate are also referred to as biased. The average for a series of measures is termed the arithmetic mean, and statistical parameters such as standard deviation and standard error describe the amount of variation about the mean.

Biologists can determine population number through a census, which is a complete count of all organisms in a population. However, censuses are rarely undertaken because of the time and expense involved. Biologists more often use surveys, which are partial counts of a population. The data from a survey can be expanded to an estimate of the total population number or simply used as an index to relative population abundance. Estimates based on survey data should be accompanied by a measure such as a confidence interval that will allow the reliability of the estimate to be assessed.

Wildlife biologists often index natality as number of young per female or number of eggs hatched per nest. Fisheries biologists who wish to assess recruitment must know the age at which mortality of young fish is similar to that of adults. Mortality rate, equal to 100% minus survival, is often determined from population age structure. Life tables can be used to calculate mortality by age class for a population.

Population structure refers to the age structure or size structure of a population. Fisheries biologists are often concerned with the quality of a fish population, which is related to its size structure. Wildlife biologists rarely manage for body size, but they do at times manage for size of structures such as antlers.

Monitoring the physiological status of individuals and populations can also provide useful vital statistics. For example, blood chemistry can reveal much about nutritional status. Body condition is also often monitored.

Food habits are the kinds and amounts of food eaten. Digestive tracts of organisms are commonly sampled to obtain food items. Numerical analysis, frequency-of-occurrence analysis, gravimetric analysis, and volumetric analysis are typically used to summarize food habits data. However, each of these methods has its own set of biases, and careful interpretation is essential.

Field biologists typically find high variation among individual samples. Thus, it is important to measure this variation, to design sampling programs carefully, and to use sufficient units of sampling effort to obtain reliable data.

Community assessment has not advanced to the same level as population assessment. However, community assessment is a subject of much current research. Expect continued improvement in community assessment tools in the future.

Models are simplified conceptual or mathematical descriptions of complex systems. The use of models allows prediction of the effects of management strategies before they are implemented. However, the reliability of models should be assessed with actual field trials. Reliability increases as more accurate data are entered into the model.

Marking, tagging, and banding can be used for procedures such as population estimates and mortality/survival estimates. Marks can be physical or chemical and are most often used for group recognition. Tags and bands are often individually numbered, allowing recognition of individual organisms. In addition, tags and bands can be attached externally or internally.

Biotelemetry allows the activity of individual organisms to be monitored. Sophisticated transmitters can even relay such information as body temperature, heart rate, and feeding activity. Most wildlife biotelemetry studies involve the use of radio transmitters. Fisheries workers also use radio systems, but ultrasonic transmitters can be more effective in some aquatic environments. The GPS technology is widely used in wildlife radio telemetry to more accurately determine the target animal's position. Technological advances in biotelemetry equipment itself are occurring rapidly and will greatly enhance research capabilities for biologists.

PRACTICE QUESTIONS

1. Why is it important to obtain some measure of variation about a sample mean?

2. What is the difference between a creel survey and a creel census? On the same body of water, which would be more expensive to obtain? Why?

3. How does a mortality rate calculated from age structure using the catch-curve method differ from mortality information generated from a life table?

4. Cutthroat trout are quite vulnerable to recreational angling and are easily overexploited. Will an overexploited cutthroat trout population have a low or high proportional size distribution? Now assume that you have found a harvest regulation tool that will improve the size structure of that overexploited population. Will the proportional size distribution now increase or decrease?

5. Ecologists have a broader suite of population assessment tools than community assessment tools. List several reasons for this difference.

6. What is the difference between numerical analysis of food habits data and frequency-of-occurrence analysis?

7. What is the difference between food habits and feeding habits?

8. What parameters would need to be available to develop a realistic computer model that would predict the age structure of a fish population? Hint: think of the three dynamic rate functions.

9. In what kinds of studies would group marking techniques be suitable and under what circumstances would individually recognizable tags be necessary?

10. Biotelemetry studies can provide much information on the behavior of individual organisms. Choose a fish or wildlife species with which you are familiar. Make a list of the different types of behaviors that could be monitored using biotelemetry.

SELECTED READINGS

Anderson, R. O., and R. M. Neumann. 1996. Length, weight, and associated structural indices. Pages 447–482 *in* B. R. Murphy and D. W. Willis, eds. *Fisheries techniques*. 2nd ed. American Fisheries Society, Bethesda, Md.

Bowen, S. H. 1996. Quantitative description of the diet. Pages 513–532 *in* B. R. Murphy and D. W. Willis, eds. *Fisheries techniques*. 2nd ed. American Fisheries Society, Bethesda, Md.

Dinsmore, S. J., and D. H. Johnson. 2005. Population analysis in wildlife biology. Pages 154–184 *in* C. E. Braun, ed. *Techniques for wildlife investigations and management*. 6th ed. The Wildlife Society, Bethesda, Md.

Fuller, M. R., J. J. Millspaugh, K. E. Church, and R. E. Kenward. 2005. Wildlife radiotelemetry. Pages 337–417 *in* C. E. Braun, ed. *Techniques for wildlife investigations and management*. 6th ed. The Wildlife Society, Bethesda, Md.

Guy, C. S., and M. L. Brown. 2007. *Analysis and interpretation of freshwater fisheries data*. American Fisheries Society, Bethesda, Md.

Hall, S. H., J. S. Collie, D. E. Duplisea, S. J. Jennings, M. B. Bravington, and J. L. Link. 2006. A length-based multispecies model for evaluating community responses to fishing. *Canadian Journal of Fisheries and Aquatic Sciences* 63:1344–1359.

Lancia, R. A., W. L. Kendall, K. H. Pollock, and J. D. Nichols. 2005. Estimating the number of animals in wildlife populations. Pages 106–153 *in* C. E. Braun, ed. *Techniques for wildlife investigations and management*. 6th ed. The Wildlife Society, Bethesda, Md.

Ney, J. J. 1999. Practical use of biological statistics. Pages 167–191 *in* C. C. Kohler and W. A. Hubert, eds. *Inland fisheries management in North America*. 2nd ed. American Fisheries Society, Bethesda, Md.

Nielsen, L. A. 1992. *Methods of marking fish and shellfish*. Special Publication 23. American Fisheries Society, Bethesda, Md.

Pimm, S. L. 1991. *The balance of nature? Ecological issues in the conservation of species and communities*. University of Chicago Press, Chicago.

Rochet, M.-J., and V. M. Trenkel. 2003. Which community indicators can measure the impact of fishing? A review and proposals. *Canadian Journal of Fisheries and Aquatic Sciences* 60:86–99.

Servello, F. A., E. C. Hellgren, and S. R. McWilliams. 2005. Techniques for wildlife nutritional ecology. Pages 554–590 *in* C. E. Braun, ed. *Techniques for wildlife investigations and management*. 6th ed. The Wildlife Society, Bethesda, Md.

Van Den Avyle, M. J. 1999. Dynamics of exploited fish populations. Pages 127–166 *in* C. C. Kohler and W. A. Hubert, eds. *Inland fisheries management in North America*. 2nd ed. American Fisheries Society, Bethesda, Md.

Winter, J. D. 1996. Advances in underwater biotelemetry. Pages 555–590 *in* B. R. Murphy and D. W. Willis, eds. *Fisheries techniques*. 2nd ed. American Fisheries Society, Bethesda, Md.

LITERATURE CITED

Akçakaya, H. R., M. A. Burgman, and L. R. Ginzburg. 1999. *Applied population ecology: Principles and computer exercises using Ramas® EcoLab 2.0*. Applied Biomathematics, Setauket, N.Y.

Anderson, R. O. 1976. Management of small warm water impoundments. *Fisheries* (Bethesda) 1(6):5–7, 26–28.

Anderson, R. O., and R. M. Neumann. 1996. Length, weight, and associated structural indices. Pages 447–482 *in* B. R. Murphy and D. W. Willis, eds. *Fisheries techniques*. 2nd ed. American Fisheries Society, Bethesda, Md.

Beever, E. A., R. K. Swihart, and B. T. Bestelmeyer. 2006. Linking the concept of scale to studies of biological diversity: Evolving approaches and tools. *Diversity and Distributions* 12:229–235.

Bethke, R., M. Taylor, S. Amstrup, and F. Messier. 1996. Population delineation of polar bears using satellite collar data. *Ecological Applications* 6:311–317.

Block, W. M., A. B. Franklin, J. P. Ward Jr., J. L. Ganey, and G. C. White. 2001. Design and implementation of monitoring studies to evaluate the success of ecological restoration on wildlife. *Restoration Ecology* 9:293–303.

Caughley, G. 1966. Mortality patterns in mammals. *Ecology* 47:906–918.

Chipps, S. R., and D. H. Bennett. 2002. Evaluation of a *Mysis* bioenergetics model. *Journal of Plankton Research* 24:77–82.

DeKeyser, E. S., D. R. Kirby, and M. J. Ell. 2003. An index of plant community integrity: Development of the methodology for assessing prairie wetland plant communities. *Ecological Indicators* 3:119–133.

Dinsmore, S. J., and D. H. Johnson. 2005. Population analysis in wildlife biology. Pages 154–184 *in* C. E. Braun, ed. *Techniques for wildlife investigations and management*. 6th ed. The Wildlife Society, Bethesda, Md.

Fausch, K. D., J. R. Karr, and P. R. Yant. 1984. Regional application of an index of biotic integrity based on stream fish communities. *Transactions of the American Fisheries Society* 113:39–55.

Gabelhouse, D. W., Jr. 1984. A length-categorization system to assess fish stocks. *North American Journal of Fisheries Management* 4:273–285.

Guy, C. S., and M. L. Brown. 2007. *Analysis and interpretation of freshwater fisheries data*. American Fisheries Society, Bethesda, Md.

Guy, C. S., R. M. Neumann, D. W. Willis, and R. O Anderson. 2007. Proportional size distribution (PSD): A further refinement of population size structure index terminology. *Fisheries* (Bethesda) 32:348.

Hanson, P. C., T. B. Johnson, D. E. Schindler, and J. F. Kitchell. 1997. *Fish bioenergetics 3.0*. WISCU-T-97-

001. Sea Grant Institute, University of Wisconsin, Madison.

Hurlbert, S. H. 1984. Pseudoreplication and the design of ecological field experiments. *Ecological Monographs* 54:187–211.

Karr, J. R. 1981. Assessment of biotic integrity using fish communities. *Fisheries* (Bethesda) 6(6):21–27.

Krebs, C. J. 1999. *Ecological methodology.* 2nd ed. Addison-Welsey Educational Publishers, Inc., Menlo, Calif.

Lancia, R. A., W. L. Kendall, K. H. Pollock, and J. D. Nichols. 2005. Estimating the number of animals in wildlife populations. Pages 106–153 *in* C. E. Braun, ed. *Techniques for wildlife investigations and management.* 6th ed. The Wildlife Society, Bethesda, Md.

Lyons, J., L. Wang, and T. D. Simonson. 1996. Development and validation of an index of biotic integrity for coldwater streams in Wisconsin. *North American Journal of Fisheries Management* 16:241–256.

Mate, B. R., R. Gisiner, and J. Mobley. 1998. Local and migratory movements of Hawaiian humpback whales tracked by satellite telemetry. *Canadian Journal of Zoology* 76:863–868.

McCoy, E. D., and H. R. Mushinsky. 2002. Measuring the success of wildlife community restoration. *Ecological Applications* 12:1861–1871.

Moore, D. S. 2003. *The basic practice of statistics.* 3rd ed. W. H. Freeman, New York.

Nichols, W. J., A. Resendiz, J. A. Seminoff, and B. Resendiz. 2000. Transpacific migration of a loggerhead turtle monitored by satellite telemetry. *Bulletin of Marine Science* 67:937–947.

Rabeni, C. F., and N. Wang. 2001. Bioassessment of streams using macroinvertebrates: Are the Chironomidae necessary? *Environmental Monitoring and Assessment* 71:177–185.

Ricker, W. E. 1975. *Computation and interpretation of biological statistics of fish populations.* Bulletin 191. Fisheries Research Board of Canada, Ottawa.

Shuman, D. A., D. W. Willis, and S. C. Krentz. 2006. Application of a length-categorization system for pallid sturgeon (*Scaphirhynchus albus*). *Journal of Freshwater Ecology* 21:71–76.

Slipke, J. W., and M. J. Maceina. 2000. Fishery Analyses and Simulation Tools (FAST). Auburn University, Auburn, Ala.

Sokal, R. R., and F. J. Rohlf. 1994. *Biometry.* 3rd ed. W. H. Freeman, San Francisco.

Thompson, D. J., and J. A. Jenks. 2005. Long-distance dispersal by a subadult male cougar from the Black Hills, South Dakota. *Journal of Wildlife Management* 69:818–820.

Van Den Avyle, M. J. 1999. Dynamics of exploited fish populations. Pages 127–166 *in* C. C. Kohler and W. A. Hubert, eds. *Inland fisheries management in North America.* 2nd ed. American Fisheries Society, Bethesda, Md.

Wall, S. S., C. R. Berry Jr., C. M. Blausey, J. A. Jenks, and C. J. Kopplin. 2004. Fish-habitat modeling for gap analysis to conserve the endangered Topeka shiner (*Notropis topeka*). *Canadian Journal of Fisheries and Aquatic Sciences* 61:954–973.

Walters, C. 1997. Challenges in adaptive management of riparian and coastal ecosystems. *Conservation Ecology* 1, no. 2, article 1, http://www.consecol.org/vol1/iss2/art1/.

White, G. C., D. R. Anderson, K. P. Burnham, and D. L. Otis. 1982. *Capture-recapture and removal methods for sampling closed populations.* LA-8787-NERP. Los Alamos National Laboratory, Los Alamos, N. Mex.

Willis, D. W. 1987. Reproduction and recruitment of gizzard shad in Kansas reservoirs. *North American Journal of Fisheries Management* 7:71–80.

CHAPTER 10

Managing the Biota

In this chapter we describe some of the common practices used by biologists to manage the biota. Although it might seem initially that a discussion of such management practices would be extensive and complicated, much of the management of biota involves management of their habitat and human users, which we will discuss in chapters 15 and 17.

Biota management techniques are often directed at the population level. For example, if Cooper's hawks have been eliminated from an area and suitable habitat is available, restoration of the population by reintroduction is an option.

Biota management techniques can also be directed at the community level. For example, a newly impounded reservoir has an existing fish community that is adapted to the environment of a river, not the environment of a reservoir. Biota management in this situation might involve the addition of a number of new fish species and other organisms to exploit the variety of newly available niches in the reservoir.

Biota management techniques can also be directed at the ecosystem level. However, at that level, biota management, habitat management, and human user management are usually combined.

Stocking has a negative connotation to some people (see sections 4.5, 4.9, 10.8, and 14.9). However, in many situations stocking is the only alternative available to the biologist. Almost all biological systems in North America have been perturbed by humans, some severely. In many cases, stocking is an effort to ameliorate those perturbations. Restoration biology (see section 2.6), for example, is based in part on stocking, as is conservation biology (see section 1.7). In North America, current population distributions of species such as white-tailed deer, wild turkeys, paddlefish, and cutthroat trout have been restored through stocking programs. Many species, both game and nongame, could be added to this list of successful restorations. Remember that the capture and transfer of wild organisms (often called trap and transfer) is a form of stocking, as is the addition of individuals from a captive-rearing program to a population or community.

This chapter will focus on partial removal and complete removal of populations, the introduction of organisms, and the culture and propagation of fish and wildlife species, including their subsequent performance upon release. In addition, a brief discussion of diseases will be presented.

10.1 Partial Population Removal

Partial population removal may involve the removal of some individuals from a population that the biologist is trying to improve, or it may involve the removal of other species that are adversely affecting the target population or populations. Removal of a portion of a population can be accomplished in numerous ways. For example, problem populations can be partially removed by selective piscicide kills, by mechanical removal, or by commercial harvest. In fisheries work, removal of a portion of a population or community is called **partial renovation** (**partial rehabilitation**). Wildlife biologists have no comparable terminology, but partial wildlife renovations are possible.

Fisheries biologists often use piscicides for the selective removal of problem species; this method is essentially unavailable to wildlife biologists. Gizzard shad, for example, often become overly abundant in reservoirs in the southeastern United States. At high population levels, gizzard shad compete with the young of other fish species for food but produce few offspring that can serve as prey for predatory fishes. Gizzard shad are sensitive to toxicants such as rotenone and antimycin at levels that will not affect most other fishes. Although the entire gizzard shad population is not eliminated with low-level toxification, a 75%–90% reduction of the adults will result in reduced competition with other fishes and increased reproduction by the surviving adult shad.

Selective removal using toxicants can also be accomplished by taking advantage of fish behavior. For example, stunted bluegill populations can be selectively controlled in this manner. A biologist can wait until the height of the bluegill spawning season and then apply piscicides to the shallow water areas used by bluegills for spawning. Most of the fishes in this habitat area at this time of year will be bluegills. Some other fishes will be killed, but the target bluegills will incur the majority of the resulting mortality.

A strategy of selectively controlling problem fish species with species-specific toxicants would be highly useful if environmentally safe, target-specific toxicants were available. Unfortunately, there are few such chemicals. The primary selective toxicants currently available are used for sea lamprey control in tributaries of the Great Lakes (Meyer and Schnick 1983; see section 7.6), and they are not completely species specific. An ecologically safe, selective toxicant for a species such as the common carp would be invaluable for both fisheries and wildlife purposes because of the environmental problems caused by this fish (see section 1.5); however, such a selective toxicant has yet to be developed (Marking 1992). Ward (2005) did find that Supaverm, which is a combination of two chemicals, selectively removed nonnative fishes (e.g., fathead minnow and largemouth bass) while not harming native fishes such as Gila chub and Gila topminnow. For wildlife biologists, before rodenticides or avicides are used to reduce abundance of problem organisms, effects on nontarget organisms must first be considered (e.g., Apa et al. 1991).

Mechanical removal can also be used to affect populations of organisms. This technique usually involves traps, seines, or other capture devices. Wildlife biologists, for example, may wish to reduce predator populations in relatively small areas in order to increase populations of ground-nesting birds. Some waterfowl species, such as mallards, gadwalls, and Canada geese, will nest in high densities on islands or in other areas where mammalian predator populations have been substantially reduced; while total elimination of predators may be a management goal, it is unlikely to be accomplished (Figure 10.1). Biologists may selectively trap in these areas to remove

FIGURE 10.1 This electric barrier was used to prevent terrestrial predators from reinvading an area after they had been removed by trapping. In this case, waterfowl nest density and nesting success increased on the peninsula protected by the barrier.

FIGURE 10.2 Merriam's wild turkeys may concentrate near farmsteads and cattle-feeding sites during late fall and winter; if the wintering flock grows too large, the turkeys can cause damage by feeding, leaving fecal droppings, and otherwise destroying cattle feeds, such as these oat bales. *[Image courtesy of C. Lehman.]*

American mink, red foxes, raccoons, striped skunks, and other predators that destroy clutches or kill nesting females. Waterfowl generally have high nesting success in these predator-reduced habitats.

High-density populations of species such as the white-tailed deer, elk, and Canada goose can cause problems in a variety of ways. For example, such populations can damage agricultural crops (Figure 10.2), natural vegetation, or urban shrubbery. Their depredations can result in substantial economic and environmental damage. Mechanical removal with a lethal method, such as shooting, can be an effective control method. However, public opinion or a dense human population in an area can preclude the use of shooting as a control measure. Under these circumstances, trapping or the use of anesthetics, followed by removal and relocation, may be used for population reduction. Such procedures can be costly, and there is also the problem of finding suitable relocation sites. At times these relocated animals die after release because of the stress involved in capture and transport, or because they are unable to compete with existing animals in the location to which they were moved.

Coyotes, especially when abundant, will prey upon domesticated livestock. Some conservation agencies hire trappers and aerial shooters that respond to such damage complaints. American beavers can cause problems when they damage trees or agricultural crops, build dams in unwanted locations, or plug culverts. Again, control methods involve substantial worker hours to trap the beavers or remove dams. In most such damage situations, removal of the specific offending individuals is more cost-effective and less environmentally damaging than indiscriminate removal of the offending species.

Most mechanical removals in fisheries are attempts to reduce the abundance of fishes such as common carp, bigmouth and smallmouth buffaloes, white suckers, various bullhead species, or stunted panfish species. Most of these species can develop high standing stocks, and they will convert a substantial proportion of the food energy available in the ecosystem to their own biomass. In such situations, recruitment of other fish species often declines. Although improvement in some fish populations has been documented after mechanical reduction of competing populations (Johnson 1977), many such attempts have been unsuccessful. Mechanical reduction of fish populations is labor intensive, and rarely is it economically feasible to continue the reduction until a sufficient biomass of the problem species has been removed. In addition, high recruitment is a common response to reduced population abundance in these fish populations. Thus, the population can quickly return to problem status unless an additional management strategy, such as predator fish protection with a length limit (see section 17.5), is used in conjunction with the removal.

Commercial harvest can be an alternative means of partial removal of organisms. It seems an attractive alternative, where and when feasible, to allow private individuals to harvest and sell the problem organisms, rather than having a conservation agency pay for mechanical removal. In wildlife, the selective trapping of furbearers is a commercial harvest method used for partial removal. There have been successful cases in which commercial fishing removed a sufficient biomass of problem species to allow increased abundance and growth of other fishes (Stephen 1986). However, it is difficult to remove enough of the problem-species population so that improvements in target populations can be achieved. Often, just when the commercial harvest is reaching a sufficient proportion of the standing stock of the problem species, catches begin to decline and the commercial harvesters want to move on to other areas where a greater profit can be made. At that time, the only alternative left to a conservation agency might be to subsidize continued commercial exploitation, and the cost–benefit ratio of such operations should be evaluated.

10.2 Complete Population and Community Removal

Fishes can be totally eliminated from a water body by means of certain toxicants or by draining. Because the use of toxicants is such a drastic and expensive measure, they are used much less frequently now than in the past. Wildlife managers essentially have no comparable methods of total removal of a population or community. Attempts can be made to use capture devices to completely remove a wildlife population, as in the example of the mammalian predators in section 10.1, but the costs are so high that only under special circumstances is this even attempted.

Total renovation (**total rehabilitation**) in fisheries work refers to the elimination of an entire fish community. *Total renovation*, like partial renovation, is a fisheries term with no equivalent wildlife term. This biota management method is most often applied in a situation in which a particular fish species is overly abundant. For example, common carp may have been introduced into a water body through an accidental inclusion with a purposeful fish stocking or as an accidental introduction with released bait from a minnow bucket; they may also have naturally dispersed into an area. When common carp reach high standing stocks, they route much of the productivity of an aquatic ecosystem into their population biomass, and densities and biomasses of other fishes generally decline. At times, the most appropriate solution to this problem may be the elimination of the entire fish community, followed by restocking of desired fish species. Similarly, panfishes can overpopulate and stunt. In some cases, total population renovation is the most realistic management alternative. Obviously, the larger the water body, the less likely it is to be toxified or drained.

Total renovations should be considered a management alternative of last resort. There are numerous negative effects associated with total renovations. If the totally renovated water body is restocked with small fishes, several years generally must pass before a fish community develops that will resemble a natural community and provide recreational fishing opportunities. If larger fishes are stocked, the restocking effort is much more expensive; such a restocking may not even be possible due to a lack of adequately sized fishes available for restocking. In addition, a biologist needs to be certain that the original problem will not simply recur in a few years. If common carp were the original problem, total renovation may be useless if there is no way to prevent the carp from reentering the water body and reestablishing a population in a few years. Any fish species upstream in the watershed from a renovated water body will eventually reach that water.

Another negative aspect of total renovations is that the problem species is not the only organism removed. Consideration should be given to restoration of other native organisms eliminated by the renovation, not just the target recreational fishes and their primary prey species. Restoration of all the native organisms eliminated would ensure the species diversity needed for the aquatic ecosystem to function correctly. Complete restoration is seldom, if ever, feasible, especially if invertebrates, key to many food webs, are included in the restoration.

Total renovations have sometimes been used by fisheries biologists to allow for the restoration of native species. For example, introduced brook, rainbow, and brown trout have often reduced or eliminated native cutthroat trout populations from Rocky Mountain streams. Some conservation agencies use a toxicant such as rotenone to eliminate the fish community from a reach of stream and then restore the native cutthroat trout. Good sites for such renovations are stream stretches above a natural barrier to fish movement, such as a waterfall. Some of the initial restorations of greenback cutthroat trout were completed using this technique (Harig et al. 2000). This subspecies of cutthroat trout, native to the eastern slope of the Rocky Mountains in Colorado, was at one time believed to be extinct. Remnant populations were found and brought into a fish hatchery. Progeny produced at the hatchery were stocked above barriers in streams that had been chemically renovated. Greenback cutthroat trout have now expanded to the point that catch-and-release fishing is allowed in some places in Colorado. Biologists in the Appalachian Mountains, where the brook trout is native, have undertaken similar renovations to remove introduced rainbow trout and restore brook trout. However, restoring native trout is not a simple process (Kruse et al. 2001). For example, disconnected, isolated populations in various headwater streams may not be able to exchange genetic material. In addition, unpredictable environmental events such as floods or volcanoes may **extirpate** (eliminate from a portion or all of their native range) such populations from entire watersheds, with no likelihood of repopulation without subsequent human intervention. Thus, biologists working on conservation restorations for native trout are engaged in an evolving science.

Depending on the size of a stream and the complexity of the native fish community in that stream, restocking of nongame fishes after total renovation may be necessary. This is especially true if renovations are done above natural barriers. When renovations are undertaken on a reach of stream with no barriers to fish movement, native fishes often recolonize the reach from above and below the treatment area.

Common carp are also viewed as a detriment by most managers of waterfowl refuges (Crivelli 1983). Common carp are well suited to the shallow, vegetated waters that refuge managers maintain for waterfowl production. When common carp populations attain high standing stocks, the quality of waterfowl habitat is reduced. Waters typically become more turbid, and stands of aquatic plants such as sago pondweed and wild celery, as well as associated aquatic invertebrates that are used as food sources by many waterfowl and other species, are reduced. Total renovation of the fish communities in waterfowl refuge pools is a tool that can be used by waterfowl biologists, especially in situations in which the water can be drained from part or all of the system.

Total renovation as a biota management tool is primarily limited to fisheries work, because fish communities are generally limited by the boundaries of a water body. One of the rare attempts at total elimination of a species in the wildlife field occurred when biologists attempted to eliminate introduced Arctic foxes from islands in the Aleutian chain (Springer et al. 1977). These islands are isolated and have distinct boundaries, which allowed this unique wildlife renovation attempt. Arctic fox predation on Aleutian cackling geese was a direct cause of the decline of the geese and their listing as an endangered species. After Arctic fox populations had been greatly reduced or eliminated, Aleutian cackling geese from captive flocks were restored to selected islands. An interesting component of this story is that biologists found it necessary to include a few adult wild birds with each release of captive-hatched goslings. The captive-hatched birds had unusually high mortality because of a tendency to migrate straight south over the Pacific Ocean. The wild birds traditionally had migrated to the continental coastline before turning south, and the adult wild birds led the captive-hatched birds along this route, perhaps because adults could recognize landscape features.

Biomanipulation, a term first coined by Shapiro et al. (1975), refers to manipulation of a fish community to reduce predation on herbivorous zooplankton, which then should be followed by an increased abundance and size structure of zooplankton, particularly *Daphnia*. In turn, this higher abundance of large-bodied zooplankton should lead to increased grazing on phytoplankton and subsequently to clearer water. The manipulation may involve complete removal of a fish community, partial removal of some of the fish community, or a strategy to favor the piscivorous fishes that will then prey upon and reduce the planktivorous fishes. This management strategy is somewhat controversial among ecologists, likely because the results of such manipulations are only sometimes successful (e.g., Hansson et al. 1998; Meijer et al. 1999; Angeler et al. 2003). Biomanipulation as a management strategy actually involves both biota management and habitat management; the biota is manipulated to produce a change in habitat.

10.3 Other Types of Population Control

Other attempts at population control have been made that do not involve killing or moving organisms. Most of these biota management techniques involve interference with reproduction or attempts at scaring or repelling organisms from an area. For example, insects, striped skunks, and coyotes have been sterilized and released into the wild. These individuals' unsuccessful attempts at reproduction can result in lower overall numbers of offspring in a population. Similarly, contraceptives can be delivered to organisms in the wild through direct injection or in baits (Turner et al. 1992). Attempts to interfere with reproduction in order to control population numbers are more likely to be successful in small populations located in confined areas. It is difficult and expensive to apply these techniques over large geographic areas. The use of contraceptives is also an area of contention with animal rights and even some animal welfare proponents (see section 16.6).

Fisheries biologists sometimes rely upon the release of polyploid or monosex fishes to prevent future reproduction by a released species (see section

4.9). Many states in the United States limit introductions of grass carp to triploid fish. Grass carp are herbivores and can be used for vegetation control. However, they are an exotic species in the United States, and their natural reproduction is not desired. Stocking of only triploid fish can ensure that these introductions do not result in the subsequent development of wild populations. Unfortunately, not all states have this requirement, and some fish sold as triploids may have been diploids. Thus, reproducing populations of grass carp now occur in many U.S. locations (Fuller et al. 1999). Triploid trout species also hold promise for recreational fisheries management. The use of these sterile fishes has the potential to protect the genetic integrity of native trout stocks while continuing to supply hatchery-reared trout for anglers (Kozfkay et al. 2006).

Another method of population control is the introduction of forms for which appropriate reproductive habitats are not available. For example, stocking a Pacific salmon species into a water body that has no salmon spawning habitat will ensure that the population can eventually be eliminated by cessation of stocking. Similarly, a wildlife species such as the chukar can be released into certain areas for recreational hunting with the knowledge that the correct habitat, including reproductive habitat, is absent and the species will not develop a self-sustaining population.

Selective control of some problem organisms may involve attempts to repel or scare them from an area. Blackbirds migrate in large, mixed-species flocks that can cause substantial economic damage to agricultural crops. Attempts to control these birds have involved repellents, fright calls, and scare devices such as shell crackers and propane exploders (Figure 10.3), but the effectiveness of these techniques is often limited. Similarly, white-tailed deer can cause economic damage to orchards, residential plantings, and other plant life. Repellents, screens, fences, scare devices, and even the spreading of human hair from a barber shop are all techniques that have been used in attempts to reduce damage. Again, the effectiveness of these techniques typically is limited.

While part of the work of a wildlife biologist involves protecting wildlife from people, it is also necessary at times to protect people from wildlife. Cougar harvest, as well as trapping and relocating, have been increased in some areas to reduce the likelihood of attacks on humans and domestic livestock. An increased number of human attack and threat situations have occurred in many North American locations as cougar populations have increased and dispersed into areas of dense human populations, and as human populations have spread into traditional cougar habitat (Kellert et al. 1996).

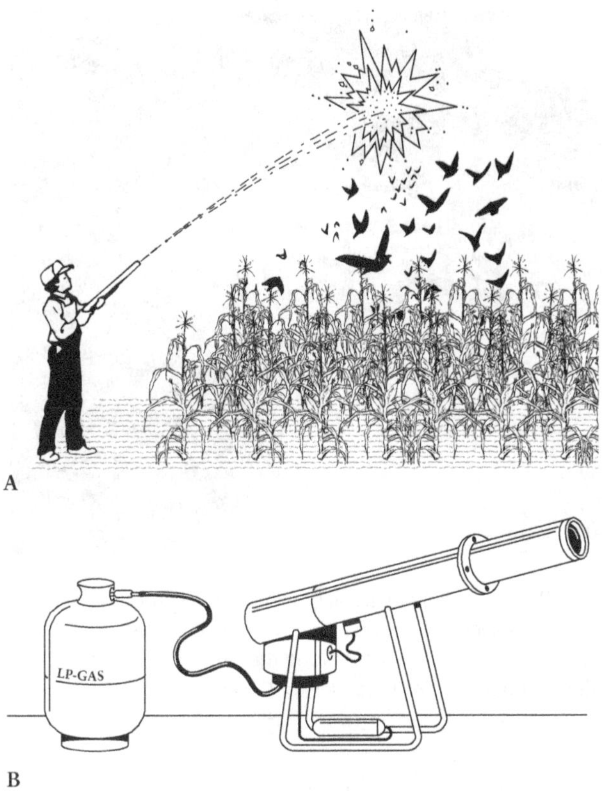

FIGURE 10.3 Devices used for scaring away problem wildlife. *(A)* Shell crackers, fired from a 12-gauge shotgun, produce an aerial explosion. *(B)* Propane cannons, also known as liquid propane gas exploders, also make loud sounds at periodic intervals.

Management techniques used to reduce human conflicts with grizzly bears include proper disposal of domesticated livestock carcasses, redistribution of **carrion** (dead-animal flesh), protection of beehives with specialized electric fence systems, education of landowners, and proper garbage storage and disposal programs. Relocation of nuisance grizzly bears to areas outside of their known home ranges has proven to be generally ineffective and no more than a temporary solution.

In fisheries work, electrical barriers (Verrill and Berry 1995) and mechanical barriers have been used to keep fishes from moving into specific areas (Figure 10.4). The long-term success of such barriers is questionable,

FIGURE 10.4 An electrical barrier used to prevent common carp movement at Heron Lake, Minnesota. The cement pad visible on the far bank runs along the stream bottom and contains electrodes. *[Image courtesy of C. Berry Jr.]*

because flooding or failure of the mechanical devices, as well as the human habit of either intentionally or unintentionally introducing fishes, can render them ineffective. However, alternatives to barriers are few, as in the case of Asian carps in the Illinois River. Lake Michigan is connected to the Illinois River through the Chicago Ship and Sanitary Canal, and an electrical barrier was constructed in an attempt to prevent movement of invasive bighead and silver carp into the Great Lakes. Whether the barrier will prevent movement is unknown at this time. Electric barriers have also been used in wildlife work to protect ground-nesting birds from predators (see Figure 10.1). In these areas, predators such as striped skunks and red foxes are removed, and the electrical barriers slow reinvasion.

10.4 Fish Culture and Stocking

Fish culture actually was the origination of the fisheries profession in the United States. Initial attempts at fisheries management were quite dependent on fish hatcheries and the culture of fishes for subsequent release into the wild (Nielsen 1999). The American Fish Culturists' Association, which evolved into the American Fisheries Society that remains highly active today, was formed in 1870. Common carp were first imported from Germany in 1877 and widely distributed across the United States. Striped bass, native to the Atlantic coast, were first shipped from New Jersey to California in 1879. Rainbow trout, native to the Pacific coast, were first distributed to the eastern United States in 1880.

Fish production methods typically are categorized as either **extensive** or **intensive fish culture** (Wedemeyer 2001). Extensive culture involves fish production in large, earthen ponds. Fish can feed upon zooplankton, insects, or other prey resources that are naturally produced in the pond. On the other extreme, intensive fish production often involves indoor facilities where water turnover rates are substantial and all food is supplied by the culturists, typically as prepared pellets. In reality, a continuum between the two production methods actually exists. Consider an earthen pond in which channel catfish are fed pelleted food. This fish production effort would be categorized between the two extremes of intensive and extensive fish culture.

Fish stockings are generally categorized as **introductory, maintenance**, or **supplemental stockings**. Introductory stockings involve the release of fishes into a new or renovated body of water or the introduction of a new fish species (either native or introduced) into waters with an existing fish community. Maintenance stockings are those made to sustain a fish population that for some reason has no, or perhaps limited, natural reproduction. Supplemental stockings are those made to augment a naturally reproducing population. Maintenance and supplemental stockings can be difficult to differentiate; in reality, there probably is a continuum across these two definitions.

Stockings of wildlife organisms can fit into the categories described for fishes; however, these terms are generally not used to describe wildlife stockings. The introduction of gray partridge into North America could be termed an introductory stocking. The restoration of wild turkeys to an area they had previously inhabited is also an example of introductory stocking. Maintenance stockings are quite rare in wildlife. However, some conservation agencies or private hunting preserves may release adults of a species that cannot reproduce in a certain area. Ring-necked pheasants, for example, typically do not recruit in the wild in the southeastern United States, but the species can be pen reared and released in a private hunting preserve. Supplemental stockings of wildlife species are also undertaken at times. Examples might include attempts to supplement existing northern bobwhite or ring-necked pheasant populations with pen-reared birds.

10.5 Introductory Stockings

One task of the biologist is to introduce native species into new or renovated water bodies or into appropriate terrestrial habitats. While quite common in fisheries work, such introductions are currently less common in wildlife. There are numerous reasons for this difference. Fisheries biologists have two tools available for the total renovation of fish communities (toxicants and draining); thus, restockings are commonly made after total renovations. Wildlife biologists have no comparable management tools. Also, fishes tend to be less mobile than most wildlife species. When a water body is newly constructed or renovated, often the only means for the reasonably rapid colonization of fishes is stocking. Conversely, when terrestrial habitat is developed, wildlife organisms are more likely to disperse naturally into the newly created habitat areas.

These differences do not mean that introduction of native species does not occur in wildlife, just that it occurs less often. For example, North American river otters and American martens have been restored by state conservation agencies in locations from which they were extirpated. The reestablishment of breeding populations of giant Canada geese in north-central states is another example of a successful restoration. This restoration program has been so successful that those states now allow regulated recreational hunting of giant Canada geese and, in many cases, have special early fall seasons to help control nuisance populations. Extensive restoration programs in various portions of their native ranges have been completed for animals such as elk, pronghorns, white-tailed deer, mountain goats, and bighorn sheep. Much of the current abundance of large mammals in the United States is the result of restoration programs that were completed earlier in this century. In a comparable management program, fisheries biologists have used cultured paddlefish to restore populations in Missouri (Graham 1997).

Peregrine falcons have been restored to many areas of the United States by a technique known as **hacking**. Offspring produced by captive birds are reared and fed with as little direct contact as possible with humans. They are then moved to the wild, fed in an elevated caged nest, again with as little human contact as possible, and then allowed to fly free. Typically, food is supplied at the hacking site until the birds have learned to fend for themselves. When wildlife biologists want to restore a northern bobwhite population, trap and transfer with wild birds is most successful (Hellickson and Radomski 1999).

Because introductory stocking of native species is a biota management tool most often used in fisheries, the remainder of this section will pertain to fishes. Introductory stockings are more common in artificial (human-created) water bodies than in natural water bodies. Natural water bodies typically contain a complement of fish species that evolved in that ecosystem. Thus, biologists generally need to do little in the way of biota management to improve the native fish communities in these waters. In fact, this natural species diversity tends to make it more difficult to affect fish communities by stocking in natural water bodies. Introductory stockings have been made in natural lakes, especially when these habitats have been modified by human activities. The introduction of some Pacific salmon species into the Great Lakes is an example of such stockings.

Artificial water bodies often have no complement of fish species that evolved in the system. Artificial water bodies exist along a continuum that ranges from highly stable waters quite similar to a natural lake to those with high rates of water exchange that may actually be little more than a wide spot in a river. The fish community that lived in a river before it was dammed is likely to contain relatively few species that will thrive in an artificial water body.

When making an introductory stocking into a new or renovated water body, the biologist must consider the available habitat and stock species best suited to the available niches. A small water body has fewer niches, and it is generally best to keep the fish community simple. Over much of the United States, largemouth bass are introduced into small water bodies as a predator and bluegills as a panfish and prey species. However, the available habitat must be considered. Largemouth bass and bluegills are appropriate species for introduction into small water bodies that have relatively clear (i.e., typically transparency of 45 centimeters or more), warm water for most of the year and contain appropriate coverage of submergent aquatic plants. If aquatic plants are overly abundant, the bluegills will have too much cover and will avoid predation by the largemouth bass, leading to a stunted bluegill population. If aquatic plants are too sparse, largemouth bass recruitment may be low at midwestern and northern locations in the United States. In these cases too few largemouth bass will be present to prey upon the bluegills, again leading to a stunted bluegill population. If a small, warm water body is constantly turbid, perhaps because of suspended colloidal clay particles, channel catfish may be a better choice for introduction. In cold water bodies, or those that have colder water in deep areas, coldwater species such as rainbow trout can be stocked.

In larger artificial water bodies, the diversity of available niches is greater and the biologist can typically introduce a wider array of fish species; some may even be transplants or exotics. One can use a large, warmwater water body as an example for this type of stocking strategy. First, consider the top-level predators available for such a system. For shallow warmwater habitat, the largemouth bass immediately comes to mind. However, both the smallmouth and spotted bass can also be used in this capacity. Largemouth bass tend to congregate in areas with aquatic plants, flooded timber, or some other type of structure, while spotted and smallmouth bass tend to use rocky habitat. However, this niche segregation is commonly not clear-cut. Biologists usually would also want to introduce open-water predators into a large water body. Some fish species commonly chosen to fill this niche include white bass, striped bass, striped bass × white bass hybrids, and walleyes. Deep-water predators also need to be considered. The flathead catfish in southern states (a warmwater species) could fill this niche.

The next category of fish species to consider for this water body is a group that is difficult to categorize; these fishes can be termed *midlevel piscivores*. Fishes such as the yellow perch, black crappie, and white crappie occur in this category. Although they do eat some prey fishes as adults, during much of their lives they feed on aquatic or terrestrial invertebrates. These panfishes are extremely important fishes in most systems. While the top-level predators are "glamour fishes," panfishes might be called the "bread-and-butter" or "family" recreational fishes. They support a tremendous portion of the recreational fish harvest across the United States and Canada because they are easily caught, can be caught in large numbers, and are low on the food chain and thus can have relatively high standing stocks. In addition, they serve as prey for the higher-level predators.

The biologist should now consider a lower trophic level, the insectivores. Some fishes obtain most of their food energy from aquatic or terrestrial insects; common examples include bluegills and rainbow trout. Actually, many fishes feed primarily on insects at certain life stages and at certain times of year. For example, during a summer hatch of a large mayfly such as *Hexagenia,* channel catfish, largemouth bass, black and white crappies, yellow perch, walleyes, and white bass will all feed on this abundant food source.

Finally, prey organisms (commonly fishes or invertebrates) need to be considered. In the past, there was a tendency for fisheries biologists to stock top-level predators without being certain that the system could provide a sufficient prey base. This rarely occurs today. The particular prey fishes used depend on the geographic location and habitat. Gizzard and threadfin shad are important prey fishes in the southeastern and midwestern United States. Many minnow species are important prey fishes. Emerald and spottail shiners serve as prey for predators in South Dakota waters, while red shiners are more likely to be common in Kansas water bodies and golden shiners are common in New England. The term **forage fish** is sometimes used synonymously with the term *prey fish*. However, animals such as cows and horses eat forage, while predators eat prey; thus the term *forage* is inappropriate.

The species stocked will vary according to water type. Obviously, warmwater species will be stocked in warm water and coldwater species in cold water. Stocking may include introduced forms if no native species can fill the available niches. **Two-story fisheries** can also exist. In large, deep water bodies that have both warm and cold, well-oxygenated water strata, a biologist can develop a warmwater or coolwater fishery in the surface waters and a coldwater fishery in the colder bottom waters. Coldwater species such as rainbow trout can be used in the coldwater habitat, while warmwater species such as largemouth bass or coolwater species such as walleyes can be used in the warmer water habitat. At northern latitudes, the construction of a water body may result in new warmwater habitat. The coldwater natives will be present to populate the bottom story of the fishery, while introduced species may be stocked into the upper, warmer water habitat.

Introductory stockings of either native or introduced fishes for food web improvement are appropriate in some circumstances. Prey fish stockings are often undertaken to improve food webs, especially if predators have already been stocked without prior consideration of the prey base. It is important to note, however, that some prey fish stockings have provided mixed results at best and that caution should be exercised. Biologists sometimes find that fish stocked as prey for adult predators actually compete with young predators. For example, threadfin shad introduced as a prey species often compete directly with young largemouth bass and black crappies.

Stockings of either native or introduced fishes can be used to develop trophy fisheries by introducing such species as muskellunge or striped bass. However, many such populations may not be sustained by natural reproduction, and maintenance stockings may be necessary.

10.6 Introductory Stockings of Exotic and Transplanted Species

You will recall that exotic species were defined as organisms introduced from another zoogeographic region. Transplants were defined as organisms moved outside their native range but within the same zoogeographic region (see section 2.6). Both exotic and transplant introductions have resulted in successes; such introductions have also resulted in ecological disasters (see section 1.5). Most of the exotic and transplant introductions that have proven to be both successful and beneficial have involved organisms introduced into areas that had been so changed by habitat modification that native species could no longer exist. Such introductions are a common form of biota management.

The Missouri River in North and South Dakota was once a relatively shallow flowing-water environment characterized by highly turbid water. The construction of a series of dams on that river resulted in a primarily nonflowing environment with deep, clear water. Many fishes native to the original river could not adjust to these habitat changes, and some new habitats were produced that no native species could inhabit. Essentially, new habitats were formed in a region that had no native fish species able to exploit those new habitats. A variety of coldwater salmonid species, including both trout and Pacific salmon, were introduced into some reservoirs on this river system, where they have been successful and beneficial. Prey species such as rainbow smelt and the invertebrate *Mysis relicta* were also introduced, because native prey species could not exploit all of the new ecological niches.

A wildlife example of biota management using exotics or transplants that has been both beneficial and successful is the introduction of exotic ring-necked pheasants. In some areas of the United States, populations of native gallinaceous birds declined in landscapes changed by human activities, primarily agriculture. New habitats and landscapes had been produced, and native forms could not fill these new ecological niches. The ring-necked pheasant could, and its introduction has been a success. The introduced gray partridge has also been successful and is even more adaptable to intensive agriculture than the ring-necked pheasant.

Introducing exotics and transplants where new habitats have not been formed can be a risky business. An example of an introduction into a habitat that had not been extensively modified and in which native species still existed took place in the Black Hills of South Dakota and Wyoming. The Black Hills had habitats that would support trout species, but trouts had apparently not dispersed naturally into that region. Brook, brown, and rainbow trout were introduced into the Black Hills, and a successful and beneficial trout fishery was produced. The success of these introduced forms, however, has resulted in population decreases of native forms, primarily minnows and suckers. Recreational fisheries advocates generally accept such decreases in the population abundance of nongame species as long as nongame species do not become threatened or endangered. However, not all agree with such stocking policies, and much greater scrutiny is now common before such introductions are made.

Other introductions of trouts in the United States have also had mixed and sometimes primarily negative effects (Figure 10.5). The brook trout native range extended from the Appalachian Mountains to the northeastern United States and southeastern Canada. Naturalized brook trout populations in the

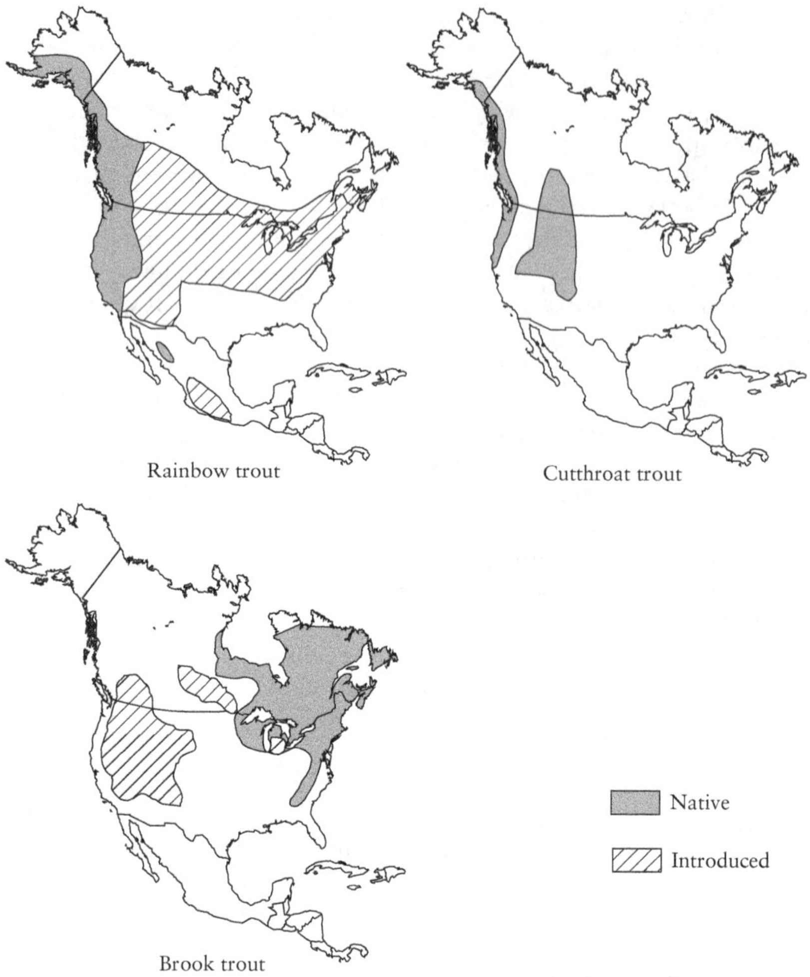

FIGURE 10.5 Distribution maps for rainbow, cutthroat, brown, and brook trout in North and Central America. Cutthroat trout have been transplanted outside of their native range, but such stockings are rare and generally localized. The brown trout is native to Europe and Asia and is an exotic species in North America.

Rocky Mountains have, through competition, reduced or eliminated native subspecies of cutthroat trout, especially in small high-elevation streams. The rainbow trout is native to the extreme western drainages along the Pacific coast of North America. These fish have become naturalized in some Rocky Mountain and Appalachian streams, where they compete with native species. The brown trout is native to Europe and thus is an exotic species in North America. Because it can tolerate warmer water temperatures than brook, cutthroat, or rainbow trout, it has been considered a desirable species for introduction in streams where water temperatures are marginal for other trouts. However, brown trout are piscivorous and can negatively affect native trouts and other species through competition and predation.

Invasive Asian carps such as the bighead carp and silver carp have rapidly increased in population abundance in the Mississippi River basin (Chick and Pegg 2001). Both the bighead and silver carp are planktivorous, and thus they have the potential to compete with native planktivores such as paddlefish and bigmouth buffalo.

European starlings and house sparrows are exotic species from the Palearctic region that have adapted to environments in the United States and

Canada. Both exotic species often outcompete native species, resulting in declining populations of what are generally considered to be more desirable species.

Many exotic ungulates have been introduced into North America and other areas of the world, and most have created problems through competition with native species, hybridization with natives, and the introduction of diseases. For example, escaped red deer have come into contact with and interbred with native elk in some areas. However, these hybrids currently have limited distribution. Barbary sheep introduced into Texas are carriers of diseases that can affect native bighorn sheep populations. Several western states have forbidden the importation of exotic ungulates.

New Zealand has incurred serious vegetation losses and erosion problems due to introductions of exotic ungulates. Likewise, native species in New Zealand face serious competition from exotic fishes and wildlife. Many islands, including Hawaii and Jamaica, have lost or are losing substantial numbers of native bird species due to competition with or predation by exotic animals.

10.7 Supplemental and Maintenance Stockings

While introductory stockings are often justifiable and successful, supplemental and maintenance stockings must be more carefully assessed. For example, a study of walleye stockings in 125 water bodies over a 100-year period indicated that 48% of introductory stockings, 32% of maintenance stockings, and 5% of supplemental stockings were successful (Laarman 1978). Stocking success depended more on environmental and biological conditions at the time of stocking than on the number and size of walleyes stocked. Two conclusions can be drawn from this study of walleye stockings. First, it appears that maintenance stocking of walleyes into waters that contain no suitable habitat for reproduction can create recreational fisheries. Second, supplemental stocking of walleyes to increase the density of a population that has some amount of natural reproduction is not likely to succeed.

Maintenance and supplemental stockings are also attempted in wildlife, although these specific terms are rarely used to describe such introductions. Private individuals commonly release 8- to 10-week-old ring-necked pheasants into areas with established wild pheasant populations. In a few cases, natural resource agencies, due to public and political pressure, have also released young pen-reared ring-necked pheasants. However, where wild ring-necked pheasants are already established, such releases have repeatedly failed to boost long-term population levels. In most release areas, the habitat is already suboptimal, or wild ring-necked pheasants would naturally be more abundant. Stocked birds are highly vulnerable to predation; however, predator control of sufficient intensity to benefit the released birds is time-consuming and expensive, and it is seldom attempted or accomplished. In addition, where intensive predator control is attempted, private individuals may illegally kill raptors and other beneficial organisms that can reduce stocking success. In recent years, public opposition to native predator control has increased. When habitat is improved, as illustrated by the substantial increases of ring-necked pheasants after the implementation of the Conservation Reserve Program (see section 1.3), wild pheasants quickly expand their numbers in response to the enhanced habitat conditions and no stocking is required. Stocking in such situations may even be harmful to existing populations because of the poor genetic fitness of the pen-reared birds and potential disease problems. However, as with Florida panthers (see section 4.8), supplemental stocking from wild-trapped animals can be attempted to increase genetic diversity in a population.

Spring release of pen-reared adult hens has also generally failed to provide appreciable increases in subsequent fall ring-necked pheasant numbers. Spring release of such birds is also extremely expensive (Hessler et al. 1970; Hill and Robertson 1988). Pheasant preserve operators generally release mature males shortly before or at regular intervals during the hunting season to minimize natural mortality of stocked birds and to increase the percentage of stocked birds that are harvested.

Supplemental and maintenance stockings usually take two forms, **put, grow, and take stockings** or **put and take stockings**. When subcatchable-sized rainbow trout are maintenance stocked into a remote mountain lake with no inlet or outlet stream, the fish use the natural productivity of the lake to grow, and then they are harvested by anglers when they reach catchable size. This is a put, grow, and take fishery and is quite different from fisheries in which catchable-sized rainbow trout are stocked into high-use streams and lakes. Catchable-sized stockings are put and take stockings; these fish are immediately available to anglers. Frequent stockings, often several per year, may be necessary to maintain populations in high-use situations in which fish harvest is important.

Many waters stocked with catchable-sized trout are quite popular with tourists. Anglers drive large, expensive recreational vehicles from many parts of the country to camp and fish in these areas. Supplemental stockings of catchable-sized fish to sustain such fisheries are often criticized. However, if the income from license sales to nonresident anglers that harvest the catchable-sized fish is greater than the cost of the culture and stocking program, the catchable-sized stockings probably will be economically justified. This is especially true if the overall economic impact of the tourism industry in these areas is considered. Other anglers may prefer to fish for wild trout, even if most or all of the fish are protected by harvest regulations and have to be released; such anglers can seek areas set aside for this type of angling experience.

Urban fisheries programs are highly reliant upon stocking. Approximately 75% of the people in the United States currently live in cities with populations of 100,000 or more. Providing fishing opportunities to these urban residents, especially those with limited financial resources and mobility, has become an important mission of many conservation agencies. These fisheries typically sustain high levels of fishing effort; as much as 30,000 hours of angling effort per hectare of water per year has been documented (Alcorn 1981). At such high levels of fishing effort, natural reproduction and growth cannot sustain fishable populations when the fish caught are removed (harvested), so stocking of catchable-sized fishes is a common management strategy. The fish species stocked is often less important to the urban angler than the opportunity to catch fish; thus, many urban programs are based on fishes such as common carp, buffaloes, or bullheads.

The terms *put and take* and *put, grow, and take* are typically not used by wildlife biologists, but these terms are applicable to wildlife stockings. Hunting preserves that release pen-reared adult birds or mammals prior to a hunt are practicing put and take stocking. Some conservation agencies sponsor such practices in areas with substantial levels of hunting effort. Put, grow, and take stockings are less often used, primarily because the release of young birds or mammals that will grow and be harvested at a later date is typically not a very successful strategy.

10.8 Philosophical Considerations for Stocking Programs

Stocking is often viewed by the public as the most important and effective management tool available to biologists. It seems to represent some form of magical solution to all fisheries and wildlife problems. If fishing or hunting success declines, the public will often request more stocking. If nongame organisms, such as the Colorado pikeminnow or black-footed ferret, decline, stocking is prominently mentioned as a potential solution. In some cases, the stocking may be politically divisive, as occurred for the wolf reintroduction to Yellowstone National Park (Smith and Ferguson 2005). It does not seem to matter if fishing has declined because a fish species has overpopulated and stunted or wildlife populations have declined because of a lack of habitat; more stockings are often requested by the public. It may take a major public relations effort by biologists to convince the public that stocking in certain situations is unnecessary, useless, or even harmful. In some cases, stocking is a valid management tool. However, it is certainly not the answer in many situations.

In many instances, stocking represents the treatment of a symptom rather than the cause of a problem. Stocking cannot address habitat degradation, overexploitation, poor population structure, the effects of introduced organisms, or other problems that affect populations and communities. Aside from the fact that stocking is often unsuccessful, it can also cause additional problems in some situations. There is increasing evidence that inappropriate stockings can harm rather than help a resource. A good case in point involves the various species of Pacific salmon. There is concern, and some evidence, that the release of hatchery-produced Pacific salmon in the native range of these fishes may actually harm native stocks, perhaps through the introduction of inferior genetic material (Hilborn 1992). Fisheries biologists in British Columbia were sufficiently concerned about the conservation of genetic material to change their method of stocking **steelheads**. The steelhead is a rainbow trout form that normally lives part of its life in marine waters. The new stocking programs in British Columbia involve the capture of small numbers of returning steelheads from each of a series of rivers, spawning and rearing those fish in individual lots, and then stocking offspring only in the stream from which their parents were collected. Pacific salmon and steelhead population declines will be more effectively addressed when their causes, such as dams that hinder migrations of adults up the river or juveniles down the river, lost or degraded spawning habitat, and other human-influenced limiting factors, are addressed. The current arguments primarily used against addressing causes rather than symptoms are economic. However, in the long term, addressing causes typically makes more economic sense than treating symptoms.

A situation similar to that of the Pacific salmon and steelhead may have been averted in the wildlife field. In the late 1980s, a plan was developed to stock massive numbers of mallards into North America. The North American population of mallards was declining, and the proposed solution was to raise large numbers of mallards in captivity for release. Fortunately, the plan has not been implemented.

A variety of problems can be caused by stocking in the wrong situation. As mentioned above and in section 4.9, the introduction of inferior genetic material into a native population may decrease the survival characteristics of that population. For example, after a strain of rainbow trout has been raised in fish hatcheries for generations, intentional and unintentional artificial selection is likely to have favored those fish adapted to hatchery existence rather than to life in the wild. If any of that strain of rainbow trout survive when stocked in the wild and reproduce with native fish, the native gene pool may be diluted. Another problem is that many introductions are made from a small gene pool. For example, if five sibling wild turkeys are trapped in one area and relocated to another to establish a new population, inbreeding will occur. The turkeys thus may not have sufficient genetic diversity to flourish in their new environment. Even if the same number of nonsibling wild turkeys were relocated to the new area, the small number of individuals might still provide minimal genetic diversity.

Another important concern about stocking programs is their influence on species diversity. For example, consider what can happen when a predatory species such as the walleye is stocked into a water body in which it was previously not present. The stocked walleyes may have an effect on other fishes and a variety of other aquatic organisms in the water body. In addition, walleyes may escape from that water body into other water bodies. Now other potential problems arise. A substantial number of large, predatory fish may have moved into an area such as a small stream in which no such top-level predators may have previously existed. What effect will these new predators have on native stream species? To add to the problem, the fish community in that small stream may have already been reduced because of habitat degradation. Thus, the influence of a stocking program on species diversity in the stream becomes a thorny issue, probably a much more important one than the effects of introduced walleyes in a human-created water body.

Introduced fishes also can affect other components of the biota in a water body; examples include amphibians (Bradford et al. 1993) as well as aquatic insects and zooplankton (Dunham et al. 2004). Once again, fisheries biologists must be concerned about ecosystem management and consider all aspects of the biotic community. The Wyoming Game and Fish Department reached a compromise position for approximately 1,200 wilderness lakes that were suitable for trout in the state (Wiley 2003). They will continue management of self-sustaining trout populations, which involved approximately half of the lakes. Trout will be allowed to gradually disappear from about 10% of the lakes in which natural reproduction and recruitment does not occur; stockings in those waters have ceased. No fish will be stocked into the remaining 40% of the lakes that were fishless; they will remain as pristine as allowed by other human activities in the wilderness.

Wildlife biologists do not receive nearly the demand from the public for stockings that fisheries biologists do. This is not to say that the wildlife field is immune to the problem. For example, there is still considerable demand for northern bobwhite and ring-necked pheasant stockings.

Biologists must carefully define the reason for a stocking program. Specific management objectives need to be developed, and those objectives need to be measurable (see section 20.6). In addition, all of the biological ramifications of the stocking must be taken into consideration. What effects will it have on other organisms? Economic issues must also be considered. What is the final cost of the organism caught or harvested? Are the benefits greater than the costs of the stocking?

Some stockings are done for political rather than biological reasons. For example, a biologist may be successful in stocking catchable-sized rainbow trout in an urban stream in the southern United States during the winter. A popular fishery may develop, but the cost of the program may outweigh the return from license sales and other economic benefits. The biologist may wish to cease the program but may find that influential political figures and the general public in the city expect the stockings to continue. The outraged biologist may approach his or her supervisors and be even further upset when an administrative decision is made to continue the program. Who is right and who is wrong in this situation? This is a difficult question to answer. Perhaps the administrators decided that bearing the cost of the stocking program was a lesser evil than bearing the costs of a political battle and negative public opinion that could carry over to larger issues.

Both the fisheries and wildlife professions have modified their activities and thought processes over the past century. Certainly, more thought is given to biological diversity and native species than in past decades. The fisheries profession as a whole has responded to this change in philosophy with a change in fish culture and fish stocking practices (Jackson et al. 2004). A survey of state and provincial management agencies in the United States and Canada indicated that both fish culture practices and fish stocking programs have changed over the past three decades. Agencies reported greater evaluation of stocking programs, increased use of native species, and greater concern over the genetic integrity of cultured fishes. In addition, the rate of increased expenditures for habitat management was double that for stocking programs.

10.9 Diseases, Pathogens, and Parasites

A **disease** is any departure from health; it is a particular destructive process in an organ or organism with a specific cause and symptoms. **Disease agents** can be physical, chemical, or biological factors that cause a disease. Disease-causing biological agents can also be called **pathogens**. This section of the text will address a few of the biological factors that result in disease. Particular pathogens cause particular diseases. Pathogens and parasites encompass a wide variety of organisms, including viruses, bacteria, protozoans, fungi, and helminths. Some are external, while others are internal. Only a few will be highlighted in this text; further information can be found in Lim and Webster (2001) and Wobeser (2006).

Diseases can be a concern in all aspects of biota management. Obviously a biologist would be concerned about disease outbreaks in wild populations. In addition, diseases can be even more of a concern when stocking or accidental introduction is involved. Any type of fish or wildlife stocking creates the opportunity for pathogens or parasites to spread to a new environment. This problem can be especially serious when exotic or transplanted organisms are involved, because they may transfer a disease against which native organisms have no defense.

The potential for the spread of pathogens or parasites is intensified in the case of stockings because most stockings use hatchery-reared fish or pen-reared wildlife. Most parasites in the wild affect only a few of the organisms in a population. In an artificially reared or held population, the crowded organisms are much more susceptible to disease outbreaks and an outbreak is likely to affect a high percentage of the population. The stress that results from crowding creates greater opportunities for large-scale eruptive disease outbreaks (**epizootics**) as well as low-level, chronic disease problems (**enzootics**).

Mammals are subject to a variety of diseases that can cause problems in wild and captive populations. American bison are known to carry the bacterial disease brucellosis. For example, American bison in Yellowstone National Park have apparently carried the disease since 1917, with little effect on the herd (Meagher 1978). The disease is also present in the Jackson Hole elk herd that ranges in the same area. Brucellosis can have serious economic effects on ranchers that raise domesticated cattle. Lawsuits have been filed over the transmission of the disease from bison to domesticated cattle in the Yellowstone Park area; the resulting decisions have basically said that transmission was probable but not proven. Most bison herds in the United States are maintained in a brucellosis-free condition.

White-tailed deer are susceptible to outbreaks of epizootic hemorrhagic disease (EHD). The causative agent for this disease is a virus, and outbreaks commonly occur in the southeastern United States and sporadically in other states and Canada. There is no effective control for this disease, but outbreaks are usually associated with high white-tailed deer densities, evoking similarities with cultured fish and pen-reared wildlife.

The possible introduction of tuberculosis into wild populations of ungulates from infected private game-farm animals is a source of concern to wildlife biologists in both the United States and Canada. It could also create a public health problem for humans. Laws regulating the transport of domesticated livestock do not necessarily apply to game-farm ungulates, and disease tests developed for domesticated livestock do not necessarily work well for other species. Another disease concern in the United States and Canada is the possible introduction of the meningeal worm into western North America. Introduction of this parasite through game-farm animals could have catastrophic effects on western elk and moose populations.

Chronic wasting disease (usually shortened to CWD) is a transmissible neurological disease that produces small lesions in brains of deer and elk. Infected animals are characterized by loss of body condition, behavioral changes, and eventual death (Williams et al. 2002). This disease actually is a transmissible spongiform encephalopathy, and it is similar to mad cow disease in cattle and scrapie in sheep. The infectious agent for chronic wasting disease is believed to be a prion, which is an infectious protein without associated nucleic acids. Although this disease is fatal for deer and elk, research suggests that humans, cattle, and other domesticated livestock are resistant to natural transmission. The possibility of human infection remains a concern, but there have been no verified cases of humans contracting chronic wasting disease. This disease can reduce the growth and size of wild deer and elk populations in areas of high disease prevalence and is of concern for wildlife biologists across North America. Chronic wasting disease has also been diagnosed in commercial game farms in several states and provinces.

Birds are also subject to a wide variety of diseases that are important in wild and captive populations. Botulism, caused by the anaerobic bacterium *Clostridium botulinum,* causes severe disease outbreaks in waterfowl. Sick birds cannot hold their necks erect, so the disease is also called limber neck (Bellrose 1976). The disease cycle commonly includes fly maggots, so it is important that carcasses of waterfowl killed by the disease be gathered and burned. Avian cholera, a bacterial disease, annually kills thousands of waterfowl during spring migrations.

Another common disease in birds is aspergillosis. This disease, which is prevalent in many birds, including gallinaceous birds and waterfowl, attacks the respiratory system. The causative agent is a fungus that often occurs on moldy grain. A viral disease that primarily attacks waterfowl is duck virus enteritis. The pathogen responsible for this disease is a herpesvirus.

West Nile virus is a mosquito-borne disease that can infect numerous organisms, including humans, horses, and birds. The virus was originally found in Africa, Asia, Europe, and Australia and was first found in New York City in 1999. The virus subsequently extended its range throughout much of the United States. Migratory birds are suspected of carrying the virus to the western hemisphere (Rappole et al. 2000). Similarly, avian influenza viruses (also known as bird flu) concern health officials because of the potential for transmission from birds to humans (Abbott and Pearson 2004).

Fishes are subject to a wide complement of diseases. Some of these occur primarily in aquaculture situations, while others can at times be found at epizootic levels even in wild populations. Among fishes, disease outbreaks are again most common under stressful or crowded conditions. In addition, poor nutrition can lead to stress, which again increases the likelihood of disease outbreaks.

Examples of bacterial diseases of fishes include bacterial kidney disease (BKD) and furunculosis, which affect salmonids, and columnaris, which commonly affects warmwater fishes such as those of the sunfish family. Viral diseases include infectious pancreatic necrosis (IPN), which affects salmonids throughout the world, channel catfish virus disease (CCVD), which obviously is of concern to the channel catfish aquaculture industry in the United States (Davis 1992), and largemouth bass virus (LMBV), which concerns both private aquaculture producers and recreational fisheries managers (Grizzle et al. 2002).

Viral hemorrhagic septicemia (VHS) was first detected in the United States during 1988, and it affected spawning salmon in the Pacific Northwest. Initially, biologists believed that the disease was primarily limited to trout and salmon. However, in 2005 biologists determined that muskellunge had been infected in Lake St. Claire, Michigan. In 2006, muskellunge in Lake Erie, Lake Ontario, and the St. Lawrence River were killed by the virus. At this time, there is substantial concern over potential epizootics in the Great Lakes, as several other fish species have been infected by the VHS virus.

The technology for diagnosing bacterial and viral fish diseases has improved substantially over the last decade, especially for early diagnosis of these diseases. However, treatment capabilities have actually declined over the same time period because of the loss of chemicals that had been approved for such uses. Two U.S. federal agencies regulate which chemicals can be used for fish disease treatment. The U.S. Food and Drug Administration is concerned with the effects of the chemicals used on human health, while the U.S. Environmental Protection Agency is concerned with the environmental effects of those chemicals. Thus, chemicals that have proven carcinogenic to humans or those that have proven harmful to the environment can no longer be used for fish disease treatment.

External parasites can also create fish health problems. Again, large outbreaks of parasites are more likely under crowded or stressful conditions. A common fish health problem in the channel catfish aquaculture industry and at times in wild populations is outbreaks of "ich" (the causative agent is the protozoan *Ichthyophtherius multifilis*). The same disease is commonly found in aquarium fishes; "ich" can actually affect a wide variety of warmwater fishes. *Trichodina* is another external parasite, and it commonly affects both salmonids and warmwater fishes. Treatment capabilities for external parasites are much better than for bacterial or viral diseases.

Some states require inspection of fishes transported across state lines in an attempt to reduce the transfer of fish diseases. However, state regulations vary widely, and no central federal authority has emerged to regulate such transport.

The effects of diseases on biota represent an important concern for wildlife and fisheries biologists. The above examples are just a few of the many that could have been mentioned. Actually, it is often difficult or impossible to identify a single causative agent for an epizootic. For example, bighorn sheep are susceptible to lungworms. Under certain conditions, the lungworm infestation can lead to an epizootic, with bighorn sheep actually dying of pneumonia, which may be caused by bacterial or viral infection. These lungworms, bacteria, and viruses have been found in apparently healthy as well as diseased bighorn sheep. The "certain conditions" that lead to a disease outbreak are complex and interrelated (Figure 10.6). Stressed animals are more likely to become sick, and sources of stress can include competition with other bighorn sheep or domesticated livestock, harassment by humans, weather, and range condition and forage quality.

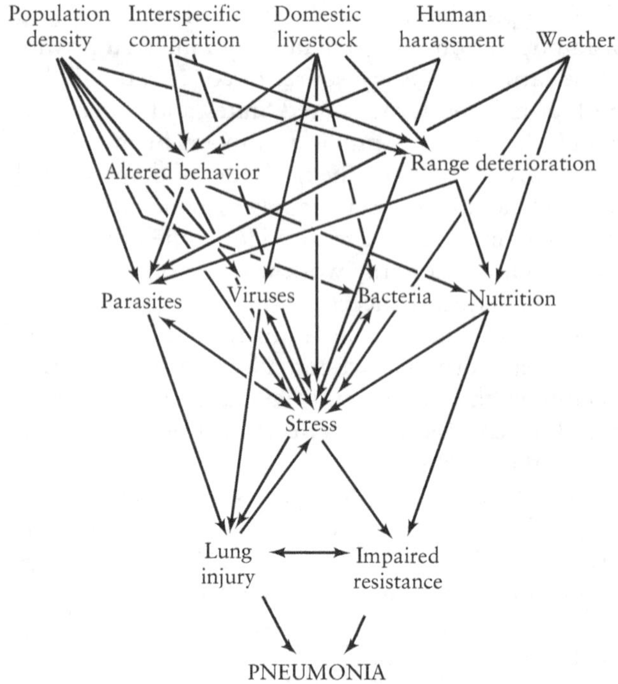

FIGURE 10.6 The causes of an epizootic in a population of wild organisms can be complex. This diagram depicts the variety of interrelated factors that can cause the lungworm–pneumonia complex in bighorn sheep. *[From Wobeser 1994.]*

SUMMARY

At times, biologists find it necessary to remove part or all of an animal community. Partial removals are often undertaken by fisheries biologists when a particular population has become overabundant, has slow growth rates, and has a poor size structure. A somewhat comparable management technique used by wildlife biologists might be extensive predator removal on a waterfowl nesting island.

Complete population renovations are attempted primarily by fisheries biologists. At times, an entire fish community may be eliminated, perhaps due to circumstances such as the overabundance of a species such as the common carp. Complete renovations have also been used to expand the population size of threatened or endangered species.

Population control has also been attempted through interference with the reproductive process by release of sterile organisms or by use of contraceptives.

Attempts to interfere with reproduction are more likely to succeed in small populations in confined areas. When fisheries biologists prefer that a species not become naturalized, they can produce and stock polyploid or monosex fish, which cannot reproduce.

Stocking of organisms can be categorized as introductory, maintenance, or supplemental. These terms are commonly used by fisheries biologists; while not used by wildlife biologists, they are still applicable to that field. Introductory stockings involve the release of organisms into a new or renovated water body or newly created terrestrial habitats or the introduction of a new species into an existing community. Maintenance stockings are made to maintain a population that for some reason has no or very limited natural reproduction. Supplemental stockings are made to augment naturally reproducing populations; they are generally less likely to be successful than introductory or maintenance stockings.

Examples of introductory stockings include the restoration of North American river otters to Missouri rivers and streams from which they had been extirpated and the introduction of brook trout to the Rocky Mountains. Maintenance stockings of walleyes are commonly made in many lakes and reservoirs. Supplemental stockings of organisms such as ring-necked pheasants generally are unsuccessful and can lead to problems of reduced genetic fitness and disease transfer. Biologists should avoid unnecessary and unwise stocking of organisms. However, stocking can be an effective tool for the management and restoration of game and nongame species.

Diseases can account for a substantial portion of the natural mortality in wild populations, and thus they are of concern to biologists. Transfer of pathogens and parasites during fish stocking or wildlife introduction may be an even greater concern, especially when exotic organisms are involved. Exotic organisms may transfer a pathogen or parasite for which native organisms have no defense.

PRACTICE QUESTIONS

1. Renovations of impoundments, such as those conducted with a fish toxicant, are less frequently undertaken by fisheries biologists today than in the past. Why might the use of this technique be declining?

2. What are some advantages and disadvantages for both mechanical removal and chemical removal undertaken for a partial rehabilitation of a fish population?

3. What are some advantages and disadvantages of having a mechanical removal of fishes completed by commercial harvest rather than by a state, provincial, or territorial conservation agency?

4. Why are supplemental stockings of ring-necked pheasants and northern bobwhites typically not used to bolster wild populations?

5. List several situations in which biologists might undertake population control to protect humans from wildlife.

6. Think of two different scenarios in which two fish species are cultured. Which of the two tends toward extensive fish culture, and which tends more toward intensive fish culture?

7. What are the differences between introductory, maintenance, and supplemental stockings? Give an example of each for both a fish and a wildlife species.

8. Why does a fisheries biologist need to consider available habitat and fish niches when deciding on the complement of fish species to stock into a new reservoir?

9. What is the difference between a put and take stocking and a put, grow, and take stocking?

10. What are some of the potential disease problems that might arise if game ranching of large ungulates is not undertaken in a careful manner?

11. If an exotic fish species were proposed for introduction, would the use of polyploid individuals provide any unique opportunities to safely study the introduction? Why?

12. In the state or province in which you live, does the state or provincial conservation agency or the agricultural agency maintain control over captive game herds? Consider the disease information provided in this chapter and indicate which agency you would prefer to be in charge.

SELECTED READINGS

Adams, L. W., L. W. VanDruff, and M. Luniak. 2005. Managing urban habitats and wildlife. Pages 714–739 *in* C. E. Braun, ed. *Techniques for wildlife investigations and management*. 6th ed. The Wildlife Society, Bethesda, Md.

Allen, L. J., ed. 1984. *Urban fishing symposium proceedings*. Fisheries Management Section and Fisheries Administrators Section, American Fisheries Society, Bethesda, Md.

Arlinghaus, R., and T. Mehner. 2004. A management-orientated comparative analysis of urban and rural anglers living in a metropolis (Berlin, Germany). *Environmental Management* 33:331–344.

Behnke, R. J. 1992. *Native trout of western North America.* Monograph 6. American Fisheries Society, Bethesda, Md.

Bolen, E. G., and W. L. Robinson. 2003. *Wildlife ecology and management.* 5th ed. Prentice Hall, Upper Saddle River, N.J.

Bryan, M. B., D. Zalinski, K. B. Filcek, S. Libants, W. Li, and K. T. Scribner. 2005. Patterns of invasion and colonization of the sea lamprey (*Petromyzon marinus*) in North America as revealed by microsatellite genotypes. *Molecular Ecology* 14:3757–3773.

Hallerman, E., D. King, and A. Kapuscinski. 1999. A decision support software for safely conducting research with genetically modified fish and shellfish. *Aquaculture* 173:309–318.

Heidinger, R. C. 1999. Stocking for sport fisheries enhancement. Pages 375–401 *in* C. C. Kohler and W. A. Hubert, eds. *Inland fisheries management in North America.* 2nd ed. American Fisheries Society, Bethesda, Md.

Hygnstrom, S. E., R. M. Timm, and G. E. Larson, eds. 1994. *Prevention and control of animal damage.* University of Nebraska, Cooperative Extension Service, Lincoln.

Kohler, C. C., and J. G. Stanley. 1984. A suggested protocol for evaluating proposed exotic fish introductions in the United States. Pages 387–406 *in* W. R. Courtenay Jr., and J. R. Stauffer Jr., eds. *Distribution, biology, and management of exotic fishes.* Johns Hopkins University Press, Baltimore, Md.

Lackey, R. T., D. H. Lach, and S. L. Duncan, eds. 2006. *Salmon 2100: The future of wild Pacific salmon.* American Fisheries Society, Bethesda, Md.

Li, H. W., and P. B. Moyle. 1999. Management of introduced fishes. Pages 345–374 *in* C. C. Kohler and W. A. Hubert, eds. *Inland fisheries management in North America.* 2nd ed. American Fisheries Society, Bethesda, Md.

Mills, L. S., J. M. Scott, K. M. Strickler, and S. A. Temple. 2005. Ecology and management of small populations. Pages 691–713 *in* C. E. Braun, ed. *Techniques for wildlife investigations and management.* 6th ed. The Wildlife Society, Bethesda, Md.

Mooney, H. A., R. N. Mack, J. A. McNeely, L. E. Neville, P. J. Schei, and J. K. Waage. 2005. *Invasive alien species: A new synthesis.* Island Press, Washington, D.C.

Noble, R. L. 1981. Management of forage fishes in impoundments of the southern United States. *Transactions of the American Fisheries Society* 110:738–750.

Pister, E. P. 2001. Wilderness and fish stocking: History and perspective. *Ecosystems* 4:279–286.

Rahel, F. J., R. T. Muth, and C. A. Carlson. 1999. Endangered species management. Pages 431–454 *in* C. C. Kohler and W. A. Hubert, eds. *Inland fisheries management in North America.* 2nd ed. American Fisheries Society, Bethesda, Md.

Reynolds, J. D., G. M. Mace, K. H. Redford, and J. G. Robinson, eds. 2001. *Conservation of exploited species.* Cambridge University Press, Cambridge, U.K.

Ruiz, G. M., and J. T. Carlton. 2003. *Invasive species: Vectors and management strategies.* Island Press, Washington, D.C.

Shivik, J. A., and D. J. Martin. 2000. Aversive and disruptive stimulus applications for managing predation. Pages 111–119 *in* M. C. Brittingham, J. Kays, and R. McPeake, eds. *The ninth wildlife damage management conference proceedings.* Internet Center for Wildlife Damage Management Conferences, University of Nebraska, Lincoln.

Teer, J. G., L. A. Renecker, and R. J. Hudson. 1993. Overview of wildlife farming and ranching in North America. *Transactions of the North American Wildlife and Natural Resources Conference* 58:448–459.

Vaske, J. J., N. R. Timmons, J. Beaman, and J. Petchenik. 2004. Chronic wasting disease in Wisconsin: Hunter behavior, perceived risk, and agency trust. *Human Dimensions of Wildlife* 9:193–209.

VerCauteren, K. C., R. A. Dolbeer, and E. M. Gese. 2005. Identification and management of wildlife damage. Pages 740-778 *in* C. E. Braun, ed. *Techniques for wildlife investigations and management.* 6th ed. The Wildlife Society, Bethesda, Md.

Wiley, R. W., and R. S. Wydoski. 1999. Management of undesirable fish species. Pages 403–454 *in* C. C. Kohler and W. A. Hubert, eds. *Inland fisheries management in North America.* 2nd ed. American Fisheries Society, Bethesda, Md.

LITERATURE CITED

Abbott, A., and H. Pearson. 2004. Fear of human pandemic grows as bird flu sweeps through Asia. *Nature* 427:472-473.

Alcorn, S. R. 1981. Fishing quality in two urban fishing lakes, St. Louis, Missouri. *North American Journal of Fisheries Management* 1:80–84.

Angeler, D. G., P. Chow-Fraser, M. A. Hanson, S. Sánchez-Carrillo, and K. D. Zimmer. 2003. Biomanipulation: A useful tool for freshwater wetland mitigation? *Freshwater Biology* 48:2203–2213.

Apa, A. D., D. W. Uresk, and R. L. Linder. 1991. Impacts of black-tailed prairie dog rodenticides on nontarget passerines. *The Great Basin Naturalist* 51:301-309.

Bellrose, F. C. 1976. *Ducks, geese, and swans of North America.* 2nd ed. Stackpole Books, Harrisburg, Pa.

Bradford, D. F., F. Tabatabai, and D. M. Graber. 1993. Isolation of remaining populations of the native frog, *Rana mucosa*, by introduced fishes in Sequoia and Kings Canyon national parks, California. *Conservation Biology* 7:882–888.

Chick, J. H., and M. A. Pegg. 2001. Invasive carp in the Mississippi River basin. *Science* 292:2250–2251.

Crivelli, A. J. 1983. The destruction of aquatic vegetation by carp. *Hydrobiologia* 106:37–41.

Davis, A. J. 1992. Channel catfish virus: A new type of herpesvirus. *Virology* 186:9–14.

Dunham, J. B., D. S. Pilliod, and M. K. Young. 2004. Assessing the consequences of nonnative trout in headwater ecosystems in western North America. *Fisheries* (Bethesda) 29(6):18–26.

Fuller, P. L., L. G. Nico, and J. D. Williams. 1999. *Nonindigenous fishes introduced into inland waters of the United States.* Special Publication 27, American Fisheries Society, Bethesda, Md.

Graham, L. K. 1997. Contemporary status of the North American paddlefish, *Polyodon spathula. Environmental Biology of Fishes* 48:279–289.

Grizzle, J. M., I. Altinok, W. A. Fraser, and R. Francis-Floyd. 2002. First isolation of largemouth bass virus. *Diseases of Aquatic Organisms* 50:233–235.

Hansson, L.-A., H. Annadotter, E. Bergman, S. F. Hamrin, E. Jeppesen, T. Kairesalo, E. Luokkanen, P. A. Nilsson, M. Søndergaard, and J. Strand. 1998. Biomanipulation as an application of food-chain theory: Constraints, synthesis, and recommendations for temperate lakes. *Ecosystems* 1:558–574.

Harig, A. L., K. D. Fausch, and M. K. Young. 2000. Factors influencing success of greenback cutthroat trout translocations. *North American Journal of Fisheries Management* 20:994–1004.

Hellickson, M., and A. Radomski. 1999. *Bobwhites of the Wild Horse Desert: Status of our knowledge.* Wildlife Management Bulletin 4, Caesar Kleberg Wildlife Research Institute, Texas A&M University, Kingsville.

Hessler, E., J. R. Tester, D. B. Siniff, and M. M. Nelson. 1970. A biotelemetry study of survival of pen-reared pheasants released in selected habitats. *The Journal of Wildlife Management* 34:267–274.

Hilborn, R. 1992. Hatcheries and the future of salmon in the northwest. *Fisheries* (Bethesda) 17(1):5–8.

Hill, D., and P. Robertson. 1988. Breeding success of wild and hand-reared ring-necked pheasants. *The Journal of Wildlife Management* 52:446–450.

Jackson, J. R., J. C. Boxrucker, and D. W. Willis. 2004. Trends in agency use of propagated fishes as a management tool in inland fisheries. Pages 121–138 *in* M. J. Nickum, P. M. Mazik, J. G. Nickum, and D. D. Mackinlay, eds. *Propagated fish in resource management.* Symposium 44, American Fisheries Society, Bethesda, Md.

Johnson, F. H. 1977. Responses of walleye *(Stizostedion vitreum vitreum)* and yellow perch *(Perca flavescens)* populations to removal of white sucker *(Catostomus commersoni)* from a Minnesota lake, 1966. *Journal of the Fisheries Research Board of Canada* 34:1633–1642.

Kellert, S. R., M. Black, C. R. Rush, and A. J. Bath. 1996. Human culture and large carnivore conservation in North America. *Conservation Biology* 10:977–990.

Kozfkay, J. R., J. C. Dillon, and D. J. Schill. 2006. Routine use of sterile fish in salmonid sport fisheries: Are we there yet? *Fisheries* (Bethesda) 31(8):392–401.

Kruse, C. G., W. A. Hubert, and F. J. Rahel. 2001. An assessment of headwater isolation as a conservation strategy for cutthroat trout in the Absaroka Mountains of Wyoming. *Northwest Science* 75:1–11.

Laarman, P. W. 1978. Case histories of stocking walleyes in inland lakes, impoundments, and the Great Lakes: 100 years with walleyes. *American Fisheries Society Special Publication* 11:254–260.

Lim, C., and C. D. Webster. 2001. *Nutrition and fish health.* Food Products Press, Binghamton, N.Y.

Marking, L. L. 1992. Evaluation of toxicants for the control of carp and other nuisance fishes. *Fisheries* (Bethesda) 17(6):6–13.

Meagher, M. M. 1978. Bison. Pages 122–133 *in* J. L. Schmidt and D. L. Gilbert, eds. *Big game of North America: Ecology and management.* Stackpole Books, Harrisburg, Penn.

Meijer, M.-L., I. de Boois, M. Scheffer, R. Portielje, and H. Hosper. 1999. Biomanipulation in shallow lakes in The Netherlands: An evaluation of 18 case studies. *Hydrobiologia* 408/409:13–30.

Meyer, F. P., and R. A. Schnick. 1983. Sea lamprey control techniques: Past, present, and future. *Journal of Great Lakes Research* 9:354–358.

Nielsen, L. A. 1999. History of inland fisheries management in North America. Pages 3–30 *in* C. C. Kohler and W. A. Hubert, eds. *Inland fisheries management in North America.* 2nd ed. American Fisheries Society, Bethesda, Md.

Rappole, J. H., S. R. Derrickson, and Z. Hubálek. 2000. Migratory birds and spread of West Nile virus in the western hemisphere. *Emerging Infectious Diseases* 6:319–328.

Shapiro, J., V. Lamarra, and M. Lynch. 1975. Biomanipulation: An ecosystem approach to lake restoration. Pages 85–96 *in* P. L. Brezonik and J. L. Fox, eds. *Proceedings of a symposium on water quality management through biological control.* University of Florida, Gainesville.

Smith, D. W., and G. Ferguson. 2005. *Decade of the wolf: Returning the wild to Yellowstone.* The Lyons Press, Guilford, Conn.

Springer, P. F., G. V. Byrd, and D. W. Woolington. 1977. Reestablishing Aleutian Canada geese. Pages 331–338 *in* S. A. Temple, ed. *Endangered birds: Management techniques for preserving threatened species.* University of Wisconsin Press, Madison.

Stephen, J. L. 1986. Effects of commercial harvest on the fish community of Lovewell Reservoir, Kansas. Pages 211–217 *in* G. E. Hall and M. J. Van Den Avyle, eds. *Reservoir fisheries management: Strategies for the 80s.* Reservoir Committee, Southern Division, American Fisheries Society, Bethesda, Md.

Turner, J. W., Jr., K. M. Liu Irwin, and J. F. Kirkpatrick. 1992. Remotely delivered immunocontraception in captive white-tailed deer. *The Journal of Wildlife Management* 56:154–157.

Verrill, D. D., and C. R. Berry Jr. 1995. Effectiveness of an electrical barrier and lake drawdown for reducing common carp and bigmouth buffalo abundances. *North American Journal of Fisheries Management* 15:137–141.

Ward, D. 2005. Selective removal of nonnative fishes using Supaverm: Toxicity screening for a candidate species-specific piscicide. *Journal of Freshwater Ecology* 20:787–789.

Wedemeyer, G. A, ed. 2001. *Fish hatchery management.* 2nd ed. American Fisheries Society, Bethesda, Md.

Wiley, R. W. 2003. Planting trout in Wyoming high-elevation wilderness waters. *Fisheries* (Bethesda) 28(1):22–27.

Williams, E. S., M. W. Miller, T. J. Kreeger, R. H. Kahn, and E. T. Thorne. 2002. Chronic wasting disease of deer and elk: A review with recommendations for management. *Journal of Wildlife Management* 66:551–563.

Wobeser, G. A. 1994. *Investigation and management of disease in wild animals.* Plenum Press, New York.

Wobeser, G. A. 2006. *Essentials of disease in wild animals.* Blackwell Publishing Professional, Ames, Iowa.

CHAPTER 11

Endangered and Threatened Species

The disappearance of species and the evolution of new species are natural processes that have occurred throughout the history of life on earth. Rates of species extinction have varied over different geologic periods, with peak extinction rates associated with five extinction spasms (Wilson 1992). Extinction rates today are being greatly accelerated by human activities and represent a sixth extinction spasm (Whitmore 1998).

It is probable that hunting by humans in North America in the last 12,000 years caused the extinction of several species of large mammals (Wilson 1992; Martin 2005). Woolly mammoths (mastodons), along with ground sloths and several other large mammals, became extinct within about 1,000 years of North American colonization by Paleo-Indian cultures 11,000 to 12,000 years ago (Wilson 1992). The specific effects of humans and climate changes in their extinction is unknown, but spear points in the bones of mastodons and other mammals, remains of large mammal bones in campfires and at campsites, and their rapid decline and extinction coincidental with human colonization all indicate a substantial human role.

The bones of many other extinct species in other geographic regions have been found in debris left by early humans. Moas, such as the slender moa and upland moa, were large **cursorial** (adapted for running) birds of New Zealand. Most species of moas were rapidly brought to extinction by the hunting activities of the Maoris after they colonized New Zealand around A.D. 1000; a few moas may have survived into the 1800s. Massive extinctions also coincided with human colonizations of Australia, Madagascar, and many Pacific islands (Wilson 1992). Other recent extinctions caused by human influences include the passenger pigeon of eastern North America, the heath hen, which lived along the east coast of North America from Massachusetts to Maryland, the blue pike of the Great Lakes region, the Steller's sea cow of the Bering Sea, and the great auk of the North Atlantic. Unfortunately, extinctions due to human influence are increasing rapidly. Many species exist that are unknown to us; they could be lost without our even being aware of their disappearance.

There are many causes of extinction, including natural causes, unregulated harvest by humans, and introduced organisms, but by far the most common cause of extinction is habitat loss or modification. Most habitat loss or modification is ultimately the result of human population growth, ever-increasing human demands for space and natural resources, and, in some cases, inappropriate uses of technology. Extinctions due to natural causes are inconsequential at this time in the history of the earth compared with human-related species losses. We seem to be rapidly increasing our ability to destroy without accompanying increases in our wisdom to preserve. Extinction is not retractable; it is permanent and final.

Human demand for natural resources, the byproducts of human life, and introduced species combine to challenge our abilities to conserve biodiversity on the planet (see section 2.1). The use of the term *biodiversity* is usually restricted to wild organisms; in a broader sense, the gene pools of all organisms, even domesticated or cultivated ones, also contribute to biodiversity.

The purposes of this chapter are to discuss endangered and threatened species, the causes of their declines, and plans for increasing their populations.

11.1 Why Be Concerned with Extinction?

To many people, the loss of little-known species, even in the great numbers currently being lost in tropical rain forests, is of little consequence. A large, familiar species, such as the bald eagle, is likely to gain greater public support and interest when its population drops to low levels than, for instance, a small, unknown insect, reptile, or plant.

There are two categories of values that can be associated with any organism. The first category includes the value of an organism to humans; that is, animals and plants that have commercial, scientific, ecological, recreational, aesthetic, and sociocultural values to humans. The second category includes values that do not relate to humans. Our anthropocentrism often causes us to believe that everything revolves around human needs. A species has value in itself; it does not need to be of direct use to humans to have value.

If the value of a species to humans were the only issue of importance, then the scientific and medical values that many species provide to humans should convince us of the benefits of preserving species. A variety of organisms have contributed greatly to humans; some of these have been quite inconspicuous. The little-known rosy periwinkle, for example, a tropical plant, has provided vinca alkaloids for much-improved treatment of Hodgkin's disease, non-Hodgkin's lymphoma, and certain cancers of the brain and lung (Hanson 2005). What if humans had caused the extinction of the rosy periwinkle before its medical value was discovered? Penicillin was derived from a mold, hardly an organism that elicits great love from humans. Physiological studies of other animals have contributed greatly to human medical science. The loss of species may destroy entire gene pools that hold the potential to unlock secrets important to our own health and survival. We know little about most species, and yet each may have value to humans.

In the development of domesticated animals and plants we have manipulated the gene pools of the original wild organisms and selected for characteristics useful to humans. Domesticated organisms have lost portions of their original gene pools that could have been important to the development of valuable characteristics in future domesticated varieties. For instance, a domesticated animal or plant may lose certain immunities to disease, adaptations to extreme heat or cold, or the ability to adapt to greater salinity levels. Once a portion of the gene pool is lost from a domesticated organism and its wild ancestor is extinct, those genetic characteristics are gone forever. Further, many wild organisms could one day provide the basis for additional valuable domesticated varieties.

Such issues demonstrate the importance of all species to humans. Additionally, all species function interconnectedly in the web of life. Life on earth is one large food web, of which humans are a part; think of it as a pyramid of cans or a house of cards. Who knows what levels above may crash if a supporting can or card is pulled? What if many cans or cards are removed? How many can be removed before the entire stack collapses?

It seems appropriate here to describe the economic values of some species that have faced extinction, are in decline, or are of unknown status. American bison provide a source of low-fat meat of increasing economic importance and also hold promise for improving grazing systems, yet bison populations once reached near-extinction levels. Aesthetics also provide economic value. The whooping crane, for example, is viewed by thousands of people at the Aransas National Wildlife Refuge in south Texas;

many of those people spend money for tourism services. Some developers are finding that property values increase as they coordinate human space needs with those of species such as nesting bald eagles or red-cockaded woodpeckers (Figure 11.1). Undisturbed areas left for these other species may ultimately add considerable value to properties that are developed for housing and other human purposes.

Wild fishes, birds, mammals, and other organisms are also important to human well-being from an aesthetic viewpoint. Millions of people are involved in observing "watchable" wildlife and fisheries resources. Underwater observation of the many colorful marine organisms found on coral reefs and in other coastal areas is becoming increasingly popular. National, state, provincial, territorial, and local parks continue to be extensively used; a primary reason for this appeal is their wild animal populations. People obtain much everyday enjoyment from the natural elements of ecosystems in which they live. We spend great sums of money on mental health care, and yet the mental health benefits of plant and animal communities that surround us are often taken for granted. How important are songbirds in a backyard or eastern fox squirrels at a feeder to senior citizens with limited mobility beyond their own homes? How much would the average citizen miss the chirping of crickets, the songs of birds, the choruses of frogs, and the rustling of cottonwood leaves if they were lost? These are the sounds of life. When birds, mammals, and other wildlife disappear from backyards, parks, or favorite picnic areas, people

FIGURE 11.1 Endangered red-cockaded woodpeckers require mature pine habitats in the southeastern United States. Pairs excavate nesting cavities in living pines that are generally over 65 years of age. *[Image courtesy of M. Lennartz. U.S. Fish and Wildlife Service.]*

quickly express their displeasure. Natural resource biologists continually receive questions and calls from the public requesting information about wild organisms. Recently, one of the most frequent questions has been, "Where are the birds?" When people do not hear or see birds in places where they expect them, they express their concern. Too often we forget how much most people really care about their natural environment.

One does not even have to see or hear wild organisms; just the knowledge that they are present has value. For example, relatively few people will ever observe firsthand the wildlife that abounds in places like the Arctic National Wildlife Refuge in Alaska or the Serengeti region of Africa, but the knowledge that those places exist has value to many people (see section 16.3). Extinctions degrade and destroy such natural environments.

An appreciation of life also engenders a respect and moral responsibility for all life. To many people, it is morally wrong to negatively influence other species to such an extent that their populations decline to dangerously low levels or reach extinction. As with all moral issues, however, what may appear to be simple is really quite complex. While people would generally agree that humans should not allow their actions to cause the extinction of species such as the peregrine falcon or the cheetah, can the same be said for a little-known subspecies of mouse, a beetle, or a bacterium that causes death or disease in humans? In the latter case, the attitude of humans toward extinction of a disease organism is understandable. Our respect for the right of most species to exist, however, should not be in question. Other species do not exist just for the well-being of humans. To quote Aldo Leopold (1949), "In short, a land ethic changes the role of *Homo sapiens* from conqueror of the land-community to plain member and citizen of it. It implies respect for his fellow-members, and also respect for the community as such."

11.2 Endangered Species Legislation and Protective Measures

In 1964, the Committee on Rare and Endangered Wildlife Species was formed in the U.S. Department of the Interior. This committee was responsible for producing the first list of species threatened with extinction; this publication was known as the *Red Book*. In 1966 the Endangered Species Preservation

Act was passed in the United States, but it included few restrictions to protect species in danger of extinction. In 1969 the Endangered Species Conservation Act was passed. This law showed promise by broadening the scope of conservation to include species throughout the world. However, the 1969 law still lacked specific measures to protect species; this legal weakness largely resulted from lobbying efforts of special-interest groups. In 1973, the current Endangered Species Act (ESA) became federal law in the United States, and with the addition of amendments every few years, it remains largely intact today (U.S. Fish and Wildlife Service 2005). The ESA includes substantial criminal penalties and fines for violations of its provisions; without penalties and fines, laws such as the ESA would be unenforceable. The ESA provides for listings of species, subspecies populations, or other distinct population segments that are **endangered** or **threatened** and specifies measures to be taken to protect and restore them. As defined in the ESA, an endangered species is a species in danger of becoming extinct throughout all or a portion of its range. A threatened species is a species that is likely to become endangered throughout all or a portion of its range. Thus, *endangered* and *threatened* are legal terms defined by the ESA and should be used only to describe species listed under that law as endangered and threatened. The secretary of the interior and the secretary of commerce act through the U.S. Fish and Wildlife Service (USFWS) and the National Oceanic and Atmospheric Administration (NOAA) Fisheries to authorize the listing of species; NOAA Fisheries is commonly known as the National Marine Fisheries Service (NMFS). The secretary of commerce is specifically authorized to list marine mammals and most marine and anadromous fish, while all other organisms are under the jurisdiction of the secretary of the interior. Additions to the lists are made through these agencies after consideration of the status of a particular species. The ESA also allows for species nominations by the public in the listing process.

The term *portion of its range* in the ESA allows for listing of populations that face local extinction, even though the species may be abundant in other parts of its range. The ESA also includes a *similarity of appearance* clause that can be used to protect an unlisted species if that species cannot be readily distinguished from a listed species in a problem area. For instance, if greater prairie-chickens were listed as endangered for a geographic area where sharp-tailed grouse also occur, the ESA could require the cessation of hunting of sharp-tailed grouse in that area because most hunters cannot readily distinguish the two species in flight.

Along with the listing of species, current rules require designation of **critical habitat** at the time a species is listed as endangered or threatened, or soon after. Critical habitat is the geographic area and ecosystem essential for the survival of the listed species; in actual practice, critical habitat has included only essential habitat that the species needs to survive and not the habitat needed to fully recover (Stanford Environmental Law Society 2001). Where critical habitat designation causes substantial economic problems, it can be delayed while still listing the species. Designation of critical habitat has proceeded more slowly than seemingly required by the ESA, and critical habitat for the majority of threatened and endangered species has not yet been designated (Cheever 2002). The ESA also provides a process, in the form of species **recovery plans**, for bringing endangered or threatened species back to healthy population levels. This process usually involves stabilizing critical habitat and, where needed, obtaining additional habitat. The ESA allows for acquisition of critical habitat using federal funds, such as those available through the Land and Water Conservation Fund Act of 1965.

Listed species released into new areas outside their existing range, such as grizzly bears in parts of Montana and Idaho, California condors in northern Arizona, and black-footed ferrets in southwestern South Dakota, are considered **experimental populations.** Such populations are listed and counted on the endangered and threatened species list. Critical habitat is not designated for experimental populations, unless the release is essential to the survival of the species. In most restoration efforts, even with species near extinction such as the black-footed ferret, experimental populations have been designated as nonessential to survival of the species. Listing a population as experimental has considerable value in working with the public and in encouraging private landowner support of restoration efforts, because regulations on experimental populations are more adaptable than those for nonexperimental populations of endangered and threatened species or subspecies.

The ESA requires other federal agencies or others involved in federally funded projects to ensure that their actions are not detrimental to endangered or threatened species. Thus, if timber sales in a national forest or development of a federally funded dam threaten the critical habitat of a threatened or endangered species, the federal agency responsible must respond to the problem or be in violation of the ESA. The line between violation of the ESA by a federal agency and its normal operating procedures in the course of such activities can at times be unclear. The USFWS and/or the NMFS are responsible for advising other federal agencies as to when they must conduct a biological assessment to evaluate the effects of their proposed actions on listed species. If the proposed actions of a federal agency are found to jeopardize a listed species or its critical habitat, even if on private land, then the agency must develop contingency plans to reduce damage. The USFWS and/or NMFS must issue a written opinion regarding agencies and federally funded projects that may affect listed species.

The ESA prohibits international trade in endangered or threatened species or parts thereof through implementation of the Convention on International Trade in Endangered Species of Wildlife Fauna and Flora (CITES). The CITES agreement also provides for regulation of importing and exporting certain species not in the endangered or threatened categories. A less well-known agreement, the Convention on Nature Protection and Wildlife Preservation in the Western Hemisphere, was also activated in part through the ESA. This convention, an agreement among various Western nations, was signed in 1940, but it was not implemented at that time. However, in 1982, Congress directed the secretary of the interior and the secretary of state to put this convention into effect, especially as it relates to migratory birds (Rohlf 1989). Other international agreements, including migratory bird treaties with Canada, Mexico, and Japan and agreements such as the International Convention for the High Seas Fisheries of the North Pacific Ocean, can also be implemented in conjunction with the ESA.

Species can be placed on the ESA list of endangered and threatened species even if they do not occur in the United States. As of June 2007, the ESA list of threatened, endangered, or experimental populations totaled 1,922 (animals: 1,175; plants: 747), of which 567 (29.5%) were from foreign countries. Approximately sixteen species in this list were counted more than once, because distinct population segments can be listed separately. The development of recovery plans and the designation of critical habitats for species in foreign countries are of lower priority than are those for organisms of special concern within the United States. However, as noted previously, just listing a foreign animal or plant provides it with legal protection in terms of importation and international trade restrictions that fall under U.S. jurisdiction. Migratory animals that spend part of the year outside the United States, such as the whooping crane, do receive high priority.

Another aspect of the ESA is that within the United States, the federal government can cooperate with state governments to develop appropriate endangered species programs and can provide federal cost sharing. Thus, most states have developed their own lists of organisms that may be threatened, endangered, or of special concern. Species or specific populations on state endangered and threatened lists are not always included on the federal listing. For example, the North American least shrew and lake sturgeon are listed as endangered in Pennsylvania but currently are not federally listed. State lists of endangered, threatened, or special-concern species can be obtained by contacting the state agency responsible for wildlife and fisheries resources or by searching their Web pages. In most cases, biologists are employed by each state to work with the endangered species program in that state.

The possibility of **incidental take** and its potential effects on an ESA-listed species is sometimes of concern to private landowners, corporations, or state and local governments that conduct various normal activities on lands supporting endangered or threatened species. *Incidental take* is defined as the taking of (killing, harming, harassing, collecting, etc.) ESA-listed species "incidental to and not the purpose of, the carrying out of an otherwise lawful activity" (Section 10, ESA). For example, a logging corporation might, through normal activities, cut down a tree with a nest cavity supporting a nesting red-cockaded woodpecker. A farmer might, in harvesting a crop, incidentally kill an endangered butterfly. Where activities that might be harmful to endangered or threatened species are undertaken by a nonfederal entity, an incidental take permit must first be obtained from the USFWS. To obtain such a permit, the applicant develops a **habitat conservation plan** (HCP) that offsets harmful effects of the proposed activity on the population in question. The HCP allows the proposed activity to continue, while promoting the recovery of the endangered or threatened population through setting aside habitat or other means. In addition to the HCP, there are several other programs administered by the USFWS that can provide funding assistance for habitat improvements to benefit endangered and threatened species on private lands. Several U.S. Department of Agriculture programs can similarly provide funding for habitat improvements that benefit ESA-listed species. The USFWS also works on a cooperative basis with Native American tribes to protect endangered and threatened species on tribal lands.

The status of federally listed species in the United States must be reviewed at 5-year intervals. Species are removed from the list if they become extinct, if they recover sufficiently to no longer fit the endangered or threatened categories, or if data demonstrate that the population is healthier than was indicated by the earlier assessment. The ultimate goal of the ESA is to remove species from the list due to population recovery.

The listing of a species as endangered or threatened is relatively difficult, even for high-profile organisms. Real problems arise when the organism is small, unknown, and seemingly valueless, such as an insect, plant, or small mammal. Public support for such organisms often diminishes when a highway project or some other human activity is negatively affected by the ESA. The federally funded Tellico Dam in Tennessee, for example, was started and nearly completed prior to the passage of the ESA in 1973. A new species of fish, the snail darter, was identified in 1973 in a portion of the Little Tennessee River that was to be inundated by the dam; no other populations of this fish were known to exist. Federal court rulings stopped dam construction for a period of years because the ESA prohibited destruction of the habitat of an endangered species. Eventually, under considerable pressure, an exemption was granted and the dam was completed. Relocated snail darters from the Little Tennessee River have become established in other waters, and additional populations have now been located.

Mandatory reductions in the harvest of old-growth forests on federal lands to protect the habitat of the threatened northern spotted owl have stirred similar levels of controversy concerning the ESA. The snail darter and northern spotted owl examples are indicative of the controversy that can develop over the ESA when its provisions have immediate economic repercussions. A delicate balance must be struck between maintaining and enforcing a strong and viable ESA and maintaining sufficient public support for its continued authorization and funding.

The ESA has other problem areas that attract complaints. Subspecies listing is a contentious issue. Some subspecies exist in very limited areas or at small population levels, even though the species as a whole has a large range and healthy populations. Another contentious issue involves small populations of a species, often at the edge of the species' range. While the species may be common in other localities, it may be found in only limited numbers at such a location.

The ESA is periodically targeted for elimination or significant change by many opposing forces, as was the case during the 2004 revision. Pressure on federal legislators can be substantial when the critical habitat segment of the ESA interferes with activities such as development on private lands, oil and gas development on public lands, and in other situations involving land use. While some modification of the legislation would not eliminate its usefulness, loss or major weakening of the ESA would reduce its utility in protecting endangered and threatened species. Unfortunately, too many humans think in terms of short-term economic gain at the expense of long-term economics and sustainability. Some groups will continue to attack legislation such as the ESA; other groups will continue to work at broadening and strengthening such legislation. There is still strong public support for the ESA.

Canada only recently (2003) passed the Species at Risk Act (SRA) to prevent wildlife species from becoming extinct and to take necessary actions to promote their recovery. Species under the SRA can be classified as extirpated, endangered, threatened, or of special concern. The act applies to Canada's extensive public lands, but not to private lands, and it applies to listed wildlife species and habitat critical to their survival. Canada's SRA is relatively weak in comparison to the ESA, but it is a good start toward recognizing and protecting wildlife diversity in Canada. There are twelve other acts, including the Canada Wildlife Act, the Migratory Birds Convention, and the Canadian Environmental Protection Act, that can apply to the conservation of threatened or endangered species in Canada. Most of the provinces in Canada have endangered species legislation; this legislation authorizes, but does not compel, action to conserve endangered species.

The International Union for Conservation of Nature (IUCN) maintains a Red List of plant and animal species throughout the world with a designation of their conservation status. This private organization is supported by various donors and conservation organizations, and it has volunteer help from authorities on various species or species groups; the IUCN is developing an increasingly rigorous approach to designation of species status. As of 2006, the IUCN was managing conservation data on approximately 41,000 species, with plans for expansion to 60,000 species by 2008. Goals of the IUCN program for listing species are to identify species in greatest need of conservation attention and to provide a global index on the state of degeneration of biodiversity. Under the IUCN program scientific information is used to list species as extinct, extinct in the wild, critically endangered, endangered, vulnerable, near threatened, and of least concern. The IUCN listings and other guidelines are available on the Internet.

FIGURE 11.2 Endangered Colorado pikeminnows have declined because of Colorado River habitat degradation and introduction of northern pike. This Colorado pikeminnow measured 60 centimeters in length; the largest specimen caught in recent years measured 125 centimeters in length. *[Image courtesy of U.S. Fish and Wildlife Service.]*

11.3 Causes of Extinction or Endangerment of Populations

Extinctions and serious population declines among wild organisms result from a variety of human influences, including destruction or alteration of habitat; addition of toxic substances to the environment; direct killing or capture of excessive numbers of organisms in an uncontrolled manner; and competition with, predation by, or hybridization with introduced or native organisms.

The causes of extinction or endangerment are highly variable, depending on the particular organism, and usually more than one factor is responsible for an organism's decline. As previously indicated, habitat destruction is the single most important cause of extinction and endangerment. For example, in coastal Texas, most of the suitable habitat within the original range of the endangered Attwater's greater prairie-chicken (a subspecies of greater prairie-chicken) has been destroyed by agricultural activities, suburban development, and brush encroachment. Black-footed ferrets are endangered because of extensive reduction of their primary prey species, the black-tailed prairie dog; tilling of their original grassland habitat; and possibly disease. Colorado pikeminnows are endangered in large part because of water projects on the Colorado River and its tributaries (Figure 11.2). However, the transplanting of northern pike into the Colorado River drainage must also be viewed as a negative influence on the Colorado pikeminnow due to competition and, for smaller pikeminnow, predation. The large and beautiful Lahontan cutthroat trout (a subspecies of cutthroat trout) of northwestern Nevada and northeastern California is threatened by dam construction and reduced water levels related to agricultural and urban water needs, competition from introduced trouts, and hybridization with transplanted rainbow trout (Gerstung 1988). Other subspecies of cutthroat trout, including the greenback cutthroat trout of the South Platte River and Arkansas River drainages of eastern Colorado (Stuber et al. 1988), are threatened for similar reasons.

The effects of introduced organisms on native species have been especially evident in island habitats. Feral goats and sheep on the Hawaiian Islands have caused difficulties for native species by destroying vegetation that native species depend upon; the list of recently extinct and endangered or threatened species for the Hawaiian Islands is long. The introduction of predators such as the small Asian mongoose on Jamaica and other islands to control introduced brown and roof rats has led to the destruction of many species of native reptiles, amphibians, and birds. The introduction of Arctic foxes to the Aleutian Islands for commercial fur farming caused the near extinction of the Aleutian cackling goose.

Another serious concern involves the introduction of closely related forms that can interbreed with species native to an area. In North America, game ranching is becoming increasingly popular, and the red deer of Europe, a subspecies of elk, is a popular large ungulate on these ranches. Game ranches can sell the meat, antlers, **velvet** (highly vascularized tissues that cover developing antlers), and hunting opportunities provided by these animals, but they are required by state laws to keep them within fenced areas. These fenced areas can include extensive ranches or smaller areas, but regardless of the situation, experience has shown that red deer do occasionally escape; native animals can also enter these fenced areas. Biologists are well aware that fences function at less than 100% efficiency in holding animals within or outside particular boundaries. When red deer escape, they may move long distances, join native elk herds, and eventually hybridize. Red deer characteristics can thus enter native elk herds and could negatively affect the gene pool of native elk (see section 4.7). Stopping this threat has proven difficult because of economic and political influences. The reality of this problem was realized in summer 2006, when more than 100 red deer of European lineage escaped a private hunting preserve in Idaho on the western edge of Yellowstone National Park; there is much concern that these escaped red deer will spread disease to wild elk and blemish the genetic purity of elk indigenous to the Yellowstone region. It would seem appropriate that exotic red deer be prohibited as a game-ranch animal in areas of North America near any population of indigenous wild elk.

Exotics and transplants are not the only threats to native organisms. Many native species have been stocked into areas without proper consideration of the genetic characteristics and genetic integrity of the original populations. Wild turkeys and largemouth bass, for example, are commonly stocked into areas that already contain those species. Stockings such as these can decrease the genetic integrity of natural populations.

Pollution can also lead to endangerment or extinction. Brown pelicans in the southeastern coastal areas of North America were brought to near extinction by environmental contamination with chlorinated hydrocarbon insecticides, such as DDT (Figure 11.3). These pesticides are concentrated through the successive trophic levels of a food web, a process known as **food chain biomagnification (biological amplification)**. In the brown pelican, a higher-level predator, pesticides reached levels that caused eggshell thinning and nesting failure. Bald eagles and peregrine falcons also suffered reproductive failure and population declines due to food chain biomagnification of chlorinated hydrocarbons, primarily DDT.

FIGURE 11.3 Brown pelicans suffered nesting failures and were reduced to dangerously low numbers prior to the banning of chlorinated hydrocarbon insecticides such as DDT. *[Image courtesy of the National Park Service.]*

Uncontrolled commercial harvest and illegal harvest have led to extinction for some species and to serious population reductions for others. The Mauritius dodo of the island of Mauritius in the Indian Ocean was extirpated by European travelers and settlers and the animals they introduced in the 1600s (Figure 11.4). Market hunting was a major compo-

FIGURE 11.4 The Mauritius dodo of the island of Mauritius became extinct in the 1600s, after Europeans settled the island, due to introduction of predators and scavengers, destruction of the forest, and human use for food.

nent in the incredible decimation and extinction of the passenger pigeon, possibly the most abundant bird in North America at one time. Hunting for feathers and meat was important in the reduction of the whooping crane. Killing of birds for their feathers also caused substantial reductions of species such as snowy and great egrets. The paddlefish, already a fish of special concern in many states, has been further reduced by illegal capture for commercial sale of their eggs. The eggs are used to make caviar. The Steller sea lion was once greatly reduced because of the value of its hide; unfortunately, after an initial recovery and for reasons not clearly understood, Steller sea lions have again declined precipitously despite being protected from harvest (Figure 11.5). The once endless herds of American bison in North America were brought close to extinction due to extensive uncontrolled slaughter. This wasteful reduction in American bison populations was purposefully carried out as government policy to bring hardships to indigenous peoples dependent on the bison for food, clothing, and shelter. American bison now appear safe from extinction, but their populations still are based on the small gene pool of those few animals that remained before restoration began. Small populations can lead to inbreeding and a lack of genetic heterozygosity, which in turn can lead to reduced adaptability (see section 4.6). Additionally, a serious challenge facing American bison is the potential loss of heritable wildness through selective breeding for increased size, nonaggressive behavior, and other characteristics desirable in domestic ungulates (see section 4.5). Domestication versus wildness of American bison is a controversy of considerable concern to many Native Americans.

FIGURE 11.5 The endangered Steller sea lion population near the Aleutian Islands has declined by more than 70% since 1970. The cause of the decline is complex, but possible factors include depletion of their food base by commercial fisheries, climate change, and illegal kill. *[Image courtesy of A. Morkill, U.S. Fish and Wildlife Service.]*

11.4 Recovery Plans and Results

Recovery plans have been written and fully or partially implemented for many ESA-listed species. These plans specify reasonable actions that can be taken to allow for the recovery or protection of listed species. Recovery plans are published by the USFWS or NMFS, but they may be written by consultants, state agencies, recovery teams (composed of a variety of professionals and agencies interested in the species), or others. For example, the USFWS published a recovery plan for the Colorado pikeminnow in 1978, and revised recovery work plans for that species are developed annually. The recovery plan includes a status report, in which the status of the Colorado pikeminnow and the reasons for its decline are reviewed, and its habitat requirements, range of distribution, limiting factors, and other aspects of its life history are addressed. The recovery plan lists objectives for specific streams or recovery areas that contain appropriate or degraded Colorado pikeminnow habitat. However, such a recovery plan does not ensure that adequate funding to implement the plan will be available.

The primary goal of any species recovery plan is to upgrade the status of the species or remove it from the endangered or threatened list. The bald eagle has recovered dramatically in North America. As of June 1995, the USFWS upgraded the status of the bald eagle from endangered to threatened in the lower 48 states. Since that time, the status of the bald eagle in those states has continued to improve, and in June 2007 the USFWS removed the bald eagle from the list of endangered and threatened species. This delisting of the bald eagle reflects the success of the recovery program and the ESA. The banning of chlorinated hydrocarbon insecticides in North America was critical to the recovery of bald eagles, brown pelicans, and other long-lived predatory birds. There has also been a concerted effort to educate the public and to protect bald and golden eagles from shooting and poisoning by vandals. Law enforcement efforts to reduce the illegal possession and sale of eagle feathers remain important to the restoration and maintenance of eagle populations. Another formerly endangered raptor, the American peregrine falcon (a subspecies), responded positively to reductions in the use of chlorinated hydrocarbon insecticides, increased protection from illegal taking, and the restoration of populations using captive breeding and release techniques; it was removed from ESA listing as of 1999. The Arctic peregrine falcon subspecies was removed from ESA listing in 1994.

Wolf restorations in the Rocky Mountain regions of Montana, Idaho, and Wyoming have proven successful after initial releases, but these populations are still classified as endangered (as of 2007) under the ESA. The continued success of these populations will likely lead to their downlisting. The western Great Lakes population of wolves, primarily located in northern Minnesota, have also been successful, and they were removed from ESA listing in 2007. Increases in the Yellowstone population of grizzly bears led to their removal from ESA listing in spring of 2007.

The whooping crane (see Figure 1.5A) is sometimes viewed as a success story, and yet the entire wild population is only about 300 birds. This is a great improvement, however, over a population that had fallen below forty individuals. The recovery plan for whooping cranes involves protection of the wild population from disturbance and habitat destruction on their nesting grounds in Wood Buffalo National Park in Canada and on their wintering grounds at Aransas National Wildlife Refuge on the Texas coast. The maintenance of multiple captive breeding flocks and restoration of additional wild flocks are also part of the recovery plan. A migrating flock of whooping cranes was established at Necedah National Wildlife Refuge in Wisconsin starting in 2001 with the release of young birds. These juveniles were trained to follow an ultralight aircraft that subsequently led them to wintering grounds at Chassahowitzka National Wildlife Refuge in Florida. Additional juveniles have since been released and trained to migrate from Wisconsin to Florida in a similar manner; experienced birds have begun migrating on their own and can lead additional released young. A pair of whooping cranes at Necedah National Wildlife Refuge successfully hatched two young in 2006; this is the first hatch of whooping cranes in the wild in the Midwest in over 100 years. In addition to the sixty-three migrating whooping cranes in Wisconsin, a resident population of sixty pen-reared birds has been released in central Florida (Kissimmee region), and some pairing and reproductive behaviors have been observed. These newly established populations in Wisconsin and Florida help to reduce the chances of extinction caused by a single catastrophic loss of whooping cranes from one population. Unfortunately, such a catastrophic loss occurred in February 2007, when seventeen whooping cranes from the Wisconsin population were killed by a violent storm on their Florida wintering grounds.

The greenback cutthroat trout is still threatened in Colorado, but restoration efforts have involved both streams and lakes, particularly in Rocky Mountain National Park. Recovery plans call for maintenance of captive breeding stocks in several locations and monitoring of the remaining populations and their habitats. While removal of introduced trout species from all of the greenback cutthroat trout's original habitat is not feasible, the recovery plan calls for total or partial renovation of some streams that this subspecies once occupied (see section 10.2). As of 2007, there were numerous self-sustaining populations of greenback cutthroat. However, not all self-sustaining populations are assured of long-term survival (Harig et al. 2000; Young and Guenther-Gloss 2004). In addition, some populations thought to be greenback cutthroat trout actually were founded from the Colorado River cutthroat trout subspecies (Metcalf et al. 2007).

The recovery plan for the threatened Ozark cavefish calls for locating existing populations and protecting cave habitats and their water recharge areas. The status of this fish is improving largely because of the discovery of new populations. Several cave sites have also been purchased to preserve remaining habitat. Primary threats to Ozark cavefish populations include groundwater pollution and illegal collecting. The endangered Pahrump poolfish of the Pahrump Valley of Nevada is also improving in status, largely resulting from transplants of the species into additional desert springs.

Endangered and threatened species, such as various populations of Pacific salmon, are often described and listed under the ESA in terms of **evolutionary significant units (ESUs)**. These units represent populations that infrequently exchange genes with other population units of the same species and represent an important component in the genetic diversity of that species (Waples 1995; Schiewe and Kareiva 2001). For these salmon, such examples are found in the various populations using different stream systems and in populations that spawn at different times of the year (see section 4.2). Spring versus fall runs of Chinook salmon in many streams or geographic areas in northwestern North America provide good examples of this concept. Spring runs of Chinook salmon on rivers that drain into the San Joaquin Delta in central California and the fall run of Chinook salmon on the Columbia River are ESUs that are listed as threatened under the ESA. One can readily see the problems introduced by indiscriminately releasing hatchery-reared juvenile salmon without regard to ESUs. Identifying a population of salmon that represents an ESU and needs protection is also highly political because of the economic costs that such a designation could mean to ongoing and future economic ventures in particular regions.

The black-footed ferret, which is native to areas containing black-tailed prairie dog colonies in the Great Plains, was thought to be extinct by the 1970s (see Figure 1.5C). Loss of black-tailed prairie dog colonies resulting from control measures and black-footed ferret disease problems both contributed to ferret declines. In the early 1980s a rancher near Meeteetse, Wyoming, identified a black-footed ferret killed by a ranch dog. This find led to the discovery of a black-footed ferret population scattered over a number of black-tailed prairie dog towns in the area. Due primarily to canine distemper, the Meeteetse population of black-footed ferrets declined a few years after it was discovered. The remaining black-footed ferrets were captured and initially held at sites in Nebraska, Wyoming, and Virginia to develop a captive breeding program. Multiple sites in separate states reduced the chances of loss of the entire species, and even the ferrets at individual sites were kept relatively separated, except for breeding purposes, to reduce the chances of the spread of disease. Reproduction has been successful in captive populations, and the number of separate captive sites has increased and now includes several zoos. The recovery plan calls for restoration into widely dispersed black-tailed prairie dog towns within the original range of the species, but suitable sites are limited (Clark and Harvey 1991). Restoration efforts have initially been successful on National Forest Service lands (Conata Basin) in southwestern South Dakota and on the Charles M. Russell National Wildlife Refuge in Montana. Sylvatic plague appears to have caused failure of at least two restoration efforts at other sites and remains a serious problem. The continued control of black-tailed prairie dog populations on private and many public lands illustrates how difficult it is to implement long-term solutions, even with a recovery plan in place.

The recovery plan for the California condor (Figure 11.6) also called for the capture of the last few individuals remaining in the wild in the late 1980s and for the development of a captive breeding program. The captively bred populations were originally divided between the San Diego and the Los Angeles zoos to reduce the chances of loss of the entire population. Captive breeding has been successful, and the species is increasing at rearing facilities. However, captive rearing of adequate numbers of California condors for restoration efforts can be risky. Four of twelve condor chicks hatched in 2006 at the World Center for Birds of Prey in Boise, Idaho, died at 3 months of age because of West Nile virus. Captively reared California condors have been released in the Sierra Nevada and coastal mountain ranges of California as well as in Arizona, Utah, and Mexico. Condors released into the wild face the problem of encroaching human populations, loss of fear of humans, and losses to toxins such as lead shot in carrion. As of June 2007, released California condors have produced five fledged young in Arizona and one in California (Peregrine Fund: World Center for Birds of Prey). On a positive note, of these six birds, four were still alive and had survived 2 years or more as of summer 2007. No California condors were fledged in the wild in 2006, but mature pairs and juveniles remain in the wild and should continue or initiate reproduction in future years.

Unfortunately, the status reports on federally listed species are seldom encouraging; most listed organisms

FIGURE 11.6 California condors are being bred in captivity and released into suitable areas to restore populations in the wild. Some reproduction by released condors has occurred in the wild in Arizona and California. *[Image courtesy of N. Synder, U.S. Fish and Wildlife Service.]*

FIGURE 11.7 Steller's eiders in Alaska are listed as threatened under the Endangered Species Act. The cause for their decline is poorly understood, but potential factors include climate-related ecosystem changes and lead poisoning from spent shot. *[Image courtesy of G. Smart, U.S. Fish and Wildlife Service.]*

are declining, stable, or of unknown status. As of June 2007, 45 listed populations had been delisted, of which 19 had recovered, 9 had become extinct, and 17 were judged as neither threatened nor endangered based on new information since the original listing. Sadly, species such as the dusky seaside sparrow have become extinct and others, such as the endangered western population of Steller sea lions, threatened Steller's eider (Alaska population), and spectacled eider, show little or no sign of recovery (Figure 11.7). On a positive note, species such as the American alligator, Columbian white-tailed deer, and Aleutian cackling goose have recovered. Funding for efforts to evaluate populations, determine the needs of listed species, and develop workable recovery plans is in short supply. While great effort and expense is being directed toward some recovery programs, much more needs to be done.

11.5 Toward a Broader Approach: Biodiversity

With large numbers of endangered and threatened species listed in the United States, a staggering list on a worldwide basis, and many species not even identified, the limited success of our efforts to save species often seems discouraging. Given our past record of recovery success, the massive destruction of tropical rain forests and other ecosystems, and the continued increase in the human population, one can understand that discouragement.

It is difficult to argue against efforts expended on the recovery of individual species listed as endangered or threatened. However, an ultimate slowing of the rate of extinctions can probably best be accomplished by broader approaches by which entire ecosystems, landscapes, and their biodiversity are conserved (Noss 1983). The Attwater's greater prairie-chicken in southeastern Texas is a case in point. Losses of native grassland to agriculture and development are the ultimate causes of the decline of this bird. The importance of native grassland regions in Texas to a variety of other grassland organisms needs emphasis. The endangered Attwater's greater prairie-chicken is really a symbol of the loss of native grasslands in Texas. The threatened marbled murrelet nests in mossy depressions on the limbs of old-growth conifers such as Douglas fir and Sitka spruce in the northwest. The cause for declining populations in marbled murrelets is an ecosystem and landscape problem. Conservation of old-growth forests along the west coast of North America is the only way to save the marbled murrelet and other troubled species dependent on that habitat. Species such as these are often referred to as **indicator species.** Indicator species are key organisms that serve as gauges of the status of ecological conditions in an area.

Ultimately, reduction in recent extinction rates can be best accomplished by maintaining adequate amounts of each ecosystem and landscape to support the native plant and animal communities of that ecosystem or landscape. Maintenance of an ecosystem can often best be accomplished through human use of that system on a truly sustainable basis. Thus, encouragement of well-managed grazing by domesticated livestock on lands containing Attwater's greater prairie-chickens could be considered one strategy for maintaining that grassland ecosystem in a form usable by those birds.

The rapid loss of species caused by the clearing of tropical rain forests can hardly be managed on a species-by-species basis. This extensive destruction of an ecosystem and the resulting broad reduction in biodiversity have repercussions yet unforeseen. Many of the species lost in tropical ecosystems are little known or even unknown to humans.

Likewise, the tilling of grasslands, especially tallgrass prairies (see section 12.1), in North America to such an extent that only remnant pieces of these habitats remain has extinguished a variety of grassland life forms from much of their original range. This, too, is a biodiversity and ecosystem problem. It is true that the planting of trees on farmsteads throughout the original tall-grass prairie has resulted in new animal species being present and has contributed to overall species diversity. However, our responsibilities in terms of conservation entail maintaining the diversity of life native to these grasslands. To maintain this biodiversity, portions of the original ecosystem need to be maintained, or in some cases restored, throughout the areas where native tall-grass prairie once occurred.

Native fishes and other aquatic life likewise can be viewed as part of the original ecosystem in which they resided. They are as integral a part of a grassland, tropical rain forest, or other ecosystem as terrestrial organisms. What happens to the upland ecosystem most certainly relates to the fate of the organisms in, or associated with, aquatic systems.

Private organizations such as The Nature Conservancy, the Sierra Club, the National Audubon Society, the National Wildlife Federation, and the Wilderness Society have made great efforts to protect ecosystems and biodiversity. These organizations often have specific policies regarding environmental issues. Their published policies on issues such as human population numbers, environmental degradation, the marine environment, wetlands, and endangered species specifically relate to ecosystems and biodiversity. The Nature Conservancy has been involved in purchasing habitat critical to a variety of species since before the ESA was passed. Its program of using private funds to purchase important habitat has been exemplary in protecting blocks of ecosystems and landscapes and the species associated with them. The IUCN has served an increasingly important role in providing guidance to governments and conservation organizations throughout the world in the conservation of species and biodiversity (see section 11.2).

Even when blocks of an ecosystem can be maintained or restored, there are many additional questions to answer. What size blocks of such habitat are needed to be truly functional? Perhaps populations of organisms known to have existed there can be restored, but what of organisms that we do not know were there? Do they even exist now? Total restoration may be impossible, but partial restoration may be accomplished in some situations.

The establishment of large blocks of land to maintain ecosystems and species is an attractive conservation measure, but political and economic realities make it a difficult and complicated one to accomplish. Private landowners often consider government purchases of land to protect species and habitat, even on a willing-seller basis, to be a threat to private property rights. Regional or local government entities may oppose purchases of land by federal government agencies because of actual or perceived losses of tax revenue. For example, both state and county governments have in several instances restricted or attempted to restrict purchases of Waterfowl Production Areas by the USFWS in states such as North Dakota. Interestingly, large federally funded projects such as dams and canals for power generation or irrigation, which also involve federal land purchases, are often strongly supported by state and local governments. In many cases biologists working for natural resource agencies are prohibited from openly opposing such projects.

Even on federal lands in the United States, such as those managed by the U.S. Forest Service and the Bureau of Land Management, protection of blocks of habitat remains difficult. For example, designation of wilderness or scenic and wild river areas, protection of remnant mature forest stands, and reduction of grazing by domesticated livestock are all measures intended to protect ecosystems. All of these actions have economic, political, social, and other implications and involve a variety of human user groups; some of those user groups have strongly opposing viewpoints. Thus, protection of portions of ecosystems on public lands can be controversial. Conflict resolution is often required if such areas are to be protected (see section 20.5).

Conservation of tropical rain forests seems especially appropriate because of the potential for losses of large numbers of species. However, it is much easier to promote this important cause than to accomplish such a task. Biologists involved in protecting tropical rain forests encounter many impediments that people in North America should readily understand. Those people currently replacing tropical rain forests with agricultural crops or harvesting rain forest timber at nonsustainable rates can easily point to the large-scale land modifications made to North American ecosystems, such as grasslands and forests, in the past. In terms of habitat and species losses, the past destruction and modification of some of our North American ecosystems have interesting similarities to the current destruction of tropical rain forests. The reasons for our large-scale ecosystem modifications were economic; the removal of tropical rain forests is also economically motivated. While North Americans are now more aware of and appreciative of the consequences of habitat loss than we were during earlier periods of our history, any effort at restoring even relatively small areas of habitat to more natural conditions is still opposed by many people.

Development of land-use practices compatible with long-term maintenance of ecosystems is critical to species protection. For example, game ranching of wild ungulates in Africa and ranching of domesticated livestock or American bison in North America both are long-term means of harvesting grass and protecting ecosystems if done on a sustainable basis. Well-managed ranchlands are remarkable examples of the potential for protecting ecosystems while reaping economic returns. Efforts to develop economic activities that are compatible with the maintenance of ecosystems offer hope for protecting those ecosystems. While it is important, for example, to protect remnant old-growth forest stands, it is also important to continue to utilize other portions of the forest for logging on a sustainable basis. The economic benefits of such logging can be an important ingredient in conserving forest ecosystems for multiple purposes. Recreational expenditures for hiking, camping, hunting, rock climbing, fishing, mountain biking, and other activities on public lands have become increasingly important to many local economies. When these benefits are recognized, they encourage public support for conservation of the ecosystems that provide them. Conservation of tropical rain forests is much enhanced when forest products, including lumber, are valuable enough to be harvested on a sustained yield basis; however, management for optimum sustained yield must include long-term consideration for unborn generations of humans and for all organisms dependent on the tropical rain forest. The rapid growth of human populations in most regions of the world and increasing use and demand for natural resources will continue to challenge our ingenuity as we attempt to protect ecosystems from degradation and subsequent catastrophes in terms of species losses.

SUMMARY

Extinctions of species have always occurred naturally, but the rate of extinctions has increased greatly because of human influences over the last 200 years. A variety of human influences cause extinctions of species, but the greatest single cause is the alteration and destruction of habitat.

Species are valuable to humans for a variety of commercial, scientific, ecological, recreational, aesthetic, and moral reasons. Wild organisms are also an integral part of the ecosystems in which they exist. Regardless of their value to humans, they are part of the working mechanisms of an ecosystem and have value in and of themselves.

Federal laws in the United States designed to protect endangered species were passed in 1966 and 1969 but were ineffective in species protection. The Endangered Species Act (ESA) of 1973 has been more effective in protecting endangered and threatened species in the United States. The listing of a species as endangered or threatened under the ESA allows for the development of a recovery plan and the designation of critical habitat. The ESA requires that the actions of federal agencies not threaten or endanger a species. International trade in endangered or threatened species or their parts is prohibited by the ESA primarily through its implementation of the Convention on International Trade in Endangered Species of Wildlife Flora and Fauna.

Recovery plans may involve the purchase of critical habitat, protection of habitats and organisms from human disturbance, release of captive-reared or wild-trapped organisms into former habitat, establishment of new populations, and other means of reducing the chances of extinction. Recovery plans are designed to upgrade the status of a species from endangered to threatened and eventually to allow the delisting of the species. There is both continued support and opposition to the ESA; most opposition is based on short-term economic issues.

Most recovery plans are species specific. In the long term, a more fruitful means of protecting the biodiversity of the earth and reducing the rate of extinctions lies in maintaining and, where necessary, restoring ecosystems and landscapes. The political and economic realities of maintaining and restoring ecosystems, even in the United States, greatly challenge our ingenuity. However, many ecosystems can be maintained by land-use practices that offer economic returns to humans while conserving the ecosystem and its unique species.

PRACTICE QUESTIONS

1. What are some of the values of wild organisms to humans?

2. What is biodiversity? How does it differ from species diversity?

3. What is an endangered species? What is a threatened species? Identify two of each. Try to use organisms not discussed in this text.

4. What federal (U.S.) agency has jurisdiction over marine mammals and most marine and anadromous fishes listed under the Endangered Species Act? Which agency has jurisdiction over all other ESA-listed organisms?

5. What is critical habitat? Provide an example for an endangered or threatened organism of your choice, preferably one other than those discussed in this text.

6. What is a recovery plan? Give an example.

7. What are some human-related causes of extinction of organisms?

8. Briefly discuss causes of extinction or endangerment for two organisms of your choice, preferably organisms other than those discussed in this text.

9. What are the benefits of an ecosystem approach to saving threatened or endangered organisms?

10. Choose two species that have been successfully removed from endangered or threatened status and explain how they recovered.

11. Provide two reasons why the black-footed ferret is in such a precarious position in terms of recovery.

SELECTED READINGS

Baur, D. C., and W. R. Irvin, eds. 2002. *Endangered Species Act: Law, policy, and perspectives.* American Bar Association Publishing, Chicago.

Ellis, R. 2003. *The empty ocean: Plundering the world's marine life.* Island Press, Washington, D.C.

Johnson, J. E. 1987. *Protected fishes of the United States and Canada.* American Fisheries Society, Bethesda, Md.

Kohm, K. A., ed. 1991. *Balancing on the brink of extinction: The Endangered Species Act and lessons for the future.* Island Press, Washington, D.C.

Montgomery, D. R. 2003. *King of fish: The thousand-year run of salmon.* Westview Press, Boulder, Colo.

Ono, R. D., J. D. Williams, and A. Wagner. 1983. *Vanishing fishes of North America.* Stone Wall Press, Washington, D.C.

Warren, M. L. Jr., B. M. Burr, S. J. Walsh, H. L. Bart Jr., R. C. Cashner, D. A. Etnier, B. J. Freeman, B. R. Kuhajda, R. L. Mayden, H. W. Robison, S. T. Ross, and W. C. Starnes. 2000. Diversity, distribution, and conservation status of the native freshwater fishes of the southern United States. *Fisheries* 25:7–31.

Wilson, E. O., ed. 1988. *Biodiversity.* National Academy Press, Washington, D.C.

Wilson, E. O. 1992. *The diversity of life.* Belknap Press, Cambridge, Mass.

Wilson, E. O. 2006. *Nature revealed: Selected writings, 1949-2006.* Johns Hopkins University Press, Baltimore.

LITERATURE CITED

Cheever, F. 2002. Endangered Species Act: Critical habitat. Pages 47–70 *in* D. C. Baur and W. R. Irvin, eds. *Endangered Species Act: Law, policy, and perspectives.* American Bar Association Publishing, Chicago.

Clark, T., and A. Harvey. 1991. Implementing recovery policy: Learning as we go? Pages 147–163 *in* K. A. Kohm, ed. *Balancing on the brink of extinction: The Endangered Species Act and lessons for the future.* Island Press, Washington, D.C.

Gerstung, E. R. 1988. Status, life history, and management of the Lahontan cutthroat trout. Pages 93–106 *in* R. E. Gresswell, ed. *Status and management of interior stocks of cutthroat trout.* American Fisheries Society, Bethesda, Md.

Hanson, B. A. 2005. *Understanding medicinal plants: Their chemistry and therapeutic action.* Haworth Press, New York.

Harig, A. L., K. D. Fausch, and M. K. Young. 2000. Factors influencing success of greenback cutthroat trout translocations. *North American Journal of Fisheries Management* 20:994–1004.

Leopold, A. 1949. *A Sand County almanac and sketches here and there*. Oxford University Press, London.

Martin, P. S., 2005. *Twilight of the mammoths: Ice age extinctions and the rewilding of America.* University of California Press, Berkeley.

Metcalf, J. L., V. L. Pritchard, S. M. Silvestri, J. B. Jenks, J. S. Wood, D. W. Cowley, R. P. Evans, D. K. Shiozawa, and A. P. Martin. 2007. Across the great divide: genetic forensics reveals misidentification of endangered cutthroat trout populations. *Molecular Ecology* 16: 4445–4454.

Noss, R. F. 1983. A regional landscape approach to maintain diversity. *BioScience* 33:700–706.

Rohlf, D. J. 1989. *The Endangered Species Act: A guide to its protections and implementation.* Stanford Environmental Law Society, Stanford Law School, Stanford, Calif.

Schiewe, M. H., and P. Kareiva. 2001. Salmon. Pages 233–243 *in* S. A. Levin, ed. *Encyclopedia of biodiversity*. vol. 5. Academic Press, Harcourt Science and Technology, San Diego.

Stanford Environmental Law Society. 2001. *The Endangered Species Act: A Stanford Environmental Law Society handbook*. Stanford University Press, Stanford, Calif.

Stuber, R. J., B. D. Rosenlund, and J. R. Bennett. 1988. Greenback cutthroat trout recovery program: Management overview. Pages 71–74 *in* R. E. Gresswell, ed. *Status and management of interior stocks of cutthroat trout.* American Fisheries Society, Bethesda, Md.

U.S. Fish and Wildlife Service. 2005. *Endangered Species Act of 1973 (as amended through December 2004).* U.S. Government Printing Office, Washington, D.C.

Waples, R. 1995. Evolutionary significant units and the conservation of biological diversity under the Endangered Species Act. Pages 8–27 *in* J. L. Nielsen, ed. Evolution and the aquatic ecosystem: Defining unique units in population conservation. Symposium 17. American Fisheries Society, Bethesda, Md.

Whitmore, T. C. 1998. *An introduction to tropical rain forests.* Oxford University Press, New York.

Wilson, E. O. 1992. *The diversity of life.* Belknap Press, Cambridge, Mass.

Young, M. K., and P. M. Guenther-Gloss. 2004. Population characteristics of greenback cutthroat trout in streams: Their relation to model predictions and recovery criteria. *North American Journal of Fisheries Management* 24:184–197.

CHAPTER 12

Habitat Types

Habitat, as defined in section 2.5, is the place where an organism lives, and it includes both biotic and abiotic components. It would be difficult to manage and conserve populations and communities effectively without a knowledge of general and specific habitat types; organisms select and require particular landscapes for various purposes. Habitat classifications can range from general categories, such as grasslands or streams, to very specialized systems, such as those describing the specific habitat needed by canvasbacks for over-water nest sites or the specific feeding site needs of rainbow darters. While a general habitat may occur within the overall range of tolerance for a species, the species may also need specific habitats that can fill particular requirements within narrower ranges of tolerance. For example, a fish species may need general habitat with water temperatures within the range of 2° to 20°C during the year, but it may also require water that is between 12° and 13°C during a specific time of the year for successful reproduction.

Terrestrial biologists most often classify habitat by vegetative characteristics; aquatic biologists rely more on structural characteristics of a water body for habitat classification. This general difference in the way wildlife and fisheries biologists view habitat may be one of the reasons that they differ somewhat in the ways in which they categorize themselves. Wildlife biologists are more apt to classify themselves by species or groups of species, for example, as elk, raptor, Neotropical migrant, or furbearer biologists. Less often will they refer to themselves as grassland or desert wildlife biologists. Conversely, fisheries biologists are more likely to identify themselves with a habitat type, for example, as a stream or lake biologist; they are less likely to categorize themselves as walleye or rainbow trout biologists. There are many exceptions to this generalization, such as wetland biologists and salmon biologists, but this difference in categorization usually holds true.

As our understanding of biological systems increases, there are fewer differences in the ways in which wildlife and fisheries biologists categorize themselves. The advent of ecosystem management, landscape ecology, and other more holistic views of habitat and organismal interaction requires having a broader outlook. It should be recognized, however, that in most cases habitat and its management account for a substantial portion of the effort expended by biologists, regardless of whether they consider themselves to be fisheries, wildlife, landscape, restoration, or conservation biologists, or any other type of natural resource-oriented biologist.

Terrestrial and aquatic habitats are not really separate systems. While they may be discussed separately for the sake of simplicity, they are closely interconnected. For example, the aquatic habitats of a stream and the terrestrial portion of the watershed of that stream are ecologically inseparable; each affects the other in myriad ways.

The primary purpose of this chapter is to identify and describe some of the habitats found in aquatic and terrestrial landscapes and ecosystems. General habitat delineations as well as more specific habitat types will be addressed.

12.1 Terrestrial Habitats

Some broad forms of classifying terrestrial habitats have already been introduced, such as biomes (section 2.1) and zoogeographic regions (section 2.6). Biomes are large geographic areas that are primarily recognizable by the type of vegetation present. Climate and soil influence vegetation type, and vegetation and physical factors influence the types of animals present. Biomes are repeated on a global basis; a biome found in North America also is often found in Europe, Asia, or other regions of the world. Animals adapted to particular biomes may be different from continent to continent, but because of convergent evolution (see section 4.4), they are often similar in form and function. Figure 12.1 depicts the distribution of major biomes of the world. There are numerous

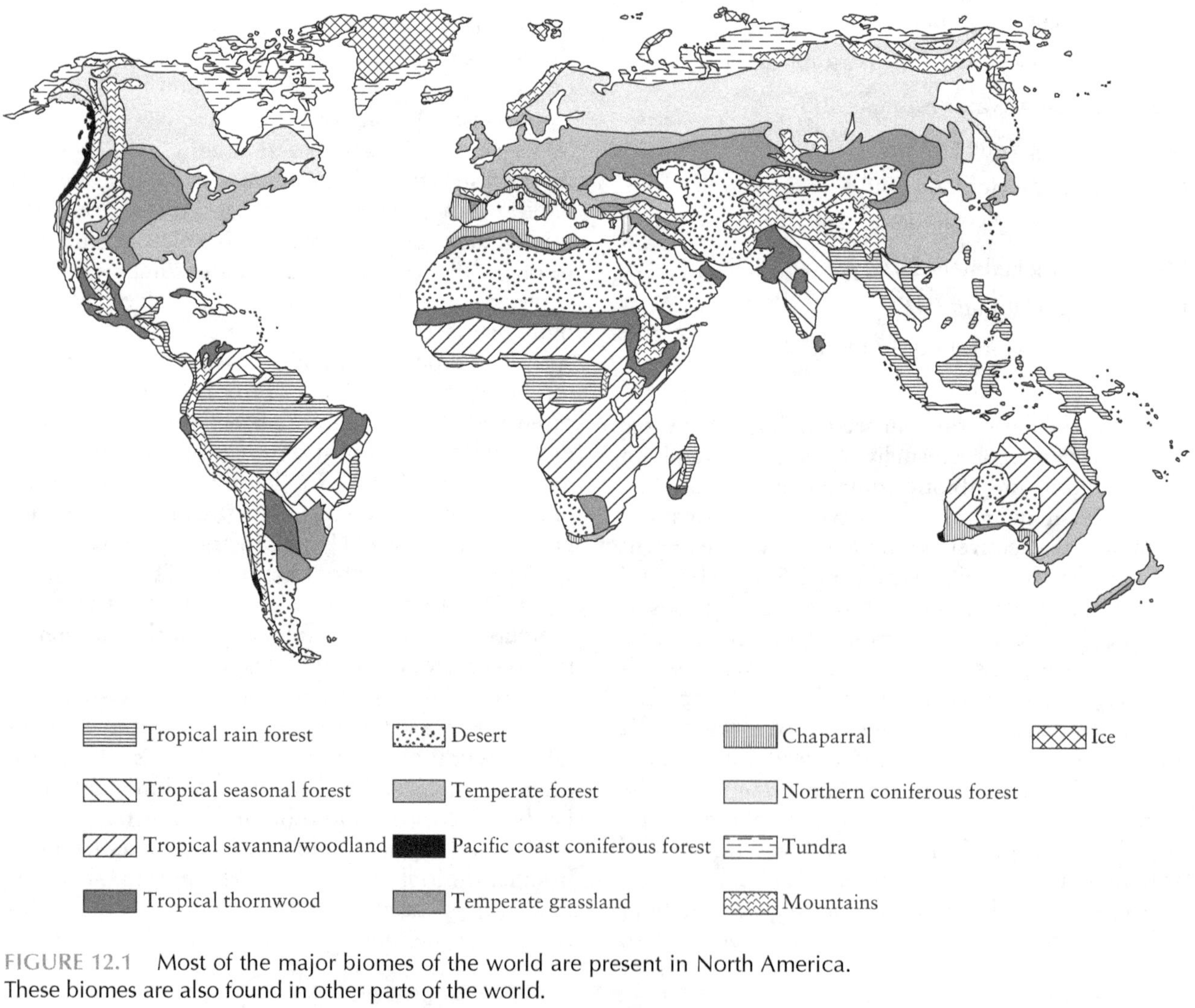

FIGURE 12.1 Most of the major biomes of the world are present in North America. These biomes are also found in other parts of the world.

FIGURE 12.2 Arctic tundra is the northernmost North American biome. *[Image courtesy of J. Flinders.]*

biome classification systems, and biome names, numbers, and descriptions vary with each system. The classification system used in this text is just one model.

In North America, **tundra**, also referred to as **Arctic tundra**, is primarily found in northern Canada and Alaska (Figure 12.2). Tundra is characterized by short, cool summers and long, cold winters. The growing season is short, and vegetation consists primarily of grasses, sedges, lichens, and low shrubby plants. A distinctive characteristic of tundra is **permafrost**, the constant presence of frozen ground, usually a short distance below the soil surface, during the entire year. Permafrost impedes tree growth by limiting root systems.

Many animals of the Arctic tundra are present only during the warmer seasons of the year. Migrating birds, such as snow geese and other Arctic geese, and many species of shorebirds, such as American golden-plovers, reproduce in this biome. There are also year-round residents, such as reindeer, Siberian brown lemmings, Arctic foxes, muskoxen, and rock ptarmigan, that are adapted for living through long, harsh winters.

Arctic tundra is a wet place, even though it has relatively low levels of precipitation. Two reasons why it remains so wet are reduced surface evaporation from the soil and reduced transpiration from plants; both result from the cool climate and short growing season. Permafrost also increases wetness by reducing moisture percolation through the soil.

Tundra can also be found at high elevations at lower latitudes. Biomes found at high elevations are called **altitudinal biomes**. The tops of high mountains in areas such as Colorado and Montana contain what is called **alpine tundra**. Alpine tundra often differs from Arctic tundra with regard to the amount of permafrost and precipitation levels. Because of high winds, wide temperature fluctuations, and high levels of ultraviolet radiation, alpine tundra may have a climate even more severe than that of Arctic tundra.

Tundra is a fragile environment because of permafrost and the short growing season. Human activity in tundra can easily cause long-term environmental damage. In addition, low species diversity contributes to the fragility of this biome. Food webs tend to be simple; thus, they are easily disturbed. The destruction or reduction of any one segment of the biota can have substantial effects on other trophic levels that depend on the lost or reduced plant or animal population.

Northern coniferous forest (**boreal forest** or **taiga**) occurs south of tundra in North America (Figure 12.3). This biome is present in broad expanses, primarily in Canada and Alaska. Northern coniferous forest overstory vegetation consists primarily of spruce, fir, and pine trees. These trees produce dense shade year-round, which often results in limited development of understory layers. This biome is warmer than tundra and usually has higher levels of precipitation. Most precipitation falls as snow. Soils of northern coniferous forests are relatively infertile and have limited use for agricultural purposes, but these forests are among the great timber-producing areas of the world.

Animals of the northern coniferous forest include both migrants and year-round residents. Large numbers of migratory birds use the area seasonally. Many birds and mammals, such as moose, Canadian lynx, spruce grouse, snowshoe hares, American three-toed woodpeckers, and wolverines are year-round residents. As in most cold areas, reptiles and amphibians are few.

FIGURE 12.3 The northern coniferous forest biome is circumpolar in the northern hemisphere. *[Image courtesy of J. Riis.]*

FIGURE 12.4 The temperate forest biome is found over most of the eastern United States. *[Image courtesy of R. Gates.]*

Northern coniferous forests are also present as altitudinal biomes. In North America, these **montane coniferous forests** can be found much farther south than Canada or Alaska; mountainous regions such as those in Washington and New Mexico and the Adirondack Mountains of New York contain montane coniferous forests.

In North America, the **temperate forest** biome occurs primarily below the northern coniferous forest biome in the eastern portion of the continent (Figure 12.4). Most of this biome is characterized by deciduous trees. Dominant trees are oaks, maples, beeches, and other hardwoods, although some areas also have large numbers of evergreen trees. Precipitation is generally higher than that in the northern coniferous forest, and the growing season is longer. These climatic conditions result in dense vegetation. The temperate forest biome is characterized by many year-round resident animal species, including white-tailed deer, wild turkeys, eastern gray squirrels, blue jays, raccoons, and various reptiles and amphibians. Human activities have eliminated or reduced numerous species, such as the American black bear, in many areas of this biome. Many bird species inhabit these forests; the majority of them migrate south during the winter. Much of the temperate forest biome in North America has been converted to agricultural cropland.

The varying climatic conditions over the range of the temperate forest biome result in different vegetation patterns and thus the presence of different animal communities. For this reason, the temperate forest biome is often divided into several subareas. For example, **southeastern mixed forest** ranges from Virginia through the Carolinas and Alabama and contains many conifers. **Outer coastal plain forest,** which ranges from Florida to Louisiana, contains vegetation such as mangroves and live oaks.

Most of the **temperate grassland** biome in North America is found west of the temperate forest and south of the northern coniferous forest (Figure 12.5). Temperate grasslands generally have relatively low precipitation compared to that of temperate forests and are characterized by wet and dry cycles. Summers are hot and winters are cold. Periodic droughts and fires retard tree growth. In contrast to the multiple vegetation layers (trees, shrubs, grasses, etc.) found in most forests, temperate grassland vegetation usually occurs in a single layer. This layer, however, is characterized by many plant species. A variety of grasses and other low-growing plants are common in this biome.

Grasslands vary depending primarily on rainfall and are often classified by grass height. **Short-grass prairies** are relatively dry and have short vegetation compared to that of **tall-grass prairies.** Intermediate between the two are **mixed-grass prairies.** Portions of eastern Colorado and eastern New Mexico represent areas of short-grass prairie. Central Kansas and most of North Dakota are mixed-grass prairie areas. Central Illinois and much of Iowa represent tall-grass prairie regions. Characteristic grassland animals include western meadowlarks, pronghorns, black-tailed prairie dogs, American badgers, burrowing owls, and greater and lesser prairie-chickens.

Because rainfall levels are low in temperate grasslands, nutrients are not easily leached out of the soil.

FIGURE 12.5 Temperate grassland covers most of central North America.

FIGURE 12.6 The desert biome is a dry area with a harsh climate. This image depicts the Sonoran desert in southern Arizona. *[Image courtesy of R. Labisky.]*

Grasslands therefore tend to have rich, productive soils and are extensively used for agricultural production. Most of the tall-grass prairie in North America has been converted to grain production; mixed- and short-grass prairies are usually used for domesticated livestock grazing or grain production.

The **desert** biome is arid, usually having less than 25 centimeters of precipitation in a year (Figure 12.6). In North America, the desert biome occurs primarily in the western and southwestern regions. Both high-altitude and low-altitude deserts are present, and each has characteristic vegetation patterns. High desert (sometimes called shrub-steppe), such as that in the Great Basin of the United States, is characterized by sagebrush, low shrubs, and intermixed grasses and forbs. The low deserts of southwestern North America have been subdivided into three subareas: the Chihuahuan, Mojave, and Sonoran deserts. These subareas vary in precipitation levels, vegetation, and animal life. For example, the Sonoran desert in southern Arizona is characterized by mesquite, creosote bush, and a variety of cacti, while the Chihuahuan desert of Mexico is rich in cacti and other plant forms that conserve water.

The desert biome has a harsh climate, and many of the plants and animals that live there have special adaptations for water conservation. Common desert animals include mule deer, black-tailed jackrabbits, numerous hawks, thrashers, and a variety of reptiles.

Some desert areas have very fertile soils. In many of these areas irrigated agriculture has been well developed. Irrigation, along with grazing of domesticated livestock and the exclusion of fire, has had substantial effects on the desert biome and on some of its native organisms.

These five biomes (tundra, northern coniferous forest, temperate forest, temperate grassland, and desert) cover most of the area of North America. However, other biomes of smaller size are also present.

One of the smaller North American biomes is the **chaparral** biome. It is characterized by a mix of trees and shrubs, often with hard, thick evergreen leaves. The coastal area of southern California is representative of this biome. Many people consider the chaparral biome to have an ideal climate. Dry, warm summers and short, mild winters are the norm, with most precipitation occurring in the cooler months. Chaparral, however, tends to burn easily; the seeds of numerous plants in this biome require fire to break the seed coat so that germination can occur. Despite this tendency and need for chaparral to burn, people continue to build homes and other structures in this biome, want the vegetation to remain the same, and then are surprised when fires occur.

Animals of the chaparral biome include black-tailed deer, California quail, and an abundance of other birds, mammals, and reptiles. Wildlife of this biome has been affected by human activities, especially fire exclusion.

Another small North American biome is **Pacific coast coniferous forest** (Figure 12.7). This biome, which is often included in the northern coniferous forest biome or even in the temperate forest biome, occurs along the Pacific coast from Alaska to northern California. A cool climate and abundant rainfall make this a highly productive timber area. Sitka spruce, giant redwoods, and Douglas firs are all commercially important trees found in this region. Because of their timber value, old-growth forests in this biome are the focal point of a major ecological controversy. Continuing concern over old-growth forests and their native biota in the Pacific Northwest represents a contentious issue that is mentioned several times in this text. Portions of this biome are wet enough to be called **temperate rain forests**. The Olympic National Forest in Washington is an example of such a rain forest.

FIGURE 12.7 The Pacific coast coniferous forest biome contains some of the largest trees in the world. *[Image courtesy of K. Jensen.]*

The southern tip of Florida represents another small North American biome, the **tropical seasonal forest**. This forest type is characterized by wet and dry seasons and by hot temperatures year-round. It contains both evergreen and deciduous tree species.

Another major biome that should be mentioned is **tropical rain forest** (Figure 12.8). Although it is not present in North America, large areas of Central America and southern Mexico consist of this forest type, as do large areas of southeast Asia and equatorial South America and Africa. The global importance of this biome necessitates its mention in this text. Tropical rain forests are characterized by heavy rainfall, usually in excess of 220 centimeters per year. While these forests appear to be exceedingly productive, most nutrients are retained in the lush vegetation. Soils, because of high rainfall, tend to be rapidly leached of nutrients. Human attempts at agriculture in tropical rain forest areas are generally short-lived, because soils are poor and their nutrients are rapidly lost.

While North America has no tropical rain forest, it is affected by human activities that are degrading this biome. The effects of this degradation impact the entire biosphere in various ways, ranging from potential climate change to reduced populations of migratory birds. In addition, tropical rain forests are a primary repository of worldwide species diversity. This biome is being lost at an alarming rate.

Biomes should not be thought of as monolithic blocks of habitat. In all of these biomes there are interspersions of a variety of habitats and landscapes. Grasslands have corridors of trees along rivers or on other wetland margins, temperate forests have grassland openings, and all biomes can have mountainous areas that contain islands of various habitat types not found in the surrounding area. In addition, human activities have modified many portions of these biomes into varied landscapes that are quite different from their original states.

In addition, biomes do not end abruptly. Ecotone areas, sometimes quite unique ones, characterize their edges. The cross timbers area of Oklahoma, where the transition between temperate grassland and temperate forest occurs, is one example. Another example is the mosaic of scrubby trees and open areas lying between the northern coniferous forest and the tundra. These edge areas can have their own unique assem-

FIGURE 12.8 Tropical rain forests are currently being lost at a rapid rate. *[Image courtesy of the U.S. Fish and Wildlife Service.]*

blages of animals and plants. Species diversity is often quite high in such edge areas because organisms from both adjacent biomes may be present, as well as organisms specifically adapted to the ecotone area.

Some species are quite adaptable and are found in numerous biomes, while other species are relegated to particular portions of certain biomes. Pronghorns, for example, are animals of open spaces. While common in some of the short- and mixed-grass prairies, they are not well adapted to tall-grass prairies. Pronghorns need to be able to see for long distances, primarily for protection from predators, and tall grass is as much a barrier to their existence as forests. Other animals, such as American crows and cougars, are very adaptable and can be found in numerous biomes.

Another method of classifying habitats is by ecoregion (see section 2.1). Ricketts et al. (1999) defined an ecoregion as "a large area of land or water that contains a geographically distinct assemblage of natural communities that (a) share a large majority of their species and ecological dynamics, (b) share similar environmental conditions, and (c) interact ecologically in ways that are critical for their long-term persistence." Ecoregions are generally much smaller units than biomes or zoogeograhic regions.

Ricketts et al. (1999) listed 116 terrestrial ecoregions in the United States and Canada. They range from Hawaiian Moist Forests to Permanent Ice. Some are quite large, such as the Northwestern Mixed Grasslands ecoregion, found in parts of five states and two Canadian provinces, and the Southeastern Mixed Forests ecoregion, found in parts of seven states. Others are quite small, such as the East Central Texas Forests ecoregion (found only in east Texas) and the California Coastal Sage and Chaparral ecoregion (found only in coastal southern California). Some states contain numerous ecoregions (Texas and Washington both have eleven); some states contain few ecoregions (Indiana has three, and Maine has two).

Ecoregions have also been delineated for aquatic habitats. Abell et al. (2000) described 76 freshwater ecoregions in North America (United States, Canada, and Mexico). These consist of lake, stream, river, and pond areas. Some are quite large, such as the Yukon freshwater ecosystem that covers most of Alaska. Others are quite small, such as the Apalachicola freshwater ecoregion that covers small parts of Florida, Georgia, and Alabama. Both the Abell et al. (2000) book and the Ricketts et al. (1999) book have excellent conservation assessments of ecoregions.

12.2 Terrestrial and Aquatic Transitional Habitats

Upland areas are terrestrial habitats, while streams and lakes are aquatic habitats. Located between the two are transitional habitats. Such transitional habitats include the riparian zones along rivers and streams, the wet margins of lakes, ponds, and other standing-water bodies, and a variety of marshes, bogs, swamps, coastal wetlands, and beach–ocean interfaces. **Marshes** are low, treeless wet areas characterized by sedges, rushes, and cattails. **Bogs** are wet areas characterized by floating spongy mats of vegetation often composed of sphagnum, sedges, and heaths. **Swamps** are wet areas that usually contain standing trees. **Coastal wetlands** are transitional areas between land and the oceans. Each of these transitional areas contains characteristic plants and animals that thrive in terrestrial–aquatic interface habitats. Transition zones are usually not abrupt but can be represented by a gradient or continuum, becoming either more terrestrial or more aquatic depending upon the distance from each of the two primary habitat types. However, some aquatic–terrestrial interfaces can be fairly abrupt, as in the case of many areas of the beach–ocean interface.

Organisms associated with these aquatic–terrestrial transition areas are many. A variety of shorebirds, such as American avocets, Baird's sandpipers, and greater yellowlegs, are dependent on transitional areas. Some mammals, such as American mink and raccoons, frequent the terrestrial–aquatic edge. Some organisms, such as American beavers and common muskrats, tend toward the aquatic end of the gradient in these habitats, as, of course, do fishes. In addition, these transitional areas often have abundant populations of amphibians and reptiles. Many animals spend the majority of their lives in these transitional areas.

Riparian zones and shorelines of lakes, ponds, and other standing-water bodies are transitional areas that are generally considered to be primarily wildlife habitats. Transitional areas such as bogs, marshes, swamps, and coastal wetlands are primarily aquatic habitats and are important to fishes, other primarily aquatic animals, and a variety of terrestrial wildlife species. Cowardin et al. (1979) divided wetland and deep-water habitats into five different systems. Their classification scheme divides these habitats into system, subsystem, and class groupings (Figure 12.9). This text will highlight the system portion of the classification. **Marine** (open-ocean areas) systems can contain transitional

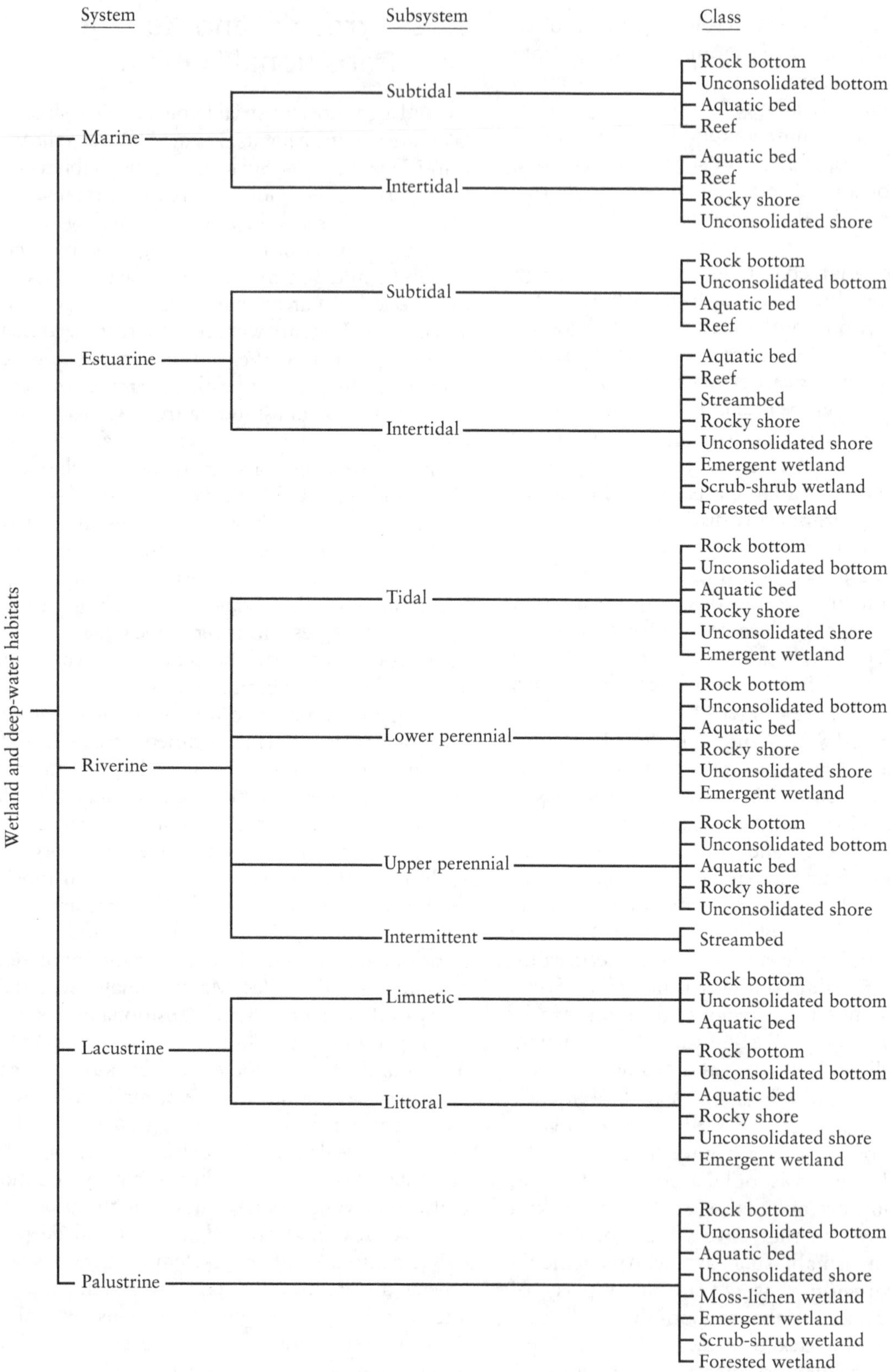
System
Subsystem
Class
Wetland and deep-water habitats
Marine
Subtidal
Rock bottom
Unconsolidated bottom
Aquatic bed
Reef
Intertidal
Aquatic bed
Reef
Rocky shore
Unconsolidated shore
Estuarine
Subtidal
Rock bottom
Unconsolidated bottom
Aquatic bed
Reef
Intertidal
Aquatic bed
Reef
Streambed
Rocky shore
Unconsolidated shore
Emergent wetland
Scrub-shrub wetland
Forested wetland
Riverine
Tidal
Rock bottom
Unconsolidated bottom
Aquatic bed
Rocky shore
Unconsolidated shore
Emergent wetland
Lower perennial
Rock bottom
Unconsolidated bottom
Aquatic bed
Rocky shore
Unconsolidated shore
Emergent wetland
Upper perennial
Rock bottom
Unconsolidated bottom
Aquatic bed
Rocky shore
Unconsolidated shore
Intermittent
Streambed
Lacustrine
Limnetic
Rock bottom
Unconsolidated bottom
Aquatic bed
Littoral
Rock bottom
Unconsolidated bottom
Aquatic bed
Rocky shore
Unconsolidated shore
Emergent wetland
Palustrine
Rock bottom
Unconsolidated bottom
Aquatic bed
Unconsolidated shore
Moss-lichen wetland
Emergent wetland
Scrub-shrub wetland
Forested wetland

habitats at the beach–ocean interface. **Estuarine** (transitional areas between marine and freshwater areas) systems will be discussed in section 12.5. The freshwater portion of the Cowardin et al. classification scheme contains three major categories: **riverine**, **lacustrine**, and **palustrine** systems. Riverine systems are associated with rivers and streams, while lacustrine systems are associated with lakes and ponds; these two categories will be discussed as aquatic habitats in sections 12.4 and 12.3, respectively. The palustrine systems (bogs, marshes, and swamps) are discussed here as transitional habitat.

While any watered area, regardless of depth or size, can be described as a wetland, biologists usually use the term **wetland** when referring to bogs, marshes, swamps, or coastal wetlands. Wetlands are generally described as areas where the water table is near ground level or just shallowly covers the land surface (Figure 12.10). Various characteristics can be indicative of a wetland area, including the presence of **hydric plants** (plants associated with water), the presence of **hydric soils** (soils developed in wet areas), or the presence of water saturation or coverage during some portion of the year. Some wetland areas have visible water at all times of the year, while others, over the course of a year or some other time period, sometimes more closely resemble a terrestrial habitat. For many wetlands, dry periods are important for maintaining productivity and vegetative characteristics. Just because a wetland is dry does not mean it is not a wetland. The fact that a wetland may contain no water for appreciable periods of time is one of the reasons why there is currently a great deal of public debate as to what constitutes a wetland. This controversy creates difficulties in efforts to protect wetlands, and finding a definition of wetlands that everyone will agree upon may not be possible.

Regardless of whether these palustrine areas contain water at all times or for only brief periods of time, their utility as habitat is unquestioned. Wetlands not only serve as habitat for aquatic-oriented organisms but also represent a landscape commonly used by many terrestrial organisms. In addition to providing habitat for wild organisms, wetlands are important in floodwater retention, nutrient cycling and trapping, groundwater recharge, microclimate modification, water purification, provision of forage and water for domesticated livestock, and other beneficial functions (see section 14.6).

FIGURE 12.9 (opposite) The Cowardin et al. classification divides wetland and deep-water habitats into different systems, subsystems, and classes. *[Adapted from Cowardin et al. 1979.]*

FIGURE 12.10 Wetland areas are characterized by the presence of hydric plants and hydric soils.

The beach–ocean interface is a transitional habitat that is quite different from most of the other areas discussed in this section. This habitat contains its own unique assemblage of organisms, ranging from marine mammals and seabirds to marine fishes and invertebrates. This habitat zone is of concern to many public agencies as well as the general public. It is an area rich in commercial products, such as fishes and shellfishes, and is heavily used for consumptive and nonconsumptive outdoor pursuits such as fishing, swimming, sunbathing, and viewing wildlife. It is biologically important because of its diverse flora and fauna.

12.3 Standing-Water Habitats

Natural lakes, ponds, and reservoirs are called standing-water, or **lentic**, habitats. However, these water bodies can have current and water movement. Water, either at the surface or underground, moving into these bodies results in current, as does wind action and the heating and cooling of water. There is, however, usually no consistent unidirectional water flow like that found in streams and rivers. Lotic systems, such as rivers and streams, have a stronger unidirectional water flow regime; these systems will be discussed in section 12.4. Within the lentic water body category, some water bodies have a considerable degree of unidirectional water flow. For example, natural water bodies can have inlets and outlets that result in constant or temporary water flows, while artificial lentic water bodies created by human activities such as damming rivers or streams often have various levels of unidirectional water flow. Lentic and lotic environments occur on a continuum; some water bodies are almost totally lentic or lotic, while others occur at various intermediate levels.

There is much variation in the terminology associated with lacustrine systems. Terms such as *pond* and *lake* are often used interchangeably and can lead to confusion. There are no universally accepted standardized definitions to describe, for example, when a pond is large enough to be called a lake. Some biologists state that for a water body to be called a lake, variations in water temperature with depth or windswept shorelines must be present. No classification system will satisfy everyone. We will describe these lacustrine systems with just one of the many classification schemes that can be used.

Natural lakes (lakes), as the name implies, are naturally formed lentic water bodies. Usually, no sizes are defined. Natural lakes can vary in size from very large water bodies, such as Lake Superior or Lake Champlain, to rather small water bodies, such as the many small natural lakes (often called ponds) found in New England. Natural lakes can be formed in numerous ways, the most common of which are glacial activity, movements of the earth's crust, and changes in river courses.

Deep natural lakes tend to be **oligotrophic**, or low in nutrients. Oligotrophic lakes are rather unproductive and usually have very transparent water. Their lack of nutrients may result from being located in unproductive areas or because they are so recent successionally that nutrient levels have not had time to accumulate. **Eutrophic** natural lakes tend to be productive and shallow. It is important to remember that the terms *oligotrophic* and *eutrophic* do not refer just to natural lakes and that the terms are used somewhat subjectively. There is a continuum of productivity levels in water bodies, and there is no set point at which the eutrophic and oligotrophic categories begin and end that everyone agrees upon. Numerous attempts at such divisions have been made, but these categorizations vary.

The remaining lentic systems are artificial (human-constructed) water bodies. Ponds and reservoirs are called **impoundments**, because they result from human activities that block natural watercourses such as rivers and streams. As mentioned above, it is difficult to categorize water bodies by size, but the following system was used by Summerfelt (1999) in relation to human-constructed water bodies. **Ponds**, the smallest impoundments, are less than 10 hectares in surface area. The previously mentioned use of the term *pond* to describe a small New England natural lake leads to a certain amount of confusion. **Large impoundments or reservoirs** have surface areas that exceed 200 hectares. Those artificial water bodies that are between 10 and 200 hectares are considered **intermediate impoundments**. Obviously, a 100,000-hectare reservoir may be quite different from a 250-hectare reservoir. What is important to remember about these impoundments is that artificial lentic systems occur in a wide variety of sizes and vary greatly with respect to the reasons for and methods of their construction, the amount of water that flows through them, and the levels of their productivity.

As already mentioned, impoundments are constructed by damming watercourses. In the case of ponds, the damming can involve small intermittent streams or small drainage areas. In the case of reservoirs, a large river such as the Columbia, Colorado, or Savannah might be dammed.

There are numerous reasons why impoundments are constructed. However, whether they are constructed for domestic or agricultural water supplies, flood control, electricity generation, navigation, recreation, or a combination of uses, they still represent aquatic habitat that can have varying degrees of utility for both aquatic and terrestrial communities.

Other artificial water bodies result from excavation activities. Gravel pit, strip mine, quarry, and borrow pit (excavations along roadways) water bodies are the result of excavation. These excavations are not purposely constructed for holding water; they are by-products of human activities usually related to mining or road construction (Figure 12.11). Some

FIGURE 12.11 Gravel pits, such as the one depicted here, are the by-product of mining activity.

FIGURE 12.12 Dugouts are constructed primarily for watering domesticated livestock but also provide aquatic habitat for some other species.

FIGURE 12.13 Ponds are small artificial water bodies that serve numerous functions.

other excavated water bodies, such as dugouts (Figure 12.12), are specifically constructed to hold water for domesticated livestock.

Ponds are the most numerous small artificial water bodies made especially for water holding (Figure 12.13). Because they are small, they tend to have a limited number of ecological niches, and therefore they have rather low species diversity. In some ways this makes them easy to manage, but in other ways their simplicity makes management more difficult. Low species diversity may simplify food webs, but this simplicity can also cause problems, because a change in one portion of the web can quickly and drastically affect other portions.

Intermediate impoundments and reservoirs are larger than ponds and usually have greater species diversity (Figure 12.14). Some are sufficiently deep enough to have water temperature gradients from the surface to the bottom. There is a great deal of variation among these water bodies in the amount of water that flows through them. Some impoundments have substantial amounts of water inflow and outflow only during brief portions of the year. Many larger reservoirs have constant inflow and outflow year-round. In some reservoirs the total volume of the water body may be exchanged in a few days; these impoundments are more lotic in nature. In other reservoirs the total exchange of water volume may take 2 or 3 years; these impoundments more closely resemble lentic systems. The rate of water exchange has important effects on the development of aquatic communities (Ploskey et al. 1984).

FIGURE 12.14 Reservoirs vary greatly in their size, depth, and rates of water volume exchange.

All lentic bodies, whether natural or artificial, can contain a variety of different zones or habitats. The number of zones depends on water depth, water body size, and numerous other physical and chemical characteristics. Terminology used to describe the various zones present in a lentic water body is quite variable. The following system is simplified, but it is representative of terminology commonly used for freshwater lentic environments (Figure 12.15).

Most lentic systems have a **littoral zone**—a peripheral shallow water area. In many cases, this zone extends from the shoreline to the limit of rooted aquatic vegetation (see Figure 12.15A). In some water bodies the entire body is shallow enough to support rooted aquatic vegetation; in this case, the entire water body consists of a littoral

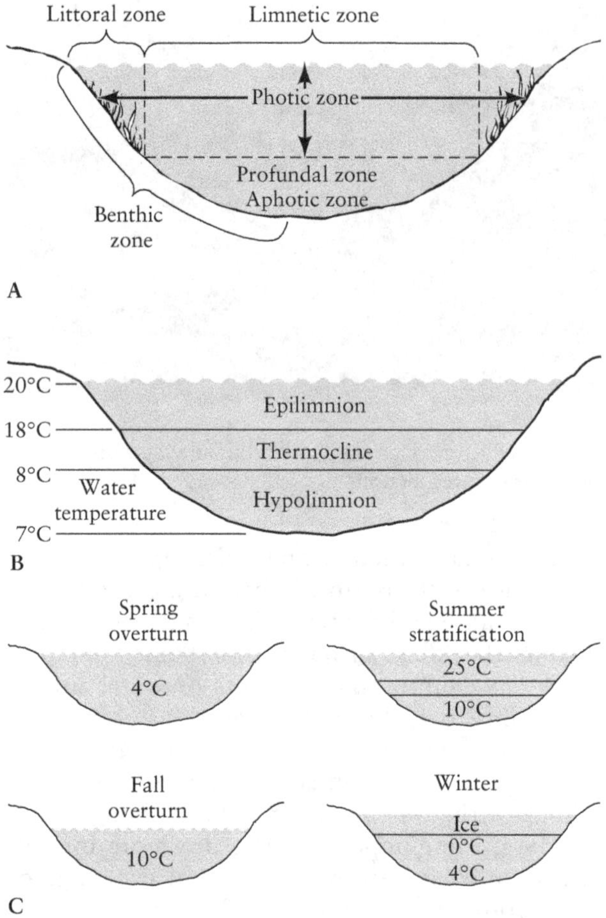

FIGURE 12.15 Various zones may be found in freshwater lentic systems. *(A)* Zones determined by water depth and sunlight penetration. *(B)* Zones determined by water temperatures. Water temperature in relation to depth in a stratified water body during summer is shown. *(C)* Seasonal patterns of thermal stratification for water bodies in northern regions that stratify.

zone. In deeper water bodies, the littoral zone is usually represented by a band of rooted aquatic vegetation along the shoreline. Using such vegetation to describe the littoral zone, however, can lead to confusion. Consider the shallow area of a water body that is too turbid or windswept to support rooted aquatic vegetation. This shallow area still represents a littoral zone, even though no rooted aquatic vegetation is present.

In lentic water bodies that have deeper water beyond the littoral zone, the open-water area is called the **limnetic zone.** The limnetic zone includes all of the water volume outside the littoral zone that is lighted. The total lighted area of a water body is called the **photic** (**euphotic** or **trophogenic**) **zone.** This lighted area serves as an environment where phytoplankton can live and photosynthesize. If a water body is deep enough to have an unlighted portion below the limnetic zone, this area is called the **profundal zone.** The profundal zone is an unlighted **aphotic** (**tropholytic**) **zone**, and it does not support photosynthesis.

In any lentic or lotic water body there is also a bottom, or **benthic zone.** This zone, depending on water depth and other characteristics, provides habitat for organisms that do not suspend in the water column.

Lentic water bodies can also be characterized by layers with different water temperatures and oxygen levels. Such **thermal stratification** occurs where there is sufficient water depth, seasonal temperature change, and protection from wind (see Figure 12.15B). In stratified water bodies, three layers are present during summer. The upper layer is called the **epilimnion**; the middle layer, where water temperature drops abruptly, is the **thermocline** (**metalimnion**); the deepest and coldest layer is the **hypolimnion.** Stratification results from the relationship of water density to water temperature. Except at near-freezing temperatures, the warmer water is, the lighter it is. Waters of different temperatures, and therefore different densities, do not easily mix. Thus, the thermocline, the area of abrupt temperature change, acts as a barrier to water mixing. The epilimnion, because it is in the photic zone and therefore has photosynthetic activity, and because it is open to wind action and other mechanical agitation, can have its oxygen supply replenished. The hypolimnion is sealed from this mixing zone by the thermocline and, depending on productivity of the water body and the relative volumes of the hypolimnion and epilimnion, can contain either adequate or inadequate amounts of dissolved oxygen for oxygen-demanding organisms. The hypolimnion is likely to retain dissolved oxygen during the entire year in oligotrophic water bodies; this is less likely to occur in eutrophic water bodies.

As autumn approaches and water temperature in the epilimnion decreases, the thermocline erodes. Eventually, water temperatures throughout the water body equalize, allowing complete mixing of the water. This phenomenon is called **fall overturn.** The water body at this time has uniform temperatures and dissolved oxygen concentrations from surface to bottom.

In cold climates, especially when ice cover is present, **inverse stratification** generally occurs in winter. Water attains its maximum density at 4°C. Below 4°C, it becomes lighter and rises to the surface. That is why ice forms on the top of a water body, not at the bottom. After ice formation, the coldest water is near the surface, and the warmest water is in the deeper areas of the water body. In spring, when ice melts and water temperatures equalize, a **spring overturn** occurs. Figure 12.15C depicts the seasonal changes that occur in a stratified water body.

The unique relationship of water density to water temperature has important ramifications for aquatic ecosystems. If water at 0°C was denser than water at 4°C, ice would form at the bottom of water bodies rather than at the top. Try to envision the effects on organisms if water bodies froze from the bottom up rather than from the top down. In fact, bottom ice can occur, and most aquatic organisms are ill prepared to handle its effects. Ice crystals from above the water surface can fall into streams and form ice masses on stream bottom rocks and other substrates. This ice, when firmly attached, is called **anchor ice.** It can dam and divert water from the streambed. Disoriented fishes and aquatic invertebrates can wander into these diverted channels and be stranded and die when the ice dams melt and the water returns to the normal stream channel. Anchor ice can also damage the benthic-community portion of the food web in a stream.

The various zones and zone characteristics present in water bodies have many ramifications for fisheries and even wildlife systems. Littoral zones serve as habitat for shallow water fish species, such as sunfishes. Limnetic zones provide habitat for open-water fishes, such as the shads. Benthic zones are inhabited by bottom-dwelling species, such as bullheads and suckers. Some wildlife, such as belted kingfishers, moose, and raccoons, feed in the littoral zone, while birds capable of diving, such as the long-tailed duck, may use the limnetic zone. Several species, including lesser snow geese and mallards, use isolated open water of the limnetic zone to escape disturbances during fall migration.

Thermal stratification can provide warmwater, coolwater, and coldwater habitats for aquatic organisms if sufficient dissolved oxygen is available. In this case, two-story fisheries, as described in section 10.5, can be present, with warmwater and coolwater fishes inhabiting one temperature layer and coldwater fishes utilizing another.

In reservoirs, thermal stratification can also have substantial effects on the organisms present downstream from the dam. The area immediately below the dam is called the **tailrace** (Figure 12.16). It can support different types of organisms, depending on the depth of the outlet and whether water is released from the warmer or the cooler layer of the reservoir. For example, trout fisheries have been developed in warm climates in tailrace areas where cold, well-oxygenated hypolimnetic water is released through a dam. Tailrace areas in colder climates may provide ice-free waters below dams. Such areas can be attractive wintering places for animals, such as Canada geese, that would normally be forced to migrate because of ice cover. However, tailrace water releases are not always beneficial. For example, if a stream or river has a native assemblage of warmwater fishes and other organisms dependent on warm water and the water released from a reservoir is cold, the native species will be reduced or eliminated.

Construction of ponds, intermediate impoundments, and reservoirs can provide habitat types not normally present in an area. Large reservoirs on the Missouri River in North Dakota, South Dakota, and Montana have provided a coldwater habitat not natural to that area. Trout and salmon, which could not otherwise live in the area, have been introduced into these new habitats. Construction of ponds and intermediate impoundments in many areas of North America has likewise provided new habitat for some species in areas where such habitat was not previously available. Numerous native and introduced fish species

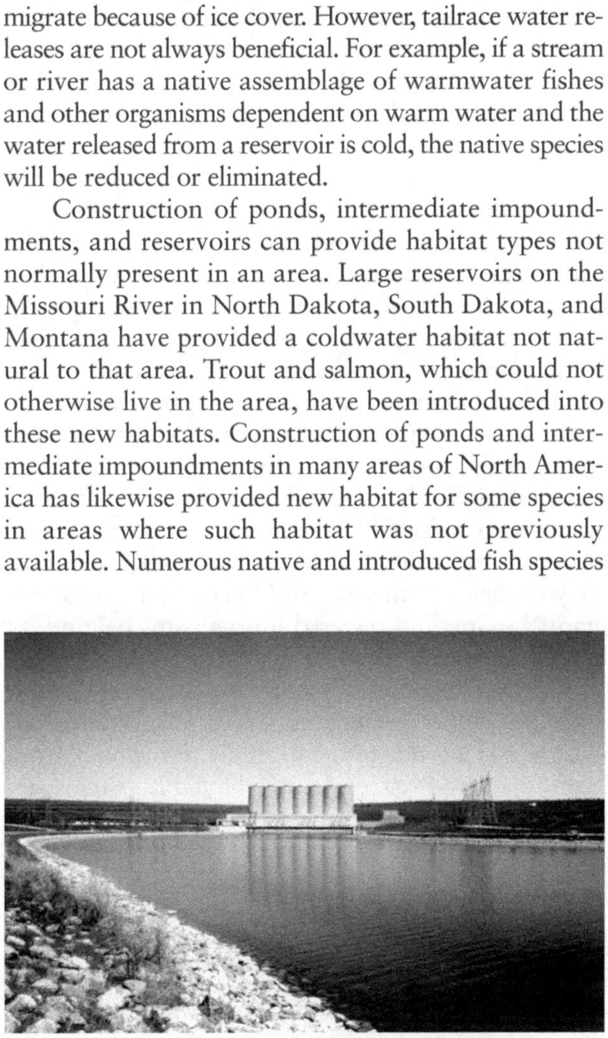

FIGURE 12.16 Tailrace areas can be ecologically very different from the lotic systems originally present.

have been placed in these artificial habitats. Impoundments have also been particularly beneficial to wildlife such as waterfowl and shorebirds, especially in areas where aquatic habitat is naturally scarce.

There are some drawbacks to these human-induced changes, however. One of these is their negative effects on native communities. For example, new habitats made available by the construction of reservoirs on the Missouri River were produced at the expense of existing habitats. What was once a free-flowing lotic system is now a series of lentic systems connected by highly modified lotic segments. Many fishes adapted to the original lotic habitat, such as paddlefish, sicklefin chubs, blue suckers, and silver chubs, have been diminished. At the locations of reservoirs, extensive areas of riparian habitat that were important to terrestrial species are also greatly modified or destroyed. When a tree-lined prairie river is impounded, for example, it loses the trees in its riparian zone. This belt of trees may have served as a corridor for movement for many species dependent on trees as well as provided habitat for resident species. In addition, the immediate area of a reservoir is not the only area affected. Modification of water flow changes downstream habitat in various ways for both terrestrial and aquatic organisms. For example, rates at which islands and sandbars are formed or eliminated are changed, scouring effects of floods are changed, seasonal fluctuations in water flows are modified, continual sediment enrichment of floodplains is ceased, and downstream habitats are affected by the modified water temperatures of reservoir outflows. Thus, ecological effects go well beyond the impoundment and can affect many aquatic and terrestrial organisms beyond its immediate locale.

Artificial aquatic systems are generally much more difficult to manage on a long-term basis than natural systems due to their rapidly changing nature. Natural aquatic systems tend to be stable systems; artificial systems generally progress rapidly through various successional stages and tend to be less stable.

12.4 Flowing-Water Habitats

Riverine habitats range from very small streams, ones that a person could step across, to rivers the size of the Mississippi. The dominant feature of this type of habitat is a fairly consistent, unidirectional water flow. Because of this water flow, lotic systems are dynamic and are constantly changing.

A stream order classification is commonly used to describe riverine systems (Figure 12.17). In this system, small unbranched tributaries are designated order-1 streams. To be an order-1 stream, a watercourse must be large enough to appear on a 7.5-minute quadrangle map. When two order-1 streams join, an order-2 stream is formed. This increasing order continues through large rivers such as the Mississippi, which may reach a stream order number as high as 7 or 8. The utility of this classification system is based on the assumption that stream order number is directly proportional not only to stream size but also to such characteristics as water discharge rate and watershed area.

Because they exhibit differences in depth, length, width, area, flow volume, shoreline characteristics, **stream gradient** (rate of descent to sea level), bottom type, and water temperature, riverine systems are quite variable. Lotic systems can experience large seasonal fluctuations in dissolved oxygen content as well as water temperature; however, compared with lentic water bodies, water temperature and dissolved oxygen concentrations in lotic systems are much more uniform from bottom to surface at any specific point in time or place. Lotic systems are also usually shallower than lentic systems. Because water is constantly moving through them, their bottom structure is less stable than that of lentic systems. In addition, nutrients in lotic systems are constantly moving downstream and must be replaced; nutrients entering a

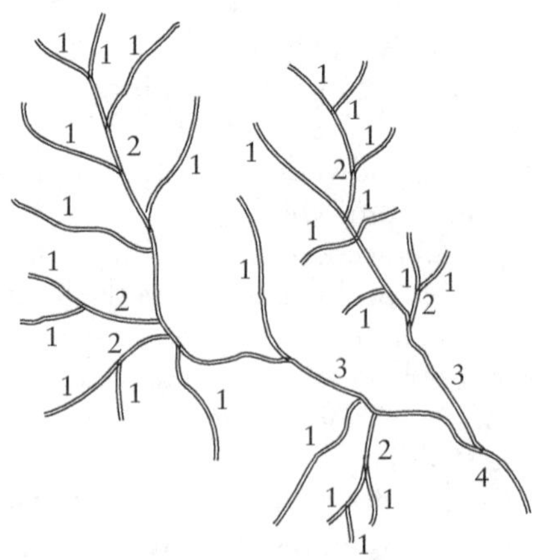

FIGURE 12.17 Stream order designations provide information on much more than stream size alone.

lentic system have more of a tendency to stay there. The productivity of lotic systems is more closely tied to the productivity levels of the surrounding terrestrial area than is that of lentic systems.

There are two broad habitat types commonly found in most headwater riverine systems: **riffles** and **pools** (Figure 12.18). Riffles are generally shallow and narrow, and their bottoms are usually characterized by gravel and large rocks. Because of their relative shallowness, current speed in riffles is greater than that in pools; this swifter current sweeps away most material of smaller particle size. The swifter the current, the larger the rocks in a riffle. Pools tend to be deeper, wider, and more soft bottomed, and they have less current flow than riffles. In some situations, pool areas can closely resemble lentic habitat. Some biologists classify the areas between the fastest portion of a riffle and a pool as **raceways, runs,** or **chutes.** These areas are intermediate in depth and current compared to riffles and pools, and they often have their own unique assemblages of plants and animals.

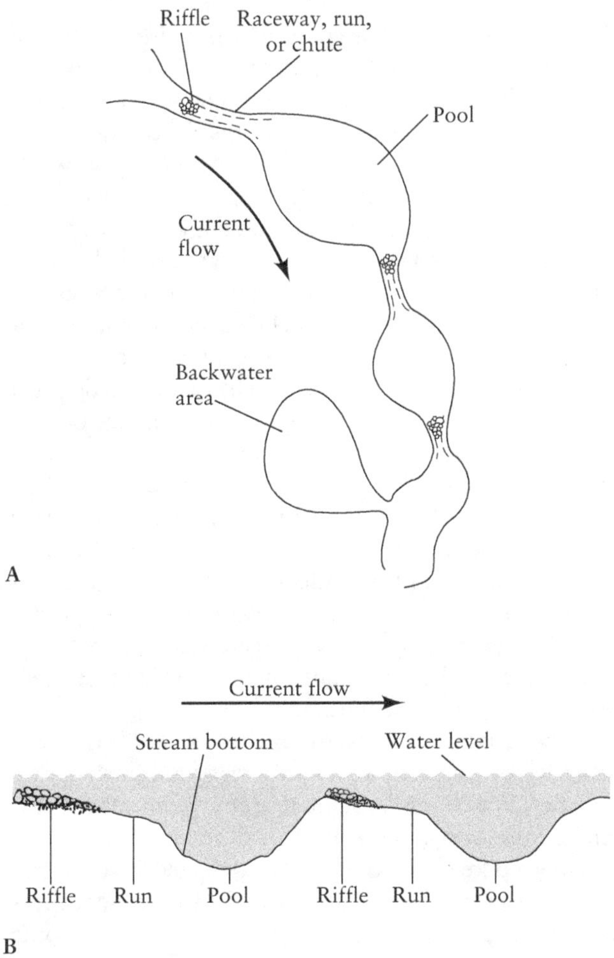

FIGURE 12.18 Lotic systems contain two broad habitat types: riffles and pools. Backwater areas may also be found in association with lotic systems. *(A)* Top view; *(B)* side view.

Because of the differences mentioned above, fishes and other organisms in a lotic system can vary greatly from pool to riffle areas. For example, fishes such as orangethroat darters, mountain madtoms, and desert suckers that require gravel or rocks for nest sites or other habitat needs are more often associated with riffle areas. Fishes that are less well adapted to current, such as green sunfish, flat bullheads, and brassy minnows, are more likely to be found in pool areas. Some fish species can be found in pool areas during one stage of life and in the riffle areas during another life stage. It is common to see quite different animal and plant communities in the two habitat types.

Lotic systems occur across a continuum, ranging from headwater mountain streams that resemble constant riffle habitat to sluggish, nearly stationary downstream rivers that are almost totally pool habitat. In addition, pool–riffle relationships are not always linear. A stream or river can have a pool area and a riffle area side by side. Backwater areas—areas connected to the lotic system, but without unidirectional current—may also be found in association with rivers and streams (see Figure 12.18). These areas have essentially no water flow, and the fishes and other biotic assemblages found in them often include species more commonly associated with lentic systems.

Although there is usually a constant unidirectional water flow in lotic systems, this flow is not uniform. As mentioned above, the shallower and narrower riffles have faster flow rates than the deeper, wider pools. Also, the closer to the bottom or sides of a lotic system, the slower the current; this pattern is a result of frictional effects. Few organisms are adapted to life in swift-current areas. Most fishes and other animals found in lotic systems associate with the bottom or with structures that reduce current flow. Most venture into the swift portion of the current only on occasional forays.

In addition, most lotic systems do not travel in a straight line; they meander in a series of curves. Wherever a stream or river curves or changes direction, one can find different current speeds and therefore different habitats. The outer segment of a curve (the **erosional zone**) tends to be deeper than the inner segment (the

depositional zone), because the swifter current in the outer segment is able to move larger bottom particles. This is also the reason why such outer segments tend to have firmer bottoms. The inner portion of the curve generally is shallower and softer bottomed, and it has a slower current speed.

12.5 Estuarine Habitats

An **estuary** is the place where a river enters the ocean. Estuarine habitats are often associated with coastal wetlands. Estuaries are areas of transition between freshwater and marine environments. Although marine habitat is not discussed in this text, the roles of estuaries, coastal wetlands, and beach–ocean interfaces as transitional zones between freshwater and marine systems and between aquatic and terrestrial habitats make their inclusion important.

An estuary can be a rather small area where a single stream enters the ocean, or it can be a sizable area, such as Chesapeake Bay, where a number of freshwater sources enter the ocean. Estuaries can experience extreme fluctuations in salinity. In addition, estuaries can be affected by tides. Salinity, and therefore density, differences between marine and freshwater systems can result in the formation of a **salt wedge**, which provides a structure similar in some ways to that of the thermal stratification found in lakes and impoundments. The salt wedge, because of its higher salinity, is heavier than the overlying fresh water and can extend upstream beneath the fresh water. This wedge can advance and retreat with tidal movement. At the interface where fresh and marine water meets, small particles and other materials are formed into larger aggregates. This nutrient-rich material contributes to the high productivity of estuaries. Also contributing to this productivity is the constant movement and flushing of water resulting from river currents and tides.

High nutrient input into estuaries generally benefits these areas and has an important effect on estuarine production. The word *generally* is used because excessive nutrient input into any water body is not beneficial (see section 14.2). Estuaries are very rich habitats, and many marine and terrestrial vertebrate and invertebrate species use them as adult habitat and as nursery areas for their young. Because of their ecological importance, the present decline in quality and quantity of estuarine habitats is of special concern to natural resource advocates. Pollution and human building activities continue to reduce the amount of estuarine area and its utility to native organisms.

Coastal wetlands, while not always associated with estuarine areas, are also important transitional zones. They are high in primary productivity, provide habitat for many year-round resident and migratory terrestrial and aquatic species, and have a natural ability to lessen flood and storm damage, minimize coastal erosion, and improve water quality. Coastal wetlands loss helped contribute to damage caused by Hurricane Katrina in 2005. They, along with estuaries, serve as spawning grounds and nurseries for over 75% of the commercially important marine fish and shellfish species.

12.6 Reproductive Habitats

Previous portions of this chapter have addressed general aquatic and terrestrial habitat types. Located within each of these general habitats are many more specific habitats that organisms use. All organisms require specific types of landscapes in order to be successful. While the previous categories, such as deserts and rivers, are general in nature, organisms require much more specific or functional habitat units. These specific habitat requirements can relate to organismal needs associated with reproduction, feeding, shelter, concealment, rest, or other characteristics that a particular species may require. In addition, habitat and landscape needs for a particular species can vary depending upon age or gender of the organism, time of year or time of day, and proximity of one type of habitat to another.

The availability of appropriate reproductive habitat is crucial for most species. Optimal reproductive habitat can provide a variety of benefits to a species. It may shield the eggs or young from predators, protect them from adverse environmental elements such as silt, sunlight, or rain, or protect them from damage by the parents.

In aquatic systems, reproductive habitat needs usually revolve around a number of habitat characteristics. For example, a suitable spawning substrate is often vital for fish reproduction. Some species, such as northern pike and brook sticklebacks, need vegetative material as spawning substrate. Central stonerollers and pink salmon require gravel substrates. Some fishes, such as the johnny darter, spawn on the undersides of rocks. Others, such as the freshwater drum, require no substrate at all and produce eggs that are free floating in the water. Water depth and current also influence the suitability of reproductive habitat. Some fishes, such as orangespotted sunfish, prefer lentic water, often less than 1 meter in depth. Others, such as lake trout, usually require lentic water and can spawn at water depths of up to 36 meters. The mountain sucker requires swift-moving shallow water, while the sauger can spawn in sluggish rivers in water more than 4 meters in depth.

To understand just how specific habitat requirements for fish reproduction may be, consider the example of the California grunion. Spawning activities of this species are closely cued to lunar and tidal cycles. They spawn at night at the peak of high tides. Both genders move onto beaches with incoming wave surges and are stranded out of water. The fish mate and the fertilized eggs are deposited in wet sand. With the next wave surge the parents are carried back to the sea. The eggs remain buried in moist sand until the next extreme high tide (usually 2 weeks later), when the fry hatch and enter the ocean (Barton 2007).

Birds also have specific needs for reproductive habitat. Western meadowlarks are grassland ground nesters, sage thrashers nest in low-lying shrubs 1.0 to 1.5 meters tall, Baltimore orioles nest in treetops, and yellow-headed blackbirds nest on emergent vegetation in wetlands. Even within a grouping such as ground-nesting birds, there is great variation: piping plovers use almost barren sandbars, common nighthawks utilize sparse ground vegetation or gravel, killdeers nest in sparse ground vegetation, and grasshopper sparrows place their nests in dense ground vegetation. Some species, such as redheads and canvasbacks, construct nests of emergent vegetation that float on water. Many birds, such as gray partridges, need solitary locations for nesting, while others nest only in colonies. Some birds, such as double-crested cormorants and great blue herons, form nesting colonies called **rookeries** and need such aggregations to be successful (see Figure 6.12).

As an example of specific nesting needs, consider belted kingfishers. Winkler (2004) described the nest tunnels of these birds. Some are in tree cavities, and others are in burrows that the birds build in earthen banks. These earthen burrows are usually near water, in steep dirt or sand banks with no vegetation on the bank face, and the banks usually have just **herbaceous plants** (nonwoody plants) on top so that the birds will not encounter tree roots when digging.

Because most mammals bear live young, they need no specific sites for laying eggs, but they still have a need for sites in which to bear young and keep them safe after birth. Southern flying squirrels need tree cavities, common muskrats need dens, Gunnison's prairie dogs need their burrow systems, and mule deer seek specific types of vegetational cover for fawning. Consider dens where kit foxes bear and raise their young. These dens usually have more entrances than dens kit foxes use for shelter during the nonbreeding season. Around nursery dens, the kit fox family also tramples dried vegetation around the den opening; this forms a disturbed apron for up to 10 meters around the den. In addition, numerous prey remains are scattered about the den opening (O'Farrell 1987).

Organisms may have other reproductive habitat needs in addition to locations in which to nest or place young. Greater and lesser prairie-chickens gather on communal areas called **booming grounds** to perform their courtship displays. Sharp-tailed grouse perform such reproductive behavior on **dancing grounds** (Figure 12.19A), while greater sage-grouse perform on **strutting grounds** (Figure 12.19B). Such places where

A

B

FIGURE 12.19 Grouse gather to perform courtship displays. *(A)* Sharp-tailed grouse on dancing grounds; *(B)* greater sage-grouse on strutting grounds. *(Part A image courtesy of J. Iron; part B image courtesy of K. Jensen.)*

these species gather to perform mating displays are sometimes collectively referred to as **leks**. Ruffed grouse need drumming logs, and western grebes need sufficient open water to perform their breeding chase. A fish such as the bluntnose minnow also needs sufficient space of the right kind to perform its reproductive behavior, as does a mammal such as the mountain goat.

12.7 Feeding Habitats

Animals may also require specific types of habitat in which particular types of food are produced. If a mammal needs specific grasses or other plants for food, habitat to support such plant life must be present. If a fish needs aquatic invertebrates that are associated with rocky substrates, rocks must be present in its feeding area. If a bird needs specific seeds or invertebrates as a food source at a particular time, habitat that can provide such food must be present. Feeding-habitat requirements may also entail vegetation or other structural elements that allow a predator to get close to or to see prey. Some raptors sit atop posts or dead trees, because these habitat features allow them to survey their feeding area more efficiently. A chain pickerel remains motionless near a vegetated area, awaiting prey species that venture away from their vegetative hiding places. In some cases feeding habitat may be used year-round, while in others such habitat may be used for only part of the year. For example, migrating birds may use a particular habitat for only a few days or hours of the year.

Requirements for feeding habitat can be quite specific. For adult shortnose sturgeon, optimal feeding habitat during summer must have, among other characteristics, water temperature between 11° and 22°C, current velocities between 15 and 45 centimeters per second, and bottom substrates ranging from vegetation to sand (Crance 1986). Habitat that falls outside of those ranges represents suboptimal habitat.

The collared peccary has dietary requirements that dictate the feeding habitat and landscapes they can inhabit. While they have a diverse diet, it is dominated in North America by succulents, such as prickly pear cactus (Hellgren and Lochmiller 2000). It is possible that cactus consumption is linked to meeting water requirements, with other dietary components supplying more nutritional foods.

Feeding-habitat requirements for birds can also be very specific. For example, Brown and Fredrickson (1986) reported that white-winged scoters feed primarily on scuds (a type of small aquatic crustacean). In their study in Saskatchewan, Brown and Fredrickson found that one species of scud provided 97% of the diet of female white-winged scoters during nesting and brood rearing. These quantities of scuds would be found only in specific types of habitat.

12.8 Cover Habitats

Habitat that is used by organisms for shelter, concealment, or escape purposes is often referred to as **cover** (Figure 12.20). Organisms use cover in numerous ways. Endotherms, such as birds and mammals, can use cover as shelter to prevent excess gain or loss of body heat. Being able to avoid wind, rain, snow, hail, and other adverse elements may be essential for survival. Fishes also use cover as shelter for energy conservation purposes. A mountain whitefish remaining behind a rock or other stream obstruction usually does so to avoid stronger water currents.

Cover is used for concealment and escape by many organisms, both aquatic and terrestrial. A small green sunfish hiding in aquatic vegetation is using that habitat to conceal itself. If it is outside this vegetation and becomes alarmed, it will seek such a place in which to escape. A white-tailed deer will use a thick stand of tim-

FIGURE 12.20 Escape cover is especially important to wild animals. The cover shown in this image could be used by a variety of species. *[Image courtesy of K. Jensen.]*

ber to conceal itself or to escape when disturbed. A ring-necked pheasant will conceal itself in heavy cover or escape to such cover when threatened.

Requirements for escape cover may be quite specific and vary by species. Pulliam and Mills (1977) described escape behavior and its relation to habitat for several bird species in Arizona. In the area they studied, a variety of birds inhabited open grassland containing scattered mesquite trees. Species such as white-crowned sparrows, chipping sparrows, rufous-crowned sparrows, and canyon towhees were highly dependent on tree or shrub cover for escape; they tended to stay close to mesquite trees. Vesper sparrows often used mesquite for escape cover, but they were able to feed far from mesquite trees and used ground cover (grass) for camouflage or escape. Grasshopper sparrows escaped to dense ground cover when necessary and used their cryptic coloration and solitary behavior to escape predators; they seldom used mesquite trees for escape cover. Chestnut-collared longspurs escaped to more open ground cover and made strong use of camouflage and aerial group flights to avoid predators.

Fishes, too, can have particular cover requirements. For example, black bullheads thrive in lotic waters where the percentage of pool and backwater areas during average summer flows is between 50% and 75% of the habitat present. This species also thrives where cover (vegetation and debris) within pools, backwaters, or littoral areas during summer represents 25% of the habitat, and where average current velocity at 0.6 meters of water depth during average summer flow is approximately 4 centimeters per second. Additionally, black bullheads do best where the littoral area represents 25% of a water body during summer (Stuber 1982).

Cover requirements for mammals can also be quite specific. For example, North American river otters use partly submerged trees and logjams as shelter. Cavities among tree roots, shrubs, and grass also provide escape cover (Melquist and Dronkert 1987).

The particular attributes needed in concealment and escape cover can vary, depending on what is being avoided. If airborne predators are a threat to a bird or mammal, some type of overhead cover is sought. In the case of fishes, all that may be needed are small crevices in rocky areas or thick vegetation that restricts predator maneuverability.

Cover needs can vary by season. Fishes often move to deeper water areas during cold periods of the year and shallow water areas during summer. In winter, deeper water areas often provide, among other things, warmer water. As mentioned earlier, water is its densest at 4°C; therefore, a water body near freezing will have its warmest water near the bottom. Elk and deer use different habitats in winter than in summer. They often migrate to higher altitude areas in summer and to lower elevations during winter. In the western mountainous regions of North America, they often select a winter range that is kept snow free by high-velocity warm winds characteristic of such areas. Daily movements among different cover habitats also occur. Any person who has angled for fishes readily learns that their habitat use can vary by time of day. For example, knowledgeable anglers trying to catch large brown trout are well aware that such fish will tend to seek cover and deeper areas during daylight and move to open, shallower areas to feed near dusk or at night.

12.9 Other Habitat Requirements

In the wildlife field, organisms are often described as using loafing or resting cover. These areas are used in a variety of activities. For example, sharp-tailed grouse use loafing cover to sun or dust themselves, while a Brandt's cormorant may rest on a rock while drying its feathers. Specific habitat requirements for other activities, such as defecation, watering, and mineral seeking, are common among wild animals. It should be noted that in many areas, sites that are particularly attractive for homesites and other human activities are the same ones that contain critical habitat components for wild animals. Deer winter ranges and riparian zones are examples of habitat types that are also popular areas for human activities. This conflict between the needs of people and animals is a problem of human management.

Biologists expend a great deal of their efforts on habitat management. Chapter 15 describes some of the procedures and techniques that they use. If one were to measure the amount of time biologists spend on various management and conservation activities, the majority of that time would be associated with habitat. This is especially true of terrestrial biologists. The current trend toward ecosystem and landscape management reinforces the precept that unless all habitat needs are met, populations and communities of aquatic and terrestrial organisms cannot reach or approach their maximal levels. In the past, wildlife and fisheries biologists have received criticism for managing for single species, usually game or commercially important organisms. To some extent this criticism was justified. However, a waterfowl biologist who manages or saves a wetland for waterfowl is in actuality also providing habitat for a large variety of other organisms. Even though such an area may be primarily managed for one species, many other species receive benefit from that habitat. Conversely, other species may be diminished by a management or conservation activity directed at a single target species or group of species. Habitat management strategies currently consider effects on a variety of organisms to a much greater extent as we move from single-species management to community, landscape, and ecosystem management.

SUMMARY

A broad form of terrestrial habitat classification is by biome. Biomes are repeated on a global basis; a biome found in North America may also be found in Europe, Asia, or other regions of the world. Important North American biomes include tundra, northern coniferous forest, temperate forest, temperate grassland, desert, chaparral, Pacific coast coniferous forest, and tropical seasonal forest. Biomes are not monolithic habitat blocks, but contain various habitat types. The edges of biomes constitute ecotone areas. Ecoregions are another way of delineating terrestrial and aquatic areas.

Riparian zones, wet margins of lakes and ponds, and a variety of marshes, bogs, swamps, coastal wetlands, and beach–ocean interfaces represent transitional areas between terrestrial and aquatic habitats. These transitional areas occur across a continuum from almost totally terrestrial to almost totally aquatic.

Natural lakes, intermediate impoundments, ponds, and reservoirs represent lentic aquatic environments. Rivers and streams represent lotic aquatic environments. Eutrophic aquatic systems are highly productive, while oligotrophic systems are relatively unproductive. Shallow water areas of a lentic system constitute the littoral zone; the limnetic and profundal zones are deeper water areas. Lentic water bodies can also be characterized by differences in water temperatures and oxygen content. Lotic systems are generally divided into riffle and pool habitats.

Estuaries, coastal wetlands, and beach–ocean interfaces represent important transitional zones. These highly productive areas contain a large number of organisms that are of great commercial and ecological importance.

While most habitat categories are general in nature, organisms also require more specific or functional habitat types. These specific habitat requirements can relate to such organismal needs as reproduction, food, shelter, concealment, and rest. Specific habitat needs for a particular species can vary depending on age or gender of the organism, time of year or time of day, or even the proximity of one type of habitat to another.

A large proportion of the efforts of a biologist involves habitat management. The trend in habitat management today is toward a more holistic system rather than single-species management.

PRACTICE QUESTIONS

1. In which biome do you live? What are some of the characteristic plants and animals of that biome? What specific habitats do each use?

2. Why is Arctic tundra such a fragile biome?

3. For each North American biome, describe as many human activities as possible that have modified that biome over the last hundred years.

4. What are the major habitat conservation issues in the area where you live?

5. In what way does the issue of wetland loss affect the region where you live? Is it even an issue? Among whom? For what reasons?

6. Describe thermal stratification in lentic systems and describe what leads to fall overturn in these systems.

7. Describe a dozen different ways in which water current affects the lives of aquatic organisms in lotic systems.

8. Based on your past observations, choose six different bird species that occur in the area where you live, then describe the differences in where they nest and the types of nests they construct.

9. How can a reservoir affect aquatic and terrestrial organisms downstream from the reservoir?

10. What are estuaries and why are they valuable to humans?

SELECTED READINGS

Abell, R. A., D. M. Olson, E. Dinerstein, P. T. Hurley, J. T. Diggs, W. Eichbaum, S. Walters, W. Wettengel, T. Allnutt, C. J. Loucks, and P. Hedao. 2000. *Freshwater ecoregions of North America: A conservation assessment.* Island Press, Washington, D.C.

Bain, M. B., and N. J. Stevenson, eds. 1999. *Aquatic habitat assessment: Common methods.* American Fisheries Society, Bethesda, Md.

Cooke, G. D., E. B. Welch, S. A. Peterson, and P. R. Newroth. 1993. *Restoration and management of lakes and reservoirs.* 2nd ed. Lewis Publishers, Boca Raton, Fla.

Farina, A. 2000. *Principles and methods in landscape ecology.* Kluwer Academic Publishers, Norwell, Mass.

Gill, F. B. 2006. *Ornithology.* 3rd ed. W. H. Freeman, New York.

Kohler, C. C., and W. A. Hubert, eds. 1999. *Inland fisheries management in North America.* 2nd ed. American Fisheries Society, Bethesda, Md.

Morrison, M. L., B. G. Marcot, and R. W. Mannan. 1992. *Wildlife-habitat relationships: Concepts and applications.* University of Wisconsin Press, Madison.

Ricketts, T. H., E. Dinerstein, D. M. Olson, C. J. Loucks, W. Eichbaum, D. DellaSala, K. Kavanagh, P. Hedao, P. T. Hurley, K. M. Carney, R. Abell, and S. Walters. 1999. *Terrestrial ecoregions of North America: A conservation assessment.* Island Press, Washington, D.C.

Smith, R. L. 1996. *Ecology and field biology.* 5th ed. HarperCollins, New York.

Vaughan, T. A., J. J. Ryan, and N. J. Czaplewski. 2000. *Mammalogy.* 4th ed. Saunders College Publishing, Philadelphia.

Wetzel, R. G. 2001. *Limnology: Lake and river ecosystems.* 3rd ed. Academic Press, San Diego.

LITERATURE CITED

Abell, R. A., D. M. Olson, E. Dinerstein, P. T. Hurley, J. T. Diggs, W. Eichbaum, S. Walters, W. Wettengel, T. Allnutt, C. J. Loucks, and P. Hedao. 2000. *Freshwater ecoregions of North America: A conservation assessment.* Island Press, Washington, D.C.

Barton, M. 2007. *Bond's biology of fishes.* 3rd ed. Thomson Brooks/Cole, Belmont, Calif.

Brown, P. W., and L. H. Fredrickson. 1986. Food habits of breeding white-winged scoters. *Canadian Journal of Zoology* 64:1652–1654.

Cowardin, L. M., V. Carter, F. C. Golet, and E. T. LaRoe. 1979. *Classification of wetlands and deepwater habitats of the United States.* FWS/OBS-79/31. U.S. Fish and Wildlife Service, Washington, D.C.

Crance, J. H. 1986. *Habitat suitability index models and instream flow suitability curves: Shortnose sturgeon.* Biological Report 82(10.129). U.S. Fish and Wildlife Service, Washington, D.C.

Hellgren, E. C., and R. L. Lochmiller. 2000. Collared peccary. Pages 429–446 *in* S. Demarais and P. R. Krausman, eds. *Ecology and management of large mammals in North America.* Prentice Hall, Upper Saddle River, N.J.

Melquist, W. E., and A. E. Dronkert. 1987. River otter. Pages 626–641 *in* M. Novak, J. A. Baker, M. E. Obbard, and B. Malloch, eds. *Wild furbearer management and conservation in North America.* Ministry of Natural Resources, Ontario, Canada.

O'Farrell, T. P. 1987. Kit fox. Pages 422–431 *in* M. Novak, J. A. Baker, M. E. Obbard, and B. Malloch, eds. *Wild furbearer management and conservation in North America.* Ministry of Natural Resources, Ontario, Canada.

Ploskey, G. R., L. R. Aggus, and J. M. Nestler. 1984. *Effects of water levels and hydrology on fisheries in hydropower storage, hydropower mainstream and flood control reservoirs.* Technical Report E-84-8. Environmental and Water Quality Operational Studies, U.S. Army Corps of Engineers, Washington, D.C.

Pulliam, H. R., and G. S. Mills. 1977. The use of space by wintering sparrows. *Ecology* 58:1393–1399.

Ricketts, T. H., E. Dinerstein, D. M. Olson, C. J. Loucks, W. Eichbaum, D. DellaSala, K. Kavanagh, P. Hedao,

P. T. Hurley, K. M. Carney, R. Abell, and S. Walters. 1999. *Terrestrial ecoregions of North America: A conservation assessment.* Island Press, Washington, D.C.

Stuber, R. J. 1982. *Habitat suitability index models: Black bullhead.* FWS/OBS-82/10.14. U.S. Fish and Wildlife Service, Washington, D.C.

Summerfelt, R. C. 1999. Lake and reservoir habitat management. Pages 285–320 *in* C. C. Kohler and W. A. Hubert, eds. *Inland fisheries management in North America.* 2nd ed. American Fisheries Society, Bethesda, Md.

Winkler, D. W. 2004. Nests, eggs and young: Breeding biology of birds. Pages 8.1–8.152 *in* S. Podulka, R. W. Rohrbaugh, and R. Bonney, eds. *Handbook of bird biology.* 2nd ed. Cornell Lab of Ornithology, Ithaca, N.Y.

CHAPTER 13

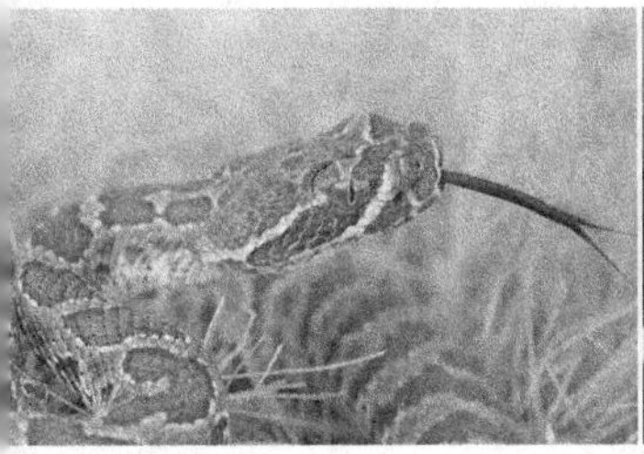

Habitat Sampling and Assessment

Biologists must be able to assess the biota and habitat components of a fishery or a wildlife system. A variety of methods used in assessing populations are described in Chapter 9; this chapter builds on those previously described assessment skills, focusing on habitat. We often assess the value of habitats by observing how and when a species, or even genders within species, use available habitats during a season, year, or other time period. Thus, animal use of an area provides information on its habitat. Biota assessment is seldom done without some form of companion assessment of habitat.

Consider a situation in which population or community assessment has indicated that an area has a particular density or number of a certain target species or species assemblage. A biologist may want to increase or decrease those numbers or determine why they are increasing or decreasing. Habitat assessment is likely to be needed to answer a variety of questions, such as, "Is a particular habitat limiting?" or "What characterizes that habitat?" or "How much of that habitat is present?" Even after habitat assessment has led to the development of a conservation strategy, habitat may need to be reassessed to determine the success or failure of any implemented management practices.

Biologists often identify habitat as the most important overall component in managing and conserving terrestrial and aquatic animals. Because of the importance of habitat, biologists need to be familiar with the methods used in its measurement and assessment. In practice, a wide variety of techniques are available for evaluating habitat. The purposes of this chapter are to introduce some of the more common procedures used for sampling and assessing habitat and to describe some of the uses of such information.

13.1 Aquatic Habitats: Water Characteristics

A variety of parameters, such as water temperature, dissolved-oxygen content, current velocity, pH, and nitrate levels, may need to be measured when assessing aquatic habitat. Various methods and devices are available for measuring these water characteristics. The devices used range from electronic instruments to rather simple field kits or tools used for obtaining approximate measurements. Some of the more elaborate electronic devices measure numerous parameters at the same time and over a long period of time. Many of these devices collect data that can be downloaded directly into a computer. These devices become more efficient with each new model.

Water Temperature

Water temperature has widespread influences on aquatic ecosystems. For example, water temperature is important in determining spawning chronology in fish communities. Water temperature also influences metabolic activity, and thus growth rates, in most aquatic organisms, as well as the dissolved-oxygen-holding capacity of water. Even rates of development and lengths of hatching periods for fish and amphibian eggs are influenced by water temperature. Many other examples of the effects of water temperature could be provided.

Water temperatures often vary with depth. Temperature variation from surface to bottom of a water body can be determined by removing water samples from various depths and measuring the temperature of each sample with a handheld thermometer or electronic temperature-measuring device. Simple devices such as **Kemmerer** and **van Dorn water samplers** (Figure 13.1) can be remotely triggered to obtain water from a particular depth. Water temperatures can also be obtained with electronic probes lowered to various depths. Obtaining temperatures at different depths is especially important in water bodies that thermally stratify. In lotic or shallow lentic water bodies where temperature stratification is lacking or poorly developed, near-surface water temperature information is usually adequate. However, in shallow water habitats, temperature fluctuation over a 24-hour period is usually greater than in deeper water bodies.

A

B

FIGURE 13.1 *(A)* The Kemmerer water sampler and *(B)* the van Dorn water sampler are used for sampling water at various depths.

Dissolved-Oxygen Levels

Knowing the concentration of dissolved oxygen in a body of water is important, because a lack of oxygen may result in stress or death for aquatic organisms. Even if low dissolved-oxygen levels do not cause death, stress from **hypoxic** (low-oxygen) conditions can reduce growth rates and increase the susceptibility of animals to diseases, pathogens, and parasites. Dissolved-oxygen concentrations of approximately 3 to 5 milligrams per liter are commonly considered adequate for warmwater fishes, while levels of 6 to 9 milligrams per liter or above are usually adequate for coldwater fishes. However, low dissolved-oxygen levels do not necessarily cause fish death (technically referred to as *fish kills*). Meyer and Herman (1990) reported that dissolved-oxygen levels of less than 1 milligram per liter, and sometimes as high as 2 milligrams per liter, are indicative of potential fish kill conditions. Some fishes, such as northern pike and fathead minnows, however, often survive the winter in shallow, ice-covered northern water bodies where dissolved-oxygen levels drop below 1 milligram per liter.

Dissolved oxygen can be measured in water samples either by titration or electronically. **Titration**, which is now seldom used, is the measurement of the quantity of a given constituent of a solution–dissolved oxygen in this instance—through the addition of reagents of known concentration until a particular chemical reaction, often indicated by a color change, is completed. Dissolved-oxygen determination is now primarily done with dissolved-oxygen probes connected to an electronic meter.

Light Penetration

Light penetration in aquatic systems varies with water depth, color, and the amounts and kinds of materials suspended in water; the amount and kind of light that penetrates to various depths helps determine primary productivity in an aquatic system. Therefore, light penetration or water transparency information has many uses. For example, if turbidity is limiting primary production, management practices to reduce turbidity can be implemented (see section 15.3).

Water transparency is sometimes estimated with a **Secchi disk** (Figure 13.2). The Secchi disk is a metal, plastic, or wooden plate approximately 20 centimeters in diameter, and it usually has white and black quarters. A line attached to the disk is marked in known length increments. The disk is lowered into the water until it disappears from view. It is then lowered another meter or so and then raised until it reappears. The average of the depths at which the disk disappears and reappears is recorded as the Secchi disk visibility. Secchi disk visibility is normally measured from the shaded side of a boat and in areas where aquatic vegetation will not hinder or obscure the disk. However, the Secchi disk does not really measure light penetration; there are a variety of photoelectric devices that measure light intensity when lowered to various depths.

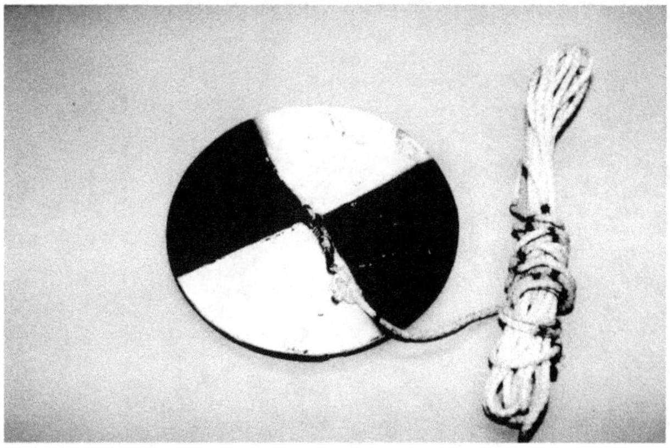

FIGURE 13.2 Water transparency is commonly approximated with a Secchi disk.

Current Velocity

Current velocity is an important variable in lotic systems. It can be measured with current meters such as the **pygmy water meter** (Figure 13.3). Water current varies with depth and position in the stream channel,

FIGURE 13.3 The pygmy water meter is used to measure current velocity in lotic systems. This photograph shows only the portion of the meter that is submerged in water when taking data.

FIGURE 13.4 Highway development can influence the channel characteristics and thus the fisheries and wildlife values of lotic and associated riparian habitats. *[Image courtesy of Brigham Young University.]*

so it is usually necessary to measure current velocity at different depths and at different locations in relation to the stream bottom and shoreline. Rapid changes in current speed over a 24-hour period are not unusual. Strip chart recorders that can be set and left unattended can obtain water current information over time.

Human construction activities can physically alter stream channel characteristics, current velocities, and current patterns. For example, the development of a highway along a river or stream may require stream channel rerouting (Figure 13.4). Biologists should obtain at least a subjective classification of flow patterns and the relative proportions of habitat types, such as pools and riffles, prior to stream modification; this information can be used in assessment of postconstruction fish habitat and in mitigation decisions.

Totally aquatic species are not the only organisms influenced by stream modification. Other organisms, such as North American river otters, American dippers, and riparian plants, are also important aspects of lotic ecosystems that are influenced by construction activities.

Other Water Characteristics

The hydrogen ion concentration in a solution is measured as **pH. Acidity** is the capacity to donate hydrogen ions. Acidic waters have a pH of less than 7. Acidic water bodies can occur naturally, such as in some areas of the Pacific Northwest and in northeastern North America, but water bodies can also become more acidic due to human activities, as discussed in section 14.1. Increased acidity in water bodies can limit or eliminate various invertebrate and vertebrate biota. Electronic meters, titration methods, and even pH-sensitive paper (which turns a specific color depending on pH) can be used to measure pH.

Alkalinity (**basicity**) is a measure of carbonates and bicarbonates in water and often reflects the carbonate content of surrounding soils. Alkalinity is usually expressed in terms of calcium carbonate equivalents. Waters with higher levels of calcium and magnesium carbonates, within limits, are more biologically productive. Alkalinity is usually measured using titration techniques.

Hardness is best defined as the total concentration of calcium and magnesium ions in water. These two elements are generally the most abundant ions in fresh water. Hardness and alkalinity are closely related, but they are measured in different ways. Water bodies with minimal concentrations of calcium and magnesium are generally lower in productivity. Hardness can be measured by titration or with electronic meters.

Conductivity (**specific conductance**) is a measure of the ability of water to convey electrical current. In general, the greater the concentration of ions in water, the greater its conductivity. Conductivity can also be used as an indication of water hardness. Conductivity is usually measured with electronic meters.

Measurement of other water quality characteristics, such as nitrogen and phosphorus content, can also be important. Causes of high nitrogen and phosphorus levels in water bodies include runoff from fertilizers placed on croplands and lawns, natural waste from domesticated livestock, and incompletely treated human sewage. Domesticated livestock holding facilities near water bodies can be a substantial source of nitrogen. High nitrogen levels, particularly in the forms of ammonia and ammonium, can limit fish productivity and may even cause mortality in some fish populations. Both nitrogen and phosphorus, especially the latter, can contribute to high productivity and excessive aquatic vegetation in water bodies.

There are many other chemical characteristics of water that can be measured; chloride, chlorine, and carbon dioxide content are examples. The water characteristics measured depend on what information is needed in a particular situation.

Procedures for water quality analysis are quite specific, because the interacting effects of turbidity, time, and temperature on water samples may influence various water quality characteristics. Specialized water quality analysis laboratories are the best source for accurate evaluation of water samples. These laboratories provide exact specifications for the collection, shipping, and care of water samples. The cost of accurate water analysis can be quite high and is often a major budgetary consideration.

13.2 Aquatic Habitats: Water Body Characteristics

Surface Area

The surface area of a water body is an important descriptor in fisheries and, to a lesser extent, wildlife work. In fisheries evaluations, standing stock is measured in weight per unit of surface area, usually in kilograms per hectare. Numbers of fish stocked are usually determined in relation to surface area. Surface area, in conjunction with productivity, is related to potential production of fishes and thus the number of angler hours of fishing a water body can support. The use of water bodies by some wildlife species is also influenced by surface area. For example, molting ducks, such as redheads and canvasbacks, traditionally gather on large water bodies; these molt gatherings may number many thousands of birds. Water bodies on which these ducks molt must be shallow and produce adequate food, but just as importantly, they must be large enough that the birds can remain far from disturbances.

Maps of water bodies are available from a variety of sources. National Wetlands Inventory maps (http://www.fws.gov/nwi/) can be obtained through the U.S. Fish and Wildlife Service; these maps provide information on both the size (surface area) and classification of water bodies. Most natural resource agencies produce maps of specific water bodies that provide surface area and other information; maps of specific lakes can often be located on the Internet by searching under the name of specific lakes or using a term such as *Wisconsin lake maps*.

If available maps do not specify surface area, it can be determined from aerial photographs of known scale, such as those available in the United States through the Consolidated Farm Service Agency. If aerial photographs are not available, it may be necessary to make a map of the water body surface. The shoreline of a water body can be walked with a handheld global positioning system (GPS) (see section 9.14) to obtain the outline of the water body. The surface area of the water body can then be obtained using geographic information system (GIS) software (Fisher and Rahel 2004) (see sections 9.9 and 13.9). There are additional means by which outlines of water body surfaces can be obtained; the use of aerial photographs and GIS are just two examples.

Depth

Depth contours are another characteristic of water bodies that are often used for conservation purposes (Figure 13.5). Depth contours of water bodies can be determined by means of weighted cables or lines with depth markings, poles or boards with depth markings, or electronic echolocators such as fish finders. Depth contours can be mapped by taking depth measurements in a systematic fashion, usually along a series of predetermined lines (**transects**). Depth information can also be entered into a computer with a coordinate system and the contours mapped using computer software.

Contour maps depicting depth can be used to evaluate fish habitat or to select sites for fishing piers, scuba-diving areas, boat docks, or fish habitat structures, such as old Christmas trees or other artificial reefs. Water depth and contour information is also used extensively in management practices involving increases or decreases in water levels.

Water level fluctuations in a water body can affect populations of many aquatic and terrestrial species (see sections 15.3 and 15.4). While most water level fluctuations influence aquatic organisms, these are not the only biota affected. Water releases from some reservoirs, for example, can periodically flood sandbars, islands, and other low areas. In some cases, it has been necessary to reduce these fluctuations during spring and early summer to protect the nests and young of animals that use these riverine areas (Figure 13.6).

Water depth and contour information also allows biologists to use water level increases or decreases as a management technique. Knowledge of depth contours can be used to answer many questions, such as, "How much water volume or surface area will be present in a water body if a certain amount of water is removed?"or "What bottom areas will be exposed

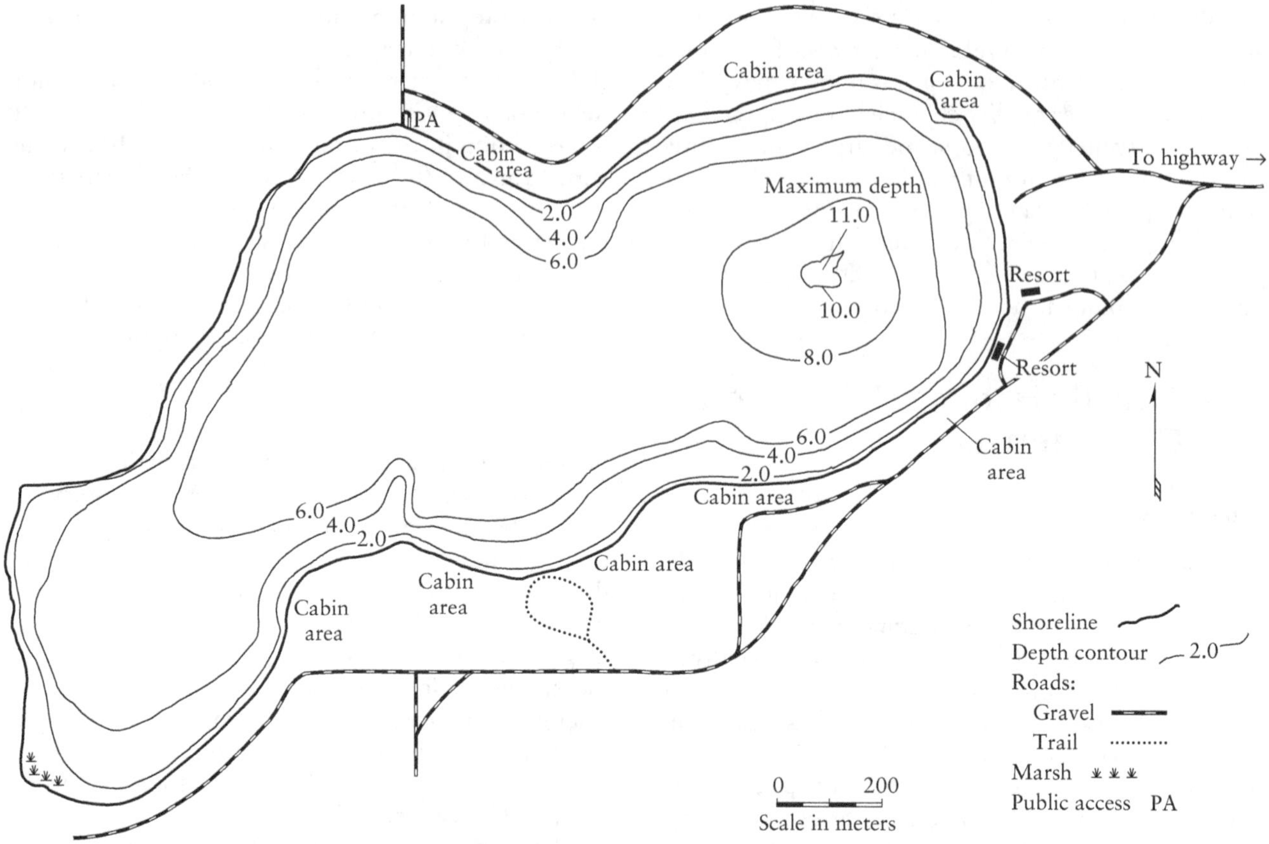

FIGURE 13.5 Depth contour maps of water bodies are important in evaluating fish habitat. The contour lines in this example are at 2-meter depth intervals.

if a certain amount of water is removed?" or "How much will the surface area increase if a certain amount of water is added?" For example, walleye, white bass, and white crappie populations did not benefit from water level management plans in Kansas reservoirs unless at least 20% of the bottom area was exposed during planned water removals. Information on depth contours would indicate how much water would need to be removed to attain 20% exposure and what areas would be exposed.

Volume

If the depth contours and area of a water body are known, water volume can be determined. In fisheries work, water volume estimates are needed for chemical treatments. For example, biologists often need to determine the amount of fertilizer that should be added to increase primary productivity or the amount of piscicide needed to eliminate a fish species or community. Water volume information can also be useful in wildlife work. Grizzly bears in Yellowstone National Park, for example, obtain fish primarily from small streams used by fish for spawning purposes. Water volume characteristics of these streams can be used to identify important feeding sites for the bears.

Shape

Another characteristic that is often measured in lentic water bodies is **shoreline development**. Shoreline development is the ratio of shoreline length to the circumference of a circle with the same area as the water body. Thus, a round water body would have a low shoreline development value, while a water body with numerous bays and arms would have a higher shoreline development value. Shoreline development is basically a measure of shoreline irregularity and

A

B

FIGURE 13.6 Piping plovers *(A)* often build their pebble- and shell-lined nests *(B)* on sites such as this sandbar along the upper Missouri River that are vulnerable to the fluctuating water levels common below many dams. *[Part A courtesy of G. Nieminen, U.S. Fish and Wildlife Service; part B courtesy of the National Biological Survey.]*

has applications in both fisheries and wildlife work. Water bodies with greater shoreline development have more areas that are protected from wind and waves and therefore provide improved habitat for some fish and wildlife species.

13.3 Aquatic Habitats: Plants and Other Substrates

Emergent and submergent aquatic plants can be identified and measured in a variety of ways. The percentage of the area of a water body that has submergent aquatic plants can be measured using echolocation. In clear, shallow water bodies, aerial photographs can reveal the extent of coverage (see section 13.4) of submergent vegetation; emergent vegetation can be photographed even if the water is turbid. Remote sensing using infrared photography can also be used for this purpose. By eliminating the blue colors and emphasizing greens, reds, and the infrared spectrum (contiguous to the red end of the visible spectrum), this technique highlights growing vegetation. Also, in clear, shallow water, floating sampling quadrats (for example, those measuring 0.5 by 0.5 meters) can be used to sample randomly or systematically. The amount of aquatic vegetation enclosed by each quadrat can be used to estimate the percentage of the water surface overlying submergent or underlying emergent aquatic plants. If individual species of plants are identified, this method can also be used to estimate relative species composition of aquatic vegetation. Aquatic plant species composition can also be assessed by setting up parallel transects in the water body and assessing the proportion of each transect intercepted by a particular plant species. Collection of aquatic plants for identification can be conducted by hand or by using simple devices such as rakes or grappling hooks. Collection of aquatic plants for determination of characteristics such as volume or weight can be done by randomly or systematically using sampling quadrats and clipping all vegetation enclosed by each quadrat.

Abundance, distribution, and types of aquatic plants in a water body can affect a variety of terrestrial species as well as fishes and other aquatic animals. For example, birds such as pied-billed and horned grebes use aquatic vegetation to form floating nests (Figure 13.7).

FIGURE 13.7 Some species, such as the pied-billed grebe, build floating nests of dead aquatic vegetation.

Yellow-headed blackbirds attach their over-water nests to emergent vegetation such as cattails. Several species of frogs and toads can lay their gelatinous strings of eggs on aquatic vegetation, as can some fishes, such as yellow perch. Aquatic vegetation provides important escape cover and food production areas for fishes. Aquatic vegetation also provides substrate and food for a variety of invertebrates that are important in the aquatic food web. Either an overabundance or a lack of emergent and submergent aquatic plants can be cause for concern to the biologist.

Long-term data describing trends in aquatic plant abundance and species composition may be valuable to biologists. Changes in aquatic vegetation may be directly related to siltation, nutrient enrichment, changes in water levels, changes in animal species composition, or other potentially human-influenced factors. Disappearance of rooted aquatic plants in a water body, for example, may indicate severe sedimentation or other pollution problems. Losses of aquatic plants or changes in plant species composition can affect other biotic components of an aquatic system. Such changes can also be important to human users, particularly as they influence water-related recreational activities. Long-term documentation of these changes is important in gaining the public support needed to correct or mitigate water quality problems.

The physical characteristics of bottom substrates are also important in describing a water body. Abiotic benthic substrates are most often categorized according to particle size. Particle sizes range from very fine silt to large boulders, with sand, gravel, and rocks of various sizes in between. Substrates consisting of small particles can be sampled with grabs (see section 7.8) or by hand. The particles can then be separated by size using a series of differently sized meshes or screens. Such measurements usually are not necessary to categorize large rocks and boulders.

Bottom particle size has a major effect on habitat suitability for many organisms. For example, silt is often a good location for aquatic vegetation but a poor location for fish egg deposition. Large rocks serve as an attachment substrate for invertebrates and as cover for fishes. In lotic habitats, fishes may seek reduced current velocities associated with large rocks. Many fish species need a specific size of gravel in which to successfully spawn. Chinook salmon, for example, tend to spawn on larger gravel than the other Pacific salmon (Scott and Crossman 1973). Most salmonids have a low tolerance for silt in intergravel spaces. Inverse relationships have been found between the quantity of sand and silt in a substrate and survival rates of emerging coho salmon and steelhead fry (Phillips et al. 1975) and pink salmon fry (Heard 1991).

Dead vegetation can also be important as an aquatic substrate. Dead trees, brush, and other nonliving vegetation can serve the same functions as abiotic substrates; they can serve as substrates for attachment or spawning, or they can moderate the effects of wind and wave action. These substrates can occur naturally or can be established by humans. In lotic systems, areas containing accumulations of dead trees and brush represent the best habitat for many fish species and other aquatic organisms. Wood ducks, green kingfishers, and various other bird species commonly perch on or feed around dead trees or brush protruding from the stream surface (Figure 13.8). Such dead vegetation accumulations are sometimes quantified in terms of number per linear distance (number per kilometer) or number per unit of surface area (number per hectare). Dead vegetative and abiotic substrates in aquatic systems have utility to more than fishes and birds; a variety of mammals, amphibians, reptiles, and invertebrates also use such structures.

Live trees and brush are also beneficial substrates. Swamps dominated by bald cypress and southern coastal areas dominated by red mangrove, for example, provide abundant structural compo-

FIGURE 13.8 This fallen tree in a river channel provides hiding cover, invertebrate prey, and reduced current velocity for fishes as well as loafing, feeding, and perching sites for birds. Various mammals, amphibians, and reptiles also use such habitat. *[Image courtesy of the National Biological Survey.]*

nents that are beneficial to a variety of aquatic and terrestrial organisms.

Geographic information system (see section 9.9 and 13.9) technology is commonly used to determine and map various habitat layers in a water body. Integrated systems can be purchased that incorporate both GPS (see section 9.14) technology and depth sounders into a single system on a boat. Such a boat can then be driven across the water body in a series of transects, or random locations can be chosen around the water body. The depth contour map in Figure 13.9A was developed for a 140-hectare lake using such integrated technology and was based on 130 random points (more points would be selected for larger water bodies). In addition, substrate and submergent vegetation presence was determined at each of those 130 points. GIS software was then used to create a bottom substrate habitat layer (Figure 13.9B) and a submergent vegetation habitat layer (Figure 13.9C). Once the substrate and vegetation layers are available, biologists often use them for additional inquiries. For example, see Paukert and Long (2004) for an example in which biologists implanted transmitters into largemouth bass and northern pike and determined fish locations at weekly intervals. These locations can then be overlaid onto various GIS layers, such as substrate or vegetation coverage, and fish habitat use can be determined.

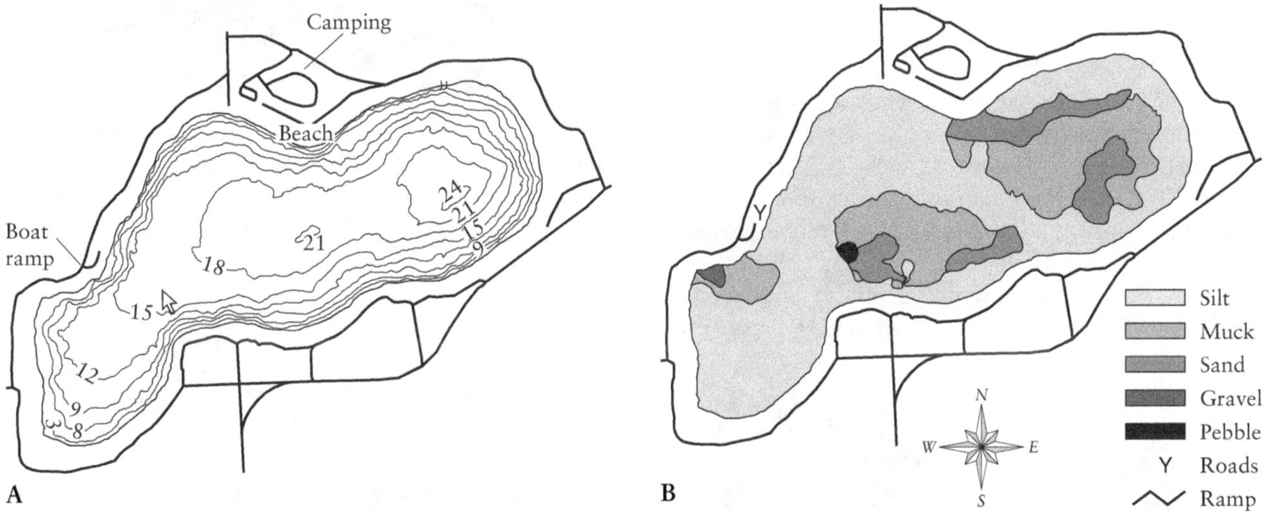

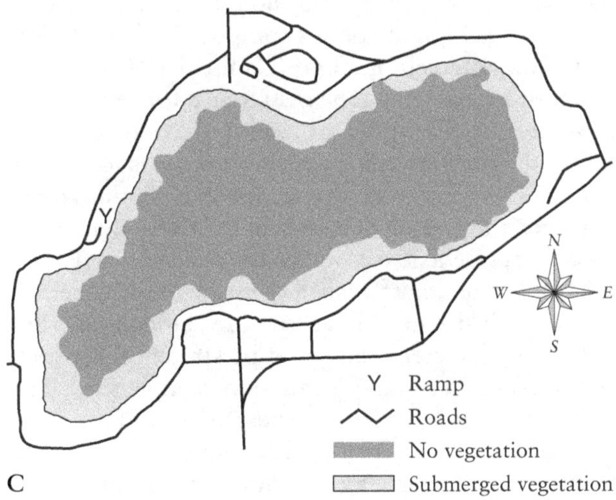

FIGURE 13.9 *(A)* This depth contour map for a 140-hectare lake was created with an integrated global positioning system (GPS) and a geographic information system (GIS). Depth contours are in feet. Prior to traveling to the lake, biologists randomly selected 130 points at which to obtain depth information; these data were then converted to the depth contour map using GIS software. *(B)* Bottom substrate map. *(C)* Submergent vegetation map.

13.4 Terrestrial Vegetation: Grasslands, Shrublands, and Other Nonforested Habitats

Several techniques exist for obtaining information concerning terrestrial vegetation. These techniques all involve taking random or systematic samples that accurately describe the habitat in question. In many studies of grassland or other low-growing vegetation, the percentage of ground obscured by grasses, forbs, and other vegetation is recorded. The percentage of the ground, water surface (cattails and other emergent plants), or other underlying surface (bottom of shallow wetland, etc.) obscured by plants as viewed from above is known as **coverage.** Daubenmire (1959) developed a method of estimating plant coverage using 20- by 50-centimeter sampling quadrats, often referred to as **Daubenmire quadrats** (Figure 13.10). The percentage of each quadrat covered by a certain plant species or group (such as forbs) is estimated and recorded in one of six coverage classes (0%–5%, 5%–25%, 25%–50%, 50%–75%, 75%–95%, and 95%–100%). For example, in an uncut alfalfa field in late spring, alfalfa would be likely to cover a high percentage of ground within a 20- by 50-centimeter quadrat and would probably be placed in the 75%–95% or 95%–100% coverage class. Frequency of occurrence, the percentage of samples in which a particular plant species or group occurs, can also be estimated using these quadrats. Other types of sampling quadrats used in a similar manner may be square or even round.

FIGURE 13.10 Plant coverage within this 20- by 50-centimeter Daubenmire quadrat is recorded in one of six coverage classes (0%–5%, 5%–25%, 25%–50%, 50%–75%, 75%–95%, and 95%–100%). The quadrat is color coded to allow more accurate estimation of coverage classes. The small accessory quadrat is used for measuring frequency of occurrence of grasses. *[Image courtesy of M. Rumble, U.S. Forest Service.]*

FIGURE 13.11 The line-intercept method works well for determining the coverage of plants such as this big sagebrush. *[Image courtesy of the Environmental Science and Research Foundation.]*

Another common technique used in assessing coverage and species composition (percentage of total coverage each plant species composes) of vegetation is the **line-intercept method** (Canfield 1941). In this method, a measuring tape is stretched for several meters between two stakes at a level above or near the top of the vegetation for which coverage is to be estimated. The percentage of the tape intercepting (passing over the top of or touching) a particular plant species represents the coverage for that species (Figure 13.11). If a biologist wished to estimate the coverage of big sagebrush in a high desert area in Idaho, for example, a 10-meter tape could be stretched between two stakes at a number of random sample sites. The biologist would compute the percentage of tape overlying or intercepting big sagebrush at each sample site [(centimeters overlying or intercepting big sagebrush/total centimeters in the 10-meter tape) × (100)],

FIGURE 13.12 Point frames are time-consuming to use, but they provide considerable accuracy in estimating coverage and species composition of grassland plants. *[Image courtesy of the Department of Animal and Range Sciences, North Dakota State University.]*

and the average of all sample sites would provide an indication of coverage of big sagebrush in the study area. The line-intercept method works best when applied to plant communities dominated by shrubs or bunchgrasses, as is characteristic of arid and semiarid lands (Higgins et al. 2005).

Point frames can be used to assess coverage of **herbaceous plants** at both the basal and canopy levels (Higgins et al. 2005). A point frame is usually a metal frame with guided holes into which metal rods (usually 10 or more) can be inserted. The frame is set above the vegetation, and the rods are lowered until they touch a stem or leaf (Figure 13.12). The first species of vegetation touched is recorded, as is bare ground if no plant is intercepted by the rod. The points where rods first touch vegetation are recorded as canopy data points; plants touched below the canopy at the soil interface are usually recorded as basal data points. Not only can a single rod touch more than one plant species, but it can also touch the same species more than once. The percentage of all rods first touching a particular species of vegetation provides the estimated canopy coverage for that species; for example, if 3 of 10 rods in a point frame first touched western wheatgrass when lowered, then the estimated canopy coverage for that grass would be 30% [(3/10) × (100)]. Actual estimates of coverage would be based on multiple point frame samples from the area being studied. Basal vegetation data points can be used in a similar manner to estimate basal coverage.

Biologists are often interested in the effectiveness of vegetative cover in concealing nesting birds, deer fawns, or other wildlife. This concealment value of vegetation can be measured by determining the degree to which a marked board or rod is obscured; marked boards or rods used for this purpose are called **visual-obstruction measurement devices (visual-obscurity measurement devices)**. One commonly used visual-obstruction measurement device is a board divided into 0.5-meter marked intervals. The board is usually 2.5 meters long, approximately 30 centimeters wide, and may be hinged in the middle so that it can be folded and carried in the field (Nudds 1977). This device is sometimes referred to as a Nudds board. The board is handheld or held in an upright position by stakes at a nest site, bedding site, or other area where a biologist wishes to obtain a quantitative measure of cover (Figure 13.13). An estimate of the percentage of each 0.5-meter section that is obscured from vision is recorded. Usually five categories, 0%–20%, 21%–40%, 41%–60%, 61%–80%, and 81%–100%, are used in estimating the coverage of each board section. The distance at which the board is viewed needs to be appropriate to the vegetation type to assure maximum variation in readings between sites. For example, if the board were viewed

FIGURE 13.13 A visual-obstruction board provides a measure of vegetation density. *[Image courtesy of the U.S. Fish and Wildlife Service.]*

FIGURE 13.14 Readings obtained from a Robel pole provide information on height and density of vegetation. *[Image courtesy of the U.S. Fish and Wildlife Service.]*

from 50 meters away at a site with very thick vegetation, 81%–100% of all the board sections probably would be obscured from view regardless of vegetation differences among sites. If the board were viewed from only 2 meters away at such a site, a high percentage of the board would be visible despite considerable differences among microhabitats. Nudds boards are most commonly viewed from 15 meters. Another method of measuring visual obstruction involves some type of background marked like a large checkerboard that is photographed to estimate the percentage obscured.

The **Robel pole**, another type of visual-obstruction measuring device, is a pole approximately 1.5 meters in height marked into 5-centimeter units that often alternate between light and dark coloration (Robel et al. 1970). The observer views the pole from a 4-meter distance and at a height of 1 meter and records the lowest 5-centimeter mark visible (Figure 13.14). The Robel pole is often read from four directions spaced 90° apart.

13.5 Terrestrial Vegetation: Forests

In forested areas, biologists can evaluate the understory for coverage, species composition, and visual obstruction using techniques similar to those discussed in the previous section. However, additional measures, such as **canopy cover, canopy volume, diameter at breast height, basal area**, tree density, and tree height, are used to assess forest habitat. Forest canopy cover, also called **canopy coverage** or **canopy closure**, is the percentage of the forest floor with tree cover directly overhead. Canopy volume is the space occupied by tree crowns (leaves and branches) per unit of area or per tree. Diameter at breast height, often referred to as DBH, is the diameter of the trunk (**bole**) of a tree at approximately 1.4 meters (approximately human breast height). Basal area is the cross-sectional area of standing tree trunks per hectare as measured at breast height.

Various types of gridded sighting tubes, such as the moosehorn densiometer, are used in estimating canopy cover in forested sites; the moosehorn densiometer provides more accurate but less precise estimates of forest canopy coverage (see section 9.1) than the spherical densiometer (Figure 13.15) (Cook et al. 1995). Estimates of canopy cover are useful to wildlife biologists in describing and predicting the potential wildlife values of timber stands. Areas with a high percentage (over 70%) of coniferous canopy coverage in cold regions are often important for wintering deer, elk, and a variety of smaller animals. Such areas provide thermal cover that reduces loss of radiant heat to the atmosphere by wintering animals. For Merriam's wild turkeys in coniferous forest habitat, areas with a high percentage of canopy coverage may provide sites where snow depth is reduced and pine seeds are available. Dense canopy coverage in the summer may provide cool areas for animals. Thus, canopy coverage has a moderating effect on temperature extremes and snow cover, especially in coniferous forest areas. Canopy coverage along streams reduces water temperature and siltation rates, and thus it may influence the suitability of a stream for species such as rainbow trout, eastern blacknose dace, or bridgelip suckers (Figure 13.16).

Diameter at breast height is measured using a DBH tape to obtain a direct reading of tree diameter by fitting the tape around the bole of the tree 1.4 meters above the ground; calipers can also be used to determine DBH. Diameter at breast height can be used to determine basal area. The basal area of trees per unit of area, along with tree density, is sometimes used by wildlife biologists to describe cover value in forested areas. Wild turkey roosting

A B

FIGURE 13.15 Canopy coverage in forests has commonly been estimated using a moosehorn densiometer (*A*), a type of gridded sighting tube, and the spherical densitometer (*B*). The type of instrument used can influence the accuracy of results (Cook et al. 1995). *[Part A courtesy of C. Lehman; part B courtesy of the U.S. Forest Service.]*

FIGURE 13.16 Canopy coverage over streams can strongly influence water temperature and thus the suitability of a stream for various fishes.

sites, for example, have commonly been described in terms of basal area of trees per hectare and tree density.

The **fixed-radius plot method** can be used for estimating tree density and basal area per hectare. A sample plot is described by a circle drawn at a measured distance from the plot center, and all trees occurring within the plot are counted. Diameter at breast height can be directly measured for all trees in the plot. Multiple sample plots can then be used to estimate the average number of trees and basal area per hectare by species and for all species combined. Fixed-radius plots provide accurate data, but this method is labor-intensive.

Basal area and density of tree species are often estimated using a **variable-radius method** that is less labor-intensive than the fixed-radius plot method. The variable-radius method is plotless and requires

FIGURE 13.17 A prism *(A)* or an angle gauge *(B)* can be used in a quick, plotless method for estimating the basal area of trees. *[Parts A and B courtesy of the U.S. Forest Service.]*

the use of a prism or angle gauge at random sampling sites in a forest to determine which trees will be included in the inventory. The observer stands at a sampling site and views the surrounding tree boles (at breast height) through the prism (Figure 13.17A) or angle gauge (Figure 13.17B). The observer counts the number of boles that appear wider than the view through the angle gauge or that are displaced to the right or left when viewed through the prism. The number of trees counted depends on the distance of the trees from the observer and the DBH of the trees. This method has also been modified for determination of canopy coverage of shrubs. The operating principles and basic mathematics behind this method are explained in Grosenbaugh (1952) and in other forest measurement texts (Bell and Dilworth 1993).

In areas where tree distribution is random, a plotless method known as the **point-centered-quarter method** is commonly used (Cottam and Curtis 1956). Sampling points can be randomly located or placed along transects. At each sampling point the distance to the nearest tree in each of four quarters is measured, and the DBH and species of the nearest tree are recorded for each quarter. Computations provide estimates of tree density per hectare by species as well as average and total basal areas. Complexities and problems with this method are discussed in greater detail in Higgins et al. (2005).

Forest composition [(number of trees of a particular species per hectare ÷ total number of trees per hectare) × 100] can be determined by recording tree species while estimating densities by plot or plotless methods. The line-intercept method can also be utilized to estimate forest composition. Forest composition is given in terms of numbers of trees, usually above some minimum size, rather than ground coverage, as was described earlier for grasses, forbs, and shrubs.

Tree heights can be accurately determined with a **clinometer** (Figure 13.18), an instrument with a scale inside the viewing area that is calibrated to read zero when the device is exactly horizontal. Each clinometer is scaled for use at a specific baseline distance from the tree to be measured, such as 20.1 or 30.5 meters. If, for example, the percentage scale reading was 69% for the top and 8% for the base, the height of the tree would be estimated by adding these two percentages, converting them to a decimal, and multiplying the result by the baseline distance (see Figure 13.18). The height in this example would be $0.77 \times 30.5 = 23.5$ meters. Tree height, along with height to the base of the crown and crown width, are used in estimating canopy volume.

Laser rangefinders are commonly used to accurately determine distance in field studies. Laser rangefinder–clinometer hybrids are available that automatically display tree height estimates after the user takes angle and distance readings by focusing on three sites: the tree top, the bole at approximately mid-height, and the trunk at the tree base. This same instrument can display slope and, depending on the model, will download directly onto a field datalogger or notebook computer.

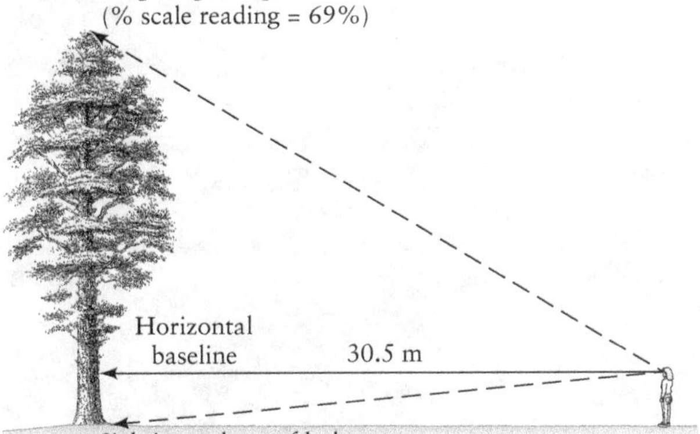

FIGURE 13.18 Biologists can use a clinometer to estimate tree height. In this example, the observer is using a clinometer scaled for a 30.5-meter baseline.

Tree height measurements could be used for such assessments as evaluating potential bald eagle or northern goshawk nesting or roosting sites. Many birds segregate horizontally in the vegetative community; that is, different species use different portions of the forest canopy and understory. Some bird species are highly specific to a particular horizontal stratum. Thus, measurements of tree height and the availability of horizontal strata could be of primary interest to a wildlife biologist.

Laser imaging detection and ranging systems or **laser altimetry** (**LIDAR**), a type of remote sensing, uses laser beams and returning scattered light to measure distances, movements, physical properties, and other characteristics of targets. Laser altimetry is becoming increasingly important in ecosystem studies, particularly in forest ecosystems (Lefsky et al. 2002). In studies of the earth surface, laser pulses are emitted and received (returning scattered light) by aircraft or satellites from downward-facing LIDAR instruments. Coupled with GPS capabilities, LIDAR technology can measure many characteristics of forest systems with considerable accuracy and can complement ground-level forestry measurements. Conventional remote sensing from high-altitude photography or from satellite imagery has yielded two-dimensional information of much value regarding the earth surface, vegetation, and other aspects of interest to ecologists. Laser altimetry is much more accurate than conventional remote sensing and also provides a depth or height perspective (three-dimensional information). Profiles of forest canopy height and cover can be rapidly and accurately generated using laser altimetry. Several studies have shown that accurate estimates of other forest characteristics, such as aboveground biomass, basal area, and foliage biomass, can be made using height and cover data from laser altimetry. In wildlife and fisheries habitat work, LIDAR appears to have considerable potential in providing data on forest canopies and forest ecosystems in general; it has also proven successful in determining biomass in ocean surface waters and may prove valuable in many other ecosystems and landscapes. Compared to high-altitude photography and satellite imagery, LIDAR is much more expensive per unit of area covered.

Anderson and Gutzwiller (2005) stress the importance of reviewing current literature before selecting methods for vegetation analysis in habitat studies. Recent advances may indicate a particular approach is more accurate than other approaches in estimating coverage, composition, canopy cover, or other vegetation characteristics in a particular plant community.

13.6 Habitat Relationships to Microclimate and Weather

In addition to sampling and assessing vegetation and other landscape features, biologists may want to evaluate the secondary effects of those features on microclimate. **Microclimate** is the climate on a very local scale and differs from the general climate of an area. The climate near the ground in tall-grass prairie, the climate in a snake denning area, and the climate under a fallen tree are all examples of microclimates. Some habitat features can moderate temperatures, wind speeds, precipitation, or even humidity. Wildlife may select habitats, at least in part, based on such microclimate characteristics.

Selection of a wintering microhabitat with reduced winds or higher temperatures, for example, allows an animal to maintain its body temperature with minimal use of additional energy to produce heat (see section 5.6). Dense stands of cattails in the northern plains of North America can reduce wind speed by as much as 95%; these areas provide protective cover that greatly reduces the wind chill factor for many wintering wildlife species. Cool temperatures and high winds can also place unusual stress on ducklings and goslings, particularly in the first week after hatch, when they are on bodies of water that lack protective emergent vegetation. Sitka black-tailed deer in Alaskan coastal areas seek old-growth forests for wintering, because their high canopy coverage and volume reduce snow accumulation on the forest floor (Kirchhoff and Schoen 1987). Warm rocks used by reptiles and wolf dens are both examples of microclimates influenced by habitat, primarily by abiotic components in these cases.

Animals may likewise seek microclimates associated with landscape features. The leeward side of a hill or of a forest patch, for example, can provide protection from wind. Reindeer (caribou) or American bison may select windswept hillsides to reduce problems with biting insects. Therefore, the secondary influence of landscape features on microclimate may need to be assessed as part of terrestrial habitat sampling and assessment.

Temperatures in terrestrial environments can be measured directly using outdoor thermometers, minimum-maximum thermometers, or continuous recording devices, or they can be obtained from local weather station records. **Anemometers** or other wind-measuring devices can provide information on wind speed. Wind speeds can also be continuously or periodically recorded with electronic equipment and later downloaded to a computer. Biologists often obtain precipitation information from local weather stations. However, precipitation gauges can be placed in study areas at little cost. Snow accumulation can be measured at various sites using poles with length markings.

In some cases, annual variations in precipitation or other weather patterns have an overriding influence on habitat. In such cases, measurements of weather variables may provide a low-cost means of predicting habitat conditions and thus survival or reproductive rates for certain wildlife populations. High reproductive success in Gambel's quail, California quail, and other quail species in the Southwest, for example, is associated with above-normal winter-spring rainfall, high soil moisture, and a resultant increase in new-growth vegetation (Figure 13.19). Conversely, unusually dry conditions can lead to extremely low reproductive success in these species. Therefore, biologists can predict habitat conditions and reproductive success in these quail species based on rainfall and soil moisture conditions. Snow goose reproductive success in the Arctic can be predicted based on satellite photographs of snow cover taken during May, when nesting begins. If snow cover is extensive in late May and early June, many pairs may forgo nesting until the following year.

FIGURE 13.19 In southwestern North America, Gambel's quail and other desert quail species experience high reproductive success in years with above-normal winter-spring precipitation and the resultant increase in new-growth herbaceous vegetation. *[Image courtesy of L. Karney, U.S. Fish and Wildlife Service.]*

13.7 Herbage, Browse, Mast, and Other Plant Foods for Wildlife

Biologists may wish to obtain information on the availability of **herbage** (grasses and forbs), **browse** (leaves, shoots, and stems of shrubs and trees), **mast**, and other plant foods utilized by wildlife species. Mast includes various fruits of trees and shrubs, especially nuts such as acorns and fleshy fruits such as blackberries and wild plums.

To measure herbage biomass, plant material can be clipped from small sample plots or quadrats in grass-forb habitat and weighed. Biomass of grasses and forbs can also be estimated using an electronic device called a **radiometer** that measures reflected light from the vegetation. The radiometer approach must be applied during midday on sunny days to equalize light conditions.

Comparitive information from cutting, clipping, and weighing herbage for the local area being evaluated is required to calibrate radiometer results with actual vegetation biomass (Higgins et al. 2005).

Dry-weight data from clipped sample plots in different areas can be compared to assess food availability and relative productivity of potential food plants. For example, herbage biomass at forest sites where ponderosa pines have been thinned to a low density and low basal area can be compared with that at denser and higher-basal-area sites to determine the influence of logging practices on herbage production. Likewise, sites protected from domesticated and wild ungulate feeding by exclosures can provide information on the combined effects on vegetation of **grazing** (feeding on herbage) and **browsing** (feeding on browse) (Figure 13.20). Mast can be sampled on trees by direct observation and counts, passively by the use of various traps to collect falling mast, or by means of sample plots located under trees.

In addition to availability information, biologists are likely to need information on the nutritional value of herbage, browse, mast, and other plant foods (Servello et al. 2005). Measurements of the crude protein, digestible protein, and digestible energy content of plant foods may prove valuable to a wildlife biologist. It should be remembered that digestible protein and digestible energy must be determined in reference to different animal groups, such as seed-eating birds or ruminants. Crude protein measurements are usually directly related to the value of plant foods to animals. However, if plants contain oils or other **secondary compounds**, such as tannins, that inhibit digestion, they can be relatively high in crude protein and still be of little value to a particular animal species. Secondary compounds are anti-herbivory types of chemicals and do not fill a primary role in plant growth and development. Digestible protein and digestible energy provide more accurate estimates of actual nutritional value of plant food sources but involve more expensive laboratory analyses than does the measurement of crude protein. The moisture content of vegetation is also important, because higher moisture content is usually associated with increased digestibility.

Biologists are interested in knowing which plants provide the highest digestible protein and digestible energy levels for wild animals; they are also interested in how these nutritional levels change with time of year and plant growth stage. Migrations of

A B

FIGURE 13.20 A comparison of vegetation inside these exclosures with that outside primarily reflects *(A)* grazing by brant near the Arctic coast in Alaska and *(B)* the influence of browsing by wintering mule deer and domesticated livestock in California. *[Part A courtesy of K. Jensen.]*

large ungulates and other animals may be directly related to the availability of high-protein forage in a particular area or habitat. The spring migration of elk to higher altitudes in the Rocky Mountains, for example, is related to the high-protein forage available at these higher altitudes. Soils usually have a strong influence on the nutrient levels and growth rates of various plant foods.

Habitats with plentiful, high-quality forage or other plant foods are able to support a more abundant and diverse wildlife assemblage than are habitats that have similar potential but are in poorer condition. For example, if a winter range for bighorn sheep is overgrazed and overbrowsed by domesticated livestock, its value to bighorn sheep, as well as to other wintering wildlife and domesticated livestock, is reduced. In such a situation, biologists would probably find reductions in the quantity and quality (digestible protein and digestible energy) of the herbage and browse compared to those of sites of similar potential that had not been mismanaged.

13.8 Prey Populations

In fisheries, prey animals such as invertebrates, frogs, or small fishes are considered to be a portion of the biota; prey is not considered to be a part of the habitat (see section 1.2). While the same is generally true in wildlife, some wildlife biologists include prey animals as a portion of the habitat. If a biologist were describing or evaluating brood habitat for wood ducks, for example, information on invertebrate populations, the primary food (prey) source for the young birds, would be critical. Even for more typical predators, such as kit foxes or diamond-backed watersnakes, an assessment of small mammals, fishes, amphibians, and other potential prey could be included in habitat assessment.

Prey availability is an important factor, whether or not it is considered to be a part of or separate from habitat. Techniques applicable to the sampling and assessment of prey populations are presented in Chapters 7 and 9.

13.9 Landscape-Level Assessment

The U.S. Fish and Wildlife Service has developed a series of **habitat suitability index models** for both fish and wildlife species (Anderson and Gutzwiller 2005). These models, generally referred to as HSI models, are based on suitability indices formulated from variables that affect the life cycle and survival rates of each species. Information from literature sources on species–habitat relationships has been consolidated to create these indices. The models are valuable to biologists both as a habitat assessment tool and for predicting the likely effects of human alterations of habitat.

Assume, for example, that an urban fisheries biologist is directing a project to improve a previously polluted stream within the city of Richmond, Virginia. The biologist is interested in determining whether the stream habitat is now appropriate for smallmouth bass, a fish species native to that area. Consulting the HSI model for smallmouth bass (Edwards et al. 1983) would allow the biologist to determine whether the stream was now suitable for that species. The habitat variables in the model include substrate type, percentage of pools, percentage of cover (stumps, boulders, and crevices), depth of pools, pH, dissolved-oxygen levels, turbidity, temperature, water level fluctuations, and gradient. These habitat variables are measured and used in a mathematical equation that will indicate the likely population abundance of smallmouth bass. The biologist must also be aware that the status of each of these stream variables is linked to the surrounding landscape. Habitat variables associated with lake-spawning smallmouth bass would, of course, differ from those for stream-spawning smallmouth bass. The HSI developed for spawning male smallmouth bass in lakes is based on depth, slope, and substrate (Clark et al. 1998).

Habitat evaluation for many species can be enhanced by or may even require a landscape-level approach (see section 2.11). For example, survival and reproduction of forest species such as the worm-eating warbler and pileated woodpecker are negatively affected by fragmentation of forest habitat, a landscape characteristic. Extensive roadless areas and intermixed meadows are important components of elk habitat. Most studies of wildlife habitat can be greatly enhanced with information on habitat characteristics at a landscape level.

There are several methods of evaluating habitat features at a landscape scale. Broad-scale habitat patterns can be mapped with satellite photography and infrared imagery. Satellite imagery provides broad-scale information on landscapes and increasingly accurate information on small units of land that may have vegetation types and other features of interest to the biologist. However, identification of microhabitats needed by animal species is often beyond the resolution available from satellite imagery. Information from the landscape-level maps produced by these methods can be digitized and entered on computers using appropriate software. Computers and landscape-level GIS provide the capacity for quick evaluations of habitat patches, corridors, wetlands, and other habitat features in terms of area and perimeter (Figure 13.21). Satellite photography and GIS technology also allow for the updating of habitat maps through time as well as the tracking of changes over time. Laser altimetry (see section 13.5) along with GIS mapping offers a three-dimensional approach to remote sensing and landscape analysis, but it is much more expensive than the use of high-altitude photography and infrared imagery.

Measures of the horizontal diversity of habitats, such as edge (see section 2.11), interspersion, and juxtaposition, can also be determined using GIS. **Interspersion** is a measure of the intermixing of different habitat types. For example, a full section of land (1.6 by 1.6 kilometers) consisting of two large crop fields, corn and soybeans, with no noncropland between the fields has very low interspersion. An area

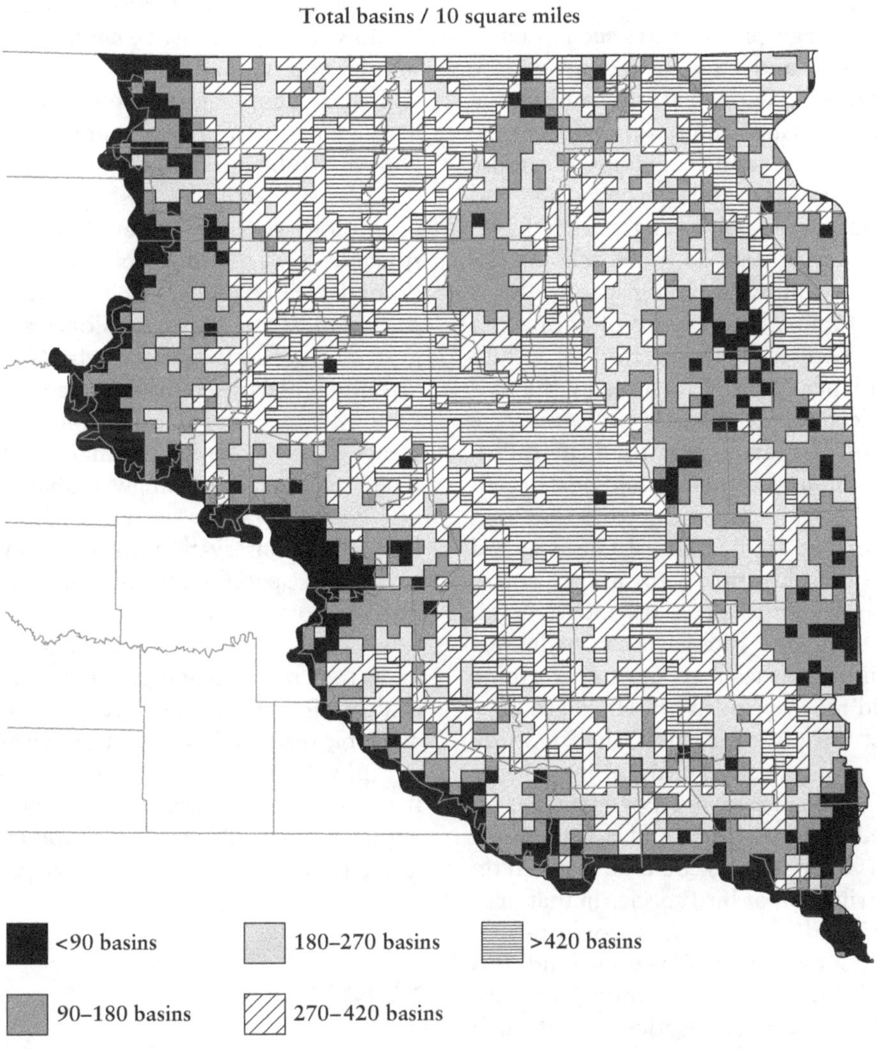

FIGURE 13.21 This GIS map codes numbers of wetland basins per unit of area (26 km^2) in eastern South Dakota to provide a landscape-level view of wetland distribution. *[Image courtesy of National Wetlands Inventory.]*

of the same size containing a small willow- and cattail-lined stream, a patch of native prairie, two shelterbelts, three ponds, a fenceline with shrub and tall-grass cover, and separate fields of sorghum, soybeans, and corn has a much higher interspersion. **Juxtaposition** is the proximity of habitats needed by an organism. For example, a common kingsnake may require sunning sites, winter denning sites, sites for depositing eggs, sites for hunting rodents, and several other habitat types within a certain area. If these needed habitats are not available within a reasonable distance of each other, the area may support no or few common kingsnakes. Techniques for quantifying interspersion and juxtaposition are described in Anderson and Gutzwiller (2005).

Information on geographic features such as habitat types, soils, roads, and land ownership can be stored in GIS databases. These features can then be viewed simultaneously (superimposed) and their relationships explored. For example, a biologist can use GIS to superimpose roads, habitat types, soils, and other characteristics on the home ranges and telemetry locations for species such as mule deer or red foxes. This capability allows a biologist to evaluate the home ranges of animals or groups of animals in relation to habitat features. Thus, selection of certain habitats and avoidance of other features by the animals could be detected by the biologist. Avoidance (or lack of avoidance) of areas near roads by radio-collared greater kudu, for example, could be detected using GIS.

Another important application of GIS is called **gap analysis**, because it identifies gaps in the protection of species and species diversity (Scott et al. 1993). Using this technique, biologists can evaluate a large area in terms of species diversity to determine which sites should receive priority in efforts to conserve the greatest variety of species. Such a study could involve one group, such as birds, or it could involve a broad array of organisms. For instance, an area of intermixed national forest and private lands containing various habitat types could be mapped on GIS; then, the distribution of bird species in that area could be superimposed. Certain areas would probably show the greatest overlap in bird species and thus might be considered unusually important for maintaining bird species diversity, particularly if such habitats were limited in area and subject to potential alteration or elimination. By superimposing land ownership patterns, biologists could then determine which or how many of these important areas were currently in public ownership. Such areas might be identified for protection from activities that could destroy their value to birds; for example, they might be designated for no logging or for some other specified logging treatment that would maintain their particular habitat.

In this manner, biologists can identify gaps in species or species diversity protection within existing habitat and land ownership patterns. If an area contains endangered or threatened species, these gaps in protection may receive special attention. Because gap analysis may identify privately owned areas in need of long-term protection from development, its possible implications in terms of public relations and landowner rights must be carefully considered. Communication with all parties involved is critical if special easements, purchases, or other long-term arrangements to provide habitat protection in these gaps are to be attempted.

Natural resource conservation benefits from GIS technology at local, regional, national, and international levels. The United Nations Environmental Program (UNEP) has developed an expanding network of data centers, such as the Global and Regional Integrated Data (GRID) program (O'Neil et al. 2005). Environmental conservation data on aspects such as biology, human impacts, hydrology, soils, climate, and ecology are being made available at GRID centers on a regional and, increasingly, global scale. Involvement of UNEP also occurs through other initiatives, such as the World Conservation Monitoring Center (WCMC), by making spatially related conservation information available on aspects such as species diversity, tropical rain forests, and other protected terrestrial and marine areas on a regional and global basis. Natural resource information from GIS technology is increasingly available through Internet sites such as those of the Geography Network and the National Biological Information Infrastructure. A more detailed review of the application of GIS spatial technologies is presented in O'Neil et al. (2005).

SUMMARY

Habitat is often considered to be the most important overall ingredient in conserving and managing populations of wild organisms. Biologists use a variety of techniques in sampling and assessing aquatic habitats. Water characteristics such as temperature, dissolved-oxygen content, pH, hardness, and nutrient levels influence the growth and survival of aquatic organisms, and measurement of these characteristics provides important information to biologists. Aquatic habitats can also be characterized by size, depth, volume, and shape. Maps depicting the shapes and depth contours of water bodies are often available from natural resource agencies; they can also be created using various mapping techniques.

Aquatic vegetation is a major habitat component for fishes and aquatic invertebrates as well as for many aquatic-oriented terrestrial species. Biologists can estimate submergent and emergent vegetation abundance using transects, quadrats, or even aerial photography. Long-term trends in aquatic vegetation can be an indicator of water quality changes. Abiotic substrates in water bodies are categorized on the basis of particle size. Live or dead trees and other woody vegetation in water bodies have value to both aquatic and terrestrial organisms.

In terrestrial ecosystems, much habitat analysis involves the sampling and assessment of vegetation. Coverage and species composition of vegetation is often estimated using quadrats or other types of sample plots. Plant coverage and composition can also be estimated with the line-intercept method or with a point frame. Vegetation density viewed horizontally can be measured with various visual-obstruction measurement devices.

In forested habitats, additional characteristics such as canopy cover, canopy volume, diameter at breast height, basal area, and tree height are often assessed. Gridded sighting tubes and the spherical densiometer have commonly been used to determine canopy cover. Tree density and basal area can be sampled in circular sample plots with fixed boundaries. Variable-radius methods requiring the use of a prism or angle gauge can also be used to estimate tree densities and basal areas in forested habitats; variable-radius methods are less labor-intensive than fixed-radius plots. The point-centered-quarter method is another plotless method commonly used to estimate tree density. Laser altimetry (LIDAR) from airplanes or satellites is becoming increasingly important in rapid determination of forest characteristics such as canopy coverage, biomass, and height.

Vegetation, rocks, and other habitat features can have a strong influence on microclimate characteristics such as wind speed, temperature, and humidity. Because of the importance of microclimates to wild animals, biologists may want to assess the influence of vegetation and landscape features on microclimates. Weather conditions in a region can sometimes be a good predictor of broad-scale habitat conditions.

Plant foods such as herbage, browse, and mast are often assessed as a measure of habitat condition. A variety of sampling methods, from traps that collect falling mast to clipping herbage, are used by biologists. Biologists may monitor the nutritional values of vegetation by assessing crude protein, digestible protein, and digestible energy.

Prey populations are not considered to be a part of the habitat component by fisheries biologists. In contrast, many wildlife biologists consider prey to be part of both the habitat and biota components of a wildlife system.

Landscape-level evaluation of habitat characteristics is often accomplished using satellite imagery and GIS technology and is important to conservation efforts at local, regional, national, and international levels.

PRACTICE QUESTIONS

1. Assume that a fisheries biologist is collecting monthly information on temperature, dissolved-oxygen concentration, and pH in a lake. For each characteristic, list two possible reasons for collecting this information.

2. Describe the use of a Secchi disk.

3. Why might information on water volume in a water body be of value to a fisheries biologist?

4. What is shoreline development and why is it an important characteristic of water bodies?

5. List three reasons why the assessment of aquatic plant abundance may be important.

6. Why would a biologist need information on substrate composition of a water body? Provide four reasons.

7. Select a method used to measure coverage of herbaceous plants or shrubs and describe the use of the method.

8. Describe the use of a visual-obstruction measurement device. Describe two situations in which such a tool might provide important information to a biologist.

9. Define the following terms: *canopy coverage, diameter at breast height*, and *basal area*.

10. What are two techniques, methods, or instruments that can be used in estimating forest characteristics, and what information does each provide?

11. What is microclimate and why is it important in animal habitat selection?

12. Name three measures of the nutritional quality of herbage, browse, and mast for wild animals.

13. What is a geographic information system (GIS)? Provide two examples of ways in which GIS could be used by a biologist.

14. What is gap analysis?

15. Define *interspersion* and relate it to current farming practices, in which field sizes are increasing.

SELECTED READINGS

Anderson, S. H., and K. J. Gutzwiller. 2005. Wildlife habitat evaluation. Pages 489–502 *in* C. E. Braun, ed. *Techniques for wildlife investigation and management.* 6th ed. The Wildlife Society, Bethesda, Md.

Bisson, P. A., J. L. Nielsen, R. A. Palmason, and L. E. Grove. 1982. A system of naming habitat types in small streams, with examples of habitat utilization by salmonids during low streamflow. Pages 62–73 *in* N. B. Armantrout, ed. *Acquisition and utilization of aquatic habitat inventory information.* Western Division, American Fisheries Society, Bethesda, Md.

Clesceri, L. S., A. E. Greenberg, and R. R. Trussell, eds. 1989. *Standard methods for the examination of water and wastewater.* 17th ed. American Public Health Association, Washington, D.C.

Higgins, K. F., K. J. Jenkins, G. K. Clambey, D. W. Uresk, D. E. Naugle, J. E. Norland, and W. T. Barker. 2005. Vegetation sampling and measurement. Pages 524–553 *in* C. E. Braun, ed. *Techniques for wildlife investigation and management.* 6th ed. The Wildlife Society, Bethesda, Md.

O'Neil, T. A., P. Bettinger, B. G. Marcot, B. W. Luscombe, G. T. Koeln, H. J. Bruner, C. Barrett, J. A. Pollock, and S. Bernatas. 2005. Application of spatial technologies in wildlife biology. Pages 418–447 *in* C. E. Braun, ed. *Techniques for wildlife investigation and management.* 6th ed. The Wildlife Society, Bethesda, Md.

Orth, D. J. 1983. Aquatic habitat measurements. Pages 61–84 *in* L. A. Nielsen and D. L. Johnson, eds. *Fisheries techniques.* American Fisheries Society, Bethesda, Md.

LITERATURE CITED

Anderson, S. H., and K. J. Gutzwiller. 2005. Habitat evaluation methods. Pages 489–502 *in* C. E. Braun, ed. *Techniques for wildlife investigation and management.* 6th ed. The Wildlife Society, Bethesda, Md.

Bell, J. F., and J. R. Dilworth. 1993. *Log scaling and timber cruising.* Oregon State University Book Stores, Inc., Corvallis.

Canfield, R. H. 1941. Application of the line interception method in sampling range vegetation. *Journal of Forestry* 39:388–394.

Clark, M. E., K. A. Rose, J. A. Chandler, T. J. Richter, D. J. Orth, and W. Van Winkle. 1998. Simulating smallmouth bass reproductive success in reservoirs subject to water level fluctuations. *Environmental Biology of Fishes* 51:161–174.

Cook, J. G., T. W. Stutzman, C. W. Bowers, K. A. Brenner, and L. L. Irwin. 1995. Spherical densiometers produce biased estimates of forest canopy cover. *Wildlife Society Bulletin* 23:711–717.

Cottam, G., and J. T. Curtis. 1956. The use of distance measures in phytosociological sampling. *Ecology* 37:451–460.

Daubenmire, R. 1959. A canopy-coverage method of vegetational analysis. *Northwest Science* 33:43–64.

Edwards, E. A., G. Gebhart, and O. E. Maughan. 1983. *Habitat suitability information: Smallmouth bass.* FWS/OBS-82/10.36. U.S. Fish and Wildlife Service, Washington, D.C.

Fisher, W. L., and F. J. Rahel, eds. 2004. *Geographic information systems in fisheries.* American Fisheries Society, Bethesda, Md.

Grosenbaugh, L. R. 1952. Plotless timber estimates: New, fast, easy. *Journal of Forestry* 50:32–37.

Heard, W. R. 1991. Life history of pink salmon (*Oncorhynchus gorbuscha*). Pages 119–230 *in* C. Groot and L. Margolis, eds. *Pacific salmon life histories.* University of British Columbia Press, Vancouver, Canada.

Higgins, K. F., K. J. Jenkins, G. K. Clambey, D. W. Uresk, D. E. Naugle, J. E. Norland, and W. T. Barker. 2005.

Vegetation sampling and measurement. Pages 524–553 *in* C. E. Braun, ed. *Techniques for wildlife investigation and management.* 6th ed. The Wildlife Society, Bethesda, Md.

Kirchhoff, M. D., and J. W. Schoen. 1987. Forest cover and snow: Implications for deer habitat in southeast Alaska. *The Journal of Wildlife Management* 51:28–33.

Lefsky, M. A., W. B. Cohen, G. G. Parker, and D. J. Harding. 2002. Lidar remote sensing for ecosystem studies. *Bioscience* 52:19–30.

Meyer, F. P., and R. L. Herman. 1990. Interpreting the scene. Pages 10–18 *in* F. P. Meyer and L. A. Barclay, eds. *Field manual for the investigation of fish kills.* Resource Publication 177. U.S. Department of the Interior, Fish and Wildlife Service, Washington, D.C.

Nudds, T. D. 1977. Quantifying the vegetative structure of wildlife cover. *Wildlife Society Bulletin* 5:113–117.

O'Neil, T. A., P. Bettinger, B. G. Marcot, B. W. Luscombe, G. T. Koeln, H. J. Bruner, C. Barrett, J. A. Pollock, and S. Bernatas. 2005. Application of spatial technologies in wildlife biology. Pages 418–447 *in* C. E. Braun, ed. *Techniques for wildlife investigation and management.* 6th ed. The Wildlife Society, Bethesda, Md.

Paukert, C. P., and J. M. Long. 2004. Geographic information systems applications in reservoir fisheries. Pages 85–111 in W. L. Fisher and F. J. Rahel, eds. *Geographic information systems in fisheries.* American Fisheries Society, Bethesda, Md.

Phillips, R. W., R. L. Lantz, E. W. Claire, and J. R. Moring. 1975. Some effects of gravel mixtures on emergence of coho salmon and steelhead trout fry. *Transactions of the American Fisheries Society* 104:461–466.

Robel, R. J., J. N. Briggs, A. D. Dayton, and L. C. Hulbert. 1970. Relationships between visual obstruction measurements and weight of grassland vegetation. *Journal of Range Management* 23:295–297.

Scott, J. M., F. Davis, B. Csuti, R. Noss, B. Butterfield, C. Groves, H. Anderson, S. Caicco, F. D'erchia, T. C. Edwards Jr., J. Ulliman, and R. G. Wright. 1993. GAP analysis: A geographic approach to protection of biological diversity. *Wildlife Monographs* 123.

Scott, W. B., and E. J. Crossman. 1973. *Freshwater fishes of Canada.* Bulletin 184. Fisheries Research Board of Canada, Ottawa.

Servello, F. A., E. C. Hellgren, and S. R. McWilliams. 2005. Techniques for wildlife nutritional ecology. Pages 554–590 *in* C. E. Braun, ed. *Techniques for wildlife investigation and management.* 6th ed. The Wildlife Society, Bethesda, Md.

CHAPTER 14

Habitat Degradation

Habitat can be altered by numerous natural and artificial means. It can be altered on a small scale that could be measured in square meters or hectares, or it can be altered on larger scales at the ecoregion, landscape, or biosphere level. Human-influenced alterations of habitat, both intentional and unintentional, can have far-reaching effects. Habitat degradation or destruction is often not as simple to define as it may appear; as in art and many other endeavors, beauty is in the eye of the beholder. The plowing of a section of grassland and its replacement with a wheat field would be viewed as habitat degradation by a biologist interested in grassland organisms, but this would not be viewed as habitat degradation by a wheat farmer. The construction of an impoundment would probably not be viewed as habitat degradation by a water skier, but it would be by someone who had enjoyed canoeing that river stretch before its impoundment. The conversion of a natural area into a disc golf course would be viewed differently by various users. However, in some cases degradation is obvious—few can find any major benefits from global climate change.

Sometimes multiple changes in habitat occur over time. A forest habitat could be changed into an agricultural crop field, eventually becoming an abandoned field of forbs, grasses, and shrubs, and then be turned into a parking lot. With each change, different people would view the resulting habitat differently.

Some habitat degradation or destruction is purposeful. For example, when humans disturb Arctic tundra or marine habitat in the Gulf of Mexico for petroleum production, we know that these areas will be degraded. There may be discussions concerning how much and what kind of damage will occur, but the activities are carried out with full knowledge that some level of degradation is going to result. Some habitat degradation or destruction is not done purposefully. For example, increased acidity resulting from the combustion of fossil fuels was not a result of human actions taken with the knowledge that this combustion would destroy some habitats; degradation was an unforeseen by-product of human activity.

Habitat degradation changes the biotic community that is present. When habitat changes, one type of habitat is destroyed but another is created. Thus, some organisms are harmed, while others benefit. By the 1960s, Lake Erie had been degraded by humans to a point at which the native lake biota had been greatly modified. Years of municipal, agricultural, and industrial pollution had resulted in dramatic shifts in the plant and animal communities of the lake. Damage to the native biota was so severe that many people pronounced Lake Erie dead. The lake was not really dead; in fact, it probably had more life than ever, just not the flora and fauna present before degradation. Efforts to reduce pollution were successful, and by the late 1980s the lake had recovered sufficiently to allow some of the native biota to rebound. Once again, Lake Erie is alive. The lake now faces another human-induced threat in the form of a variety of invasive species. Only time will tell what effects these introduced species will have on the native biota of the lake.

Degradation of habitat is not a process accomplished only by humans. When an American beaver dams a stream to ensure itself a home, when a tree produces chemicals that inhibit the growth of nearby plants, or when a white-tailed prairie dog clips grass around its burrow so that it will be less vulnerable to predation, those organisms are modifying habitat to their benefit. Each modification degrades that habitat for some species and improves it for others. It is impossible for organisms, humans included, to exist without in some way changing or degrading the habitat in which they live. However, humans are different in that they can make conscious choices. Even if the final choice is to degrade an area, humans often have some idea of the effects they will have. The degree to which destruction or degradation occurs can often be reduced or corrected with proper management. There is a field of study called **restoration ecology** that is directed toward proper management; it involves the study of recuperating degraded, damaged, or destroyed ecosystems through active human intervention.

This chapter describes some of the ways in which habitats are degraded or destroyed by humans. Habitat degradation or habitat destruction will be interpreted in this text as human-induced changes that most biologists consider to be negative in relation to native or desirable plant and animal populations and communities.

14.1 Air Pollution

Air pollution in its various forms has many implications with regard to habitat degradation. While many facets of air pollution could be discussed, three particular problems will be highlighted: global climate change, acidic deposition, and ozone depletion.

Global climate change is probably the result of human actions and has important implications for fisheries and wildlife resources. "Global climate change could have a greater potential to change life on the planet than anything else we could do outside of nuclear war" (Cunningham and Saigo 2001). Increases in **greenhouse gases** that result from human activities are thought to be the cause of global climate change. Greenhouse gases include carbon dioxide (CO_2), water vapor, ozone, methane, nitrous oxide, and chlorofluorocarbons, with CO_2 being the most important. These gases, when released into the atmosphere, may act somewhat like the glass in a greenhouse. The glass lets in light from the sun, but it keeps some of the resulting infrared radiation (heat) from escaping; the resulting heat buildup raises the ambient temperature in the greenhouse. Human activities, such as fossil fuel combustion, release large amounts of CO_2 into the atmosphere (see section 2.2). Deforestation, especially in tropical rain forest areas of the world, also contributes to the greenhouse effect, because when vegetation is removed, plants that would use CO_2 are eliminated.

World climate change is not new; climatic conditions on the earth have changed often throughout geologic time. These past changes, however, usually occurred at a relatively slow pace. Gradual climatic modifications that affect ecosystems and their biota allow time for many organisms to either adapt or move. Rapid, human-induced climatic change may not allow the time necessary for organisms to adapt to the new conditions or disperse to areas with more appropriate climates. Habitat fragmentation (see section 14.7) adds to this problem by restricting the movements of some animals.

There seems to be sufficient evidence that global climate change is occurring. There also seems to be sufficient evidence that it is the result of human activities, particularly fossil fuel use and deforestation. This issue has attracted much political rhetoric, primarily because of potential short-term economic effects caused by the need to change some of our energy-use and land-use practices. The rhetoric usually revolves around atmospheric scientists not having proven that human activities are the cause of this climate change. However, recall from section 1.6 that the scientific method can only reject or not reject hypotheses; it cannot prove anything.

The deleterious effects of global climate change far exceed any small benefits that humans would accrue from this change. It would not eliminate life on the planet, it would just greatly change it. A few of the major changes include the following: many plant and animal extinctions; rising sea levels that would inundate excellent crop production areas and coastal cities; increases in estuary salinities; displacement of agricultural production; increases in tropical diseases in northern areas; increases in severe storm activity; greatly altered weather patterns; and saltwater intrusion into freshwater aquifers. In addition, the negative effects of world CO_2 increases in oceans and resultant marine acidification problems are just starting to be understood. Changes will be on a global scale and will affect all locations on the planet and most of its inhabitants.

Acidic deposition, which is often inappropriately called **acid rain** or **acid precipitation**, results primarily from fossil fuel combustion. When rain, snow, fog, dew, or solid airborne particles contain high levels of acidity, acidic deposition occurs. Rain is naturally somewhat acidic; however, human activities can increase its acidity. As fossil fuels such as coal and oil are burned, sulfur dioxide and nitrogen oxides are released into the atmosphere. These react with water and oxygen to produce compounds of low pH, primarily sulfuric acid and nitric acid. When deposited on land or in water, these acids are harmful to many organisms.

The level of acidity or basicity is usually expressed as pH, a measure of hydrogen ion concentration (Figure 14.1) (see section 13.1). A neutral solution has a pH of 7. If a solution has a pH above 7, it is basic, or alkaline, and if it has a pH below 7, it is acidic. Each whole number on the pH scale from 1 to 14 represents a tenfold difference in hydrogen ion concentration. For example, a solution with a pH of 6 is ten times more acidic than a pH 7 solution, and a pH 4 solution is one hundred times more acidic than a pH 6 solution.

The environments most immediately affected by acidic deposition are those with low alkalinity levels. For example, an area with soil pH of 7.2 would be more susceptible to acidic deposition problems than an

	pH	Substance
	14	← NaOH solution (lye), 4%
		← Oven cleaner
	13	← Limewater
		← Hair remover
	12	← Household ammonia
		← Washing soda (about 1%)
	11	
		← Soap solutions
↑	10	
	9	← Baking soda
Basic	8	← Bicarbonate of soda (about 1%)
		← Blood
Neutral	7	← Pure water
		← Cow's milk
Acidic	6	
		← Unpolluted rainwater
	5	← Squash, pumpkin
↓	4	← Tomatoes
		← Oranges
	3	← Vinegar; soft drinks
	2	← Limes
		← Normal stomach acidity (pH 1.0-3.0)
	1	← Acid stomach

FIGURE 14.1 The pH scale. Unpolluted rainwater is already acidic, but acidic deposition events generally involve much lower pH values. The pH scale runs from 1 to 14, with 7.0 being neutral, below 7 being acidic, and above 7 being alkaline.

area with soil pH of 8.1. The more basic an environment, the more hydroxide ions it has; these hydroxide ions neutralize hydrogen ions. **Buffering capacity** is the ability of water or soil to react with and neutralize acids. Areas with good buffering capacity are not immune from acidic deposition, but they are less likely to exhibit immediate negative effects.

With respect to habitat of wild organisms, the effects of acidic deposition can be manifested in numerous ways. By harming particular plants and animals, acidic deposition can substantially alter habitats and affect organisms dependent on those habitats. In aquatic systems, acidic deposition can result in mortality among fishes, aquatic plants, and a variety of other aquatic organisms (Chiras and Reganold 2005). Acidic deposition may also occur at levels that do not kill organisms but instead reduce their ability to reproduce or to survive early developmental stages. Adult brook trout, for example, can survive acidity levels that their eggs and young cannot tolerate. Because aquatic invertebrate numbers are reduced by acidic deposition, all organisms, including fishes, birds, and mammals that depend on aquatic invertebrates as a food source, are negatively affected. In addition, all organisms dependent on these fishes, birds, and mammals are negatively affected. Acidic deposition can also affect the cycling of elements such as nitrogen or act to leach elements such as aluminum, lead, mercury, and cadmium from soil; these can be toxic to aquatic organisms when moved into water bodies.

Acidic deposition also has the capacity to affect terrestrial ecosystems. One well-established effect is that of acidic deposition on soil chemistry and the concomitant effects on terrestrial plant roots (Raven et al. 1993). The development of roots, as well as their uptake of nutrients and water from the soil, are affected. Essential minerals, such as calcium and magnesium, readily wash out of acidic soils, while heavy metals, such as manganese and aluminum, dissolve in acidic soils and become available for absorption by roots in toxic amounts.

The above are not the only negative effects of acidic deposition on terrestrial habitats and ecosystems. It inhibits nitrogen-fixing bacteria and has numerous other effects, such as reducing seed germination, which results in reduced growth or death of agricultural crops and forests. In some areas of North America and Europe, entire forests have perished because of acidic deposition.

To date, parts of the northeastern United States and southeastern Canada have been the areas most severely affected by acidic deposition in North America (Figure 14.2). Large areas of land in other parts of the world, especially in Scandinavia, have also been negatively affected by acidic deposition, and still other areas throughout the world are at risk. There is little question that acidic deposition has had and will continue to have substantial negative effects on aquatic and terrestrial resources. In addition to its negative effects on fisheries and wildlife resources, economic costs resulting from the destruction of timber, decreased agricultural productivity, damage to human structures, and damage to domesticated livestock are large.

New technology has helped reduce emissions that lead to acidic deposition, and many people think that this problem has essentially been solved. This is incorrect. The effects of acidic deposition are worse than originally thought, and our efforts to control this problem have proven inadequate (Withgott and Brennan 2007).

Another result of air pollution that has important implications for fisheries and wildlife resources is the depletion of the atmospheric ozone layer that protects the earth from ultraviolet radiation. Ozone is a form of pollution at low altitudes, but it is a very beneficial gas at higher altitudes. Most of the potential negative effects of ozone depletion at higher altitudes are usually mentioned in relation to human problems, such as skin cancer, eye cataracts, severe sunburn, suppression of the immune system, or reductions in agricultural production. However, ozone depletion affects many nonhuman organisms as well. Ozone depletion can result in negative occurrences such as retardation of the growth of plants, the death of larval organisms such as shrimp and crabs, and the depletion of plankton. Think back to the discussion in section 2.3 of trophic levels, food webs, and pyramids of energy and biomass. A reduction in plankton would reduce the bases of aquatic food webs and therefore negatively affect all higher trophic levels, including humans.

Ozone depletion has been associated with the use of chlorofluorocarbons (CFCs). Natural processes, such as volcanic eruptions, also affect ozone levels; however, there is considerable evidence that CFCs are the primary cause of higher-altitude ozone depletion. As CFCs are released from air conditioners and refrigerators (where they function as coolants), aerosol cans (where they function as propellants), and other sources, they enter the atmosphere. The CFCs move upward to the stratosphere, where, under the influence of ultraviolet radiation, they release chlorine atoms, which speed the breakdown of ozone (O_3) into oxygen (O_2). With less ozone in the stratosphere, more ultraviolet radiation can reach the surface of the earth.

While CFC usage has been greatly reduced and human actions to address this problem seem to be having success, this form of pollution is still a major concern. Its effects are global in scope.

The thin envelope of air surrounding our planet is essential to all living organisms. While various forms of air pollution can harm fisheries and wildlife resources, the broader ways in which such pollution affects all organisms, including humans, should be of paramount concern.

FIGURE 14.2 Soil sensitivity to acidic deposition varies depending on the natural buffering capacity of an area. A large part of North America is sensitive to acidic deposition.

14.2 Water Pollution

There are numerous materials that pollute water, some of which have more negative effects than others. Important aquatic pollutants include oxygen-demanding wastes, sediments, organic chemicals, inorganic chemicals, and heat. Aquatic pollutants can affect not only those organisms that live in water but also those organisms that use water for drinking or other purposes. Thus, both aquatic and terrestrial species are affected by water pollution. Water pollution can impact small localized areas, or it can affect large areas; the dead zone in the Gulf of Mexico, caused by agricultural pollution, is a prime example of a large-scale problem. The effects of pollution on surface water will be emphasized in this text, but pollutants also damage groundwater.

Oxygen-demanding wastes include domestic sewage, domesticated livestock wastes, and a host of other biodegradable organic wastes that cause water to be depleted of dissolved oxygen. Because the majority of aquatic organisms obtain their oxygen from that dissolved in water, such depletions can have severe negative effects. When aquatic systems receive large amounts of oxygen-demanding wastes, decomposers that utilize these wastes increase in numbers. Many of these organisms (**aerobic decomposers**) use dissolved oxygen during their activities. Other decomposers (**anaerobic decomposers**) do not use dissolved oxygen but can produce toxic substances, such as ammonia, through their anaerobic processes. As the amount of dissolved oxygen decreases or the amount of toxic substances increases, aquatic organisms are affected. These effects can be lethal, or they can act to reduce growth or reproduction or cause organisms to be more susceptible to parasites, pathogens, and diseases. There is also a potential for the direct spread of diseases by this pollution source.

Sediments are water-insoluble particles, primarily soil, that erode into surface waters from surrounding land. Some sediments come from natural erosional processes, while others result from human activities such as agriculture, mining, logging, or construction. Large sediment particles settle to the bottom of a water body. Smaller sediment particles can remain suspended in the water column for longer periods of time. As sediments settle to the bottom, they can smother fish eggs or young, invertebrates, aquatic plants, and a variety of other organisms. The smaller sediment particles that remain suspended can reduce the efficiency of sight-feeding organisms, clog the gills of fishes or the filtering systems of filter feeders such as many molluscans, and, by lowering light levels, reduce primary production. Sediments can also act as vehicles for the transport of nutrients, toxic metals, pesticides, herbicides, and other substances from terrestrial to aquatic systems.

Organic chemicals, such as oil, pesticides, detergents, and a wide array of other contaminants, can also have negative effects on aquatic biota. Some are cancer or tumor causing in humans and other organisms, some result in added nutrient input, some are toxic, and some cause damage by collecting on the outer surfaces of organisms. All can reduce the utility of water bodies as habitat for plants and animals. For example, federal, state, provincial, and territorial governments often issue advisories concerning water quality. These may take the form of fish consumption advisories or even advisories against water contact or use.

Oil contamination originates from various sources, ranging from oil drained from a lawn mower to massive marine oil spills. The *Exxon Valdez* spill off the coast of Alaska illustrated the extensive damage and costs that can result from such contamination. That spill occurred in 1989, and its effects (and lawsuits) are still present. Included in this source of pollution are all of the gasoline and similar hydrocarbons that evaporate into the atmosphere and eventually make their way into aquatic systems. Oil pollution from a variety of sources continues to be a major problem, and much more could be said about the negative effects of this form of contamination on aquatic and terrestrial systems.

Inorganic chemicals and nutrients such as toxic metals, acids, salts, nitrogen, and phosphorus can also be damaging to aquatic habitats. Acids released by mining activities, industry, or acidic deposition can degrade habitat. Salts from irrigation runoff, urban storm drains, or industry can negatively affect aquatic systems. Toxic substances such as mercury, cyanide, arsenic, and chromium can toxify aquatic ecosystems. Nitrogen and phosphorus can be discharged into aquatic systems from a variety of sources. Large amounts of these nutrients result in excessive algal growth or dense stands of other unwanted aquatic vegetation (Figure 14.3). Such excessive aquatic plant growth can block light needed by more desirable vegetation, can cause noxious odors, and can reduce the aesthetics of aquatic systems. Such growth can also interfere with a variety of recreational activities. In addition, when the excess vegetation dies, its decomposition can lead to decreases in the amount of dissolved oxygen and increases in the amounts of toxic substances.

FIGURE 14.3 Excessive plant growth, such as this dense stand of water fern in a Texas water body, not only can result in oxygen depletion but also can produce toxic substances and reduce the recreational utility of a water body. *[Image courtesy of B. Blackwell.]*

Heat can also be a water pollutant when water is removed from a water body and is used to cool power plants or other industrial facilities. When the heated water is returned to the water body, it changes the aquatic habitat. Native plants and animals adapted to lower water temperatures may be unable to exist at the new higher water temperatures.

Another aquatic pollutant that is recently gaining more attention is the problem of **endocrine disruptors** (**hormone disruptors**). These are toxicants that interfere with the endocrine (hormone) system of animals, including humans. Sources of this pollutant are many and include releases from paper pulp mills, human birth control pills and patches, antibiotics, and some agricultural chemicals. These toxicants affect reproduction and other processes controlled by hormones. The endocrine systems of animals respond to exceedingly small amounts of chemicals, so even highly diluted amounts of these substances pose a problem. We have much to learn about the effects of these disruptors on humans and other animals.

14.3 Urbanization

As the human population increases, the rate of urbanization also increases. Even if human population levels were stable, urbanization would probably continue. For all of previous human history the majority of the population, sometimes a large majority, lived in rural areas. That scenario has recently changed, and the human urban population of the world now exceeds the rural population (Withgott and Brennan 2007). The growth of cities and suburbs has many ramifications for fisheries and wildlife resources. Its most obvious effect is habitat loss.

As land-use patterns change, the types of organisms present also change. Those species having specific habitat needs not met in human-developed areas will experience population reductions, while other species that can use these habitats will increase. The species aided may be desirable or undesirable ones. For example, house sparrows do well in urban areas, as do Canada geese. Most people consider Canada geese (in acceptable numbers) to be desirable and house sparrows to be undesirable. White-tailed deer and raccoons can also thrive in areas with large human populations. Many other species are generally intolerant of either the habitat changes resulting from urbanization or the increased presence of humans. The increased number of domesticated cats in urban areas, for example, has a negative effect on wildlife, especially on songbird numbers. Even if natural habitats remain within urban areas, these areas often exist as habitat fragments within the urban setting; the utility of these small islands of habitat to many terrestrial organisms is marginal (see section 14.7).

The effects of urbanization on aquatic organisms are usually even more profound. Natural and artificial water bodies are often drained and used as construction sites when urbanization occurs. Even if they are not totally drained, they are often reduced in size or in some other way negatively modified. If they cannot be drained, concrete-lined "rivers" are often the result; such rivers are the ultimate in habitat degradation. They have little utility for aquatic or riparian life forms. However, there has been a positive trend toward protection and creation of aquatic habitats in urban areas—many new housing and business developments now include artificial lakes and wetlands.

An additional effect of urbanization is increased interaction between humans and wild animals, which often leads to a variety of problems. Conflicts between people and animals are increasing as humans encroach on areas traditionally used by wildlife. For example, the building of a housing development in an area historically used by deer as a winter range can result in the deer causing damage to landscape plantings; it can also lead to reduced survival and poor health for the deer. As human populations expand or move into new areas and as cougar numbers increase, attacks on pets and people occur with greater frequency. Houses, sheds, and other structures can provide excellent habitat for animals such as striped skunks, and the edible refuse associated with human habitations is an attractant to bears and raccoons. Many species, however, cannot exist in habitats intermixed with lawns or parking lots; the habitat needed by these species is completely lost when urbanization occurs. Try to envision a grizzly bear living in an urban setting; it would neither do well nor be accepted as a neighbor.

While contact with wild organisms, both plant and animal, is desired by many people, problems do arise. By their nature, most nondomesticated creatures are wild, and they do not react to humans in the ways depicted in cartoon fantasies. Human tolerance tends to wane when an eastern fox squirrel gets into an attic and shreds the quilt great-grandmother hauled across the Oregon Trail, or when deer eat prize roses, or when an American alligator eats the family dog. Additionally, some animals may lose their innate wildness in urban settings, becoming caricatures of their wild counterparts; the way we perceive them also changes. The wild and wily mallard, white-tailed deer, or yellow-bellied marmot may become a popcorn-eating creature that is quite different from its conspecifics in the wild. In many people's view, this tends to degrade the organisms. Some animals, such as coyotes, have been able to adapt to new surroundings while still remaining in an essentially wild state. In many urban areas, coyote populations have been able to expand, but they differ from their rural counterparts in being rather nonvocal in their behavior. Similarly, the American crow was in the past a crafty bird of rural areas that disdained places of human activity; it is now commonly found in urban areas and is less wary of humans.

While urbanization benefits some aquatic and terrestrial species, it generally results in habitat degradation for most wildlife and fish species. The field of urban wildlife and fisheries is a rapidly growing area of study. Based upon world human population growth projections and an increasing urbanization of that population, we can expect more human–wildlife interactions.

14.4 Agriculture

Of all human activities, none has directly modified large areas of natural landscapes more than agriculture. Vast expanses of land, especially in the temperate grassland and temperate forest biomes, have been changed by this activity. As human populations have increased, their needs for food, natural fibers, and other agricultural products have increased as well. As agricultural technology has changed, the effects of agriculture on habitat have also changed. Some of these effects have been essentially unavoidable as the need for land devoted to agriculture has increased. Some agricultural activities, however, have had unnecessarily negative effects because of the manner in which they have been and are being conducted.

North American agriculture, more so today than ever, is influenced by a wide array of economic and social forces. Capital outlay has increased as land prices have increased; as equipment has gotten larger, more complicated, and more costly; and as herbicide, pesticide, and fertilizer use and costs have increased. It has become increasingly difficult for the agricultural community to spend money or forgo income to benefit wildlife and fisheries resources. However, practices such as lax pollution controls and agricultural subsidies, which shift costs from agricultural producers to the general public, are being scrutinized more now than at any time in the past. The costs of agricultural activities that degrade the environment, which were once passed on to the commons and to other private property, are now more often being viewed as costs that should be borne by those who cause the degradation. The **commons** are those resources such as air, water, wildlife, fish, and publicly owned land that are owned by everyone but by no individual in particular. More will be said about the commons in section 14.10.

Agricultural activities have changed terrestrial and aquatic habitats in a variety of ways. Some of these changes can be viewed as beneficial, but many have been primarily negative. Generally positive changes include pond construction and the production of food for wildlife. Irrigated farmland in arid regions has provided abundant food and water for a variety of wildlife species. Some domesticated animal grazing systems benefit both wildlife and domesticated livestock. Agricultural practices that clear woodlands for small fields have been beneficial to such species as the northern bobwhite and eastern cottontail, although they have not been beneficial to other species, such as the American redstart and American black bear. Some of these habitat changes were and are essentially unavoidable as humans change landscapes in their use of land.

Some other agricultural activities have been unnecessarily damaging. Some of these practices promote accelerated soil erosion, which clogs water bodies with sediments. Increasingly efficient methods of draining lowlands have encouraged the spread of agricultural crops into marginal areas. Increased use of fertilizers and pesticides has negatively affected terrestrial and aquatic wildlife. The change from smaller to larger field sizes has negatively affected wildlife by reducing interspersion and edge habitat. The removal of obstructions such as trees or fencelines, which hinder the movement of irrigation equipment or increasingly large tractors and farm implements, has also had negative effects. Fence-to-fence agriculture, and in some cases the total removal of fences and their associated habitat, has been harmful

FIGURE 14.4 Little habitat for wild organisms remains in areas where intensive agriculture is practiced. Species diversity is a casualty in such areas. *[Image courtesy of R. Meeks.]*

(Figure 14.4). The use of irrigation in agriculture has also had some widespread negative effects on both terrestrial and aquatic systems. Irrigation has allowed the movement of agriculture into areas previously too dry for such pursuits, with concomitant negative effects on most of the native biota of such areas. In addition, at this time we know little about how the use of genetically engineered agricultural crops, a practice that is increasing rapidly, will affect ecosystems and landscapes.

Another negative aspect of agriculture has been its tendency toward monoculture systems. **Monoculture** is the growing of a single species, such as corn or wheat, over a large area. Not only does it reduce habitat diversity, but a monoculture also provides excellent habitat for certain plant pathogens and undesirable plant species that thrive in the monocultured crop. Damage caused by these organisms elicits an agricultural response in the form of herbicides and pesticides that can cause additional environmental damage. Monoculture is a good example of how changing a habitat reduces some species but aids others. What more could a corn rootworm, a pest on that crop, want than field after field planted with corn?

Agricultural practices that involve domesticated livestock also have caused some unnecessary problems. When livestock grazing is properly managed, it can benefit vegetation and wild animals. However, incorrect grazing practices can be damaging to vegetation as well as to wildlife and fisheries resources, and they will eventually be costly to the landowner as well. If too many cattle, sheep, or goats are placed into too small an area for too long a time, habitat degradation occurs. Domesticated livestock-grazing activities can reduce the numbers and biomass of desirable grasses (called **decreaser species** because they decrease under grazing pressure) and increase the numbers of less desirable grasses. These less desirable grasses can be **increaser species** (which increase under grazing pressure) or **invader species** (which invade from other areas into heavily grazed areas). These changes can have profound effects on wildlife using the area. Mechanical damage to streambeds and riparian zones is another common result of domesticated livestock activity. Riparian zones are attractive feeding and watering places for domesticated livestock, and this often leads to overuse of such areas by the animals (Figure 14.5).

In the United States today, there are a number of contentious issues concerning domesticated livestock grazing, as well as mining and other consumptive pursuits, on federally owned land by private citizens and companies. There are questions about whether current levels of grazing can be sustained on a long-term basis and how much of a negative effect grazing has on wild animals and their habitat. There is also concern about whether other uses of federal land, such as for camping, fishing, hiking, and wildlife viewing, are receiving adequate attention and prioritization. To further fuel this issue, the grazing fees currently charged on federally owned land are well below the fees charged on similar private land. Federal laws require that federally owned lands be managed for multiple purposes (see section 18.8), but there are questions concerning the effectiveness of such laws. Because federal lands cover such a large area, especially in western states, the way grazing and other activities are managed on these lands is of great importance.

Desertification is a major problem in some parts of the world, has substantial negative effects on ecosystems and landscapes, and is closely tied to inappropriate agricultural practices. Desertification is land degradation caused by natural (primarily climatic change) or human-induced factors in dry areas. It is characterized by devegetation, groundwater depletion, severe erosion, and other negative occurrences. It is usually caused by overgrazing, overcutting, inappropriate irrigation, and cultivation of unsuitable land (Chiras and Reganold 2005). Some areas of the southwestern United States have been greatly affected, as

A

B

FIGURE 14.5 Riparian zone degradation due to domesticated livestock grazing affects both terrestrial and aquatic organisms. The damage can be very obvious *(A)* or more subtle *(B)*. The scene depicted in *(B)* may look fairly normal, but the degradation reflected by sloughing stream banks, a lack of shoreline vegetation and undercut banks, and other problems indicates poor grazing management. *[Images courtesy of W. Platts.]*

have large areas in Africa, the Middle East, and central Asia. Such areas can be rehabilitated or at least stabilized, but costs are substantial.

Positive changes are occurring in agriculture, as evidenced by increases in minimum-till and no-till farming, more efficient use of fertilizers and pesticides, removal of some marginal land from crop production, and other improvements that are reducing environmental damage. Regardless of the environmental effects of agriculture, one aspect cannot be changed: Humans will always need food and agriculture will always change landscapes, which will be beneficial to some species and harmful to others.

14.5 Logging

Products made from wood have long been used by humans in a variety of forms, as has wood as a fuel source. To obtain wood, trees must be cut. In some situations tree cutting has minimal or positive effects on wild animal populations, while in other situations such cutting can be devastating. The effects are dictated by the scope and kind of logging practiced and the wild organisms in question.

In many forest management activities, humans tend to replace multispecies forests with monoculture forests. This type of ecosystem simplification causes problems similar to those described for agricultural crops by reducing habitat diversity and, in turn, the species diversity of organisms that inhabit the forest. Mature natural forests generally have several layers of understory and overstory, including shrubs, smaller tree species, larger tree species, and other layers. They also tend to have an interspersion of differently aged vegetation. Monoculture reduces these layers by encouraging not only a single species of tree but also a single year class of trees. As vegetation diversity is reduced, so are food supplies, nesting areas, and other niche characteristics used by various animals. Such monoculture practices also promote larger-scale problems with diseases, parasites, pathogens, and undesirable species. The natural system of checks and balances inherent with species diversity is lost.

FIGURE 14.6 Clear cutting of large areas can reduce the utility of forested areas to both terrestrial and aquatic organisms. Severe erosion is often associated with this forest management practice. *[Image courtesy of K. Jenkins.]*

When extensive logging occurs, habitat fragmentation (see section 14.7) results. Logging, particularly where **clear cutting** (removal of all trees) occurs (Figure 14.6), can also lead to accelerated soil erosion and stream warming that negatively affects organisms living in water bodies within the watershed. Erosion and warming affect downstream areas as well. Pacific salmon fisheries, for example, are affected by timber management practices not only in the streams where the fishes reproduce but also in downstream areas through which they migrate.

One of the primary characteristics of logging activities is the road network needed for log removal. Such roads usually damage stream habitat through the resulting erosion and vehicular traffic. Additionally, a newly constructed logging road network opens an area to increased human use for other activities. Although this can be beneficial for some recreational activities, areas previously accessible only to foot traffic are opened to a variety of motorized vehicles, even after logging operations are completed. Areas that once were remote and received little human disturbance become more accessible; this increased accessibility can have negative effects on some species. A road placed too close to a bald eagle or northern goshawk nest, for example, may cause the birds to abandon the nest or to avoid further use of the nest site. Even without roads, continued development of the variety and capabilities of off-highway vehicles (OHVs) has resulted in increased human access to previously remote areas.

Destructive forest wildfires throughout North America in the late 1800s and early 1900s were followed by increased fire-suppression activities. The natural burning patterns of forests were disrupted by fire suppression, and this resulted in habitat changes, often to the detriment of species diversity. Forest fires that occur in areas subjected to frequent burning are less destructive than fires that occur in unburned forests with a great deal of fire fuel buildup. However, logging does not always lead to fire suppression; the burning of unusable parts of trees from logging operations during droughts has led to wildfires and loss of human life (Biswell 1989).

Not all logging activities need be destructive. **Selective cutting** of specific trees or **block cutting** (clear cutting of small areas) can usually be accomplished with less damage, especially if care is taken in log removal and road construction. In some cases block cutting can even serve as a partial substitute for fire. Some wildlife species can be benefited in numerous ways by correct logging practices. Ruffed grouse, for example, use aspen trees of a variety of sizes. Managing aspen stands so as to maintain a combination of recently cut areas, young-tree areas, and mature-tree areas will fill a variety of ruffed grouse habitat needs. Similarly, wild turkeys, elk, and deer can benefit from the creation of forest openings. However, these forest openings may also increase populations of other less desirable species, such as the parasitic brown-headed cowbird. This species, which lays its eggs in the nests of other species, prefers forests with openings. By creating habitat for the brown-headed cowbird, logging can reduce the nesting success of other birds.

Fires, floods, storms, and diseases at times kill and weaken large tracts of forest. **Salvage logging** is the removal of these dead or diseased trees to improve forest health. However, such logging is coming under increased public scrutiny. It has been practiced under the premise that it reduces future fires, increases forest health, and utilizes otherwise wasted timber resources. However, cost analysis of such logging is being questioned, as are its benefits to improving forest health.

In the last 30 to 40 years, many areas of public land in the western United States have been logged, even when this resulted in below-cost timber sales (a form of subsidized logging). The usual arguments in favor of this practice concern jobs and the control of construction material costs. This practice has, however, disrupted much forest habitat in the western United States. Logging practices on federally owned land, including salvage logging, will continue to receive much attention from a variety of groups, such as logging companies, politicians, advocacy groups representing a variety of interests, biologists, people who make their living by logging, and others.

Deforestation, especially of tropical rain forests, is a global issue of great importance to wildlife and fisheries resources. These forests are the world's great repository of species diversity. Approximately half of all tropical rain forests are believed to have already been cut (an area about half the size of the United States). Even when these cut areas revert back to forest habitat, the resulting forest is not the same as the original multilayered, diverse habitat. Although we know of the contributions of tropical rain forests to the functioning of the earth's biosphere, cutting continues at a rapid rate.

Humans will continue to use wood for a variety of purposes. When logging is correctly managed, the damage it causes can be minimized. The potential for short-term and long-term damage to aquatic and terrestrial populations, like the value of the timber itself, is an economic factor that should be considered when logging is planned. All too often in the past, only the obvious monetary value of the timber removed was considered in economic analyses. When decisions are made concerning location, extent, or kind of logging to be carried out, they should be based on full economic considerations. Timber is not the only resource involved.

14.6 Wetland Loss

Many people have often equated wetlands with wasteland. Coastal wetlands, floodplain forests, bogs, marshes, and other wetland types have generally been viewed as places to drain, fill, or in some other way make dry. Over the last 200 years, housing construction, harbor building, agricultural encroachment, the diking of rivers, and a variety of other human activities have led to extensive losses of wetlands (Figure 14.7). California, for example, has lost 91% of its wetlands, Ohio 90%, Iowa 89%, Indiana 87%, and Missouri 87%. Twenty-two states have lost over one-half of their wetlands (Dahl and Johnson 1991). However, recent federal administrations have followed a no-net-loss policy on wetlands, and through restoration and creation activities, for the first time wetland gains have exceeded losses (Dahl 2006).

The value of wetlands to both terrestrial and aquatic organisms, including humans, is great. Wetlands represent a habitat type that is exceptionally productive. A large percentage of commercially important marine species live a portion of their lives in coastal wetlands and estuaries. North American waterfowl populations are dependent on wetlands, as are vast numbers of nongame organisms, ranging from salamanders to butterflies. Wetlands also contain large numbers of plant species dependent on this type of habitat. In short, wetlands have excellent species diversity.

Wetland inhabitants are adversely affected by wetland loss in numerous ways in addition to the actual loss of their habitat. For example, as wetlands are lost, waterfowl and shorebirds are forced into smaller and smaller areas. Such concentrations increase the likelihood of disease outbreaks. Great numbers of ducks and geese concentrate in the late winter and early spring in the Rainwater Basin in south central Nebraska. This is an area where over 90% of the natural wetlands have been drained. Thousands of waterfowl often die during this concentration period because of avian cholera.

Wetlands serve as natural water-purifying systems, and as those systems are reduced, water pollution problems are aggravated. The microclimate of

FIGURE 14.7 Wetland drainage not only results in decreased wildlife and fisheries resources but also negatively affects many of the other benefits of wetlands to humans. *[Image courtesy of the U.S. Fish and Wildlife Service.]*

an area is modified when water is artificially removed through wetland loss. This climate modification affects all organisms, including humans and their agricultural crops and domesticated livestock. Numerous other direct and indirect negative consequences can be attributed to wetland loss.

Wetlands are currently one of the focal points in an often heated discussion concerning habitat maintenance versus personal property rights (see section 14.10). The benefits of wetlands extend to all people, but wetlands are mainly the property of individuals. This debate is an excellent example of the way biological, social, economic, and other issues can interact to create a complex problem. It is sometimes argued that wetlands are only good for ducks, mosquitoes, snakes, or whatever other wild organisms directly use the area. Those who support wetland drainage argue that these areas are being made useful to humans when they are drained and used for direct human purposes such as agricultural crops or buildings. A person making such a statement is demonstrating a lack of knowledge concerning the negative ecological and economic ramifications of wetland loss to humans, as well as the negative effects of that loss on wild organisms. Humans are a part of the ecosystems in which they live, not separate from or above them.

14.7 Habitat Fragmentation

Clear-cut forests, power line rights of way, series of dams and reservoirs on rivers, agricultural fields, all types of roads, urban sprawl, and a wide variety of other human actions have two things in common: all are the result of human activities, and all result in habitat fragmentation (see section 2.11). The isolated bits of habitat resulting from these activities do not function ecologically in the same way as comparably sized undivided habitat blocks. Such altered landscapes can affect some animal migration patterns, nesting or roosting patterns, feeding behavior, and a myriad of other activities pursued by organisms. In addition, some species simply need undisturbed areas. Roads and accompanying traffic, activities associated with power lines, and recreational activity on water bodies may prevent some animal species from using an area.

Habitat fragmentation is a rather insidious process; it is similar to death by a thousand cuts. When viewed from a close perspective it looks innocuous—a new trail here, a new house there, a new agricultural field over there. Until one views an area from a landscape level, one cannot see the scope of fragmentation over large geographic areas.

There are some wildlife benefits from habitat fragmentation, such as increased ecotone areas and the creation of openings in large blocks of timberland, but these benefits must be weighed against drawbacks. One of the most pressing problems in both wildlife and fisheries work is finding ways to ameliorate the negative effects of continued habitat fragmentation.

14.8 Channelization

Stream channelization and associated activities, such as dredging (primarily for deepening water bodies) and snag clearing (removal of dead trees), are forms of habitat degradation that affect both aquatic and terrestrial biota. In our efforts to improve shipping channels or to move water more rapidly out of an area by straightening or deepening rivers and removing flow obstructions, humans change the characteristics of lotic systems and the biota that can inhabit them. The ability of lotic systems to release energy is modified, and this energy is often expended in ways that harm the general environment and the works of humans. In many situations our human purposes seem contradictory. On the one hand we view water as an important economic resource to be saved, protected, and stockpiled, while on the other hand we try to move it out of an area as rapidly as possible so that we can plant crops (which need water) or construct buildings in floodplains.

The initial problem with channelization is that stream or river length is reduced. Mark Twain, in his book *Life on the Mississippi*, described such a loss of river length on the lower Mississippi River between Cairo, Illinois, and New Orleans, Louisiana. In his time as a riverboat pilot on the Mississippi in the mid-1800s, the distance between the two towns decreased by 67 miles because of cutoffs that changed the river channel; these human-influenced cutoffs decreased river length from 1,040 miles to 973 miles. He calculated that the river was being shortened by 1.3 miles per year and that at that rate of loss, in 742 years the Mississippi River between Cairo and New Orleans would be just 1.75 miles long. The point is that as a river is straightened, river length (and therefore habitat) is lost, if perhaps not to the extent that Mark Twain facetiously predicted.

Even after rivers are shortened, further degradation occurs. Dredging to deepen channels and snag clearing to remove obstructions, two other methods of speeding water flow, are also destructive to habitat. Dredging disturbs the development of benthic communities. In addition, materials dredged from such areas need to be placed somewhere; this placement can cause further habitat damage. Snag clearing removes one of the most important habitat features of a lotic system. The areas around snags have been shown to be one of the most attractive locations for fishes and other aquatic organisms. Even organisms such as birds, reptiles, and amphibians utilize these snags.

Most human efforts to move water more rapidly out of an area cause further problems. That water must go somewhere; usually it becomes a problem for other humans downstream.

Another form of river and stream degradation that should be mentioned is dam construction. Impoundment of a free-flowing river changes that water body from a lotic into a primarily lentic environment, thereby destroying lotic habitat (see section 12.3). Construction of impoundments also establishes barriers to the movement of organisms, especially fishes. These barriers have their greatest effects on migratory species. Many efforts, usually only partially successful, have been made to assist various fish species around or over these obstructions. The series of fish ladders and other structures built into dams on Pacific Northwest rivers to assist Pacific salmon are an example.

14.9 Introduced Organisms and Competition

The previous sections of this chapter have highlighted habitat degradation or destruction resulting from direct physical changes in habitat caused by humans. Biological changes that result from competition with or activities of other organisms can be just as destructive as the previously mentioned forms of degradation. Some of these changes occur through natural expansion in the geographic range of an organism, while others are obviously induced by humans through the accidental or purposeful introduction of organisms. An example of the latter would be habitat degradation resulting from the grazing of domesticated livestock (see section 14.4).

Introduced organisms other than domesticated livestock can also result in degraded habitats. The exotic common carp, for example, has caused widespread aquatic habitat damage that has contributed to the reduction of numerous native organisms in North America. Activities of this species, among other effects, increase water turbidity and decrease the abundance of rooted aquatic plants. Common carp have negatively affected populations of terrestrial organisms associated with water bodies as well as native aquatic organisms. The exotic house sparrow competes for nest sites with native birds, such as purple martins and eastern bluebirds. Kudzu is an introduced plant that has degraded large terrestrial areas in the southeastern United States. Purple loosestrife has had and is having negative effects on native plant species in wetland areas. There are a multitude of other examples of introduced forms that have caused habitat degradation. One added species or one subtracted species can make the difference between a correctly functioning ecosystem and one that is degraded and unstable.

14.10 Habitat Degradation, the Commons, and the Takings Issue

Pollution and other forms of degradation result in habitat modification. The commons, as already described, represent resources such as air, water, wildlife, fish, and public land that are owned by everyone but by no specific individual. The takings issue involves a response by some people to protect what they believe to be their rights as private property owners. They believe that they should not be inhibited from pursuing activities involving their private property. If regulations or laws intended to protect the commons and other private property reduce their ability to use their property in some way, they believe that this represents a taking of their property rights. In some cases, when they believe that takings have occurred, these private property owners believe that they should be monetarily compensated. In the United States, such compensation has been sought through legal action for a variety of perceived takings, including civil rights laws, public health and safety laws, conservation measures, zoning actions at various levels, and others.

The takings issue has far-reaching implications for natural resource conservation and management with regard to regulatory activities such as clean water and endangered species legislation and the decisions of land-use planning boards and commissions. The following are sample scenarios in which some people believe takings are occurring. A property owner wants to construct an apartment building but cannot, because the area is zoned for single-family dwellings; this would represent a taking of the ability of that property owner to collect rent from multiple apartment units rather than from one house. A farmer wants to plant an agricultural crop within a few meters of a lake, but a law or regulation prohibits farming within 30 meters of that lake; those 30 meters would represent a taking, and compensation would be due for the loss of income from the crops that would have been planted on that strip of land.

To better understand the relationships among pollution, habitat degradation, the commons, private property, and the takings issue, one must first briefly review the history of pollution control in the United States. Initially, pollution control was directed at **point-source pollution**. Point-source pollution emanates from single, identifiable, concentrated sources, such as factory smokestacks or drainpipes from municipal sewage plants. These sources release pollutants into the environment at a specific location and represent the most easily identifiable pollution sources. Laws and regulations were promulgated that required some point polluters to reduce pollution from these sources. One of the rationales underlying such regulations or laws involves costs that were being passed on by polluters to the commons and to other private property owners. For example, when pollution emanating from a sewage effluent degrades a resource such as a river or lake, that degradation reduces the value of that resource. Someone will eventually have to pay to correct the action of the polluter. It might be a local government that incurs costs to improve the water body for recreation, it might be another industry that incurs costs to make the polluted water usable in its industrial process, or it might be an individual landowner who wants to use the water for some purpose but cannot do so until it is purified. Controls on point-source pollution shifted the cost burden, at least partially, from other private property owners and the commons to the polluter. Initially there was resistance to this shift in costs, but the shift is now generally well established in the mind of the public. However, there are some people who would like to see such laws and regulations reduced or eliminated; to them, such laws and regulations represent takings and a threat to economic development.

The next phase in pollution control, which is at least partially responsible for fueling the takings issue, involves **nonpoint-source pollution**. This term refers to more diffuse and dispersed forms of pollution that are not released at a single identifiable point. For example, fertilizer runoff from a dozen farms is nonpoint-source pollution. It does not enter a pipe and then flow into a lake or onto a neighbor's land at one point; it enters at multiple points. Efforts to control nonpoint-source pollution and to partially shift its costs to the nonpoint polluters have met with resistance. This resistance to increased regulation of activities conducted on private property is just one facet of the takings issue response by some people.

The concept of the right of an individual to do what he or she wants to with property that he or she owns may seem like a cornerstone of the free-enterprise system. Usually property rights and takings arguments are based on the Fifth Amendment of the U.S. Constitution, which states that private property "shall not be taken for public use without just compensation." However, there is a long legal and judicial history that addresses what represents a taking, and there are many laws and regulations that limit the rights of property owners. For example, there are laws and regulations that prohibit one property owner from draining water from his or her land onto the property of another person. There are rules that keep someone from opening a manufacturing plant or a cocktail lounge next to your home. There are regulations that prohibit people from having noxious plants, such as Canada thistle, on their land. A listing of all the controls on what individuals can or cannot do on or to their own property would be lengthy. All are legal safeguards that protect the personal property rights of others, the health and safety of the public, the civil rights of all people, and the commons. The rules or regulations mentioned above, and many others, do represent the limitation of some property rights, but such limitations are generally not supported in court cases as infringements of Fifth Amendment rights, because they protect the rights of others. Such rules and regulations are generally accepted by most people, even most advocates of the takings issue.

Most of the takings issue problems involving habitat degradation occur in the course of attempts either to shift some pollution costs back to the degraders or to modify land-use practices away from activities that property owners have traditionally pursued with little or no regulatory interference. Unfortunately, the effect of what someone does on or to his or her own property does not stop at their property line, nor does that effect remain a problem only for the commons. Pollution and habitat degradation caused by airborne soil particles resulting from improper agricultural practices can be used as an example. Consider a few of the locations where these soil particles come to rest. They enter the lungs and eyes of people, domesticated livestock, and wild animals. They end up in automobile, tractor, and airplane engines. They fill lakes, ponds, and drainage ditches. They invade homes, factories, restaurants, and grocery stores. It is obvious that these airborne soil particles affect many people in many ways. In addition, the loss of topsoil eventually results in losses to the polluting landowner through reduced crop production.

The question is, at what point do the rights of the private property owner supersede the rights of other people? In the United States, 200 years of laws and court decisions have upheld the tenet that no one has the absolute right to use property in a way that may harm the public health or welfare or that may damage the quality of life for neighboring property owners or for the community as a whole. However, a stream of national and state legislation that addresses the takings issue is continuously being introduced, and in some cases passed. Such legislation either provides for monetary compensation for perceived takings or imposes increased paperwork requirements on agencies or boards that serve to impede perceived takings (red-tape legislation).

The takings issue will continue to be a divisive one. Because of its complexity, it will not be easily resolved. It represents an area in which continued efforts at conflict resolution are needed. Conflict resolution with regard to the takings issue has and probably will continue to be primarily political and legal in nature. Section 20.5 further addresses the topic of conflict resolution.

SUMMARY

Habitat can be degraded by numerous natural and artificial means. Human-influenced degradation can be intentional or unintentional. As habitats are degraded, they become more usable to some organisms and less usable to others.

Three types of air pollution problems that have substantial effects on wild organisms are global climate change, acidic deposition, and ozone depletion. Global climate changes may result from increased greenhouse gases in the atmosphere and represent an extraordinary challenge, because their effects are on a truly global scale. Environmental changes in pH affect both aquatic and terrestrial organisms. Ozone depletion results primarily from the use of chlorofluorocarbons.

Important water pollutants include oxygen-demanding wastes, sediments, organic chemicals, inorganic chemicals, and heat. Oxygen-demanding wastes reduce habitat usability for organisms that obtain their oxygen from that dissolved in water. The issue of endocrine disrupters is an important problem area that is now receiving much more attention.

Urbanization results in habitat loss and increased interactions between people and wildlife. The field of urban wildlife and fisheries is a growing area of study.

Agricultural activities modify large areas of natural landscapes. The increased use of herbicides, pesticides, and fertilizers has had dramatic effects on wildlife and fisheries resources, as have increased field size and monoculture. Domesticated livestock can also contribute to habitat degradation. Desertification, primarily caused by improper agricultural practices, is a substantial problem in some regions of the world.

Timber management practices vary in their effects on wildlife and fisheries resources. The cutting of trees represents an agent of habitat change, as do other aspects of forestry practices, such as monoculture and road construction. All too often, economic analyses of logging revolve around timber value alone. The timber is not the only resource involved. Deforestation, especially of tropical rain forests, remains an issue of global importance.

Wetlands are all too often viewed as wastelands that need to be used for the obvious, immediate economic benefit of humans. The economic benefits of wetlands go far beyond their values associated with conversion to agricultural crops or housing developments, or even as wildlife and fisheries resources. They are an integral part of the landscape that contributes many economic, ecological, and aesthetic benefits.

Habitat fragmentation, such as that caused by forest clear cutting, power line rights of way, and road construction, reduces the utility of habitat to most wild organisms. Isolated bits of habitat do not function ecologically in the same manner as comparably sized unfragmented habitat blocks.

Channelization and associated activities, such as dredging and snag clearing, are forms of habitat degradation that affect lotic systems. River length is shortened and specific habitats are reduced. Dams serve as barriers to migratory animals and also change lotic water bodies into lentic habitats.

Purposefully and accidentally introduced organisms can degrade physical habitat or compete with native organisms. As the number of introduced organisms increases, problems associated with them increase.

Pollution, habitat degradation, the commons, private property, and the takings issue are all interrelated. The takings issue is primarily the result of attempts to have degraders of natural resources pay more of the costs of their polluting activities and to modify habitat-degrading land-use practices that property owners have traditionally pursued with little or no regulatory interference.

PRACTICE QUESTIONS

1. What is restoration ecology?

2. Describe the process that results in acidic deposition.

3. List some human activities that result in the release of greenhouse gases and describe how these gases may lead to global climate change.

4. How do oxygen-demanding wastes lead to aquatic habitat degradation? What are some sources of oxygen-demanding wastes?

5. What are the primary causes of desertification?

6. List a dozen different ways in which agriculture can negatively affect wild organisms.

7. Define decreaser, increaser, and invader species, and describe some ecological and economic ramifications associated with decreased populations of decreaser species.

8. What is meant by forest clear cutting, and what other less ecologically destructive forest management practices can be used?

9. List some benefits of wetlands other than their benefits to wildlife and fisheries resources.

10. What is habitat fragmentation? List a number of human activities taking place where you live that contribute to habitat fragmentation.

11. What is involved in the takings issue? Describe four situations in the area where you live in which takings could be an issue.

SELECTED READINGS

Chiras, D. D. 2006. *Environmental science.* 7th ed. Jones and Bartlett Publishers, Sudbury, Mass.

Chiras, D. D., and J. P. Reganold. 2005. *Natural resource conservation: Management for a sustainable future.* 9th ed. Pearson Prentice Hall, Upper Saddle River, N.J.

Dessler, A. E., and E. A. Parson. 2006. *The science and politics of global climate change: A guide to the debate.* Cambridge University Press, Cambridge, U.K.

Ehrlich, P. R., and A. H. Ehrlich. 1991. *Healing the planet: Strategies for resolving the environmental crisis.* Addison-Wesley, Reading, Mass.

Gergel, S. E., and M. G. Turner, eds. 2002. *Learning landscape ecology: A practical guide to concepts and techniques.* Springer-Verlag, New York.

Groom, M. J., G. K. Meffe, and C. R. Carroll. 2006. *Principles of conservation biology.* 3rd ed. Sinauer Associates, Inc., Sunderland, Mass.

Hill, M. K. 2004. *Understanding environmental pollution: A primer.* Cambridge University Press, Cambridge, U.K.

Laws, E. A. 2000. *Aquatic pollution: An introductory text.* 3rd ed. John Wiley & Sons, New York.

Van Andel, J., and J. Aronson, eds. 2006. *Restoration ecology: The new frontier.* Blackwell Publishing, Malden, Mass.

LITERATURE CITED

Biswell, H. H. 1989. *Prescribed burning in California wildlands vegetation management*. University of California Press, Berkeley.

Chiras, D. D., and J. P. Reganold. 2005. *Natural resource conservation: Management for a sustainable future.* 9th ed. Pearson Prentice Hall, Upper Saddle River, N.J.

Cunningham, W. P., and B. W. Saigo. 2001. *Environmental science: A global concern.* McGraw Hill, New York.

Dahl, T. E. 2006. *Status and trends of wetlands in the conterminous United States 1998 to 2004.* U.S. Department of the Interior, Fish and Wildlife Service,Washington, D.C.

Dahl, T. E., and C. E. Johnson. 1991. *Status and trends of wetlands in the conterminous United States, mid-1970s to mid-1980s.* U.S. Department of the Interior, Fish and Wildlife Service, Washington, D.C.

Raven, P. H., L. R. Berg, and G. B. Johnson. 1993. *Environment.* Saunders College Publishing, Fort Worth, Tex.

Withgott, J., and S. Brennan. 2007. *Environment: The science behind the stories.* 2nd ed. Pearson Benjamin Cummings, San Francisco.

CHAPTER 15

Habitat Management

A large proportion of the strategies used to conserve wildlife and fisheries resources involve the management of habitat. Habitat is often manipulated to improve or degrade the ecological characteristics required by a particular species or group of species. Such habitat manipulations also have positive and negative effects on nontarget organisms, and these effects should be taken into consideration when implementing any habitat management technique. A species or community may simply be more or less desirable to humans in that particular place or at that particular time. The desirability of a species or community depends on the management objectives in that specific situation. For example, improving habitat for Canada geese may be an objective in some settings, but degrading habitat for that species may be an objective where too many geese are a problem, such as on a golf course or in a park.

In some instances, the management strategy implemented may address multiple needs of a population or community. For example, controlled burning of an area may be done to improve food supply, escape cover, nesting cover, and other habitat characteristics for grassland songbirds. The management practice implemented may also address a single need. For example, the construction and placement of American kestrel nesting boxes is directed only at the reproductive requirements of that species.

Habitat management can take many forms. It can involve the addition of natural structures, such as gravel and rock to encourage walleye spawning. It can involve the removal of natural structures, such as trees, if grassland species are the target organisms. It can involve the addition of artificial objects, such as eastern bluebird nesting boxes, or the removal of artificial objects, such as old buildings to reduce striped skunk populations. Habitat management can also be aimed at modifying the ecological succession that would naturally occur if no action were taken. For example, the water level in some wetlands can be raised to inhibit cattail growth.

Most habitat management activities in the past have been directed primarily at single species of recreational or commercial importance. There is, however, an increasing tendency to direct habitat management efforts at multispecies combinations and at larger land areas such as watersheds. In addition, more effort is being directed at habitat management for native and nongame species now than in the past.

Habitat management has always helped more species than just the target organisms; most habitat management efforts are beneficial to a variety of species.

However, the recent trend toward the promotion of species diversity requires that biologists take more of a landscape and ecosystem approach to habitat manipulation, rather than just directing activities at a single species. In reality, such a broader approach can usually benefit recreational or commercial species as well as the other organisms present.

There is a growing trend to attempt to manage the habitats of larger and larger areas. As methods of landscape and ecosystem management advance, biologists will become more proficient at these practices. However, as with most endeavors, scale is an issue. We currently are much more effective at habitat management of smaller areas and much less proficient at managing larger landscapes. Often, current large-parcel management efforts are just a number of smaller-scale activities designed to work in concert at a larger scale.

The purpose of this chapter is to provide an overview of some of the most commonly used habitat management techniques. This discussion will follow the order of the habitat types presented in Chapter 12, with additional information on the management of watersheds.

15.1 Management of Terrestrial Habitats

Management techniques employed in terrestrial habitats are highly diverse. Common techniques include efforts at improving food production, cover, and water availability. As previously mentioned, some techniques are intended to benefit particular species or communities, while other techniques are primarily intended to reduce the utility of (degrade) habitat for some species or communities.

Food Production

Food production techniques in terrestrial habitats typically concentrate on mast, browse, or grasses and forbs, including agricultural crops. As mentioned in section 1.3, most managed wildlife species are primarily herbivorous; therefore, this discussion primarily involves vegetation management. Even if carnivorous wildlife species, such as cougars or northern harriers, are the target species, most of the habitat management techniques implemented still involve vegetation management, because predator population abundance is related to food availability and predators often prey on herbivorous species. For example, if vegetation is managed for herbivorous prey, populations of carnivores will be affected. If insectivorous birds, such as Brewer's sparrows or Blackburnian warblers, are the target species, then food supply protection is often a concern. In such situations, biologists may monitor pesticide use and runoff from agricultural and urban areas to ensure that insect populations are not depleted. Because these insect prey species are often herbivores, habitat management is still involved.

Three primary methods used to improve food production are propagation, release, and protection. **Propagation** involves planting seeds or transplanting young trees or shrubs. Shelterbelt planting, for example, is a type of propagation. Plant species such as red cedar and wild plum are sometimes included in shelterbelt plantings because the mast they produce attracts game and nongame wildlife species, although they are considered nuisance species in some locations. Propagation is often used in urban settings to provide specific types of foods for songbirds. Much information is available on habitats, such as mast-producing plants, selected by various birds (Cody 1985; Marzluff et al. 2001; Odell et al. 2003).

Browse is typically used as food by a variety of animals, including mule deer, black-tailed jackrabbits, North American porcupines, and ruffed grouse. As with mast, browse can be propagated. Planting is a common method of increasing browse.

There are few wildlife species that do not include in their diets at least some grasses or forbs, or their seeds, or the insects that feed on these plants. A common propagation practice that allows the establishment of grasses and forbs is range reseeding. In general, sowing seeds, either by hand or mechanically, without seedbed preparation is unlikely to be successful. In most cases, it is best to prepare a seedbed of exposed soil so that newly planted seeds can germinate and grow with reduced competition. The species planted will vary by geographic location. Similarly, plantings of cereal crops in food plots for wildlife can provide both food and cover for many game and nongame species. The crops planted vary by region, but such plots most commonly contain corn, millet, sorghum, or a combination of these plants. Plots of tall varieties of sorghum, sometimes mixed with corn or shorter sorghum varieties, can provide important winter cover and food in northern areas. Such plots are particularly important in areas of intensive agriculture. They need to be large enough so that drifting snow will not destroy their cover and food values in severe winters. In urban settings, feeding stations with different types of seeds or other foods can be provided for a variety of animals as a substitute for propagation.

Mention should also be made here of the ever-increasing efforts to restore communities of native grasses and forbs by reseeding. At one time, the cost of this practice was prohibitive because seeds for these native plants were at a premium. Seed costs have been reduced as demand has resulted in increased seed production. Increased native plant reseeding has resulted in an economically important form of alternative agriculture.

Release refers to the reduction of competition from other, less desired plant species. If competing plants are removed from the area surrounding a mast-producing tree, for example, more mast production can be expected. Conversely, release can also occur after a reduction in the target plant species, as lower abundance may result in a greater number of healthy plants. Release is an important food management tool, and it can be accomplished by mechanical methods, chemical methods, or burning.

Mechanical release methods can include pulling a chain or cable through vegetation to break and uproot undesirable plants, using conventional agricultural tillage equipment such as disks or plows, or even hand pulling. Another mechanical release technique is the use of domesticated livestock, whose grazing and trampling can stimulate or depress the growth of various plants. In urban settings, mechanical methods such as mowing or pruning can result in the release of targeted plants.

Timber management can also be used by biologists for improving food production (Yahner et al. 2005). Thinning of tree stands allows light penetration and increased production in the understory. Clear cutting can also be used as a management tool. Clear cutting has a bad reputation because of the environmental damage caused by clear cuts over huge tracts of land (see section 14.5). However, timber stand improvement using small clear cuts or block cuts can provide open areas in which grasses, forbs, and young trees can sprout, providing much wildlife food. Selective cuts can also promote forest rejuvenation.

Management for uneven-aged timber stands can be especially important for some animals. For example, biologists manage ruffed grouse in northern states such as Minnesota and Michigan by managing aspen stands. Different portions of the stands are cut at approximately 10-year intervals such that four stages of aspen are always available, from newly sprouting aspens in recently cut areas to mature aspens that have not been cut for 30 or more years. First-stage areas (most recently cut) are important feeding areas for young ruffed grouse. Adults feed and roost most often in third-stage areas (cut 20 years ago). Second (cut 10 years ago) and fourth (cut 30 years ago) stages are used by the adults during breeding. This type of habitat management is also useful in urban environments, where the promotion of different layers in a forested area can be used to promote species diversity. This practice can be applied to everything from small backyard areas to larger, more natural areas in the urban landscape.

Chemical control of competing plants can also be used for improving food production, but many biologists are concerned about the long-term and sublethal effects of chemicals. Perhaps the best advice, if chemical control is deemed to be the most effective release method in a particular situation, is to consider the use of small handheld sprayers that will allow highly selective placement of herbicides.

Fire is a natural force in many landscapes and ecosystems. Whether intentionally set or naturally occurring, fires have influenced both forest and grassland habitats in North America for thousands of years. For example, estimates of historic, lightning-set fires in the northern Great Plains range as high as 92 fires per 10,000 square kilometers per year (Higgins 1984). Fires deliberately set by native peoples were apparently even more frequent on North American prairies than lightning-set fires (Higgins et al. 1989).

Fire as a release mechanism can remove dead vegetation that hinders new growth, free nutrients to enrich the soil, reduce invader species and encourage native species, and in general create and maintain habitats attractive to wildlife. Fire can also result in earlier vegetative growth in the spring. A **prescribed burn** (**controlled burn**) is a human-induced fire that mimics the course of nature. Prescribed burns can be categorized as **reclamation burns**, which are used on mistreated and unmanaged lands, or **maintenance burns**, which are set regularly on a particular area as a management practice. **Backfires**, which are set upwind from a barrier such as a road or a cleared section of land, burn against the wind. Such burns tend to move slowly and burn thoroughly, so they are often used for reclamation burns. A backfire may also be set in an area where a biologist wants a headfire to stop burning. **Headfires**, those set to move with the wind, tend to burn more rapidly and less thoroughly and are more often used for maintenance burns.

Prescribed burns generally should be undertaken before the growing season, and certainly before seed heads of desired grasses begin to develop. The actual timing of prescribed burns varies with latitude. Prescribed burns may be done in February or March in Texas or Florida grasslands, but not until April or May in North Dakota grasslands. The timing of a prescribed burn also varies with the life cycle of the plant species being managed. For example, biologists in northern regions who want a prescribed burn to favor a cool-season grass, such as tall or intermediate wheatgrass, would burn between mid-April and mid-May. However, if they want the prescribed burn to favor a warm-season grass, such as big bluestem or Indian grass, the burn would be delayed until early June. Prescribed burns can also be done during the fall, again with the actual date varying by latitude. Burning at inappropriate times of year (Bleich et al. 2005) and in vegetation such as big sagebrush (Connelly et al. 2000) can produce negative effects on wildlife populations.

Woodlands can also be burned to improve food production. Reclamation burns are likely to be hot and intense, because unmanaged woodlands often contain a substantial amount of fuel. Maintenance burns are higher in frequency and lower in intensity, as the usual management objective is to decrease litter and allow understory plants such as grasses and forbs to develop. Thus, maintenance burns are generally set during cooler times of the day when humidity levels are high.

Protection, as the word implies, refers to attempts to protect plants until they mature and produce the desired food. Protection may involve screening or fencing the plants. Like propagation, protection is a common habitat management strategy in urban settings. Mast, browse, grasses, and forbs can be enhanced by protection. Saplings of quaking aspen are an example of browse species that can respond rapidly and positively to reduction in browsing from species such as cattle and elk.

Cover Development

Cover development is an important habitat management tool, because wildlife species depend on a variety of cover types. Often it is difficult to distinguish between management for cover and management for food, because both aspects of a habitat may be improved by the same management techniques.

Cover can be provided by such features as hedgerows, dense timber stands, shelterbelts, brush piles, and natural or artificial roosts or nests. Maintaining undisturbed cover along fences, in railroad rights of way, at roadsides, and in backyards can encourage many wildlife species. An effective management strategy for species that nest along roads is to delay roadside mowing until most nesting has been completed. In some areas, such roadside habitat management is a policy. Dense nesting cover, often a mixture of grasses and legumes, such as alfalfa, can be planted to enhance nesting success, while in other situations, native grasses are used for such cover development.

Some wildlife species, such as many of the squirrels and woodpeckers, rely on den or nesting trees. Protection of such habitat is important for these wildlife species. The U.S. Forest Service identifies and clearly marks den and nesting trees on some of its lands. These trees are protected, even in areas where logging is occurring. Artificial nest structures are built for many wildlife species. Some waterfowl species, such as Canada geese and mallards, will nest in artificial structures such as nest baskets (Figure 15.1). Nest boxes can

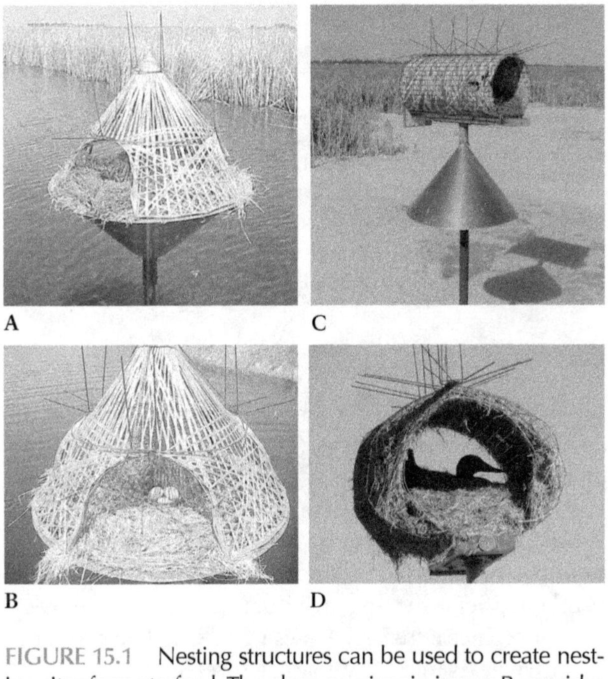

FIGURE 15.1 Nesting structures can be used to create nesting sites for waterfowl. The close-up view in image B provides a view of eggs in the nesting basket in image A. The metal cone below the nesting structure in image C prevents access by mammalian predators. A hen mallard is on a nest inside the structure in image D; the wire on top of the structure helps prevent perching by raptors. *[Images A and B courtesy of P. Mammenga; images C and D courtesy of M. Grovijahn.]*

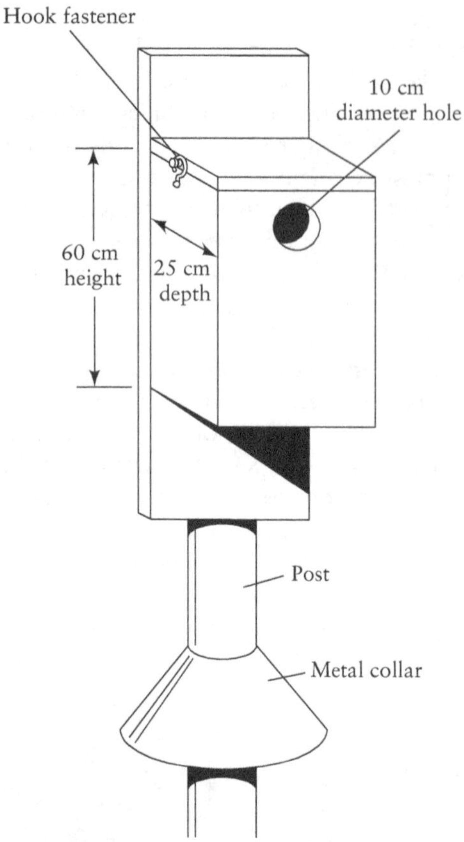

FIGURE 15.2 A nest box constructed to encourage wood duck nesting. Several centimeters of wood shavings are placed in the box, which should be cleaned annually. The box can be mounted on a post or attached to a tree at a height of 3–10 meters above the ground or water.

be used for animals such as squirrels, American kestrels, and wood ducks (Figure 15.2), and nesting platforms are commonly used for ospreys (Ewins 1994). Bat boxes can be constructed to provide roosting locations, and butterfly houses can be constructed to allow predator avoidance and thermal regulation. While artificial nesting structures can be useful, they cannot replace extensive losses of natural habitats such as native grasslands and mature forests.

The reproduction and survival of some species can be promoted through other cover development techniques. The building of islands in water bodies, for example, can provide areas safe from nonaerial predators. Islands can be purposefully built during the construction of artificial water bodies, formed by excavating material from the neck of a peninsula on an existing water body, or added to existing water bodies by placing materials in the water at particular locations (Figure 15.3).

Nesting by undesirable species can be discouraged with cover management techniques. Areas on buildings that are attractive to house sparrows or rock pigeons for nesting and roosting can be reduced by eliminating ledges, openings accessible to birds, and small projections from building designs. Squirrels can be discouraged from nesting in house attics by effectively screening small openings. Spotted and striped skunks and rodents can be discouraged from using the undersides of buildings by adequately sealing such areas to prevent their entry.

A variety of other devices and techniques can be used to control animal access to areas needing protection. For example, areas can be fenced to prevent or control access by domesticated livestock and to allow the development of better wildlife cover. Interstate highway fences can be constructed in urban areas to prevent deer or other large ungulates from gaining access to areas of high traffic. In western areas of North America, consideration must be given to the migrations of species such as mule deer. Highway underpasses are sometimes constructed to allow the movement of these animals. Orchards may need to be fenced to prevent access by deer, which feed on fruit or may damage

FIGURE 15.3 Waterfowl nesting islands can be created by excavating earth across a peninsula to form an island or by hauling sufficient fill to an area to create an island. This rectangular nesting island was built by Ducks Unlimited on the Harris Waterfowl Production Area in Sheridan County, North Dakota. *[Image courtesy of R. Warhurst.]*

trees. Predator exclosures constructed of wire fencing have been used to increase the nest success of piping plovers and least terns, two nongame bird species of concern (Kruse et al. 2002). Nesting baskets can be placed over water to reduce predation by terrestrial animals. Metal collars can be placed around trees with natural nest sites, on trees and poles supporting artificial nesting structures (see Figure 15.2), or on birdhouses and bird feeders in backyard urban settings to reduce predator access. The sheet metal around the bases of telephone poles seen in many areas of the eastern and central United States is used to reduce eastern fox squirrel and raccoon access; power outages can occur if these animals gain access to electrical wires. Electric barriers are used to reduce nonaerial predator access to peninsulas as a means of protecting nesting waterfowl and other ground-nesting birds (see Figure 10.1).

Many research studies involve the placement of exclosures that prevent animal access to a certain area (see Figure 13.20). These control areas are then compared to areas where access is allowed. For example, fences that prevent elk access to riparian habitat in wintering areas can provide information on the effects of elk on willows and other vegetation.

Biologists must be aware of habitat needs on a broad landscape level, particularly with the increasing emphasis on ecosystem management. Juxtaposition of habitat types is important to both terrestrial and aquatic organisms. Northern bobwhites, for example, require food, cover, and other habitat types in close proximity so that the needs of a covey or pair can be met within a relatively small home range. Species such as the northern bobwhite that require a variety of habitats in close proximity are sometimes referred to as **edge species.**

Other species, referred to as **interior species,** require large patches of a relatively homogeneous landscape, such as tall-grass prairie or temperate forest. These species are negatively affected by management practices that fragment larger patches of habitat into smaller patches with numerous ecotone areas. The northern spotted owl, which uses old-growth forests in the Pacific Northwest, is an example of an interior species. Similarly, many songbirds in the eastern United States are dependent on large patches of woodland habitat. For example, Robbins et al. (1989) found that the likelihood of occurrence for Neotropical migrants such as red-eyed vireos, wood thrushes, scarlet tanagers, and ovenbirds increased as the area of forest increased to 100 hectares or more.

Water Development

The need for development of water sources varies depending on the abundance of water in a given area and the water needs of target organisms. Water development might include improvement of natural water sources such as springs, seeps, and water holes, the construction of **guzzlers,** or the placement of backyard birdbaths. Guzzlers, also known as gallinaceous guzzlers because they are particularly useful to upland game birds, are often constructed in arid regions to provide water for wildlife (Figure 15.4). Guzzlers typically are watertight

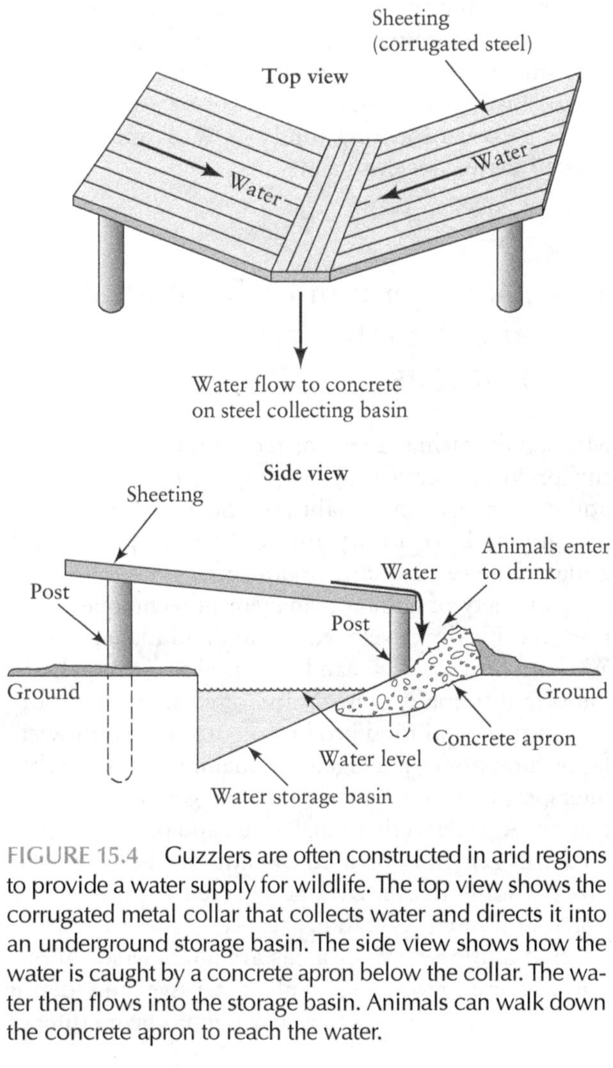

FIGURE 15.4 Guzzlers are often constructed in arid regions to provide a water supply for wildlife. The top view shows the corrugated metal collar that collects water and directs it into an underground storage basin. The side view shows how the water is caught by a concrete apron below the collar. The water then flows into the storage basin. Animals can walk down the concrete apron to reach the water.

tanks set into the ground and filled by means of a rain-collecting collar. Rainwater collected is stored underground to reduce evaporation. Birds and mammals walk down a sloping ramp to the water. There are numerous guzzler designs.

In some situations, water development for terrestrial species may involve the construction of artificial water bodies. Small farm and ranch ponds, for example, are particularly useful in some regions where natural watering places for wildlife are a limiting factor.

In urban areas, natural water bodies are often degraded, with resulting decreases in water availability. In these circumstances, something as simple as a ceramic birdbath or a rock with a depression that will hold water can be useful to wildlife.

Numerous terrestrial wildlife species obtain sufficient water from dew, leafy vegetation, insects, and metabolic water and may not need drinking-water sources; some desert species such as kangaroo rats survive on metabolic water alone (see section 5.10). Biologists should be aware of actual water needs before allocating funds on what may be unneeded water developments.

15.2 Management of Terrestrial and Aquatic Transitional Habitats

Many habitat management techniques discussed in the previous section also apply to terrestrial and aquatic transitional habitats. Such management techniques have utility for terrestrial species and aquatic and semiaquatic organisms.

A variety of habitat management techniques have been specifically developed for wetland management. Wetland areas, which can be natural or artificial, are important to many animals and often serve as important nursery and food production areas. Natural wetlands are often managed to maintain a desirable interspersion of open water and emergent aquatic vegetation. A wetland that is half open and half plant covered will generally be used by more bird species and other wildlife than a wetland covered with emergent vegetation such as cattails, especially if the plant-covered and open-water areas are interspersed about the wetland. However, while extensive and dense growth of emergent vegetation can limit the usefulness of wetlands to some species, it can be beneficial to others. In areas with severe winters, such vegetation can provide important winter cover for animals such as sharp-tailed grouse, ring-necked pheasants, white-tailed deer, and a variety of nongame species. Wetland vegetation can be manipulated by mechanical or chemical means or controlled biologically, as with common muskrats. If too few muskrats are present, cattails and other emergent vegetation may completely cover a shallow wetland. If muskrats are overly abundant, they may reduce the vegetation far more than would be desirable. Domesticated livestock can also be used to reduce emergent vegetation in wetland habitats through grazing and trampling. Burning can also be used as a management tool for wetlands. Vegetation can also be burned over ice or over water to remove dead plant material and increase nutrient cycling.

If a natural wetland is too uniformly shallow, techniques are available to develop deeper areas that are more likely to provide open water or to provide some water in drought years. These deeper areas can be produced in a variety of ways. Earth-moving equipment can be used to excavate some areas. This excavation is usually done in the form of **level ditching** (Figure 15.5), in which ditches are dug in the wet-

FIGURE 15.5 The level ditch shown in this aerial photograph was created by excavation. In this wetland, the ditch holds water even during dry periods when the remainder of the wetland is dry. During wet periods, the ditch provides an area of deeper water. *[Image courtesy of M. Kjellsen.]*

FIGURE 15.6 Explosives can be used to create new wetlands or modify existing wetlands. While this technique was used historically, it has rarely been used in the past two decades given concern over availability of explosives.

land, often in a zig-zag pattern. Another means by which deeper areas can be formed in wetlands is blasting with ammonium nitrate, dynamite, or other explosives (Figure 15.6). Care must be taken during blasting or level ditching to ensure that too large or deep an area is not made, because smaller wetlands may be dewatered.

Any type of excavation or artificial manipulation of natural wetlands should be subjected to careful scrutiny because of its possible influences on the appearance and function of the wetland. However, in many areas where wetlands have become choked with dense emergent vegetation, such wetlands have already been greatly modified by agricultural and other human land-use practices.

Water level manipulation is another management tool for both natural and artificial wetlands. Natural wetlands can sometimes be enhanced by installing water level management devices such as outlet control structures. If wetland areas are interconnected, various portions of these wetlands can be kept in different successional stages with such control structures (Figure 15.7). Some natural wetland areas, such as floodplain riparian forests in the southern United States, can be managed by water level manipulation during summer and winter for a variety of wildlife species. Wintering mallards and breeding and wintering wood ducks are often targeted in the management of these riparian wetlands. Water level determines the availability of invertebrates as well as mast, such as acorns, needed by waterfowl. Management for hydric soils, hydric plants and their seeds, detritus, and the resulting invertebrate production in these riparian forests can be important to a variety of terrestrial and aquatic organisms.

There are efforts under way to produce totally artificial wetlands to mitigate losses of natural wetlands. This may seem like a simple procedure; flood an area and it becomes a wetland. In reality, it is not so simple. The aquatic plants and animals normally found in a wetland often need to be introduced.

Artificial wetlands can be created in many areas, generally by building a dam or by intentionally flooding an area using dike systems. Water levels can be manipulated in these areas, especially if sufficient water is available for pumping and subsequent flooding of wetland areas at various times of the year. Many biologists flood artificial wetlands during spring and fall migration periods for shorebirds and waterfowl. During midsummer, water levels can be lowered, and either natural hydric plants or planted crops can be allowed to grow in the basin. Reflooding will then provide highly attractive habitat for various wildlife species.

FIGURE 15.7 The water level in this restored wetland is maintained at the height of the water control structure. *[Image courtesy of the U.S. Fish and Wildlife Service.]*

There is also growing interest in restoring wetland areas, such as marshes and swamps, that were drained at some time in the past. Often, all that is required is the filling or plugging of the drainage ditches initially dug to drain the wetland (see Figures 15.7 and 20.1). However, if drainage tile was put in place to drain an area, the tiles may need to be removed or broken. In some cases, even if drained for many years, reflooded wetlands still contain a bank of wetland plant seeds just awaiting the correct conditions to germinate and grow. In other cases, reflooding alone may not result in reestablishment of plants needed for complete restoration.

Coastal wetlands in North America include tidal salt marshes, tidal freshwater marshes, and mangrove wetlands. Such wetlands are particularly prone to damage caused by urban development and pollution because of the high human population densities in many of these areas. Losses of the protective islands and shallow water areas that tend to reduce wave action in coastal wetlands are also reasons why these habitats have been degraded. Current management techniques for these wetlands include laws and regulations intended to prevent their degradation.

Beach–ocean interfaces are also often protected by laws and regulations that prevent their degradation. In addition, beach stabilization structures, the addition of sand, and other measures are employed to manage this habitat.

15.3 Management of Standing-Water Habitats

The manageability of lentic habitats varies by water body type. For example, smaller water bodies are easier to manipulate than larger water bodies. This section is divided into subsections concerning dissolved-oxygen levels, water levels, water access, aquatic vegetation, terrestrial landscaping, fertilization, and artificial structures.

Dissolved-Oxygen Levels

Some lentic systems can undergo **winterkill** if dissolved oxygen is depleted under the ice. Additionally, naturally produced toxic substances, such as hydrogen sulfide and ammonia, can increase under ice cover. Both low dissolved-oxygen levels and toxic substances can negatively affect fishes and other aquatic organisms (Figure 15.8). The loss of oxygen occurs when respiration (use of oxygen) by plants, animals, and bacteria exceeds the production of oxygen by plants during photosynthesis. Winterkill is therefore more likely to occur if a water body is shallow and contains large amounts of decaying organic material, especially algae or other aquatic plants. Such conditions are more likely in eutrophic waters. The probability of winterkill also increases with duration of ice cover, thickness of the ice, depth of snow on the ice, and ice opaqueness. Clear ice allows light penetration, and photosynthesis can occur under such ice. However, snow on the ice will block light penetration. Under winterkill situations, water bodies first lose their oxygen near the bottom, and oxygen depletion then progresses upward toward the ice.

FIGURE 15.8 Results of a winterkill are highly visible just as the ice leaves a water body in the spring. Risk of winterkill in shallow water bodies increases with nutrient loading, thickness of ice, depth of snow that covers the ice, and duration of ice cover. *[Image courtesy of B. Graeb.]*

Water bodies that are prone to winterkill are generally managed in one of three ways: the entire system of water and surrounding land (the watershed) can be managed to reduce nutrient and sediment inputs, aeration systems can be installed, or such waters can simply be managed as marginal waters.

Water bodies can be prone to winterkill because of incorrect land management in the watershed. Reducing the inflow of nutrients, organic material, and sediments into a water body can help to reduce or avoid winterkill. Such inflows can come from industrial, residential, agricultural, or other sources. Controlling these inflows requires control of land use, which often involves private property, so a variety of social, legal, and economic issues involving property rights must be considered (see section 14.10). This is the primary reason why stopping or reducing nutrient and sediment inflow is difficult to accomplish. Thus, biologists are often forced to treat symptoms rather than the actual causes of this problem.

Aeration systems prevent winterkill by maintaining dissolved-oxygen levels and providing an open-water area for the dispersion of potentially harmful toxic gases. One type of aeration system is illustrated in Figure 15.9. In this system, an air compressor or windmill is placed on the shore and an air line is extended into the water body. As bubbles rise from the air line, a current is created that carries the warmer (typically near 4°C), deeper water to the surface, and an opening is maintained in the ice. Because these systems are called aeration systems, most people assume that the air bubbles produced are the primary source of oxygen input into the water body. In reality, little oxygen transfer occurs by this means. Most of the added oxygen is produced by phytoplankton during photosynthesis because the open area in the ice allows sunlight penetration. In addition, just maintaining an exposed water surface allows for a small amount of oxygen addition by surface mixing.

FIGURE 15.9 An aeration system that uses a compressor on shore pumps air to a diffuser on the pond bottom. The rising bubbles carry warmer (~4°C) water from the bottom to the surface, and an opening remains in the ice during winter. *[Image courtesy of T. Lea.]*

If aeration systems are not correctly managed and maintained, aquatic organism death can be caused by the system itself. For example, if the system is not started until oxygen levels are depressed and toxic gases such as hydrogen sulfide are present, it can cause warmer unoxygenated water from the bottom of the water body to mix with oxygenated water above it. As a result, the average oxygen availability is reduced and winterkill may occur because the toxic gases are also present. In addition, too much circulation induced by too big an aeration system may cool the entire water column, which could lead to fish deaths in geographic locations with long winters and periods of extended ice cover.

The cost-effectiveness of using an aeration system needs to be closely evaluated. There are costs involved in the construction, maintenance, and operation of such systems (for example, electricity to run the compressor). A conservation agency would be more likely to invest money in an aeration system in an area that had few fishing waters. If other, higher-quality water bodies were present in the same locale, aeration probably would not be undertaken. Aeration systems are popular with lake associations and other citizen groups and private individuals, and they commonly request that the state install a system or allow them to install it.

In much of North and South Dakota, Minnesota, Wisconsin, Michigan, and southern Canada, numerous shallow, eutrophic water bodies are naturally prone to winterkill. Conservation agencies in many of these areas manage these water bodies as marginal waters, meaning that they are managed for short-term production of recreational fishes, recognizing that the frequency of winterkill will be high. For example, perhaps winterkill can be expected once every 3 years. Such marginal water bodies can still provide many hours of recreational angling. They tend to be quite productive, and fishes that are stocked after a winterkill often have rapid growth rates. In North and South Dakota and Canada, fishes such as the northern pike, yellow perch, and even rainbow trout are commonly stocked into these waters. Some of these marginal waters can also be used for aquaculture purposes. Bait fishes such as fathead minnows can be raised, and sport fishes such as walleyes and yellow perch can be reared to advanced sizes, captured, and then stocked elsewhere.

Summerkill, like winterkill, is a result of oxygen deprivation. Summerkill generally occurs in eutrophic waters. As with winterkill, summerkill can be associated with incorrect land-use practices in the watershed. Consider one of the marginal, eutrophic water bodies discussed in the previous paragraphs on winterkill and the following conditions. During midsummer, hot and sunny weather dominates for a 2-week period, and a large **algal bloom** develops. *Bloom* is a commonly used term that denotes a rapid increase in algal populations. The weather then changes, and a 5-day period of cool, cloudy, calm weather occurs. The clouds block some sunlight, some of the algae die, and the calm weather allows for little replenishment of oxygen by wind action. Just before sunrise, after several days of this weather pattern, respiration by plants and animals, combined with the extra respiration caused by increased numbers of bacteria feeding on the decaying algae, may reach a point at which oxygen supplies are depleted. Summerkill occurs if the dissolved-oxygen level drops too low to support some or all of the fishes present.

Summerkill can also occur in a stratified water body. In eutrophic waters, the hypolimnion often loses its oxygen. The hypolimnion is the zone of decomposition, and its oxygen cannot be replenished because photosynthesis does not occur at that depth. Hypolimnetic water cannot mix with better-oxygenated water in the epilimnion because of water temperature differences (see section 12.3). High winds can cause destratification, however, allowing complete mixing, and the combination of low oxygen levels, warm water temperatures, and the presence of other dissolved gases such as hydrogen sulfide can lead to a partial or complete summerkill in the entire water body. Even without total mixing, a partial summerkill can occur; fishes restricted to the hypolimnion by temperature requirements may be the only ones to die if that layer loses its oxygen.

Aeration systems can be used to decrease the likelihood of summerkill. In small water bodies that can become stratified, aeration systems are used to help prevent stratification. Thus, the entire water body circulates and oxygen is available throughout. In addition, summer aeration is at times used in oligotrophic water bodies, where the hypolimnion can act as a nutrient trap. Destratification will allow the nutrients to cycle throughout the water body for use by plants.

Fishes vary in their sensitivity to low dissolved-oxygen levels and thus in their susceptibility to winterkill or summerkill. Most fish species can live at lowered oxygen levels if they are slowly acclimated to reduced concentrations. However, many sport fishes, such as walleye, trouts, and basses, are the first to die when dissolved-oxygen levels are reduced. Fishes such as northern pike, bullheads, and common carp are more tolerant of low dissolved-oxygen levels. While many fishes require at least 4–5 milligrams of dissolved oxygen per liter of water, these tolerant fish species can exist at oxygen levels below 1 milligram per liter. Partial kills can be a problem for fisheries biologists. Biologists responsible for waters at risk of winterkill or summerkill would usually prefer a complete kill rather than a partial kill that allows less desirable fishes to survive and reproduce the next spring. However, complete kills seldom occur. Little is known about tolerances and effects of winterkill and summerkill on most nongame fish species and other aquatic organisms.

Water Level Manipulation

Planned water level manipulation can be a useful tool for biologists who manage water bodies. A newly constructed water body generally has an initial period of high productivity, because as it fills, nutrients are continuously being added to the system. Water level manipulation is sometimes used to mimic this initial period of high productivity.

Decreases in water level (**drawdowns**) concentrate predators and prey; many biologists use drawdowns to force members of a stunted species, such as bluegills, out of cover so that predators, such as largemouth bass, can reduce their populations. This management practice improves habitat for the predator species while degrading habitat for other species, in this case the stunted panfishes. Increased mortality of the small panfishes will result in faster growth rates and improved size structures for both predator and prey species. Drawdowns expose the bottom of a water body to oxidation, and nutrients are thus released from bottom soils. Drawdowns also allow for compaction of silt, resulting in improved water clarity after reflooding. Drawdowns may also affect nontarget organisms, either positively or negatively.

Drawdowns allow the growth of terrestrial vegetation on exposed shoreline areas. Such vegetation may be natural (for example, smartweed and many sedges will commonly grow in such environments) or may result from human seeding of plants such as millet. When reflooded, this vegetation results in increased populations of bacteria, phytoplankton, zooplankton, and aquatic insects. Obviously, this increased overall food supply results in increased productivity for fishes; wildlife that feed on invertebrates, such as shorebirds and many duck species, also benefit. Some fishes may also increase recruitment as a result of the flooded vegetation. Largemouth bass recruitment, for example, may be low in turbid water bodies that lack submerged aquatic plants. Flooding of terrestrial vegetation results in improved habitat for largemouth bass, and their recruitment temporarily increases. Flooded terrestrial vegetation is also attractive to waterfowl because it provides seeds as a food source.

Flooded vegetation can help increase water transparency. Some waters remain turbid because they contain suspended clay particles. These particles have a negative charge, and therefore they repel each other. The material remains in suspension because the particles cannot clump and settle to the bottom. Decaying aquatic vegetation provides an attachment site for these clay particles; thus, biologists can expect improved water transparency in such waters after terrestrial vegetation has been flooded. Increased water clarity is beneficial to some species and reduces populations of other species. Sight-feeding predators may become more effective in clearer water, while prey species may be reduced in numbers because of increased vulnerability to predation. Increased water clarity can also lead to increased productivity due to increased photosynthetic activity that results from greater light penetration.

Water level management does not always include the addition or removal of water; it may instead involve the modification of the bottom of a water body. Dredging to deepen a water body functions in much the same manner as adding water. Dredging is a costly procedure in existing water bodies, and deepening is much more likely to be economically feasible if the excavation occurs when artificial water bodies are being constructed.

While water levels can be increased or decreased by pumping, siphoning, or controlling water releases, water level management is most often done in water bodies that have a preexisting means of water retention and release, such as impoundments or reservoirs. Many such water bodies are multiple-use impoundments, meaning that they were constructed for water supplies, irrigation, flood control, electricity generation, recreation, navigation, and other purposes. Because these multiple uses often conflict, biologists find it difficult to implement a water level manipulation plan that will satisfy all users. Conflicts among multiple uses, even just recreational uses, can complicate water level management. Most recreational facilities are designed for a certain water level. Water levels above this elevation can cause flooding and shoreline erosion from waves, and water levels below this elevation can result in the facilities being difficult to use; for example, boat ramps may be out of the water, or camping and picnicking areas may be too far from the water. Water level manipulation intended to improve a fishery can have substantial negative effects on other organisms. For example, island-nesting birds can be reduced if the bottom of the water body around the island is exposed. Despite these problems, water level management can be one of the most successful tools for fisheries management in impoundments.

Water Access

Aquatic habitat management in urban settings is often less habitat management than access management. The management emphasis is less upon habitat and more upon making space available for large numbers of people. Urban water bodies are often made more accessible to anglers and other users by providing access along open shorelines, fishing piers, and rock jetties. In most of these situations, angling effort is so high that habitat management within the water body could not possibly provide fish populations of sufficient density, so stocking is sometimes a common practice (see section 10.7). At the extreme end of the urban habitat spectrum are very artificial fisheries, often placed in permanent or temporary water bodies such as swimming pools, which primarily serve youth, the elderly, and the physically disadvantaged. Such fisheries are simply based on water-holding mechanisms to support put-and-take fisheries, and little habitat management is involved.

Water access management is also important in nonurban settings. Road networks, boat-launching areas, and parking lots are important means of access. In addition, amenities such as fish-cleaning stations, restrooms, picnic areas, and even food, bait, or fuel concessions may also be needed. Access management is usually based on the amount of fishing or on other public uses at a particular water body.

Aquatic Vegetation

Fisheries biologists are commonly called upon to address problems associated with aquatic vegetation in water bodies. Most human users tend to view aquatic vegetation as a problem that needs control. It can make fishing, as well as water skiing and swimming, more difficult and can impede the use of outboard motors. Overly abundant vegetation can allow some panfish species to avoid predators and overpopulate, and it can increase the likelihood of winterkill or summerkill. Some aquatic vegetation, primarily some blue-green algae species, also produces substances that can be toxic to aquatic and terrestrial animals.

Although excess vegetation can be a problem, aquatic vegetation has many positive attributes. Aquatic vegetation produces tremendous amounts of food for aquatic and terrestrial organisms. Farm ponds constructed in grassland ecosystems have proved particularly valuable for nesting ducks. Nesting hens and their ducklings find an abundant supply of aquatic invertebrates associated with the submergent vegetation in these ponds. Further, emergent vegetation is important escape cover for ducklings and escape and nesting cover for a variety of other animals.

Many aquatic stages of insects cannot exist without vegetation. Vegetation buffers the effects of wind action and therefore serves to make water clearer and to reduce wave height in large water bodies. Vegetation provides cover for young fishes, both for escape from predators and for protection from wind and wave action. In reality, most aquatic communities benefit far more from aquatic vegetation than anglers, boaters, swimmers, and other recreational resource users realize. For example, various studies have indicated that largemouth bass/bluegill communities function best when submerged aquatic plants cover 20%–40% of the water body surface area. Therefore, if vegetation is a problem, the best means of vegetation management may be a combination of controlled and uncontrolled areas.

Aquatic vegetation control methods can be categorized as mechanical, chemical, or biological. Mechanical control methods can range from simply using a rake to keep a fishing lane or swimming area open to expensive, boat-mounted mechanical vegetation harvesters that cut and remove vegetation. Dredging of shallow water areas can result in deeper water and reduced vegetation, but this method is expensive. Sometimes opaque plastic sheeting used to cover small areas of bottom can provide an effective control. Drawdowns compact bottom soils and generally result in reduced rooted aquatic plant abundance for a few years. A drawdown over winter can be especially effective, because it ensures freezing and desiccation of soils and plant material as well as compaction of soils.

Chemical control involves the use of herbicides to reduce vegetation. It is important that a biologist using chemical control methods be able to identify the aquatic plant species causing the problem, because many chemicals are selective. Chemicals that are effective on algae, for example, are often ineffective on vascular aquatic plants. In the United States, both the Environmental Protection Agency and the Food and Drug Administration control the types of herbicides that can be used in aquatic habitats. The effects of chemical control tend to be temporary; plants typically regrow quickly after the herbicide has been applied. Thus, application at regular intervals may be needed, which can become quite expensive. In addition, many chemicals are designed to eliminate plants completely, which can be quite detrimental to aquatic communities, especially if the entire water body is treated.

One method of biological control is the introduction of grass carp. This exotic species, a native of Asia, is herbivorous as an adult. However, its introduction into North America was quite controversial, given the problems created by its close relative, the common carp, and other exotic introductions. Unfortunately, grass carp are now reproducing in some of the large river systems of the United States. While triploid grass carp, which are sterile, can be used to avoid this problem, grass carp commonly eliminate nearly all vegetation in a water body; it is difficult to obtain partial vegetation control with this method. In addition, grass carp are long lived and can affect a water body into which they have been introduced for 10 or more years. Fisheries biologists are therefore often unwilling to use this control method in water bodies where the fish species present rely on aquatic vegetation or food organisms associated with it, because complete vegetation removal could harm the structure and dynamics of fish communities (Bettoli et al. 1993). Similarly, wildlife biologists are concerned with grass carp effects on vegetation that is needed by waterfowl and other organisms.

It should be obvious that the various mechanical, chemical, and biological methods for controlling excess aquatic vegetation most often treat symptoms rather than causes. Large expenditures can be made with very transitory results. The best method of controlling excess vegetation is avoiding the causes. Correct planning and construction of new artificial water bodies can reduce the problem; for example, large, shallow expanses of water should be avoided. For already constructed artificial water bodies or natural water bodies, correct watershed management will reduce the problem. By controlling erosion and nutrient input, excess vegetation problems can usually be minimized.

Terrestrial Landscaping

Aquatic habitats, both lentic and lotic, can often benefit from terrestrial landscaping. Properly placed trees and shrubs can reduce the effects of wind on water bodies, help to reduce water temperatures, and provide an aesthetically pleasing setting. They also can produce quiet water areas for the enjoyment of bird watchers, anglers, swimmers, and other human users. In addition, maintaining shoreline areas in trees, grasses, or forbs improves water quality by reducing erosion. Such terrestrial landscaping can have negative effects by providing raptor perches and increasing predation.

Fertilization

Fertilization is sometimes used as an aquatic habitat management tool, most often to increase carrying capacity for fishes in a water body. This tool should be used only in areas that have infertile soils and resulting infertile waters, as commonly occurs in the southeastern United States, where soils are highly leached. In established waters, phosphorus is typically the limiting nutrient and often can be the sole ingredient used in a fertilization program. An infertile 2-hectare pond in Alabama, for example, may contain a standing stock of only 15 kilograms per hectare of largemouth bass and 40 kilograms per hectare of bluegills, whereas a typical 2-hectare pond in the Midwest may contain 40 kilograms per hectare of largemouth bass and 150 kilograms per hectare of bluegills. More fertile waters and water bodies in the northern part of North America should not be fertilized because of the increased risk of winterkill.

While not a form of fertilization, the addition of lime is sometimes used as a means of counteracting environmental damage resulting from acidic deposition or acid mine drainage. Liming neutralizes acidity directly, and it also increases the buffering capacity of a water body. Liming is usually not a cost-effective habitat management tool except under certain conditions, and it is most commonly used for ponds and other impoundments.

Artificial Structures

Fisheries biologists sometimes place artificial structures in lentic water bodies to attract fishes. A variety of materials can be used, including brush piles or Christmas trees weighted by cement blocks (Figure 15.10), automobile tires bundled into pyramids,

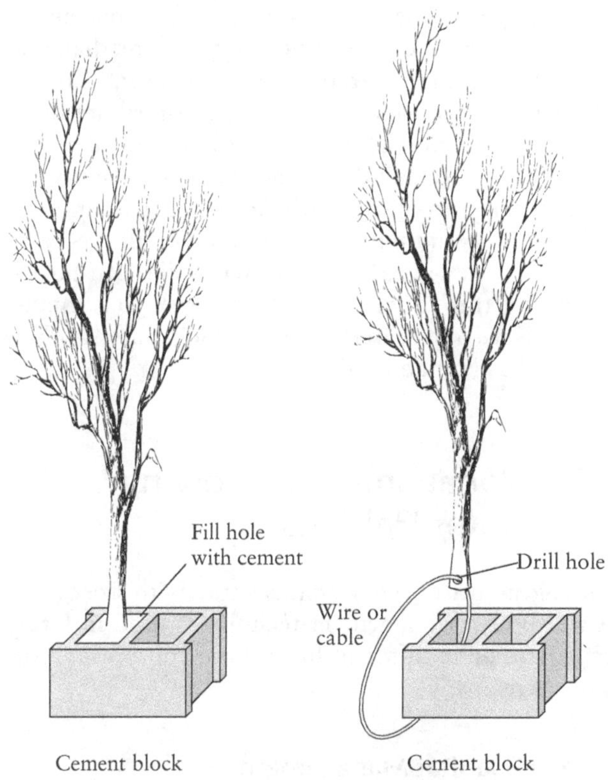

FIGURE 15.10 Fish attractors can be constructed from trees or brush and cement blocks.

concrete blocks and rubble piles, and old car or bus bodies (primarily used in the ocean). In some marine situations, old oil well drilling rigs and old ships are sunk to serve as artificial reefs. Most biologists use artificial structures as fish concentrators and do not expect an increase in fish productivity (Pardue and Nielsen 1979). However, extremely large artificial reefs can apparently result in increased productivity (Prince et al. 1979).

Biologists involved in preconstruction planning for a new impoundment should attempt to retain as much of the standing timber and natural structure as possible. Standing timber will concentrate and provide habitat for many fish species, and it is much easier to retain the timber than to add artificial structures at a later date. Unfortunately, after a number of years, flooded timber begins to decompose, and its usefulness declines. It is important to note that decomposition of organic materials in water bodies has at times created concerns over water quality, especially in human water supply impoundments. For example, some regulatory agencies are concerned that decomposition of standing timber in reservoirs or brush piles added to a reservoir as fish concentrators will cause increased levels of trihalomethane (a suspected human carcinogen). However, Layher (1984) reported that such practices contribute little to the degradation of water quality compared to inputs from the remainder of the watershed. Decaying vegetation does allow methylation of mercury, resulting in contamination of aquatic life. Newly flooded or newly expanded water bodies can be affected by this process, which is a risk to humans when fishes and waterfowl from such water bodies are used as human food.

15.4 Management of Flowing-Water Habitats

Manageability of lotic habitats varies with size of the water body. Management techniques used in large rivers are quite different from those appropriate for small streams.

Stream Habitat Management

Stream habitat management is a common practice in coldwater streams; warmwater and coolwater streams are less often managed, although more attention has been paid to them over the last two decades. Much of stream habitat management is undertaken to improve degraded sections of streams. Stream habitat improvement has been an important part of coldwater stream management in the Rocky Mountains, Appalachian Mountains, and upper Midwest (Michigan, Minnesota, and Wisconsin) for many years.

Human activities such as domesticated livestock grazing (see Figure 14.6) and other agricultural practices, building and road construction (see Figure 13.4), forest cutting (see Figure 14.7), and urbanization can result in losses of streamside vegetation and overhanging banks as well as a wider and shallower stream basin. A shallower stream basin without overhanging vegetation can lead to increased water temperatures, which can result in the loss of coldwater fish habitat if the water warms too much. Various habitat management practices can be used to alleviate such problems.

One of the best methods of stream habitat improvement is the preservation of natural conditions. Therefore, fencing can be an important tool. Simply by fencing a stream to exclude domesticated livestock or vehicles, biologists can allow the stream to recover its original vegetation and channel. Fencing can be expensive, and the cost-effectiveness of such a strategy must be considered.

Many artificial stream habitat improvement structures can be built in the streambed to improve the quality of habitat. **Current deflectors** and **bank covers** (Figure 15.11) are used to narrow streams and provide overhead cover along the erosional zone or straight reaches. Narrowing a stream increases its current velocity, which can result in lower water temperatures and increased scouring effects of the current on the streambed. Overhead cover provides shade and can also result in lower water temperatures. Many fishes seek the cover provided by a simulated overhanging bank.

Current deflectors can be constructed from large rocks or logs. If logs are used, they must be dug into the stream bank to prevent their loss during times of high water flow. Deflectors can also be constructed with rock-filled wire baskets called **gabions**. Often, current deflectors are used to direct stream current into a stream bank or tree roots, creating a natural overhanging bank (Figure 15.12).

The type of bank cover used depends on the stream bottom (Hunt 1988). In streams with sandy

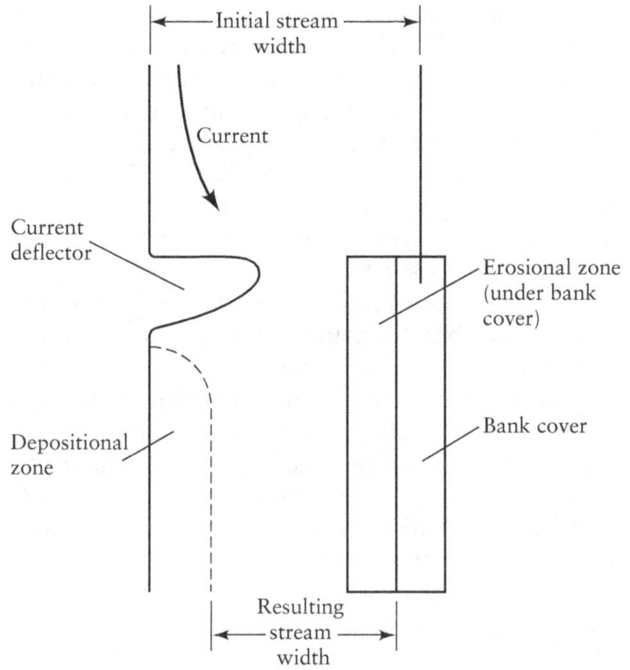

FIGURE 15.11 Current deflectors (such as rocks or logs) can be used to direct current. In this case, current is directed under an artificial bank cover (see Figure 15.14). Consequent increases in current velocity prevent silt from accumulating under the structure, resulting in an overhanging bank. The area of the bank behind the current deflector will become a depositional zone. Thus, stream width will be narrowed and current velocity increased by use of these habitat improvement techniques.

bottoms, pilings can be driven into the stream bottom to support overhead covers. Where bottoms are hard, biologists often use "skyhook" or "lunker" structures (Figure 15.13). Wire, rocks, and boards can all be used to create these structures. After soil is placed on the top and vegetation allowed to regrow, users such as anglers and picnickers can rarely tell that the structures are anything other than natural.

Brush bundles can be placed on the depositional zones found on inside bends of streams (Figure 15.14), where they collect silt and eventually help to narrow the streambed. **Half-logs** (Figure 15.15) can be used to provide overhead cover in the middle of a stream. Although half-logs are occasionally used in coldwater streams, they are more commonly used in warmer streams that contain smallmouth bass.

Dams, generally small ones, are also used to improve stream habitat. Generally, a deeper water area, or plunge pool, forms below the dam. Dams are typically used only on steep-gradient streams. In shallow-gradient streams, a dam slows water flow, opens the overhead canopy, and can allow too much stream warming. Similarly, removal of vegetation along small coldwater streams, for example, at high altitudes in the Rocky Mountains, allows greater sunlight penetration, thereby warming the waters, which increases productivity. However, in some coldwater streams, temperatures during midsummer are often near the upper limit tolerated by coldwater fishes. Shade from vegetation is essential for keeping water temperatures cool in such streams in the upper Midwest and in low-elevation mountain streams. Stream habitat improvement experts always stress that proper techniques must be selected on a case-by-case basis. There is no standard procedure that can be applied to all stream types.

Impoundment construction can also be used in areas where streams are naturally intermittent. If water is collected in an impoundment during high runoff periods, a minimum stream flow can be provided over the entire year by means of planned water releases. A stream that has little utility for constant aquatic life can thus be transformed into one with continuous populations of aquatic organisms.

Poor land-use practices in a watershed can lead to high sedimentation rates in a stream. Such sediments

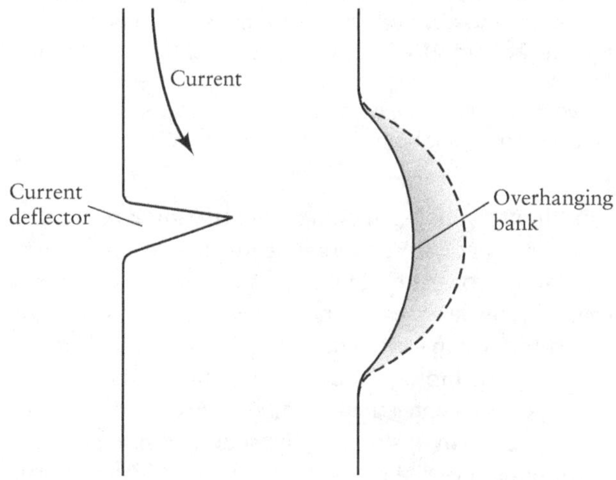

FIGURE 15.12 A current deflector being used to create a natural overhanging bank. This method can be especially effective when current is deflected toward the roots of a large tree.

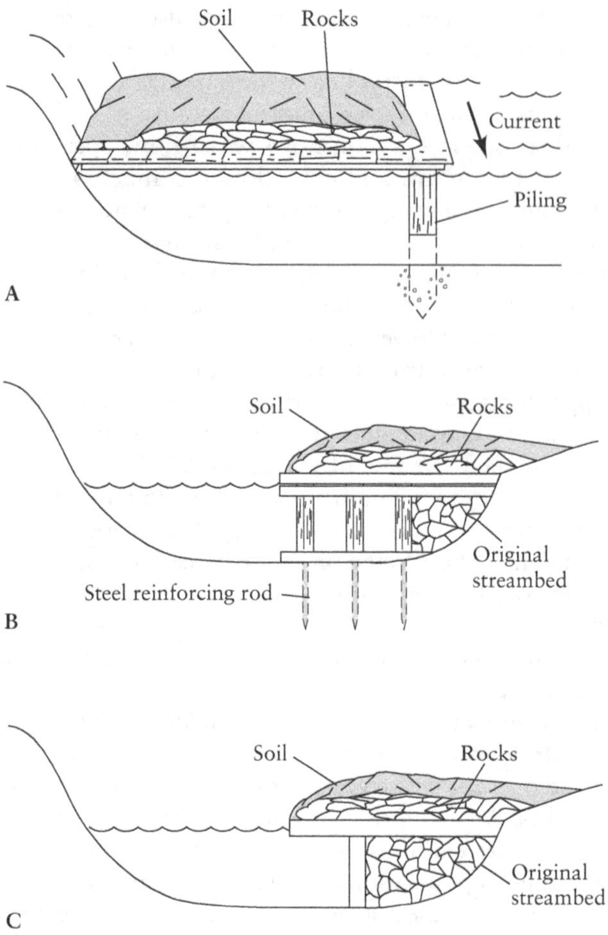

FIGURE 15.13 Various types of bank covers can be used to simulate overhanging banks. *(A)* A structure built directly on pilings and covered with rocks and soil can be used on a sandy bottom. *(B)* A "lunker" structure; steel reinforcing rods can be used if the stream bottom is sufficiently soft. *(C)* A "skyhook" structure held in place with rocks and soil can be used on a rock-bottomed stream.

can fill interstitial spaces in gravel that may be important for aquatic invertebrate production or as spawning habitat for fishes. Habitat improvement structures that narrow a stream and increase its velocity can sometimes clean sediment from a stream reach. In extreme cases, biologists can add sediment traps to a stream or even enter the streambed and clean the gravel mechanically by dislodging the sediment and actually vacuuming it out of the water (Sigler and Sigler 1990).

Special note should be made here of the challenges of stream habitat management in urban environments. Some of this habitat management involves the construction of habitats that closely resemble those found in more natural settings. For example, Rapid Creek flows through Rapid City, South Dakota. By placing large numbers of stream habitat improvement structures in this stream, biologists have been able to maintain a large population of wild brown trout. Population quality of wild fishes in an urban environment often can be maintained only through the use of harvest regulations (see Chapter 17).

Maintenance of **minimum instream flows** is another stream habitat management practice that has become increasingly important. To maintain populations of aquatic organisms, every stream must have some minimum amount of water flow, and at least that minimum must be maintained throughout the year, not just for 8 months, 10 months, or 11.5 months. Maintaining such minimum flows is usually a matter of human user management, as individuals, companies, and government agencies may all have the legal right to remove water from streams. Usually such removal is for industrial, agricultural, or municipal purposes. It should also be noted that removal of groundwater can affect surface water levels. Only relatively recently has the need for water rights for aquatic organisms become an important issue.

The problem of maintaining minimum instream flows is most prevalent in western North America, where water is scarce. The problem is often exacerbated by increasing human populations. Attempts to maintain minimum instream flows in arid areas can

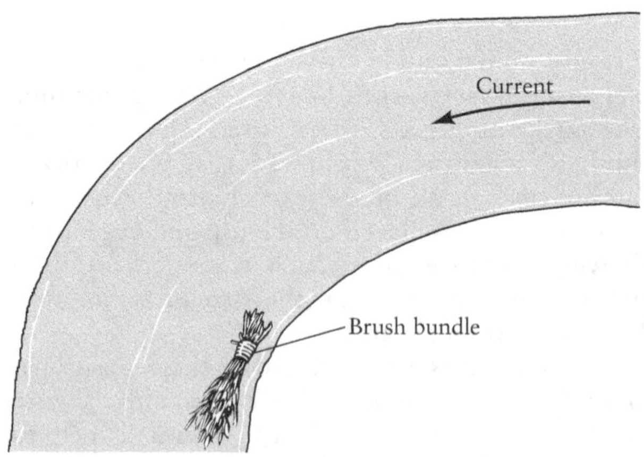

FIGURE 15.14 A brush bundle can be staked on the inside bend of a stream. The bundle can simply be a number of tree branches tied together. Silt will collect around the bundle, resulting in a narrowing of the streambed.

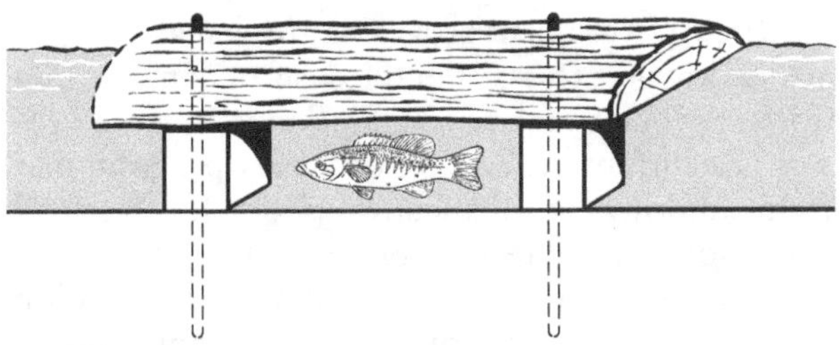

FIGURE 15.15 Half-logs can be suspended near the surface in the middle of a stream to provide overhead cover for fishes.

result in conflict among the various users of stream water. A farmer may need water to irrigate orchards or crops, a city may need water for domestic use, a factory may need water for its manufacturing process, and the aquatic organisms in the stream also need water. Satisfying all users can be a problem, especially if all of these water needs added together exceed the total flow of a river or stream; this situation is not uncommon. Efforts to provide minimum instream flows are being made in many areas, and often conflict resolution must occur where the issue is contentious. Much more will be said about conflict resolution in section 20.5.

River Habitat Management

Most artificial structures placed in large rivers are there for reasons other than fish habitat improvement. For example, the channelized sections of rivers such as the Missouri and Mississippi are used for barge traffic. Numerous bank stabilization structures have been built along these rivers. **Wing dikes** are bank stabilization structures that are usually covered with rip-rap (large rocks) and are used to concentrate the current in the middle of the river channel (Figure 15.16), thereby keeping the channel sufficiently deep for barge passage. While such structures can be quite attractive to many aquatic species, they do not compensate for the loss of habitat diversity resulting from the channelization (see section 14.8). One habitat management practice for rivers that benefits fish and wildlife is the protection of natural snags (trees, shrubs, or portions thereof that fall into rivers and streams). A more common procedure has been to remove the snags to allow for more rapid water flow. Protection of these naturally occurring structures can provide habitat for a variety of aquatic organisms; portions of fallen tree trunks and limbs protruding above the stream surface may also be used by wood ducks, belted kingfishers, common mergansers, perching songbirds, various turtles, and other wildlife.

Fishways are artificial structures that are designed to allow fish passage over or around a barrier. Natural barriers in a river, for example, might be modified by using explosives to blast a series of pools or steps in rock. Fish ladders and other mechanisms to allow fish passage are often built over or around dams. Such devices are commonly used on dams constructed on rivers that serve as migratory routes for anadromous species such as the Pacific salmon on the west coast of North America and the Atlantic salmon on the east coast, often with limited success.

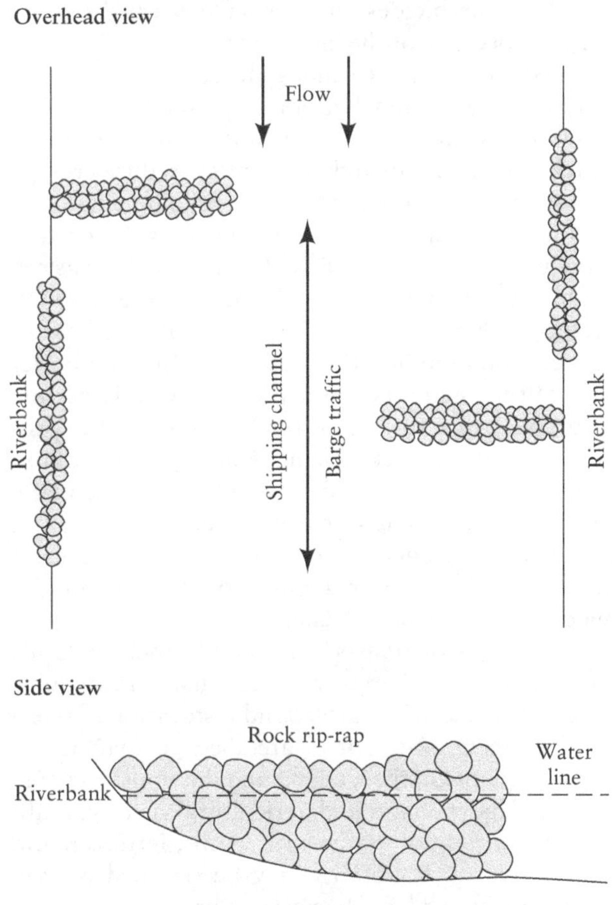

FIGURE 15.16 Wing dikes, covered with large rock rip-rap, are often constructed in large rivers. The primary function of these devices is to concentrate river flow in a center channel to provide deep water for barges. Because of turbulence caused by wing dikes, the opposite shore is typically protected with rip-rap as well.

15.5 Watershed Management

For both lentic and lotic systems, water quality is closely linked to land-use practices within the watershed (Brooks et al. 2003). Water quality in all surface waters is affected by alterations in the landscape, often as a result of agricultural practices or industrial and urban runoff. Thus, all aspects of a landscape must be considered. Modifications to land draining into lakes and streams can occur far from an affected body of water. Point-source pollution, such as sewage effluents or drain pipes, is easier to identify and control. Control of nonpoint-source pollution, such as sediment (mostly topsoil), fertilizers (especially nitrogen and phosphorus), animal wastes, and chemicals (e.g., herbicides and pesticides), can be more difficult because of the diffuse sources.

Sediment management can be attempted in many ways. Conservation buffers of planted vegetation slow sediment and filter runoff water before it reaches a stream. In addition, buffers reduce erosion from wind and provide habitat for wildlife. Examples include grassed waterways within crop fields that slow water, guide it off the field, and reduce erosion. Stabilizing stream banks with rocks (rip-rap), grass, trees, or other cover helps to reduce erosion, filter nutrients, and minimize flood damage. Riparian buffer strips of planted vegetation also reduce undercutting and collapsing of stream banks, especially if protected from livestock grazing. Planting or maintaining vegetation on lake shores will similarly control erosion in a standing water body. Sediment traps can be constructed above water bodies to trap sediment from their drainage area and also add recreational, wildlife, and aesthetic benefits.

Wet soils of riparian areas and other wetlands help maintain and improve water quality in streams, rivers, and lakes. Thus, wetland restoration is an effective technique for watershed management. Aquatic plants in the wetland and the transition area between the wetland and agricultural fields can absorb potential pollutants such as silt, fertilizer, and animal waste. Pollutants may be retained by wet soils long enough for bacteria to break them down into harmless compounds. Wet soils also store excess water, releasing it slowly over time into the downstream watershed. Prevention of flood damage can result in considerable savings that help justify wetland restoration costs.

Often, government programs are available to private landowners to assist in the costs of conservation efforts. As one example, the Wetlands Reserve Program is a current (i.e., 2007) voluntary program that offers landowners financial support for wetland restoration and protection projects. The program is administered by the Natural Resources Conservation Service. The federal government obtains conservation easements from landowners and provides cost-share payments for wetland rehabilitation practices. The program encourages tree plantings, but it primarily focuses on wetland rehabilitation. Such locations often are in river corridors and benefit the river as well as provide flood control downstream by creating areas where rivers can flood.

15.6 Management of Estuarine Habitats

Much of the management of estuarine habitat is focused on protection of this very productive habitat type. The rate of loss of estuarine habitat is high along both the Atlantic and Pacific coasts of North America and along the Gulf of Mexico. Estuaries are essential reproductive and nursery habitat for many marine fishes, crustaceans, molluscans, and other aquatic species. In addition, many birds and mammals use this habitat type for all or part of their lives.

Deepening channels for navigation and filling wet areas to form solid land causes substantial losses of estuaries and estuarine wetlands in the United States. In addition, wave action from ship traffic causes erosion of estuary margins. Contamination of estuaries with toxic substances has also affected fisheries and wildlife resources, and it has affected humans that are part of the food webs involving organisms from contaminated areas.

The value of estuarine habitat, both in economic and aesthetic terms, is finally being recognized in the United States, and protection and restoration efforts are being made. For example, the U.S. Army Corps of Engineers diverted fresh water from the Mississippi River into Louisiana coastal marshes. Intrusion of salt water has greatly altered these wetlands, and the return of fresh water, which in the past was provided by spring floods, will facilitate recovery of this habitat. Similarly, potential future increases in sea level may substantially affect coastal estuaries. Thus, suspected global climate change may directly affect estuaries, which will then affect aspects of the marine biotic community that are dependent on estuaries for some life stage (Herke and Rogers 1999).

SUMMARY

Terrestrial habitat management is typically focused on food production, cover development, and water availability. Most recreational and commercial wildlife species are primarily herbivorous, and even most managed carnivores rely on herbivorous prey; therefore, management of plants is an important part of wildlife conservation. Food production and cover development can be managed through propagation, release, or protection.

Fire is an important terrestrial management tool that is used for both grasslands and woodlands. Fire removes dead vegetation that hinders new growth and releases nutrients to enrich the soil. Prescribed burning is an artificial technique that mimics the course of nature.

Cover management is important to many terrestrial animals. Many wild animals nest on the ground and can be managed by techniques such as delaying the mowing of roadside ditches or planting dense nesting cover. Some wildlife species rely on den or nesting trees, so such trees are often protected from disturbance during timber harvest in public woodlands. Nest structures or nesting islands are built for many wildlife species. Use of particular areas by certain animals can also be prevented with habitat management techniques.

Many habitat management techniques used for terrestrial habitats can also be used for aquatic and terrestrial transitional habitats. Natural wetlands are often managed to produce an interspersion of open water and emergent aquatic vegetation. Vegetation can be manipulated through mechanical removal, chemical control, or biological control. If natural wetlands are too uniform in depth, level ditches can be excavated throughout the wetland, or explosives can be used to form deeper areas. Water level manipulation in natural or artificial wetlands can also be used to maintain wetland vegetation in the desired successional stage.

Management of lentic water habitats for fisheries often requires consideration of oxygen levels. Eutrophic waters at northern latitudes are susceptible to winterkill caused by under-ice oxygen depletion. Aeration systems can be placed in such waters to prevent winterkill, but cost-effectiveness of such a strategy must be assessed. Eutrophic waters at all latitudes are susceptible to summerkill. Oxygen levels may drop below those needed to support fishes and other aquatic life, especially just prior to dawn.

Planned water level manipulation is an important habitat management tool for impoundments. Newly flooded impoundments typically have an initial flush of productivity. Many water level management plans are designed to mimic this period of high initial productivity.

Control of overly abundant aquatic vegetation is another habitat management technique used in lentic waters. Control can be accomplished by mechanical, chemical, or biological methods. Biologists are cautious about vegetation control, because removal of too much vegetation can harm aquatic communities.

Fertilization is a management technique that can be applied to infertile water bodies, such as those found in the southeastern United States, to increase their carrying capacity.

Artificial structures are often added to lentic water habitats. Where natural structure is limited, structures such as brush piles will attract many fish species. Generally, concentration of fishes for recreational angler harvest is the goal of biologists who install these artificial structures.

Habitat management in lotic waters is most commonly attempted in smaller coldwater streams. Much of this management is undertaken to improve degraded sections of streams. Sometimes habitat improvement simply involves the reestablishment of natural conditions. In other cases, instream structures are constructed. Experts stress that stream habitat improvement techniques must be chosen on a case-by-case basis. Another major concern with regard to stream habitat management is maintenance of minimum instream flows.

A common trend in habitat management today is management, or at least consideration, of entire watersheds. Expensive habitat renovations in an impoundment are not a wise investment unless the watershed has been treated to reduce the original problem. It does little good to remove the sediment from an impoundment if land management practices within the watershed still allow high rates of soil erosion.

PRACTICE QUESTIONS

1. Why is it difficult to distinguish between the use of fire as a habitat management technique to improve food production and its use to improve cover? Provide an example of a prescribed burn applied to grasslands and a burn applied to woodlands that would improve both food and cover.

2. In the future, clean air standards may reduce the ability of wildlife biologists to undertake prescribed burns in some areas. Name some alternative techniques that could be used instead of burning.

3. In which of the North American biomes would it be worthwhile to construct guzzlers? What animal species might benefit from guzzlers in each of these biomes?

4. What is dense nesting cover? List some animal species that might benefit from the use of this habitat management technique. How would they benefit?

5. What are several differences between a restored wetland and an artificial wetland?

6. How does an aeration system maintain an opening in the ice in a lentic water body?

7. How does planned water level manipulation in an older reservoir mimic the conditions that occur in a newly constructed reservoir? Is this desirable? Why?

8. Aquatic vegetation control is a two-edged sword. Too much vegetation can impair the functioning of a fish community, but so can too little vegetation. What problems are associated with each of these two scenarios?

9. Small dams are sometimes constructed as habitat improvement structures in small streams. However, they can be appropriate for some environmental conditions and inappropriate for others. Assume that you are managing for brook trout. Would you use a small dam on a high-elevation stream in the Rocky Mountains in which water temperatures typically did not exceed 13°C during midsummer? Would you use a small dam on a low-elevation stream in Minnesota or Wisconsin in which summer water temperatures sometimes approached 21°C? In both cases, why or why not?

10. Think of a small water body located near your home. How might management of the watershed improve the fish community and the wildlife community in that watershed?

11. What is the status of estuarine habitats in the United States?

12. Are minimum instream flows a concern in the area where you live? If so, describe the various human users involved.

SELECTED READINGS

Adams, L. W., L. W. VanDruff, and M. Luniak. 2005. Managing urban habitats and wildlife. Pages 714–739 *in* C. E. Braun, ed. *Techniques for wildlife investigations and management.* 6th ed. The Wildlife Society, Bethesda, Md.

Bleich, V. C., J. G. Kie, E. R. Loft, T. R. Stephenson, M. W. Oehler Sr., and A. L. Medina. 2005. Managing rangelands for wildlife. Pages 873–897 *in* C. E. Braun, ed. *Techniques for wildlife investigations and management.* 6th ed. The Wildlife Society, Bethesda, Md.

Carpenter, S. R., and J. F. Kitchell. 1993. *The trophic cascade in lakes.* Cambridge University Press, Cambridge, U.K.

Herke, W. H., and B. D. Rogers. 1999. Maintenance of the estuarine environment. Pages 321–344 *in* C. C. Kohler and W. A. Hubert, eds. *Inland fisheries management in North America.* 2nd ed. American Fisheries Society, Bethesda, Md.

Johnson, D. L., and R. A. Stein, eds. 1979. *Response of fish to habitat structure in standing water.* Special Publication 6. North Central Division, American Fisheries Society, Bethesda, Md.

Laubhan, M. K., S. L. King, and L. H. Fredrickson. 2005. Managing inland wetlands for wildlife. Pages 797–838 *in* C. E. Braun, ed. *Techniques for wildlife investigations and management.* 6th ed. The Wildlife Society, Bethesda, Md.

McKinstry, M. C., W. A. Hubert, and S. H. Anderson. 2004. *Wetland and riparian areas of the intermountain west: Ecology and management.* University of Texas Press, Austin.

Meehan, W. R., ed. 1991. *Influences of forest and rangeland management on salmonid fishes and their habitats*. Special Publication 19. American Fisheries Society, Bethesda, Md.

Orth, D. J., and R. J. White. 1999. Stream habitat management. Pages 249–284 *in* C. C. Kohler and W. A. Hubert, eds. *Inland fisheries management in North America*. 2nd ed. American Fisheries Society, Bethesda, Md.

Perry, R. W., R. E. Thill, D. G. Peitz, and P. A. Tappe. 1999. Effects of different silvicultural systems on initial soft mast production. *Wildlife Society Bulletin* 27:915–923.

Ploskey, G. R. 1986. Effects of water-level changes on reservoir ecosystems, with implications for fisheries management. Pages 86–97 *in* G. E. Hall and M. J. Van Den Avyle, eds. *Reservoir fisheries management: Strategies for the 80s*. Reservoir Committee, Southern Division, American Fisheries Society, Bethesda, Md.

Resiner, M. 1993. *Cadillac desert: The American West and its disappearing water (revised and updated edition)*. Penguin Books, New York.

Rodiek, J. E., and E. G. Bolen. 1991. *Wildlife and habitats in managed landscapes*. Island Press, Washington, D.C.

Schneberger, E., ed. 1970. *A symposium on the management of midwestern winterkill lakes*. North Central Division, American Fisheries Society, Bethesda, Md.

Summerfelt, R. C. 1999. Lake and reservoir habitat management. Pages 285–320 *in* C. C. Kohler and W. A. Hubert, eds. *Inland fisheries management in North America*. 2nd ed. American Fisheries Society, Bethesda, Md.

Warner, R. E., J. W. Walk, and C. L. Hoffman. 2005. Managing farmlands for wildlife. Pages 861–872 *in* C. E. Braun, ed. *Techniques for wildlife investigations and management*. 6th ed. The Wildlife Society, Bethesda, Md.

Wesche, T. A., and D. J. Isaak. 1999. Watershed management and land-use practices. Pages 217–248 *in* C. C. Kohler and W. A. Hubert, eds. *Inland fisheries management in North America*. 2nd ed. American Fisheries Society, Bethesda, Md.

Williams, J. E., C. A. Wood, and M. P. Dombeck, eds. 1997. *Watershed restoration: Principles and practices*. American Fisheries Society, Bethesda, Md.

Yahner, R. H., C. G. Mahan, and A. D. Rodewald. 2005. Managing forestlands for wildlife. Pages 898–919 *in* C. E. Braun, ed. *Techniques for wildlife investigations and management*. 6th ed. The Wildlife Society, Bethesda, Md.

LITERATURE CITED

Bettoli, P. W., M. J. Maceina, R. L. Noble, and R. K. Betsill. 1993. Response of a reservoir fish community to aquatic vegetation removal. *North American Journal of Fisheries Management* 13:110–124.

Bleich, V. C., J. G. Kie, E. R. Loft, T. R. Stephenson, M. W. Oehler Sr., and A. L. Medina. 2005. Managing rangelands for wildlife. Pages 873–897 *in* C. E. Braun, ed. *Techniques for wildlife investigations and management*. 6th ed. The Wildlife Society, Bethesda, Md.

Brooks, K. N., P. F. Ffolliott, H. M. Gregersen, and L. F. DeBano. 2003. *Hydrology and the management of watersheds*. 3rd ed. Iowa State Press, Ames.

Cody, M. L., ed. 1985. *Habitat selection in birds*. Academic Press, San Diego.

Connelly, J. W., K. P. Reese, R. A. Fischer, and W. L. Wakkinen. 2000. Response of a sage grouse breeding population to fire in southeastern Idaho. *Wildlife Society Bulletin* 28:90–96.

Ewins, P. J. 1994. *Artificial nest structures for ospreys: A construction manual*. Canadian Wildlife Service, Toronto, Ontario.

Herke, W. H., and B. D. Rogers. 1999. Maintenance of the estuarine environment. Pages 321–341 *in* C. C. Kohler and W. A. Hubert, eds. *Inland fisheries management in North America*. 2nd ed. American Fisheries Society, Bethesda, Md.

Higgins, K. F. 1984. Lightning fires in North Dakota grasslands and in pine-savanna lands of South Dakota and Montana. *Journal of Range Management* 37:100–103.

Higgins, K. F., A. D. Crossway, and J. L. Piehl. 1989. *Prescribed burning guidelines in the northern Great Plains*. EC 760. South Dakota Cooperative Extension Service, Brookings.

Hunt, R. L. 1988. *A compendium of 45 trout stream habitat development evaluations in Wisconsin during 1953–1985*. Technical Bulletin no. 162. Wisconsin Department of Natural Resources, Madison.

Kruse, C. D., K. F. Higgins, and B. A. Vander Lee. 2002. Influence of predation on piping plover, *Charadrius melodus*, and least tern, *Sterna antillarum*, productivity along the Missouri River in South Dakota. *Canadian Field-Naturalist* 115:480–486.

Layher, W. G. 1984. Compatibility of multiple uses: Potable water supplies and fisheries. *Fisheries* (Bethesda) 9(6): 2–11.

Marzluff, J. M., R. Bowman, and R. Donnelly, eds. 2001. *Avian ecology and conservation in an urbanizing world*. Kluwer Academic Publishers, Norwell, Mass.

Odell, E. A., D. M. Theobald, and R. L. Knight. 2003. Incorporating ecology into land use planning—the songbirds' case for clustered development. *Journal of the American Planning Association* 69:72–82.

Pardue, G. B., and L. A. Nielsen. 1979. Invertebrate biomass and fish production in ponds with added attachment surface. Pages 34–37 *in* D. L. Johnson and R. A. Stein, eds. *Response of fish to habitat structure in standing water.* Special Publication 6. North Central Division, American Fisheries Society, Bethesda, Md.

Prince, E. D., O. E. Maughan, D. H. Bennett, G. M. Simmons Jr., J. Stauffer Jr., and R. J. Strange. 1979. Trophic dynamics of a freshwater artificial tire reef. Pages 459–473 *in* H. Clepper, ed. *Predator–prey systems in fisheries management.* Sport Fishing Institute, Washington, D.C.

Robbins, C. S., D. K. Dawson, and B. A. Dowell. 1989. Habitat area requirements of breeding forest birds of the middle Atlantic states. *Wildlife Monographs* 103. The Wildlife Society, Bethesda, Md.

Sigler, W. F., and J. W. Sigler. 1990. *Recreational fisheries: Management, theory, and application.* University of Nevada Press, Reno.

Yahner, R. H., C. G. Mahan, and A. D. Rodewald. 2005. Managing forestlands for wildlife. Pages 898–919 *in* C. E. Braun, ed. *Techniques for wildlife investigations and management.* 6th ed. The Wildlife Society, Bethesda, Md.

CHAPTER 16

Assessment of Human Users

Biological dimensions have always represented a primary aspect of what is done in wildlife and fisheries. It was not until the mid-1970s that human dimensions aspects of wildlife and fisheries gained stature as a recognized field of study (Decker et al. 2001). There were good reasons for this emergence.

To thoroughly understand a fishery or a wildlife system, it is essential to understand the human dimensions component. Biologists can obtain a wealth of information from and about the individuals and groups who directly and indirectly use wildlife and fisheries resources. Currently, more information is available on consumptive resource users than on nonconsumptive users; however, assessment of nonconsumptive users continues to improve as the importance of these users becomes more apparent.

Human dimensions assessment may involve a variety of types of information collected in numerous ways. Some of the information may be biological in nature, such as species taken, species seen, or the size of organisms taken. Other information may be demographic in nature, such as the age or gender of the people who utilize the resource. Still other information may concern human attitudes and values on a wide array of natural resource issues. Information on items such as species taken and size has been gathered by biologists for many years—attitude and value information is a more recent development (the last 30 years or so).

Economic value is another type of information that is often needed concerning wildlife and fisheries resources. Economic theory and practice are human endeavors; thus, economic assessment is included in this chapter. The economic values of natural resources can be difficult to determine accurately. For example, the economic value of taking a fishing trip or viewing songbirds is not measured solely by the amount of money spent in either of these activities. They provide satisfaction beyond the aspects that require monetary expenditures. In addition, because such assessment involves human ethics and values, the economic value placed on a natural resource can vary from one individual or group to the next. Like some economic values, human ethics and values can be difficult to determine.

The initial portion of this chapter highlights some ways in which natural resource information is gathered from humans and the reasons why such information is collected. That is followed by a brief discussion of natural resource economics. Human ethics and values are then addressed, because they are logical outgrowths of any discussion of human dimensions assessment and natural resource economics.

16.1 Surveys and Censuses

This discussion pertains primarily to the collection of biological and demographic data from resource users. Survey and census information gathering for the purpose of wild organism population assessment was introduced in section 9.2. This chapter builds on that previous information concerning censuses and surveys.

Surveys and censuses can be conducted on-site (in the field) as natural resource agency personnel meet directly with the public; they can involve observation of users by resource personnel, telephone or mail data collection, or some form of group discussion. At times combinations of formats are used. Common examples of on-site data collection formats are fisheries creel surveys and wildlife roadside bag checks (Figure 16.1). Data collection can also involve information on nonconsumptive users such as hikers, campers, or bird-watchers. Information gathered can include a wide variety of data, depending on the purposes of the survey or census.

FIGURE 16.1 A great deal of useful information can be obtained in face-to-face contact with resource users, as in this bag check of big-game hunters. *[Image courtesy of L. Parker.]*

By conducting creel surveys or censuses, fisheries personnel can obtain biological information such as the size, species, and number of fishes being caught. If those data are obtained in conjunction with information on time spent fishing, harvest and success rate estimates can be calculated. Determination of hours spent angling is one of the more common and important objectives of such creel surveys or censuses. Natural resource agencies often set management objectives for particular water bodies in terms of expected catch per angler-hour of fishing effort for a particular species (see section 20.6). Determination of the success or failure of management strategies is often impossible without data on catch (or harvest) and effort (angler-hours). Such information can also be used to formulate new regulations or other corrective measures if management objectives are not being met.

Biological data can also be obtained through bag surveys or censuses of wildlife species. Roadside bag surveys may be used to determine age, gender, and number of deer or other game species harvested by hunters. This information can be used to assess success or failure of management objectives set in terms of hunter success rates, male–female ratios, or adult–young ratios. Face-to-face meetings between biologists and fisheries and wildlife resource users also have utility for law enforcement and public relations purposes.

Censuses and surveys need not target only game organisms. For example, the National Audubon Society annually conducts a Christmastime bird count. This effort is national in scope, carried out by thousands of volunteers, and provides numbers and distributions for all birds observed. This information is useful in determining long-term trends in bird species distribution and abundance.

Most biological data are collected from users by means of surveys (partial counts); censuses are seldom done, because they require complete counts. When surveys are performed, specific sampling techniques must be used so that data obtained can be expanded to an estimate of the whole (see section 9.2). Specifics of how, when, and where surveys need to be conducted are beyond the scope of this text, but an example will demonstrate the difficulties that can be associated with obtaining reliable survey data.

Assume that angler-hours per year (annual total fishing effort) must be determined to assess resource use on a particular water body. One cannot sample every day, so which days are chosen? Think of how the data would be biased if only weekends or holidays were sampled. How are night anglers surveyed compared to day anglers, or boat anglers compared to shore anglers? If ice fishing occurs in enclosed fish houses, how are people and numbers of lines being used counted? These and other variables combine to make accurate survey data difficult to obtain, which places a premium on planning and survey design.

Some information obtained may not be biological in nature but instead is intended to profile the users. Demographic data such as user age, gender, distance traveled to utilize a resource, personal income level, education level, and harvest method are common components of surveys or censuses. This information helps biologists determine who is using a resource, how, and possibly why. Such information may affect the way the resource is managed.

Some data can best be obtained by face-to-face contact with users. In fisheries work, resource personnel can move from one user or group of users to another to obtain information. One can also wait at an access point, such as a boat dock or fishing pier, for users. For wildlife bag surveys or censuses, biologists can use primary roads that funnel users to particular locations or travel around an area, checking hunters that they encounter. For surveys of campers, various campsites can be visited. For hikers, resource personnel can gather information at various points on a trail system.

There are a number of distinct advantages to direct personal contact. First, the information received is more likely to be accurate. For example, which length measurement of a fish do you think would probably be more accurate, one made by a biologist or one made by an angler? Wildlife resource personnel at a roadside bag check can see a doe or a fawn deer that might have been reported as an antlered buck if some other type of survey or census had been used. An estimate of time spent bird-watching is likely to be more accurate if resource personnel can discuss starting and finishing times with a bird-watcher who is unsure of when he or she started, ate lunch, or finished the activity.

Another advantage of personal contact is the public relations work that can be accomplished. Users often have questions about laws or regulations, where and how to be successful in their recreational activity, and other information. Personal contact allows them access to natural resource personnel who can answer such questions.

There are also drawbacks to personal contact, especially if it is conducted improperly. Personal contact is expensive, and most natural resource agencies have limited funds. Another drawback of personal contact is the possibility of inaccurate results if the questions and information obtained are not standardized; information obtained should be the same regardless of who performs the interviews. An additional drawback is that resource users who are particularly satisfied or dissatisfied are often more talkative and cooperative than users who are more neutral in their success or opinion.

Personal interviews involve interactions between people and thus require special care. The interviewer may be interrupting the recreational activity of the user, or the user may be finished and want to return home. In addition, the interviewer represents authority, and each person reacts differently to authority. These and other factors, some of them very psychological in nature, combine to result in a need for proper interviewing techniques. The interviewer should be courteous and informative and should attempt to develop trust. Resource personnel should also convey to the user the reasons for the survey or census so that he or she understands that the information requested is important enough to justify the interruption.

Gathering on-site field information does not always require actual personal contact with users. For example, aerial surveys or censuses can be conducted from a variety of types of aircraft, vehicle-counting devices can be placed at resource use access points, or a biologist can obtain estimates of use just by viewing an area.

There are other means of obtaining information from users without face-to-face contact. Survey or census instruments, such as questionnaires, can be provided at particular locations. They can be completed by the user and left at that location or returned by mail. In some situations, specific users may be surveyed or censused by mail. It is a common practice for many natural resource agencies to obtain data by requesting that users provide information concerning catch or harvest success, area used, time spent on the recreational activity, and other data. In some cases users are requested to maintain detailed activity logs; this is usually done on a voluntary basis. The U.S.

Fish and Wildlife Service annually requests information from waterfowl hunters by having them report birds taken by species, date, and location. Wings and tail feathers from birds taken are also requested so that information on species and age structure can be obtained. In some fisheries surveys, anglers may be asked to provide a variety of detailed information as well as scales from fishes caught.

Telephone surveys or censuses are also used to gather information. It allows for an exchange of questions and answers, which can increase both accuracy and public relations benefits, though it is not as personal as face-to-face interviews.

Focus groups can also be used to develop information. Such groups are often used prior to an activity, such as a mail or telephone survey, to explore a topic before a survey is conducted. These groups provide an added perspective in survey planning and also help generate ideas.

Most people will readily cooperate with resource personnel in data collection because they are interested in natural resources and their successful conservation and management. They are often interested in the results obtained and will usually respond positively when asked if they would like information on the results of the survey or census. Supplying such information provides educational benefits and is also a useful public relations tool. Remember, however, that data gathering should be done with specific purposes in mind. It is too costly to collect data without a good reason, and users' time is valuable to them and should not be wasted.

User census and survey information is voluminous and is available from a variety of federal, state, provincial, territorial, and local sources. In the United States, the most extensive survey of natural resource users is the *National Survey of Fishing, Hunting, and Wildlife-Associated Recreation* (U.S. Department of the Interior 2002). This survey has been conducted approximately every 5 years since 1955. Initially data were collected at the national and regional levels, but the survey was later expanded to include information at the state level. Over time it has also expanded to include information on nonconsumptive wildlife activities, such as wildlife observation and photography. Survey results provide a wealth of information and are available in formats with combined national data and individual state data. The Canadian Wildlife Service started producing a similar national survey in 1981 and now does so every 5 years.

16.2 Attitude Assessment

Wildlife and fisheries biologists have always been adept at obtaining biological information from resource users, but they have often been much less effective in assessing user attitudes. As mentioned in section 1.2 of this text, the biota and habitat aspects of a fishery or a wildlife system have received the most attention from biologists; the human user components of these three-part systems, especially attitudes, have received less attention. Many biologists, while well educated in the biota and habitat aspects of a fishery or a wildlife system, often lack educational depth in the human dimensions aspect. Current curriculum additions in the higher education of future natural resource biologists is correcting this deficiency. While all biologists cannot become human dimensions experts, they should be knowledgeable about basic aspects of human dimensions. They should also be able to work effectively with people specifically educated in the human dimensions area. This increased attention to human dimensions, especially attitudes of resource users and how such attitudes can be assessed, is resulting in the development of more appropriate natural resource conservation planning.

Attitude assessment can include direct, indirect, consumptive, and nonconsumptive users of a resource and also segments of the general public not included in those categories. Because wildlife and fisheries resources in the United States belong to the public as common resources, all people have some stake in how those resources are managed and conserved. In addition, attitude assessment goes far beyond whether someone is satisfied with a particular law or regulation, believes that boat access areas are adequate, has been successful in bagging a trophy deer, or has added three new species to the list of birds observed. Each fishery or wildlife resource use provides multiple sources of satisfaction with a whole series of potential recreational outcomes that can accrue to users. All of these outcomes must be understood if accurate attitude assessments are to be obtained.

Biologists need to understand how to measure, evaluate, and integrate the socioeconomic, cultural, and political elements of the conservation environment (Krueger et al. 1986). Typically, attitude assessment relates to the beliefs, values, behaviors, and socioeconomic and demographic characteristics of user groups or publics and how those characteristics can be incorporated into an overall natural resource conservation scheme (Gigliotti and Decker 1992).

Most human dimensions theorists agree that attitudes can be divided into two separate components: affect, the emotional component, and belief, the perception (cognitive) component. (*Affect* is used as a noun by human dimensions specialists and should not be confused with the words *affect* [verb] and *effect* [noun] as they are commonly used.) Many also add a behavioral (intentions and actions) component (conation) to the model (Fishbein and Ajzen 1975). If one is to understand public preferences and attitudes toward natural resource conservation and management actions and policies, then it is necessary to examine user beliefs.

Beliefs represent the information that a person has about an object and what links that object has to some attribute. For example, the belief that "red-cockaded woodpeckers need protection" links the object red-cockaded woodpecker to the attribute protection. The object of a belief may be an organism, person, group, behavior, policy, or event, and the associated attribute may be any object, trait, property, quality, characteristic, outcome, or event. Beliefs can vary in strength from person to person; that is, people usually differ in terms of the ways in which an object is associated with the attribute in question. Beliefs about an object can be positive or negative, and the attitude of a person is viewed as corresponding to the total affect associated with his or her belief. For example, the evaluation of red-cockaded woodpeckers by a business executive will be positive or negative based on the peron's belief concerning a statement that "red-cockaded woodpeckers need protection" and the sum of other beliefs he or she holds about these birds. If he or she values these birds, then the belief that they need protection will be a positive belief.

Pierce (1979) defined values as "preferred outcomes or preferred ways of achieving those outcomes." In the above example, if the business executive values red-cockaded woodpeckers, then holding that value would probably lead to a positive rating for the belief that they need protection. Conversely, if a person such as a landowner has contact with these birds only in the context of their being on his or her land and possibly impacting choices in land use, the belief that these birds need protection might receive a negative rating. Understanding the attitude a person has toward something requires learning about the belief and value systems of that person. Understanding the diverse reasons for the attitudes of people who participate in wildlife- and fisheries-related activities, what they expect from these activities, and their sources of satisfaction and dissatisfaction can allow a biologist to better incorporate the human dimensions component into conservation and management decisions. As mentioned, the diverse reasons for the attitudes of consumptive, nonconsumptive, indirect, and direct users all require assessment.

The most common method currently used to obtain attitude information is surveying people by means of questionnaires. When such surveys are properly conducted, a representative sample of people and attitudes can be obtained. In a representative sample all are represented, not just those with a strong interest one way or the other.

Public input into natural resource decisions is often dominated by vocal minorities. Assessment based on just those individuals is usually inaccurate because those who feel strongest are more apt to speak out than others, thereby skewing results. For example, voluntary response questionnaires, such as tear-out questionnaires found in magazines or those left on location, although popular methods of collecting public input, are not valid survey methods and often obtain the opinions of only the vocal minority. Opinion call-ins or online voting are not valid survey methods, nor are compendia of letters of complaint or support, nor are surveys performed at public meetings. Such surveys lack control over and knowledge about the sample.

A disadvantage of even a valid questionnaire format used as a public participation tool is that it does not allow for two-way communication. This disadvantage can be overcome by linking questionnaires with other public involvement strategies, such as meetings or interviews, that involve two-way communication.

The process of developing an attitude survey instrument or questionnaire does not begin with writing questions. The initial step is to develop specific goals, objectives, and strategies with regard to information needed. Section 20.6 describes the planning process, including the development of goals, objectives, and strategies. The question, "What management decisions do we intend to make?" must be answered first. The answer to that question is used to develop goal statements. The question, "What information is needed to make these management decisions?" leads to development of objectives. The question, "How, specifically, can we obtain this information?" leads to strategy statements. Specific questions, such as whom to sample and how many people to sample, must also be answered, and other aspects of the survey strategy determined. Only after all of these issues are settled can question writing begin. The development of survey goals, objectives, and strategies—indeed, the entire planning process—works most effectively if a team of people is involved.

Question wording is crucial in obtaining valid information. If questions are vague, incomplete, ambiguous, or open to different interpretations by respondents, the survey results will be of little use. Consider the following pitfalls in question writing.

The question, "Do you want a quality songbird management program for this area?" is imprecise. A songbird biologist or any number of bird-watchers would interpret the term *quality* quite differently. Another example of an imprecise question would be, "Did you fish during the last year?" Does this mean during the last 365 days, this calendar year, last calendar year, or during the year shown on your fishing license?

Avoid double-barreled questions. These are questions with two or more parts; for example, "Were you satisfied with your ring-necked pheasant harvest and pheasant hunting regulations this year?" Such questions can confuse respondents, because they may have different answers for each part. It is safest to avoid lengthy and complex questions. Use the simplest wording possible that conveys the intended meaning.

Avoid irrelevant questions that surveyed people are not really in a position to answer. For example, one would not ask a hunter, "Do you think the deer harvest this year exceeded management objectives of harvesting 50,000 deer for the state?" Few hunters could give a meaningful response.

Another serious mistake is the use of loaded questions. These are questions that channel the response in a particular direction. "Don't you think the mourning dove season should be eliminated?" is an example. Such wording implies that the season should indeed be eliminated. "What is your opinion about the mourning dove season?" or "Do you support or oppose the mourning dove season?" are more appropriately worded questions. Another type of loaded question results from using an unequal set of response choices. Giving the respondent a choice among "superior," "excellent," and "good" does not provide sufficient choices. There should be a full range of choices, and none should be obviously different from the rest; for example, "How would you rate the State Game and Fish Department—superior, excellent, good, fair, poor, not worth a cent?" The

Here are some statements that deal with your feelings about catching trout. Please indicate the extent to which you agree or disagree with each of the following statements. Please circle one number for each item.

	Strongly disagree	Slightly disagree	Neutral	Slightly agree	Strongly agree
1. A fishing trip can be successful to me even if I don't catch trout.	1	2	3	4	5
2. The bigger the trout I catch, the better the fishing trip.	1	2	3	4	5
3. When I go fishing, I am satisfied only when I catch some trout.	1	2	3	4	5
4. Catching a "trophy" trout is the biggest reward for me.	1	2	3	4	5
5. It does not matter to me what type of trout I catch.	1	2	3	4	5
6. How I catch a trout is as important to me as actually catching one.	1	2	3	4	5
7. If I thought I would not catch trout, I would not go fishing.	1	2	3	4	5
8. The more trout I catch, the happier I am.	1	2	3	4	5

FIGURE 16.2 The Likert scale method is an attitude survey method that provides information on both direction and intensity of affect. *[Courtesy of L. Cigliotti.]*

How would you describe your treatment by the conservation officer for your most recent contact?

1. Fair	1	2	3	4	5	Unfair
2. Rough	1	2	3	4	5	Kind
3. Professional	1	2	3	4	5	Unprofessional
4. Good	1	2	3	4	5	Bad
5. Unfriendly	1	2	3	4	5	Friendly

FIGURE 16.3 The semantic differential method of attitude assessment uses a scale with each endpoint anchored by an adjective or phrase. *[Courtesy of L. Gigliotti.]*

last response is very different from the others, and the question is loaded by its implication that the Game and Fish Department is so bad that a special response category is needed.

Most biologists are not well qualified to write survey questions. Assistance from experienced human dimensions professionals is needed; that is why the team approach produces excellent results. The biologist knows what information is desired, and the human dimensions expert knows how to word questions to obtain accurate information.

Biological data tend to be quantitative in nature; for example, how many of a particular species were taken or seen, whether they were male or female, or what their lengths were. Attitude information tends to be qualitative; rather than "yes" or "no" or a specific number, an attitude assessment question has a range of potential responses. For this reason, specific instruments for attitude measurement have been developed. Their intent is to obtain qualitative information that can be put into a quantifiable format.

Attitudes can be measured with a survey tool called the **Likert scale method** (Figure 16.2). Attitudes have both direction of affect (positive, negative) and intensity of affect; a good attitude scale will measure both of these components. Attitude statements are developed that relate to the object of concern, and respondents are asked to respond to the statements on a scale ranging from "strongly agree" to "strongly disagree." The survey instrument depicted in Figure 16.2 measures different aspects of the importance of catching trout to trout anglers. The total of all responses results in an index measuring the importance of catching trout, while the individual items explore the component parts of this concept.

Another means of measuring attitudes is the **semantic differential method**. This method rates an object or concept on a scale with each endpoint anchored by an adjective or phrase. Figure 16.3 is an example of the use of this method to evaluate the experience of a person being checked by a conservation officer. Note that it also measures both direction and intensity of affect. In the **ranking scale method** (Figure 16.4), the respondent is asked to order a list of items according to personal preference or some other standard. The **frequency scale method** (Figure 16.5) asks a question about the likelihood of a certain event or action. The **checklist scale method** (Figure 16.6) provides a list of items and asks the respondent to indicate whether each

Please rank (using the numbers 1 through 6) the following methods of sewage sludge management according to your overall preference. Use 1 for your first choice.

_____ Burying in landfills
_____ Incineration (burning)
_____ Application to agricultural land for production of human food
_____ Application to agricultural land for production of animal feeds
_____ Use as lawn and garden fertilizer or soil conditioner
_____ Application to forest lands

FIGURE 16.4 The ranking scale method requires the respondent to rank a list of items. *[Courtesy of L. Gigliotti.]*

You find a raptor with a broken wing. How likely might you be to contact a raptor rehabilitation center concerning the bird? Please circle only one number for your response.

Never								Very likely
1	2	3	4	5	6	7	8	9

FIGURE 16.5 The frequency scale method requires the respondent to indicate the likelihood of an event. *[Courtesy of L. Gigliotti.]*

applies in a certain context. With the **magnitude scale method**, an attempt is made to build a scale (Figure 16.7). The respondent is given a short list of items and asked to allocate a given number of points among them to represent their importance or value.

Forcing respondents to indicate the presence of an attitude when none exists can result in inaccurate information. To avoid this problem, "no opinion," "do not know," or other neutral responses can be added to the response categories. These are valid responses, and including them can have an important effect on attitude assessment accuracy. Which of these neutral response categories to include and when to include them depend largely on the specific situation.

Attitude assessment methods in Figures 16.2–16.7 are all examples of closed-ended formats; that is, respondents choose one or more appropriate responses from a given set of alternatives. Advantages of closed-ended questions are that they are quick and easy for the respondent to use and that they readily lend themselves to quantitative analysis. A disadvantage of closed-ended formats is that the validity of results depends on the adequacy of the given set of alternative responses as well as on the wording of the question. Therefore, designing closed-ended questions requires in-depth knowledge of the survey topic.

By contrast, open-ended questions, such as, "What would you like done in this area that would improve your whale-watching experience?" do not provide alternative responses from which to choose but instead provide for a free and potentially more informative response. Open-ended questions are best used to generate ideas and to explore issues; they are also used when adequate knowledge for the development of a list of appropriate response categories is lacking. They are more difficult for the respondent to answer, and the responses are more difficult to quantify. Open-ended questions work best in personal interviews. In mailed surveys, open-ended questions are often left blank or lead to an increase in nonresponses to the survey.

Please rate from zero (0) to nine (9) the importance of each reason for why you fish. Please circle one number for each reason.

	Not a reason									A very important reason
1. To catch fish to eat	0	1	2	3	4	5	6	7	8	9
2. To catch fish for fun and excitement	0	1	2	3	4	5	6	7	8	9
3. For companionship (friends and/or family)	0	1	2	3	4	5	6	7	8	9
4. To get away and relax	0	1	2	3	4	5	6	7	8	9
5. To enjoy nature	0	1	2	3	4	5	6	7	8	9
6. To enjoy my fishing equipment	0	1	2	3	4	5	6	7	8	9
7. To catch trophy-sized fish	0	1	2	3	4	5	6	7	8	9

FIGURE 16.6 The checklist scale method of attitude assessment provides a list of items and asks the respondent to indicate whether each applies in a certain context. *[Courtesy of L. Gigliotti.]*

To show how much consideration you believe should be given to each of these categories by planners when considering development in your area, divide 100 points among the four categories below. The higher the number of points given to a category, the more importance you feel it should have compared with the other categories. Remember that the total must equal 100.

_____	Human health	_____	Environmental quality
_____	Economics (costs)	_____	Aesthetics (beauty) of the area

Total = 100

FIGURE 16.7 In the magnitude scale method, the respondent is provided with a list of items and asked to allocate points among them. *[Courtesy of L. Gigliotti.]*

These and any other attitude assessment tools can reveal much information about user attitudes, including those that may conflict with what the biologist believes is biologically correct and what different portions of the public may want; all of these preferences must be considered. What should a biologist do, for example, if he or she is involved in an effort to restore an extirpated native species, such as the wolf, to an area and assessment of the human dimension indicates that this is the course of action desired by the general public but not by the people in the area of the restoration? Assessing attitudes does not necessarily solve problems.

16.3 Economic Assessment

Natural resource economics is a field of study in itself, and it is too involved to address here except in a cursory fashion. **Natural resources** are things such as land, water, and wildlife that are useful and valuable in the condition in which they are found. **Economics** is the study of the allocation of resources. The human social aspects of natural resource use constitute an important element of resource value. Assessment of the economic values of wildlife and fisheries resources may take into consideration a wide variety of economic activities associated with these resources.

It is important to have information on economic values associated with particular natural resources, and such information is becoming more critical than ever. Without knowledge of their economic value, it is difficult to assess the importance of natural resources relative to that of other economic factors. For example, it is likely that for wetlands conservation to be ultimately successful, the total economic value of wetlands will need to be judged against the economic values of competing activities, such as agricultural pursuits and commercial and residential land development. When economics is the prime criterion for decisions on whether a particular area or habitat type is worth conserving instead of developing, adequate and correct economic information must be available.

Unfortunately, it is difficult to obtain adequate economic information concerning natural resources such as a fishery or a wildlife system. They are inherently much more difficult to assess than some other human undertakings, such as agriculture, forest crops, or shopping malls, because a good portion of their value is attitudinal, and attitudes vary from person to person. Economic assessment of wildlife and fisheries resources may involve direct, indirect, consumptive, or nonconsumptive uses or users, or any combination thereof. The best current economic data available relate to publicly owned resources used in a consumptive manner. It should be noted, however, that the nonconsumptive use of natural resources is rapidly growing; it is also an area in which better economic information is needed.

Economic assessment of wildlife and fisheries resources may include the economics of hunting, fishing, trapping, and other consumptive wildlife and fisheries activities. It may also include activities such as game farming, game ranching, and aquaculture, as well as **fee hunting** or **fee fishing**, which involve charging fees for the right to trespass and take either publicly or privately owned organisms on private land. **Ecotourism** is an economic activity associated primarily with the nonconsumptive use of natural resources. Ecotourism is tourism primarily directed at viewing specific habitats or organisms in their wild state. Examples include bald eagle watching tours, trips to view tropical rain forests, desert wildflower viewing trips, tours to view autumn foliage in temperate forests, and a wide variety of other similar activities.

What is the economic value of a particular natural resource, what are the problems associated with its assessment, and how is an assessment made? A good method of starting to answer such questions is to provide some examples of the problems involved in the economic assessment of fisheries and wildlife resources. Many large reservoirs in North America are managed by the U.S. Army Corps of Engineers (COE). When each reservoir was constructed, the COE assigned particular economic values to a variety of uses of that reservoir. These economic values to a large degree dictated how each reservoir was managed. By definition, multiple-use reservoirs have multiple functions, such as power generation, irrigation, flood control, navigation, and recreation. However, the economic values of these uses, often determined many years ago when the reservoirs were constructed, may differ from their current economic values; these values may also be quite different in the future. Conflicts have arisen because the original economic assessments placed more emphasis on the nonrecreational values than on their recreational values. The original emphases were also generally on consumptive uses, such as irrigation and power generation, rather than nonconsumptive uses, such as bird-watching, boating, and hiking. Use patterns have changed, and so have the economics of the situation, but the management of these reservoirs has been slow to change. The COE has tried, usually unsuccessfully, to address the multiple and often conflicting and changing uses of the reservoir system. The conflicts should be obvious. At what reservoir level is power production greatest? At what reservoir level is flood control most effective? How is downstream navigation best protected? At what water level is the fisheries resource best served? When do water discharges need to be reduced or stabilized to protect downstream nests of endangered or threatened bird species? How much water can be appropriated for irrigation, and when and where? How much concern should there be for nonconsumptive uses? These issues are essentially economic in nature; however, biological, social, and political issues must be considered as well.

Another example of how natural resource economics and biological, social, and political considerations interact is the issue of old-growth forests in the Pacific Northwest. The lightning rod of this issue has been the effect of timber harvest on the northern spotted owl, but there are many other issues to address. What of the economic, aesthetic, and biological values of the myriad other animal and plant species dependent on those old-growth forests? What economic and social factors affect the people and towns dependent on that diminishing forest resource? What are the economic and social effects on the commercial and recreational salmon fisheries that are dependent on salmonid spawning habitat protected by old-growth forests? How does one modify the time frame (1 year, 2 years, 30 years) in which the economic assessments are made? These and other issues require, among other things, adequate economic assessment before educated decisions can be made.

Economic issues are directly related to human dimensions, because economics is a human undertaking. Identifiable economic benefits from wildlife and fisheries resources accrue in numerous ways. Steinhoff et al. (1987) classified economic values of wildlife (and fisheries) into two types, **exercised values** and **option values**. Exercised values represent benefits that are currently accruing to society. Option values pertain to maintaining an option to use, experience, or enjoy something in the future.

Within exercised values, Brown et al. (2001) listed two categories: **direct benefits** and **indirect benefits**. Direct benefits accrue from personal interaction with animals. Activities such as fishing, bird-watching, or wildlife photography represent activities associated with direct benefits. Indirect benefits include activities such as reading or watching television programs about wildlife and fisheries, in which the user is not in direct contact with the resources.

Option values include a variety of human attitudes. One value is a **bequest value**. It is the value of knowing that wild organisms and their habitats will be available for future human generations. Another example of an option value is the willingness of a person to pay an annual fee to keep a natural area open.

When assigning an economic value to the worth of something, some measure must be used. Economists most frequently use dollars as a quantitative measure of value. Value estimations are relatively easy to quantify in some cases and rather difficult in other situations.

Direct expenditures involve all expenditures a recreationist makes on items such as food, lodging, and goods related to the activity pursued. **Indirect**

expenditures are generated as the monies from direct expenditures are respent to pay salaries or to purchase new supplies or services. Indirect expenditures multiply their value as they are spent and respent both inside and outside the area of the original direct expenditure.

In some cases dollar expenditures are insufficient for determining total value; that is, dollar expenditures do not always cover the total value of a recreational activity. Value over and above expenditures is known as **consumer surplus (net willingness to pay)** (Brown et al. 2001). Economists use two methods of estimating consumer surplus. The **contingent valuation approach** is designed to estimate the value of changes in the quality of an activity by asking about the willingness of someone to pay for those changes. For example, how much more would one pay for increased fishing opportunities or permits to watch wild brown bears. The objective is to determine the most that people would pay for an activity. The second method of estimating consumer surplus is the **travel cost method**. It represents a method of estimating benefits associated with specific recreation sites, based on the distance traveled to reach that location. The objective is to develop a demand curve for the site to estimate who would visit the area at various costs of travel.

An additional economic value that is often overlooked is the value of resources that are harvested by consumptive users such as hunters, anglers, or someone gathering wild plants to dry and arrange. These plants and animals have value to the user, but again, such values can be difficult to determine. A pen-raised ring-necked pheasant in a grocery store may cost $10. A dinner of ring-necked pheasant in a restaurant may cost anywhere from $15 to much more in very fashionable restaurants. Some species used for recreational purposes are not commercially raised. How can their economic value be estimated? The consumptive user who harvests such an organism and consumes it is obviously obtaining a benefit. Its cost may be great, based on expenditures made to harvest the organism, but some value is being obtained. There are no specific mechanisms available to make such value estimates, but such values can be determined in other ways. For example, Southwick and Loftus (2003) produced a publication intended to assist in assessing the values of various fish species killed accidentally or purposefully; this listing can also be used for other economic assessment purposes. It provides values of fishes by length and weight for a variety of species. This is just one tool of many available that can be used to assess the economic value of individual organisms.

A problem involved with the economic assessment of natural resources is the fact that estimates are based only on current values and knowledge. What is the economic value of some undescribed or nondescript insect before and after it is learned that the insect contains the key ingredient needed to cure arthritis? What is an Atlantic salmon worth today compared to its economic value in the year 2096? In most cases, we do not know the answers to such questions, but we still try to place an economic value on an item. Changes with time and knowledge are factors that can cause a natural resource to be overvalued or undervalued when assessments are projected into the future.

It is important to reiterate the difficulty of obtaining a total assessment of the monetary value of a fishery or a wildlife system. There are many benefits that are difficult to include in economic models. Activities such as fishing, bird-watching, visiting national parks, and other recreational pursuits that utilize fisheries and wildlife resources are activities with multiple sources of satisfaction. What is the value of the social well-being one obtains from a natural resource? What are the physiological benefits? What are the psychological benefits? What are the cultural benefits? These hard-to-measure benefits make it difficult to put dollar values on such activities.

16.4 Human Ethics and Values

While not a form of resource-user assessment, consideration of human ethics and values can be important in managing and conserving a fishery or a wildlife system. Ethics and values are directly related to human attitudes and to numerous other aspects of natural resource valuation. The ethics and values of a person or group can greatly influence the worth they place on a fishery or a wildlife system. Many topical areas could be covered in this portion of the text, but comments will be restricted to three areas: ethics and values in general, conservation ethics and values, and animal rights and animal welfare ethics and values. Different belief systems are represented in each of these areas. A further discussion of ethics and values is presented in section 20.5. That discussion revolves around biopolitics and conflict resolution.

Ethics (standards of conduct and moral philosophy) and **values** (the qualities of a thing according to which it is thought of as being more or less desirable, useful, or important) are obviously dependent on the individual. While the ethics and values of some people would allow them to follow wildlife and fisheries regulations minimally, others would go further. For example, while regulations might allow for the harvest of walleyes from deep water, some people would not do so if they believed that the fish would die when released. While some people think nothing of damaging fragile habitat with an off-highway vehicle, others would not consider such an activity.

Even when trying to be ethical, humans can cause problems. A number of years ago, during a study of piping plovers (endangered in some parts of North America and threatened in others) and least terns (an endangered species), a sandbar area on the Missouri River was posted as a no-recreational-use zone because the plovers and terns were nesting there. River sandbars in the region are popular recreational spots. Some users entered the restricted area, removed all the eggs from the nests, played volleyball, and then returned the eggs to the nests. They were trying to be ethical as they viewed it, but their actions were still harmful to the birds. In other cases, people entered such restricted areas and purposefully destroyed nests. Those people may even have been attracted by the signs; this would be indicative of a totally different set of ethics and values.

Ethics and values need not be viewed only as a consumptive-use issue. Examples like that of the volleyball players above demonstrate that nonconsumptive users are also involved in this issue. Even indirect users are involved. The recycling of paper, aluminum, and other recyclables, for example, is done for more than just economic reasons. Recycling is seen as ethical by many people because of the trees and habitat it saves from cutting and the reduction in environmental damage resulting from mining and manufacturing. The increased interest of the general public in protecting organisms and habitat is a result of decisions based on ethics and values by those who support conservation; those who do not support such conservation have also made a decision based on ethics and values. Two people can have diametrically opposed opinions, and both consider their particular viewpoint to be ethical.

Another example of the ambiguity of what is or is not ethical is the debate over saving individual organisms harmed by oil spills. Large amounts of time and money are spent in trying to save a relatively few animals each time an oil spill coats the bodies of birds or mammals. Many animals are damaged and killed as a result of this pollution source (Figure 16.8). Some people would contend that it is ethical to save these few animals; others would contend that it is unethical to expend all of that effort on a few, when similar efforts or expenditures directed at general research or management would do more good.

Human-induced problems do not provide the only examples of ethical conflicts. It seems that almost annually massive efforts are made to save whales that become stranded on beaches. While this problem makes for good news coverage and seems ethical to some, others believe that it is unethical to expend more time and resources on a few individual stranded whales than is annually expended on research to benefit all whales.

There are ongoing debates concerning the ethics of recreational hunting and fishing and indeed the use of any organism by humans for any purpose, even as pets. These are individual decisions with no answers that are universally right or wrong. What may seem perfectly right and ethical to one person, even what is allowable by law, may seem unethical or wrong to

FIGURE 16.8 The ethics of saving a few individual animals harmed by an oil spill is often questioned when similar time and monetary expenditures could assist greater numbers of animals in other ways. It is probable that this American coot could have been saved if found alive and washed thoroughly to remove the oil that caused its death.

another. This dichotomy of opinion on issues such as hunting, fishing, trapping, animal experimentation, keeping animals as pets, human consumption of animals, and many other similar activities will never be resolved until all people believe the same thing, and the chance of that ever occurring is remote.

Ethics and values also change with time. One hundred years ago few people would have thought it unethical to attempt to exterminate all wolves; that is not the case today. Changing attitudes result in changing values and ethical positions. Attitudes can be affected based on gender, age, occupation, income, religion, education, culture, urban versus rural settings, and any number of other characteristics that may influence how people think.

Whatever action is taken by a person or group, it is usually done with the belief that their position is more ethical than that of their opposition. Many solutions to problems in a democratic society are arrived at by compromise, but in some situations compromise is not considered to be an option because of ethical considerations.

16.5 Conservation Ethics and Values

There is a wide range in belief systems that relate to how the environment and the fisheries and wildlife resources it supports should be used or protected. The term *conservation*, for example, has been defined in many ways. One simply stated definition is "the effective management and use of natural resources." However, the concept of "effective management and use" varies depending on the person or group of people concerned. To some, conservation is akin to **preservation**, which is leaving natural systems the way they are. For example, those who advocate the preservation position would support the total closure of many areas to human activity. To others, conservation is management and use for the benefit of both humans and wild organisms. In this latter group, management and use can be directed primarily at benefiting either humans or wild organisms or at some point in between.

Because belief systems with regard to conservation vary to such an extent, confusion often enters the picture. For example, some people advocate resource management and use that primarily favors activities such as grazing, logging, and mineral extraction. While it is the ethical belief of these people that their position represents the best form of conservation, other people with a different set of conservation ethics and values would disagree with them. Some people primarily support resource management and use for short-term economic benefits at the expense of the long-term viability and economic benefits of the resources. They seldom question whether the short-term utilization they support is sustainable over a longer time frame. Those who disagree with them support a more balanced, long-term approach to resource management and use, an approach that more equitably considers both economics and the wild organisms involved.

16.6 Animal Rights and Animal Welfare

The last segment of this chapter concerns two other belief systems, **animal rights** and **animal welfare**. These belief systems are also rooted in ethics and values. This topic is so contentious that not all people would even agree on the following definitions and descriptions.

Animal rights involves a belief that all other animals have rights equal to those of humans. While these rights are extended primarily to other mammals, advocates of this philosophy often extend such rights to other animals as well. They believe that any form of management or use by humans interferes with these rights. Animal rights advocates are not necessarily supportive of animal welfare. For example, if a deer population were starving or dying from some disease, animal rights supporters would advocate letting the deer fend for themselves. They believe that humans should not intervene. Obviously, any consumptive use, experimentation, or other similar activity using animals would be opposed. It should also be obvious that animal rights advocates can believe in different levels of animal rights.

Animal welfare is a different ethical belief system. Correct handling and humane treatment of animals by humans are the primary goals of animal welfare advocates. Within the animal welfare philosophy, activities such as consumptive use and experimentation are allowable if they are conducted in as humane a manner as possible and if they do not threaten populations or species. The majority of biologists believe in animal welfare and thus are primarily concerned with the well-being of populations and species.

There are a variety of federal, state, provincial, territorial, and local laws and regulations that protect animals. Many of these are quite specific as to what is and is not acceptable in relation to what humans can do to other animals. Wildlife and fisheries biologists have developed additional guidelines for the handling and use of the organisms with which they work. Some organizations involved with the development of such guidelines are the American Society of Ichthyologists and Herpetologists, the American Fisheries Society, the American Institute of Fisheries Research Biologists, the American Society of Mammalogists, the American Ornithologists' Union, the Cooper Ornithological Society, the Wilson Ornithological Society, the Association for the Study of Animal Behaviour, and the Animal Behavior Society. There are also other groups that have developed additional guidelines for the care and treatment of animals. However, because biologists are also humans, there is individual variation in their commitment to animal welfare and their adherence to these guidelines.

Animal welfare supporters do not include just biologists. Consumptive and direct users, such as trappers, hunters, anglers, agriculturists, and pet owners, should all be concerned with the humane treatment of the animals that they use or keep. Numerous precautions, such as shooting only at animals that are within range, checking trap lines often, using proper fish release methods, keeping farm animals in well-ventilated buildings, and remembering to feed and water the dog or cat, contribute to proper concern for and care of animals. Even nonconsumptive and indirect users are involved with, and should be concerned about, the welfare of animals. For example, concerned bird- watchers do not disturb nesting birds, concerned snowmobilers do not chase deer, and concerned farmers may delay hay cutting during the height of the ring-necked pheasant nesting season. In most cases, the humane treatment of animals primarily requires a feeling of concern and the use of common sense.

SUMMARY

Surveys and censuses are means of gathering information. They can be conducted on-site, through the observations of biologists, or through telephone or mail contact. Data on species, sizes, numbers, sex ratios, and other biological information can be collected. Demographic information on users can also be gathered. Personal contact during surveys and censuses can result in more accurate information and also serves a variety of public relations and law enforcement functions. Telephone and mail censuses and surveys lack the degree of personal contact available in face-to-face encounters but still have much utility. Focus groups are often used during the planning stages of surveys.

Attitude assessment planning usually requires the interaction of biologists and human dimensions experts. Assessment methods usually measure both the direction and intensity of attitudes. Proper question wording is critical if accurate information is to be obtained. A variety of methods are used to measure attitudes. Some involve closed-ended questions, while others involve open-ended questions. Closed-ended questions allow for the collection of data that are more easily quantified.

The total economic values of wildlife and fisheries resources can be difficult to determine. Individuals or groups of people can place different values on outdoor activities, and such activities provide multiple sources of satisfaction. Some economic values, such as direct benefits and indirect benefits, are more easily determined from a monetary standpoint. Others, such as bequest values, are more difficult to evaluate in terms of dollars. An additional benefit is the value of natural resources that are harvested by consumptive users.

Ethics and values vary depending on the individual or group involved. What is ethical to some may be unethical to others. Ethics and values concern consumptive, nonconsumptive, direct, and indirect users. Ethics and values vary depending on whether an individual or group defines conservation as management and use directed primarily at the well-being of the biota and habitat or management and use directed primarily at the economic benefits of that biota or habitat to humans. Positions on animal rights and animal welfare also vary with individual or group ethics and values.

PRACTICE QUESTIONS

1. List ten different types of biological information that can be obtained from fisheries creel surveys and wildlife bag checks.

2. List six different types of demographic data that can be obtained in wildlife and fisheries surveys and censuses.

3. What are some advantages of direct personal contact with users while performing a survey or census?

4. Human attitudes can generally be divided into two separate components. What are they?

5. Why are public meetings, tear-out questionnaires, opinion call-ins, and online voting of little value as accurate survey methods?

6. What is wrong with the following attitude assessment question? "The Wildlife Division is an excellent organization that is interested in direct user needs and indirect users as well. Agree or disagree?"

7. Describe five activities or resources upon which you place an option value.

8. Select a natural resource–related activity in which you participate and list all the kinds of satisfaction that you derive from that activity.

9. Define *conservation*. Define *preservation*.

10. Describe the major differences between the animal rights and animal welfare philosophies.

SELECTED READINGS

Decker, D. J., T. L. Brown, and W. F. Siemer, eds. 2001. *Human dimensions of wildlife management in North America*. The Wildlife Society, Bethesda, Md.

Elliot, H. 2005. *Ethics for a finite world: An essay concerning a sustainable future*. Fulcrum Publishing, Golden, Colo.

Field, B. C. 2005. *Natural resource economics: An introduction*. Waveland Press Incorporated, Long Grove, Ill.

Gannon, W. L., S. Barnack, M. S. Burt, B. Danielson, J. W. Dragoo, C. W. Edwards, T. Findley, M. R. Gannon, E. J. Gentz, V. L. Jackson, A. D. Krevitz, K. D. Lundy, S. A. Macdonald, D. K. Odell, A. J. Pinter, R. S. Sikes, and B. R. Tallent. 1998. *Animal care and use guidelines*. Available at the American Society of Mammalogists Web site.

Gaunt, A. S., and L. W. Oring, eds. 1997. *Guidelines to the use of wild birds in research*. Available at the Ornithological Council Web site.

Grafton, R. Q., J. Kirkley, T. Kompas, and D. Squires. 2006. *Economics for fisheries management*. Ashgate Publishing Company, Burlington, Vt.

Knuth, B. A., and S. L. McMullin. 1996. Measuring the human dimensions of recreational fisheries. Pages 651–684 *in* B. R. Murphy and D. W. Willis, eds. *Fisheries techniques*. 2nd ed. American Fisheries Society, Bethesda, Md.

Nickum, J. G., H. L. Bart, P. R. Bowser, I. E. Greer, C. Hubbs, J. A. Jenkins, J. R. MacMillan, J. W. Rachlin, J. D. Rose, P. W. Sorensen, and J. R. Tomasso. 2004. *Guidelines for the use of fishes in research*. Available at the American Fisheries Society Web site.

Pitcher, T. J., and C. Hollingworth, eds. 2002. *Recreational fisheries ecological, economic, and social evaluation*. Blackwell Sciences Limited, Iowa State University Press, Ames.

Posewitz, J. 1994. *Beyond fair chase: The ethic and tradition of hunting*. Falcon Publishing Company, Helena, Mont.

Tisdell, C. A., ed. 2002. *The economics of conserving wildlife and natural areas*. Edward Elgar Publishing, Cheltenham, U.K.

Weithman, A. S. 1999. Socioeconomic benefits of fisheries. Pages 193–213 *in* C. C. Kohler and W. A. Hubert, eds. *Inland fisheries management in North America*. 2nd ed. American Fisheries Society, Bethesda, Md.

LITERATURE CITED

Brown, T. L., D. J. Decker, and W. F. Siemer. 2001. Valuing wildlife: Economic perspectives. Pages 57–73 *in* D. J. Decker, T. L. Brown, and W. F. Siemer, eds. *Human dimensions of wildlife management in North America*. The Wildlife Society, Bethesda, Md.

Decker, D. J., T. L. Brown, and W. F. Siemer, eds. 2001. *Human dimensions of wildlife management in North America*. The Wildlife Society, Bethesda, Md.

Fishbein, M., and I. Ajzen. 1975. *Belief, attitude, intention and behavior: An introduction to theory and research*. Addison-Wesley, Reading, Mass.

Gigliotti, L. M., and D. J. Decker. 1992. Human dimensions in wildlife management education: Preservice opportunities and in-service needs. *Wildlife Society Bulletin* 20:8–14.

Krueger, C. C., D. J. Decker, and T. A. Gavin. 1986. A concept of natural resource management: An application to unicorns. *Transactions of the Northeast Section of The Wildlife Society* 43:50–56.

Pierce, J. C. 1979. Water resource preservation: Personal values and public support. *Environment and Behavior* 11:147–161.

Southwick, R. I., and A. J. Loftus, eds. 2003. *Investigation and monetary values of fish and freshwater mussel kills.* Special Publication 30. American Fisheries Society, Bethesda, Md.

Steinhoff, H. W., R. G. Walsh, T. J. Peterle, and J. M. Petulla. 1987. Evolution of the valuation of wildlife. Pages 34–48 *in* D. J. Decker and G. R. Goff, eds. *Valuing wildlife: Economic and social perspectives.* Westview, Boulder, Colo.

U.S. Department of the Interior, Fish and Wildlife Service, and U.S. Department of Commerce, Bureau of the Census. 2002. *2001 National survey of fishing, hunting, and wildlife-associated recreation.* U.S. Government Printing Office, Washington, D.C.

CHAPTER 17

Management of Human Users

Most individuals interested in wildlife and fisheries choose this field of study because they enjoy being outdoors; utilize wild organisms for recreational purposes such as hunting, fishing, or observing; obtain satisfaction from working with animals; have a concern for these resources; or any combination of the above. However, the world of the wildlife or fisheries biologist is not simply one of biota and habitat. In fact, many of the activities of a fisheries or wildlife biologist revolve around managing people who use these resources. Biologists often need to manage human activities to conserve the animal and plant communities for which they are responsible.

Most wildlife and fisheries regulations concern direct consumptive users of a resource, such as individuals who hunt, fish, or trap. However, nonconsumptive users are also regulated. For example, people wishing to view bald eagles may need to be regulated by location, number, and activity so as not to unduly disturb the birds, especially if the eagles are nesting. Even indirect users are regulated. For example, regulations commonly limit the amount of water that can be removed from a stream by industry, municipalities, or agriculture.

This chapter will primarily address direct consumptive users, because they are the ones most closely regulated. Regulation of consumptive users tends to be more complicated and addresses more conservation issues than does regulation of nonconsumptive users. For example, consider the variety of regulations that could be directed at two fish species, largemouth bass and riffle sculpins. Because the largemouth bass is pursued over a wide geographic area for recreational purposes, a long list of regulations, such as how many can be taken and what size they must be to be harvestable, is promulgated to control its use. The riffle sculpin is not a game species. Few people have heard of or seen this small stream fish, which is found primarily in coastal streams of Washington and Oregon. Regulation pertaining to this fish may consist of a protective rule that simply states that the species cannot be captured. Regulation may also be absent because no one uses this species consumptively. The riffle sculpin is obviously not a fish that would attract limits on size, number, or methods of capture. Rules that protect riffle sculpins may more commonly involve regulation of indirect users, often through environmental protection laws such as clean water legislation. The degree of regulation or nonregulation is not necessarily indicative of the biological importance of a species but instead is based on the organism in question.

Regulations generally address one or more of the following aspects of consumptive use: "who" could pertain to items such as user age and residence, "what" to species or size, "when" to time of year or time of day, "where" to protected areas or portions of those areas, and "how" to type of bait or harvest method.

Management of human users is essential to the proper conservation of natural resources. People can often be the overriding factor in conservation decisions and in the likelihood of their subsequent success or failure. The purpose of this chapter is to discuss some aspects of human user management in relation to wildlife and fisheries resources.

17.1 Purposes of Regulations

Controlling **overexploitation** (**overharvest**) is a common purpose of regulations. Overexploitation, however, can be a rather nebulous concept. Taken literally, overexploitation might be defined as the point at which harvest mortality causes undesirable reductions in fish, wildlife, or other populations (see section 3.3). However, no single definition of overexploitation exists. Overexploitation sometimes suggests such extensive harvest that insufficient adult organisms remain to replace themselves through reproduction. In other situations, the term is simply used to imply that the quality of a population has been impaired by some level of harvest. The distinction in terms of quality is similar to that made in comparing maximum and optimum sustained yields (see section 3.6).

Commercial fisheries provide numerous examples of overexploitation. Commonly, a new resource is found, harvest increases over time, overexploitation eventually occurs, the adult stock rapidly declines, and commercial harvest plummets. A typical example was described by Talbot (1973) for the Pacific sardine fishery in the Pacific Ocean off the California coast (Figure 17.1). Overexploitation in this context does not mean that the resource or species is extirpated; it is simply reduced below the level needed to support a commercially viable venture.

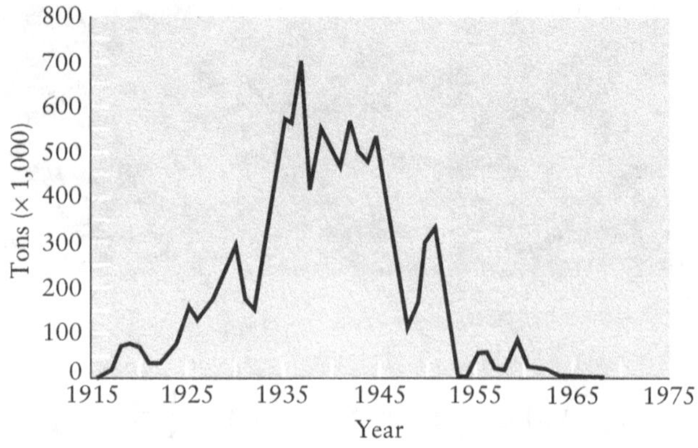

FIGURE 17.1 Catches of Pacific sardines in the Pacific Ocean off California, 1917–1968. Harvest increased gradually, and then dramatically, until the stock was overexploited. *[From Talbot 1973.]*

In inland recreational fisheries, regulated angler harvest typically does not reduce an adult stock to a point at which it could not replace itself. The reasons for this are quite understandable. Most biologists believe that recreational fishing becomes increasingly poor as the adult stock is reduced, and that most anglers quit fishing before the adult stock is reduced to the point at which insufficient adults remain for reproductive purposes. However, overexploitation of recreational fisheries can result in a decrease in the quality or size structure of the fish population. For example, consider a largemouth bass population in a water body with a surface area of approximately 60 hectares that is located near a city with a population of 100,000 people. Angling effort at such a water body typically could be about 500 hours per hectare per year. Given this scenario and no regulation of largemouth bass harvest other than a five-fish-per-day **creel limit** (the number of fish that one person can harvest and remove in a day), one would expect to find few bass longer than 30 centimeters in the water body. There would be enough adult largemouth bass to reproduce and replace themselves, but the level of harvest would preclude many fish from reaching larger sizes.

Another reason for regulation may be to achieve a particular management goal. For example, the Colorado Division of Wildlife might wish to manage a specific elk population for the quality of harvestable bulls. The management objective might be to produce an elk population in which 40% of the bulls have four or more points on each antler. Such a population generally cannot exist without special harvest regulations.

Regulations may also be implemented to make a resource available to more users. For example, while a daily creel limit or **bag limit** (the number of individual wildlife organisms and sometimes fish that one person can harvest and remove in a day) or **possession limit** (the total number of individual organisms that can be possessed by one person at any one time) can be implemented primarily to attain a biological management objective, it can also be intended to ensure that a publicly owned resource is distributed among as many users as possible. Without limits, someone who is very adept at catching steelhead or shooting ruffed grouse may be able to take more than his or her share of these resources.

Purposes for regulation can be deeply rooted in psychology. Some individuals who use a resource may not be satisfied with a recreational experience unless they attain the limit. While some would consider such an attitude unethical, it can still be a factor in recreational outcomes. If the limit is six and that is what some users get, they are satisfied; if the limit is four and that is what they get, they are satisfied; if the limit is six and they get four, they may be dissatisfied. The particular number can be less important than the goal. The psychology of such situations can create unanticipated results. For example, consider a situation in which biologists wish to reduce harvest and therefore reduce the creel limit on rainbow trout from eight to four. Because of the psychology of limits, and therefore human goals, it is possible that harvest could actually increase with such a creel limit reduction. If the limit is eight and an angler has three fish, the angler may stop fishing; if the limit is four and the angler has three fish, that person may continue fishing until a fourth fish is caught.

Regulation may also be applied for reasons that are primarily social or political in nature. For example, while a management objective for watchable wildlife might call for increased populations of Canada geese in a metropolitan area, golf course groundskeepers, airport superintendents, or the general public might not desire such increases; in fact, they might even want decreases. The length of the hunting season for ring-necked pheasants in midwestern states is another example. Biologically, a male-only season could be opened during the fall and run through late winter (perhaps mid-October through February) with no negative effects on the ring-necked pheasant population. Ring-necked pheasants are polygynous, and they require as few as one male to every ten females for successful reproduction. Even with an extremely long season, the males would not be harvested to a sufficient extent to reach such a skewed ratio of males to females. Why, then, do many natural resource agencies set season lengths of 1 or 2 months? The answer may vary from state to state, but such season length limitations are often the result of social concerns. A commonly cited reason for shorter seasons is landowner tolerance. Even though wildlife species are a part of the commons, much hunting necessarily occurs on privately owned land. The commissions or boards that preside over most natural resource agency activities (see section 20.4) often express concern that landowners will not tolerate the bother and interruptions that would accompany a longer hunting season.

Are socially and politically based regulations wrong? There is no definitive answer to this question; the answer varies among individuals. An enthusiastic biologist may insist that resource decisions, whether on regulations or other conservation strategies, should be based only on biology. Another person, such as an agency administrator, may have a different viewpoint. Very often administrators have more experience in biopolitical activities and recognize that a political dispute over a minor regulation proposal may incur a cost that carries over to larger, perhaps more important, matters. To administrators, a socially based regulation may be highly appropriate. Who is right? We certainly cannot say. However, it is important that people be aware that day-to-day decisions made by natural resource agencies are influenced by social and political considerations as well as by biological ones (see section 20.5). In the last two decades, there has been increasing input on regulations by the general public. Many people have different opinions on natural resource management and conservation, and they readily voice these opinions.

17.2 History of Harvest Regulations

The earliest fish harvest regulations implemented in the United States eliminated the use of fish traps, nets, and seines, particularly near the mouths of streams and during spawning seasons (Redmond

1986). By the early 1900s, increasingly restrictive regulations had progressed to minimum length limits and closed spring seasons to allow fish to spawn at least once before being caught. Minimum length limits were often very low, such as 15 centimeters for trout species and 20 centimeters for largemouth or smallmouth bass.

During the first half of the twentieth century, maximum sustained yield (MSY) (see section 3.6) was the common management strategy. Biologists strove to maximize fish harvest without harming subsequent generations of a fish population. Regulations were liberalized after studies in the 1940s indicated that fishing could be allowed in the spring without affecting reproduction or numbers of fish harvested in following years. Studies in the 1940s and 1950s continued to support increased recreational fish harvest. By the 1960s, however, investigations indicated that, in some small water bodies, predator fishes could be overharvested, upsetting the balance in the fish community and seriously affecting future fishing quality of both predators and panfishes. The imposition of effective **size limits** (the size of organisms that can legally be harvested) and **harvest quotas** (the specific number or total weight of organisms that can be harvested) often prevented predator overharvest and maintained balanced fish communities that provided good fishing. With the advent of refined fish tagging and recovery methods, biologists demonstrated that high exploitation rates of **black basses** (largemouth, smallmouth, and spotted bass and other members of the genus *Micropterus*) and white crappies also occurred in large water bodies. Harvest restrictions applied to large reservoirs in the 1970s demonstrated that practical, properly applied, and well-enforced size limits could maintain recreational fishing and a harvest of desirable quality. These management practices reflected an optimum sustained yield (OSY) strategy.

Redmond (1986) concluded that an important challenge for the future is to increase public acceptance of harvest restrictions to achieve the necessary compliance. Compliance will be necessary to maintain high-quality fishing for a growing number of more knowledgeable and better-equipped recreational anglers. Compliance is typically accomplished through both enforcement and education.

Shaw (1985) summarized the history of wildlife regulation in the United States. The "Era of Abundance" lasted from 1600 through 1849. Wildlife was abundant, and human effects on wildlife populations were minimal. The "Era of Overexploitation" began in 1850 and lasted until 1899. Shaw used the destruction of the vast herds of American bison as the best example of this overexploitation. This decline in wildlife abundance was accelerated by settlement of wild lands and by technological improvements that allowed increased harvest without any accompanying increase in conservation measures. During both of the above periods, regulation was infrequent. The "Era of Protection," 1900–1929, was the immediate response to the Era of Overexploitation. Formal regulation of wildlife harvest became well established during this time period.

The wildlife profession evolved during the "Era of Game Management," 1930–1965. While legal protection was an important part of wildlife conservation, it was apparent that biological information was also needed so that biologists could apply conservation strategies that would be effective for protecting particular species. Biology of an organism had to be understood before proper regulations could be applied. Wildlife conservation efforts during this era were largely limited to game species.

Finally, we have reached the "Era of Environmental Management." Conservation measures have spread beyond game species to reach all types of organisms and their accompanying habitats, with concomitant increases in the complexity of laws and regulations.

17.3 Contrast between Wildlife and Fisheries Regulations

Fisheries harvest regulations commonly include both number and size (length or weight) limitations. Harvest regulations for recreational fisheries are based on the fact that many fish species can be caught and released with a reasonable expectation of survival. Hooking mortality does occur, and it is likely to be highest when live bait is used or when water temperatures are relatively warm. Where substantial, hooking mortality can be minimized by artificial-lure-only requirements and by careful handling techniques. Wildlife regulations, in contrast, most often limit the number of organisms harvested; size regulations are less frequently used. Obviously, it is not possible to release a common snipe or a mountain goat once it has been harvested. This single difference greatly influences how a fishery can be regulated compared to regulation of most wildlife resources.

The numbers of harvestable species and individuals (density) are usually lower for wildlife in a particular area than for fishes in a water body; this difference also contributes to the differences between fisheries and wildlife regulations. Another difference relates to differing growth patterns. Fishes, because they exhibit indeterminate growth, have populations that can generally be characterized by a wider variety of individual fish sizes.

In fisheries, it is possible for nearby waters to be managed differently. Two water bodies within close proximity can be managed for entirely different sizes of fishes or for entirely different fish species. Such water-body-specific conservation has few equivalents in wildlife. Larger management units are typical in wildlife, because most wildlife species are less limited in their movements than fishes.

For nongame animals, there are also differences among the regulations aimed at protecting fishes, mammals, and birds. Most fish and mammal nongame species either are not protected by law or are protected by not being included on lists of organisms that can be harvested. Exceptions typically occur when a species is threatened or endangered. In contrast, most nongame bird species are protected because they are included in international agreements on migratory species and because the public has requested their protection.

17.4 Regulating "Who"

Harvest of wildlife and fisheries organisms can be regulated simply by determining who can legally take these resources. For example, a state, province, or territory that has a limited allowable harvest for a particular wildlife species may set a regulation decreeing that only residents can hunt that species; nonresident hunting may not be allowed. Similarly, if a state, province, or territory allows a limited harvest of a fish species that is not particularly abundant, nonresident anglers may be prohibited from fishing.

In the United States, the user's level of skill, interest, and knowledge can also be used to limit participation in a recreational activity. A common example of this strategy is the hunter education programs that are mandatory in most states. Until individuals have demonstrated an appropriate level of knowledge by passing an examination, they are prohibited from purchasing a hunting license. There are calls for even more rigorous testing in some circumstances. For example, some people believe that proficiency should be tested before individuals are allowed to purchase archery licenses for big-game species. In most European countries, licensing requirements for hunting and fishing are much more stringent. For example, German anglers can obtain a government fishing license only after passing an examination that includes knowledge of fish biology, aquatic habitat, fishing techniques, and proper fish handling information (von Lukowicz 1998).

Most natural resource agencies allow minors to hunt, fish, or trap at reduced license fees. This is usually done to encourage the entry of new participants into these activities and with the belief that these individuals are inexperienced and therefore less likely to be proficient. At times, reduced fee structures are provided for senior citizens. This is often considered justifiable because senior citizens may be on a fixed income.

Many natural resource agencies provide special considerations for landowners. Landowners often can obtain licenses at no or reduced cost, and they may receive a substantial portion of the licenses when a lottery or drawing system is used. Other benefits for landowners include exemptions from license requirements for hunting or fishing for some species on their own land. The general philosophy underlying such benefits is that landowners are providing habitat for organisms and thus are already contributing to the maintenance of the resource.

Individuals that have a physical disability, former prisoners of war, veterans, military personnel not living in their normal state of residence, and nonresident college and university students often receive special consideration for obtaining licenses. People with a physical disability may be allowed to harvest wildlife with equipment that cannot be used by the general public. Nonresidents in the military and attending colleges and universities often are allowed to purchase lower-cost resident licenses rather than more expensive nonresident licenses.

Regulating who can harvest a resource can also be accomplished by limiting the number of individuals that can participate, as further discussed in section 17.5. For example, limiting the number of licenses or permits can limit the number of participants.

17.5 Regulating "What"

Regulating what can be harvested most often involves regulation by species, number, gender, or size. Each of these will be discussed separately.

Regulation by Species

In any particular area, there are many species that cannot be harvested. Some may be endangered or threatened, some may exist in low numbers in that particular area, and others may not be legally classified as game species (sometimes termed *sport species*). Regulations on the harvest or nonharvest of particular species can be highly variable from location to location. For example, the mourning dove is classified as a game bird in some areas but as a songbird in others. In some states, provinces, and territories, fish species are classified as game or nongame based on the taxonomic group, such as family, to which they belong. Thus, in some areas, darters, a type of fish too small to be a recreationally important fish, are legally classified as game species and are subject to protection because they belong to the same family (Percidae) as walleyes. In other areas, it is legal to take darters because they are not specifically listed as game species. Game species are often regulated in numerous ways; nongame species are often regulated only with regard to prohibition of harvest.

Regulation by Number

The number of individuals to be harvested can be managed in a variety of ways. Numbers of permits to be sold for big-game species are often set by geographic unit. When demand is greater than the number of permits available for a unit, some type of lottery or drawing is generally used to determine successful applicants.

Quotas that allow the harvest of a certain proportion of a population are a theoretically attractive means of preventing overharvest of some species. Quotas can be set for wildlife, fishes, or any other species if there is sufficient knowledge of the numbers of that particular species and how many are needed to maintain viable and quality populations. Determining such quotas and monitoring the harvest, however, can entail substantial economic costs. Consider the largemouth bass as an example. Graham (1974) found that harvest of more than 40% of the largemouth bass in Missouri impoundments resulted in a decreased size structure and less effective predation on panfishes. Based on this finding, a biologist in charge of a 20-hectare impoundment might decide to set a quota under which 40% of the largemouth bass in that population could be harvested. When that quota was reached, harvest would stop. However, it would be expensive and time-consuming to obtain a population estimate for largemouth bass in that impoundment, and even after a reliable estimate was made, a thorough, expensive creel survey or census would still be needed to monitor the amount of harvest. The quota system could certainly work, but it would be difficult to manage very many water bodies using this technique because of economic limitations. Quotas are commonly used for commercial fisheries and are also used to manage paddlefish in some Missouri River states.

Quotas are sometimes applied to the harvest of wildlife species. For example, geese that congregate on waterfowl refuges during migration or wintering may be subject to high levels of exploitation on surrounding lands. Biologists may attempt to manage specific populations of geese by setting quotas on the number that can be harvested. These quotas are designed to assure that breeding stock from a particular area is not depleted. Consider a hypothetical population of Canada geese that nests near Hudson Bay and spends much of late October and early November on a state waterfowl refuge in Minnesota. The Canada geese rest on large water bodies within the refuge and feed primarily on surrounding private lands. While moving to feeding areas or returning to the refuge, the Canada geese are hunted, both on portions of the refuge open to goose hunting and on private lands. Assume that breeding ground surveys and fall migration counts indicate that the fall population of these Canada geese has declined to 80,000 from about 140,000 over 10 years, and that the management objective for this population has been set at 135,000 geese. Band returns indicate that 60% of all hunting mortality in this population is occurring on or near the state refuge in Minnesota. Biologists may place harvest quotas on this population to reduce mortality and allow the population to reach management objectives. When the harvest quota is reached, the season on Canada geese in that area is closed, at least until that population has migrated from the refuge. If daily harvest is high, the season may last less than a week; if harvest is low, the season may be allowed to continue much longer. Such regulatory practices sound ideal, but they can be difficult in terms of public relations. Consider the pressures brought upon those making the decision to close a season early because quotas have been reached.

Quotas are actually quite common in wildlife regulation, although the term is seldom used. For example, when the number of big-game licenses to be sold is determined for a geographic unit, that number is a reflection of the estimated number of animals present and the population level that biologists wish to maintain in that area. Knowing these numbers and having a historical indication of what hunter success rates will be allows biologists to determine the number of animals that can be taken and therefore the number of licenses to be issued.

Creel limits in fisheries and bag limits for wildlife also regulate what is taken. If a fish population is overharvested, common sense seems to dictate a reduction in the number of fish that each angler can harvest. Despite the apparent attractiveness of this response, fisheries biologists have found that creel limits are usually an ineffective tool for preventing overharvest. To effectively control overharvest, creel limits would typically have to be so low that most anglers would consider them unacceptable. Consider the following case history. In 1982, there was no creel limit for white crappies in Kansas, and anglers were allowed to keep all the crappies they caught. At about this time, there were concerns that overharvest might be negatively affecting some white crappie populations in the state. The solution proposed by most anglers was to enact a statewide creel limit for crappies. When questioned about the appropriate number, most people recommended a limit somewhere between twenty-five and fifty. A creel survey was undertaken that year at Clinton Reservoir in the northeastern part of the state. Based on the creel survey results, estimates were made of the reduction in harvest that would have occurred had various creel limits been in effect (Figure 17.2). A creel limit of thirty would have reduced total harvest by 1%, and a creel limit of ten (much too restrictive for most anglers) would have reduced harvest by only 11%. It would have taken a creel limit of five to result in a harvest reduction of 26%, which may or may not have been sufficient to improve the quality (i.e., size structure) of the white crappie population. Where overharvest truly is a problem, creel limits generally

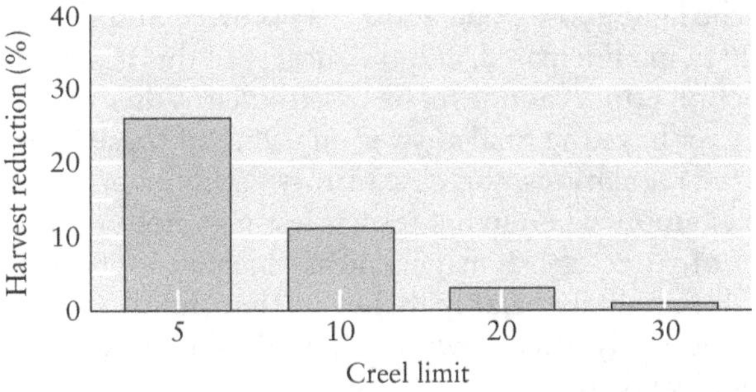

FIGURE 17.2 Reductions in the white crappie harvest that would have resulted if various creel limits had been in effect for Clinton Reservoir, Kansas, during the 1982 fishing season. No creel limit was actually in effect that year.

have to be very restrictive to sufficiently reduce harvest. As a result, creel or bag limits are often applied for social or political, rather than biological, reasons.

In both Chapters 3 and 10, we indicated that biologists at times employ a management strategy designed to decrease population size for a particular organism. Such efforts may or may not be successful. Regulations for lesser snow geese and Ross's geese were liberalized in the 1990s in the central portion of the United States in an attempt to control populations; the purpose of this was to reduce habitat destruction on the breeding grounds caused by too many geese (Kerbes et al. 1990; Ganter et al. 1995). Weight declines in hatching-year (HY) and after-hatching-year (AHY) (see section 8.1) lesser snow geese along with habitat destruction originally alerted biologists to this problem. Daily bag limits of five per day were common during fall seasons in the late 1980s, but they commonly increased to daily limits of twenty by 2005. Spring harvest was also allowed in the 1990s (Light Goose Conservation Order, U.S. Fish and Wildlife Service), and regulations also were eased in aspects such as allowing the use of electronic calls to enhance the likelihood of harvest. Harvest of light geese (i.e., blue, Ross's, and snow geese) approximately doubled after the initial conservation order for a spring season (beginning in 1996) and has remained at that level through spring of 2006. The number of geese harvested initially increased at a fast rate until approximately 1998–2000, when harvest stabilized or declined slightly. Wintering ground counts indicate that light goose populations for the midcontinent region have declined from an estimated 3 million in 1998 to 2.5 million in 2006; reduction of the light goose population has occurred at a slower rate than anticipated, and the success of this goose reduction effort cannot yet be determined. While we do not yet have the final answer on utility of these liberalized regulations, this case history certainly provides an example wherein hunter harvest may not be sufficient to accomplish management objectives. The general public invariably believes that hunting will undoubtedly affect a wildlife population, and this is not always true.

Some unique species are managed with a lifetime limit. For example, a limited number of permits are available each year for bighorn sheep in the Badlands of North Dakota. Applicants can obtain only one license in a lifetime. The license is good for one hunting season, and the permittee cannot apply for another license in subsequent years.

A creel or bag limit may serve to distribute harvest more equitably among a variety of users over a longer period of time. In goose-hunting areas with quotas, for example, low daily bag limits allow more hunters to pursue geese over a longer period of time. Restrictive creel or bag limits also help to reduce the waste of harvested animals that may occur when anglers or hunters harvest a large number of a species in a single day. At the least, a user arriving one day and not catching many fish or seeing much game does not hear stories about massive numbers of animals that were taken and removed the previous day.

Wild animals are limited resources, and creel and bag limits add perceived value to them. If a natural resource agency sets no creel or bag limit for a species, users may believe that it has little value. The same philosophy applies to possession limits. Possession limits are often set at two to five times the daily bag or creel limit. However, possession limits are difficult to enforce; a search warrant would have to be obtained to check home freezers and commercial freezer lockers to determine whether an individual had violated the possession limit.

Regulation by Gender

Regulation by gender of organisms harvested can also be used to modify harvest. This is a common practice in wildlife, but it is seldom used in fisheries. Sexual dimorphism or dichromatism in the animal to be regulated is an obvious prerequisite. Polygynous species are often regulated in this manner. Ring-necked pheasant harvest, for example, can be regulated by allowing harvest of males only, as described in section 17.1. The presence of fewer males will have no effect on reproductive success the following spring. There is even some question as to whether, biologically, a bag limit on males is needed at all. Few hunters would still be afield in an area where males composed only 10% of a population. Remember, however, that there are reasons for regulations other than biological ones.

Similar arguments can be made for some mammalian species in which gender can be readily determined, such as white-tailed deer. Allowing harvest of antlered deer only is one way to maintain large harvests without negatively affecting population numbers. In the case of deer, there is, however, another consideration. When hunters harvest male ring-necked pheasants, they do not generally differentiate between young and old males; this cannot be said for harvest of large ungulates, in which antler or horn size is a consideration. Unlimited buck harvest can lead to a shift toward a younger population age structure (see section 3.7). Many other species that can be readily sexed before harvest are regulated by gender. Examples include bull versus cow and calf elk, gobbler versus hen wild turkeys, and ram versus ewe and lamb bighorn sheep.

Monogamous species or species in which gender cannot be readily differentiated before harvest must be managed with more care to ensure that population numbers are not, over time, adversely affected by harvest. Examples of such species are northern bobwhites, American woodcocks, and white-winged doves.

Regulation by Size

Size limits are commonly used as a fisheries management technique, and they are sometimes used in wildlife management. Wildlife size limits usually pertain to structures such as antlers or horns rather than body size, and they are commonly termed *appearance restrictions*. Such a regulation might restrict harvest of mule deer or elk to those individuals with four or more antler points per side. Similarly, bighorn sheep might be regulated with a requirement for a full horn curl, or perhaps a three-quarter curl (Figure 17.3). Whether or not such appearance regulations actually increase the proportion of large or trophy animals in a population is debatable (Connelly et al. 2005). For example, regulation of elk harvest using point

FIGURE 17.3 Some big-game species are managed with appearance regulations. This bighorn ram has horns that exceed a half curl but have not yet reached a three-quarter curl. Full curl is attained when the tip of the horn reaches the base of the horn, above the eye. *[Image courtesy of South Dakota Tourism.]*

restrictions may result in increased kill of illegal (having insufficient points) bulls and decreased overall yield (Boyd and Lipscomb 1976). The genetic implications of such a strategy are also not clear. Thelen (1991) suggested that harvest of only bulls with high numbers of antler points could be genetically counterproductive. Another part of the decision process for a trophy bull elk management program would involve hunter preferences. For example, Manfredo et al. (2004) found that the majority of Colorado resident and nonresident elk hunters preferred to maximize hunting frequency; they preferred frequent hunts for moderate-size animals over alternatives that favored trophy hunting.

While size implies both length and weight, the majority of size regulations in fisheries work are based on organism length. Length is easy to measure, and enforcement on this basis is easier. A fish can be eviscerated but transported with the head intact to allow measurement by enforcement officers. Common length limit regulations include **minimum length limits,** under which all fish less than a specified length must be released and fish longer than the specified length may be kept; **protected slot length limits,** under which all fish within a specified length range must be released, while those outside the slot length range may be kept; **reverse slot length limits,** under which all fish within a specified length range may be kept, while those outside the slot length range must be released; and **maximum length limits,** under which fish up to a maximum length may be kept but all longer fish must be released. We will further distinguish these types of length limits using largemouth bass examples for minimum and slot length limits and a cutthroat trout example for maximum length limits.

Consider a fisheries biologist responsible for a 30-hectare impoundment in the midwestern United States that has appropriate habitat for largemouth bass. Water transparency in the impoundment exceeds 45 centimeters for most of the year, and submerged aquatic vegetation covers approximately 25% of the surface area during midsummer. A sample of largemouth bass is obtained by electrofishing at night during the spring. The size structure of 20-centimeter and longer largemouth bass in the population sample is depicted in Figure 17.4. It should be immediately apparent that the quality of the population is low. Of the eighty fish (20 centimeters and longer) collected, only nine (12.7%) were 30 centimeters or longer. An appropriate management objective in this case might be 40%–70% of these fish being 30 centimeters or longer. Before a strategy can be implemented to attain this management objective, the cause of the poor size structure must be understood.

Poor size structure for largemouth bass in an impoundment can result from either overpopulation or overharvest. Overpopulation is characterized by high density, slow growth, and poor body condition. As a result, few largemouth bass grow to sizes longer than 30 centimeters. Overharvested populations are characterized by lower density, faster growth, and good body condition, but few fish exceed 30 centimeters,

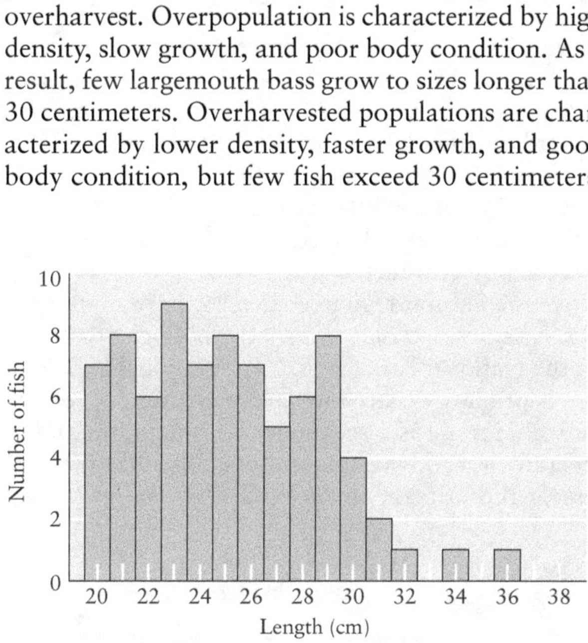

FIGURE 17.4 Size structure for a sample of 20-centimeter (cm) and longer largemouth bass collected by night electrofishing during the spring in a small midwestern impoundment.

because most are harvested by the time they reach this length. Once a biologist has determined which of these problems is occurring, the appropriate length limit type can be selected.

The size structure of an overharvested largemouth bass population that has moderate to fast growth rates and that has excessive angling mortality can be improved with a minimum length limit. If such a description fit the population depicted in Figure 17.4, one would expect to see a substantial increase in the number of fish that exceeded 30 centimeters within 1 or 2 years after imposition of a 38-centimeter minimum length limit.

Conversely, if the largemouth bass population depicted in Figure 17.4 was near carrying capacity and exhibited slow growth and poor body condition, then a protected slot length limit could be used to improve its size structure. A slot length limit commonly used in midwestern states protects largemouth bass from 30 to 38 centimeters long. Harvest of smaller fish is allowed, and is necessary, to reduce population density. If density is reduced, growth rates will increase, and fish will begin to grow past 30 centimeters. Fish are then protected until they reach 38 centimeters, and fish above the slot length range can again be harvested by anglers. A 30–38-centimeter protected slot length limit was imposed for largemouth bass in Murdo Lake, a 19-hectare impoundment in central South Dakota (Figure 17.5). In the year 1 spring electrofishing sample, only 22% of the 20-centimeter and longer largemouth bass exceeded 30 centimeters. The slot length limit was imposed that spring. By year 3, nearly 60% of the 20-centimeter and longer largemouth bass exceeded 30 centimeters, and over 25% exceeded 38 centimeters.

Some natural resource agencies have quite specific missions. Consider a fisheries biologist working for the National Park Service. The National Park Service is primarily geared toward protection of the resource. Harvest is a secondary consideration and is generally not allowed unless it will not affect the success of the primary mission. Gresswell and Varley

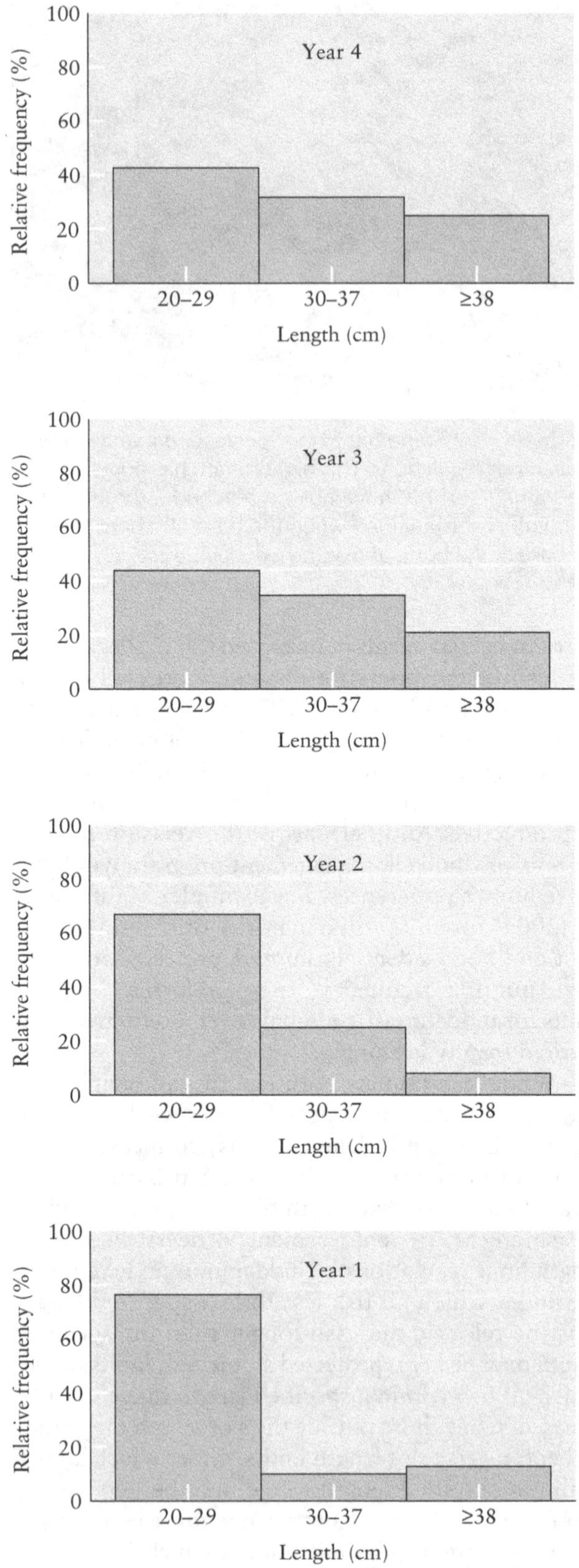

FIGURE 17.5 Size structure for 20-centimeter (cm) and longer largemouth bass collected by night electrofishing during the spring over a 4-year period in Murdo Lake, South Dakota. A 30–38-centimeter protected slot length limit was imposed in year 1.

(1988) used a 33-centimeter maximum length limit to manage the Yellowstone cutthroat trout population in Yellowstone Lake in Yellowstone National Park. The lake received high levels of angling effort. The maximum length limit allowed fish less than 33 centimeters in length to be harvested, but all longer fish had to be immediately released. After this regulation was implemented in conjunction with an artificial-lures-only requirement, 5- to 10-year-old fish reappeared in the lake, 45-centimeter fish were abundant, 51-centimeter fish were present, and anglers caught and released five large (illegal size) fish for every legal-length fish they caught. In reality, this regulation returned the Yellowstone cutthroat trout population to a size structure that was more characteristic of the lake before human impacts. The spawning runs of these large Yellowstone cutthroat trout out of Yellowstone Lake then became energetically profitable for grizzly bears to fish. Thus, the food base for a threatened species, the grizzly bear, was increased. Unfortunately, the case history does not end here. Lake trout were discovered in Yellowstone Lake in approximately 1994 (Kaeding et al. 1996). Despite attempted control measures, the lake trout population is increasing. Biologists fear that increased abundance of large, piscivorous lake trout could then reduce abundance and thus the spawning runs of cutthroat trout, negatively affecting the grizzly bear (Reinhart et al. 2001; Koel et al. 2005).

Most fisheries biologists find length limits to be the best way to manage fish populations when overharvest is a problem. Given sufficient enforcement efforts to ensure compliance, length limits decrease harvest but not fishing opportunity. When angling effort is sufficiently high, few fish will exist in a population that are longer than the minimum length limit or the upper end of a protected slot length limit. However, that does not mean that fishing must cease. Anglers have the opportunity to continue fishing, although they rarely have the opportunity to legally harvest a fish.

At the far end of the spectrum of length limit regulations are total **catch-and-release fisheries**. Under catch-and-release regulations, it is legal to capture a species, but any fish caught must be released. Some areas are managed as trophy fisheries with these regulations. Brown, cutthroat, and rainbow trout under catch-and-release regulations commonly reach large sizes. However, such changes occur only where habitat is appropriate. Such regulations are not a panacea that guarantees production of large fish. Catch-and-release fishing is growing in popularity with anglers, even where it is not a regulation.

17.6 Regulating "When"

Regulations specifying when organisms can be taken usually involve either time of year or time of day. Time restrictions were some of the earliest harvest regulations used in the United States. Many early regulations attempted to limit harvest to a single time of year and prohibited harvest during breeding seasons. Limiting season length can regulate the amount of harvest that actually occurs. Especially in hunting, seasons are generally open during the time of year when most populations are at peak density and a harvestable surplus typically is available. Most bird and mammal hunting seasons, for example, are in the fall in the northern hemisphere. Exceptions do occur in which hunting is allowed at other times of year, such as spring wild turkey seasons, but in this case, only males are legal game.

Another tool of the biologist is manipulation of the time of day when a species can be taken. Half-day hunting can be used to reduce harvest, or it can simply be a tool to reduce harassment of migratory birds and thus the likelihood of premature migration. Duck harvest can be limited by opening hunting time at or after sunrise rather than before sunrise. The movement and vulnerability to hunting of most ducks is greatest in the crepuscular periods prior to sunrise and after sunset. Ducks are so vulnerable to hunting after sunset that a sunset daily closure is mandatory in the United States.

Most hunting is a daylight activity. Exceptions, such as nighttime raccoon hunting, are rare. Fishing can be much more of a nighttime activity than hunting, but night fishing is prohibited in some situations by some natural resource agencies.

The use of closed seasons often presents a dilemma for fisheries biologists. Most fishing season closures occur during the late winter and early spring. The idea of disallowing recreational harvest during a time period that includes the spawning season for many species is attractive to anglers. If overharvest is suspected of reducing fish population quality, people often are quick to request a closed season. However, if a fish population truly is being overharvested, then the season must be closed at the time of year when most harvest typically occurs if sufficient harvest reduction is to be accomplished. For example, Redmond (1974) summarized case histories for largemouth bass harvest in five Missouri impoundments that ranged in size from 9 to 85 hectares. In the first 4 days after the impoundments were opened to fishing, 40%–69% of the largemouth bass were harvested by anglers. To prevent overharvest of largemouth bass in these impoundments, a closed season would obviously have to last 11 months and 26 days, or perhaps longer. Therefore, fisheries biologists generally prefer to find alternatives to closed seasons to prevent overharvest in situations in which exploitation is substantial.

Much tradition often surrounds opening day, and excitement can build as that day approaches. However, we certainly do not mean to imply that there are not social and biological justifications for closed seasons in fisheries. Closed fishing seasons are still used in some northern states, though they are uncommon in southern states. Kiefer et al. (1995) and Suski et al. (2003) reported that even catch-and-release fishing of male smallmouth bass during the spawning season may negatively affect reproductive success, which could justify a closed season at this time of year. Biologists believe that smallmouth bass populations in low-productivity waters in the northern United States and Canada are most vulnerable to such angling activities. Similar concerns have been expressed for largemouth bass (e.g., Philipp et al. 1997). While negative effects on individual largemouth bass have been documented, no negative effects of angling have been documented at the population level (i.e., bass recruitment and subsequent population abundance) (see section 3.3), although research continues (see Siepker et al. 2007).

17.7 Regulating "Where"

The locations where hunting or fishing is allowed are commonly regulated. Refuges and sanctuaries are provided for some wildlife species, with a common example being waterfowl refuges. Waterfowl refuges may provide a sanctuary from hunting in some cases, while in others they may attract waterfowl to an undisturbed area but actually increase hunter harvest. For example, as in the case described in section 17.5, a refuge may allow geese to rest undisturbed between feeding flights to lands surrounding the refuge. However, the birds may be subject to hunting off the refuge and may even be overexploited without some type of control on total harvest.

Sanctuaries have rarely been used as a freshwater fisheries management tool, although they are now common in commercial marine fisheries. A sanctuary of sufficient size could be used to prevent overexploitation of a freshwater fish species. Overcoming the potential opposition to such a closure could be difficult, however, especially if the area had traditionally been a popular fishing location. There are cases where sanctuaries have been provided in freshwater fisheries. For example, the Wisconsin Department of Natural Resources maintains sanctuaries for lake trout in the Wisconsin waters of Lake Superior. These sanctuaries typically are primary spawning grounds for the lake trout, and no recreational or commercial harvest is allowed there.

Geographic hunting units are often delineated by natural resource agencies, especially where licenses are limited in number and granted by lottery. Thus, harvest can be regulated according to these units. Harvest can be higher in areas where population density of a species is high than in areas with lower population densities. Although they are less common, similar differentiations can be made in different fishing waters.

Another important factor determining where people can hunt, fish, watch birds, hike, canoe, or pursue any other recreational activity involves trespass laws. Landowners, in conjunction with trespass laws, determine how access is obtained. This issue may initially seem to be a simple one—that is, it might be assumed that if land is privately owned, access is under the control of the landowner. In reality the situation is not so simple. In some areas, for example, hunting is allowed on private land unless that land has been posted against hunting or trespassing. Another example involves laws designating waterway ownership. Some streams and lakes are publicly owned, while others are in private ownership. In some situations, a stream or river is open to public access, but the banks are not. In other cases, a stream or river is open to public access, but users cannot touch the banks or the bottom of the watercourse. Where individuals can go and what they can do varies a great deal depending on regional laws.

There are also a wide variety of other laws and regulations that determine where activities can be pursued. Laws and regulations controlling activities in road rights of way, near dwellings, and in proximity to domesticated livestock all control "where" aspects of resource use.

17.8 Regulating "How"

Regulating methods that can be used to take organisms certainly influences eventual harvest. For example, restricting anglers to shore-only fishing and prohibiting angling from boats would reduce harvest of fishes commonly found in limnetic areas, such as walleyes, white bass, and striped bass. A commonly used fisheries regulation is a restriction on the type of **terminal tackle** (**terminal gear**) that can be used. Terminal tackle is the fishing equipment at the end of a fishing line, such as live bait or artificial lures. The characteristics of the angling public and the amount of harvest can be influenced by whether anglers are restricted to flies only or artificial lures only or whether live bait is allowed. For example, a fly-fishing-only regulation typically results in a substantial drop in angling effort for that water.

There are a number of other laws and restrictions regulating how fishes and other aquatic organisms can be taken. Common restrictions control the number of lines and hooks that can be used, whether fishes can be snagged (hooked somewhere in the body other than the mouth), whether animals can be speared, whether lines must be tended, whether artificial light can be used, and whether and what kinds of traps, nets, or seines can be used. In most cases, the use of electricity, piscicides, or explosives is illegal except with special permits.

Similarly, wildlife biologists can regulate hunting weapons to influence harvest. Big-game hunters using rifles, slugs in shotguns, primitive firearms (black powder), and archery all have different efficiencies. Even within these groups there can be further differentiation; for example, rifle caliber or power, the minimum draw weight of a bow, or the number of shells or cartridges that a weapon can hold or a hunter can carry while hunting can be regulated. The use of telescopic sights is often debated for black-powder rifles. How organisms are taken can also relate to nonhunting mortality. For example, lead shot can no longer be used for hunting waterfowl because of its toxicity (Bellrose 1982). Nontoxic shot, such as steel, must be used. There are many other potential restrictions on how wildlife can be taken, including those on the use of dogs, artificial light, baits, calls, and a wide variety of other techniques or devices.

Even regulating how access can be gained can influence hunter or angler use and the characteristics of the users that are present. Access by walking only, for example, will certainly draw a different clientele than access by motorized vehicle. In addition, there is likely to be less overall effort and harvest where motorized vehicles are not allowed.

17.9 Regulations for Nonharvested Species

Many people enjoy viewing terrestrial and aquatic organisms; such nonconsumptive use may still require regulation. For example, if sufficient numbers of bird-watchers seek a rare bird species such as the California clapper rail, even simply for viewing, they may damage salt-marsh habitat. Large numbers of people wishing to view breeding and nesting ospreys or bald eagles can disturb the birds to the point at which they are not successful in their reproductive attempts. Protected (no trespassing) areas are sometimes delineated around bald eagle or osprey nests in the lower forty-eight states to prevent viewers and others from disturbing the breeding and nesting process. Posting of no-trespassing signs on islands and sandbars has been used to protect nesting piping plovers and least terns from nonconsumptive disturbance (see section 16.4). At times, protection may be needed for sensitive habitats, such as sand dunes or wildflower areas.

Many fishes are also considered watchable wildlife in appropriate habitat conditions, such as in freshwater and marine environments where water clarity is high. Swimmers, snorkelers, and scuba divers desiring to view the underwater world are increasing in numbers. In some areas, numbers had increased to the point where environmental damage was occurring, and diver numbers had to be limited. Similarly, people commonly gather to watch the spectacle of migrating salmonids as they leap to pass over barriers.

Species that are considered game animals may be important watchable organisms in some locations. Regulations that prohibit angling or hunting may be implemented to maintain high population abundance for those species in those locales. The organisms may be in a refuge situation, or they may be in a location that for human safety purposes precludes harvest. Waterfowl refuges attract many people who wish to view waterfowl; often such users view the animals before, during, and after hunting seasons.

17.10 Public Relations and Regulation

In the past, biologists often determined what techniques would be applied to address a conservation problem and then informed the public of what they intended to do. Those days are gone, and for more than one reason. Today, public input into management decisions is essential. Natural resource agencies are cognizant of this change; in fact, many are hiring human dimensions specialists to address this facet of resource conservation. In addition, conservation programs today need to be geared toward satisfying the needs of many user groups, not just the anglers and hunters who were traditionally interested in natural resources. For agencies to be successful in maintaining wild populations of plants and animals, the support of the general public is needed. Most biologists are aware that the future welfare of many wild organisms depends on broad-based public support.

Public involvement in the past was usually in the form of public meetings. These were often attended by a very few people who had rather narrow interests. Those who attended were also usually willing to accept only those options provided by wildlife and fisheries professionals. Biologists can no longer present a short, selected list of already-planned conservation alternatives. There are many more people and groups interested in a much wider range of conservation options. It is also important to schedule public meetings sufficiently early in the process so that public input can be obtained on various potential strategies. Such meetings also have to be made available to a broad spectrum of the public.

As wildlife and fisheries professionals have evolved in their knowledge of the human dimensions, it has become apparent that public meetings are not the only means through which public opinion should be obtained. In fact, a few loud voices, often termed a vocal minority (see section 16.2), at a public meeting can actually have far more influence than their opinions should carry. Thus, many natural resource agencies are beginning to collect broader data on public opinions and attitudes. Instead of simply holding a public meeting and gaining the views of fifty people concerned with an issue, an agency might mail a questionnaire to all individuals who use a resource (or a statistically valid sample), as well as to other citizens who do not use the resource. More valid public opinion information is collected by such means. Many natural resource agencies survey citizens at regular intervals to obtain information on their preferences and attitudes. Such data can then be used to direct the future efforts of the agency.

17.11 The Regulation Process

The regulation process is becoming increasingly associated with and influenced by politics (see Chapter 20). Final decisions on regulations (as well as other

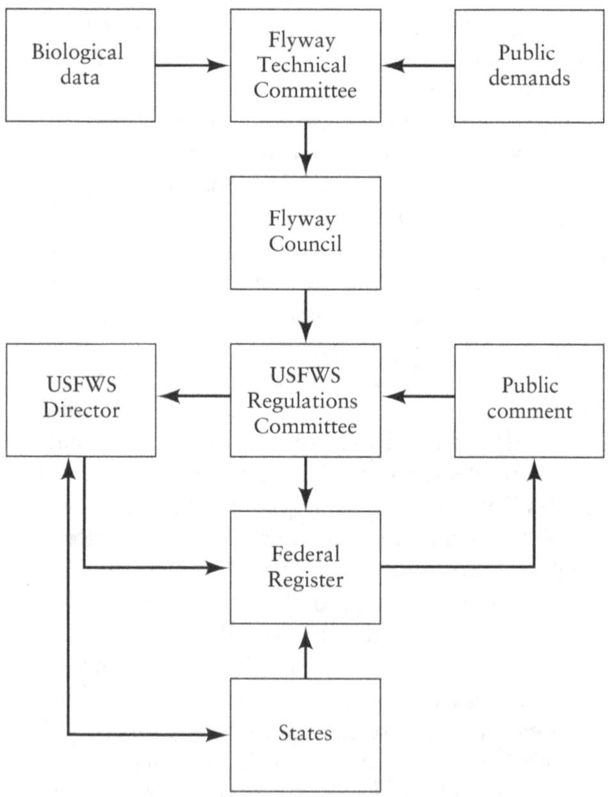

FIGURE 17.6 The regulation development system for the harvest of migratory waterfowl in the United States. (USFWS is the U.S. Fish and Wildlife Service.)

FIGURE 17.7 In North America, migratory waterfowl are managed by flyway. There are four such flyways.

conservation decisions) usually are not made by professional biologists.

When a state resource in the United States, such as a nonmigratory bird species, is involved, the citizens of the state control regulation. However, such decisions can still be questioned. Legal action may be taken by nonresidents to try to obtain some measure of input into state decisions. Federal agencies may become involved in interstate aspects of regulations affecting a variety of aquatic and terrestrial organisms. Regulations affecting endangered and threatened species, migratory birds, and marine fisheries require federal action, often in conjunction with other countries.

For example, the process for setting regulations on the harvest of North American waterfowl is intricate. These migratory birds are managed by international treaty. Initially it might seem like a simple procedure to determine when and how long waterfowl seasons will be, how many birds can be harvested, and which species can be harvested, but the determination procedure is more complicated than most people realize. Figure 17.6 depicts the steps in this process.

The initial step involves the collection of biological data on waterfowl population trends. This information is then assessed by a Flyway Technical Committee. There are four of these committees, one for each waterfowl flyway in North America (Pacific, Central, Mississippi, and Atlantic) (Figure 17.7). The committees are composed of waterfowl biologists from each state, province, or territory in the flyway. This phase of the process addresses the biology of the situation.

Each Flyway Technical Committee submits its recommendations to its respective Flyway Council. The council is composed of one person from each state, province, or territory in the flyway, usually the chief natural resource agency administrator. This administrator may or may not be educated specifically in natural resource conservation. Based on input from the Flyway Technical Committee and other information received, the council makes a recommendation.

In the United States, each council's recommendation goes to the U.S. Fish and Wildlife Service (USFWS) Regulations Committee. This committee publishes the proposals in the *Federal Register* (a publication that contains in printed form all activities of the U.S. federal government) and solicits public comment. The public comment portion of the process involves public meetings at which any and all stakeholders can comment about the proposed regulations. Testimony can be oral or written; all input is entered into a subsequent issue of the *Federal Register.*

After all testimony is received and considered, the USFWS Regulations Committee sets a framework for what the states can do with regard to seasons, bag limits, and other aspects of waterfowl harvest regulation. Each state then can begin its decision-making process to set regulations in that individual state. These decisions are usually made by the citizen commission or board in the state, again with biological and public input. After the state decisions are made, the state must notify the federal government of how it has decided to work within the guidelines provided, and a final total season framework, by state and flyway, is published in the *Federal Register.* As you can see, there are many opportunities for public input into this process.

As the science of wildlife and fisheries progresses and improves, biologists have more and better information that can influence and improve their decisions. Very often, this knowledge has led to specific strategies for fishes that vary by water body. Similarly, wildlife regulations commonly vary by geographic unit. The negative aspect of this improved management has been an increase in the complexity of regulations. Some wildlife and fisheries professionals are concerned that complex regulations could prevent participation in recreational activities by new and marginally interested individuals. Participation by such individuals is deemed to be desirable.

17.12 Information and Education

Regulations are not the only means by which biologists can modify human behavior and attitudes. Increasingly, biologists are providing educational opportunities to the public through dissemination of information. The formats used are highly diverse: articles in popular and natural resource agency magazines, pamphlets to be distributed to the public (Figure 17.8), news releases, public meetings, call-in talk shows, and public service announcements on radio and television have all been used. All agencies now have Internet Web pages, and many have forums that allow questions and answers to be discussed or even debated. Similar forums also are hosted by private individuals. While the Internet has greatly increased access to information, the sources of information must be considered, as misinformation is common.

Most natural resource agencies have an information and education staff to help with information dissemination. The primary responsibilities of these people lie in providing as much information as possible to the widest possible audience. Their education and background can be invaluable during the production of informational leaflets, booklets, and other communication forms. For example, fisheries biologists working for a natural resource agency have the technical expertise to provide information for a booklet that describes fish management in private ponds. However, the information and education personnel have the expertise to ensure that the appropriate writing style and presentation format will be used for the booklet.

FIGURE 17.8 Many informational publications are produced for individuals interested in natural resources.

Professional societies, such as The Wildlife Society and the American Fisheries Society, are also involved in information and education efforts. Their activities range from work at the local level to work at the federal and even international levels.

Much education can occur through informal contacts as well. A biologist who encounters anglers at a boat ramp or hunters at a bag check station can provide them with valuable information not only on local fish and wildlife communities but also on larger issues, such as the Clean Water Act.

Education cannot take the place of regulations and enforcement of regulations; it is simply an additional tool. Enforcement of regulations is essential for their success. However, public relations and information dissemination techniques can be used to educate the users who must comply with the regulations. If the majority of users understand and accept a regulation, it is more likely to succeed, and enforcement will certainly be easier to accomplish.

Education can also be used to modify human behavior without regulation. Many natural resource agencies devote substantial amounts of time and money to maintaining good relationships between resource users and landowners. In some areas, public service announcements remind users that they are guests on private lands and should act responsibly and courteously. Many agencies are encouraging users to be more nonconsumptive, for example, by promoting catch-and-release fishing without specific regulation. Fisheries and wildlife resources are renewable, but they should be considered limited. Educational efforts can be directed at users to convince them that there are many recreational benefits to be obtained from each natural resource–related activity, and that reaching a goal such as a bag or creel limit is not the only benchmark for a successful recreational experience.

SUMMARY

Biologists need to manage human activities in order to conserve the natural resources for which they are responsible. Management of the human user component of fisheries and wildlife systems, while always important, has received ever-increasing attention within the profession. Regulations on human activities generally affect who can use resources, what species or sizes can be taken, when the activities can occur, where the activities are allowed, and how the activities can be accomplished.

Regulations are commonly imposed to prevent overexploitation of a resource; however, this is not the only justification for regulation. Specific regulations may be used to achieve a specific management goal. Regulations may also be implemented to divide a resource more equitably among users. Finally, regulations may be applied for reasons that are primarily social or political.

Both wildlife and fisheries regulations and laws have progressed through a series of changes over time in North America. Early settlers found abundant resources, which were exploited without regulation. Through time, as fisheries and wildlife resources declined, conservation philosophies changed, and regulations and laws were altered.

The primary difference between wildlife and fisheries regulations lies in the fact that anglers often are able to release the fish that they catch. Hunters, in contrast, cannot release an animal once it has been harvested. In addition, fish movement is limited by the boundaries of a water body; wildlife species are more mobile.

Regulating who can use wildlife and fisheries resources can simply involve a determination of who can legally harvest organisms. Regulation of what can be harvested generally involves restrictions on the species, number, gender, or size of organisms that can be taken. Regulation of when use can occur generally involves either time of year or time of day. The location where recreational activities can occur is commonly regulated as well. Finally, natural resource users are often restricted in how they can legally pursue a recreational activity.

Regulations are also needed to manage nongame or nonharvested species. The disturbance caused when a large number of wildlife or fisheries watchers converge on a site can negatively affect the organisms. Thus, disturbance may need to be eliminated by means of no-trespassing regulations or regulation of the number of people allowed to participate.

In the past, biologists often determined appropriate conservation strategies and informed the public of their plans. Today, public input into such decisions is essential. The support of the general public, including those who do not consumptively use resources, is especially important.

Enforcement of regulations is an essential component of wildlife and fisheries management, but information and education programs can be equally important. An informed public is likely to be more supportive of regulations and less likely to violate them. The Internet has certainly provided an excellent central resource for the public, but incorrect information available at some sites not hosted by agencies is also a concern.

PRACTICE QUESTIONS

1. What are the three components of a fishery or a wildlife system? If necessary, refer back to section 1.2.

2. Why is human user management such an important component of wildlife and fisheries conservation?

3. Describe a situation where you live in which a wildlife or fisheries regulation was imposed primarily for social or political, rather than biological, reasons.

4. What roles do law enforcement and education programs play in determining success of regulations?

5. If your objective for a mule deer population was to produce a buck harvest that included a substantial proportion of individuals with four or more points, what regulations might you use to reach that objective?

6. Your objective for largemouth bass in a 30-hectare impoundment is to produce a population in which 40%–70% of the 20-centimeter and longer bass are also longer than 30 centimeters. Currently, only 5% of the fish exceed 30 centimeters. What information is needed to select the appropriate length limit to attain your objective?

7. Make a list of some bird and mammal species that can be managed with separate regulations for each gender. What characteristics of each species make such regulations possible?

8. Choose a nongame species in your area and describe potential human activities that might affect this species and require regulation.

9. Why do fisheries biologists consider daily creel limits to be relatively ineffective for prevention of fish population overharvest?

10. Discuss a situation in which establishment of a sanctuary might be an appropriate conservation strategy for a wildlife population and for a fish population. Include a management objective for each population in your discussion.

11. Go to the home page for your state natural resource agency and determine the number of different length limits used for fishes in your state.

SELECTED READINGS

Connelly, J. W., J. H. Gammonley, and J. M. Peek. 2005. Harvest management. Pages 658–690 *in* C. E. Braun, ed. *Techniques for wildlife investigations and management*. 6th ed. The Wildlife Society, Bethesda, Md.

Decker, D. J., and C. C. Krueger. 1999. Communication for effective fisheries management. Pages 61–82 *in* C. C. Kohler and W. A. Hubert, eds. *Inland fisheries management in North America*. 2nd ed. American Fisheries Society, Bethesda, Md.

Denney, R. N. 1978. Managing the harvest. Pages 395–408 *in* J. L. Schmidt and D. L. Gilbert, eds. *Big game of North America: Ecology and management.* Stackpole Books, Harrisburg, Pa.

Hawkins, A. S., R. C. Hanson, H. K. Nelson, and H. M. Reeves, eds. 1984. *Flyways: Pioneering waterfowl management in North America.* U.S. Fish and Wildlife Service 1984–667–157. U.S. Government Printing Office, Washington, D.C.

Noble, R. L., and T. W. Jones. 1999. Managing fisheries with regulations. Pages 455–480 *in* C. C. Kohler and W. A. Hubert, eds. *Inland fisheries management in North America*. 2nd ed. American Fisheries Society, Bethesda, Md.

Pollock, K. H., C. M. Jones, and T. L. Brown. 1994. *Angler survey methods and their application in fisheries management*. Special Publication 25. American Fisheries Society, Bethesda, Md.

Walters, C. J., and S. J. D. Martell. 2004. *Fisheries ecology and management*. Princeton University Press, Princeton, N.J.

LITERATURE CITED

Bellrose, F. 1982. Impact of ingested lead pellets on waterfowl. Pages 633–641 *in* J. T. Ratti, L. D. Flake, and W. A. Wentz, eds. *Waterfowl ecology and management: Selected readings*. The Wildlife Society, Bethesda, Md.

Boyd, R. J., and J. F. Lipscomb. 1976. An evaluation of yearling bull elk hunting restrictions in Colorado. *Wildlife Society Bulletin* 4:3–10.

Connelly, J. W., J. H. Gammonley, and J. M. Peek. 2005. Harvest management. Pages 658–690 *in* C. E. Braun, ed. *Techniques for wildlife investigations and management*. 6th ed. The Wildlife Society, Bethesda, Md.

Ganter, B., F. Cooke, and P. Mineau. 1995. Long-term vegetation changes in a snow goose nesting habitat. *Canadian Journal of Zoology* 74:965–969.

Graham, L. K. 1974. Effects of four harvest rates on pond fish populations. Pages 29–38 *in* J. L. Funk, ed. *Symposium on overharvest and management of largemouth bass in small impoundments*. Special Publication no. 3. North Central Division, American Fisheries Society, Bethesda, Md.

Gresswell, R. E., and J. D. Varley. 1988. Effects of a century of human influence on the cutthroat trout of Yellowstone Lake. Pages 45–52 *in* R. E. Gresswell, ed. *Status and management of interior stocks of cutthroat trout*. Symposium 4. American Fisheries Society, Bethesda, Md.

Kaeding, L. R., G. D. Boltz, and D. G. Carty. 1996. Lake trout discovered in Yellowstone Lake threaten native cutthroat trout. *Fisheries* (Bethesda) 21(3):16–20.

Kerbes, R. H., P. M. Kotanen, and R. L. Jefferies. 1990. Destruction of wetland habitats by lesser snow geese: A keystone species on the West Coast of Hudson Bay. *Journal of Applied Ecology* 27:242–258.

Kiefer, J. D., M. R. Kubacki, F. J. S. Phelan, D. P. Philipp, and B. L. Tufts. 1995. Effects of catch-and-release angling on nesting male smallmouth bass. *Transactions of the American Fisheries Society* 124:70–76.

Koel, T. M., P. E. Bigelow, P. D. Doepke, B. D. Ertel, and D. L. Mahony. 2005. Nonnative lake trout result in Yellowstone cutthroat trout decline and impacts to bears and anglers. *Fisheries* (Bethesda) 30(11):10–19.

Manfredo, M. J., P. J. Fix, T. L. Teel, J. Smeltzer, and R. Kahn. 2004. Assessing demand for big-game hunting opportunities: Applying the multiple-satisfaction concept. *Wildlife Society Bulletin* 32:1147–1155.

Philipp, D. P., C. A. Toline, M. F. Kubacki, D. B. F. Philipp, and F. J. S. Phelan. 1997. The impact of catch-and-release angling on the reproductive success of smallmouth bass and largemouth bass. *North American Journal of Fisheries Management* 17:557–567.

Redmond, L. C. 1974. Prevention of overharvest of largemouth bass in Missouri impoundments. Pages 54–68 *in* J. L. Funk, ed. *Symposium on overharvest and management of largemouth bass in small impoundments*. Special Publication no. 3. North Central Division, American Fisheries Society, Bethesda, Md.

Redmond, L. C. 1986. The history and development of warmwater fish harvest regulations. Pages 186–195 *in* G. E. Hall and M. J. Van Den Avyle, eds. *Reservoir fisheries management: Strategies for the 80s*. Reservoir Committee, Southern Division, American Fisheries Society, Bethesda, Md.

Reinhart, D. P., M. A. Haroldson, D. J. Mattson, and K. A. Gunther. 2001. Effects of exotic species on Yellowstone's grizzly bears. *Western North American Naturalist* 61:277–288.

Shaw, J. H. 1985. *Introduction to wildlife management*. McGraw-Hill, New York.

Siepker, M. J., K. G. Ostrand, S. J. Cooke, D. P. Philipp, and D. H. Wahl. 2007. A review of the effects of catch-and-release angling on black bass, *Micropterus* spp.: Implications for conservation and management of populations. *Fisheries Management and Ecology* 14:91–101.

Suski, C. D., J. H. Svec, J. B. Ludden, F. J. S. Phelan, and D. P. Philipp. 2003. The effect of catch-and-release angling on the parental care behavior of male smallmouth bass. *Transactions of the American Fisheries Society* 132:210–218.

Talbot, G. B. 1973. The California sardine-anchovy fisheries. *Transactions of the American Fisheries Society* 102:178–187.

Thelen, T. H. 1991. Effects of harvest on antlers of simulated populations of elk. *The Journal of Wildlife Management* 55:243–249.

von Lukowicz, M. 1998. Education and training in recreational fishery in Germany. Pages 287–293 *in* P. Hickley and H. Tompkins, eds. *Recreational fisheries: Social, economic, and management aspects*. Blackwell Science, Malden, Mass.

CHAPTER 18

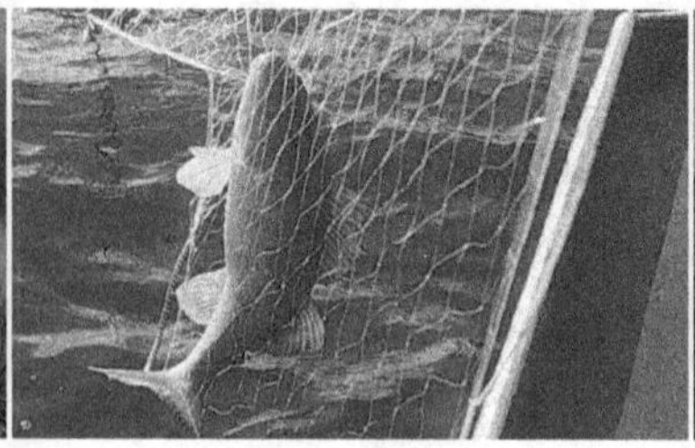

Wildlife and Fisheries Legislation

Wildlife and fisheries legislation in the United States and Canada reflects over a century of efforts by individuals and groups to conserve natural resources for the good of all citizens. Early colonists and other settlers in North America found an abundance of wildlife and fisheries resources that seemed beyond depletion. With the settling of North America and associated increases in human population and industrialization, negative effects of uncontrolled use of common resources soon became evident. To some, the future lay in dividing and depleting wildlife and fisheries resources with no thought to their future availability; to others, with greater vision, the potential for management of wildlife and fisheries resources as renewable resources was evident.

The wisdom of those who advocated conservation of our natural resources led eventually to the establishment of a North American heritage in wildlife and fisheries legislation. The common historical backgrounds, stable governments, and democratic natures of the United States and Canada were important factors in the evolution of the existing framework of wildlife and fisheries legislation in these two societies. Other nations with stable governments have also developed important foundations in wildlife and fisheries legislation, but none more strongly reflects the values of democratic societies than the legislation found in North America.

Extensive wildlife and fisheries legislation exists within individual states, provinces, and territories and within the federal governments of the United States and Canada. It would be impossible to review all such legislation in this text. The primary emphasis of this chapter is on U.S. federal legislation as it affects natural resources on a broad geographic scale. Some comparisons with Canadian legislation are also provided. An attempt is made to interweave the state, provincial, territorial, and federal perspectives with regard to jurisdiction over wildlife and fisheries resources.

Natural resource legislation in Mexico is relatively recent compared to that of the United States and Canada. Federal responsibility for wildlife and fisheries in Mexico has shifted continuously among agencies. Funding, professional personnel, and enforcement have all been lacking, and state natural resource agencies (31 states) are practically nonexistent. Faced with an extensive federal deficit and social problems, wildlife and fisheries legislation and accompanying management have not been a priority. A recent review of wildlife and fisheries laws, policy, administration, and conservation challenges in Mexico was presented in Valdez et al. (2006). Because of the above issues, legislation in Mexico will seldom be mentioned in this chapter.

18.1 States' Rights Doctrine

Wildlife and fisheries resources in any nation are dependent on legislation as well as on the traditions that led to that legislation. In Canada and the United States, immigrants found vast expanses of land with abundant wildlife, fisheries, and other natural resources. Most early settlers in North America were from the British Isles, France, and other European countries and had lived under laws and traditions that largely reserved hunting and fishing rights for royalty, landowners, and the privileged few. These immigrants shared in the abundant wealth of fish and wildlife resources in North America and rejected the European system. Early tradition, as well as court decisions in the United States, confirmed that the colonies and then the states, representing the people, assumed ownership of and responsibility for wildlife and fisheries resources. This meant that these resources belonged to everyone, not just the privileged class. In the United States, the right of individual states to control wild animal populations within their borders falls under the realm of the **states' rights doctrine**.

The states' rights doctrine, as it relates to fisheries and wildlife resources, remained largely intact through the 1800s. The U.S. Supreme Court acknowledged state ownership of wildlife and fisheries resources in two particularly important cases. The first case, *McCready v. Virginia*, in 1876, confirmed the right of the Commonwealth of Virginia to limit access to oysters within state waters. In the case of *Geer v. Connecticut* in 1896, Edgar M. Geer had attempted to transport American woodcocks, ruffed grouse, and northern bobwhite, legally taken in Connecticut, across the state line. The court ruled that Connecticut had the right to limit transport of legally taken game from the state for sale or use outside state borders. The *Geer* decision is viewed as the landmark states' rights decision and is commonly referred to in discussions and court decisions involving state versus federal jurisdiction over fisheries and wildlife resources.

In the United States prior to 1900, any federal involvement in conservation of wildlife and fisheries resources or their control through the legal system was considered the usurping of rights that constitutionally belonged to the states. These rights are based on the Tenth Amendment to the U.S. Constitution, which states, "The powers not delegated to the United States by the Constitution, nor prohibited by it to the States, are reserved to the States respectively, or to the people."

Actual colonial and state legislation limiting the taking of wildlife and fisheries resources by restricting time of year harvest could occur, requiring licenses, and even closing seasons occurred early in the history of the United States. Prior to independence, twelve of the thirteen colonies had already enacted closed seasons on deer, and Massachusetts actually closed deer hunting for several years starting in 1718 (Leopold 1933). New York required the first hunting license in 1864. By 1880, all states had game laws, although enforcement was often limited and sporadic. As early as 1652, the colony of Massachusetts closed fishing during the spawning period for species such as Atlantic cod, haddock, pollock, and Atlantic mackerel. By 1877, twenty-seven states had outlawed nets and seines for capturing fishes (Redmond 1986). Additional information concerning the history of fisheries and wildlife regulations appeared in section 17.2.

The Canadian Constitution Act (1867) provided for federal regulation of "sea coast and inland fisheries." The system that currently exists, as in the United States, recognizes provincial and territorial responsibilities as well as selected federal (migratory birds and coastal fisheries) responsibilities for wildlife and fisheries resources. Most Crown lands (public lands) in Canada are under provincial or territorial rather than federal control, a major difference from the United States.

18.2 The Federal Legislative Process in the United States

Prior to discussing wildlife- and fisheries-related legislation, a brief description of the U.S. federal legislative process is appropriate. Legislative proposals can originate in a variety of ways—with individuals, groups, members of Congress, or the executive branch of the federal government—but all proposed legislation must be introduced by a member of Congress. Several members of the House of Representatives and Senate can cosponsor the same bill. The typical route by which proposed legislation becomes federal law in the United States is illustrated in Figure 18.1.

Important legislation, such as the Endangered Species Act or Clean Water Act, must periodically be reauthorized by Congress. Thus, passage of legislation and years of successful implementation do not guarantee the future status of that legislation. Wildlife and fisheries biologists, conservation groups, other advocacy groups, and the general public can

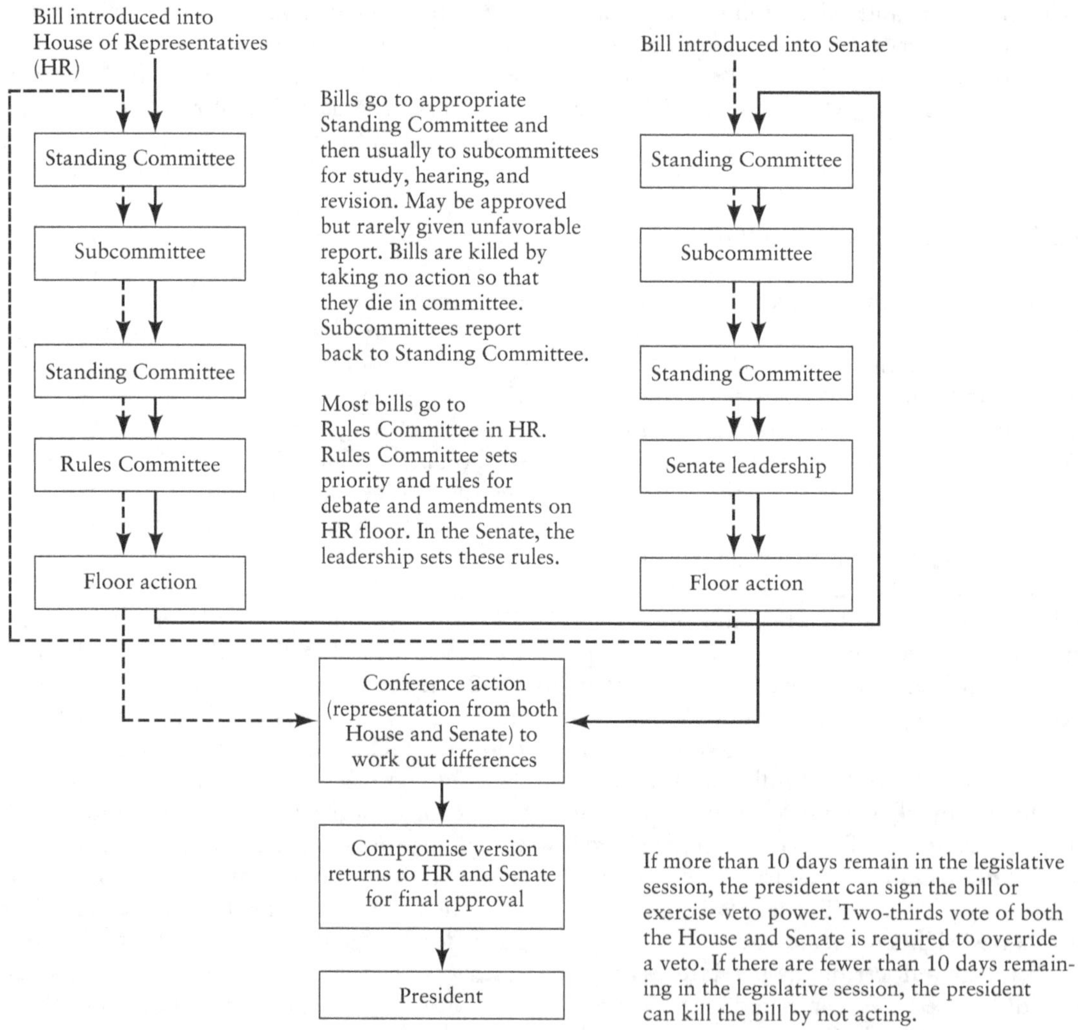

FIGURE 18.1 A diagram of the federal legislative process in the United States. The solid line indicates a bill introduced in the House of Representatives, while a dashed line indicates a bill introduced in the Senate.

have an important influence on reauthorization of legislation, development of amendments to legislation, introduction of new legislation, and opposition to legislation they are against. The legislative process involves a good deal of biopolitical interaction (see section 20.5).

18.3 Early Federal Legislation in the United States: The Lacey Act

The U.S. federal government became directly involved in wildlife legislation with the passage of the Lacey Act, which is also called the Game and Wild Bird Preservation Act of 1900. The Lacey Act, most importantly, regulates the interstate shipment of illegally killed animals. The Lacey Act was viewed as helping state law enforcement and conservation efforts and for that reason was not considered an infringement on the states' rights doctrine. The U.S. government is empowered by the Constitution to regulate interstate commerce, and thus its involvement in wildlife and fisheries conservation through the Lacey Act is constitutionally protected. Legislation similar to the Lacey Act, the Black Bass Act of 1926, was originally passed to control the interstate shipment of illegally taken largemouth bass. This legislation was later amended to include other freshwater and marine fishes and even shellfishes. In 1981, amendments to the Lacey Act repealed the Black Bass Act and provided for a single comprehensive law. Wildlife, fishes, shellfishes, some plants, and other protected organisms are now covered by the Lacey Act, as are parts or eggs of wildlife, fishes, and other protected organisms. If someone were to illegally kill a greater sage-grouse in eastern Oregon and transport that bird, or even its leg or some other body part, across the state line to Idaho, that person would be in violation of the Lacey Act. Likewise, an angler from Alabama fishing illegally in Mississippi and transporting a catch of channel catfish and largemouth bass across the border to Alabama would be in violation of the Lacey Act. It is also a violation of the Lacey Act for a nontribal member to illegally take wildlife, fish, or plants on a tribal reservation and transport them any distance, even without crossing reservation lines.

The Lacey Act also regulates international commerce in protected organisms. Anyone who transports, ships, or sells protected species contrary to foreign law or U.S. federal statutes, as well as state law, is in violation of the Lacey Act. Thus, if the horn of the protected black rhinoceros from Africa or illegally taken eggs of greater or lesser rheas from Argentina were shipped into the United States, those shipments would constitute a Lacey Act violation. As another example, assume that someone was illegally netting white sturgeons in the Snake River in Washington and then selling the eggs to a second party, who transported them to Europe for sale as caviar. Both the people illegally capturing the fish and those purchasing and transporting the illegally taken fish or eggs would be in violation of the Lacey Act. Any third party who knew that the eggs were illegally taken and was storing them prior to shipment would also be in violation.

The Lacey Act also prohibits the importation of certain species into portions of the United States. Importation of any species of mongoose, for example, is prohibited by the Lacey Act. Prohibition of the introduction of injurious species was originally a feature added in the interest of agriculture and horticulture (Bean and Rowland 1997). This prohibition has been broadened to incorporate a wide array of organisms, including some plants, because of their possible negative effects on natural ecosystems. Two other federal acts, the Non-Indigenous Aquatic Nuisance Prevention and Control Act (1990) and the Alien Species Prevention and Enforcement Act (1992), reflect further attempts to stop introduction of harmful species into the United States (Musgrave et al. 1998).

In Canada, the Wild Animal and Plant Protection and Regulation of International and Interprovincial Trade Act was passed in 1992; this act has many similarities to the Lacey Act. It prohibits interprovincial, interterritorial, and international commerce in illegally taken wild organisms or their parts, and it also prohibits introduction of species potentially harmful to ecosystems within Canadian federal, provincial, or territorial borders.

18.4 Migratory Bird Treaties and Legislation

Early in the 1900s, Canada and the United States recognized that migratory birds, such as waterfowl, needed protection within an international framework to assure long-term conservation of this resource. In 1913, the U.S. Congress passed the Migratory Bird Act, which declared that migratory and insectivorous birds were under the custody and protection of the federal government. This 1913 legislation was challenged in two federal district courts and was judged invalid on the basis of the earlier *Geer* decision. Even though the U.S. federal government recognized the opposition of many states to any federal regulation of wildlife, the federal government exercised its constitutional right to sign treaties with foreign nations by signing a migratory bird treaty with Canada in 1916. Canada, in supporting and signing the Migratory Bird Treaty with the United States, likewise acknowledged the need for an international approach to conservation of migratory birds.

Following the signing of the Migratory Bird Treaty, the Migratory Bird Treaty Act of 1918 was enacted to implement the treaty in the United States; in Canada the treaty was implemented by the Migratory Bird Convention Act of 1917 (Gilbert and Dodds 2001). In the United States, passage of the Migratory Bird Treaty Act meant that the Supreme Court no longer needed to decide the constitutionality of the 1913 Migratory Bird Act. The Migratory Bird Treaty Act in the United States and the Migratory Bird Convention Act in Canada made it illegal in both countries, except as permitted through regulations set by federal authority, to kill, possess, sell, purchase, or otherwise deal in migratory birds or their parts, eggs, or nests. States, provinces, and territories may make and enforce migratory bird laws and regulations if they are consistent with or more restrictive than those allowed under federal jurisdiction.

The constitutionality of the Migratory Bird Treaty Act was challenged in the U.S. Supreme Court in 1920 in the case of *Missouri v. Holland*. Ray Holland was a U.S. game warden enforcing the Migratory Bird Treaty Act in Missouri. The state of Missouri contended that enforcement of the Migratory Bird Treaty Act was a violation of constitutional rights reserved to the states under the Tenth Amendment. The authority of the U.S. government to use police powers to arrest persons violating the Migratory Bird Treaty Act was upheld in this important decision. The power of the U.S. government to enforce this legislation is derived from its constitutional powers to make and uphold treaties with foreign nations. Interestingly and wisely, the Supreme Court noted that migratory birds are transient between states and nations and that without the treaty, there might soon be no migratory birds left for any power to regulate. Upholding the Migratory Bird Treaty Act was seen as a defeat to the states' rights view of wildlife jurisdiction.

In the United States, the Migratory Bird Treaty Act did not authorize the acquisition of migratory bird habitat. The passage of the Migratory Bird Conservation Act (1929) resulted in the formation of the Migratory Bird Conservation Commission and allowed the secretary of the interior to purchase and lease migratory bird refuges. The Migratory Bird Hunting Stamp Act (1934) provided funding for the purchase and administration of **Waterfowl Production Areas** and purchase of **wetland easements** (in later years). Under this act, hunters in the United States pursuing waterfowl and some other migratory birds, such as sandhill cranes, must purchase a Migratory Bird Hunting and Conservation Stamp (Figure 18.2). This legislation is now more appropriately titled the Migratory Bird Hunting and Conservation Stamp Act, and the stamp itself is commonly referred to as a duck stamp.

Funds from the sale of duck stamps in the United States have provided substantial benefits in terms of migratory bird habitat purchases and easements. Preservation of these habitats has also benefited many other nonmigratory species of animals as well as plants. Funds from the stamp were inadequate for acquisition of sufficient wetlands by fee title and permanent easements, and it was necessary to borrow against future duck stamp income. Under the Wetlands Loan Act (1961), funds were made available for the purchase of wetlands or wetland easements over a 7-year period. Funding for the Wetlands Loan Act was later advanced for additional years and has recently been based on 1-year extensions. These borrowed funds were to be repaid by income from the Migratory Bird Hunting and Conservation Stamp Act. However, funding advanced for wetland acquisition has in large part been forgiven by Congress and does not need to be paid back from duck stamp funds. Payment of part of the cost of preserving wetlands from the general fund (funds provided by all taxpayers) seems reasonable, considering the variety of benefits to society resulting from wetland preservation. Reauthorization of the Wetlands Loan Act (or a similar act) will hopefully continue into the future. Canada passed a Migratory Bird Hunting Permit Act in 1966 that requires waterfowl hunters to purchase a federal migratory bird-hunting license. Funds from the purchase of this federal license go into the general fund, with no stipulation on their use. However, in 1984, Canada implemented a Wildlife Habitat Conservation Stamp Program to accompany the federal migratory bird-hunting license as a vehicle for raising money for habitat conservation projects throughout the nation. Funds from the purchase of the Wildlife

A

B

FIGURE 18.2 Funds from the sale of the Migratory Bird Hunting and Conservation Stamp *(A)* in the United States are primarily used to purchase or otherwise protect wetlands and associated upland habitat, such as Waterfowl Production Areas *(B)*. *[Image A courtesy of the U.S. Fish and Wildlife Service; image B courtesy of K. Jensen]*

Habitat Conservation Stamp are transferred to and administered by Wildlife Habitat Canada, a nonprofit foundation. Although waterfowl and wetland projects have been supported by these funds since inception, the funds are not exclusively dedicated to this purpose.

In 1936, a treaty for the conservation of migratory birds (and migratory game mammals) was signed by the United States and Mexico. Another treaty was signed by the United States and Japan in 1972 to cover birds migrating between the United States, Japan, and their territories. Canada does not have migratory bird treaties with countries other than the United States.

In 1986, Canada and the United States jointly signed the North American Waterfowl Management Plan (NAWMP), an ambitious plan to provide for conservation of waterfowl habitat in North America; in 1994, Mexico also became a partner in NAWMP. Goals of NAWMP involve protection and enhancement of waterfowl habitat throughout North America, with the objective of restoring ducks and geese to 1970s population levels. Funding for NAWMP comes from the three federal governments along with partnerships with state, provincial, tribal, private, and corporate sponsors. The North American Wetlands Conservation Act (1989) provides funding assistance and administrative direction from the United States for implementation of NAWMP throughout the continent.

18.5 Funding: A Carrot-and-Stick Approach

In 1937 the Pittman-Robertson Act (PR Act), or Federal Aid in Wildlife Restoration Act, was passed in the United States. This important legislation enacted an 11% tax on arms and ammunition and provided that proceeds be held in the federal treasury for use in wildlife restoration and management by the states. The act takes a carrot-and-stick approach by encouraging states to reinvest their hunting license income in wildlife resources. Funds available to states through the PR Act, commonly referred to as **PR funds** or **federal aid funds**, represent the carrot. These funds are available to a state only if its income from the sale of hunting licenses goes directly to the state agency responsible for wildlife and fisheries resources. If license money is taken away from that agency, the state loses its PR funds; this represents the stick. Wise federal legislators saw that income from hunting licenses would be a great temptation to state governments that might want to divert it to a variety of non-wildlife-related projects and expenses. Efforts are constantly made at both the federal and state levels to divert PR funds to purposes other than those intended by the act. Only a watchful and well-informed public interested in wildlife resources can ensure that these user-fee monies continue to be directed to their intended purposes.

Pittman-Robertson funds are apportioned to states on the basis of total state area and number of paid hunting license holders, each composing 50% of the apportioning formula. No state is to receive less than 0.5% and no state more than 5% of the total annual PR funds available. Pittman-Robertson funds pay up to 75% of the cost of state projects that are submitted to the U.S. Fish and Wildlife Service and accepted for federal aid participation. The state must provide 25% in nonfederal matching funds. State management and research projects involving the acquisition, lease, restoration, rehabilitation, and improvement of wildlife habitat are eligible for PR funding.

In 1970 an amendment to the PR Act directed that monies from an existing federal excise tax on pistols and revolvers be placed in the PR fund. Since 1975, half of the federal tax on bows and arrows also goes into the PR fund. Part of the annual revenue from these taxes is divided among the states for hunter safety programs, including construction, operation, and maintenance of outdoor target ranges.

The Dingell-Johnson Act (1950) (DJ Act), or Federal Aid in Sport Fish Restoration Act, was modeled after the PR Act and contains similar benefits, except they apply to fisheries resources. The DJ Act provides for proceeds of a tax on fishing tackle to be held in the federal treasury and granted to the states for restoration and management of state fisheries resources. Funds from the DJ Act are often referred to as **DJ funds** or federal aid funds. As with PR funds, DJ funds are available to states only if their income from fishing licenses is restricted to use by the state agency responsible for fisheries and wildlife resources. Also, as with PR funds, constant attempts are made at both the state and federal levels to divert such funding away from fisheries. Distribution of DJ funds to states is based 40% on land area and 60% on fishing licenses sold. As with the PR Act, states must contribute a 25% nonfederal match in order to use DJ funds.

In 1984 the **Wallop-Breaux Amendment** to the Federal Aid in Sport Fish Restoration Act extended the federal excise tax to include additional fishing gear. An excise tax on electric motors and sonar devices was also added, as was a federal tax on fuel for motor boats. The Wallop-Breaux Amendment also added import duties on foreign-manufactured fishing tackle and boats. The Wallop-Breaux Amendment has more than tripled the federal aid funds available to states for fisheries work.

The Pittman-Robertson and Dingell-Johnson acts have greatly benefited wildlife and fisheries resources in the United States. We are aware of no other country with similar legislation. Wording of these legislative acts has also successfully protected state hunting and fishing license fees from diversion into other programs.

18.6 Other Federal Wildlife and Fisheries Legislation

A variety of other U.S. federal legislation affects wildlife and fisheries resources; a few important examples follow. The Fish and Wildlife Coordination Act of 1934 (last amended in 1995) signaled the intention of the U.S. government to coordinate all activities of federal agencies that affect wildlife and fisheries resources. Under this act, federally funded projects influencing fisheries or wildlife resources must be coordinated with the U.S. Fish and Wildlife Service (USFWS). For example, if a major water development project using federal funds, such as a dam with hydroelectric facilities, were planned, the USFWS would have to be consulted and effects of the project on fisheries and wildlife resources given equal consideration (1958 amendment) with its other aspects. Equal consideration for fisheries and wildlife resources may require **mitigation** for lost habitat. For example, the influence of a dam on wildlife is likely to be negative because of the loss of riparian and other bottomland habitat associated with the original river or stream. Mitigation might involve the purchase of private land with riparian and bottomland habitat values in another part of the state in which the project is developed. Even if such areas are purchased and put under public ownership and management, it is still difficult to mitigate for extensive riparian habitat destruction caused by many water projects. In general, the mitigation process has not always been successful. Until recently, the U.S. Congress delayed appropriation of funding for habitat mitigation until after dams and other projects were completed; the process was slow, sometimes requiring several decades before reasonable mitigation for habitat loss could be accomplished. Habitat mitigation costs are now included as part of the initial appropriations needed before construction begins on a federal project.

The Bald Eagle Protection Act (1940), amended in 1962 to include the golden eagle, provides for complete protection for both species in the United States. Until being delisted in 2007 (see section 11.4), the bald eagle in the lower 48 states was also protected by the Endangered Species Act. The U.S. Endangered Species Act was discussed in Chapter 11 and thus is not included here. In Canada, endangered species and threatened species such as the whooping crane and swift fox are protected through the Species at Risk Act (2003) (see section 11.2)

The Marine Mammal Protection Act (1972) in the United States addressed a variety of marine mammal issues. States lost their authority to manage marine mammals through this legislation, but they could regain some of that authority by instituting conservation programs that meet federal standards. The Marine Mammal Protection Act places restrictions on harvest of marine mammals. Limited exceptions are allowed for native peoples and for incidental kills from legal commercial operations. The act also allows for harvest of selected marine mammals if populations are kept at optimum carrying capacity, a term not clearly defined in the legislation. The act must be considered in international treaties and agreements between the United States, Japan, Russia, and Canada. A related act, the Fur Seal Act of 1966, prohibits taking of northern fur seals by persons or vessels under U.S. jurisdiction in the North Pacific; exceptions are provided for Indians, Aleuts, and Eskimos who traditionally depended on these seals for sustenance.

The Magnuson Fishery Conservation and Management Act (1976) is the primary act governing marine fisheries in the U. S. coastal waters and has been amended multiple times through 2006. In 1996, this much-amended act was renamed the Magnuson-Stevens Fishery Conservation and Management Act and is now commonly referred to as the Sustainable Fisheries Act. The primary goal of the Sustainable Fisheries Act is to restore marine fisheries by eliminating overfishing and rebuilding overfished stocks important to commercial, recreational, and subsistence fisheries. Many marine fishes and shellfishes have been depleted to levels that can no longer support substantial commercial harvest. This act governs the conservation and management of fish and shellfish in coastal and continental shelf waters and also governs harvest related to salmon and other anadromous fishes that spawn in rivers in the United States.

The Fish and Wildlife Conservation Act (1980), commonly known as the Nongame Act, provided a framework for the conservation of the many nongame vertebrates in the United States. Nongame species are often overlooked unless they are endangered or threatened. Unfortunately, Congress did not provide an adequate funding mechanism for the Fish and Wildlife Conservation Act. Since passage of the Nongame Act, there has been and continues to be efforts to develop a dependable and long-range funding source. However, beginning in 2000, states became eligible for federal funding for part of their nongame needs under the State Wildlife Grants Program. From inception through 2006, this program has provided a total of \$350 million in congressional appropriations to be apportioned to states. Depending on the nature of the project, 25%–50% of the funding from this source must be matched by state or private funds. States must obtain prior approval from the U.S. Fish and Wildlife Service on planned expenditures of these State Wildlife Grant funds and, as of 2005, must have submitted a Comprehensive Wildlife Conservation Strategy to qualify.

The U.S. government recognized the importance of international wetland protection for conserving migratory birds and other species in the North American Wetlands Conservation Act of 1989. Under this legislation, funds can be spent for wetland acquisition and protection in Canada and Mexico as well as in the United States. Funding for wetland protection in Canada and Mexico is particularly important to the success of the North American Waterfowl Management Plan (see section 18.4). Interest earned on Pittman-Robertson funds prior to their distribution to the states is now reserved for funding of wetland acquisition and improvement under this legislation. These interest-generated funds previously went into the federal general fund and had long been targeted by conservation interests; their use in helping to fund the North American Wetlands Conservation Act seems appropriate. Additional funds come from fines collected by the Department of Justice for violations of the Migratory Bird Treaty Act and from federal appropriations.

There is much additional U.S. legislation dealing with fish and wildlife on an international scale. These are well illustrated by legislation on oceanic fisheries and anadromous fishes. For example, federal fisheries legislation in the United States such as the Atlantic Salmon Convention Act (1982), the Pacific Salmon Treaty Act (1985), the Atlantic Tunas Convention Act (1975), and the Northern Pacific Halibut Act of (1982) all address important national and international resource issues that require cooperative efforts with other nations (Musgrave et al. 1998).

18.7 General Environmental Legislation

While many of the following pieces of legislation are usually described as environmental legislation, they go far beyond that. All are important to public health, safety, and welfare as well as to wildlife and fisheries resources.

The National Environmental Policy Act (1969) in the United States, popularly known as NEPA, is broad in its implications, because it established the Council on Environmental Quality as well as the Environmental Protection Agency. In addition, through NEPA, **environmental impact statements** (EIS) are required for all projects receiving federal funds that substantially affect the quality of human life. Construction of a dam or a highway using federal funds, for example, requires preparation of an EIS that includes the possible effects of the project on wildlife and fisheries resources. The Council on Environmental Quality must assess the EIS before allowing a project to proceed. The Council on Environmental Quality was mandated by executive order in 1979 to also evaluate effects of major federal actions abroad. In addition, the United States signed a Convention on Environmental Impact Assessment with several European countries in 1991 requiring the signatories to consult one another regarding transboundary environmental impacts.

Environmental impact assessment in Canada is mandated through the Canadian Environmental Assessment Act (1992). Projects using federal funds, on Crown lands, or being undertaken by federal agencies require environmental impact assessment under this decision. Assessment is conducted by a federally appointed review panel.

Regulation of various toxic substances released into the environment can be important to wildlife and fisheries resources. By their nature, pesticides are designed to kill certain organisms or groups of organisms that damage crops, threaten human health, or in some other way cause economic damage. Pesticide control in the United States was initially under the Federal Insecticide, Fungicide, and Rodenticide Act of 1947 (FIFRA), which was later amended and replaced by the Federal Environmental Pesticide Control Act of 1972 (FEPCA). Under FEPCA, pesticides were placed into two categories, general use and restricted use. Restricted-use pesticides require certified applicators. The intent of the legislation is to stop applications that involve unreasonable adverse effects on the environment. Human health is the primary concern, with wildlife and fisheries considerations secondary. However, prohibition of use due to wildlife and fisheries concerns can occur, as in the cases of strychnine and compound 1080 (sodium monofluoroacetate), which were removed from use as coyote poisons in 1972. The Toxic Substances Control Act of 1976 extended control to substances other than pesticides. Federal authority for control of toxic substances and other types of pollution in Canada is provided by the Canadian Environmental Protection Act (1999).

Despite pesticide control legislation, numerous instances of direct mortality and secondary effects on wild organisms due to pesticides still occur. Some of these incidents cause considerable mortality in vertebrate and invertebrate species. In addition, pesticide effects that do not kill organisms outright often remain undetected (Figure 18.3). For example, sublethal effects, such as reduced growth rates due to reductions in the invertebrate foods available to young ducks and gallinaceous birds, are particularly difficult to identify (Grue et al. 1988). Similarly, sublethal effects of chemical pollutants have been linked to declines in eel populations in the United States and Europe (Robinet and Feunteun 2002).

Water quality affects aquatic plants and animals as well as terrestrial animals associated with aquatic

FIGURE 18.3 Agricultural pesticides sometimes kill invertebrates in wetlands, resulting in reduced growth rates or even death for ducklings such as these blue-winged teal. Some pesticides may also directly kill birds through contact or ingestion.

systems. Thus, the Clean Water Act of 1977 is important to wildlife and fisheries resources. This legislation directs the Environmental Protection Agency to set standards that limit pollution in lakes, streams, and other surface waters.

18.8 Legislation Directed at Management of Federal Lands

Legislation influencing the management of federal lands in the United States is particularly important to wildlife and fisheries resources, especially in the western half of the United States, where these lands cover an extensive area. The Taylor Grazing Act of 1934 was the beginning of control over grazing by domesticated livestock on vacant, unappropriated, and unreserved federal lands under the jurisdiction of the Bureau of Land Management (BLM). The Taylor Grazing Act clearly states that nothing in the act should be construed so as to restrict hunting and fishing on these lands. It also notes that grazing permittees have no right to interfere with hunting, fishing, and many other recreational activities on these lands. Without this stipulation, public recreation on BLM lands could have been curtailed by grazing or other commercial interests.

The Multiple Use-Sustained Yield Act (1960) is the legal basis for multiple use of national forests in the United States. The Classification and Multiple Use Act of 1964 provided a similar policy on BLM lands. These legislative acts state that national forests and BLM lands will be managed on a multiple-use basis, with fish and wildlife given consideration along with other uses, such as timber, range, watershed, and outdoor recreation (Figure 18.4). Despite these legislative actions, timber, livestock, and mining interests have generally received higher priority than wildlife and fisheries resources on federal lands. The two multiple-use acts described above are still important, because they give legal impetus to conservation of wild animals and their habitats on federal land.

Further support for the multiple-use concept was provided through the Public Rangeland Improvement Act of 1978; this act required improvement of federal lands that have been degraded by domesticated livestock grazing. This legislation, which imposed higher grazing fees and grazing restrictions, set forth demands for the transfer of federal lands to private ownership. Many ranchers, however, recognized that federal land, if sold, would go to the highest bidder, and that they could lose the grazing leases they controlled while the land was in federal ownership. In addition, if grazing lessees bought the land, they would also need to begin paying state property taxes on land they formerly leased at low rates without paying taxes. Economic considerations prevailed, and demand for a massive sale of federal lands quieted. Resistance to federally mandated rangeland improvement efforts continues at county levels and through local legislation designed to challenge federal authority and to counter federal efforts at public land management (Musgrave et al. 1998). There is always a threat of the loss of federal lands through sale to private individuals or corporations. The potential ceding of federal lands to individual states is also a concern, as states could subsequently sell the land to private owners or in other ways exclude the public. Additionally, states might also be more likely to favor timber, mining, and grazing interests. It is clear from past and current efforts by some special interest groups that the existence of federal land in the United States and the right of public access to this land cannot be taken for granted.

FIGURE 18.4 The Chugach National Forest in Alaska must be managed for multiple uses, including wildlife and fisheries resources, under the Multiple Use-Sustained Yield Act. *[Image courtesy of K. Jensen.]*

The National Forest Management Act (NFMA) of 1976 required federal forest management plans to comply with the National Environmental Policy Act. This legislation specifically directs the Forest

Service to maintain the biological diversity of native vertebrate species within national forests. The contemporary Federal Land Policy and Management Act of 1976 (FLPMA) had many similarities to NFMA, but it was directed at BLM lands. Both pieces of legislation enhanced the multiple-use concept on federal lands.

The Wilderness Act of 1964 mandated preservation in a natural state of large tracts of national forests, national parks, and other federal lands in remote areas; 769,000 hectares of remote federal land was originally established in the National Wilderness Preservation System. Areas are set aside in which roads, timber harvests, commercial development, and buildings are prohibited. Prior mining laws allow mining activity in areas where valid claims were filed before 1983. The Endangered American Wilderness Act of 1978 brought an additional 526,000 hectares of unprotected national forest into the National Wilderness Preservation System. Passage of the Alaska National Interest Lands Conservation Act (ANILCA) in 1980 also established large tracts of wilderness; these Alaskan areas constitute over two-thirds of the National Wilderness Preservation System in the United States. The designation of extensive areas in Alaska as part of the National Wilderness Preservation System has been important in helping preserve these pristine ecosystems by preventing logging, road development, oil and gas development, and other economic activities. However, ANILCA has sometimes been controversial, particularly with some Alaskan residents. Economic pressure continues for development of oil and gas reserves on the vast and biologically rich coastal plain of the Arctic National Wildlife Refuge in Alaska; this portion of the refuge was set aside for special status through passage of ANILCA but did not receive protection as national wilderness and can be opened to oil and gas development through an act of Congress.

Protection of ecosystems on federal wildlife refuges was the impetus behind passage of the National Wildlife Refuge System Improvement Act of 1997. Wildlife and fisheries conservation on all of these federal lands was enhanced through the Sikes Act (1960). This act authorized the secretaries of the departments of Defense, Agriculture, and Interior to develop cooperative plans for wildlife and fisheries conservation and habitat restoration on lands under their jurisdiction.

18.9 Federal Legislation Influencing Fish and Wildlife Habitat on Private Lands

Some of the legislation that most influences the environment as well as public health and welfare involves U.S. Department of Agriculture (USDA) programs. In 1956, Congress passed the Soil Bank Act, which, at its peak, diverted almost 12 million hectares of land from crop production and placed it in protective vegetative cover for a period of 3 to 10 years. The highest enrollments in the Soil Bank program were in the Plains states from Texas to North Dakota. The Soil Bank provided habitat for a wide variety of wildlife species.

The **Conservation Reserve Program** (CRP) of the 1985 Food Security Act (Farm Act or Farm Bill) and the conservation compliance elements of the **swampbuster provisions** of this 1985 act have provided important benefits to the environment and to a wide variety of wild organisms; the **sodbuster provisions,** as will be explained, have been disappointing in terms of protecting wildlife and fisheries resources. The CRP and the swampbuster and sodbuster provisions were continued through 1990 and 1996 reauthorizations of the Farm Act and, most recently, through the Farm Security and Rural Investment Act (Farm Act) of 2002. The swampbuster provisions include the loss of certain federal payments for landowners receiving federal crop subsidies through the USDA if they produce crops such as corn or soybeans on wetlands drained or otherwise converted to farmland after December 1985. After the passage of the 1990 Farm Act, any wetland conversion, even without planting a crop, became a violation subject to loss of USDA subsidies.

The sodbuster provisions require landowners plowing highly erodible grasslands to implement a soil conservation plan to keep soil erosion within limits specified or be subject to loss of certain USDA subsidies. Sodbuster has proven to be of little value in keeping remaining areas of native prairie from being plowed. The sodbuster provisions only apply to land defined as highly erodible, and even on those lands, the landowner only needs to comply with a soil conservation plan before plowing native prairie. Sodbuster needs to be strengthened with crop subsidy penalties tied to further destruction of native prairies. The continued destruction of our native prairies is most discouraging. Development of drought-resistant soybeans led to the conversion of approximately 70,000 hectares of native prairie in central South Dakota alone between 1987 and 2002 (Higgins et al. 2002). Populations of many wildlife species dependent on prairies for nesting, such as the long-billed curlew and northern pintail, are placed in an increasingly precarious position by continued destruction of remaining prairies.

The Conservation Reserve Program provides payments to landowners who take marginal cropland out of production. Under CRP, extensive amounts of marginal cropland have been idled and seeded to grass or grass–legume cover for a period of 10 to 15 years. The primary goals of the original CRP legislation were to reduce farm surpluses, control erosion, and improve water quality. Enrollment in CRP reached over 13 million hectares under the 2002 Farm Act (Brady 2005). In the northern Plains states, where much of the CRP land is located, these lands provide extensive areas of excellent nesting and other habitat for a variety of game and nongame wildlife (Figure 18.5). Fisheries resources have also benefited from the CRP through reduction of siltation and chemical pollution of aquatic habitats.

While payments to landowners for idling CRP lands are substantial, they are often lower than other payments that would be made through the USDA for other programs, such as crop subsidies, crop deficiency payments, annual set-aside payments, and disaster payments, if CRP were not in existence. Thus, the CRP provides both producer benefits and conservation benefits at a lower price. CRP lands are only to be grazed or mowed for weed control or habitat management purposes. However, pressure from CRP enrollees, especially during drought when there is high demand and good prices for hay, has often led to the allowance of additional amounts of mowing and grazing on CRP land without a reduction in CRP payments.

FIGURE 18.5 This Conservation Reserve Program (CRP) field in the north-central United States is dominated by intermediate and tall wheatgrass with inclusions of smooth bromegrass and provides excellent habitat for a variety of wildlife species. Loss of CRP would cause declines in many upland nesting bird species, such as mallards, bobolinks, ring-necked pheasants, and sharp-tailed grouse. *[Image courtesy of K. Jensen.]*

Unfortunately, land retirement programs such as CRP sometimes penalize those who have practiced good land stewardship by not plowing marginal ground. Many landowners who plowed areas of short-grass or mixed-grass prairie to plant wheat in areas such as eastern Colorado were rewarded with CRP contracts. Those landowners who had continued to graze livestock rather than plowing these lands, despite the higher immediate economic returns from wheat, were unjustly penalized by not receiving CRP contracts. Anticipation of soil retirement programs can even encourage plowing of marginal areas so that the landowner can qualify for the program. Penalization of good land stewardship needs to be corrected in future land retirement programs.

Two other USDA programs valuable to wildlife include the **Wetlands Reserve Program** (see section 15.5), first established by the 1990 Farm Act, and the **Grassland Reserve Program**, authorized in the 2002 Farm Act. The Wetlands Reserve Program provides technical and financial assistance to landowners primarily for restoration of drained or agriculturally degraded wetlands. More than 647,000 hectares of degraded or drained wetlands had been restored by 2005 (Rewa 2005). The Grassland Reserve Program provides technical and financial assistance to landowners in conserving, managing, and improving grasslands while allowing grazing and other compatible uses (Wood and Williams 2005). This program has been valuable in protecting over 200,000 hectares of grasslands from conversion to croplands.

Soil Bank and CRP, while accomplishing many goals and providing many benefits to wildlife and fisheries resources, were both initiated as temporary or short-term programs. Land retirement and other conservation programs should be of a longer-term nature, with permanent vegetative cover required and with mowing, haying or grazing, and fire used sparingly and only to enhance the quality and wildlife value of these lands. All current USDA programs valuable to wildlife such as CRP, swampbuster, and the Wetlands Reserve Program must be reauthorized every few years and could be dropped, greatly reduced, or enhanced each time the Farm Act is reauthorized. Continued reauthorization of these programs requires broad expressions of support to federal legislators from farmers, ranchers, conservation groups, and the interested public.

18.10 Treaties, Tribes, and Wildlife and Fisheries Resources

In the United States, the right of indigenous peoples to hunt and fish on their own tribal lands, and to set their own seasons and manage their own wildlife and fisheries resources, generally is not challenged by states. However, off-reservation tribal hunting and fishing rights are often contested. Some aspects of hunting and fishing rights of indigenous peoples have been highly controversial and are still being litigated; differing interpretations of treaty wording often cause such litigation.

The U.S. government signed numerous treaties with indigenous tribes during the period when territories were being settled. Specifications in these treaties are as legal as those in treaties signed with foreign nations. Treaties conferred many advantages on settlers, military establishments, railroads, and other nonindigenous users of land, and indigenous tribes gave up substantial areas of land in most treaties. In exchange for giving up land, tribes were sometimes promised, in very specific terms, certain rights to fish and hunt. These treaties often may seem to conflict with the tenet that states have the authority to regulate wildlife and fisheries resources within their borders. However, rights guaranteed to tribes by federal treaty supersede state rights to wildlife and fisheries resources.

Rights of indigenous peoples to hunt or fish, especially off reservations, under some treaties have been a source of controversy. In parts of Wisconsin, for example, spring spearing of walleyes by tribal members on traditional but off-reservation sites has caused much controversy. Similar treaty rights are being exercised in the hunting of large ungulates on some off-reservation sites in states such as Idaho and Washington. The U.S. government attempts to enforce hunting and fishing rights specified in these treaties because they are legal and lawful agreements.

Indigenous peoples exercising off-reservation fishing rights that are guaranteed by treaty cannot be required to purchase a state license, as their rights are more extensive than those of other citizens of the state (*United States v. Winnans*, 1905) (Sigler 1995). However, the state may restrict fishing by indigenous peoples to the extent that it is necessary to conserve a particular stock or run of fish (*Puyallup Tribe, Inc. v. Washington Department of Game*, 1968). Where a stock of fish is being depleted, the state must first restrict anglers and commercial operations not covered by treaty rights.

Under treaty rights, tribes in the state of Washington may take salmon "at usual and accustomed places." An important U.S. District Court decision in the state of Washington in 1975, popularly called the Boldt Decision (*United States v. Washington*), provided a clear interpretation of the quotas of anadromous fishes allowed to indigenous peoples. Judge Boldt declared that tribes could take up to 50% of the annual harvest at traditional fishing sites (Gilbert and Dodds 2001). The total tribal and nontribal harvest must allow for adequate escapement of spawning fishes for conservation purposes. The Supreme Court upheld the Boldt Decision in *Washington v. Washington State Commercial Passenger Fishing Vessel Association* (1979), but it did stipulate that fishes taken on reservations be counted toward the 50% of the harvest allotted to the tribes (Bean and Rowland 1997). The Boldt Decision had important implications in the allocation of fish resources in the northwestern United States, especially that of Pacific salmon and steelheads.

Reservation authorities have the right to allow or refuse hunting or fishing rights to people who are not tribal members. They can also require and enforce the purchase of a tribal hunting or fishing license by nonmembers. In some cases, state natural resource agencies also require a state hunting or fishing license for people who are not tribal members but who hunt or fish on reservation lands. The matter of state jurisdiction over hunting and fishing on reservation lands by nontribal members is still unresolved in many situations and has been the subject of several court cases. Aspects such as the degree to which a state is involved in fisheries or wildlife conservation on the reservation—for example, in stocking fish or management of migratory wild ungulates—are important in this jurisdictional question (Sigler 1995). Jurisdiction over wildlife and fisheries resources on nontribal land (owned by nonindigenous peoples) within historical reservation boundaries has also been controversial in some states.

Most larger reservations now have their own natural resource agencies, sell licenses to nonmembers and members of the tribe, and in many cases employ law enforcement personnel. Some of these tribes, such as the Yakama Indian Nation in Washington and the White Mountain Apaches in Arizona, also employ university-educated biologists to assist in managing their wildlife and fisheries resources. Such tribes have had much success in natural resource management. Funds from the sale of licenses, including fees of $5,000 to $15,000 for a trophy bull elk, can result in substantial income to the tribe.

Many aspects of the hunting and fishing rights of indigenous peoples in the United States still remain unclear. The rights of states, tribes, and all citizens involved will eventually be clarified. Unfortunately, much time and money is being expended in litigation over these rights. In the meantime, many cases of cooperation among states, tribes, and the federal government are providing hope for the natural resources, average citizens, and tribes. Wildlife and fisheries conservation on tribal lands has improved greatly in the past few decades, and new natural resource opportunities for the public are developing. Tribal resources can be an important part of the natural resource base of a state, attracting hunters, anglers, and others, and benefiting the overall economy (Figure 18.6).

In Canada, the rights of indigenous or aboriginal people are broader than those in the United States. Basically, indigenous peoples under federal treaty may hunt for food during all seasons of the year on all unoccupied Crown lands, such as the Northern Provincial Forest, and on any lands to which they have "acquired a right of access." Courts have determined that all wildlife management units, wildlife refuges, and parks where an open hunting season has been held are open to year-round hunting by indigenous people. Private land may also be hunted outside of the regulated game season if the indigenous people have been granted permission by the landowner.

The *Sparrow* decision (1990; *Regina v. Sparrow*) in Canada had a major effect on resource use. Basically, the decision states that in the case of federally governed species, such as waterfowl and some fishes, the hunting, fishing, and commercial fishing rights of nonindigenous people are secondary to the food-gathering rights of indigenous people. If harvest restrictions are required, they are to be applied first to the nonindigenous commercial and recreational users.

FIGURE 18.6 Some tribal reservations have excellent wildlife and fisheries resources and are managing those resources to provide income for the tribe and recreational opportunities for tribal members and nonmembers. Here, a tribal biologist and tribal personnel are conducting a stream survey on the Fort Belknap Reservation in Montana. *[Image courtesy of W. Stancill.]*

18.11 Federal Jurisdiction and Status of States' Rights

As noted earlier, provincial or territorial governments in Canada have jurisdiction over most Crown lands. The Canadian federal government (Canadian Wildlife Service) maintains jurisdiction over species at risk, international trade in endangered species, migratory birds, marine mammals, and all animals within national parks, wildlife refuges, or other nationally significant habitats. It also has jurisdiction over mammals, such as reindeer (caribou), that migrate over provincial-territorial boundaries (Gilbert and Dodds 2001).

In the United States, federal holdings, such as national forests and BLM lands, remain under the jurisdiction of the federal government. The states, however, have much control over wildlife and fish populations on those federal lands (see section 20.4). For example, elk hunters in Idaho or Montana hunt primarily on federal lands. The states are responsible for elk-hunting regulations, hunting licenses, and law enforcement, and their jurisdiction over elk has for the most part been unchallenged. Likewise, fishing on most federal lands is regulated by the individual states.

Federal rights to regulate wildlife and fisheries in the United States are found in the following constitutional powers: federal treaty-making power, federal property power, federal commerce power, and federal authority to regulate land under the commerce clause (Bean and Rowland 1997). The increase over the past century in federal legislation related to wildlife resources, public lands, and environmental issues was discussed in the preceding sections. Federal authority over wildlife, fisheries, and lands seems particularly far-reaching through the commerce clause of the Constitution. Nevertheless, in a major boost for states' rights in managing their wildlife and fisheries resources, a 2005 decision (*State of Minnesota v. Governor of State of North Dakota*) by the U.S. District Court supported North Dakota's rights to differentiate between residents and nonresidents in restricting waterfowl hunting in the state. In this decision, restrictions on out-of-state hunting were judged to not "run afoul" of the commerce nor privileges and immunities clauses in the U.S. Constitution. A similar U.S. District Court decision in 2006 (*Taulman v. Hayden*) supported the right of the Kansas Department of Wildlife and Parks to restrict nonresident hunting and fishing, even for nonresidents owning considerable land in Kansas.

The U.S. Fish and Wildlife Service has increasingly shared federally mandated authority with the states in regard to several aspects, such as management of endangered species and migratory birds. As of 2006, the rights of the states to manage resident wildlife and fish populations (even on most federal lands) and control hunting and fishing regulations, including exclusionary policies on nonresidents, remains firmly entrenched. The rights of both states and of the federal government in management of wildlife and fish continue to be important; a long-term balance between state and federal authority along with cooperative efforts seems most beneficial in management of these resources.

SUMMARY

A strong tradition of legislation relating directly to wildlife and fisheries resources exists in the United States and Canada at federal, state, provincial, and territorial levels. Wildlife and fisheries legislation in North America reflects the traditions and democratic natures of the people that colonized the continent from Europe. In the United States, jurisdiction over wildlife and fisheries in the early years clearly fell to the states. Federal wildlife and fisheries legislation in the United States began with the Lacey Act in 1900, in which the federal government exercised its constitutional power to regulate interstate commerce. Federal wildlife and fisheries legislation in the United States has also resulted from the government's constitutional power to make treaties with foreign nations and its right to protect its own property (property clause).

In both Canada and the United States, the need to manage migratory birds, especially waterfowl, on a national and international basis was recognized in the early 1900s. The federal governments of both countries established such jurisdiction by signing a migratory bird treaty.

The Federal Aid in Wildlife Restoration Act (PR Act) of 1937 and the Federal Aid in Sport Fish Restoration Act (DJ Act) of 1950 in the United States provided funding to states for many activities important to wildlife and fisheries resources. Funds are collected by the federal government from taxes on firearms, ammunition, fishing gear, and other selected sporting goods used by hunters and anglers. A state can receive such funds only if state hunting and fishing license fees remain with the state natural resource agency for reinvestment in wildlife and fisheries resources.

Environmental legislation in the United States, such as the National Environmental Policy Act, the Clean Water Act, the Federal Environmental Pesticide Control Act, and the Toxic Substances Control Act, has been important in terms of the welfare of wildlife and fisheries resources. This legislation also greatly affects public health, safety, and welfare. Other legislation directed at U.S. federal lands, such as the two multiple-use acts, the National Forest Management Act and the Federal Land Policy and Management Act, provides a legal basis for the conservation of wildlife and fisheries resources on federal lands. Wildlife and fisheries resources in the United States have also benefited from U.S. Department of Agriculture programs designed to retire cropland, reduce crop surpluses, and conserve soil. Programs such as the Soil Bank and Conservation Reserve Program provided for extensive habitat development on private lands. Additional programs or provisions in the Farm Act such as the

Wetland Reserve Program, swampbuster provisions, and the Grassland Reserve Program have been valuable in preserving and creating habitat.

Government treaties with indigenous peoples in North America often granted specific rights to harvest wildlife and fisheries resources in exchange for advantages given to nonindigenous peoples. In the United States, these treaty-given rights generally supersede the right of a state to manage wildlife and fisheries within its borders. Many treaties grant rights that extend to usual and accustomed hunting and fishing areas outside reservation boundaries; these rights have been the subject of controversy. In Canada, indigenous people have broad treaty rights to gather food by hunting and fishing on public lands throughout the year.

Most Crown lands and resident wildlife and fish in Canada are under provincial control. In the United States, the states have traditionally been responsible for harvest of resident game and fish species on federal and nonfederal lands. Federal agencies are responsible for management of wildlife and fish habitat in national forests, on national grasslands, and on other federal lands. Even though migratory birds are under federal jurisdiction, considerable cooperation occurs among the states or provinces and their respective federal agencies.

Legislation, jurisdiction, and legal precedents have strong influences on wildlife and fisheries resources. Interested parties can and should have an important influence on legislation related to wildlife and fisheries.

PRACTICE QUESTIONS

1. What is the states' rights doctrine as it relates to wildlife and fisheries?

2. Why is the Lacey Act important? What are its essential features?

3. How did the U.S. and Canadian federal governments successfully gain jurisdiction over migratory birds? Why might some states, provinces, and territories have opposed federal jurisdiction over migratory birds?

4. Why is the Migratory Bird Hunting Stamp Act (1934) so important to waterfowl conservation efforts in the United States?

5. What is meant by the carrot-and-stick approach as it relates to Pittman-Robertson (PR) and Dingell-Johnson (DJ) legislation?

6. Why is the National Environmental Policy Act (1969) important to wildlife and fisheries resources in the United States?

7. What is the legal basis for multiple-use management on national forest lands? On Bureau of Land Management lands? How are the National Forest Management Act and the Federal Land Policy and Management Act related to earlier legislation on multiple use?

8. Describe the swampbuster provision of the Farm Act and its importance to wildlife and fisheries resources. Why is the sodbuster provision of the Farm Act of minimal value to wildlife and fisheries resources?

9. What were the three primary goals of the initial Conservation Reserve Program?

10. What is the legal basis for allowing off-reservation hunting or fishing rights to some tribes?

11. Who has jurisdiction over most Crown lands in Canada? Compare this situation with jurisdiction over federal lands in the United States.

SELECTED READINGS

Bean, M. J., and M. J. Rowland. 1997. *The evolution of national wildlife law.* 3rd ed. Praeger Publishers, Westport, Conn.

Gilbert, F. F., and D. G. Dodds. 2001. *The philosophy and practice of wildlife management.* 3rd ed. Krieger Publishing, Malabar, Fla.

Haufler, J. B., ed. 2005. *Fish and wildlife benefits of Farm Bill conservation programs: 2000-2005 update.* Technical Review 05-2. The Wildlife Society, Bethesda, Md.

LITERATURE CITED

Bean, M. J., and M. J. Rowland. 1997. *The evolution of national wildlife law.* 3rd ed. Praeger Publishers, Westport, Conn.

Brady, S. J. 2005. Highly erodible land and swampbuster provisions of the 2002 Farm Act. Pages 5–15 *in* J. B. Haufler, ed. *Fish and wildlife benefits of Farm Bill conservation programs: 2000-2005 update.* Technical Review 05-2. The Wildlife Society, Bethesda, Md.

Gilbert, F. F., and D. G. Dodds. 2001. *The philosophy and practice of wildlife management.* 3rd ed. Krieger Publishing, Malabar, Fla.

Grue, C. E., M. W. Tome, G. A. Swanson, S. M. Borthwick, and L. R. DeWeese. 1988. Agricultural chemicals and the quality of prairie-pothole wetlands for adult and juvenile waterfowl: What are the concerns? Pages 55–64 *in* P. J. Stuber, coord. ed. *Proceedings of the National Symposium on Protection of Wetlands from Agricultural Impacts.* Biological Report 88(16). U.S. Fish and Wildlife Service, Washington, D.C.

Higgins, K. F., D. E. Naugle, and K. J. Forman. 2002. A case study of changing land use practices in the northern Great Plains, U.S.A.: An uncertain future for waterbird conservation. *Waterbirds* 25 (Special Publication 2):42–50.

Leopold, A. 1933. *Game management.* Scribners, New York.

Musgrave, R. S., J. A. Flynn-O'Brien, P. A. Lambert, A. A. Smith, and Y. D. Marinakis. 1998. *Federal wildlife laws handbook with related laws.* Government Institutes, Inc., Rockville, Md.

Redmond, L. C. 1986. The history and development of warmwater fish harvest regulations. Pages 186–195 *in* G. E. Hall and M. J. Van Den Avyle, eds. *Reservoir fisheries management: Strategies for the 80s.* Reservoir Committee, Southern Division, American Fisheries Society, Bethesda, Md.

Rewa, C. A. 2005. Wildlife benefits of the Wetlands Reserve Program. Pages 133–145 *in* J. B. Haufler, ed. *Fish and wildlife benefits of Farm Bill conservation programs: 2000-2005 update.* Technical Review 05-2. The Wildlife Society, Bethesda, Md.

Robinet, T. T., and E. E. Feunteun. 2002. Sublethal effects of exposure to chemical compounds: A cause for the decline in Atlantic eels? *Ecotoxicology* 11:265–277.

Sigler, W. F. 1995. *Wildlife law enforcement.* 4th ed. Wm. C. Brown Co., Dubuque, Iowa.

Valdez, R., J. C. Guzman-Aranda, F. J. Abarca, L. A. Tarango-Arambula, and F. C. Sanchez. 2006. Wildlife conservation and management in Mexico. *Wildlife Society Bulletin* 34:270–282.

Wood, F., and J. Williams. 2005. The grassland reserve program: New opportunities to benefit grassland wildlife. Pages 147–153 *in* J. B. Haufler, ed. *Fish and wildlife benefits of Farm Bill conservation programs: 2000-2005 update.* Technical Review 05-2. The Wildlife Society, Bethesda, Md.

CHAPTER 19

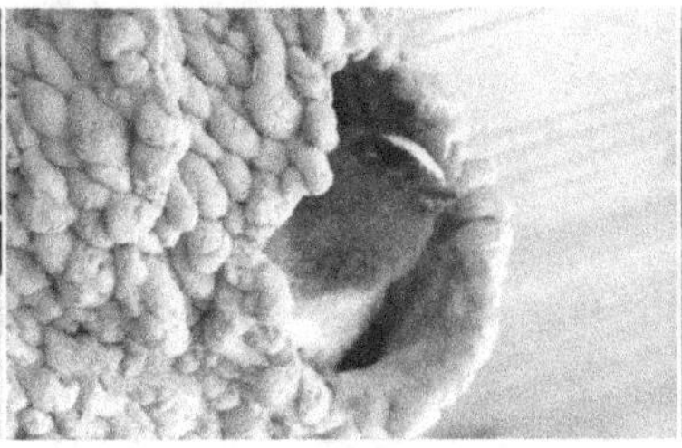

Wildlife and Fisheries Law Enforcement

Laws and regulations, as discussed in Chapters 11, 17, and 18, are important tools in wildlife and fisheries conservation; they are a facet of conservation of natural resources directed at human users. Such laws and regulations have little value, however, unless they are enforced by law enforcement personnel as well as in the court system. Some enforcement involves species taken for consumptive purposes, such as by hunters or anglers. Other enforcement involves species that are nonconsumptively important, as in the case of laws and regulations concerning nongame or endangered and threatened species. Enforcement may also involve laws and regulations relating to habitat, such as those associated with wetland conservation, clean water, or timber harvest. While a particular law or regulation may relate to any of the above natural resources, the enforcement procedures are usually quite similar.

A substantial number of wildlife and fisheries professionals spend part or all of their time on law enforcement. The importance of law enforcement in the total picture of natural resource conservation and management is sometimes given insufficient attention by other wildlife and fisheries professionals and by the public. This chapter is intended to place these activities in perspective with regard to their position in natural resource conservation and management. Other purposes of this chapter are to discuss some basic concepts of wildlife and fisheries law enforcement and to provide insights into the nature of this work. To avoid repetition of the lengthy term *wildlife and fisheries law enforcement*, the term *enforcement* or *law enforcement* will often be used. We intend the reader to assume that this enforcement involves wildlife and fisheries issues.

19.1 The Need for Law Enforcement in Wildlife and Fisheries

Wildlife and fisheries professionals may be involved in different facets of research, management, education, administration, or any combination of these areas; they may also be involved in law enforcement. People with expertise in all of these areas are needed in the overall effort to conserve and manage wildlife and fisheries resources.

Wildlife, fish, and other wild organisms are managed in part through **laws** and **regulations**. By dictionary definition, laws and regulations are essentially the same. However, in wildlife and fisheries, as in many other spheres, the terms *law* and *regulation* often have different meanings. Laws can be defined as rules enacted by an elected governing authority, such as a federal, state, provincial, territorial, or local government. Regulations can be defined as rules made by nonelected government units, such as state or provincial wildlife and fisheries commissions or boards, or by various units of a federal government, such as the U.S. Department of the Interior or the U.S. Department of Agriculture. Planning boards, park boards, and other local nonelected entities may also have regulatory responsibilities. The term *code* will not be used in this chapter, although it is often used interchangeably with the terms *law* and *regulation*. Both laws and regulations are enforceable under the court system.

While a law may be broad in scope, regulations are usually quite specific in their application. Regulations are often needed to effectively implement a law. For example, a federal law might require the U.S. Forest Service to implement a management plan in a particular area. The U.S. Forest Service would then need to promulgate regulations to optimize activities such as timber harvest, hiking, camping, hunting, rock climbing, or fishing in that area to implement the plan. Such regulations could relate to consumptive, nonconsumptive, direct, and indirect resource use activities.

It is appropriate here to provide some specific examples of the potential scope of law and regulation enforcement. In the case of golden and bald eagles, U.S. federal laws (Bald Eagle Protection Act and Migratory Bird Treaty Act) state that these species may not be taken by any means, with the possible exception of certain religious and ceremonial uses by indigenous peoples. Eagle feathers are illegal to possess, even if one is found lying on the ground as a result of molting. Because of their monetary value, eagle feathers and eagle talons, as well as headdresses, necklaces, and other items made from these parts, may be illegally sold on the black market to collectors. Without law enforcement, laws protecting eagles would be meaningless in the face of the market value of their feathers and other parts (Figure 19.1). In specific situations regulations may also control the distance from which people can view nesting bald or golden eagles at a particular time and place, or when and what types of motorized vehicles or boats can be used in an area where eagles are nesting. For any of these rules to be effective in protecting eagles, enforcement must occur.

FIGURE 19.1 Eagle feathers, as in this American Indian headdress and dance bustle (worn on the lower back), have a substantial value to collectors, but their possession is legal only for approved religious purposes by indigenous peoples in North America. *[Image courtesy of South Dakota Tourism.]*

In many trout streams in North America, Europe, New Zealand, and other countries, very specific regulations have been set to avoid depletion of fisheries in terms of both abundance and quality. For example, in a portion of the Snake River below Jackson Lake in Wyoming, there are specific regulations for cutthroat trout, allowing only catch-and-release fishing. The biological and social reasons for such regulations were discussed in Chapter 17. Without enforcement efforts, however, some people fishing these waters would keep substantial numbers of cutthroat trout, and the social and biological benefits of these regulations would be negated. The same can be said for all regulations that help to conserve wild organisms.

Much of what administrators, research biologists, managers, and those involved in the rule-making process attempt to accomplish would be negated without law enforcement. Laws and regulations are ineffective unless enforcement has a substantial presence or visibility. Thus, law enforcement is a vital and necessary part of natural resource conservation and management.

19.2 Types of Enforcement Officers—Their Education and Training

Most of the discussion about enforcement in this text is specific to wildlife and fisheries personnel in the United States at the federal and state levels. However, most of the points discussed in this chapter also apply to wildlife and fisheries enforcement personnel in other nations that have similar enforcement programs. The enforcement of international agreements restricting trade in protected species or their parts requires a cooperative effort among nations.

A variety of personnel, such as park rangers, foresters, customs agents, and others, can have enforcement responsibilities that involve fisheries and wildlife resources; however, in this text we will limit our discussion to those people whose enforcement responsibilities are tied primarily to wildlife and fisheries. Personnel involved in wildlife and fisheries law enforcement with state, provincial, or territorial natural resource agencies have a wide range of titles and responsibilities depending on the agency involved. **Conservation officer** is a title often used for personnel in these agencies who have wildlife and fisheries law enforcement responsibilities in addition to many other duties. The titles **game and fish warden,** game warden, or fish warden are sometimes used for resource personnel with strictly enforcement responsibilities. Other titles for positions with enforcement responsibilities include fish and wildlife officer, wildlife conservation officer, conservation enforcement officer, conservation ranger, natural resource officer, and renewable resource officer. For these positions, some natural resource agencies require only an interest in law enforcement with no specific academic training in the biological sciences; a bachelors' degree from a 4-year college or university is, however, frequently required. Individuals with a college degree in wildlife and fisheries or a related area are generally more successful in the competition for available enforcement positions in natural resource agencies. Persons with little or no education in wildlife and fisheries or the biological sciences are less likely to understand the biological rationale for regulations and law enforcement efforts. Moreover, even if their responsibilities are strictly related to law enforcement, their ability to interact effectively with other wildlife and fisheries professionals and the public can be reduced if they possess little biological knowledge.

Enforcement personnel with the U.S. Fish and Wildlife Service Division of Law Enforcement include **special agents** or, at a higher level of experience and responsibility, **resident agents in charge.** These nonuniformed law enforcement officers are normally armed and often become involved in undercover operations designed to investigate criminal activity related to wild animals and plants. Their education is highly variable, but a college degree is normally required. Degrees in criminology, criminal justice, police science, and wildlife and fisheries are all appropriate for these federal positions. Some education in wildlife and fisheries is highly recommended. Special agents and resident agents in charge with degrees in subjects closely related to law enforcement, and who also have a substantial interest in wildlife and fisheries work, have made effective professionals.

Uniformed **wildlife inspectors** and their supervisors working for the U.S. Fish and Wildlife Service (Division of Law Enforcement) assist in monitoring and helping enforce the Convention on International Trade in Endangered Species (CITES) (see section 11.2), as well as other related national and international laws. Wildlife inspectors work cooperatively with other inspectors from agencies such as the Department of Agriculture and are stationed at various ports of departure and entry, such as air and sea ports and border crossings. Wildlife inspectors are involved in law enforcement, but they are unarmed. Most violations detected by wildlife inspectors do not involve criminal intent. Wildlife inspectors call upon special agents or resident agents in charge when criminal intent, such as smuggling of endangered species or their parts, is involved.

Canada, like the United States, has a Wildlife Enforcement Division within the Canadian Wildlife Service. Federal wildlife officers enforce laws protecting migratory birds, species at risk, and nationally significant habitat and often work cooperatively with provincial conservation officers. The Wildlife Enforcement Division is organized into two specialized sections: investigations and inspections. Knowledge about plant and animal identification and a general background in wildlife ecology or a related biological field are important in preparing for a wildlife officer position in Canada. The Royal Canadian Mounted Police can enforce federal wildlife laws such as the Migratory Bird Convention Act but are primarily involved with other law enforcement duties.

FIGURE 19.2 Contact with landowners, conservation groups, and other members of the public is an important part of the work of a conservation officer. The impression left by this conservation officer may influence the attitude of this landowner toward wildlife and fisheries professionals, the employing agency, and natural resources. *[Image courtesy of the Idaho Department of Fish and Game.]*

Increasingly, Indian reservations have their own natural resource agencies, which serve a broad public and provide employment opportunities for wildlife and fisheries law enforcement personnel. Law enforcement personnel for natural resource agencies on reservations are usually titled conservation officers or fish and game wardens, and their responsibilities are similar to those of nontribal enforcement personnel.

Persons seeking careers in law enforcement need to develop effective communication skills and be able to positively interact with people. Conservation officers and other personnel involved in law enforcement are often the primary contact between the public and a natural resource agency (Figure 19.2). Thus, the demeanor and communication skills of these law enforcement professionals greatly influence public attitude toward natural resources, as well as natural resource agencies and their personnel. This need for effective communication skills cannot be overstated. Most natural resource agencies are finding it increasingly difficult to find candidates with acceptable levels of oral and written communication skills.

Individuals entering positions with wildlife and fisheries law enforcement responsibilities are often required to obtain considerable additional training at special law enforcement schools. This training is generally broad, covering areas such as the rights of individuals, search and seizure, types of violations, interrogation, evidence, record keeping, use of firearms, court procedures, and other legal and procedural matters that often confront law enforcement personnel. Such law enforcement training is extremely important. Many new officers have 4 or 5 years of college but little law enforcement experience or knowledge. Enforcement is the area in which mistakes can be most costly to the individual and the natural resource agency. Poor preparation and training can lead to civil lawsuits against the agency and the officer or even to the officer's injury or death.

Most new nonfederal natural resource personnel with enforcement responsibilities attend a law enforcement training program along with new officers from other enforcement agencies, such as police and sheriff's departments. In addition, semiannual training and recertification in the use of firearms, defensive tactics, and legal issues is often required. Resource personnel with law enforcement responsibilities also receive training in cardiopulmonary resuscitation techniques and other emergency medical procedures, because these people are often the first to respond to an emergency situation.

Enforcement personnel with the U.S. Fish and Wildlife Service attend a 16-week training course at the

Federal Law Enforcement Training Center in Glynco, Georgia. After completion of this training, new officers are assigned to a resident agent in charge for 1 year of on-the-job training. After that experience, these officers qualify for a single-person duty station as a special agent. These agents annually receive 40 hours of in-service training that includes legal updates on search and seizure, evidence handling, forensics, intelligence on wildlife and fisheries violation trends, physical training, firearms training, and policy.

Enforcement of wildlife and fisheries laws and regulations can be dangerous for an officer; violators cause death or injury to these officers every year. Violators are particularly dangerous when under the influence of alcohol or a controlled substance. Persons with a previous record of arrests (see section 19.5) and convictions also represent an increased threat to enforcement officers.

For persons planning a career in wildlife and fisheries law enforcement, additional education in criminal justice, criminology, and law enforcement are all desirable and can provide a competitive edge. Experience with law enforcement at a city, state, campus, or military level can also be valuable. For example, an undergraduate student in this field might want to complete a degree in wildlife and fisheries sciences with a minor in criminal justice. While working on the degree, a few semesters of employment as a campus police or security officer is usually possible. During the summer the student might work as a park ranger in a state, provincial, territorial, or federal park where the duties involve some enforcement responsibilities.

19.3 Types of Violators

Violators of wildlife and fisheries laws and regulations vary greatly in terms of their intent and the degree to which they are conscientious and law-abiding. Three basic categories of violators are recognized by enforcement personnel: **accidental**, **opportunist**, and **intentional violator**.

The accidental violator is a person whom law enforcement personnel generally dislike citing for a violation—for example, the person who attempts to fill an antlerless white-tailed deer license but accidentally shoots a buck with very small antlers. Other examples might be an angler who misreads the regulations and is fishing with bait in an area restricted to fly fishing, or an out-of-state hiker who did not know that campfires were allowed only in specific locations. Officers must at times write such citations despite their great reluctance to do so.

Opportunist violators are people who had no real intention of breaking a law or regulation, but the opportunity to do so presented itself and they could not resist. Examples might be a hiker who comes across a fossil bone in a national park and tucks it into his or her backpack, or a wildflower enthusiast who happens to find a particularly beautiful example of a protected plant and collects it. Another example might be a quail hunter who is having too much success. This person shoots a bag limit of birds, but on the way back to his or her vehicle cannot resist shooting an additional bird when it flushes. These violations were not planned, but the opportunity was too much for them to resist. Many people have at some time, either accidentally or because of opportunity, violated wildlife or fisheries laws and regulations.

The intentional violator starts a particular activity with the intention of breaking the law. Someone who sets illegal gill nets for steelheads or paddlefish, someone who poisons golden eagles, someone who shoots white-tailed deer by night spotlighting, or someone who captures a protected foreign species for import purposes is among the ranks of intentional violators. Intentional violators, especially those seeking large economic returns from illegal natural resource exploitation, are a special problem for enforcement personnel. Such persons may be dangerous, and they are likely to be very difficult to reform. They are involved in serious criminal activity and should not be considered normal resource users; they are criminals in much the same way that a person robbing a bank or burglarizing a home is a criminal. In addition, they are often involved in some other illegal activity. For example, one group involved in the smuggling of illegally taken walrus tusks in Alaska was found to be smuggling drugs as well; this group was considered to be very dangerous (see section 19.8). One wildlife case in the Midwest ended up involving illicit drugs and prostitution. A stern educational talk or even the embarrassment of apprehension is not likely to deter such people or change their behavior.

A common example of an intentional violator is the person who purchases a big-game license for a nonhunting family member and then fills that license himself. These people are not usually dangerous and may change their behavior after being cited for a violation or confronted by social pressure from friends who avoid such a practice. Such persons are often honest in their dealings with others and may have high moral values, but for some reason they believe that breaking certain wildlife and fisheries laws is not morally wrong. This type of violation is planned, is often repeated, and occurs for a variety of harvested species. Section 16.4 contains a discussion of human ethics and values.

Some individuals intending to harvest, capture, or kill wildlife, fishes, or other wild organisms illegally for their own personal use can be dangerous and can represent a serious source of depletion of wildlife and fisheries resources. Law enforcement officers must remain alert and cautious in approaching or apprehending any violator.

19.4 Wildlife and Fisheries Violations

Violations of wildlife and fisheries laws and regulations can be placed in nine major categories, as illustrated in Table 19.1. In addition to these violations, there are other subcategories of violations with which law enforcement officers need to be familiar. The major classes of violations have generally been tested and upheld by the U.S. Supreme Court, as reviewed by Sigler (1995).

A person who shoots an eastern cottontail out of season in the United States has broken a state wildlife regulation. Likewise, a person hunting waterfowl in the United States who forgets to sign the Migratory Bird Hunting and Conservation Stamp (i.e., duck stamp) on his or her license has broken a federal wildlife regulation. Actions such as these are illegal because they violate laws and regulations meant to assist in the conservation and management of terrestrial and aquatic organisms. Other violators break these laws or regulations but also transgress moral principles. A person who legally shoots five snow geese and then decides to dispose of them in a road ditch because they are too difficult to dress has violated a law or regulation and has also transgressed standards of acceptable human conduct. Such acts of wanton waste are considered morally wrong in most societies regardless of their legal status in terms of actual laws and regulations. This distinction is important in considering the rehabilitation of violators, because it is much more difficult to alter the behavior of those individuals or groups who violate the basic laws of human conduct.

Violations can be categorized as **felonies, high-grade misdemeanors** (sometimes called **class I** or **class A misdemeanors**), and **low-grade misdemeanors** (sometimes called **class II** or **class B misdemeanors**). In some situations additional classes of misdemeanors may be recognized; to avoid confusion, the terms *high-grade misdemeanor* and *low-grade misdemeanor* will be used in this book.

The seriousness of violations and their classification as felonies, high-grade misdemeanors, or low-grade misdemeanors varies greatly. A felony is a grave crime, ranging from burglary to robbery or even murder. An example of a felony natural resources violation in some states would be a second conviction for killing a deer during a closed season. Such a violation is not at all equivalent in seriousness to murder, but it is similar to a burglary or robbery in that these animals are being taken from the state and its people. Felony convictions are most common for violators breaking federal laws that protect wildlife and fisheries resources. Smuggling of endangered species or their parts, for example, would generally be a felony; this is a federal offense. Felony convictions for wildlife and fisheries violations are relatively uncommon.

Misdemeanor means misbehavior or misdeed and refers to violations that are less serious than felonies. In some states, a first offense for killing a deer or elk out of season would be a high-grade misdemeanor. Violations such as lacking an appropriate hunting or fishing license, fishing with bait in an area open only to artificial lures, or illegally killing a female ring-necked pheasant during the pheasant season would be low-grade misdemeanors in most states. In wildlife and fisheries law enforcement, most violations are low-grade misdemeanors.

19.5 Individual Rights

Law enforcement personnel in wildlife and fisheries must be aware of the legal rights of the individual. These fundamental rights are contained in the U.S. Constitution for specific reasons related to the democratic nature of our government. Courts have decided on the interpretation of the U.S. Constitution in various cases brought before them. A natural resource law enforcement officer must be aware of the rights of individuals as interpreted in certain important court cases (Sigler 1995; Rush 2000).

TABLE 19.1 Primary types of wildlife and fisheries law and regulation violations.

Type of violation[a]	Examples or further description
Taking or attempting to take animals out of season	Hunting, fishing, or otherwise taking species of wild animals when the season is closed, including after-hours violations
Taking or attempting to take animals or plants in an illegal area	Includes any area closed to hunting, fishing, or taking of the species sought; posted private land for which the violator does not have permission is included here
Taking or attempting to take animals with an improper license or no license	Illegal purchase of resident license by nonresident; unsigned hunting stamp for waterfowl or other species requiring a signed stamp; no license
Taking or attempting to take animals by illegal methods	Use of a shotgun for waterfowl hunting that holds more than three shells; use of a small-caliber rifle (such as a .22 caliber) to hunt large ungulates; use of worms on a stream limited to artificial lures; use of gill nets to capture fish
Illegally possessing wild animals or plants or their parts	Exceeding possession limits on legally taken game or fish (a person with too many trout or ducks in the freezer would fit this description); possessing game, fish, or other species illegally, even if it cannot be shown that the defendant actually killed or captured the animal
Following illegal procedures	Failure to properly tag a legally taken animal requiring tagging before transportation (most natural resource agencies require that a signed and dated tag be placed on large ungulates, usually around the leg, before placement in a vehicle or building or when the animal is brought into camp)
Illegally importing or exporting wild animals or plants or their parts	Many species are protected, but some, such as endangered or threatened species, are of particular concern; importing species that could cause agricultural or ecological damage without a permit is also restricted
Taking, possessing, or endangering nongame or endangered species	Killing, capturing, or possessing nongame vertebrates such as yellow warblers; many other organisms such as mollusks, butterflies, and plants are also included; even harassment of an animal or destruction of habitat could fall under this category, particularly for endangered species
Selling or attempting to sell wildlife, fishes, or other wild animals or plants in violation of state, provincial, territorial, or federal law	Sale of the parts or whole bodies of migratory bird species in North America is illegal with few exceptions; many other wild species are also protected from sale

Source: Adapted from Sigler 1995.
[a]A single violation may fit into more than one category.

Every law enforcement officer in the United States, including those in wildlife and fisheries, should be aware of protection provided to the individual by the Fourth Amendment to the U.S. Constitution. The Fourth Amendment protects individuals from search and seizure without **probable cause**. The specific wording of this amendment is as follows: "The right of the people to be secure in their persons, houses, papers, and effects, against unreasonable searches and seizures, shall not be violated, and no Warrants shall issue, but upon probable cause, supported by Oath or affirmation, and particularly describing the place to be searched, and the persons or things to be seized." Enforcement officers must be aware of these rights as they have been defined by the courts; they must also be cognizant of exceptions to search-and-seizure protection. Violation of the Fourth Amendment rights of a person may result in loss of evidence because of the **exclusionary rule**, which states that evidence obtained through violation of the Fourth Amendment rights of a person cannot be used in court. In addition, the officer may be held liable and sued in civil court by the suspected offender for the violation of his or her rights. The exclusionary rule is important in deterring illegal conduct by enforcement officers in the course of gathering evidence.

An **arrest** is generally considered to be the taking of a person into custody by a legal authority. During an arrest an officer can search the immediate person of the arrestee and the possessions within immediate control (**grabbing distance**) of that person without a search warrant. This ruling gives the officer access to items that might otherwise be destroyed or that might be a danger to the officer at the time of the arrest.

The search of houses, vehicles, and businesses is limited under the Fourth Amendment. Any searches of premises or belongings beyond the immediate reach of the arrestee should be made only with a search warrant. Search by consent is allowable in some situations, but the courts prefer the use of a search warrant. In many cases officers now electronically record conversations with suspected violators for possible use in court in case search by consent is allowed.

Vehicles present a special case in terms of search and seizure because of their mobility. In an important decision (*Carroll v. United States*, 1925), the U.S. Supreme Court upheld the actions of enforcement officers who stopped and searched a vehicle based on its being heavily loaded and driven by two persons suspected (based on undercover work) of smuggling liquor. Liquor was also visible in the passenger compartment when the officers approached the stopped vehicle. Based on the *Carroll* decision, a vehicle can be stopped and searched without a search warrant if there is probable cause to believe that a crime has been committed. If enforcement personnel have facts and circumstances within their knowledge that would cause a person of reasonable caution to believe that wrongdoing has occurred or is occurring, then probable cause exists. For instance, if a conservation officer watched an individual shoot a duck out of season and place the duck in the trunk of his or her car, this would be probable cause to stop and search the vehicle. The officer(s) involved would have probable cause to believe that **poaching** (illegal taking of wild animals) had occurred and that seizable evidence was in the vehicle. Probable cause must be based on more than just suspicion of the officer.

Vehicles can also be stopped but not searched on the basis of **reasonable suspicion**, a legal standard somewhat less stringent than probable cause but representing more than just a hunch by the officer. An example of reasonable suspicion would be the stopping of a particular vehicle on a lonely gravel road at 2:00 a.m. based on a vehicle description given by a landowner who reported spotlighting and rifle shots in the area 30 minutes earlier. Another example of reasonable suspicion might be stopping a pickup truck containing anglers who had bragged of overbagging American shad to a service station attendant, who then reported the vehicle's license number, description, and other information to appropriate authorities. If, upon stopping these vehicles on reasonable suspicion, the officer observed deer hair and fresh blood on the trunk and door handle in the spotlighting case or found large numbers of American shad visible in the bed of the pickup, probable cause for further search would then exist. Once probable cause is established, hidden areas such as the trunk of a car must be opened upon officer request; if not opened on request, the vehicle can be impounded and a search warrant obtained, or the trunk could be forcibly opened.

Some states have inspection statutes that allow officers to stop vehicles carrying visible game or fish. In such cases the officers can then further inspect the vehicle for harvested game or fish, check licenses, and otherwise determine if a violation has occurred. Officers at roadside check stations can inspect vehicles with visible game or fish but cannot inspect the vehicle without permission if no harvested species are visible.

Under the Fifth Amendment to the U.S. Constitution, a person has the right to remain silent if his or her testimony might be self-incriminating. The Fifth Amendment reads as follows: "No person shall be held to answer for a capital, or otherwise infamous crime, unless on a presentment or indictment of a Grand Jury, except in cases arising in the land or naval forces, or in the Militia, when in actual service in time of War or public danger; nor shall any person be subject for the same offence to be twice put in jeopardy of life or limb; nor shall be compelled in any criminal case to be a witness against himself, nor be deprived of life, liberty, or property, without due process of law; nor shall private property be taken for public use, without just compensation." Because of language in the Fifth Amendment, a series of rights must be read by an arresting officer when the **aura of custody** exists. Aura of custody is generally defined as any circumstance where a reasonable person detained by enforcement officers would believe that he or she is not free to leave. A **Miranda warning** is required whenever a person is taken into custody or is detained by an officer to a point at which the aura of custody exists. Persons in custody, prior to questioning, must be informed that they have the right to remain silent, that any statement they make could be used as evidence against them in a court of law, and that they have the right to the presence of an attorney (the right to counsel is covered under the Sixth Amendment). The attorney can be retained by the defendant or appointed by the state. The defendant can waive these rights if desired.

19.6 Arrest and Interrogation

Perhaps the first action that an officer should take when approaching a suspected violator is to clearly identify himself or herself as an enforcement officer. Most violations in wildlife and fisheries are handled on a **citation** basis. The citation directs the violator to appear before a **magistrate** (a judge having jurisdiction over minor cases) at some later date or, in some cases, simply pay a set fine for the violation if such is agreeable. Many states consider a citation a type of arrest. After writing a citation, however, the officer can direct his or her attention to other responsibilities without having to take the violator directly to a magistrate.

In most states, if the officer observes a violation directly, he or she can arrest or cite the violator without a warrant; citations commonly require no court appearance. Also, if there is probable cause to suspect that a felony or high-grade misdemeanor has been committed, the officer can arrest without a warrant. Another type of arrest is a warrant arrest, in which the officer obtains or is given a warrant from a legal magistrate. The warrant may be the result of a complaint filed by the officer or by other people who identified the individual and the violation. Thus, if the officer has good evidence that a person has committed a violation, an arrest can be accomplished through a warrant. The magistrate providing the warrant will require witnesses or other evidence to indicate probable cause. Evidence is usually provided in the form of an affidavit that lists probable cause. Once the violator is located, the officer must present the warrant and then make the arrest.

Good judgment is required with regard to the intentions and character of the individual to be arrested (including citations). Because officers must be acutely aware of dangers to their own lives, they will in most instances need to be able to make some type of character judgment. A nighttime arrest of a person or persons poaching deer or elk is a dangerous situation that must be approached with particular care; this is considered a high-risk contact. Nighttime arrests of armed violators are particularly hazardous. If the officer is alone, it may even be necessary to hold the suspects under spotlight and loudspeaker while calling for backup support from other officers. At a minimum, the officer should provide information on the location, suspect vehicle license number, and other information associated with the arrest by radio prior to approaching the vehicle.

The arrest of someone approached for fishing without a license, cutting wood where doing so is not allowed, or illegally digging wildflowers is considered an unknown-risk contact. To begin with, usually no weapons are involved. However, even these situations can turn hostile. In a tragic situation that occurred in the southeastern United States, a conservation officer approached a man to check his fishing license and catch. The officer cited the angler for fishing without a license. As the officer turned to walk away after writing the citation, the man shot and killed the officer with a small-caliber rifle. The officer may not have even suspected a violation of fishing laws when he first approached this person; he certainly did not suspect any danger to his own life. Thus, even in a situation that appears safe, the officer must remain alert. The officer cannot use deadly force on a person resisting arrest for a misdemeanor unless the life of the officer or another innocent subject is threatened.

In writing citations or making other arrests, enforcement officers need to use good judgment. That good judgment comes from having common sense, being alert to all situations, knowing how to handle potentially dangerous situations in the safest manner possible, knowing the legal rights of individuals, and understanding the laws and regulations relating to wildlife and fisheries resources. The officer, in working with people who have violated the law, is repeatedly faced with decisions about whether or not to write a citation for various low-grade misdemeanors. The effective officer is able to make decisions regarding the issuance of citations with fairness and with a degree of leniency appropriate to the situation (Eliason 2003).

Some enforcement officers are highly trained in interrogation techniques. In-depth coverage of interrogation is beyond the scope of this book; however, it should be evident that such training is a valuable tool for obtaining evidence and confessions from persons suspected of a violation. Officers trained in interrogation techniques know how to ask questions, how to read body language (eye contact, expressions of nervousness, and other actions), where and how far away to stand or sit in relation to the person being interrogated, and how to make it easier to obtain an admission of guilt. The skilled interrogator has both an innate potential for this work and the training and experience to further develop that skill.

19.7 Evidence

An enforcement officer investigating infractions or reports of infractions of wildlife and fisheries laws or regulations faces a wide variety of situations in terms of gathering evidence. Observation of a violation by a conservation officer is considered **direct evidence** because it directly proves or disproves a fact (Sigler 1995). Even for such encounters, however, the officer needs to maintain an accurate written record of the occurrence; audio- or video-taped evidence could also be kept to support the written record. Normally these records are adequately covered in the citation itself, but the officer should keep personal records containing other information about the citation. The date, time, nature of the infraction, and a description of vehicle and site, as well as other general information, should be recorded. Citation situations could still go to court, in which case accurate records will be needed. If a case reaches a judicial court, the officer may be carefully cross-examined as to the details of the incident; any written record would be valuable in this situation, especially because the time between the infraction and the court case can be lengthy.

In many cases, the infraction is not directly observed by an officer. Testimony from others, evidence from the site of the violation, or even information about a vehicle involved in transportation of the violators may be critical in these circumstances. Such evidence is considered **indirect** or **circumstantial evidence**. Indirect evidence, either alone or in combination with other evidence or facts, would allow a court or judge to reasonably infer guilt or innocence of an individual. The officer may visit the violation site to locate evidence. Photographs of the site should be taken for use in court if necessary. Animal hair, body parts, blood, spent shells, soil samples, prints from boots, tire tread prints, and other evidence may be collected. If possible, a cast or photograph of any tire tread and boot or shoe prints found at the site should be taken (Figure 19.3).

Even observations of vegetation may prove worthwhile. For example, suppose that enforcement officers receive information that a trophy white-tailed deer buck that is on display as a taxidermy mount was illegally taken in an area quite distant from the area for which the hunter held a permit. The vegeta-

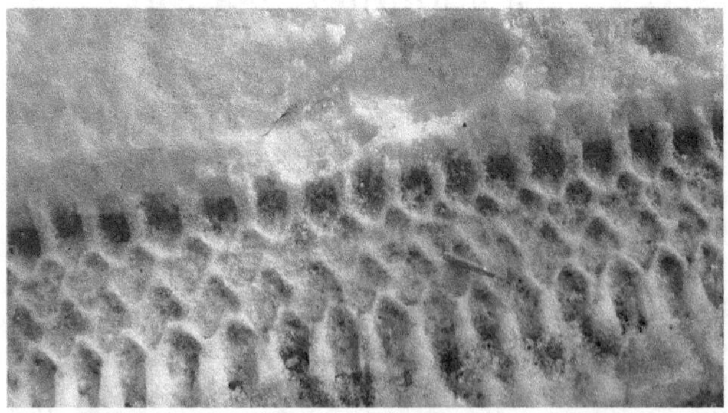

FIGURE 19.3 Images or casts of tire tread patterns, footprints, and other evidence from the site where a white-tailed deer was poached could serve as evidence in court. Evidence such as spent rifle cartridges should be collected without destroying fingerprints.

tion in the suspected area of the violation and the area for which the hunter had a hunting permit may differ greatly. Conservation officers might discover seeds or other plant parts in the hair of the mount that could not have come from the claimed kill area; this indirect evidence could result in arrest and conviction, and, in some cases, will cause the suspected individual or individuals to admit guilt.

The officer, if successful in locating a suspect vehicle after it has been driven from the violation site, will first look for evidence on the outside of the vehicle or in the visible areas of the passenger compartment. Tire tracks from the suspect vehicle can be compared to those found at the violation site. If boot or shoe prints are found near the suspect vehicle, these too can be compared to those from the violation site.

If a case goes to court, the time lapse between the infraction and the court appearance can be many months. For this reason, law enforcement officers should be concerned about how they label and store evidence. Evidence must be clearly labeled, stored in an organized fashion, and protected so that it will not be destroyed by accident or be accessible to unauthorized individuals. Small items of evidence can be sealed in an envelope and signed by a witness and the officer. Supporting photographs of physical evidence should be kept, preferably in a separate location.

Physical evidence that can decompose should be placed in a locked freezer with access limited to authorized individuals. Freezers containing evidence should be checked periodically to guard against accidental defrosting and loss of evidence. Evidence that cannot be preserved should be carefully documented using photographs and written records. Any transfer of evidence should be carefully recorded on a receipt indicating the reason for the transfer as well as the date and time. If it can be shown that unauthorized individuals have had access to physical evidence in a case, the evidence may be dismissed in court.

Use of DNA (see section 4.8) analysis in wildlife or fisheries investigations is becoming more common as techniques improve and cost declines. A recent review of DNA use in wildlife law enforcement is presented in Cassidy and Gonzales (2005). An interesting example of DNA use involved an individual charged with killing a male (buck) white-tailed deer with a firearm during a bow season. Forensic analysis indicated that the claimed arrow hole in the cape from the taxidermy mount had metal residue from a bullet. The suspect claimed the cape used with the antlers was not the one submitted to the taxidermist, but DNA analysis showed the cape and antlers were from the same deer (Cassidy and Gonzales 2005). Similarly, identification of endangered species contraband using DNA analysis is becoming an increasingly important tool in law enforcement. A recent study used DNA analysis to determine the species origin of turtle meat in markets in the southeastern United States; most of the turtle meat was from the snapping turtle, but approximately 25% was from the alligator snapping turtle, a species protected in all but one state (Roman and Bowen 2000). Such techniques have also been used to determine the origin and legality of whale meat sold in various countries. Forensic analysis using DNA will become increasingly important, particularly in relation to threatened and endangered species.

The Clark Bavin National Fish and Wildlife Forensics Laboratory in Ashland, Oregon, operated by the U.S. Fish and Wildlife Service, is the only laboratory of its kind and serves a worldwide constituency. The laboratory provides assistance to state wildlife and fisheries agencies and to federal agents and inspectors in the United States; more than 150 nations that have signed the CITES agreement can use this laboratory in endangered species investigations. Prior to 1988, law enforcement officers in wildlife and fisheries had little or no forensics support, while many forensics laboratories were available for other types of law enforcement. Personnel at the National Fish and Wildlife Forensics Laboratory have expertise in a wide range of forensics specialties. The laboratory has several sections and is able to integrate information dealing with morphology, criminalistics, serology, DNA analysis, pathology, analytical chemistry, and other areas in evaluation of various pieces of evidence related to wildlife and fisheries violations. Evidence sent to the laboratory includes materials such as teeth, hair, feathers, scales, blood, powdered rhinoceros horn, sea turtle oil, or suspected poisons. Evidence may be examined by complex procedures such as electrophoresis, immunoassays, and DNA sequencing to identify species origin of blood and tissue. Analysis using DNA may also be important in identifying hair or other tissues left at crime sites by individuals involved in higher-profile violations, such as illegal taking of endangered species or killing of elk in a national park.

19.8 Undercover Operations and Illegal Commercialization of Wildlife and Fisheries Resources

Direct evidence of illegal activities is sometimes obtained by natural resource agencies through the use of undercover agents. These agents might go hunting with guides suspected of infractions; purchase illegally taken plants, animals, or their parts from violators; or otherwise interact and work directly with those involved in illegal trafficking of wildlife and fisheries resources. For example, special agents for the U.S. Fish and Wildlife Service posed as clientele to hunt waterfowl in Texas with hunting guides suspected of repeated **overkill** and illegal **baiting**. Overkill is killing more than the legal daily limit or accumulating more than the allowed possession limit for the season. Baiting involves hunting in an area where some type of material has been scattered to attract animals, such as grain to attract waterfowl. In baiting cases, prior knowledge of the baiting by those hunting the baited area does not need to be proven for conviction on a baiting violation. In the case described above, the agents observed severe overkilling as well as hunting on or above an area that had been baited; several of these guides were later taken to court and convicted.

In a midwestern state, federal and state officers worked cooperatively to obtain evidence leading to the conviction of a guide and out-of-state clients who were illegally hunting white-tailed deer and ring-necked pheasants. Two enforcement officers posed as out-of-state hunters and were able to hunt with this guide and other clients. The undercover officers observed multiple violations committed by the clients and the guide. After a few days of hunting, the undercover officers allowed the violating clients to leave the site and arrested them as they crossed the state border with illegal game, thus ensuring a Lacey Act conviction (see section 18.3). The guide was also arrested.

Federal and sometimes state law enforcement officials may become involved in uncovering illegal international trade in wildlife and fisheries resources. An extensive international market exists for illegal animal and plant parts, whole specimens, and live organisms. For example, the gallbladders of bears have a high economic value in some cultures in which they are used as medicine. The value of gallbladders on the black market leads to illegal bear killing in North America and other areas of the world. Similar illegal traffic in animal parts such as walrus tusks, eagle feathers, and antlers also poses serious problems. Off the coast of Alaska, for instance, special agents uncovered a major walrus tusk operation in which the animals were being indiscriminately killed and only the tusks removed for sale on the black market. The chain of illegal activity stretched across the Pacific to Hawaii and to Asia and Europe; illegal drugs were also involved. Some of the people who were involved in the purchase and foreign trafficking of the walrus tusks were considered especially dangerous.

Illegal trophy hunting within the borders of national parks in the United States has become a serious problem. Clients will pay large sums of money to hunt and kill trophy elk, mule deer, bighorn sheep, or other animals that are present in these national parks (Figure 19.4). In many cases trophy animals are shot and only selected parts, such as the antlers and canine teeth (bugler or whistler teeth) of elk, are taken. Guides involved in these activities are aware of the movements of park rangers and other park personnel and may even monitor their radio conversations. Most U.S. national parks are difficult to patrol adequately because of the extensive land areas involved and the inadequate numbers of enforcement personnel. Investigations of these illegal trophy hunts have required intricate undercover operations involving long-term commitments from special agents or other enforcement officers that may extend beyond 1 or even 2 years.

In fisheries, the illegal harvest of freshwater mussels for sale overseas is an area of concern; one reason that mussels are valuable is that the shells can be ground for use in pearl culture. At present, many states, especially inland states, have little knowledge concerning their shellfish resources. Pacific salmon, because of their high market values, have also been the target of poaching rings in several instances. In fact, many fishes, including walleyes and channel catfish, have market values that make them attractive to poachers; illegal nets are commonly used in these types of illicit fishing activities (Figure 19.5). The commercial value of sturgeon and paddlefish eggs on international markets has encouraged illegal fishing operations in various areas of North America. The eggs are primarily destined for use as caviar. The eggs from a single fish can bring hundreds or even thousands of dollars.

Illegal imports or exports of live animals, including various parrot species and other birds, exotic fishes, butterflies, reptiles, amphibians, and mammals, are all within the enforcement responsibilities of wildlife and fisheries officers. Parrots, for example, may be smuggled into the United States alive and then sold in legitimate pet stores as captive-reared animals. Illegally taken feathers, hides, ivory, internal organs, and other animal parts can also bring substantial income to violators. Not even plants are immune from illegal harvest. In the southwestern United States, for example, illegal trade in plants such as saguaro cacti can result in large monetary gains to people who traffic in these organisms. Such sales can constitute a particular threat to endangered and threatened organisms. International, federal, state, provincial, territorial, and local cooperation is often involved in enforcement efforts to control illegal imports and exports (see the CITES and the Lacey Act, sections 11.2 and 18.3, respectively).

Undercover operations can be extremely dangerous if the identity of the undercover agent is discovered. People involved in illegal activities often check the backgrounds of clients and test them in other ways to ensure they are not law enforcement personnel working undercover. Officers in undercover operations are often placed in compromising situations in which they themselves may be forced to break the law to maintain their cover. Officers in most undercover operations also must work with armed suspects. Nevertheless, undercover enforcement activities are critical to uncovering and eliminating illegal operations.

Unfortunately, the seriousness of infractions of wildlife and fisheries laws and regulations is at times poorly recognized by the court system. For persons in illegal operations involving large cash benefits, the penalty for capture, prosecution, and conviction may

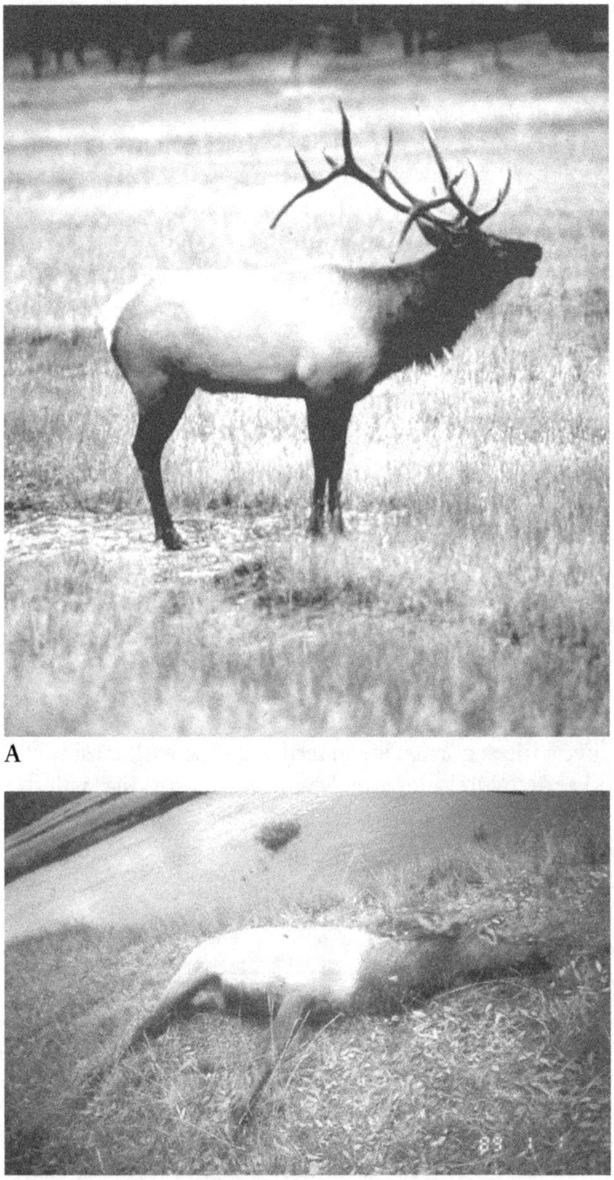

A

B

FIGURE 19.4 This mature male elk *(A)* was illegally killed and its trophy antlers taken *(B)* within the borders of Yellowstone National Park. *[Image A courtesy of B. Lindholm; image B courtesy of the National Park Service.]*

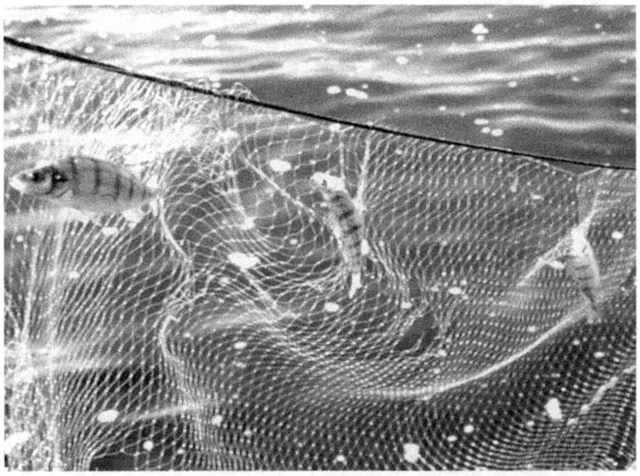

FIGURE 19.5 Gill nets may be set by poachers, thus robbing the public of a valuable resource. *[Image courtesy of J. Lott.]*

not be sufficient to deter the activity. If a guide receives $15,000 per customer for illegally taken trophy elk on federal lands in the United States, is a single fine of $10,000 likely to deter such an activity? Fortunately, there is an increasing recognition by the general public and the court system of the seriousness of illegal activities involving a wide variety of natural resources.

19.9 Trends and Changes in Wildlife and Fisheries Law Enforcement

The backgrounds of most law enforcement professionals in wildlife and fisheries have traditionally included an interest in hunting and fishing and in the out-of-doors; most persons entering the field have been Caucasian males. There is still limited diversity in race and gender of persons hired in wildlife law enforcement, although some progress has occurred in recent years.

A career in wildlife and fisheries law enforcement at state, provincial, or federal levels has traditionally offered many benefits, such as lack of constant supervision, variety in daily tasks, autonomy of work, opportunities to meet people, and often the chance to work out-of-doors (Eliason 2006). For many officers, the challenge of apprehending intentional and repeated law violators remains an important and rewarding aspect of enforcement. These positive aspects have been and continue to be important in attracting new enforcement officers and in helping to maintain morale. Responsibilities of enforcement personnel, especially at state or provincial levels, have expanded in recent decades to include a wide variety of conservation-related responsibilities as well as other policing duties. The term *conservation police* is sometimes used to identify wildlife and fisheries enforcement officers because of their broadening responsibilities (Falcone 2004). The ratio of wildlife and fisheries enforcement personnel to the overall population continues to be very low in comparison to other types of police officers. In the United States, the ratio of police officers to the overall population is approximately 24:10,000, while that of conservation officers is close to 1:10,000 (Falcone 2004). This lack of personnel, along with increased tasks and responsibilities, can lead to stressful work conditions and can negatively influence morale. In many states, conservation officers are now empowered with full or partial police powers for matters seemingly unrelated or remotely related to wildlife and fisheries work. State conservation officers often oversee or sometimes assist other police agencies with operations related to illegal drugs, enforcement of off-highway-vehicle laws, alcohol misuse on public lands, vandalism on public lands, boating laws, laws against driving while intoxicated, trespass, and an array of other societal problems. State, provincial, and federal wildlife enforcement officers have also found that required reports and other paperwork have greatly increased, demanding an important part of their time.

The involvement of the public in outdoor recreational activities in North America has broadened as a by-product of increased affluence and ability to travel. This increased participation in outdoor activities, increasing populations, drug problems, and increasingly complex bureaucracies, laws, and paperwork are all changes that will continue to influence the tasks, responsibilities, and stress placed on enforcement personnel (Falcone 2004). Still, enforcing wildlife and fisheries regulations, working in the out-of-doors, meeting the public, educating youth and adults, and many other challenging and enjoyable activities remain attractive aspects of a career in law enforcement. In addition, state and federal work laws no longer allow dedicated law enforcement officers to average more than 40 hours per week on the job without overtime compensation; working 60 to 80 hours without extra compensation was not uncommon in earlier years. The change in work hours per week has been positive in that it allows officers to have a more normal life in terms of time with family and other activities. Individuals who enjoy working with the public, have a strong interest in natural resources, and obtain an appropriate educational background will always be needed in law enforcement.

SUMMARY

A variety of federal, state, provincial, territorial, and local laws and regulations exist to assist in the conservation and management of wildlife and fisheries resources, but without adequate enforcement these laws and regulations cannot be effective. Enforcement personnel in wildlife and fisheries come from a variety of backgrounds, but a college degree in wildlife and fisheries or a related area is highly desirable. Education and training in criminology, criminal justice, and law enforcement can also be valuable. In the United States, most states as well as the federal government require special training in law enforcement for new officers as well as additional on-the-job and annual in-service training.

Violators of wildlife and fisheries laws and regulations are often categorized as accidental, opportunist, or intentional violators. The accidental violator breaks laws or regulations unintentionally. The opportunist violator does not plan to break the rules but does so when an opportunity is too tempting to resist. Many people who use natural resources consumptively, nonconsumptively, directly, or indirectly have violated a law or regulation in one of these ways. The intentional violator starts an activity intending to break the rules.

In most cases, wildlife and fisheries violations are low-grade misdemeanors. More serious violations can be high-grade misdemeanors or, in a few cases, even felonies. An example of a possible felony would be the smuggling of endangered species or their parts.

The Fourth Amendment to the U.S. Constitution protects individuals from unreasonable search and seizure. Violation of the Fourth Amendment rights of a person by law enforcement officers can lead to the exclusion of evidence in later court proceedings. In general, the search of areas beyond the immediate reach of an arrested individual, including a house or place of business, is allowed only with a search warrant. A search warrant is granted based on probable cause that a serious violation has occurred and that evidence relating to that violation will be found through the search. Vehicles are an exception to the rule because of their mobility, and in certain situations they can be searched without a warrant if probable cause is present. When persons are taken into custody, the officer must inform them of their rights through what is known as the Miranda warning.

Most violations in wildlife and fisheries are handled by issuing a citation rather than making an arrest. The cited individual often pays a set fine or may appear before a magistrate. Officers can make arrests or write citations upon directly observing a violation. If they have evidence of a violation but did not observe the incident, they can make an arrest with a warrant from a legal magistrate. Good judgment and common sense are required to be successful in law enforcement.

Wildlife and fisheries personnel involved in law enforcement need to keep accurate records of evidence and events. Photographs and other items of evidence are recommended in more serious cases. Court cases often occur months after the infraction, and documented evidence can be especially valuable in the officer's testimony.

Undercover agents are sometimes used to obtain direct evidence of wildlife and fisheries violations of the most serious nature. Undercover work can be dangerous to the officers involved, but it is sometimes needed to obtain evidence against violators. People involved in the smuggling of animals, plants, or their parts for commercial profit may also be associated with other illegal activities, such as drug trafficking.

Despite the increased responsibilities and paperwork involved in wildlife and fisheries law enforcement positions, the work still has many enjoyable and challenging aspects for those with a strong interest in natural resources, an appreciation for working out-of-doors, and an ability to work with people.

PRACTICE QUESTIONS

1. Give examples of four laws and four regulations pertaining to natural resource use.

2. What are the general in-service training requirements for wildlife and fisheries personnel with law enforcement responsibilities?

3. List the three types of violators of wildlife and fisheries laws and regulations and give an example of each.

4. List four examples of wildlife or fisheries violations involving nonconsumptive users of resources.

5. What protection does the Fourth Amendment to the U.S. Constitution provide for individuals?

6. What is the exclusionary rule?

7. What is the *Carroll* decision, and how is it related to wildlife and fisheries law enforcement?

8. What is the Miranda warning?

9. When does an officer need a warrant to make an arrest or issue a citation?

10. Assume that you are an officer called by a landowner to the site of a recent deer poaching. What types of evidence would you attempt to collect?

11. What kinds of illegal commercial operations relating to wildlife and fisheries might require undercover investigation? Try to use examples other than those provided in the text.

12. What are three advantages and three disadvantages of a career with an emphasis in wildlife and fisheries law enforcement? These may be very specific to your own life goals.

SELECTED READINGS

Grosz, T. 1999. *Wildlife wars: The life and times of a fish and game warden*. Johnson Books, Boulder, Colo.

Muth, R. M., and J. F. Bowe, Jr. 1998. Illegal harvest of renewable natural resources in North America: Toward a typology of the motivations for poaching. *Society and Natural Resources* 11:9–24.

Reisner, M. P. 1991. *Game wars: The undercover pursuit of wildlife poachers*. Viking Penguin, New York.

Schroeder, L. D. 2006. *A hunt for justice: The true story of a woman undercover wildlife agent*. Lyons Press, Guilford, Conn.

Sigler, W. F. 1995. *Wildlife law enforcement*. 4th ed. Wm. C. Brown Co., Dubuque, Iowa.

LITERATURE CITED

Cassidy, B. G., and R. A. Gonzales. 2005. DNA testing in animal forensics. *Journal of Wildlife Management* 69:1454–1462.

Eliason, S. L. 2003. Throwing the book versus cutting some slack: Factors influencing the use of discretion by game wardens in Kentucky. *Deviant Behavior: An Interdisciplinary Journal* 24:129–152.

Eliason, S. L. 2006. Factors influencing job satisfaction among state conservation officers. *Policing: An International Journal of Police Strategies and Management* 29:6–18.

Falcone, D. 2004. America's conservation police: Agencies in transition. *Policing: An International Journal of Police Strategies and Management* 27:56–66.

Roman, J., and B. W. Bowen. 2000. The mock turtle syndrome: Genetic identification of turtle meat purchased in the south-eastern United States of America. *Animal Conservation* 3:61–65.

Rush, G. E. 2000. *The dictionary of criminal justice*. 5th ed. Dushkin/McGraw Hill, New York.

Sigler, W. F. 1995. *Wildlife law enforcement*. 4th ed. Wm. C. Brown Co., Dubuque, Iowa.

CHAPTER 20

Organizations and Agencies, Biopolitics and Conflict Resolution, and Planning

Organizations and agencies that interact directly or indirectly with fisheries and wildlife resources are numerous. One can go to the Internet and find information on any number of organizations, groups, agencies, and others. Never before has so much up-to-date information on such entities been so readily available. If one does not know where to go on the Web, the National Wildlife Federation publishes the *Conservation Directory*, which lists most entities that affect wildlife and fisheries resources. This publication, which exceeds 630 pages, represents a valuable source of information. The *Conservation Directory* can be purchased in hard copy or searched on the federation's Web page. It contains material on international, national, and regional natural resource organizations; universities with wildlife and fisheries academic programs; state, provincial, territorial, and federal agencies and commissions; professional societies and organizations; and a wide array of other natural resource–related entities. Even many indirect-user groups, ranging from farm organizations to public health associations, are listed and described. Knowledge concerning the roles of these entities and how they interact is needed if one is to fully understand the fields of fisheries and wildlife.

The various agencies, people, and other entities that are involved in some way with wildlife and fisheries often interact in the biopolitical arena. **Biopolitics** involves the interaction of the biological aspects of natural resource conservation and management with a variety of political, legal, ethical, and social factors. Biopolitical activity can be divisive or can be positively channeled toward efforts at conflict resolution.

For a biologist, organization, or entity to be effective and successful in conserving and managing natural resources, planning is necessary. The planning process helps to ensure that vision and direction are maintained. Both strategic and operational planning skills are essential.

The intent of this concluding chapter is to describe a variety of agencies and organizations that are involved with natural resources and also to provide examples of ways in which they and individuals can interact in biopolitics and conflict resolution. The chapter concludes with a brief description of the planning process.

20.1 Professional Societies and Organizations

There are many professional societies and organizations devoted to various aspects of fisheries and wildlife. Some are international in scope and may have a variety of subdivisions at local, regional, state, provincial, or territorial levels. Others operate primarily at the national or international level. Most professional societies and organizations hold meetings at least once a year and produce scientific publications that communicate knowledge about natural resources. Entities described in this chapter represent some of the larger professional societies and organizations involved specifically with North American wildlife and fisheries resources; there are many other professional societies and organizations that are involved with these resources that cannot be mentioned in this text because of space limitations.

The American Fisheries Society (AFS) is a professional scientific society formed in 1870 that is dedicated to conservation of fisheries resources and professional growth of its members. The society was originally formed by North American fish culturists. Those early culturists recognized that fisheries resources were being depleted, and through their action federal fisheries and wildlife resource agencies were initiated in the United States. The AFS is international in scope (United States, Canada, Mexico, and other countries) and has numerous subdivisions. These subdivisions range from regional divisions to state and provincial chapters. There are also numerous special interest sections, such as those concerned with fish culture, fisheries management, education, socioeconomics, and bioengineering. The AFS has a certification program for professionals educated in fisheries science as well as a code of ethics. It publishes several scientific journals, including *Transactions of the American Fisheries Society,* the *Journal of Aquatic Animal Health,* the *North American Journal of Fisheries Management, Fisheries*, and the *North American Journal of Aquaculture.* It also publishes numerous books and conducts meetings at the state, provincial, regional, and international levels.

The Wildlife Society was founded in 1937. It is the wildlife counterpart to the American Fisheries Society. The society is an international scientific and educational organization of professionals and students engaged in wildlife research, management, education, and administration that is dedicated to the sound stewardship of wildlife resources and the environments upon which wildlife and humans depend. It has interest-specific working groups, such as biometrics and wildlife damage management, as well as regional and state subdivisions. The society has a certification program for biologists educated in the wildlife field and a code of ethics. The Wildlife Society publishes *The Journal of Wildlife Management*, *Wildlife Monographs*, and *The Wildlife Professional.* It conducts annual professional meetings at the state, provincial, regional, and international levels and also publishes books.

The American Society of Ichthyologists and Herpetologists, established in 1913, promotes the scientific study of fishes, amphibians, and reptiles. Its primary publication is *Copeia.* The American Society of Limnology and Oceanography, established in 1936, promotes the advancement of various aquatic science disciplines through scientific and technical symposia, colloquia, and meetings, and it promotes scientific research. Its major publication is *Limnology and Oceanography.* The North American Benthological Society, founded in 1953, is an international scientific organization that promotes a better understanding of the biotic communities of lake and stream bottoms and their role in aquatic ecosystems. One of its publications is the *Journal of the North American Benthological Society.*

Professional mammalogists may belong to the American Society of Mammalogists, which was established in 1919. The society encourages research and learning in all phases of mammalogy, and it publishes the *Journal of Mammalogy.*

There are four primary ornithological societies, all of which promote ornithological science and conservation through publications, annual meetings, and membership. These organizations and the research journals they publish are as follows: the American Ornithologists' Union (*The Auk* and *Studies in Avian Biology*), the Cooper Ornithological Society (*The Condor*), the Wilson Ornithological Society (*The Wilson Journal of Ornithology*), and the Association of Field Ornithologists (*Journal of Field Ornithology*). The American Ornithologists' Union also publishes the *Check-list of North American Birds,* an important and much-used listing of the common and scientific names of birds.

The Native American Fish and Wildlife Society services the needs of fish, wildlife, and natural resources on tribal lands across the United States. It is comprised of professional and technical personnel associated with Native American natural resource programs and hosts annual meetings.

The Ecological Society of America, established in 1915, encourages the study of organisms in relation to their environment and promotes the exchange of ideas among those interested in ecology. The society has a certification program for professional ecologists and a code of ethics. It publishes *Ecology, Ecological Applications, Ecological Monographs,* and other publications. The Society for Conservation Biology was established in 1985; it is dedicated to providing the scientific information and expertise required to protect biological diversity in the world. Its publications are *Conservation Biology* and *Conservation in Practice.*

The Association of Fish and Wildlife Agencies is an organization that represents all of North America's state fish and wildlife agencies. It promotes sound management and conservation and speaks with a unified voice on important fish and wildlife issues.

Range scientists are represented professionally by the Society for Range Management, a group established in 1948. The mission of the society is to promote and enhance the stewardship of rangelands to meet human needs based on science and sound policy. Its publications include *Rangeland Ecology and Management* and *Rangelands.* The Society of American Foresters, founded in 1900, represents all segments of the forestry profession and is the accreditation authority for professional forestry education in the United States. Its mission is to advance the science, education, technology, and practice of professional forestry. Its publications include the *Journal of Forestry* and *Forest Science.*

The Organization of Wildlife Planners was founded in 1978. This group consists of professional state and federal fisheries and wildlife resource planners, natural resource educators, and professional conservationists who are dedicated to improving, through education and training, the quality of resource management and planning.

There are three other professional organizations that should be mentioned because their activities can have substantial effects on both university students and natural resource issues. The National Association of University Fisheries and Wildlife Programs (NAUFWP) fosters improved communication among its members and other agencies, organizations, and the general public. The National Association of University Forestry Resources Programs serves the same function as NAUFWP, but for forestry programs. The National Association of State Universities and Land-Grant Colleges is an umbrella organization that involves many disciplines, including wildlife and fisheries.

There are many other organizations and societies that are concerned with scientific or professional activities in the fisheries and wildlife fields. Some have very specific interests, such as the American Cetacean Society and the Desert Fishes Council, while others are quite general, such as the Southeastern Association of Fish and Wildlife Agencies. All are involved in some phase of professional or scientific enhancement of fisheries or wildlife resources.

20.2 Advocacy Organizations

Advocacy organizations are groups that support particular views and function as advocates for those views. Their advocacy usually takes the form of legal, political, or public information activities, but some are also involved in resource management and conservation. While professional societies and government agencies and commissions sometimes function in advocacy roles, that is not their primary purpose; organizations described in this section usually have advocacy as one of their primary roles. There is a bewildering array of advocacy groups; only a few will be mentioned in this text.

It is difficult to categorize the many advocacy organizations that have wildlife and fisheries resources as one of their major concerns, but some grouping is beneficial for discussion purposes. Advocacy organizations described in this chapter will be divided into six categories. The first category contains groups that are best described as general resource conservation organizations. They pursue a broad array of activities directed at natural resource advocacy. The second category contains general environmental action organizations. Their advocacy activities encompass areas that often go far beyond wildlife and fisheries resources. The third grouping consists of animal rights and animal welfare organizations. These two types of organizations are quite different, and each has its own agenda for advocacy. The fourth grouping has as its major focal point the promotion of hunting, fishing, and other outdoor activities. The fifth category contains organizations that support single species or groups of species. Some of the organizations included in this group go far beyond support of a single species or group of species, but their primary activity is focused on the organism or organisms identified in the name of the group (for example, The Wolf Group or Trout Unlimited). The last category consists of advocacy groups that support particular types of habitats or natural areas. Groups in all six categories interact with professional societies, government agencies and commissions, and other entities in the area of biopolitics.

Keep in mind that this is only a partial listing of natural resource advocacy organizations and that the classification scheme used is subjective. The description of what each group stands for is taken from its own description as provided on its Web page or in the *Conservation Directory*. In some cases, further information is provided, especially with regard to the group's position on the direct consumptive use of fisheries and wildlife resources.

General Resource Conservation Organizations

The Wildlife Management Institute promotes professional management of natural resources for the benefit of those resources. The institute has been the longtime sponsor of an important annual meeting, the North American Wildlife and Natural Resources Conference. A large proportion of its activities are directed at habitat loss problems. The institute considers activities such as recreational hunting to be an essential element of wildlife conservation. It supports and encourages recreational hunting that is conducted legally, safely, and ethically.

Many professional biologists and even greater numbers of nonprofessionals with an interest in natural resources belong to the National Wildlife Federation. This large organization, founded in 1936, has national, state, and local affiliates. Its mission is to educate, inspire, and assist individuals and organizations to conserve wildlife and other natural resources and to protect the environment. Among its other positions, the National Wildlife Federation supports hunting because, under professional regulation, wildlife populations are a renewable natural resource that can safely sustain taking.

The Izaak Walton League of America promotes the education of the public to conserve, maintain, protect, and restore soil, forest, water, air, and other natural resources. The league also promotes the enjoyment and wholesome utilization of those resources. The league believes that hunting should be considered a valuable management tool where it is compatible with other resource uses and purposes. It also believes that hunting and fishing seasons should be based upon scientific and biological data rather than political pressure.

General Environmental Action Organizations

Many people, including numerous amateur and professional ornithologists, belong to the National Audubon Society, an organization that was founded in 1905 and uses solid science, policy research, lobbying, and citizen science and action to protect the land and habitat needed for a healthy planet. It has an extensive network of chapters and field offices in the United States and Latin America. The society has never been opposed to hunting of game species if it is done ethically and in accordance with laws and regulations designed to prevent depletion of wildlife resources.

The Nature Conservancy, founded in 1951, is an organization committed to preserving the plants, animals, and natural communities that represent the diversity of life on earth. The conservancy works with states to identify ecologically significant natural areas. As a portion of its advocacy program, the conservancy directly purchases lands in need of protection using private, corporate, and other donations. The conservancy has protected more than 116 million acres of valuable lands and waters worldwide.

The objectives of the Sierra Club, founded in 1892, are to explore, enjoy, and protect the wild places of the earth and to practice and promote the responsible use of ecosystems and resources. Sierra Club advocacy work includes lobbying, litigation, public information, and publishing. The club has numerous field offices, chapters, and groups. The Sierra Club's philosophy for more than 100 years has been that responsible use of the earth's resources can include hunting and fishing when appropriately regulated.

Wilderness is synonymous with The Wilderness Society, an organization founded in 1935. Its mission is to deliver to future generations an unspoiled legacy of wild places, biological diversity, clean air and water, towering forests, rushing rivers, and silent deserts. The society has numerous field offices across the United States. The Wilderness Society views hunting as a legitimate use of wilderness areas, national forests, and other areas, subject to appropriate regulation for species protection.

The Worldwatch Institute, founded in 1974, is quite different from the previously mentioned general environmental action organizations. It attempts to inform policy makers and the public about emerging global problems. The institute stresses public education, and it produces an annual publication entitled *State of the World* and various other publications with which anyone interested in natural resources should be familiar.

Greenpeace, founded in 1971, is an organization dedicated to preserving the earth and the life it supports. Among other things, Greenpeace seeks to protect biodiversity, prevent pollution, and end all nuclear threats. This group has been involved in some highly confrontational and visible actions.

The World Wildlife Fund, founded in 1961, attempts to protect endangered species and their habitats. The fund works in over 100 countries to conserve diversity. It is the world's largest privately financed conservation organization. The fund insists that hunting and trapping be regulated so that the survival of any species not be threatened. It opposes hunting where it might adversely affect threatened or endangered species.

Animal Rights and Animal Welfare Organizations

A large number of local humane organizations are encompassed within the American Humane Association. It provides training and resources to animal care and control agencies. This organization, founded in 1877, is opposed to the hunting of any living creature for fun, trophy, or sport.

The Humane Society of the United States, founded in 1954, is dedicated to the protection of animals, both domestic and wild. The society is involved in humane and environmental education, federal and state legislative activities, laboratory animal welfare, wildlife and habitat protection, and a number of other areas. The society is opposed to hunting for fun, trophy, or sport.

Defenders of Wildlife, founded in 1947, advocates for wildlife, endangered species, and habitat. This group is neither anti-hunting nor pro-hunting. Its goal is sustained populations of all native wildlife species.

Friends of Animals, established in 1957, is an international organization that works to protect animals from cruelty, abuse, and institutional exploitation. It strongly opposes hunting and the management of habitat to promote game animals. This group is primarily interested in animal rights.

Similarly, The Fund for Animals, founded in 1967, is unalterably opposed to the recreational killing of wildlife and can best be described as an animal rights group. Preserving wildlife and promoting humane treatment for all animals are its main goals.

People for the Ethical Treatment of Animals is also an animal rights organization. This group opposes the use of animals for hunting or fishing, domestication, food, experimentation, or other purposes. According to their philosophy, keeping animals as pets is unethical.

Organizations That Promote Outdoor Activities

A wide variety of programs to create a better understanding of and more active participation in the shooting sports and practical conservation are sponsored by the National Shooting Sports Foundation, which was founded in 1960. One of the publications that the foundation produced, *What They Say About Hunting,* has been liberally used in this chapter as a source for the positions of various organizations on issues such as hunting, wildlife management, and consumptive use of wildlife and fisheries resources. The position statements included in the publication are those made by the organizations themselves. The foundation supports hunting as a needed tool of management. It has financed conservation activities involving both game and nongame wildlife.

The Boone and Crockett Club was organized in 1887; it promotes hunting ethics and helps to establish wildlife conservation practices that will lead to the recovery of big-game animals in North America. The club also maintains size records of North American big-game animals.

The National Rifle Association of America, founded in 1871, has a variety of missions, including promotion of public safety and the safe handling of small arms. It believes that well-regulated hunting is a beneficial use of renewable natural resources. One high-profile activity of this organization is its support of the right to possess and use firearms.

U.S. Sportsmen's Alliance provides direct lobbying and grassroots coalition support to protect and advance the rights of hunters, anglers, trappers, and scientific management professionals. It has a foundation that is responsible for public education, legal defense, and research.

Organizations That Support Single Species or Groups of Species

Ducks Unlimited, founded in 1937, is involved with much more than just ducks. It has a large field staff and many chapters across the United States and Canada. The goal of Ducks Unlimited is to conserve, restore, and manage wetlands and associated habitats for North America's waterfowl. This large organization has been responsible for many important and beneficial wetland projects (Figure 20.1). Ducks Unlimited supports hunting as an important recreational activity, a tool for wildlife management, and a wholesome family activity.

The Foundation for North American Wild Sheep, founded in 1977, promotes the scientific management of all species of wild sheep, the protection of their populations and habitats, and the reestablishment of their populations in available habitats. The foundation funds research on wild sheep, other wildlife studies, habitat improvement, and sheep transplants, and it supports hunting and game management policies.

The National Wild Turkey Federation, founded in 1973, is dedicated to the conservation of wild turkeys and the preservation of the hunting tradition. It has many state and local chapters, and it supports a research grant program.

Pheasants Forever, founded in 1982, is a conservation organization formed to protect and enhance pheasant and other wildlife populations through habitat restoration, public awareness and education, and land and water management policies. This group has many chapters spread across North America and also has numerous field representatives.

The Rocky Mountain Elk Foundation, founded in 1984, has the mission of ensuring the future of elk, other wildlife, and their habitats. The foundation funds projects involving land protection, habitat enhancement, management, research, conservation education, and hunting heritage. A field staff and local chapters characterize this group.

Trout Unlimited is an international conservation organization dedicated to conservation, protection, and restoration of coldwater fishes and their water-

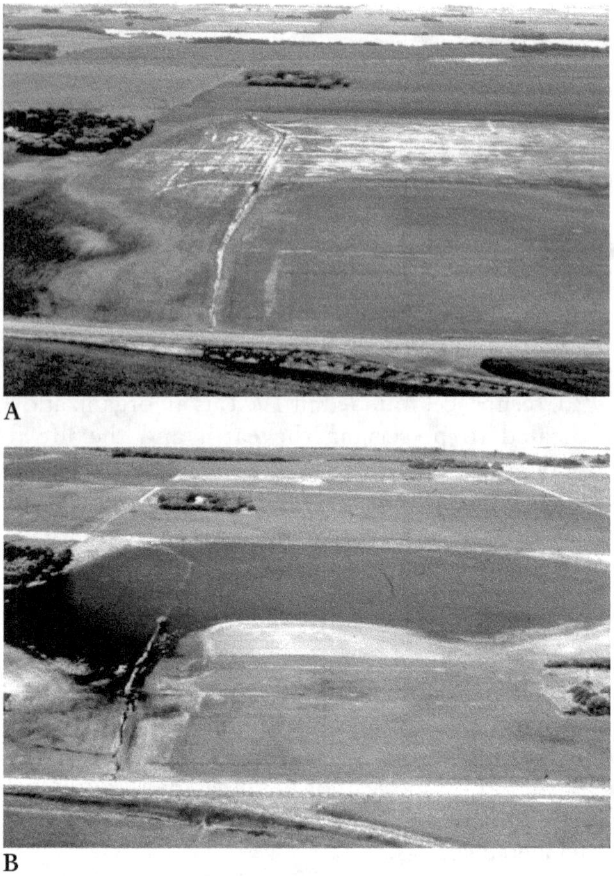

FIGURE 20.1 Ducks Unlimited conducts many wetland restoration and conservation projects. The two photos show a wetland *(A)* before and *(B)* after a Ducks Unlimited–sponsored restoration project. By plugging the drainage ditch shown in *(A)*, the wetland was restored. *[Images courtesy of R. Meeks.]*

sheds. This group, founded in 1959, has affiliates in countries such as Canada, New Zealand, and Australia.

Whitetails Unlimited, founded in 1982, supports programs that ensure the present and future well-being of white-tailed deer and its habitat. It also educates the general public on the importance of sound conservation practices, expends resources on habitat, promotes the hunting tradition, and supports research on white-tailed deer.

The International Osprey Foundation, founded in 1981, is dedicated to studying the problem of restoring osprey populations to stable numbers, making recommendations to enhance the continued survival of the osprey, and initiating educational programs. The North American Bluebird Society, founded in 1978, is involved in research and education to promote the recovery of bluebirds and other native, cavity-nesting species.

The Wolf Group was founded in 1996 to promote wolf reestablishment in historic ranges. In 1975, the Desert Tortoise Council was formed to assure the continued survival of viable populations of the desert tortoise. The Save the Manatee Club, founded in 1981, is concerned with public awareness and education, research, and legal action for the endangered West Indian manatee and its habitat. Concern for the future of wild Atlantic salmon and its habitat provided the impetus for founding the Atlantic Salmon Federation in 1982. Bat Conservation International, founded in 1982, attempts to document and publicize the value of bats and promote their conservation and management.

Organizations That Support Particular Habitats or Natural Areas

The Grassland Heritage Foundation, founded in 1976, is dedicated to prairie preservation and education. The Rainforest Alliance, founded in 1986, is dedicated to the protection of ecosystems and the people and wildlife that depend on them. Interest in preservation of the Indiana Dunes National Lakeshore for public use and enjoyment brought about the organization of the Save the Dunes Council. The council, founded in 1952, is concerned with protecting the ecological values of the dunes region, preserving Lake Michigan, and combating air, water, and hazardous waste pollution. The Adirondack Mountain Club, founded in 1922, is dedicated to the protection and responsible recreational use of the New York State Forest Preserve, parks, and other wild lands and waters. Conservation of caves, a unique habitat type, provided the basis for founding a national organization, the American Cave Conservation Association, in 1977. The Jackson Hole Conservation Alliance, founded in 1979, is dedicated to responsible land stewardship in Jackson Hole, Wyoming.

20.3 Federal Agencies and Commissions

The number of federal agencies and commissions that in some way affect fisheries and wildlife resources is larger than most people assume. While the fisheries and wildlife resource responsibilities of agencies such as the U.S. Fish and Wildlife Service and the Canadian Wildlife Service are obvious, many other agencies and commissions are also involved with these resources. It is not possible to describe all of them here, but most of the more important entities in the United States and Canada are listed and briefly described.

The Department of the Interior contains numerous important natural resource agencies. The Bureau of Land Management manages millions of hectares of public land. These lands are managed under multiple-use principles and provide outdoor recreation, watershed protection, and fish and wildlife habitat. The Bureau of Reclamation is also part of the Department of the Interior and is involved in areas such as reservoir management. Recreational opportunities and fish and wildlife enhancement are included in its duties.

The National Park Service (NPS) is also part of the Department of the Interior. The NPS administers national parks and seashores, national monuments, and other properties of national significance for their recreational, historic, and natural values. It also coordinates the wild and scenic river and national trail systems.

Another Interior Department agency is the U.S. Fish and Wildlife Service (USFWS). It is the lead federal agency in the preservation of migratory birds, threatened and endangered wildlife, certain mammals, and sport fishes. The USFWS provides technical assistance, manages refuges and fish hatcheries, conducts research, and educates the public about fish and wildlife resources.

Many students in university fisheries and wildlife programs in the United States will become familiar with the Cooperative Fish and Wildlife Research Unit program in the U.S. Geological Survey portion of the Department of the Interior. The program, which is unique and successful, represents a cooperative effort among the Department of the Interior, universities, state natural resource agencies, and the Wildlife Management Institute. The missions of this program are graduate education, research, and service.

Within the U.S. Department of Agriculture (USDA), one natural resource agency is the Forest Service. It administers programs for applying sound conservation and utilization practices to natural resources of national forests and national grasslands. The Natural Resources Conservation Service is also in the USDA. It is the primary federal agency that works with private landowners to help them conserve, maintain, and improve their natural resources.

The Cooperative State Research, Education, and Extension Service (CSREES) is another portion of the USDA that has many responsibilities important to wildlife and fisheries resources. Its extension activities include coordination of the fisheries and wildlife extension specialists found at universities that receive federal extension funding. Its research responsibilities include agricultural research conducted at state agricultural experiment stations, of which wildlife and fisheries research represents a part. The CSREES is also the federal entity that conducts wildlife and fisheries academic program reviews at **land-grant universities**. Land-grant universities are certain institutions for higher education in the United States receiving federal aid under the Morrill Acts of 1862 and 1890.

Another USDA agency that directly affects fisheries and wildlife resources is the Animal and Plant Health Inspection Service (APHIS). One of its responsibilities is to resolve conflicts involving wildlife deemed injurious to agriculture, animal husbandry, forest and range resources, or other wildlife. A portion of what APHIS does is associated with control of damage to agricultural crops by wild animals.

The National Oceanic and Atmospheric Administration (NOAA) is part of the Department of Commerce. A portion of its responsibilities involve oceanic research and management of the marine environment. Within NOAA is the National Marine Fisheries Service, which provides management, research, and services for the protection and rational use of living marine resources. Coastal and marine resources are managed by NOAA through the National Ocean Service.

Within the Department of the Army, the Corps of Engineers (COE) is involved with water management and waterway research. Many large reservoirs are managed by the COE, and a part of its responsibilities lies in determining how natural resources are affected by its actions. The COE also regulates dumping of any type of fill into natural wetlands that are navigable or contribute water to navigable waterways through the Section 404 permit system derived from the Clean Water Act. Service branches, such as the Army, are also responsible for environmental concerns and natural resource management on military lands.

The Department of State is involved with foreign relations, and this at times encompasses some aspects of natural resource management. An entity in the Department of State concerned with fisheries and wildlife resources is the Bureau of Oceans and International Environmental and Scientific Affairs. This bureau deals with global issues such as trade and environment, biodiversity, global climate change, ocean policy, and fisheries and marine conservation.

There are also numerous independent agencies in the U.S. government that affect fisheries and wildlife resources. The Environmental Protection Agency, the National Science Foundation, and the Tennessee Valley Authority are examples.

The U.S. federal government is also a participant in a variety of international, national, and regional commissions that are involved with fisheries and wildlife resources. The responsibilities of these commissions are fairly specific. A good example is the Great Lakes Fishery Commission, which, among other activities, advises the U.S. and Canadian governments on fisheries management, including sea lamprey control. Some other examples of such commissions are the International Whaling Commission, the Migratory Bird Conservation Commission, the North East Atlantic Fisheries Commission, the Marine Mammal Commission, the Pacific Salmon Commission, the Gulf States Marine Fisheries Commission, and the International Pacific Halibut Commission.

The Canadian federal government also has agencies that affect fisheries and wildlife resources. The Canadian Forest Service promotes the sustainable development of Canada's forests. The Canadian Wildlife Service is devoted to the protection and management of migratory birds and nationally important wildlife habitat, endangered species, research on nationally important wildlife issues, control of international trade in endangered species, and international treaties. The service is a part of Environment Canada, the department of the Canadian government responsible for coordinating environmental policies and programs as well as preserving and enhancing the natural environment and conservation of wildlife. Environment Canada is divided into geographic regions and also includes Parks Canada, which manages the Canadian National Parks system. The Canadian Environmental Assessment Agency is also a part of Environment Canada, as is the National Water Research Institute.

20.4 State, Provincial, and Territorial Agencies

Every state, province, and territory has at least one lead agency specifically responsible for wildlife and fisheries resources. While these entities may have different designations, such as *division*, *department*, *commission*, or *agency*, and their titles may vary to include terms such as *fish*, *wildlife*, *game*, *parks*, *natural resources*, *aquatic resources*, *conservation*, *lands*, *environment*, *forests*, or *renewable resources*, all are involved with the conservation of wildlife and fisheries resources. Examples are the Arizona Game and Fish Department, the Iowa Department of Natural Resources, the New Hampshire Fish and Game Department, the Louisiana Department of Wildlife and Fisheries, and the Nova Scotia Department of Natural Resources. Some of these agencies are headed by people educated in wildlife and fisheries, while others are headed by political appointees.

Some unique features characterize many state fisheries and wildlife agencies in the United States. In most states, citizen boards or commissions oversee these agencies (Figure 20.2). In many cases, these boards or commissions make the final decisions about natural resource conservation and management; in some states, they function only in an advisory role. Where the board or commission is the final authority, professional biologists employed by the agency may recommend various actions, but decisions are made by the citizen board or commission. These boards or commissions are often required by law to include a variety of types of citizen representation. Their composition may involve variation in such things as political party affiliation, regional representation, rural versus urban orientation, and occupation. Such variation in board or commission composition is mandated in an attempt to make these bodies as nonpartisan and representative as possible.

An additional unique characteristic of state fisheries and wildlife agencies is that they generally are funded differently than most other state agencies. While most other state agencies are primarily funded through the direct appropriation of general funds (monies collected from all state citizens), most state fisheries and wildlife agencies receive at most only a small percentage of their funding from general funds. Their primary sources of funds are license sales (hunting, fishing, trapping, and others) and excise taxes on fishing and hunting equipment (see section 18.5). Thus, these fisheries and wildlife agencies are primarily funded only by those people in a state who are direct consumptive users of its fisheries and wildlife resources (mainly anglers and hunters). Some states have unique funding sources that include nonconsumptive-user contributions. Missouri, for example, has a portion of its sales tax specifically directed toward natural resources.

FIGURE 20.2 Decisions affecting wildlife and fisheries resources are made in most states by natural resource commissions or boards. *[Image courtesy of K. Moum.]*

It is also important to remember that even though most state wildlife and fisheries agencies are funded almost exclusively by direct consumptive users, the agencies are usually responsible for all wild organisms, not just those involved with recreational hunting, fishing, or trapping. This can pose problems for fisheries and wildlife agency commissions and boards, because they are answerable to the direct consumptive users for how funding is expended; some direct consumptive users are not always supportive of expenditures such as those for nongame species.

Another unique feature of these state agencies is the resources that they control. In the United States, Supreme Court decisions have given control of fisheries and wildlife resources to the citizens of each state (see section 18.1). Except for some organisms, such as migratory birds, endangered species, wild horses and asses (burros), and marine mammals, which are under federal jurisdiction, each state has authority over and ownership of aquatic and terrestrial organisms within its borders. Even in the case of organisms such as migratory birds and endangered species, states may assume many conservation and management responsibilities within federal guidelines. On lands that are under jurisdiction of a federal agency, such as the Forest Service or the Bureau of Land Management, the federal agency manages or controls the habitat, not the wild animals on those lands.

There are other state, provincial, and territorial agencies aside from those primarily responsible for fisheries and wildlife resources that can have substantial effects on those resources. State, provincial, and territorial agencies involved with agriculture, environmental protection, water and natural resource development, and other issues can at times be in agreement with or in conflict with those agencies responsible for fisheries and wildlife resources.

20.5 Biopolitics and Conflict Resolution

An early lesson that any person being educated in wildlife and fisheries learns is that these natural resources are not necessarily managed based on the biology of the situation. Biopolitics influences all decisions, and it influences some decisions more than others. Biopolitical interactions ensure that economic, political, cultural, ethical, and other factors will be considered in the decision-making process. As the process of decision making becomes more difficult and complex, effective methods of conflict resolution become more important.

Depending on the scope and nature of a natural resource issue, a variety of entities may become involved in biopolitical and conflict resolution activities: federal agencies and commissions; state, provincial, or territorial agencies; directly or indirectly related advocacy groups; professional societies and organizations; and individual private citizens. The term **stakeholder** is commonly applied to those individuals or groups who have something at stake or an interest in a particular decision or resource. Different stakeholders will have different degrees of interest in any biopolitical issue; that is, for some, the issue may be of paramount importance, while for others, interest may be only peripheral.

All stakeholders can be considered to function, at least to some degree, in an advocacy role, but motivations are quite variable from one stakeholder to the next. Most stakeholders, especially the nongovernmental and nonscientific ones, desire to appear representative of general public opinion. In an attempt to ensure that they are viewed as representing large numbers of people, they often use the term *grassroots support* or *grassroots group*. Some groups are truly grassroots in nature and represent many people, while others have few actual supporters and only cloak their activities in the guise of grassroots support. Many of these latter groups are motivated primarily by self-interest and short-term economic gain. Their primary support generally comes in the form of monetary contributions from a few individuals, groups, or businesses that do not really represent public sentiment.

Biopolitical activities can take many forms, but generally they revolve around public opinion modification, political action, legal action, and conflict resolution. Depending on the power of each stakeholder to influence public opinion and political action, all will have some effect on the eventual outcome or decision. Biopolitics can involve attempts to sway public opinion through various means. Providing information to the public is usually an ongoing activity for all groups, including advocacy groups. This information can range from the totally factual to the untrue. At times, the discussion of an issue digresses to the point at which only emotionalism is used in an attempt to gain public support. There are other situations in which actions taken by advocacy groups to sway public opinion may be or border on the illegal. Activities such as group demonstrations or harassment of various types of users are often directed at swaying public opinion, but they can also be used in efforts to intimidate or ridicule others who hold different opinions. Such actions seldom result in solutions and often serve to further polarize opposing sides on an issue.

Biopolitics can also involve political action by stakeholders as a means of attaining their goals. These attempts to influence the political process through elected officials can be aimed at issues ranging from the local to the national and even international level. One method by which stakeholders can influence the positions of elected officials is making campaign contributions to politicians who agree with their position. Another method is providing information to political decision makers concerning issues in which stakeholders are interested. Legal action can be taken in situations in which legitimate differences of opinion occur. Lawsuits intended to stop or modify a decision or to affect the decision-making process are common. While legal action is not necessarily negative, there can be problems associated with it. One legal method used is represented by lawsuits directed at preventing stakeholders, often individuals, from speaking out on controversial issues. These lawsuits are called **SLAPP suits**, which is an acronym for **Strategic Lawsuits Against Public Participation**. The intent of such lawsuits is not necessarily to win a court case against people or groups who speak out on an issue, but instead is to stifle expression by costing those people or groups time, effort, and money to defend themselves. They are a form of intimidation. Such suits are usually initiated by stakeholders who have large monetary resources and large economic interests at stake in a decision.

There are alternative conflict resolution techniques that allow for more reasonable solutions to biopolitical issues. One method of conflict resolution involves the **coordinated resource management** (CRM) process. This process has at times proven successful in solving conflicts that have arisen in natural resource conservation. It is primarily effective in solving problems related to publicly owned resources designed for multiple use, but other types of conflict can also be addressed with this process. An example might be a situation in which different individuals or groups have conflicting desires with regard to how a piece of public grassland should be managed. For example, some users might think that domesticated livestock grazing should be the primary consideration; others might think that soil conservation is being ignored; others might believe that wildlife and fisheries resources are not being maintained correctly; others might think that their hiking, camping, or bird-watching opportunities are being diminished. Such an issue has economic, social, biological, and ethical aspects. In the CRM process, the significant stakeholders involved are all represented in an effort to develop a compromise management plan that has consensus support. The process is monitored and led by a nonpartisan facilitator. Note that compromise is involved; not all stakeholders get everything they want. Also note the term *consensus*. Consensus means that all must agree, or at least agree not to disagree; this is not a majority decision process. If all cannot reach agreement, then the process has not worked. Even failure to reach a consensus is generally not a total loss, however, because each stakeholder has the opportunity to learn the concerns of others. To some people, the CRM process makes sense; to others, it means that their ethics would have to be compromised and is therefore unacceptable.

While the CRM process is a method of solving an existing conflict, **integrated natural resource management** is a process designed to address areas of potential conflict before conflict occurs. It is a procedure for organizing different human users of a natural resource to produce the greatest value of goods and services from that resource over a given period of time. For the process to be successful, "a comprehensive plan or accounting of the direct or intended resource effects plus all the indirect or unintended resource effects must be described, and where possible, quantified and put into common units that allow comparisons between alternative uses of natural resources" (Loomis 1993).

There are a multitude of specific examples that could demonstrate or depict wildlife- and fisheries-related biopolitics and conflict resolution. Three examples are given here. The first is a hypothetical situation in which state biologists recommend an increase in the ring-necked pheasant bag limit; the second addresses marine fisheries resources; and the third involves chronic wasting disease in Wisconsin deer.

Ring-Necked Pheasant Bag Limits

Let us assume a hypothetical situation in which state natural resource agency biologists propose to their commission or board an increase in the daily bag limit of ring-necked pheasants from two birds to three birds. What stakeholders might become involved in such a decision? Obviously biologists are involved. They have used biological information available to them, which indicated that the resource could accommodate the additional harvest. The state citizen commission or board is involved because, in many states, it will make the final decision. Numerous other stakeholders may take an interest in the decision and attempt to influence its outcome.

Tourism or economic development entities in the state may support an increased bag limit so that they can use it as a selling point in their promotional activities. People or organizations representing service industries such as motels, service stations, restaurants, and sporting goods stores may think the increase is a good idea and be supportive of it. Organizations that represent hunters may be supportive because their members welcome the increase.

It is probable that there would be others who would not be supportive of the increase. Organizations representing landowners may oppose an increased bag limit because of the added bother imposed by more hunters. They may also want a greater voice in future decisions on issues such as this one. Wildlife—the wild ring-necked pheasant in this example—is in the unique situation of being a publicly owned resource that often lives on privately owned land. Landowners do not own the birds, but they control access to many of them.

An increase in bag limits on ring-necked pheasants may be opposed by animal rights groups. They may argue that a zero-bird limit should be considered. Some groups that represent hunters may be opposed because of the possible reduction in hunting opportunities for residents if the number of nonresident hunters increases due to increased limits. Hunting preserve operators may be opposed if they perceive an increased bag limit as having a negative effect on their business; they may also be supportive if they believe such a change is economically beneficial to them. Some people may oppose the increase because they believe that the added hunting pressure would harm the resource. Some may be opposed or in favor just because some other stakeholder they oppose or support is involved in the process. Depending on the situation, there would probably be numerous other people and groups with an interest in this issue. However, it should be apparent that much more is involved than just the issue of whether the daily bag limit should be two birds or three birds.

A decision would still have to be made, and all stakeholders would work to have that decision made in a manner that they favor. While the decision is in the process of being made, political pressure can be brought to bear, either for or against increased bag limits, by state legislators, county commissioners, and others. Even after the decision is made, attempts to change the decision, regardless of its outcome, can be made through legislative or legal means. This complex of interactions can make what initially appears to be a simple situation into a complicated one. To further exacerbate the situation, the final decision can affect future related or even unrelated situations. If some stakeholders perceive the decision as a loss or a win, it could affect their future actions. In the end, the issue of how the ring-necked pheasant resource might be affected by the increased bag limits may have little influence on the outcome.

Marine Fisheries

Fisheries resources are not immune to biopolitical influences; conflict resolution is also needed to solve problems in this area. For example, marine fisheries stocks can be a point of contention. This is becoming an even more contentious issue as wild marine fisheries resources diminish. These fisheries stocks may be found in the open ocean or in coastal waters. They may therefore exist in waters controlled by one or more countries, or they may exist in international waters well beyond political boundaries (commons of the world). In the United States, one of the obvious indicators that marine fisheries resources represent a biopolitical issue is the placement of primary governmental responsibility for those resources. Rather than being controlled by a natural resource agency, most marine fisheries resources are under the control of the Department of Commerce and the Department of State.

Special commissions, sometimes representing many countries, have been established to help in the conservation and management of marine resources. These commissions often act to resolve conflicts concerning the use of a particular marine resource. A few examples can best demonstrate the magnitude and scope of such commissions. The International Pacific Halibut Commission was organized in 1923 and represents an agreement between the governments of the United States and Canada to investigate and manage the stocks of Pacific halibut within the waters of each country. The International Whaling Commission was established in 1946 under international convention to provide for proper conservation of whale stocks and thus make possible the orderly development of the whaling industry. Over seventy countries are members of this whaling convention. One problem associated with such conventions is that membership is voluntary. Countries do not have to belong, and those that do not belong are not bound by commission decisions.

Not all such commissions are international. For example, the Atlantic States Marine Fisheries Commission, organized in 1942, is the result of a compact among states to promote the better utilization of the fisheries resource of the fifteen Atlantic seaboard states (Maine to Florida).

There are many functions for which such commissions may be responsible. They may have research responsibilities, or they may be responsible for determining the condition or viability of a particular fishery stock. Often a substantial portion of their efforts is directed toward conflict resolution. One example of such an activity is **resource allocation.** If a particular marine fisheries resource cannot be sustained under the harvest pressure of competing resource users, such a commission may need to allocate portions of that resource to particular users. It may allocate resources between recreational and commercial fishing interests, among different countries, between indigenous people and other users, or among any other number and kind of users. Whenever a fisheries resource is in high demand, conflict will almost invariably result from resource allocation.

The habitat that supports a fisheries resource may also support human economic interests other than fisheries; this situation results in a whole new set of problems and conflict. What happens if a particular shellfish fishery is negatively affected by electric power generation, shipping corridor activities, logging, or a host of other alternative uses? To the extent that these alternative uses adversely affect production of the shellfish fishery, an allocation of a formerly harvestable portion of the shellfish stock must be made to cover these non-fishery-induced losses. Only a certain proportion of the stock can be harvested while still maintaining an optimum sustained yield.

There are other issues that marine fisheries commissions may need to address. One of these is **incidental catches (bycatches)**, in which the utilization of one fishery resource affects another through unintentional capture. For example, when tuna are purse seined, incidental catches of porpoises and dolphins may occur. Even seabirds, such as horned puffins and marbled murrelets, can become a bycatch when caught in gill nets. Such conflicts must be resolved. Another issue that often needs resolution by such commissions is **ghost fishing.** Ghost fishing refers to the capacity of a variety of different fishing gears, such as gill nets and traps, to continue to capture organisms even after the gear is lost or abandoned.

The above are only a few of the issues that may cause conflict leading to biopolitical action in marine fisheries situations. It should be obvious that conflict resolution is essential if these fisheries resources are to be maintained. It should also be obvious that different stakeholders have different agendas with regard to how a resource is managed.

Chronic Wasting Disease in Wisconsin

Chronic wasting disease (CWD) (see section 10.9) is a fatal disease in deer and elk. It was first discovered in Colorado in the 1960s. While it was a concern in the western United States, it had not been found east of the Mississippi River until 2002, in Wisconsin. Although numerous CWD outbreaks have been attributed to infection from captive deer or elk herds, no one knows how the Wisconsin outbreak started.

The Wisconsin Department of Natural Resources (DNR) had instituted a random disease surveillance program in the late 1990s. Its purpose was to look for deer with tuberculosis, which had established itself in nearby Michigan. In February 2002, three samples from deer killed in a small area of southwestern Wisconsin tested positive for CWD. That discovery triggered what was probably the most expensive, intense, and controversial wildlife management program in Wisconsin history.

Immediately as the news hit the popular media, stakeholders began to surface, including those involved with human health, deer hunters, landowners, animal agriculturists, wildlife managers, legislators, and others. The DNR adopted an intensive model for disease control. Within days, aerial surveys were underway, a disease management area of approximately 500 square miles was created (centered on the three positive samples), a spring hunt was authorized to collect more samples, public forums were scheduled, and interagency teams for science and conservation were created. The DNR also declared a total eradication program for all deer in the designated area. The DNR response precluded substantive stakeholder input because of its speed. The DNR also lacked any management precedent for the situation.

Superficially, the goal of eradication seemed straightforward and perhaps possible—with the support of hunters and landowners. However, there were problems. The disease management area has a landscape dominated by agriculture and recreational woodlands; 95% of the land is in private ownership. Deer densities were high (forty to fifty/square mile) and deer hunting in the area is popular (there are approximately 700,000 Wisconsin deer hunters). In addition, deer hunting opportunity was the primary reason for ownership for many landowners.

Initially people were frightened and confused, but as time passed the initial shock subsided. Pressure to control the disease from hunters, landowners, and legislators never materialized in a substantial way. The uncertainty of the science, origin, and distribution of the disease, alternative theories about the problem, and landowner realization that the proposed cure (deer eradication) might be worse than the disease combined to thwart the needed level of cooperation. As soon as human and livestock health concerns faded, their stakeholders went on to other issues. Even the animal rights interests were largely silent.

Since CWD discovery, the DNR has tried just about everything feasible to drastically reduce the deer herd. While densities have been reduced, total eradication has not occurred and even a reconsidered level of five deer/square mile has not been reached. Some landowners protecting deer on their land have been pitted against their neighbors who want the deer shot. Two citizen opposition groups formed soon after the DNR announced its eradication plan and were very hostile to the program.

A fundamental problem remains: the deer population goals that the DNR believes are necessary do not allow for a huntable deer population. While most landowners recognize that there are too many deer and are willing to reduce the herd, many are unwilling to reduce it to five or fewer deer/square mile.

As of the 2006 deer season in Wisconsin, CWD has spread across most of the southern portion of the state and into northern Illinois. Management efforts are still focused on surveillance, research, and herd reductions, but herd reduction remains a slow process. Communication and cooperation with stakeholders has expanded. Stakeholders have asked for more traditional hunting seasons, and many seem resigned to living with CWD. Only time will tell what could have been done or should have been done.

In November 2006 the Wisconsin Legislative Audit Bureau evaluated the DNR response to CWD. Its report indicated that over $32 million had been expended to date on the issue, efforts to eradicate CWD have been ineffective, and the DNR approach to CWD should be reevaluated.

The biopolitical and conflict resolution activities described in the ring-necked pheasant, marine fisheries, and Wisconsin deer examples are not necessarily the most complex examples that could have been provided. When a bill such as the Food Security Act (commonly referred to as the Farm Bill), which affects a wide variety of people and groups, including those interested in wildlife and fisheries resources, is introduced in the U.S. Congress, an extremely complex interaction is set in motion. Other legislative actions, such as those dealing with endangered species or clean air or water, similarly increase the complexity of biopolitical interactions.

There are many issues and problem areas other than the ones mentioned in this chapter that involve, to varying degrees, fisheries and wildlife biopolitics and conflict resolution. Some contemporary issues facing wildlife and fisheries include restoration efforts for endangered species, genetic engineering, multiple uses on federal lands, effectiveness and ecological appropriateness of many stocking programs, responsibility for nongame organisms, ecosystem management as it affects target species management, preservation versus conservation, maintenance of genetically unique populations such as some Pacific salmon stocks, effects of climate change and habitat fragmentation, animal damage concerns, negative effects of commercial aquaculture on the environment, limitation of some nonconsumptive uses such as national park visitation, declines of world marine fisheries, game ranching, ocean acidification, maintenance of minimum instream flows, biodiversity efforts and their effects on economically important species, consumptive versus nonconsumptive use of some species, and the ecological appropriateness of exotic and transplanted organisms. Most of these issues have been mentioned in other portions of this text; others have not and are worthy of further inquiry.

Biopolitics and conflict resolution can be an exasperating area for biologists, but biologists must be involved in these processes. In fact, biologists are becoming much more adept at successful and positive interaction in the biopolitical and conflict resolution arenas. A good deal of their success can be attributed to the increased education and training they are receiving in human dimensions aspects. Success comes from learning the art of the possible, that is, knowing that any problem is solvable if correctly approached. Biologists must continue to improve their ability to communicate with individuals and groups in a positive way to resolve differences. With this skill, biologists can serve as facilitators in bringing people, knowledge, and money together for the benefit of natural resources. As human populations increase and natural resources come under more stress, biologists will need to become even more involved and proficient in the biopolitical and conflict resolution arenas, or they will fail to attain their biological goals.

20.6 Planning

It may seem incongruous to place planning at the end of this text, given that planning is usually one of the first undertakings of any successful action. However, good planning is not possible unless a previous knowledge base has been obtained. The entire text before this section and information you already possessed represent a start toward that knowledge base.

Eventual success or failure of any undertaking is usually a result of the effectiveness of its planning. Those responsible for wildlife and fisheries resources, or for that matter, any public, private, or individual undertaking, must develop plans to determine what results they want to achieve and what methods they will use to attain those results. Such plans must then be implemented and evaluated.

There are a number of different planning modes and management systems, such as **management by objective** (**MBO**) and **total quality management** (**TQM**). While planning and management methods often have substantive differences in their approaches, most have some basic elements of similarity. Following are some basic elements that are present in most planning methods. Figure 20.3 depicts steps in the planning process.

Any agency, department, or individual should first be aware of what one's mission entails. This can be accomplished by developing a single mission statement or, if an entity has numerous missions, a set of mission statements. A **mission statement** identifies the purpose or reason for the existence of an entity; that is, why the entity exists, whom it serves, and its ultimate aims. Correct determination of a mission can be much more difficult than one would initially think. There are a multitude of examples to illustrate how an incorrect perception of mission can lead to problems. Railroad companies in North America represent one such example. What was once a preeminent mode of transportation has fallen on rather hard times, and many railroad companies have suffered as a result. Many of their problems stem from a time when they planned and managed for being in the railroad business, while in reality their mission should have been defined as the transportation business. Similarly, ranchers who raise domesticated livestock by grazing are not just in the cattle or sheep business; they are also, and maybe more importantly, in the grass business. In any situation it is necessary to have or develop a mission statement, because without an accurate and clearly defined mission, planning is unlikely to be effective.

FIGURE 20.3 Planning is a continuous process; it really has no beginning or end, except for specific aspects that are completed during the process.

The following is an example of a mission statement for a state natural resource agency.

> The Division of Wildlife and Fisheries will manage the wildlife and fisheries resources of the state and their associated habitats for their sustained and equitable use, and for the benefit, welfare, and enjoyment of the citizens of this state and its visitors.

The next step is the development of goals and objectives. The number and direction of these goals and objectives will vary depending upon what one expects or is expected to accomplish toward the mission. The mission statement is usually relatively static and unchanging compared to the goals and objectives.

Goals are broad categories that are specific but do not identify measurable outcomes. Goals represent outcomes that the entity desires to reach in relation to its mission. For example, two goals (of many) for the natural resource agency described by the previously defined mission statement might be as follows.

> To conserve and enhance stream resources of the state and to increase public knowledge about them.
>
> To manage wild turkeys in the state so as to maximize user opportunity while maintaining populations consistent with the ecological, social, aesthetic, and economic values of the people of the state.

Objectives need to be specific, measurable, output or benefit focused, realistic, and have a specific time frame. They are directed at attaining a specific goal; they are what one specifically wants to accomplish. For example, some objectives directed at the previously described goal of conserving and enhancing stream resources might be as follows.

> Develop a watershed-based aquatic resource management policy for streams by 2011.
>
> Provide an annual minimum of 800,000 angler days of sustainable fishing on state streams by 2014.
>
> Conduct stream preservation and restoration projects at the rate of at least 10 kilometers of stream annually through 2014.

The next planning step is the development of **strategies** to meet particular objectives. Strategies are actions that can be undertaken to ensure that objectives will be met; they represent how one will attempt to accomplish something. To accomplish the above-described objective of 800,000 angler days, some strategies might be as follows.

> Inventory existing aquatic uses and preferences through appropriate survey mechanisms.
>
> Inventory existing aquatic resources in state streams.
>
> Develop and initiate watershed-scale management plans for sustainable recreational fishing.
>
> Develop and initiate a stream corridor access program to enhance stream user opportunities.

Development of missions, goals, objectives, and strategies represents the **strategic planning** phase of planning. After that phase is completed, the plan must be put into operation. This next phase is the **operational planning** phase. The operational planning process has a number of key components. Goals, objectives, or strategies require monetary resources; thus, planning for expenditures (budget development) is required. This step may involve the allocation of available funding or requests for additional funding. Goals, objectives, or strategies require human resources, as well as planning with regard to who will do what. All persons involved with the plan may not contribute to each strategy or even to each goal or objective. Operational planning requires a delineation of what responsibilities will be addressed by particular persons.

Development of missions, goals, objectives, and strategies and implementation of an operational plan have little utility if results are not evaluated. Not only should end results be evaluated but progress should also be periodically assessed. Results of periodic evaluations may dictate development of new or revised strategies to meet objectives.

SUMMARY

Information concerning groups and organizations that impact wildlife and fisheries can be obtained on the Internet. In addition, the *Conservation Directory*, published by the National Wildlife Federation, provides information on a wide array of entities that affect wildlife and fisheries resources. Entities that affect wildlife and fisheries can be classified as professional societies and organizations; advocacy organizations; federal agencies and commissions; and state, provincial, and territorial agencies. Included among advocacy organizations are groups concerned with general resource conservation, general environmental action, animal rights and welfare, outdoor activities, single species or groups of species, and particular habitats or natural areas.

An early lesson for any person being educated in wildlife and fisheries is that natural resources are not necessarily managed based on the biology of a situation. Biopolitics is a major factor in conservation decisions and ensures that economic, political, cultural, ethical, and other factors will be considered in the decision-making process. Conflict resolution must be undertaken before many biological actions are implemented. Two methods of conflict resolution are coordinated resource management and integrated natural resource management. There are a wide array of issues around which various stakeholders—such as professional societies; advocacy groups; federal agencies and commissions; state, provincial, and territorial agencies; local entities; and individuals—can interact. These interactions can be complex and contentious. Biologists need to continue to improve their biopolitical and conflict resolution skills. These skills will become even more important in the future.

The planning process is a continuing activity. A mission statement, goals, objectives, and strategies must first be developed; this strategic planning phase is followed by operational planning. Periodic and final evaluations of the results of the plan are also necessary. If a natural resources–oriented entity is to be successful, effective planning is essential.

PRACTICE QUESTIONS

1. Which of the described professional societies were you aware of before reading this text? Do you belong to any? Which ones?

2. Name two advocacy groups with which you are familiar. Are they among the ones mentioned in this text? If they were not described, what are their missions?

3. List eight different federal agencies in the United States that affect wildlife and fisheries resources. Describe their responsibilities.

4. What are the unique characteristics of most state wildlife and fisheries agencies with regard to how they are funded, who usually makes the final decisions on natural resource questions, and what resources they are responsible for managing?

5. Biopolitics often involves public opinion modification, political action, legal action, and conflict resolution. Provide an example of a resource conservation issue in the area in which you live and describe how each of these facets of biopolitics interacts with that issue.

6. What is a SLAPP suit?

7. With regard to your own future, develop a mission statement, goals, objectives, and strategies.

SELECTED READINGS

Carpenter, S. L., and W. J. D. Kennedy. 2001. *Managing public disputes: A practical guide for professionals in government, business, and citizen's groups*. 2nd ed. John Wiley & Sons, New York.

Conover, M. 2002. *Resolving human–wildlife conflicts: The science of wildlife damage management*. Lewis Publishers, Boca Raton, Fla.

Decker, D. J., and C. C. Krueger. 1999. Communication for effective fisheries management. Pages 61–81 *in* C. C. Kohler and W. A. Hubert, eds. *Inland fisheries management in North America*. 2nd ed. American Fisheries Society, Bethesda, Md.

d'Estree, T. P., and B. G. Colby. 2004. *Braving the currents: Evaluating environmental conflict resolution in the river basins of the American west*. Kluwer Academic Publishers, Norwell, Mass.

Fazio, J. R., and D. L. Gilbert. 2000. *Public relations and communications for natural resource managers*. 3rd ed. Kendall/Hunt Publishing Company, Dubuque, Iowa.

Jacobson, S. K. 2005. Communications for wildlife professionals. Pages 24–42 *in* C. E. Braun, ed. *Techniques for wildlife investigations and management*. 6th ed. The Wildlife Society, Bethesda, Md.

Krueger, C. C., and D. J. Decker. 1999. The process of fisheries management. Pages 31–59 *in* C. C. Kohler and W. A. Hubert, eds. *Inland fisheries management in North America*. 2nd ed. American Fisheries Society, Bethesda, Md.

National Wildlife Federation. 2005. *Conservation directory*. 50th ed. Island Press, Washington, D.C.

LITERATURE CITED

Loomis, J. B. 1993. *Integrated public lands management*. Columbia University Press, New York.

GLOSSARY

abiotic. Nonliving, as the abiotic portion of an ecosystem.
abomasum. The fourth and last chamber of a ruminant stomach.
abundance. A quantitative measure of a resource, usually expressed in biomass or numbers per unit of area.
accidental violator. A person who accidentally violates a wildlife or fisheries law or regulation.
accuracy. The closeness of a measure to the true value.
acidic deposition. The deposition of rain, snow, fog, dew, or solid airborne particles having high levels of acidity. This phenomenon is often referred to as acid rain or acid precipitation, but acidic deposition is a more accurate term.
acidity. The capacity to donate hydrogen ions.
acid precipitation. See **acidic deposition.**
acid rain. See **acidic deposition.**
active capture device. A capture device that requires human or mechanical action to move the device in order to capture the target organism.
active transport. The use of body energy to transport salts or other ions across a semipermeable membrane against a concentration gradient.
adaptation. Changes in structure, physiology, behavior, or life history that allow species to adjust to their environment. From a genetics standpoint, this is a process in which allele frequencies of a population are subjected to selection by environmental factors, thereby increasing the ability of the population to survive and reproduce in that environment.
adaptive management. A structured process of learning by doing, typically based on substantial pre-management information review and model development.
additive mortality. Mortality that exceeds the total mortality rate that would have been expected due to natural mortality alone. This term is typically used to refer to the possible added mortality resulting from excessive harvest by humans.
adult. An animal that has reached normal breeding age or size.
advocacy organization. A group that supports particular views and functions as an advocate for those views. Its advocacy usually takes the form of legal, political, or public information activities.
aerial. Having flight capabilities; occurring in flight.
aerobic. Living in the presence of free molecular oxygen.
aerobic decomposer. A decomposer that uses oxygen during its activities.
after-hatching year (AHY). An age classification used to denote a bird that is 1 year of age or older.
age class. A classification for wildlife species by chronological age in years or by some other age or life stage grouping. This term is increasingly being used by fisheries biologists.
age group. A classification for fishes based on the number of years a fish has lived. Fish in their first year of life are denoted as age group 0, fish in their second year as age group 1, and so on. Often the word *group* is deleted, for example, age 0. *Age class* is increasingly being used as a synonym for *age group* by fisheries biologists.
age pyramid. A graphic representation of the age structure of a population, beginning with the youngest age as the pyramid base.
age structure. The age distribution of a population.
agonistic behavior. Behavior involving aggression or threat displays.
algal bloom. A rapid increase in algal populations.
alkalinity. A measure of the carbonates and bicarbonates present in a solution.
alleles. The alternative forms of a gene.
Allen's rule. The tendency of ears, bills, legs, tails, and other extremities to be shorter in endotherms found in the colder portions of the range of a species.
allopatric. Having nonoverlapping ranges of distribution.
allopatric speciation. Speciation occurring when populations are spatially separated.
alpha cell. A specialized cell in the pancreas that produces glucagon.
alpine tundra. An altitudinal biome similar in its characteristics to the tundra biome.
altitudinal biome. A biome resulting from changes in altitude rather than changes in latitude.
altricial. Poorly developed and relatively helpless at hatching or birth. This term is not commonly applied to fishes; however, a fish such as the northern pike that is poorly developed at the time of hatching may be considered altricial.
altruism. Behavior involving aid giving that has a cost to the aid-giving individual.

alveoli (singular: alveolus). Small dead-end sacs in mammalian lungs where gas exchange with the blood occurs.
ambient temperature. The temperature of the environment.
ampullary organ. A specialized structure found in some fish species that can function as an electroreceptor; can be sensitive to pressure, touch, and temperature; or can function as a chemical stimulus receptor.
anabolic process. A metabolic process that builds up or maintains body tissue and stores energy.
anabolize. To assemble proteins or store energy by anabolic processes.
anadromous. Reproducing in fresh water but living most of the life cycle in a marine environment.
anaerobic. Living in the absence of free molecular oxygen.
anaerobic decomposer. A decomposer that does not use oxygen during its activities.
anchor ice. Ice formed on rocks and other streambed substrates resulting from ice crystals falling into the water from above the stream.
androgenesis. Reproduction without fertilization that produces viable young that have only the genetic material of the father.
anemometer. A mechanical device used to measure wind velocity.
animal rights. A philosophy based on the belief that other individual animals have rights equal to those of humans.
animal welfare. A philosophy advocating the correct handling and humane treatment of nonhuman animals.
anion. A negatively charged ion.
annulus (plural: annuli). An annual mark found on fish scales and bones, mammal teeth, and other hard structures.
anthropocentric. Believing that humans are the center of the universe.
anthropogenic. Human-induced; often relating to the impact of humans on nature.
anthropomorphism. The attribution of human shape or characters to, among other things, other animals.
antimycin. An antibiotic that is used as a general-purpose piscicide.
aphotic zone. The portion of a freshwater ecosystem that is unlighted and in which photosynthesis does not occur. This zone is generally considered to be the portion that receives less than 1% of the surface light intensity.
aquaculture. The culture and husbandry of aquatic organisms.
aquifer. An underground water body in a permeable formation of sand, gravel, or rock.
arboreal. Living in trees.
Arctic tundra. See **tundra**.
arithmetic mean. A statistical parameter obtained by summing all observations, then dividing by the number of observations.
arrest. The taking of a person into custody by a legal authority.
artificial selection. Selection by humans of the parents of each generation based on desired traits.
aura of custody. Any circumstance where a reasonable person detained by law enforcement officers would believe he or she is not free to leave.
Australian region. The zoogeographic region comprising Australia, New Zealand, and New Guinea.
autonomic response. An innate behavior that acts independently of volition.
autotroph. An organism, such as a plant, that can produce its own food.

backfire. A fire purposefully set upwind from a barrier such as a road or a cleared area. Backfires are sometimes set to stop future fires from passing the backfired area. They move slowly and burn thoroughly.
baculum (plural: bacula). The bony inclusion found in the penis of some mammals.
bag census. A complete count of animals taken by hunters. This term is also used occasionally in relation to fisheries.
bag limit. The number of a wildlife species that a person may legally harvest in a day. This term is sometimes used in reference to fishes.
bag seine. A seine with an added bag of mesh that increases capture efficiency. The bag is usually located midway between the two ends of the seine.
bag survey. A partial count of animals taken by hunters. Such a survey is often expanded to an estimate of the total harvest. This term is also used occasionally in relation to fisheries.
baiting. Hunting over an area where some type of material has been placed to attract animals.
bal-chatri trap. A passive capture device, usually used for raptors, consisting of a number of monofilament

loops that ensnare the feet of a bird landing on the trap.

ballast water. Water carried in the hulls of ships to provide stability.

bank cover. An artificial instream device used to provide overhead cover.

basal area. The cross-sectional area of standing tree trunks at breast height; often determined on a per-hectare basis.

basicity. See **alkalinity.**

Bayluscide. A selective piscicide used primarily on sea lampreys.

beach seine. See **seine.**

beam. The main trunk of an animal antler.

beam trawl. An active capture device for aquatic organisms consisting of a net, the mouth of which is held open by a rigid beam.

beard. A specialized group of hairlike feathers such as those found on the breast of a male wild turkey.

benthic. Associated with the bottom of an aquatic ecosystem.

benthic zone. The zone in aquatic ecosystems associated with the bottom of a water body.

bequest value. The value of knowing that wild organisms and their habitats will be available to future generations.

Bergmann's rule. The tendency of endotherms living in colder climates to have larger body sizes than subspecies or closely related species living in warmer regions.

beta cell. A specialized cell in the pancreas that produces insulin.

bias. The distance of an estimated value from the actual value or true parameter.

big game. Large animals, primarily mammals, that are hunted for recreational purposes.

biocentric. A philosophy in which all creatures in an ecosystem are viewed as being equally important. A more biology-centered philosophy than an anthropocentric viewpoint.

biodiversity. The variety and variability among living organisms and the ecological complexes in which they occur.

biogeochemical cycle. The cycling through an ecosystem of an element that is essential for growth of living cells.

biological amplification. See **food chain biomagnification.**

biological diversity. See **biodiversity.**

biology. The study of living things.

biomass. The weight or mass of a particular species in some specific unit of area; for example, the total weight of all the American robins in a square kilometer area. This term can also refer to the total weight or mass of all organisms in a specific unit of area.

biome. A major terrestrial community or major ecosystem that recurs on the various continents, classified by the dominant vegetation present.

biopolitics. The interaction of the biological aspects of natural resource conservation and management with a variety of political, economic, legal, ethical, and social factors.

biosphere. The layer around the earth in which all living things occur.

biota. All of the animals and some of the plants involved in a fishery or a wildlife system. *Biota* is most often defined as all living organisms.

biotelemetry. The use of devices placed on or in animals to relay information about the animals to a biologist; these devices usually produce radio or ultrasonic signals.

biotic. Living, as the biotic portion of an ecosystem.

biotic community. The interacting assemblage of living organisms in an area; at times, this assemblage is subdivided into plant and animal communities. Some scientists now use the term *assemblage* in place of *community*.

biotic potential. The maximum rate of population increase under ideal conditions; reproductive potential.

black basses. Largemouth bass, smallmouth bass, spotted bass, and other members of the genus *Micropterus*.

block cutting. A forestry practice that involves the clear cutting of small areas of timber.

bog. A wet area characterized by floating spongy mats of vegetation often composed of sphagnum, sedges, and heaths.

bole. The trunk of a tree.

booming ground. A lek used by greater or lesser prairie-chickens.

boreal forest. See **northern coniferous forest.**

bottleneck effect. Loss of genetic variability resulting from a restricted number of individuals reproducing in a population.

bottom-up control. Occurs when numbers or biomass of organisms in higher trophic levels are limited primarily by availability of food resources in lower trophic levels.

box trap. A passive capture device used for both birds and mammals in which the entry of an animal triggers a mechanism that closes the trap entrance.

brail. A pole or rod placed on each end of a seine that allows a worker to hold onto and pull the seine.

branchiostegal ray. A bony portion of the hyoid (throat) support system in fishes that is sometimes used in age determination.

breeding tubercle. An epidermally derived skin structure on fishes that is usually used during reproductive behavior.

browse. The leaves, shoots, and stems of trees and shrubs. This term is generally used in reference to food production for terrestrial animals.

browsing. The act of feeding on the leaves, shoots, and stems of trees and shrubs.

brush bundle. An artificial instream device that collects silt and is most often used to narrow a streambed.

buffering capacity. The ability of water or soil to react with and neutralize acids.

Burlese-Tullgren funnel. A sampling device primarily used for terrestrial invertebrates. The medium containing the animals is placed in the funnel, and the animals are forced out of the medium into a collecting jar.

bursa of Fabricius. A dorsal pocket in the cloaca that is present in young birds but disappears or becomes shallow in adult birds.

bycatch. See **incidental catch**.

caeca (singular: caecum; alternative spellings: ceca/cecum). Outpocketings from the intestine containing microbial populations that break down complex plant matter and improve digestive efficiency; found in birds and mammals.

calamus. The bare shaft of a feather below the feathering.

cannon net. An active capture device primarily used on mammals and birds, consisting of a lightweight net that is carried over a group of animals by mortar-type projectiles.

canopy. See **overstory**.

canopy closure. See **canopy cover**.

canopy cover. Primarily used in forests to describe the percentage of the forest floor with tree cover directly overhead; also used in relation to shrubs and occasionally herbaceous vegetation.

canopy coverage. See **canopy cover**.

canopy volume. The space occupied by tree crowns per unit of area or per tree.

carnivore. An organism that primarily feeds on heterotrophs.

carrion. Dead animal flesh; the primary food source for animals such as California condors and black vultures.

carrying capacity. The maximum biomass of a population that can be sustained within a defined area throughout a specified period of time. In wildlife, the definition usually includes an added factor such as "without causing damage such as that from overbrowsing."

catabolic process. A metabolic process that breaks down substances and releases energy.

catabolize. To release energy through the chemical breakdown of substances.

catadromous. Reproducing in the ocean but living most of the life cycle in a freshwater environment.

catch-and-release fishery. A fishery in which regulations require the live release of all fishes that are captured.

catch-curve method. A method of determining the mortality rate in a fish population based on the declining abundance of successive age groups in a single sample.

cation. A positively charged ion.

cementum annuli (singular: cementum annulus). Annual deposits on the teeth of some mammals that can be used for age determination.

census. A complete count of organisms present or harvested.

chaparral. A biome characterized by a mix of trees and shrubs, often with hard, thick evergreen leaves. The coastal area of southern California is representative of this biome.

checklist scale method. A survey or census method used to measure human attitudes in which the respondent is provided with a list of items and is asked to indicate whether each applies in a certain context.

chemical imprinting. Imprinting of a young animal on the particular chemical characteristics of its natal area.

chloride cell. A specialized cell in fishes used to obtain or excrete salts against an osmotic gradient.

chromosome. A structure composed of DNA that carries genes; it is found in the nuclei of cells.

chromosome aberration. Genetic alteration resulting from changes in chromosomes, such as deletion, duplication, inversion, and translocation.
chute. An intermediate area located between the pools and riffles of a headwater lotic freshwater system.
cichlid. A member of the fish family Cichlidae, which includes fishes such as the Nile tilapia, the blue tilapia, and the Jack Dempsey.
circadian behavior. Activity patterns or behaviors that change over the course of the day.
circannual behavior. Activity patterns or behaviors that change over the course of the year.
circuli (singular: circulus). Concentric markings around the focus of a fish scale that result from growth of the fish.
circumstantial evidence. See **indirect evidence.**
citation. A document that directs the violator of a law or regulation to appear at some later date in court or, in some cases, simply pay a set fine for the violation.
Clarke-Bumpus sampler. A specialized capture net used primarily to sample plankton.
class A misdemeanor. See **high-grade misdemeanor.**
class B misdemeanor. See **low-grade misdemeanor.**
class I misdemeanor. See **high-grade misdemeanor.**
class II misdemeanor. See **low-grade misdemeanor.**
clear cutting. A forestry practice that involves the removal of all trees from a large area of forest in one cutting operation.
cleithra (singular: cleithrum). Paired bones that provide support to the pectoral fin of a fish that can be used to determine fish age.
clinometer. A device used to estimate angles; commonly used to determine tree height or the angle of a cable leading to a trawl.
cloaca. The common chamber in birds and some other animals into which the intestinal, kidney, and reproductive canals empty.
Clover trap. A passive capture device used primarily to capture deer in which the entry of an animal activates a device that closes the trap entrance (and exit).
clumped dispersion. A dispersion pattern in which members of a species are found in aggregations; clumped dispersions usually reflect the distribution of habitats as well as social tendencies.
clutch. A group of eggs.
coastal wetland. A transitional area between land and ocean.
codominance. The genetic condition in which both alleles at a locus are expressed phenotypically to the same degree.
cohort. A group of individuals in a population born or hatched during a particular time period, such as a year. The term also represents a rank of taxonomic hierarchy below class and above order.
coldwater species. Freshwater fish species that survive, grow, and reproduce best in cold water. They generally spawn at water temperatures below 13°C (55°F).
commons. Those resources such as air, water, wildlife, fish, and publicly owned land that are owned by everyone but by no individual in particular.
community. See **biotic community.**
community structure. The species present and their relative composition in an assemblage of organisms.
compensatory mortality. An increase or decrease in one form of mortality that compensates for an increase or decrease in another form of mortality. Compensatory mortality occurs, for example, if increased mortality from predation results in decreased mortality from disease, and total mortality remains stable.
competition. The relationship between two or more organisms living in the same area that have overlapping niche requirements for a resource that is in limited supply.
competitive exclusion principle. The principle that no two species can simultaneously and completely occupy the same niche for an indefinite period of time; also called Gause's principle.
concentrate selector. A ruminant that is capable of digesting only vegetation with low fiber concentrations.
condition. The relative plumpness of a fish; various indices are used to quantify fish condition.
conductivity. A measure of the ability of water to convey electrical current. In general, the greater the concentration of ions, the greater the conductivity.
confidence interval (CI). A measure of precision associated with statistical methods; typically, the interval within which a parameter has a certain probability of occurring.
connectivity. The continuity or connectedness of a habitat or vegetation type across the landscape.
conservation. The effective management and use of natural resources.
conservation biology. A multidisciplinary science seeking to investigate human effects on biological di-

versity and to develop practical approaches to prevent extinction of species.

conservation genetics. Devising ways to maintain as high a degree of genetic variability as possible in populations, especially those with low numbers of individuals.

conservation officer. A title often used for a state, provincial, or territorial natural resource worker who has both law enforcement and other responsibilities.

Conservation Reserve Program. A program of the 1985, 1990, 1996, and 2002 Farm Act (Farm Security and Rural Investment Act) that provides financial assistance to farmers in converting highly erodible croplands and some other sensitive areas to perennial vegetation such as native grasses, trees, or riparian buffer strips.

conspecific. A member of the same species.

consumer surplus. The value of something over and above expenditures.

consumptive user. A person who harvests a resource, such as a deer, a trout, a tree, or a flower.

contest competition. Competition in which some individuals compete better for a resource at the expense of other competing individuals.

contingent valuation approach. Estimation of the value of changes in the quality of an activity by asking about the willingness of someone to pay for those changes.

controlled burn. See **prescribed burn.**

convergent evolution. A process by which species with different origins and histories develop similar traits and characteristics and fill similar biological niches.

coolwater species. Freshwater fish species that survive, grow, and reproduce best in cool water. They generally spawn at water temperatures between 4° and 16°C (40° and 60°F).

coordinated resource management (CRM). A conflict resolution technique that attempts to get all stakeholders interested in a resource conservation issue to agree on a particular management plan or course of action.

coprophagia. Eating feces to recover nutrients that escaped initial digestion.

corpus luteum (plural: corpora lutea). A hormone-secreting body in the ovary that develops during pregnancy in mammals.

corral trap. A passive capture device used for a variety of terrestrial animals in which the animals are usually driven into the trap and the entrance (exit) is closed behind them.

corridor. A narrow strip of land that differs, usually in terms of dominant vegetation, from the surrounding areas. This term can also refer to aquatic corridors such as rivers or streams.

countercurrent exchange. The exchange of gases or heat by means of an arrangement of closely associated blood vessels that facilitates more effective transfer.

cover. Habitat that is specifically used by animals for activities such as shelter, concealment, or escape.

coverage. Percentage of the ground, water surface (cattails and other emergent plants), or other underlying surface (e.g., bottom of shallow wetland) obscured by plants as viewed from above.

coverts. Small overlying feathers on the wings and tails of birds.

creel census. A complete count of fishes caught by anglers.

creel limit. The number of a fish species that a person may legally catch and harvest in a day.

creel survey. A partial count of fishes caught by anglers. Such a survey is typically expanded into an estimate of the total harvest.

crepuscular. Of or pertaining to the twilight hours at dawn and dusk.

critical habitat. The geographic area or ecosystem essential for the survival of a threatened or endangered species. This term is specifically related to the U.S. Endangered Species Act.

crop. An expanded area in the esophagus of some birds where food can be stored until space is available in the stomach.

current deflector. An instream device most often used to narrow a stream by redirecting water flow.

cursorial. Having a highly developed ability to run.

cyclic population. A population in which abundance increases and decreases on a regular multiyear or other periodic basis.

dancing ground. A lek used by sharp-tailed grouse.

dart gun. An air or powder-charged gun that shoots a syringelike projectile carrying an anesthetic, used primarily to capture large mammals.

Daubenmire quadrat. A 20- × 50-centimeter plot used to estimate plant coverage; usually used in the assessment of grasses and forbs.

deamination. Removal of an amine group from an amino acid.
decomposers. Organisms, such as bacteria and fungi, that reduce dead organic matter into elemental components.
decreaser species. A plant species, usually a grass of high forage quality, that decreases in number or biomass with increased grazing intensity.
deletion. A form of genetic change that results when a chromosome breaks, then reunites, but loses a segment in the process.
density-dependent factor. An environmental factor that affects a population in proportion to its density; that is, its effect increases or decreases as population size increases or decreases.
density-independent factor. An environmental factor that acts independently of population density; that is, its effect is constant regardless of population size.
dentine lake. An area in a mammal tooth where surface enamel has been worn away, exposing the dentine layer; dentine lakes can be used to estimate age.
depositional zone. The inner portion of a curve in a bending or turning lotic system. Current flow and water depth are less here than in the erosional zone on the outside bend.
desert. An arid biome usually having less than 25 centimeters of precipitation in a year. In North America this biome is primarily found in the western and southwestern regions.
desertification. Land degradation caused by natural factors or human-induced activities in dry areas.
determinate growth. A growth pattern in which an animal reaches some general adult size and essentially stops growing. Mammals and birds exhibit this growth pattern.
detritivore. An organism that feeds primarily on detritus.
detritus. Dead organic matter.
dewlap. A fold of skin on the neck of an animal such as that found on some lizards.
diadromous. Migrating between freshwater and marine environments. Anadromous and catadromous fishes are diadromous.
diameter at breast height (DBH). The diameter of a tree trunk approximately 1.4 meters above the ground.
diaphysis. The shaft of a long bone.
diel behavior. See **circadian behavior.**
diploid. Having two complete sets of chromosomes.
direct benefit. Benefits that accrue to users from a personal interaction with resources; economic benefits from activities such as hunting, fishing, and bird watching.
direct evidence. Evidence that directly proves or disproves a fact. Direct observation of a person committing a wildlife violation by a wildlife conservation officer would be direct evidence.
direct expenditure. An expenditure on a recreational activity involving items such as food, lodging, or goods.
direct user. A person who contributes to, uses, or directly benefits from a wildlife system or a fishery.
disclimax. An ecosystem maintained by continual disturbances, such as fires, such that the shifting-mosaic steady state community or later seral stages of ecological succession are not allowed to occur.
disease. Any departure from health; a destructive process in an organ or organism with a specific cause and symptoms.
disease agent. A physical, chemical, or biological factor that causes a disease.
dispersal. The outward movement of individuals away from their established areas of activity.
dispersion. The spatial pattern of individuals in a population; pairs or groups of animals can also demonstrate spatial patterning.
dispersive process. A genetic process that causes random change in allelic frequencies and a loss of genetic diversity in populations over time.
diurnal. Of or pertaining to the daytime; occurring during daylight hours.
DJ funds (Dingell-Johnson). Funds derived from a U.S. federal excise tax on a variety of fishing gears and equipment. These funds are available to states for the acquisition, restoration, rehabilitation, and improvement of fish habitat and for fisheries conservation and research.
DNA (deoxyribonucleic acid). The chemical used to store genetic information in most organisms.
domestication selection. Selection that operates during artificial propagation to produce adaptations to the captive environment.
dominant allele. An allele that is typically expressed over a recessive allele.
drag line. A rope or cable that is pulled by two people or two vehicles across grassland cover to flush nesting birds.

drawdown. A water level management practice involving the removal of water from a water body. In relation to wildlife, it can also refer to reduced water levels resulting from evaporation.
dredge. An active capture device used in fisheries work that is dragged over the bottom of a water body to capture benthic organisms. A dredge usually has blades, rakelike teeth, or other structures that enhance capture efficiency.
drive net. A passive capture device used for some mammals and birds that entangles the target organisms in a single panel of netting.
drop net. An active capture device primarily used for birds and mammals consisting of a net that is suspended above the ground; when target animals move under the net, it is dropped on them.
duplication. A form of genetic change that results when a segment of a chromosome is duplicated; that section appears twice in the genetic composition of the organism.
D-vac sampler. A large field vacuum cleaner used for capturing terrestrial invertebrates.

ecological pyramid. A graphic representation of the relationships among the various trophic levels in terms of numbers, biomass, or energy flow, in which the greatest quantity is represented by the base of the pyramid.
ecological succession. The process that occurs as an ecosystem changes from an early, immature stage to a stable final stage.
ecology. The study of the interrelationships of organisms with other organisms and the environment.
economics. The study of the allocation of resources.
ecoregion. An area of land or water that contains a geographically distinct assemblage of natural communities that share a large majority of their species and ecological dynamics, share similar environmental conditions, and interact ecologically in ways that are critical for long-term persistence.
ecosystem. An interacting system of biotic and abiotic components in a particular area or place.
ecosystem function. The way an ecosystem works, particularly the way in which nutrients and energy move through and within the system.
ecosystem structure. The composition of an ecosystem with regard to the species of plants, animals, and microbes present, the nutrients available, and the availability of water, or, more generally, a grocery list of what is present.
ecotone. The transitional area between two different ecosystems.
ecotourism. Tourism associated primarily with the nonconsumptive use of natural resources.
ectothermic. Having a body temperature that is generally the same as the ambient temperature. Fishes, reptiles, and amphibians are generally ectotherms; their primary source of heat comes from outside their bodies.
edge. See **ecotone.**
edge species. A species (or subspecies) that requires a variety of habitats in close proximity to one another. This term is most commonly used for birds, less often used for mammals, and not used for fishes.
Ekman grab. A sampling device used to obtain both organisms and substrate from the benthic zone of an aquatic habitat.
emergence trap. A device used to capture aquatic invertebrates as they emerge from their aquatic life stages into their terrestrial life forms.
emergent vegetation. Rooted aquatic plants that grow through the water column and above its surface.
emigration. Migration of an animal out of an area.
emulsify. To finely divide globules of fat.
endangered species. A species that is in danger of becoming extinct throughout all or a portion of its range. This term commonly pertains to the U.S. Endangered Species Act and to similar acts in some other countries, although other entities, such as states, provinces, and territories, can also develop endangered species lists.
endocrine disruptor. A toxicant that interferes with the endocrine (hormone) system of an organism.
endothermic. Maintaining body temperature at some specific level regardless of ambient temperature. Birds and mammals are endotherms; their primary source of heat comes from within their bodies.
energy flow. The movement of energy through an ecosystem.
environment. The total surroundings of an organism, including both biotic and abiotic components.
environmental impact statement. A legally required assessment of the effects of federally funded projects or activities in the United States that can be expected to have a substantial effect on the quality of human life. Under the National Environmental Policy Act,

the effects of such a project or activity on fisheries and wildlife resources must be considered.

environmental resistance. The environmental factors that interact to limit the abundance of a population; examples include predation and competition.

enzootic. A low-level chronic disease problem.

epilimnion. The uppermost water layer in a freshwater lentic system that is thermally stratified.

epiphyseal cartilage. The cartilage separating the epiphysis and the diaphysis of a long bone.

epiphysis. The end of a long bone.

epizootic. A large-scale eruptive disease outbreak.

erosional zone. The outer portion of a curve in a bending or turning lotic system. Current flow and water depth are greater here than in the depositional zone.

esophagus. That portion of the digestive tract of an animal located between the pharynx and the stomach.

essential amino acid. An amino acid that an organism must obtain in its diet, either because it cannot manufacture that amino acid or because the quantities it can manufacture are insufficient to provide for its needs.

estivation. Hypothermy during the summer.

estuarine. Of or pertaining to an estuary.

estuary. A transitional area where a freshwater system enters the ocean.

ethics. Standards of conduct and moral philosophy.

Ethiopian region. The zoogeographic region comprising Africa south of the Sahara, southern Arabia, and Madagascar.

euphotic zone. See **photic zone.**

eutrophic. Rich in nutrients. A eutrophic water body is generally characterized by reduced water clarity.

evolution. The theory pertaining to the process of continuous and gradual transformation of lines of descent from a common ancestor; the changes in the gene pool of a population over time.

evolutionary significant unit (ESU). Population units that infrequently exchange genes with other population units of the same species and represent an important component in the genetic diversity of that species.

exclusionary rule. The rule that evidence obtained through the violation of the Fourth Amendment rights of a suspect cannot be used in court.

exercised value. A benefit currently accruing to society; involves economic values of natural systems.

exotic. An organism introduced from another zoogeographic region.

experimental gill net. A gill net with numerous sections having different-sized meshes.

experimental populations. Populations of species listed under the U.S. Endangered Species Act that are released into new habitat outside their existing range and not in contact with nonexperimental populations. Critical habitat is generally not designated for experimental populations unless they are considered essential to survival of the species.

exploitation. Harvest of a resource by humans; usually conveyed as a rate or a percentage.

exploitative competition. Competition in which competing individuals divide the resource somewhat equally and all suffer similar effects from resource shortages.

extensive fish culture. Production method typified by fish reared in earthern ponds and feeding on natural prey such as zooplankton, insects, or prey fishes.

extirpate. To eliminate an organism from a portion or all of its native range.

fall overturn. An event that occurs during fall in a thermally stratified lentic system in which water temperature throughout the water body is equalized and the entire water body mixes.

false annuli (singular: false annulus). Marks formed on hard body parts, such as scales or bones, which appear to be annuli but are not; these marks can be caused by growth variation during periods of stress.

fecal pellets. The droppings of some animals; also called pellets or scats. These terms are usually used in reference to mammals.

fecundity. The potential number of reproductive products or units that can be produced by an organism.

federal aid funds. See **PR funds; DJ funds.**

Federal Register. A publication that contains in printed form all activities of the U.S. federal government.

feeding habits. Feeding behavior; when, where, and how an organism obtains food.

fee fishing. The payment of fees for the right to trespass and take either publicly or privately owned fisheries resources on private land.

fee hunting. The payment of fees for the right to trespass and take either publicly or privately owned wildlife resources on private land.

felony. A grave crime, such as burglary or murder. Some serious violations of wildlife and fisheries laws and regulations fall into in this category.
fen. A peat-accumulating wetland that receives some drainage from surrounding mineral soil and usually supports marshlike vegetation.
feral animal. An animal that has reverted to the wild state after having been domesticated.
filial imprinting. The process by which a young animal becomes behaviorally attached to a parent.
fingerling. The life stage of a young fish between 25 millimeters in length and the length at age 1.
first law of thermodynamics. In any ordinary chemical or physical change, energy is neither created nor destroyed but merely changed from one form to another.
fish. A single fish, or more than one representative of the same species; one largemouth bass or 200 largemouth bass are fish. Can also be used as the plural form.
fisheries. More than one fishery.
fisheries biology. The biological study of a fishery.
fisheries conservation. The effective management and use of fisheries resources.
fisheries ecology. The study of the interrelationships of organisms that are considered to be a part of the fisheries realm with other organisms and the environment.
fisheries management. The art and science of manipulating the biota, habitat, or human users of a fishery to produce some desired end result.
fisheries science. The process of obtaining knowledge about and studying fisheries resources.
fishery. A system composed of three interacting components: the biota, the habitat, and the human users. The term *fishery* can relate to a single target species or multiple target species, their habitat, and human effects on both organisms and the habitat.
fishes. One or more representatives from at least two fish species; one northern pike and one muskellunge are fishes. Can also be used as the plural of more than one fish of a species.
fishway. A structure designed to allow fish passage over or around a barrier.
fitness. A measure of the survival and reproductive success of individuals or populations; the product of genes and the environment interacting throughout the lifetime of an organism.
fixed-radius plot method. A method used to estimate tree density and basal area in which the plot radius is a specific measured distance from the plot center.
fledge. To grow feathers necessary for flight; to reach the life stage at which flight is possible.
flight-intercept trap. A device used to capture and hold both terrestrial and emerged aquatic flying invertebrates.
float line. A buoyant line attached to the top of a net used to capture aquatic organisms.
focus. The center or origin of a fish scale.
food chain. A simple, linear, graphic representation of energy flow through various trophic levels; for example, from the sun to clover to a white-tailed deer to a gray wolf.
food chain biomagnification. The concentration of a substance, such as a pesticide, at increasing levels in animals higher in a food web.
food habits. What an organism eats.
food web. A graphic representation of the complex interrelationships that occur as energy flows through the various trophic levels of an ecosystem.
forage fish. See **prey**.
forb. A nonwoody plant that is not a grass or grasslike.
foregut fermenter. A mammal with a ruminant stomach.
fossorial. Burrowing into the earth.
founder effect. The effect of inbreeding that occurs when a few individuals leave a larger population to establish a new population.
frequency scale method. A survey or census method used to measure human attitudes in which the respondent indicates the likelihood of a particular event or action.
fry. The life stage of a fish from the time it hatches until it reaches 25 millimeters in length.
fundamental niche. The niche occupied by an organism when there is no competition from other species.
funnel trap. A passive capture device, used for birds and mammals, that involves an arrangement of netting or wire mesh such that animals can readily find their way into the trap, but not out of it.
furbearer. A mammal commonly harvested for its hide.
fyke net. A passive capture device in which a lead net directs fishes into a trap that contains funnels. A fish can readily find its way into the trap through the funnels, but not out of it.

gabion. An instream device consisting of a rock-filled wire basket that can be used as a current deflector or to stabilize banks.
gallinaceous. Of or pertaining to the order Galliformes, which includes birds such as gray partridge and spruce grouse.
game and fish warden. A title often used for a state, provincial, or territorial natural resource worker who has only law enforcement responsibilities.
game animal. An animal species that is harvested for recreational purposes; these species are usually birds, mammals, or fishes. At times commercially harvested animals are also placed in this category.
game farming. The equivalent of game ranching, except that smaller organisms, such as game birds, are used. See **game ranching.**
game ranching. The husbandry of native animals, usually large ungulates, for the production of meat and other products. Also, a pay-for-hunting enterprise in which animals, primarily exotic ungulates, are harvested for sport purposes.
gap analysis. A landscape-level analysis of gaps in the protection of species and species diversity.
gas bladder. An organ found in fishes in which a fish can increase or decrease internal gas content and thus its body density.
gastric lavage. Flushing contents from the stomach of an organism with pumped water.
Gause's principle. See **competitive exclusion principle.**
gear. A fisheries term for equipment used to capture or sample aquatic organisms.
gene pool. All of the alleles present within the individuals constituting a population.
generalist. An organism that can utilize a wide variety of resources or habitats.
genes. Segments of hereditary material that are positioned on chromosomes within cells, are heritable, and determine the characteristics of an organism. A segment of DNA that codes formation of a particular protein.
genetic bottleneck. See **bottleneck effect.**
genetic diversity. All of the genetic variation within a species, including both within and between population components.
genetic drift. Random changes in allelic frequencies due to natural sampling errors that occur in each generation.
genetic forensics. The use of genetics to make inferences regarding past events.
genetic integrity. The genetic characteristics of a group of organisms that distinguish it from other groups or populations of that species.
genetic marking. Purposefully changing allele frequencies in a group of organisms in order to identify them or their descendants.
gene transfer. The artificial insertion of genetic information from one species into another.
genital papillae (singular: genital papilla). External, fleshy protuberances through which fish eggs and sperm are discharged; can be used to differentiate the sexes in some species.
genotype. The specific genetic information possessed by an organism.
geographic information system (GIS). A system for the computerized mapping and analysis of geographic features using digitized data.
geographic range. The geographic area within which a species occurs.
geophagia. The ingestion of soil by animals to obtain minerals.
gestation. The growth of an embryo in the uterus. This term is primarily applied to mammals.
gestation period. The period of time during which gestation occurs.
ghost fishing. The capacity of some fishing gears, such as gill nets, to continue capturing organisms even after they are lost or abandoned.
gill net. A passive capture device used in fisheries work in which fishes are captured by becoming wedged or tangled in a single panel of netting.
gizzard. A portion of the digestive tract of some animals that is muscularized for food-grinding purposes.
glandular stomach. The anterior stomach in birds, the simple stomach in nonruminant mammals, and the abomasum in ruminant mammals.
global positioning system (GPS). A device that uses earth-orbiting satellites to locate a specific position on the surface of the earth.
glomeruli (singular: glomerulus). Conglomerations of blood vessels associated with the functional units (nephrons) of a kidney.
glucagon. A hormone that causes stored polysaccharides and amino acids in the liver to be converted to glucose.
gluconeogenesis. The process of glucose synthesis from noncarbohydrates such as fats and proteins.
goal. A broad, specific, but nonmeasurable outcome desired by an entity in relation to its mission.

gonadosomatic index (GSI). An index used to determine the reproductive status of a fish; obtained by dividing gonadal weight by body weight and multiplying by 100.
grabbing distance. The area within the immediate control of a suspect; this area is generally searchable for evidence without a search warrant.
Grassland Reserve Program. USDA program first authorized in the 2002 Farm Act (Farm Security and Rural Investment Act) that provides technical and financial assistance to landowners in conserving, managing, and improving their grasslands while still allowing grazing and other compatible uses.
grass/roughage eater. A ruminant that is capable of digesting coarse and fibrous foods.
grazing. The act of feeding on grasses and forbs.
greater secondary coverts. The largest covert feathers overlying the secondary feathers on a bird wing.
greenhouse gas. A gas, such as carbon dioxide, water vapor, ozone, methane, nitrous oxide, or chlorofluorocarbons, that lets visible light from the sun reach the surface of the earth but inhibits infrared radiation from escaping.
guild. A group of organisms that uses environmental resources in a similar manner.
gular sac. A distended membrane in the throat area of some animals.
guzzler. An artificially constructed watering area for terrestrial animals. Guzzlers are often constructed for wild animals in arid regions.
gynogenesis. Reproduction without fertilization that produces viable young that have only the genetic material of the mother.

habitat. The specific set of environmental conditions under which an individual, species, or community exists.
habitat conservation plan (HCP). A habitat conservation plan by a nonfederal landowner or business that allows the continuation of activities that are likely to result in incidental take (death, injury, disruption) of organisms listed as endangered or threatened under the U.S. Endangered Species Act. The plan must be accepted by the U.S Fish and Wildlife Service.
habitat fragmentation. The breaking of larger patches of habitat into smaller patches.
habitat suitability index (HSI) model. A model based on suitability indices formulated from variables that affect the life cycle and survival of a species.
habituation learning. Learning not to respond to a meaningless stimulus.
hacking. A wildlife term usually used to denote the methods involved in raising captive raptors and releasing them into the wild.
half-log. An instream device consisting of a split portion of a log that is staked in a streambed to provide overhead cover.
hardness. A water quality parameter that is governed by the content of calcium and magnesium salts in water, largely combined with bicarbonate and carbonate and with sulfates, chlorides, and other anions of mineral acids.
harvestable surplus. The surplus production that can be removed without adversely affecting a population.
harvest mortality. Mortality resulting from human activities directed at taking organisms for food or recreational purposes.
harvest quota. The specific number or total weight of a species that may be legally harvested; for example, the biomass of black basses from a lake or the number of Canada geese from a population that may be taken during a given time period.
hatching curve. A plot of the number of bird clutches hatched in relation to the progression of the reproductive season.
hatching year (HY). An age classification used to denote a bird that is less than 1 year old.
headfire. A fire purposefully set to move downwind. Headfires burn more rapidly but less thoroughly than backfires.
herbaceous cover. See **herbage.**
herbaceous plants. Nonwoody plants, such as grasses or forbs.
herbage. Vegetation consisting of grasses, sedges, and forbs.
herbivore. An organism that eats autotrophs; also called a primary consumer.
heterosis. See **hybrid vigor.**
heterotherm. An ectothermic animal that maintains a body temperature above the ambient temperature.
heterotroph. An organism that feeds on autotrophs or other heterotrophs.
heterozygosity. The condition of having one or more pairs of dissimilar alleles at one or more loci in homologous chromosome segments.
heterozygous. Having two dissimilar alleles at a particular locus.

hibernation. Hypothermy during the winter, during which the body temperature drops to approximately ambient temperature.
hierarchal order. The order of dominance in an animal social group.
high-grade misdemeanor. A serious violation of a law or regulation, but one less serious than a felony.
hindgut fermenter. A species in which fermentation occurs primarily in the caecum and intestine.
homeostasis. The tendency to maintain relatively stable conditions within the body.
homeothermic. See **endothermic.**
home range. The area within which an individual animal normally travels in the course of its daily activities.
homing. The capacity of an animal to return to a familiar site that is outside the range of direct senses such as sight and hearing.
homologous chromosomes. The two chromosomes that carry the same set of genes.
homozygosity. The condition of having identical alleles at one or more loci in homologous chromosome segments.
homozygous. Having two copies of the same allele at a particular locus.
hoop net. A passive capture device consisting of mesh-covered hoops and funnel-shaped throats for capturing fishes. A fish can easily find its way into the net through the funnels, but not out of it.
hormone disruptor. See **endocrine disruptor.**
human dimensions. How people value wildlife and fisheries, how they want these resources managed, and how they affect or are affected by wildlife and fisheries management decisions.
human users. Direct, indirect, consumptive, and nonconsumptive users of fisheries or wildlife resources; the human component of a fishery or a wildlife system.
hybrid. An offspring resulting from the mating of parents that are genetically unlike. In most cases hybrids represent crosses between different species, subspecies, or strains.
hybrid vigor. An increase in some phenotypic value of a hybrid relative to its parents; also called heterosis.
hydric plant. A plant associated with a wetland.
hydric soil. A soil developed in a wetland.
hydroacoustics. A technique involving the use of sonar that is used in fisheries work to locate, count, or study fishes.
hydrologic cycle. The cycling of water through the biosphere.
hydroxyapatite matrix. A mineral matrix of calcium phosphate of which vertebrate bones and teeth are primarily composed.
hyperosmotic. Having a greater ion concentration than its surroundings; freshwater bony fishes are hyperosmotic to their environment.
hyperphagy. Rapid and lengthy feeding.
hypolimnion. The lowermost water layer in a freshwater lentic system that is thermally stratified.
hyposmotic. Having a lower ion concentration than its surroundings; marine bony fishes are hyposmotic to their environment.
hypothermy. The lowering of body temperature in endotherms to save energy during periods of stress.
hypothesis. A proposition or supposition regarding a theory that is tentatively accepted to explain known phenomena or to provide a basis for further investigation.
hypoxic. Having a deficiency or lack of oxygen in body tissues.

ichthyocide. See **piscicide.**
immature. An organism that is too young to breed and can be distinguished from an adult based on external characteristics; also called a juvenile. Can also be used to denote fishes that are juveniles and have not reached sexual maturity.
immigration. Migration of an animal into an area.
impoundment. A water body formed by blocking a natural watercourse such as a river or stream.
inborn behavior. See **innate behavior.**
inbreeding. Mating between closely related individuals.
inbreeding depression. A reduction in fitness or vigor due to increased homozygosity resulting from inbreeding.
incidental catch. Those organisms unintentionally captured in addition to the target organisms or species.
incidental take. The taking (killing, harming, harassing, collecting, etc.) of U.S. Endangered Species Act–listed species "incidental to, and not the purpose of, the carrying out of an otherwise lawful activity."
incomplete dominance. Partial dominance by one allele such that both alleles are expressed phenotypically, but to different degrees.
increaser species. A plant species, usually a grass of lower forage quality than decreaser species, that in-

creases in number or biomass in an area with increased grazing intensity.
indeterminate growth. A growth pattern in which an animal essentially continues to grow throughout its life; fishes generally exhibit this growth pattern.
index. A value, often a ratio, used in lieu of the actual number; for example, the number of fish caught per hour of electrofishing is sometimes used as an index to population density.
indicator species. A key organism that serves as an indicator of the status of ecological conditions in an area.
indigenous. Originally found in or native to an area.
indirect benefit. Benefits that accrue from situations where the user is not in direct contact with a resource; economic benefits from activities such as reading about or watching wildlife and fish.
indirect evidence. Evidence that, either alone or in combination with other evidence or facts, would allow a court or judge to reasonably infer guilt or innocence of an individual.
indirect expenditure. An expenditure generated when the money from a direct expenditure is respent to pay salaries or to purchase new supplies or services.
indirect user. A person who uses or manages some aspect of a natural resource and in doing so affects some other aspect of the resource. For example, someone who harvests trees and thus has an effect on wildlife and fisheries resources.
individual distance. The minimum distance maintained by an animal between it and another animal, usually in group situations.
inguinal canal. The canal in some mammals through which the testes descend into the scrotum.
innate behavior. A behavior pattern with which an animal is born.
inorganic. Of or pertaining to minerals or other materials made of noncarbon compounds.
insectivore. An organism that eats insects.
insight learning. Learning that requires an animal to have the ability to perceive relationships.
instinct. An innate stereotyped behavior that is characteristic of a given species.
insulin. A hormone produced by beta cells in the pancreas that stimulates liver, muscle, and fat tissues to take up glucose.
integrated natural resource management. A conflict resolution technique designed to address areas of potential conflict in natural resource conservation before conflict occurs.
intensive fish culture. Production method typified by fish rearing in confined tanks or raceways, with high water volume exchanges and using pelleted food.
intentional violator. A person who starts a particular activity with the intention of violating a wildlife or fisheries law or regulation.
interference competition. See **contest competition.**
interior species. A species (or subspecies) that requires an extensive area of relatively homogeneous habitat. This term is commonly used for birds, less often used for mammals, and not used for fishes.
intermediate impoundment. A constructed lentic water body that is between 10 and 200 hectares in surface area and intermediate in size between a pond and a reservoir.
interspecific. Between or among species.
interspersion. A measure of the intermixing of different habitat types. It is also a measure of the horizontal diversity of habitats.
intraspecific. Within a species.
introduced organism. An exotic or transplanted organism.
introduction. See **introduced organism.**
introductory stocking. The release of a fish species into a new or renovated body of water, or the introduction of a new fish species (either native or introduced) into a water body with an existing fish community. This term can also be used in relation to wildlife species.
introgression. The process by which the phenotype of one group of organisms becomes dominant after several generations of interbreeding.
invader species. A plant species that invades from other areas into heavily grazed areas. These plants often reduce the carrying capacity of an area for domesticated livestock and wildlife.
invasive species. An introduced organism that rapidly expands its range.
inverse stratification. Stratification that occurs in freshwater lentic systems under ice cover in which the warmest (and most dense) water (4°C) is at the bottom of the water column.
inversion. A form of genetic change that results when the sequence of genes on a chromosome becomes inverted.
inversity. An inverse (negative) relationship between population density and natality or recruitment.

irruptive population. A population that increases or decreases dramatically at irregular intervals.
isolating mechanism. A mechanism that tends to keep organisms from interbreeding.
isosmotic. Having an osmotic pressure similar to that of the surrounding environment.

J-curve. A curve that describes population growth when a population in a new and favorable environment increases at an exponential rate.
Juday net. A specialized capture net used primarily to sample plankton.
juvenal plumage. The plumage following natal down in birds.
juvenile. See **immature.**
juxtaposition. A measure of the proximity of habitats needed by an animal. It is also a measure of the horizontal diversity of habitats.

Kemmerer water sampler. A metal device that can be lowered into a water body, remotely triggered to close, and brought to the surface with a water sample from a particular depth.
ketogenesis. A catabolic process involving the hydrolysis of fat.
kin selection. Natural selection operating on the interactions between closely related cooperating individuals.
***K*-selected species.** A species with a reproductive strategy that involves the production of small numbers of offspring and high levels of parental care.
kype. The hooked jaws found on some male salmonids during the reproductive season.

lactation. Milk secretion in mammals.
lacustrine. Of or pertaining to lentic systems.
lake. See **natural lake.**
lampricide. A selective piscicide that primarily affects or kills lampreys.
land-grant university. An institution of higher education in the United States receiving federal aid under the Morrill Acts of 1862 and 1890.
landscape. A heterogeneous land area composed of interacting ecosystems that are repeated in similar patterns throughout an area or region. A landscape can include aquatic systems.
landscape ecology. The study of structure, function, and change in landscapes.
large impoundment. See **reservoir.**
larval fish. A newly hatched fish.
laser altimetry. See **laser imaging detection and ranging systems.**
laser imaging detection and ranging systems (LIDAR). A remote sensing technology using laser beams and returning scattered light to measure distances, movements, physical properties, and other characteristics of target areas.
laser rangefinder. A device that uses laser technology to determine distances to opaque objects.
lateral line system. A sound reception system found in fishes and some amphibians that is used to perceive mechanical sound.
law. A rule enacted by an elected governing authority, such as a federal, state, provincial, territorial, or local government.
law of tolerance. The principle that the absence or failure of an organism in a particular habitat is controlled by a qualitative or quantitative deficiency or excess with respect to any one of several factors that may approach the limits of tolerance for that organism.
lead line. A weighted line attached to the bottom of a net used to capture aquatic organisms. (In this term the pronunciation of *lead* is like that of *pencil lead.*)
lead net. A net radiating from a passive capture device that directs animals into the device. (In this term the pronunciation of *lead* is like that of *lead dog.*)
learned behavior. Behavior that is not innate but instead results from remembered experience.
leghold trap. A passive capture device, primarily used on mammals, in which an animal trips a device that then captures the animal by its leg.
lek. A term that is sometimes used to describe an area where animals gather to perform courtship displays.
length-frequency analysis. A method used to age a fish population sample, based on the principle that fish of similar age, especially younger fish, in a particular body of water tend to be of approximately the same length.
lentic. Of or pertaining to standing water, as in a pond or lake.
lesser secondary coverts. Covert feathers located immediately above the middle secondary coverts on a bird wing.
level ditching. Earth-moving activity done to modify a wetland so that deeper water areas are present; of-

ten done to ensure that some areas retain water even during low-precipitation periods.

LIDAR. See **laser imaging detection and ranging systems.**

life table. In wildlife, a summary of survivorship or mortality in a population from one age class to the next.

Likert scale method. A survey or census method used to measure human attitudes by which both direction of affect (positive or negative) and intensity of affect can be measured.

limiting factor. A single environmental component that limits population growth.

limiting resource. See **limiting factor.**

limnetic zone. The lighted portion of the deeper water zone beyond the littoral zone in a freshwater lentic water body.

Lincoln method. In wildlife, a method of estimating population number by means of a single mark-and-recapture period.

line-intercept method. A technique used to assess coverage or species composition of plants in which a tape is stretched above the vegetation and the percentage of the tape overlying plants is determined.

littoral zone. The shallow water zone in freshwater lentic systems, generally characterized by rooted aquatic vegetation.

live trap. See **box trap.**

locus (plural: loci). The location of a particular gene on a chromosome.

logistic growth curve. See **S-curve.**

lotic. Of or pertaining to flowing water, as in a river or stream.

lower critical temperature. The ambient temperature below the thermoneutral zone of an animal at which increased energy consumption must occur for the animal to maintain temperature homeostasis.

lower zone of metabolic compensation. The temperature zone below the lower critical temperature within which an animal can maintain temperature homeostasis by expending additional energy.

low-grade misdemeanor. A low-level violation of a law or regulation.

macroelement. An inorganic element required in relatively large amounts by an organism.

magistrate. An official having jurisdiction over minor court cases.

magnitude scale method. A survey or census method used to measure human attitudes in which respondents are given a list of items and asked to allocate points based on importance or value to them.

maintenance burn. One of a series of prescribed burns set regularly to maintain conditions resulting from previous burns.

maintenance stocking. The release of a fish species into a population that has no or very limited natural reproduction. This term can also be used in relation to wildlife species.

management by objective (**MBO**). A planning or management process that involves a systems approach to managing an organization.

marginal coverts. Covert feathers above the lesser secondary coverts on a bird wing (toward the leading edge of the wing).

marine. Of or pertaining to the ocean.

marsh. A low, treeless wet area characterized by plants such as sedges, rushes, and cattails.

mast. Fruit produced by trees and shrubs, such as acorns and wild plums. This term is generally used in reference to food production for terrestrial animals.

maximum length limit. A regulation under which all fish of a particular species that reach or exceed a specified length must be released.

maximum sustained yield (**MSY**). A traditional management philosophy with the goal of achieving the maximum yield that can be sustained over a series of years from a population; the maximum biomass of a population that can theoretically be harvested without affecting future harvest.

megaherbivore. An herbivore that weighs over 1,000 kilograms.

metabolic rate. The rate at which chemical reactions occur in an organism.

metabolic water. A by-product, along with carbon dioxide, of the oxidation process in cells (respiration) wherein carbohydrates, lipids, or proteins are metabolized.

metalimnion. See **thermocline.**

metapopulation. A population existing as a set of geographically separate subpopulations, but with some genetic interchange.

meter net. A device commonly used to capture fish eggs and larvae. The opening of the net is 1 meter in diameter.

microbivore. An organism that feeds on the nutrients and energy in bacteria and fungi.

microclimate. The climate on a very local scale. Microclimate differs from the general climate of an area.
microelement. An inorganic element required in trace amounts by an organism.
microflora. Small organisms such as bacteria or small fungi.
microhabitat. A small area of intensive use within the overall habitat of an organism.
middle secondary coverts. Covert feathers between the greater secondary coverts and lesser secondary coverts on a bird wing.
migration. A two-way movement of an animal to and from an area with characteristic regularity or with changes in life history stage.
milk teeth. The first set of teeth in mammals.
milt. The sperm-bearing fluid of fishes.
mineralization. The release of inorganic nutrients that have been immobilized or kept out of nutrient cycles because of their inclusion in something such as the body tissues of an organism.
minimum instream flow. The minimum flow needed to ensure sufficient water for aquatic organisms at all times of the year.
minimum length limit. A regulation under which all fish of a particular species that are less than a specified length must be released, while those exceeding the length may be kept.
Miranda warning. A procedure for informing a suspect of his or her constitutional rights.
mission statement. A statement identifying the purpose or reason for the existence of an entity, who it serves, and its ultimate aims.
mist net. A passive capture device used on birds that entangles them in a single panel of netting.
mitigation. The legally required replacement of certain habitats, such as wetlands or riparian areas, that are lost due to projects that receive federal funds. Mitigation may involve the purchase of private lands in some other area of a state to replace habitat that is lost to project construction. Mitigation can also be required for some nonfederally funded activities.
mixed-grass prairie. That portion of the temperate grassland biome characterized by grasses that are intermediate in stature and by precipitation levels that are intermediate between those of short-grass and tall-grass prairies.
model. A conceptual or mathematical description used to portray a complex system.
modified-fyke net. A passive capture device in which a lead net directs fishes into a trap containing funnels. A fish can readily find its way into the trap through the funnels, but not out of it.
molt. In birds, the loss of feathers in the process of renewal. Many other animals also molt various body parts.
monoculture. The growing or raising of a single species over a large area.
monogamous. Of or pertaining to a mating system characterized by pair bonding between a single male and a single female.
monosaccharide. A simple sugar such as glucose.
monosex population. A population that consists of only one sex.
montane coniferous forest. An altitudinal biome similar in characteristics to the northern coniferous forest biome.
mortality. The death of an individual. Mortality rate is commonly expressed as the percentage of the population that dies in a year. In addition, mortality rates can be determined for age groups, year classes, or age classes.
movement. A change in the position or posture of an animal; these changes are not necessarily regular in nature as would occur in migrations.
multidimensional niche. The variety of niche components that constitute the total niche of an organism.
mutation. A change in the DNA of a gene.

natal area. The place of birth or hatching for an animal.
natal down. The short, fluffy feathering on newly hatched birds.
natality. The birth or hatching of an animal. Natality rate is commonly expressed as the number of individuals born or hatched within a specified time period.
native range. The geographic area in which a species is indigenous.
naturalized organism. A naturally occurring offspring produced by exotic or transplanted organisms.
natural lake. A naturally formed lentic water body.
natural mortality. Mortality caused by predation, starvation, disease, accidents, or other natural causes.
natural resource. Something such as land, water, or wildlife that is useful and valuable in the condition in which it is found.
natural resource economics. The study of the allocation of natural resources.

natural selection. A theory that attempts to account for the adaptation of organisms to their environment, which states that such adaptation results from organisms of different genotypes in a population contributing differently to the gene pool of succeeding generations.
Nearctic region. The zoogeographic region comprising North America and Greenland.
Neotropical migrant. A bird that migrates between the Nearctic and Neotropical zoogeographic regions. These birds would more appropriately be called Nearctic-Neotropical migrants.
Neotropical region. The zoogeographic region comprising South America, Central America, and the West Indies.
net gun. A gun that fires projectiles that propel a small net away from the gun. This device is primarily used to capture flying birds.
net willingness to pay. See **consumer surplus.**
neuromast. A specialized sensory structure associated with the lateral line system of fishes that can detect the displacement of water.
niche. The role or function of an organism in a biotic community.
niche segregation. The tendency for two species that live in the same area and require similar resources to have niche requirements that differ in one or more dimensions.
night-vision scope. See **starlight scope.**
nitrogen fixation. Conversion of inert, gaseous nitrogen to ammonia, nitrites, or nitrates.
nocturnal. Of or pertaining to the night; active during periods of darkness.
nonconsumptive user. A person who does not harvest a resource but does use it for some purpose.
nonessential amino acid. An amino acid that an organism can manufacture in sufficient quantities within its own body.
nongame animal. An animal species that is not harvested for recreational or commercial purposes; an animal that is legally described as a nongame species.
nonhomologous chromosomes. Chromosomes that carry different sets of genes.
nonpoint-source pollution. Pollution emanating from diffuse or dispersed sources.
northern coniferous forest. A biome characterized by evergreen trees and cool climate. In North America this biome is located south of and adjacent to the tundra biome.
Nudds board. A 2.5-meter-long, 30-centimeter-wide, marked board used to determine the horizontal concealment or cover value of vegetation.
nutrient cycle. See **biogeochemical cycle.**
nutrient immobilization. The keeping of nutrients out of circulation, as in the case of those in the body tissues of an organism.

objective. A specific, measurable, output- or benefit-focused, and realistic outcome desired by an entity within a certain time frame in relation to attaining a specific goal.
olfactory. Of or pertaining to the sense of smell.
oligosaccharide. A short chain of two or more simple sugars.
oligotrophic. Low in nutrients. An oligotrophic water body is generally characterized by high water clarity.
omasum. The third chamber of a ruminant stomach.
omnivore. An animal that eats both autotrophs and heterotrophs.
operational planning. Planning for the implementation of a strategic plan.
opportunist violator. A person who had no intention of violating a wildlife or fisheries law or regulation but did so when the opportunity occurred.
optimal foraging theory. A theory based on the premise that an animal should not expend more energy than necessary to obtain food.
optimum sustained yield (OSY). A management philosophy that considers ecological and socioeconomic factors.
option value. The value of maintaining an option to use, experience, or enjoy something in the future; involves economic values of natural resources.
organic. Of or pertaining to carbon-containing compounds, including living or dead organisms.
Oriental region. The zoogeographic region comprising India, Myanmar, Indochina, Malaysia, Sumatra, Java, Borneo, and the Philippines.
osmotic balance. Water and mineral balance in an organism.
osmotic pressure. The pressure difference resulting from differing water and ionic concentrations on two sides of a semipermeable membrane.
otolith. An earstone found in the inner ear of animals. In fishes, otoliths are commonly used in age determination.
otter trawl. An active capture device for aquatic organisms in which the mouth of the trawl is maintained

in an open position by weighted boards called otter boards.
outbreeding. Mating between distantly related individuals.
outer coastal plain forest. A portion of the temperate forest biome that ranges from Florida to Louisiana, characterized by mangroves and live oaks.
ovarian follicle. A structure in the ovary that may develop into an ovum.
overexploitation. Variously defined as the point at which harvest mortality causes undesirable reductions in fish, wildlife, or other populations; the level of harvest at which adult organisms cannot replace themselves; or the level of harvest at which population quality is impaired.
overharvest. See **overexploitation.**
overkill. The harvesting of more than the legal daily limit or the accumulation of more than the possession limit for the season.
overstory. The topmost vegetation in an area. This term is most often used in reference to tree crowns in forests, which are also known as the canopy.
oxygen-demanding wastes. Pollutants such as domestic sewage, domesticated livestock wastes, and other biodegradable organic wastes that cause water to be depleted of dissolved oxygen.

Pacific coast coniferous forest. A small North American biome extending from Alaska to northern California that is characterized by a cool climate and abundant rainfall. Douglas firs, giant redwoods, and Sitka spruce are the dominant trees in this biome.
Pacific salmon. A group of salmonid fishes that includes pink, chum, coho, sockeye, and Chinook salmon.
Palearctic region. The zoogeographic region comprising Europe and Asia north of the Himalayas.
palustrine. Of or pertaining to bogs, marshes, and swamps.
panfish. A small game fish such as a bluegill, white or black crappie, white perch, or yellow perch.
parabronchi. The site of gas exchange in bird lungs.
parasite. An organism that obtains energy by living in or on another organism.
parr. The life stage of a salmonid fish from the time of initial feeding until the fish develops sufficient pigmentation to obliterate the parr marks.
parr mark. A vertical pigmentation band on the sides of young salmonids and certain other fishes. Numerous parr marks are usually present.
partial rehabilitation. See **partial renovation.**
partial renovation. A term used in fisheries to denote the removal of a portion of a population or community. This term can also be used in relation to wildlife species.
passerine bird. A member of the largest order of birds, the Passeriformes.
passive capture device. A device that is put in place, essentially remaining stationary, such that the movement or behavior of the target animal results in its capture.
passive entanglement device. A device that is put in place, essentially remaining stationary, and the target animal when encountering the device becomes ensnarled.
passive entrapment device. A device that captures animals that move into an enclosed area and are retained in that area.
patch. A continuous, nonlinear (not long and narrow) surface area within a landscape that differs from surrounding areas.
patch size. The size of a particular patch within a landscape; for example, the surface area of a forest surrounded by urban habitat.
pathogen. A disease-causing biological agent.
pecking order. See **hierarchal order.**
pellets. See **fecal pellets.**
peptide group. A combination of two or more amino acids.
permafrost. The presence of frozen ground, usually a short distance below the soil surface, during the entire year.
Petersen grab. A sampling device used to obtain organisms and substrate from the benthic zone of an aquatic habitat.
Petersen method. In fisheries, a method of estimating population number from a single mark-and-recapture period.
pH. A measure of the hydrogen ion concentration in a solution.
pharynx. That portion of the digestive tract of an animal between the oral cavity and the esophagus.
phenotype. The actual physical expression of genetic traits. Phenotype is a product of the genotype and the environment.

pheromone. A chemical cue used in animal communication.
photic zone. The portion of a freshwater ecosystem that is lighted and in which photosynthesis occurs. The photic zone is generally considered to be the portion that receives 1% or more of the surface light intensity.
photoperiod. The proportion of light and dark in a 24-hour period.
photosynthesis. Energy fixation by autotrophs using sunlight as a source of energy.
physoclistous fishes. Fishes that as adults do not have a duct connecting the digestive tract and the gas bladder.
physostomous fishes. Fishes that as adults have a duct connecting the digestive tract and the gas bladder.
phytoplankton. Tiny aquatic plants that drift or float in water.
pigeon milk. A rich milk-like material produced in the crops of reproductively active doves and pigeons.
piloerection. Erection of hair in mammals as commonly used in temperature control or in various behaviors.
pineal gland. A structure located in the brain that can function in light sensitivity. Other potential functions are also attributed to this endocrine gland and structures derived from it.
pineal organ. See **pineal gland.**
pinna (plural: pinnae). The external ear structure in mammals that directs sound into the middle ear.
pinnae feathers (singular: pinna feather). Elongated feathers on the necks of some birds, such as greater prairie-chickens.
pipping. Breaking of the eggshell during hatching in birds.
piscicide. A toxicant or poison that affects fishes.
piscivorous. Feeding on fishes.
pit trap. A passive capture device used on a wide variety of animals in which the target animal falls into a pit or hole from which it cannot escape.
poaching. The illegal taking of a wild organism.
poikilothermic. See **ectothermic.**
point-centered-quarter method. A method used to determine tree density in which the distance from a sampling point to the nearest tree is measured in each of four quarters.
point frame. A metal frame into which metal rods are inserted that is used to determine basal and canopy coverage of plants, usually grasses and forbs.
point-source pollution. Pollution emanating from a specific, identifiable location.
polyandrous. Of or pertaining to a polygamous mating system in which one female mates with multiple males.
polygamous. Of or pertaining to a mating system in which a single individual of one sex mates with multiple individuals of the opposite sex.
polygenic trait. A genetic trait controlled by multiple genes.
polygynous. Of or relating to a polygamous mating system in which one male mates with multiple females.
polyploidy. The genetic condition of having more than two sets of chromosomes.
polysaccharide. A straight or branched chain of hundreds or thousands of sugar molecules.
pond. A lentic water body that is less than 10 hectares in surface area. A pond is usually constructed by blocking a small intermittent watercourse.
pool. That area of a headwater freshwater lotic system characterized by being deep, wide, and soft bottomed. Current flow in pools is less than that in riffles because of the greater volume of space present.
population. All the individuals of one species within a specified area at a given time. From a genetics standpoint, these are groups within a species that actively interbreed and are adapted to similar environments.
population density. The number of organisms of one particular species present in a specific unit of area at a given time.
population dynamics. The study of changes in numbers or biomass of organisms in populations and the factors (natality or recruitment, growth, and mortality) that influence those changes.
population estimate. The expansion of a sample into an estimate of the total population number.
population form. See **population structure.**
population function. See **population dynamics.**
population structure. The size structure, age structure, or sex ratio of a population. The three dynamic rate functions (natality or recruitment, growth, and mortality) interact and result in population structure.
population viability. The ability of a population to sustain itself demographically.
possession limit. The total number of a species that one person may legally possess at any one time. Possession limits are commonly two to five times the daily creel or bag limit for a species.

postbreeding. An age classification term used in wildlife to denote animals that have bred.
prebreeding. An age classification term used in wildlife to denote animals that have not bred.
precision. The closeness of repeated measures to one another.
precocial. Well developed, alert, and mobile soon after birth or hatching. This term is not generally applied to fishes; however, a fish such as the walleye that is well developed at the time of hatching may be considered precocial.
predator. An animal that attacks, kills, and eats other animals.
prescribed burn. An intentional and controlled burning of an area for habitat management purposes.
preservation. A conservation philosophy that involves leaving natural systems as they are.
prey. An animal that a predator eats.
PR funds (Pittman-Robertson). Funds derived from a U.S. federal excise tax on such items as arms and ammunition. These funds are available to states for the acquisition, restoration, rehabilitation, and improvement of wildlife habitat and for wildlife conservation and research.
primaries. See **primary feathers.**
primary consumer. An organism that primarily eats autotrophs; also called an herbivore.
primary coverts. Covert feathers overlying the primary feathers on a bird wing.
primary feathers. Large feathers arising on the posterior portion of a bird wing beyond the "wrist."
primary producer. An autotroph or plant; an organism that can produce its own food.
primary succession. Ecological succession occurring on a site unchanged by living organisms, such as a new volcanic island that appears above the surface of the ocean.
probable cause. The situation in which enforcement personnel have facts and circumstances within their knowledge that would cause a person of reasonable caution to believe that wrongdoing has occurred or is occurring.
production. The biomass accumulated by a population during a year or some other specific period, including both living organisms and those that died during that period.
productivity. The rate at which energy is captured and stored in an ecosystem.
profundal zone. The unlighted portion of a freshwater ecosystem located below the limnetic zone.
promiscuous. Of or pertaining to a mating system that is not characterized by male-to-female pair bonding.
propagation. The planting of seeds or transplanting of trees and shrubs.
Property Clause. A clause in the U.S. Constitution that allows the federal government jurisdiction over its property.
proportional size distribution (PSD). The percentage of stock-length fish in a sample that are also of quality length; formerly known as proportional stock density.
protected slot length limit. A regulation under which all fish of a particular species that are within a specified length range must be released, while those outside that length range may be kept.
protection. The prevention of damage to plants until they mature.
proventriculus. The glandular stomach in birds.
proximal. Near the body, as opposed to distal (away from the body).
pseudoreplication. Treating replicate samples as if they represent replicate experimental units.
ptiloerection. Erection of feathers in birds as commonly used in temperature control or in various behaviors.
purse seine. An active capture device used in fisheries work in which a purse line on the net bottom, when drawn together, prevents fishes from escaping.
put and take stocking. The stocking of fish of catchable size for immediate angler use. This term could also be used in relation to wildlife species.
put, grow, and take stocking. The stocking of small fish into a water body where they will grow to catchable size and become available to anglers. This term could also be used in relation to wildlife species.
pygmy water meter. A meter used to determine water current velocity.
pyloric caeca (singular: pyloric caecum; alternative spellings: ceca/cecum). In fishes, structures found between the stomach and intestine that probably function in both digestion and absorption.

race. See **subspecies.**
raceway. An intermediate area located between the pools and riffles of a headwater freshwater lotic system. This term is also applied to rectangular rearing units, often made of cement, used in aquaculture.

radiometer. An electronic device that can be used to estimate the biomass of grasses and forbs.
raghorn. A male elk approximately 2.5 years of age.
random dispersion. A dispersion pattern in which the position of an organism is independent of the positions of its conspecifics; this pattern occurs when a species lacks social tendencies toward clumping and where there is little variation in the environment.
random-stratified sample. A survey in which random samples are collected from predetermined categories (stratifications). A random-stratified creel survey might use boat anglers and shore anglers as stratifications but would require random samples of each angler type.
range. See **geographic range** in reference to organism distribution. In reference to statistics, range is the difference between the largest and smallest observations in a sample.
ranking scale method. A survey or census method used to measure human attitudes in which the respondent is asked to order a list of items according to personal preference or some other standard.
raptor. A predatory bird that has feet with sharp talons or claws adapted for seizing prey and a hooked beak for tearing flesh. Hawks, eagles, and owls are examples of raptors.
realized niche. The niche occupied by an organism when some competition with other species is occurring.
reasonable suspicion. A legal standard somewhat less stringent than probable cause but representing more than just a hunch of a law enforcement officer.
recessive allele. An allele that is typically not expressed in the presence of a dominant allele.
reclamation burn. A prescribed burn conducted on mistreated or unmanaged land.
recovery plan. A plan developed to bring an endangered or threatened species back to healthy population levels. This term is specifically related to the U.S. Endangered Species Act.
recruitment. A term denoting the number of individuals hatched or born in any year that survive to reproductive size. Some alternative definitions include that number of individuals that reach harvestable size, a particular size or age, or a size captured by a particular sampling gear.
rectrices (singular: rectrix). The relatively large, conspicuous tail feathers on birds.
reflex. A simple innate behavior exhibited by an animal that is an autonomic response to a stimulus.
regulation. A rule enacted by a nonelected government entity such as a state fisheries and wildlife commission, a federal agency, or a local board.
regulatory gene. A gene that controls the expression of a structural gene during development or among different tissues.
relative abundance. Information obtained from samples or observations and used as an index to population abundance; often based on catch per unit of sampling effort in fisheries.
relative species composition. The proportional (percentage) numerical abundance of a species within a collection of species.
relative weight. An index of condition in fishes in which the actual weight of a fish is divided by a standard weight selected to represent an optimal weight for that species at that length, then multiplied by 100.
release. The result of reduced competition from other, less desired plant species. Release can also occur after a reduction in the numbers of the target species.
remote sensing. The measurement of or acquisition of information about some property of an object or phenomenon by means of a recording device that is not in physical or intimate contact with the object or phenomenon under study. Aerial photography, for example, is a form of remote sensing.
renewable resource. Resources such as animals, plants, and solar energy that replenish naturally through some biological or geophysical process.
reservoir. An impoundment that exceeds 200 hectares in surface area. Some reservoirs can be quite lotic in nature, while others are primarily lentic.
resident agent in charge. A title used for a law enforcement officer employed by an agency such as the U.S. Fish and Wildlife Service who has more experience and responsibility than a special agent.
resource allocation. The allocation of portions of a particular resource, such as a particular fishery, to various resource users.
restoration. The act of reestablishing a native population that has been extirpated from a portion or all of its native range.
restoration ecology. The study of recuperating degraded, damaged, or destroyed ecosystems through active human intervention.
restored species. A species that has been extirpated from a portion or all of its native range and then successfully reestablished in the vacated area.

rete mirabile. A specialized netlike assemblage of blood vessels that brings large numbers of small arteries and veins into close contact to increase countercurrent exchange.
reticulum. The second chamber of a ruminant stomach.
reverse slot length limit. A regulation under which all fish of a particular species that are within a specified length range may be kept, while all those outside that length range must be released.
rhythmic behavior. An activity pattern or behavior of an animal that is associated with daily or seasonal changes.
riffle. That area of a headwater freshwater lotic system characterized by being shallow, narrow, and rock bottomed. Current flow in riffles is stronger than that in pools because of the reduced volume of space present.
riparian zone. The habitat associated with the edges of rivers and streams.
riverine. Of or pertaining to lotic systems.
***r–K* continuum.** The range of reproductive strategies exhibited by animals, from producing large numbers of offspring and providing no parental care *(r)* to producing small numbers of offspring and providing substantial parental care *(K)*.
Robel pole. A marked pole used to determine the horizontal concealment or cover value of vegetation.
rocket net. A lightweight net that is carried over a group of animals by small rocket-type projectiles.
rookery. A nesting colony, usually of birds.
rotenone. A type of general-use piscicide.
***r*-selected species.** A species with a reproductive strategy that involves production of large numbers of offspring and little parental care.
rumen. The first chamber of a ruminant stomach.
ruminant. An ungulate with an even number of toes that regurgitates and chews its food repeatedly.
run. An intermediate area located between the pools and riffles of a headwater lotic freshwater system. This term can also be used in reference to a particular stock or stocks of fish.
rut. The period associated with male breeding activity in some ungulates.

sagebrush rebellion. A movement on the part of some people, primarily those holding leases on U.S. federal grazing lands, in opposition to the Public Rangeland Improvement Act of 1978.
salmonid. A member of the fish family Salmonidae (trout and salmon).
salt gland. A gland used for the active transport of salts out of the bloodstream.
salt wedge. Water with a higher salt content that, because of its higher density, is located under water with less salinity. Salt wedges are often associated with estuarine areas.
salvage logging. The logging of dead or diseased trees to improve forest health.
sample. In reference to animal populations, a sample is a subset or portion of the total number of organisms in a population. In a broader sense, a sample represents any subset or portion of the whole.
sample replication. Repeating a sample multiple times.
saprophyte. A nonanimal, such as a fungus, that feeds mainly on dead plant material.
scale (as used in landscape ecology). The spatial and temporal dimensions involved in landscape or other ecological studies.
scale annuli (singular: scale annulus). Annual growth rings found on fish scales.
scats. See **fecal pellets**.
scavenger. An animal that eats dead animals.
S-curve. A J-curve that has flattened at the upper end because the population is being restricted by one or more limiting factors.
seabirds. Birds that are associated with oceans and, at times, their shorelines.
Secchi disk. A metal, wooden, or plastic disk 20 centimeters in diameter that is used to estimate water clarity. Opposite quarters of the disk are usually colored white and black.
secondaries. See **secondary feathers**.
secondary compounds. Anti-herbivory types of chemicals that do not fill a primary role in plant growth and development.
secondary consumer. An organism that feeds on primary consumers.
secondary feathers. Large flight feathers on the posterior margin of a bird wing arising between the "wrist" and "elbow."
secondary succession. Ecological succession occurring on a site that was previously occupied by other living organisms, such as a burned forest.
secondary tree cavity. A cavity that has already been constructed by another animal.

second law of thermodynamics. When energy is transferred or transformed, much is lost or dispersed in unusable forms, such as heat.
seine. An active capture device used in fisheries work consisting of a panel of netting that is pulled through the water to capture the target animals.
selective cutting. A forestry practice that involves cutting only specific kinds and sizes of trees in an area so that some trees remain in the cut area.
semantic differential method. A survey or census method used to measure human attitudes in which an object or concept must be rated by the respondent on a scale, with each endpoint anchored by an adjective or phrase.
semicircular canals. A portion of the inner ear of vertebrates that contributes to the sense of balance and equilibrium.
seral stage. One of the transient but recognizable stages in an ecological succession.
sere. The entire sequence of seral stages in an ecological succession.
sexual dichromatism. Differences in the color, pattern, hue, or shade of animals due to gender.
sexual dimorphism. Differences in the shape or form of animals due to gender.
sexual segregation. Differential use of space by genders within a species during portions of the year.
shifting-mosaic steady state. Mature and relatively stable end stage of ecological succession; previously referred to as climax community.
shorebirds. A group consisting primarily of small wading birds, most of which feed along shores and mudflats.
shoreline development. The ratio of shoreline length to the circumference of a circle with the same area as the water body. Shoreline development is basically a measure of shoreline irregularity.
short-grass prairie. That portion of the temperate grassland biome characterized by grasses that are short in stature; the driest of the prairie types.
size limit. The size of an animal that can be legally harvested.
size structure. The size distribution of a population; for example, the relative proportions of small, medium-sized, and large individuals in a population.
SLAPP suit. A lawsuit intended to stifle expression by costing people or groups time, effort, or money to defend themselves. The intent of such a lawsuit is not to win a legal decision but to intimidate someone into silence.
smolt. The life stage of a salmonid fish at the time of physiological adaptation to life in the marine environment. The process is termed *smoltification* and occurs even in forms that spend their entire lives in fresh water.
snap trap. A passive capture device, usually used on mammals, such as the familiar mousetrap.
snare. A passive capture device, most often used to capture mammals, that consists of a wire or rope noose that captures the animal by the leg or neck.
sociality. The tendency to gather in organized groups.
sociobiology. The study of the biological basis of social behavior and the organization of societies.
sodbuster provisions. Provisions in the 1985, 1990, 1996, and 2002 Farm Act (Farm Security and Rural Investment Act) that require landowners who convert grasslands to croplands and who receive federal crop supports to comply with certain soil conservation measures. These provisions, formally known as the highly erodible land provisions, have been of little value in protecting native prairies from destruction.
songbird. A bird species that produces long vocal displays consisting of various series of complex notes and repeated patterns.
southeastern mixed forest. A portion of the temperate forest biome that ranges from Virginia through the Carolinas and Alabama and contains numerous evergreen trees.
spaced dispersion. See **uniform dispersion.**
spawn. In fishes, to release the male and female reproductive products. The term can also be used as a noun to denote either the male or female reproductive products.
spawning check. A false annulus on a fish scale (or other hard body part) caused by growth pattern anomalies at the time of spawning. See **false annuli.**
special agent. A title used for a law enforcement officer employed by an agency such as the U.S. Fish and Wildlife Service.
specialist. An organism that can utilize only a narrow range of resources or habitats.
speciation. The process of species formation.
species (plural: species). A naturally occurring group of organisms that can successfully interbreed and produce fertile offspring.
species diversity. A measure that includes both species richness and species equity or evenness. Most often,

only certain portions of a community are assessed at one time, such as the trees in a forest.

species equity. A measure of the evenness of abundance of the various species in a community or portion of a community. Maximum equity occurs when all species in a portion of a community—birds, for example—are equally abundant.

species evenness. See **species equity.**

species richness. The number of species in a community or portion of a community.

specific conductance. See **conductivity.**

speculum. A colored, sometimes iridescent, patch on the secondary feathers of some bird wings.

spherical densiometer. A device used to determine tree canopy coverage that consists of a concave, mirrored surface marked with a grid.

spot-mapping. A noncapture population estimation technique commonly used for songbirds that sing or call within established territories that involves plotting the locations of individual birds on a gridded map.

spring overturn. An event that occurs during spring in a thermally stratified lentic system in which water temperature throughout the water body is equalized and the entire water body mixes.

stakeholder. An individual or group that has something at stake in or an interest in a particular decision or resource.

standard deviation (SD). A statistical term denoting the average deviation between multiple observations and the mean.

standard error (SE). A statistical measure of variation; the standard error of the mean is obtained by dividing the standard deviation by the square root of the sample size.

standardized sampling. Sampling with effective devices using the same device at the same sites at the same time of year and from year to year. Standardized sampling is a common procedure when the purpose of sampling is to monitor long-term trends in a population.

standing crop. See **standing stock.**

standing stock. The abundance (number or biomass) of organisms present at a given time in a given area.

starlight scope. A type of optical equipment that gathers light and can be used to observe animals at night.

states' rights doctrine. In the United States, the power of individual states to control the wildlife and fisheries resources within their borders, derived from the Tenth Amendment to the Constitution.

steelhead. A rainbow trout form that normally spends a portion of its life in marine waters.

stock. A group of animals with a common ancestry or parentage that is adapted to a particular environment. When used as a verb, the term denotes the release of organisms, as in fish stocking. Also see **population.**

stock–recruitment relationship. A relationship that demonstrates the influence of parental abundance on subsequent population recruitment. Theoretically, the production of recruits is low when parental abundance is very low or very high and is generally high at moderate levels of parental abundance.

strain. Cultured populations that reliably express a suite of characteristics.

strategic planning. The development of missions, goals, objectives, and strategies.

Strategic Lawsuits Against Public Participation. See **SLAPP suit.**

strategy. An action undertaken to ensure that an objective will be met.

stream gradient. Rate of descent to sea level.

stress. The result of biotic and abiotic forces that extend the internal stabilizing mechanisms of an organism beyond its ability to maintain homeostasis.

structural gene. A gene that encodes an amino acid sequence or the structure of a protein.

strutting grounds. A lek used by sage grouse.

stunted population. A fisheries term denoting a population characterized by high density and slow growth.

subadult. An age classification used in wildlife to denote an individual that is too young to breed but is externally indistinguishable from an adult.

submergent vegetation. Aquatic vegetation that grows in the water column but generally does not extend above the water surface.

subspecies. A formally recognized geographic subdivision of a species. The members of a subspecies resemble one another to some extent and differ from other subdivisions in some recognizable way. Where ranges of subspecies overlap, they regularly interbreed to produce fertile offspring.

subyearling. See **young-of-the-year.**

suction trap. A trap that draws in air; used primarily to capture aerial insects.

summerkill. The death of aquatic organisms during summer resulting from dissolved oxygen depletion.
sunfish. A member of the fish family Centrarchidae. Most often the term is used in reference to smaller members of the family, such as bluegills and rock bass, but even largemouth bass are members of the sunfish family.
supplemental stocking. The release of fish into a population to augment natural reproduction. The term can also be used in reference to wildlife species.
Surber stream sampler. A device used to capture larger, primarily benthic, aquatic organisms, such as insects, in lotic habitats.
surplus production. That portion of production that can be removed from a population by natural causes or human harvest without adversely affecting future population levels.
survey. A partial count or sample of organisms present or harvested.
swamp. A wetland that usually contains standing trees.
swampbuster provisions. Provisions in the 1985, 1990, 1996, and 2002 Farm Act (Farm Security and Rural Investment Act) that require landowners who receive federal funding to follow certain regulations with regard to wetland conservation.
sweep net. The familiar butterfly net used primarily to capture terrestrial invertebrates.
symbiotic. Of or pertaining to a relationship in which interacting species are closely and permanently dependent on each other.
sympatric. Having overlapping ranges of distribution.
sympatric speciation. Speciation occurring when populations are not spatially separated.
systematic process. A genetic process that changes the allelic frequencies of a population in some predictable, nonrandom fashion.

taiga. See **northern coniferous forest.**
tailrace. The area immediately below a dam that receives water released from an impoundment.
tall-grass prairie. That portion of the temperate grassland biome characterized by grasses that are tall in stature; the wettest of the prairie types.
taxis. The oriented movement of an organism toward (positive taxis) or away from (negative taxis) a stimulus.
taxon (plural: taxa). The general term for a taxonomic group, whatever its rank.
temperate forest. A biome generally characterized by deciduous hardwood trees and moderate precipitation. In eastern North America it is found south of the northern coniferous forest biome.
temperate grassland. A biome characterized by the dominance of grass as a vegetation type. In North America it is primarily located west of the temperate forest biome and south of the northern coniferous forest biome.
temperate rain forest. A portion of the Pacific coast coniferous forest biome characterized by exceptionally high precipitation levels.
terminal gear. See **terminal tackle.**
terminal tackle. Fishing equipment located at the end of a fishing line.
territory. The portion of the home range of an animal that it defends against others of the same or sometimes closely related species.
tertiary consumer. An organism that primarily feeds on secondary consumers.
TFM (3-trifluoromethyl-4-nitrophenol). A selective piscicide used primarily on sea lampreys.
thermal stratification. The seasonal presence of layers with different temperatures and dissolved-oxygen levels in freshwater lentic water bodies.
thermocline. The middle temperature zone in a lentic water body that is thermally stratified; the thermocline is the zone of greatest temperature variation.
thermoneutral zone. The range of ambient temperatures within which an animal requires no additional energy beyond that needed for normal body metabolism to maintain temperature homeostasis.
threatened species. A species that is likely to become endangered throughout all or a portion of its range. This term commonly relates to the U.S. Endangered Species Act, although other entities, such as some countries, states, provinces, and territories, can also develop threatened species lists.
timberline. The highest altitude at which trees can grow in a particular area.
tines. The points on animal antlers.
titration. The measurement of the quantity of a given constituent of a solution through the addition of reagents of known concentration until a particular chemical reaction is completed. Water quality characteristics, such as dissolved-oxygen content, can be determined by titration.
top-down control. Dynamic in which predators have a primary controlling influence on numbers or

biomass of organisms in lower trophic levels or even on smaller predators in the same trophic level.

torpor. A period of lethargy or sluggishness. See **hypothermy.**

total mortality. The sum of natural and harvest mortality. Total mortality is commonly expressed as the overall death rate of a population in a 1-year period.

total rehabilitation. See **total renovation.**

total renovation. A term used in fisheries to denote the complete removal of a fish community. This term can also be used in relation to wildlife species.

total quality management (TQM). A planning or management process developed by Dr. W. Edwards Deming, who stressed employee empowerment to promote product quality.

trace element. See **microelement.**

trammel net. A passive capture device used in fisheries that entangles organisms in two or three panels of netting.

transect. A straight line along which samples are taken. Transects can be used in terrestrial or aquatic situations and are variable in length, depending on what is being sampled.

transgenic organism. An organism whose genetic makeup has been subject to artificial modification.

translocation. A form of genetic change that results when a segment from one chromosome becomes attached to a homologous chromosome.

transplant. An organism moved outside its native range but within the same zoogeographic region.

trap net. A passive capture device used in fisheries that contains either rectangular frames that stabilize the net or floats and stakes that hold the net open. The netting is arranged so that a fish can readily find its way into the trap but not out of it. Lead nets are usually present.

travel cost method. A method of estimating benefits associated with specific recreation sites based on the distance traveled to reach that location.

trial-and-error learning. Learning by means of sorting through several nonrewarding actions to find an action that will provide a reward or avoid a punishment.

triploidy. The genetic condition of having three sets of chromosomes.

trophic cascade. Indirect influence of consumer–resource interactions on other trophic levels in the community.

trophic level. The position of an organism in a food web or food chain; for example, the first trophic level consists of primary producers.

trophogenic zone. See **photic zone.**

tropholytic zone. See **aphotic zone.**

tropical rain forest. A biome characterized by high precipitation levels, hot climatic conditions, and lush vegetation. Large portions of equatorial South America, Southeast Asia, and the Malayan archipelago, and to a lesser extent Africa, consist of this biome.

tropical seasonal forest. A biome characterized by wet and dry seasons and hot temperatures throughout the year. The southern tip of Florida is representative of this biome.

tundra. A biome characterized by short, cool summers; long, cold winters; and permafrost. Large areas of northern Canada and Alaska consist of this biome.

turbidity. Water opaqueness due to suspended particulate matter.

two-story fishery. A fishery in which warmwater and/or coolwater fishes are present in one temperature stratum and coldwater fishes are present in another temperature stratum.

ultratrace element. An inorganic element required in extremely small amounts by an organism and generally only lacking in some sterile captive situations.

ultraviolet survey trap. A device used to capture night-flying insects, both terrestrial and aquatic.

understory. Vegetation that lies beneath and is shaded by a canopy or overstory.

ungulate. A hoofed animal.

uniform dispersion. A dispersion pattern in which organisms are regularly or systematically spaced; often related to territoriality or strong intraspecific competition.

upland game. A category that includes most birds and small mammals that are used for recreational hunting and are found in terrestrial systems.

upper critical temperature. The ambient temperature above the thermoneutral zone of an animal at which increased energy consumption must occur for the animal to maintain temperature homeostasis.

upper zone of metabolic compensation. The temperature zone above the upper critical temperature within which an animal can maintain temperature homeostasis by expending additional energy.

values. The qualities of a thing according to which it is thought of as being more or less desirable, useful, or important.
van Dorn water sampler. A plastic device that can be lowered into a water body, remotely triggered to close, and brought to the surface with a water sample from a particular depth.
variable-radius method. A method of estimating tree density and basal area in which a prism or angle gauge is used to determine which trees are included in the inventory.
velvet. The soft, highly vascularized tissues that cover developing antlers.
visual-obscurity measurement device. See **visual-obstruction measurement device.**
visual-obstruction measurement device. A marked board or rod used to determine the horizontal concealment or cover value of vegetation.
volatilize. To readily vaporize at a relatively low temperature.

Wallop-Breaux Amendment. An amendment to the Dingell-Johnson Act that places an excise tax on a variety of fishing gears and equipment not covered in the initial act.
warmwater species. Freshwater fish species that survive, grow, and reproduce best in warm water. They generally spawn at water temperatures above 16°C (60°F).
waterfowl. A bird group consisting of ducks, geese, and swans.
Waterfowl Production Areas. Federally owned areas in the United States purchased with funding provided by the Migratory Bird Hunting Stamp Act (1934) and other sources to provide habitat primarily for migratory bird species. Many other wild organisms are benefited by these areas.
wattle. A fleshy lobe on the throat of a bird, such as that found on the male wild turkey.
wetland. An area where the water table is near ground level or just shallowly covers the land surface, such as a bog, marsh, swamp, or coastal wetland. This term can also be used to denote any area with surface water regardless of water depth or size.
wetland easement. A federally owned easement on private land in the United States purchased with funding provided by the Migratory Bird Hunting Stamp Act (1934) and other sources to provide habitat primarily for migratory bird species.
Wetlands Reserve Program. U.S. Department of Agriculture program authorized in the 1990, 1996, and 2002 Farm Act (Farm Security and Rural Investment Act) that provides technical and financial assistance to landowners, primarily for restoration of drained or agriculturally degraded wetlands.
wild animals. Animals that are legally defined as wild by state, provincial, territorial, or federal governments.
wildlife. Wild terrestrial and partly terrestrial vertebrate animals. In a broader sense, this term can refer to all wild terrestrial, partly terrestrial, and aquatic organisms.
wildlife biology. The biological study of wildlife.
wildlife conservation. The effective management and use of wildlife resources.
wildlife ecology. The study of the interrelationships of organisms that are considered to be a part of the wildlife realm with other organisms and the environment.
wildlife inspectors. U.S. Fish and Wildlife Service uniformed but unarmed law enforcement officers stationed at ports of entry and departure in the United States; these officers inspect luggage, persons, and other materials for violation of CITES or other national or international wildlife laws. Similar inspector positions exist in Canada.
wildlife management. The art and science of manipulating the biota, habitat, or human users of a wildlife system to produce some desired end result.
wildlife science. The process of obtaining knowledge about and studying wildlife resources.
wing dike. A bank stabilization structure that is used to concentrate current into the middle of a river channel.
winterkill. The death of aquatic organisms primarily resulting from dissolved-oxygen depletion and increased toxic gases under ice cover.
Wisconsin plankton net. A specialized capture net used primarily to sample plankton.

year class. The year in which a fish was hatched or born; for example, a fish hatched in 2008 will always be a member of the 2008 year class. This term also has some wildlife usage.
year-class strength. The abundance of a particular year class. Year-class strength is typically high for years when environmental conditions are appropriate for reproduction and recruitment and poor for years when they are not. This term is most commonly used in fisheries.

yearling. A term commonly used for an ungulate in its second summer and fall after birth. In fisheries, this term refers to an age-1 fish.

yield. That portion of a population harvested by humans.

young-of-the-year. A fish at age 0. This term is also used to identify the young of wildlife species.

zooplankton. Tiny aquatic animals that primarily float or drift in water.

APPENDIX: Vertebrates Used as Examples in this Book

FISHES

Scientific (genus and species) and common names for North American fishes were obtained from the following source: J. S. Nelson, E. J. Crossman, H. Espinosa-Perez, L. T. Findley, C. R. Gilbert, R. N. Lea, and J. D. Williams. 2004. *Common and scientific names of fishes from the United States, Canada, and Mexico.* American Fisheries Society, Special Publication 29, Bethesda, Md.

Common name	Scientific name
albacore	*Thunnus alalunga*
alligator gar	*Atractosteus spatula*
American eel	*Anguilla rostrata*
American shad	*Alosa sapidissima*
Arctic char	*Salvelinus alpinus*
Arctic grayling	*Thymallus arcticus*
Asian carp (in North America)	
bighead carp	see bighead carp
black carp	*Mylopharyngodon piceus*
common carp	see common carp
grass carp	see grass carp
silver carp	see silver carp
Atlantic cod	*Gadus morhua*
Atlantic mackerel	*Scomber scombrus*
Atlantic salmon	*Salmo salar*
banded pygmy sunfish	*Elassoma zonatum*
bighead carp	*Hypophthalmichthys nobilis*
bigmouth buffalo	*Ictiobus cyprinellus*
black bullhead	*Ameiurus melas*
black crappie	*Pomoxis nigromaculatus*
black hagfish	*Eptatretus deani*
blacktail shiner	*Cyprinella venusta*
bleeding shiner	*Luxilus zonatus*
blue catfish	*Ictalurus furcatus*
bluegill	*Lepomis macrochirus*
blue pike	see walleye
blue sucker	*Cycleptus elongatus*
blue tilapia	*Oreochromis aureus*
bluntnose minnow	*Pimephales notatus*
bowfin	*Amia calva*
brassy minnow	*Hybognathus hankinsoni*
bridgelip sucker	*Catostomus columbianus*
brindled madtom	*Noturus miurus*
brook silverside	*Labidesthes sicculus*
brook stickleback	*Culaea inconstans*
brook trout	*Salvelinus fontinalis*
brown bullhead	*Ameiurus nebulosus*
brown trout	*Salmo trutta*
bull trout	*Salvelinus confluentus*
California grunion	*Leuresthes tenuis*
Cape Fear shiner	*Notropis mekistocholas*
central stoneroller	*Campostoma anomalum*
chain pickerel	*Esox niger*
channel catfish	*Ictalurus punctatus*
Chinook salmon	*Oncorhynchus tshawytscha*
chum salmon	*Oncorhynchus keta*
coho salmon	*Oncorhynchus kisutch*
Colorado pikeminnow	*Ptychocheilus lucius*
Colorado River cutthroat trout	see cutthroat trout
common carp	*Cyprinus carpio*
creek chub	*Semotilus atromaculatus*
cutthroat trout	*Oncorhynchus clarkii*
Colorado River cutthroat trout	*O. c. pleuriticus*
greenback cutthroat trout	*O. c. stomias*
Lahontan cutthroat trout	*O. c. henshawi*
Rio Grande cutthroat trout	*O. c. virginalis*
Yellowstone cutthroat trout	*O. c. bouvieri*
desert sucker	*Catostomus clarkii*
dollar sunfish	*Lepomis marginatus*
eastern blacknose dace	*Rhinichthys atratulus*
emerald shiner	*Notropis atherinoides*

fantail darter	*Etheostoma flabellare*
fathead minnow	*Pimephales promelas*
firemouth cichlid	*Cichlasoma meeki*
flat bullhead	*Ameiurus platycephalus*
flathead catfish	*Pylodictis olivaris*
freshwater drum	*Aplodinotus grunniens*
Gila chub	*Gila intermedia*
Gila topminnow	*Poeciliopsis occidentalis*
gizzard shad	*Dorosoma cepedianum*
golden shiner	*Notemigonus crysoleucas*
golden trout	see rainbow trout
grass carp	*Ctenopharyngodon idella*
great barracuda	*Sphyraena barracuda*
greenback cutthroat trout	see cutthroat trout
green sunfish	*Lepomis cyanellus*
guppy	*Poecilia reticulata*
haddock	*Melanogrammus aeglefinus*
hornyhead chub	*Nocomis biguttatus*
Jack Dempsey	*Cichlasoma octofasciatum*
johnny darter	*Etheostoma nigrum*
Lahontan cutthroat trout	see cutthroat trout
lake chub	*Couesius plumbeus*
lake sturgeon	*Acipenser fulvescens*
lake trout	*Salvelinus namaycush*
largemouth bass	*Micropterus salmoides*
longnose dace	*Rhinichthys cataractae*
longnose gar	*Lepisosteus osseus*
longnose sucker	*Catostomus catostomus*
Mexican tetra	*Astyanax mexicanus*
mountain madtom	*Noturus eleutherus*
mountain sucker	*Catostomus platyrhynchus*
mountain whitefish	*Prosopium williamsoni*
muskellunge	*Esox masquinongy*
Nile tilapia	*Tilapia nilotica*
northern hog sucker	*Hypentelium nigricans*
northern pike	*Esox lucius*

orangebelly darter	*Etheostoma radiosum*
orangespotted sunfish	*Lepomis humilis*
orangethroat darter	*Etheostoma spectabile*
Ozark cavefish	*Amblyopsis rosae*
Pacific barracuda	*Sphyraena argentea*
Pacific hagfish	*Eptatretus stouti*
Pacific halibut	*Hippoglossus stenolepis*
Pacific sardine	*Sardinops sagax*
paddlefish	*Polyodon spathula*
Pahrump poolfish	*Empetrichthys latos*
pallid sturgeon	*Scaphirhynchus albus*
pink salmon	*Oncorhynchus gorbuscha*
plains minnow	*Hybognathus placitus*
pollock	*Pollachius virens*
pumpkinseed	*Lepomis gibbosus*
rainbow darter	*Etheostoma caeruleum*
rainbow smelt	*Osmerus mordax*
rainbow trout	*Oncorhynchus mykiss*
golden trout	*O. m. aquabonita*
razorback sucker	*Xyrauchen texanus*
redbreast sunfish	*Lepomis auritus*
redear sunfish	*Lepomis microlophus*
Red River pupfish	*Cyprinodon rubrofluviatilis*
red shiner	*Cyprinella lutrensis*
riffle sculpin	*Cottus gulosus*
Rio Grande cutthroat trout	see cutthroat trout
river carpsucker	*Carpiodes carpio*
rock bass	*Ambloplites rupestris*
round goby	*Neogobius melanostomus*
sand shiner	*Notropis stramineus*
sauger	*Sander canadensis*
sea lamprey	*Petromyzon marinus*
shorthead redhorse	*Moxostoma macrolepidotum*
shortnose sturgeon	*Acipenser brevirostrum*
shovelnose sturgeon	*Scaphirhynchus platorynchus*
sicklefin chub	*Macrhybopsis meeki*
silver carp	*Hypophthalmichthys molitrix*
silver chub	*Macrhybopsis storeriana*

silver lamprey	*Ichthyomyzon unicuspis*
skipjack herring	*Alosa chrysochloris*
smallmouth bass	*Micropterus dolomieu*
smallmouth buffalo	*Ictiobus bubalus*
snail darter	*Percina tanasi*
snakeheads (in North America)	numerous species possible
sockeye salmon	*Oncorhynchus nerka*
southern cavefish	*Typhlichthys subterraneus*
spoonhead sculpin	*Cottus ricei*
spottail shiner	*Notropis hudsonius*
spotted bass	*Micropterus punctulatus*
spotted gar	*Lepisosteus oculatus*
spotted sucker	*Minytrema melanops*
starhead topminnow	*Fundulus dispar*
steelhead	see rainbow trout
striped bass	*Morone saxatilis*
sturgeon chub	*Macrhybopsis gelida*
suckermouth minnow	*Phenacobius mirabilis*
threadfin shad	*Dorosoma petenense*
thresher shark	*Alopias vulpinus*
Topeka shiner	*Notropis topeka*
Utah chub	*Gila atraria*
walleye	*Sander vitreus*
blue pike	variously regarded as a color phase of the walleye, a subspecies *S. v. glaucum,* or a species *S. glaucum*
wedgespot shiner	*Notropis greenei*
western mosquitofish	*Gambusia affinis*
white bass	*Morone chrysops*
white catfish	*Ameiurus catus*
white crappie	*Pomoxis annularis*
white perch	*Morone americana*
white shark	*Carcharodon carcharias*
white sturgeon	*Acipenser transmontanus*
white sucker	*Catostomus commersonii*
yellow bullhead	*Ameiurus natalis*
yellow perch	*Perca flavescens*
Yellowstone cutthroat trout	see cutthroat trout

AMPHIBIANS AND REPTILES

Scientific (genus, species, and subspecies) and common names for North American amphibians and reptiles were obtained from the following: B. I. Crother, Chair. 2000. *Scientific and standard English names of amphibians and reptiles of North America north of Mexico, with comments regarding confidence in our understanding*. Society for the Study of Amphibians and Reptiles, Herpetelogical Circular No. 29.

alligator snapping turtle	*Macrochelys temminckii*
American alligator	*Alligator mississipiensis*
American bullfrog	*Rana catesbeiana*
bullsnake	*Pituophis catenifer sayi*
common kingsnake	*Lampropeltis getula*
Cope's gray treefrog	*Hyla chrysoscelis*
desert tortoise	*Gopherus agassizii*
diamond-backed watersnake	*Nerodia rhombifera*
eastern collared lizard	*Crotaphytus collaris*
eastern diamond-backed rattlesnake	*Crotalus adamanteus*
loggerhead sea turtle	*Caretta caretta*
long-nosed leopard lizard	*Gambelia wislizenii*
marine iguana	*Amblyrhynchus cristatus*
plains gartersnake	*Thamnophis radix*
prairie rattlesnake	*Crotalus viridis viridis*
smooth greensnake	*Opheodrys vernalis*
snapping turtle	*Chelydra serpentina*
southern leopard frog	*Rana sphenocephala*
southern watersnake	*Nerodia fasciata*

tiger salamander	*Ambystoma tigrinum*
western fence lizard	*Sceloporus occidentalis*
western hog-nosed snake	*Heterodon nasicus*

BIRDS

Scientific (genus and species) and common names for North American birds were obtained from the following: American Ornithologists' Union. 1998. *Check-list of North American birds.* 7th ed. American Ornithologists' Union, Washington, DC. Updates to that list from the 42nd to 47th Supplements to the Check-list as published in *The Auk* for each of those years also were used.

acorn woodpecker	*Melanerpes formicivorus*
Aleutian cackling goose	see cackling goose
American avocet	*Recurvirostra americana*
American black duck	*Anas rubripes*
American coot	*Fulica americana*
American crow	*Corvus brachyrhynchos*
American dipper	*Cinclus mexicanus*
American golden-plover	*Pluvialis dominica*
American goldfinch	*Carduelis tristis*
American kestrel	*Falco sparverius*
American redstart	*Setophaga ruticilla*
American robin	*Turdus migratorius*
American three-toed woodpecker	*Picoides dorsalis*
American white pelican	*Pelecanus erythrorhynchos*
American wigeon	*Anas americana*
American woodcock	*Scolopax minor*
Andean condor	*Vultur gryphus*
Arctic tern	*Sterna paradisaea*
Attwater's greater prairie-chicken	see greater prairie-chicken
Baird's sandpiper	*Calidris bairdii*
Baird's sparrow	*Ammodramus bairdii*
bald eagle	*Haliaeetus leucocephalus*
Baltimore oriole	*Icterus galbula*
bank swallow	*Riparia riparia*
barn owl	*Tyto alba*
barn swallow	*Hirundo rustica*
belted kingfisher	*Ceryle alcyon*
black-and-white warbler	*Mniotilta varia*
black-backed woodpecker	*Picoides arcticus*
black-billed magpie	*Pica hudsonia*
Blackburnian warbler	*Dendroica fusca*
black-faced sandgrouse	*Pterocles decoratus*
black phoebe	*Sayornis nigricans*
black vulture	*Coragyps atratus*
blue jay	*Cyanocitta cristata*
blue-winged teal	*Anas discors*
bobolink	*Dolichonyx oryzivorus*
Bohemian waxwing	*Bombycilla garrulus*
Brandt's cormorant	*Phalacrocorax penicillatus*
brant	*Branta bernicla*
Brewer's sparrow	*Spizella breweri*
brown-headed cowbird	*Molothrus ater*
brown kiwi	*Apteryx australis*
brown pelican	*Pelecanus occidentalis*
burrowing owl	*Athene cunicularia*
cackling goose	*Branta hutchinsonii*
Aleutian cackling goose	*B. h. leucopareia*
cactus wren	*Campylorhynchus brunneicapillus*
California clapper rail	see clapper rail
California condor	*Gymnogyps californianus*
California quail	*Callipepla californica*
Canada goose	*Branta canadensis*
giant Canada goose	*B. c. maxima*
canvasback	*Aythya valisineria*
canyon towhee	*Pipilo fuscus*
Carolina parakeet	*Conuropsis carolinensis*
chestnut-bellied sandgrouse	*Pterocles exustus*
chestnut-collared longspur	*Calcarius ornatus*
chicken (domestic fowl)	*Gallus gallus domesticus*
chimney swift	*Chaetura pelagica*
chipping sparrow	*Spizella passerina*
chukar	*Alectoris chukar*
clapper rail	*Rallus longirostris*
California clapper rail	*R. l. obsoletus*

cliff swallow *Petrochelidon pyrrhonota*
common eider *Somateria mollissima*
common grackle *Quiscalus quiscula*
common merganser *Mergus merganser*
common murre *Uria aalge*
common nighthawk *Chordeiles minor*
common poorwill *Phalaenoptilus nuttallii*
common snipe *Gallinago gallinago*
common tern *Sterna hirundo*
Cooper's hawk *Accipiter cooperii*

double-crested cormorant *Phalacrocorax auritus*
downy woodpecker *Picoides pubescens*
dusky grouse *Dendragapus obscurus*
dusky seaside sparrow see seaside sparrow
dwarf cassowary *Casuarius bennetti*

eastern bluebird *Sialia sialis*
eastern kingbird *Tyrannus tyrannus*
eastern meadowlark *Sturnella magna*
emperor penguin *Aptenodytes forsteri*
emu *Dromaius novaehollandiae*
European starling *Sturnus vulgaris*

ferruginous hawk *Buteo regalis*
field sparrow *Spizella pusilla*
Florida scrub jay *Aphelocoma coerulescens*

gadwall *Anas strepera*
Gambel's quail *Callipepla gambelii*
giant Canada goose see Canada goose
glaucous gull *Larus hyperboreus*
golden eagle *Aquila chrysaetos*
grasshopper sparrow *Ammodramus savannarum*
gray-cheeked thrush *Catharus minimus*
gray partridge *Perdix perdix*
great auk *Pinguinus impennis*
great blue heron *Ardea herodias*
great egret *Ardea alba*
greater prairie-chicken *Tympanuchus cupido*
  Attwater's greater prairie-chicken *T. c. attwateri*
  heath hen *T. c. cupido*
greater rhea *Rhea americana*
greater sage-grouse *Centrocercus urophasianus*
greater white-fronted goose *Anser albifrons*
greater yellowlegs *Tringa melanoleuca*
great horned owl *Bubo virginianus*
green kingfisher *Chloroceryle americana*
green-winged teal *Anas crecca*

heath hen see greater prairie-chicken
Heermann's gull *Larus heermanni*
horned grebe *Podiceps auritus*
horned lark *Eremophila alpestris*
horned puffin *Fratercula corniculata*
house finch *Carpodacus mexicanus*
house sparrow *Passer domesticus*

Iceland gull *Larus glaucoides*
ivory gull *Pagophila eburnea*

Japanese quail *Coturnix japonica*

killdeer *Charadrius vociferus*
king penguin *Aptenodytes patagonicus*
Kirtland's warbler *Dendroica kirtlandii*

least tern *Sternula antillarum*
lesser prairie-chicken *Tympanuchus pallidicinctus*
lesser rhea *Pterocnemia pennata*
lesser snow goose see snow goose
little blue heron *Egretta caerulea*
long-billed curlew *Numenius americanus*
long-billed dowitcher *Limnodromus scolopaceus*
long-tailed duck *Clangula hyemalis*

mallard *Anas platyrhynchos*
marbled murrelet *Brachyramphus marmoratus*
Mauritius dodo *Raphus cucullatus*
Merriam's wild turkey see wild turkey
mountain bluebird *Sialia currucoides*
mourning dove *Zenaida macroura*

northern bobwhite	*Colinus virginianus*
northern flicker	*Colaptes auratus*
red-shafted flicker	a race of northern flicker
yellow-shafted flicker	a race of northern flicker
northern goshawk	*Accipiter gentilis*
northern harrier	*Circus cyaneus*
northern hawk owl	*Surnia ulula*
northern mockingbird	*Mimus polyglottos*
northern pintail	*Anas acuta*
northern spotted owl	see spotted owl
osprey	*Pandion haliaetus*
ostrich	*Struthio camelus*
ovenbird	*Seiurus aurocapillus*
Pacific black duck	*Anas superciliosa*
passenger pigeon	*Ectopistes migratorius*
pectoral sandpiper	*Calidris melanotos*
peregrine falcon	*Falco peregrinus*
American peregrine falcon	*F. p. anatum*
Arctic peregrine falcon	*F. p. tundrius*
petrels	there are 18 North American species
pied-billed grebe	*Podilymbus podiceps*
pied currawong	*Strepera graculina*
pigeon	see rock pigeon
pileated woodpecker	*Dryocopus pileatus*
piping plover	*Charadrius melodus*
purple martin	*Progne subis*
razorbill	*Alca torda*
red-cockaded woodpecker	*Picoides borealis*
red-eyed vireo	*Vireo olivaceus*
redhead	*Aythya americana*
red-necked phalarope	*Phalaropus lobatus*
red-shafted flicker	see northern flicker
red-tailed hawk	*Buteo jamaicensis*
red-winged blackbird	*Agelaius phoeniceus*
ring-billed gull	*Larus delawarensis*
ring-necked pheasant	*Phasianus colchicus*
rock pigeon	*Columb429a livia*
rock ptarmigan	*Lagopus muta*
rock sandpiper	*Calidris ptilocnemis*
rose-breasted grosbeak	*Pheucticus ludovicianus*
Ross's goose	*Chen rossii*
ruby-throated hummingbird	*Archilochus colubris*
ruddy duck	*Oxyura jamaicensis*
ruffed grouse	*Bonasa umbellus*
rufous-crowned sparrow	*Aimophila ruficeps*
sage grouse	*Centrocercus urophasianus*
sage thrasher	*Oreoscoptes montanus*
sandhill crane	*Grus canadensis*
scaled quail	*Callipepla squamata*
scarlet tanager	*Piranga olivacea*
scissor-tailed flycatcher	*Tyrannus forficatus*
seaside sparrow	*Ammodramus maritimus*
dusky seaside sparrow	*A. m. nigrescens*
sharp-shinned hawk	*Accipiter striatus*
sharp-tailed grouse	*Tympanuchus phasianellus*
slender moa	*Dinornis torosus*
snow goose	*Chen caerulescens*
lesser snow goose (blue and white phases)	*C. c. caerulescens*
snowy egret	*Egretta thula*
snowy owl	*Bubo scandiacus*
song sparrow	*Melospiza melodia*
sooty grouse	*Dendragapus fuliginosus*
spectacled eider	*Somateria fischeri*
spotted owl	*Strix occidentalis*
northern spotted owl	*S. o. caurina*
spruce grouse	*Falcipennis canadensis*
Steller's eider	*Polysticta stelleri*
Swainson's hawk	*Buteo swainsoni*
trumpeter swan	*Cygnus buccinator*
turkey (domesticated)	see wild turkey
turkey vulture	*Cathartes aura*
upland moa	*Megalapteryx didinus*
upland sandpiper	*Bartramia longicauda*

vesper sparrow	*Pooecetes gramineus*
western grebe	*Aechmophorus occidentalis*
western kingbird	*Tyrannus verticalis*
western meadowlark	*Sturnella neglecta*
western tanager	*Piranga ludoviciana*
white-breasted nuthatch	*Sitta carolinensis*
white-crowned sparrow	*Zonotrichia leucophrys*
white-winged dove	*Zenaida asiatica*
white-winged scoter	*Melanitta fusca*
whooping crane	*Grus americana*
wild turkey	*Meleagris gallopavo*
Merriam's wild turkey	*M. g. merriami*
Wilson's phalarope	*Phalaropus tricolor*
wood duck	*Aix sponsa*
wood thrush	*Hylocichla mustelina*
worm-eating warbler	*Helmitheros vermivorum*
yellow-headed blackbird	*Xanthocephalus xanthocephalus*
yellow-shafted flicker	see northern flicker
yellow warbler	*Dendroica petechia*

MAMMALS

Scientific (genus and species) and common names for mammals were obtained from the following: D. E. Wilson and D. M. Reeder, eds. 2005. *Mammal species of the world.* 3rd ed., vol. 1 and 2. The Johns Hopkins University Press, Baltimore, Md.

African elephant	*Loxodonta africana* (African bush elephant) *Loxodonta cyclotis* (African forest elephant)
American badger	*Taxidea taxus*
American beaver	*Castor canadensis*
American bison	*Bison bison*
American black bear	*Ursus americanus*
American marten	*Martes americana*
American mink	*Neovison vison*
Arctic fox	*Vulpes lagopus*
Arctic ground squirrel	*Spermophilus parryii*
Arizona pocket mouse	*Perognathus amplus*
Asian elephant	*Elephas maximus*
ass (domesticated)	*Equus asinus*
Audubon bighorn sheep	see bighorn sheep
aurochs	*Bos taurus*
barbary sheep	*Ammotragus lervia*
big brown bat	*Eptesicus fuscus*
bighorn sheep	*Ovis canadensis*
Audubon bighorn sheep	*O. c. auduboni*
black-eared deermouse	*Peromyscus melanotis*
black-footed ferret	*Mustela nigripes*
black rhinoceros	*Diceros bicornis*
black-tailed deer	see mule deer
black-tailed jackrabbit	*Lepus californicus*
black-tailed prairie dog	*Cynomys ludovicianus*
blue wildebeest	*Connochaetes taurinus*
bobcat	*Lynx rufus*
bottlenose dolphin	*Tursiops truncatus*
brown bear	*Ursus arctos*
brown rat	*Rattus norvegicus*
buffalo	see American bison
burro	see ass
Canadian lynx	*Lynx canadensis*
caribou	see reindeer
cat (domesticated and feral)	see wildcat
catamount	see cougar
cattle (domesticated)	see aurochs
chamois	*Rupicapra rupicapra* (alpine chamois) *Rupicapra pyrenaica* (Pyrenean chamois)
cheetah	*Acinonyx jubatus*
collared peccary	*Pecari tajacu*
Columbian white-tailed deer	see white-tailed deer
common chimpanzee	*Pan troglodytes*
common muskrat	*Ondatra zibethicus*
cougar	*Puma concolor*
coyote	*Canis latrans*
coypu	*Myocastor coypus*

Dall's sheep *Ovis dalli*
desert cottontail *Sylvilagus audubonii*
dog (domesticated) *Canis familiaris* (probably not a distinct species but instead descended from various species of *Canis*)
domesticated guinea pig *Cavia porcellus*

eastern cottontail *Sylvilagus floridanus*
eastern fox squirrel *Sciurus niger*
eastern gorilla *Gorilla beringei*
eastern gray squirrel *Sciurus carolinensis*
eastern mole *Scalopus aquaticus*
elk see red deer
European bison *Bison bonasus*
European roe *Capreolus capreolus*

fallow deer *Dama dama*
feral pig see wild boar
Florida panther see cougar
Florida manatee see West Indian manatee
Franklin's ground squirrel *Spermophilus franklinii*

giant kangaroo rat *Dipodomys ingens*
giraffe *Giraffa camelopardalis*
goat (domesticated and feral) *Capra hircus*
golden-mantled ground squirrel *Spermophilus lateralis*
greater kudu *Tragelaphus strepsiceros*
grizzly bear see brown bear
Gunnison's prairie dog *Cynomys gunnisoni*

Hamadryas baboon *Papio hamadryas*
Harris's antelope squirrel *Ammospermophilus harrisii*
hoary marmot *Marmota caligata*
hooded seal *Cystophora cristata*
horse (domesticated and feral) *Equus caballus*
house cat see wildcat
house mouse *Mus musculus*
human *Homo sapiens*

ibexes
Alpine ibex *Capra ibex*
Nubian ibex *C. nubiana*
Siberian ibex *C. sibirica*
Spanish ibex *C. pyrenaica*

killer whale *Orcinus orca*
kit fox *Vulpes macrotis*

little brown myotis *Myotis lucifugus*
little pocket mouse *Perognathus longimembris*
lodgepole chipmunk *Tamias speciosus*
long-tailed weasel *Mustela frenata*

mandrill *Mandrillus sphinx*
marsh rabbit *Sylvilagus palustris*
meadow vole *Microtus pennsylvanicus*
Mexican long-nosed bat *Leptonycteris nivalis*
Mexican vole *Microtus mexicanus*
montane vole *Microtus montanus*
moose *Alces americanus*
mouflon *Ovis musimon*
mountain cottontail *Sylvilagus nuttallii*
mountain goat *Oreamnos americanus*
mountain lion see cougar
mule deer *Odocoileus hemionus*
Sitka black-tailed deer *O. h. sitkensis*
muskox *Ovibos moschatus*

North American deermouse *Peromyscus maniculatus*
North American least shrew *Cryptotis parva*
North American porcupine *Erethizon dorsatum*
North American river otter *Lontra canadensis*
northern fur seal *Callorhinus ursinus*
northern grasshopper mouse *Onychomys leucogaster*
northern pocket gopher *Thomomys talpoides*
nutria see coypu

Ord's kangaroo rat	*Dipodomys ordii*
Pacific jumping mouse	*Zapus trinotatus*
pallid bat	*Antrozous pallidus*
panther	see cougar
pig (domesticated and feral)	see wild boar
platypus	*Ornithorhynchus anatinus*
polar bear	*Ursus maritimus*
pronghorn	*Antilocapra americana*
puma	see cougar
pygmy rabbit	*Brachylagus idahoensis*
raccoon	*Procyon lotor*
rat	see roof rat
red deer (numerous subspecies)	*Cervus elaphus*
red fox	*Vulpes vulpes*
red squirrel	*Tamiasciurus hudsonicus*
reindeer (numerous subspecies)	*Rangifer tarandus*
roof rat	*Rattus rattus*
sea otter	*Enhydra lutris*
sheep (domesticated)	see red sheep
short-beaked echidna	*Tachyglossus aculeatus*
short-finned pilot whale	*Globicephala macrorhynchus*
Siberian brown lemming	*Lemmus sibiricus*
Sitka black-tailed deer	see mule deer
small Asian mongoose	*Herpestes javanicus*
snowshoe hare	*Lepus americanus*
southern flying squirrel	*Glaucomys volans*
sperm whale	*Physeter catodon*
Steller sea lion	*Eumetopias jubatus*
Steller's sea cow	*Hydrodamalis gigas*
striped skunk	*Mephitis mephitis*
swift fox	*Vulpes velox*
thirteen-lined ground squirrel	*Spermophilus tridecemlineatus*
tiger	*Panthera tigris*
walrus	*Odobenus rosmarus*
West Indian manatee	*Trichechus manatus*
Florida manatee	*T. m. latirostris*
white-footed deermouse	*Peromyscus leucopus*
white rhinoceros	*Ceratotherium simum*
white-tailed deer	*Odocoileus virginianus*
Columbian white-tailed deer	*O. v. leucurus*
white-tailed jackrabbit	*Lepus townsendii*
white-tailed prairie dog	*Cynomys leucurus*
wild boar	*Sus scrofa*
wildcat	*Felis silvestris*
wisent	see European bison
wolf	*Canis lupus*
wolverine	*Gulo gulo*
woodchuck	*Marmota monax*
yellow-bellied marmot	*Marmota flaviventris*
yellow-nosed cotton rat	*Sigmodon ochrognathus*

CHAPTER OPENING PHOTO CREDITS

CHAPTER 1, *p. 1, left to right:* brown bear eating salmon, S. Hillebrand/U.S. Fish and Wildlife Service; red-winged blackbird, U.S. Fish and Wildlife Service; destruction of native prairie, Boyd Schulz; Parr marks on Rio Grande rainbow trout, U.S. Fish and Wildlife Service.

CHAPTER 2, *p. 22, left to right:* polar bear and cubs, S. Miller/U.S. Fish and Wildlife Service; blue-winged teal, D. Menke/U.S. Fish and Wildlife Service; stream habitat, Fall River, Colorado, D. Willis; pallid sturgeon, G. Wanner.

CHAPTER 3, *p. 52, left to right:* bull elk, G. Zahm/ U.S. Fish and Wildlife Service; white-crowned sparrow, D. Dewhurst/U.S. Fish and Wildlife Service; electrofishing for sauger, D. Willis; largemouth bass, D. Walter.

CHAPTER 4, *p. 66, left to right:* black-footed ferrets, U.S. Fish and Wildlife Service; Arctic peregrine falcon and young, U.S. Fish and Wildlife Service; fish hatchery, M. Barnes; 3-day-old walleye, M. Brown.

CHAPTER 5, *p. 79, left to right:* giant kangaroo rat, G. Harrison/U.S. Fish and Wildlife Service; bullsnake, U.S. Fish and Wildlife Service; sub-Arctic tundra near Churchill, Manitoba, L. Flake; channel catfish, E. Engbretson/U.S. Fish and Wildlife Service.

CHAPTER 6, *p. 105, left to right:* muskoxen on Nunivak Island, U.S. Fish and Wildlife Service; bohemian waxwings, J. Iron; measuring opening to American black bear den, U.S. Fish and Wildlife Service; male sockeye salmon at spawning time, L. Flake.

CHAPTER 7, *p. 124, left to right:* American badger, U.S. Fish and Wildlife Service; acorn woodpecker, G. Kramer/U.S. Fish and Wildlife Service; boat and trap net, D. Willis; paddlefish, S. Fisher.

CHAPTER 8, *p. 144, left to right:* adult male Dall's sheep, L. Flake; Merriam's wild turkey gobblers in snow, C. Lehman; otolith, S. Sammons; measuring walleye, B. Blackwell.

CHAPTER 9, *p. 164, left to right:* black-tailed prairie dogs, R. Singer/U.S. Fish and Wildlife Service; American bullfrog, D. Backlund; walleye tagging, M. Brown; gastric lavage, T. DeBates.

CHAPTER 10, *p. 188, left to right:* Cope's gray treefrog, D. Backlund; Arctic tern, D. Dewhurst/U.S. Fish and Wildlife Service; releasing wild turkeys, R. Schauer/South Dakota Department of Game, Fish and Parks; redear sunfish, B. Condello.

CHAPTER 11, *p. 209, left to right:* wolf tracks, U.S. Fish and Wildlife Service; recently extinct dusky seaside sparrow, P. Sykes/U.S. Fish and Wildlife Service; sandhill cranes along the Platte River, D. Menke/U.S. Fish and Wildlife Service; Chinook salmon, U.S. Fish and Wildlife Service.

CHAPTER 12, *p. 225, left to right:* Florida manatee, J. Reid/U.S. Fish and Wildlife Service; common eider nest, U.S. Fish and Wildlife Service; water control structure at Bear River National Wildlife Refuge, U.S. Fish and Wildlife Service; brown trout, E. Engbretson/U.S. Fish and Wildlife Service.

CHAPTER 13, *p. 247, left to right:* prairie rattlesnake, D. Backlund; northern spotted owl, J. and K. Hollingsworth/U.S. Fish and Wildlife Service; browse line on quaking aspen caused by elk and deer, C. Lehman; northern pike, T. DeBates.

CHAPTER 14, *p. 270, left to right:* Hoary marmot, B. Hickey/U.S. Fish and Wildlife Service; red-necked phalarope, U.S. Fish and Wildlife Service; destructive effects of cattle in river, T. St. Sauver; bluegill, Nebraska Game and Parks Commission.

CHAPTER 15, *p. 288, left to right:* moose in Anchorage, L. Flake; bald eagle, D. Menke/U.S. Fish and Wildlife Service; Mt. Rainier National Park, H. Thonet; black crappie, E. Engbretson/U.S. Fish and Wildlife Service.

CHAPTER 16, *p. 311, left to right:* raccoon, D. Menke/U.S. Fish and Wildlife Service; plains garter snake, D. Backlund; children fishing, D. Lucchesi; rainbow trout and proud fisherman, M. Cornwell/U.S. Fish and Wildlife Service.

CHAPTER 17, *p. 327, left to right:* red squirrel, D. Dewhurst/U.S. Fish and Wildlife Service; Heermann's gull, L. Flake; Alaska Peninsula National Wildlife Refuge, M. Emery/U.S. Fish and Wildlife Service; young angler with yellow perch, B. Blackwell.

CHAPTER 18, *p. 346, left to right:* collared lizard, U.S. Fish and Widlife Service; American avocet, U.S. Fish and Wildlife Service; walleye captured in gill net, D. Willis; smallmouth bass and angler, B. Graeb.

CHAPTER 19, *p. 363, left to right:* wolf, T. Brooks/U.S. Fish and Wildlife Service; cliff swallow and nest, D. Dewhurst/U.S. Fish and Wildlife Service; fly fishing for salmon on the Kenai River, Alaska, L. Flake; research biologist with walleye, B. Graeb.

CHAPTER 20, *p. 379, left to right:* white-tailed deer with collar and radio transmitter, J. Jenks; black-billed magpie, D. Menke/U.S. Fish and Wildlife Service; radio-tracking American black bear, J. and K. Hollingsworth/U.S. Fish and Wildlife Service; Colorado River cutthroat, Wyoming Game and Fish Department.

INDEX

Note: Page numbers followed by f indicate figures; those followed by t indicate tables.

CPSIA information can be obtained
at www.ICGtesting.com
Printed in the USA
LVHW02s2131040818
585985LV00002B/4/P

9 781464 109133